Solutions Manual

Wackerly/Mendenhall/Scheaffer's

Mathematical Statistics with Applications

SIXTH EDITION

Brian Caffo · Galin Jones

University of Florida

DUXBURY

THOMSON LEARNING™

Australia • Canada • Mexico • Singapore • Spain • United Kingdom • United States

For more information about this or any other Brooks/Cole products, contact:
BROOKS/COLE
511 Forest Lodge Road
Pacific Grove, CA 93950 USA
www.brookscole.com
1-800-423-0563 (Thomson Learning Academic Resource Center)

Printed in the United States of America

10 9 8 7 6 5 4 3 2 1

ISBN: 0-534-38237-1

CONTENTS

CHAPTER 1 WHAT IS STATISTICS?

1.1

a. Population: All generation-X age US citizens. To be more technical, the population would be the collection of 0's and 1's, with each corresponding to a generation-X age US citizen (0 those who do not want to start their own business, 1 otherwise).
Objective: To estimate the proportion of 1's in the population (the proportion of generation-X age US citizens who want to start their own business).

b. Population: The population would be the collection of healthy adults in the US To be more technical the population of interest is the collection of temperatures of all healthy adults in the US.
Objective: To estimate the true mean temperature.

c. Population: Single family dwelling units in the city. More specifically the collection of weekly water consumption's for single family dwelling units in the city.
Objective: To estimate the true mean water consumption.

d. Population: Assign the value 0 if a tire manufactured by the company in the specific year has a safe tread; assign the value 1 if the tire has an unsafe tread. The population consists of the set of 0's and 1's, each corresponding to one of the tires manufactured during the year.
Objective: To estimate the proportion of 1's in the population.

e. Population: A set of 0's and 1's, each corresponding to an adult resident of the state. Assign the value 1 or 0 according as the resident does or does not favor a unicameral legislature.
Objective: To estimate p, the proportion of 1's in the population.

f. Population: Times until recurrence for all people who have had a particular disease.
Objective: To estimate μ, the average time until recurrence.

1.2

a. Each student will obtain a slightly different histogram. Excluding the two extremely large measurements (for Alaska and Hawaii), the range of the other 48 measurements is $1713 - 565 = 1148$. To obtain approximately eight classes, the length of each subinterval should be approximately $\frac{1148}{8} = 143.5$, which we round to 150. The first subinterval is chosen as 500.5–650.5, so that the smallest measurement can be included, and the tally is shown below.

Class i	Class Boundaries	Tally	Frequency	Relative Frequency
1	500.5– 650.5	1	1	.02
2	650.5– 800.5	1	1	.02
3	800.5– 950.5	111	3	.06
4	950.5–1100.5	1111 1111 111	13	.26
5	1100.5–1250.5	1111 1111 11	12	.24
6	1250.5–1400.5	1111 1111	9	.18
7	1400.5–1550.5	1111	4	.08
8	1550.5–1700.5	1111	4	.08
9	1700.5–1850.5	1	1	.02
	2300.5–2450.5	1	1	.02
	3050.5–3200.5	1	1	.02

The relative frequency histogram is shown in Figure 1.1. Notice the two extremely large values that make the histogram awkward to graph.

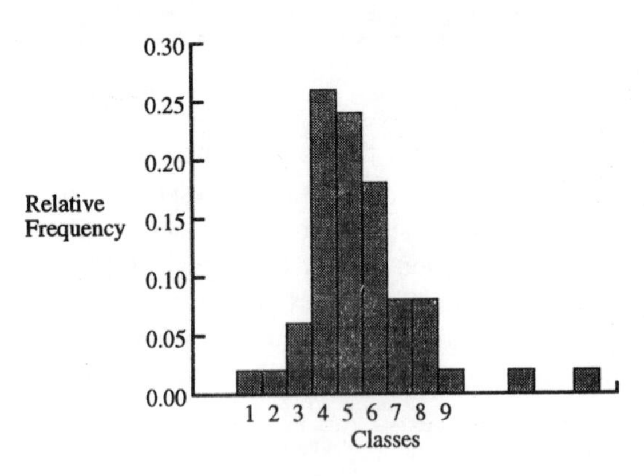

Figure 1.1

1.3 Similar to Exercise 1.2. We chose seven intervals of length 2.

Class Boundaries	Tally	Frequency	Relative Frequency
.005– 2.005	1111 1111 11	12	.48
2.005– 4.005	1111 1	6	.24
4.005– 6.005	111	3	.12
6.005– 8.005	1	1	.04
8.005–10.005	11	2	.08
10.005–12.005		0	.00
12.005–14.005	1	1	.04
		25	1.00

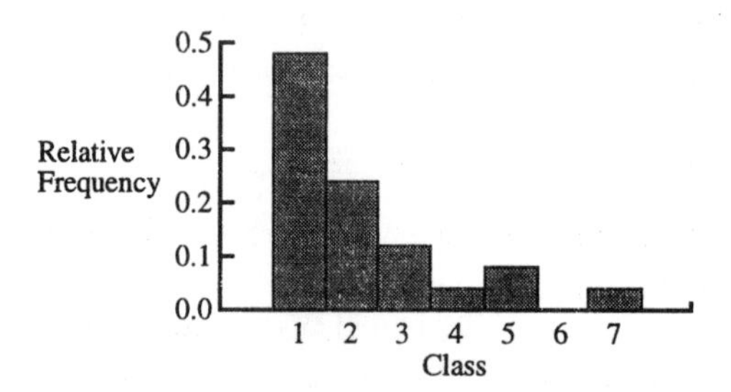

Figure 1.2

1.4 **a.** The range of the data is $38.3 - 1.8 = 36.5$. We choose to use seven class intervals of length 6 ($\frac{36.5}{7} = 5.2$, which when rounded to the next largest integer is 6). The subintervals 1.75–7.75, 7.75–13.75, 13.75–19.75, and so on, are convenient and the tally is shown below.

Class i	Class Boundaries	Tally	Frequency	Relative Frequency
1	1.75– 7.75	1111 1111 1111 1111 11	22	.88
2	7.75–13.75	1	1	.04
3	13.75–19.75	1	1	.04
4	19.75–25.75		0	0
5	25.75–31.75		0	0
6	31.75–37.75		0	0
7	37.75–43.75	1	1	.04

The relative frequency histogram is shown in Figure 1.3.

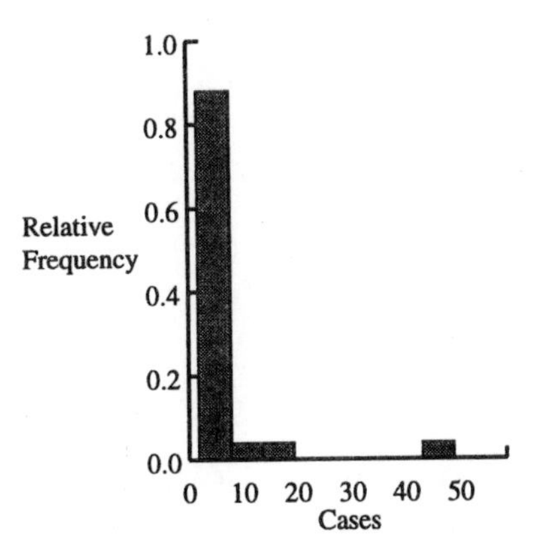

Figure 1.3

b. Use the original data, which is ranked in order of magnitude. The proportion of cities with greater than 10,000 cases is $\frac{3}{25} = .12$.

c. The proportion of cities reporting less than 3000 cases of AIDS in 1992 is $\frac{11}{25} = .44$. Therefore, the probability that a randomly selected city from the given 25 cities is 0.44.

1.5

a. The categories with the largest grouping of students are 2.45 to 2.65 and 2.65 to 2.85. Each of these categories contains 7 of the 30 polled students.

b. $7/30 = .23$

c. $7/30 + 3/30 + 3/30 + 3/30 = 16/30 = .53$

1.6

a. The modal category in this case is 2 (quarts of milk). About 36% (9 people) of the 25 sampled fell into this category.

b. The proportion of people who purchased 3, 4 and 5 quarts of milk are .2, .12 and .04 respectively. Therefore the answer is .2 +. 12 + .04 $=$ .36.

c. Note that 8% of the people purchased 0 while 4% purchased 5. Thus a total of 8% + 4% $=$ 12% purchased 0 or 5. Therefore 1-.12 = .88 (88%) of the people purchased between 1 and 4 quarts of milk.

1.7

a. Note that $9.7 = 12 - (1)2.3$ and $14.3 = 12 + (1)\,2.3$. Therefore the interval $(9.7, 14.3)$ represents breathing rates within 1 standard deviation of the mean. According to the empirical rule approximately 68% of college students should have breathing rates in this interval.

b. Note that $7.4 = 12 - (2)\,2.3$ and $16.6 = 12 + (2)\,2.3$ therefore we are now interested in the percentage of college students with breathing rates within 2 standard deviations of the mean. According to the empirical rule this percentage should be around 95%.

c. We know that 68% of students should have breathing rates between 9.7 and 14.3 (by part **a**). We also know 95% of students should have breathing rates between 7.4 and 16.6 (by part **b**). This leaves (95 - 68) = 27 to lie between both 14.3 and 16.6 and 9.7 and 7.4. By symmetry then 13.5% = 27/2 should lie between 14.3 and 16.6. Therefore 68+13.5=81.5% of college students should have breathing rates between 9.7 and 16.6.

d. Note that $5.1 = 12 - 3(2.3)$ and $18.9 = 12 + 3(2.3)$ therefore we are interested in the proportion of college students that have breathing rates outside of 3 standard deviations of the mean. According to the empirical rule this should be approximately 0.

1.8

a. $14 - (1)\,17 = -3$.

b. As 68% of observations from a normal distribution should lie within 1 standard deviation from the mean $100 - 68 = 32\%$ should lie outside. This implies 16% should lie below 1 standard deviation from the mean.

c. If the population in question were normal then approximately 16% of people would spend less than -3 hours on the Internet. This obviously cannot happen and therefore the population cannot be normal

1.9

a. $\sum_{i=1}^{n} c = c + c + c + \ldots + c$, where the sum involves n elements. Hence $\sum_{i=1}^{n} c = nc$.

b. $\sum_{i=1}^{n} cy_i = cy_1 + cy_2 + cy_3 + \ldots + cy_n = c(y_1 + y_2 + y_3 + \ldots + y_n) = c \sum_{i=1}^{n} y_i$.

c. $\sum_{i=1}^{n} (x_i + y_i) = x_1 + y_1 + x_2 + y_2 + x_3 + y_3 + \ldots + x_n + y_n$

$$= (x_1 + x_2 + x_3 + \ldots + x_n) + (y_1 + y_2 + y_3 + \ldots + y_n) = \sum_{i=1}^{n} x_i + \sum_{i=1}^{n} y_i.$$

Consider the numerator of s^2, which is $\sum_{i=1}^{n} (y_i - \overline{y})^2$.

$$\sum_{i=1}^{n} (y_i - \overline{y})^2 = \sum_{i=1}^{n} (y_i^2 - 2y_i + \overline{y})^2 = \sum_{i=1}^{n} y_i^2 - \sum_{i=1}^{n} 2y_i\overline{y} + \sum_{i=1}^{n} \overline{y}^2$$

$\overline{y}$ and $\overline{y}^2$ are constants with respect to the variable of summation (i). Hence

$$\sum_{i=1}^{n} (y_i - \overline{y})^2 = \sum_{i=1}^{n} y_i^2 - 2\overline{y} \sum_{i=1}^{n} y_i + n\overline{y}^2 = \sum_{i=1}^{n} y_i^2 - 2\overline{y}\,(n\overline{y}) + n\overline{y}^2 = \sum_{i=1}^{n} y_i^2 - n\overline{y}^2$$

with the second equality following from the fact that $\sum_{i=1}^{n} y_i = n\overline{y}$.

Thus, $s^2 = \frac{1}{n-1}\left[\sum_{i=1}^{n} y_i^2 - n\overline{y}^2\right]$, or, since $\overline{y}^2 = \frac{1}{n^2}\left[\sum_{i=1}^{n} y_i\right]^2$,

$$s^2 = \frac{1}{n-1}\left[\sum_{i=1}^{n} y_i^2 - \frac{1}{n}\left(\sum_{i=1}^{n} y_i\right)^2\right]$$

1.10

We know that $n = 6$ and that the values of y_i are 1, 4, 2, 1, 3, and 3. The most efficient method of solution is obtained by creating a table of values as follows:

y_i	y_i^2
1	1
4	16
2	4
1	1
3	9
3	9
14	40

The two necessary sums are $\sum_{i=1}^{n} y_i = 14$ and $\sum_{i=1}^{n} y_i^2 = 40$. Then

$$s^2 = \frac{\Sigma y_i^2 - \left[\frac{(\Sigma y_i)^2}{n}\right]}{n-1} = \frac{40 - \left(\frac{14^2}{6}\right)}{6-1} = \frac{7.333}{5} = 1.47$$

$$s = \sqrt{s^2} = \sqrt{1.47} = 1.21$$

1.11

a. The two necessary sums for calculating $\overline{y}$ and s are

$$\sum_{i=1}^{50} y_i = 62{,}622.0 \text{ and } \sum_{i=1}^{50} y_i^2 = 86{,}027{,}202.0.$$

Then $\overline{y} = \frac{\sum_{i=1}^{n} y_i}{n} = \frac{62{,}622.0}{50} = 1252.44$

and $s^2 = \frac{1}{n-1}\left[\sum_{i=1}^{n} y_i^2 - \frac{1}{n}\left(\sum_{i=1}^{n} y_i\right)^2\right]$

$= \frac{1}{49}\left[86{,}027{,}202.0 - \frac{1}{50}(62{,}622.0)^2\right]$

$= 155{,}038.86$

$s = \sqrt{s^2} = \sqrt{155{,}038.86} = 393.75$

b. The table below shows the intervals, the counts, and the expected counts for each interval.

k	$\overline{y} \pm ks$	Interval Boundaries	Frequency	Expected Frequency
1	1252.44 ± 393.75	858.69 to 1646.19	45	34
2	1252.44 ± 787.50	464.94 to 2039.94	48	47.5
3	1252.44 ± 1181.25	71.19 to 2433.69	49	50.0

There are many more values within one standard deviation than expected because the two extremely large values give a large standard deviation.

1.12 a. Calculate $\Sigma y_i = 80.63$ and $\Sigma y_i^2 = 500.7459$. Then $\overline{y} = \frac{\Sigma y_i}{n} = \frac{80.63}{25} = 3.23$

$s^2 = \frac{1}{24}(500.7459 - 260.04788) = 10.03$

$s = \sqrt{10.03} = 3.17$

k	$\overline{y} \pm ks$	Interval Boundaries	Frequency	Expected Frequency
1	3.23 ± 3.1669	0.063 to 6.397	21	17
2	3.23 ± 6.3338	−3.104 to 9.564	23	23.75
3	3.23 ± 9.5007	−6.271 to 12.731	25	25

1.13 a. Calculate $\sum_{i=1}^{25} y_i = 141.9$ and $\sum_{i=1}^{25} y_i^2 = 2147.61$. Then $\overline{y} = \frac{1}{n}\sum_{i=1}^{n} y_i = \frac{141.9}{25} = 5.676$

and $s^2 = \frac{1}{n-1}\left[\sum_{i=1}^{n} y_i^2 - \frac{1}{n}\left(\sum_{i=1}^{n} y_i\right)^2\right]$

$= \frac{1}{24}\left[2147.61 - \frac{1}{25}(141.9)^2\right]$

$= 55.9244$

$s = \sqrt{s^2} = \sqrt{55.9244} = 7.48$

b.

k	$\overline{y} \pm ks$	Interval Boundaries	Frequency	Expected Frequency
1	5.676 ± 7.48	−1.80 to 13.15	23	17
2	5.676 ± 14.96	−9.28 to 20.63	24	23.75
3	5.676 ± 22.43	−16.75 to 28.11	24	25

The number of measurements falling within each interval does not compare well with the expected number because of the highly skewed nature of the data. The empirical rule works for mound-shaped data. See Exercise 1.10.

1.14 a. First calculate $\sum_{i=1}^{24} y_i = 103.6$ and $\sum_{i=1}^{24} y_i^2 = 680.72$. Then $\overline{y} = \frac{1}{n}\sum_{i=1}^{n} y_i = \frac{103.6}{24} = 4.32$

and $s^2 = \frac{1}{23}\left[680.72 - \frac{1}{24}(103.6)^2\right]$

$= 10.15,$

$s = 3.19$

b.

k	$\overline{y} \pm ks$	Interval Boundaries	Frequency	Expected Frequency
1	4.32 ± 3.19	1.13 to 7.51	22	16.32
2	4.32 ± 6.37	−2.05 to 10.69	22	22.80
3	4.32 ± 9.56	−5.24 to 13.88	23	24

Removal of the large value decreases $\overline{y}$. s is half of what it was in Exercise 1.9.

Even though the intervals contain a number of values similar to before, they are much narrower.

1.15 For Exercise 1.2, the approximation is

$\frac{\text{range}}{4} = \frac{3168-565}{4} = 650.75$, while $s = 393.75$.

Note the poor approximation due to the extreme values.

For Exercise 1.3, the approximation is

$\frac{\text{range}}{4} = \frac{12.48-.32}{4} = 3.04$, while $s = 3.17$.

For Exercise 1.4, the approximation is

$\frac{\text{range}}{4} = \frac{38.3-1.8}{4} = 9.125$, while $s = 7.48$.

1.16 The approximation is range/4 $= (800 - 200)/4 = 150$.

1.17 Notice that 1 standard deviation below the mean is $34 - (1)53 = -19$. The empirical rule suggest 16% of the observations from a population that is normally distributed should lie 1 standard deviation below the mean (see 1.8). As we cannot have negative amounts of chloroform in the water this population cannot be normally distributed.

1.18 By the empirical rule approximately 68% of the measurements fall in the interval $\mu \pm \sigma = 420 \pm 30$, or 390 to 450. Hence $(100 - 68)\% = 32\%$ fall either above 450 or below 390. Since the normal distribution is symmetric, the fraction above 450 will equal the fraction below 390 (see Figure 1.4), so that $\left(\frac{1}{2}\right)(.32) = .16$ is the approximate probability of exceeding $450.

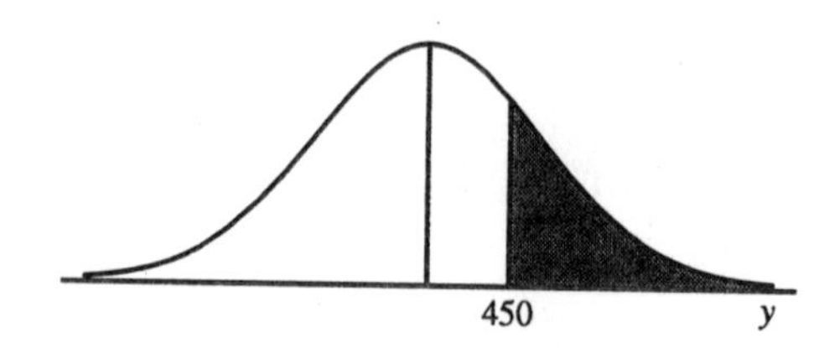

Figure 1.4

1.19 Similar to exercise 1.18. The point representing a gain of 20 pounds is shown in Figure 1.5 to be one standard deviation below the mean. Using the same reasoning as in Exercise 1.12, 16% of the measurements fall below 20, and hence $100 - 16 = 84\%$ fall above 20. Hence .84 is the approximate probability that a weight gain exceeds 20, and the manufacturer is probably correct.

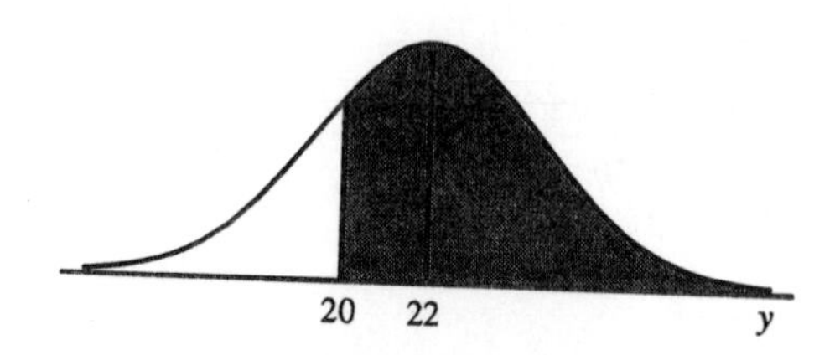

Figure 1.5

1.20 $\sum_{i=1}^{n}(y_i - \overline{y}) = \sum_{i=1}^{n} y_i - n\overline{y} = \sum_{i=1}^{n} y_i - \frac{n\left[\sum_{i=1}^{n} y_i\right]}{n} = \sum_{i=1}^{n} y_i - \sum_{i=1}^{n} y_i = 0.$

1.21 **a.** We assume that the set of 1521 games is the population. Then $\mu = 143$ and $\sigma = 26$. $169 = \mu + \sigma$, and by the empirical rule approximately 32% should be outside the interval $(\mu - \sigma, \mu + \sigma)$. Then half or approximately 16% should be greater than 169.

b. $117 = \mu - \sigma$ and half of approximately 68%, 32%, of the games should be between $\mu - \sigma$ and μ. $195 = \mu + 2\sigma$ and half of approximately 95%, or 47.5%, should be between μ and $\mu + 2\sigma$. Now, 32% + 47.5% = 81.5%, so that approximately 81.5% of the games should have ended with a total score between 117 and 195 points.

c. No. A score of 225 is greater than $\mu + 3\sigma$. Such a score is very unlikely according to the empirical rule.

1.22 **a.** $s \approx \frac{\text{range}}{4} = \frac{112-78}{4} = 8.5$

b. Each student will obtain a slightly different frequency histogram. As an example, choose five intervals of length 7.

Class Boundaries	Tally	Frequency	Relative Frequency
77.5– 84.5	1111	4	.21
84.5– 91.5	111	2	.11
91.5– 98.5	1111 1111	9	.47
98.5–105.5	1	1	.05
105.5–112.5	111	3	.16

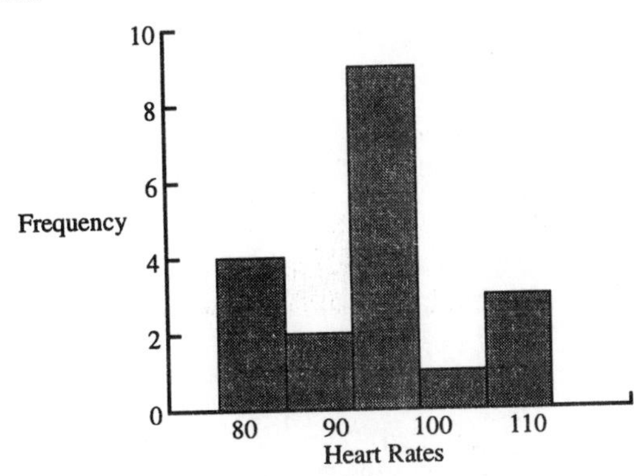

Figure 1.6

From the histogram, $\overline{y}$ appears to be about "95" and s appears to be about 10.

c. Calculate, first, $\sum_{i=1}^{20} y_i = 1874.0$ and $\sum_{i=1}^{20} y_i^2 = 117{,}328.0$. Then $\overline{y} = 93.7$ and $s = 9.55$.

d.

k	$\overline{y} \pm ks$	Interval Boundaries	Frequency	Expected Frequency
1	93.7 ± 9.55	84.1 to 103.2	13	0.65
2	93.7 ± 19.11	74.6 to 112.8	20	1.00
3	93.7 ± 28.66	65.0 to 122.4	20	1.00

These results are reasonably consistent with the empirical rule.

1.23 **a.** $s = \frac{\text{range}}{4} = \frac{716-8}{4} = 177$

b. We chose to use 15 classes of length 50.

Class Boundaries	Tally	Frequency	Relative Frequency
.5– 50.5	1111 1111 1	11	.125
50.5–100.5	1111 1111 1111 1	16	.182
100.5–150.5	1111 111	8	.091
150.5–200.5	1111 1111 1111	15	.170
200.5–250.5	1111 1111	10	.114
250.5–300.5	1111 1	6	.068
300.5–350.5	1111	5	.057
350.5–400.5	1111 1	6	.068
400.5–450.5	111	3	.034
450.5–500.5	11	2	.023
500.5–550.5	1	1	.011
550.5–600.5	1	2	.023
600.5–650.5	1	1	.011
650.5–700.5	1	1	.011
700.5–750.5	1	1	.011
		88	.999

The last column should sum to 1.000. The value .999 is due to rounding off the relative frequencies. The figure is omitted.

c. Calculate $\Sigma y_i = 18{,}550$ and $\Sigma y_i^2 = 6{,}198{,}356$. Then
$\overline{y} = \frac{18{,}550}{88} = 210.80$, $s^2 = \frac{1}{87}(6{,}198{,}356 - 3{,}910{,}255.7) = 26{,}300.00$, and $s = 162.17$.

d.

k	$\overline{y} \pm ks$	Interval Boundaries	Frequency	Expected Frequency
1	210.8 ± 162.17	48.6 to 373.0	63	.72
2	210.8 ± 324.34	−113.5 to 535.1	82	.93
3	210.8 ± 486.52	−275.7 to 697.3	87	.99

1.24 For the data in Exercise 1.10, the range is 3 and the ratio of range to standard deviation is

$$\frac{\text{range}}{s} = \frac{3}{1.211} = 2.477.$$

In Exercise 1.16, with $n = 20$, the ratio is

$$\frac{\text{range}}{s} = \frac{34}{9.5537} = 3.559.$$

Finally, in Exercise 1.17, with $n = 88$, the ratio is

$$\frac{\text{range}}{s} = \frac{708}{162.17} = 4.365.$$

As we might expect, for these three samples the ratio increases as n increases.

1.25 Assuming that the distribution of scores is bell-shaped, the empirical rule provides a means for describing the variability of the data. The results of the empirical rule are shown below.

k	$\overline{y} \pm ks'$	Interval Boundaries	Percentage of Measurements Within the Interval
1	72 ± 8	64 to 80	Approximately 68%
2	72 ± 16	56 to 88	Approximately 95%
3	72 ± 24	48 to 96	Nearly 100%

Hence one would expect 68% of the 340 scores, that is, $.68(340) = 231.2$ or 231 scores, to fall in the interval 64 to 80. Similarly, 95% of the scores, that is, $.95(340) = 323$ scores, should fall in the interval 56 to 88.

1.26 We know $\mu = 27$ and $\sigma = 14$. For this normally distributed population, we expect 68.9% of the days to have a daily discharge between $\mu - \sigma = 13$ and $\mu + \sigma = 41$. Thus, we expect 16%, or half of 32%, to be less than 13 mg/ℓ.

1.27 It is given that the distribution of bearing diameters has a mean of 3.00 and a standard deviation of .01. We wish to calculate the fraction of this machine's production that will fail to meet specifications, given that the only bearings that will meet specifications are those whose diameters lie in the interval 2.98 to 3.02. Notice that this interval contains the values within two standard deviations of the mean. That is,

$$\overline{y} \pm 2s' = 3.00 \pm 2(.01) = 3.00 \pm .02.$$

Hence, if we assume that the distribution of bearing diameters is approximately bell-shaped, the empirical rule states that the fraction of acceptable bearings will be approximately .95. Consequently, the fraction of this machine's production that will fail to meet specifications is approximately $1 - .95 = .05$.

1.28 We suppose $\mu = 0$ and $\sigma = 1.2$. We expect 34% of the differences to be between 0 and $\mu + \sigma = 1.2$. We also expect 47.5% of the differences to be between 0 and $\mu + 2\sigma = 2.4$. Hence, we expect $47.5 - 34 = 13.5\%$ of the differences to be between 1.2 and 2.4.

1.29 Assuming the distribution of yields is mound-shaped, the empirical rule states that approximately 95% of all measurements fall in the interval $\mu + 2\sigma = 60 \pm 20$, or 40 to 80. Only 5% fall outside this interval and, by symmetry, .025 is the fraction below 40. If the yield should fall below 40 (an unlikely event, having probability .025), one would suspect an abnormality in the process.

1.30 We have a set of n measurements, denoted by $y_1, y_2, \ldots, y_n$. For each of these measurements the quantity $|y_i - \overline{y}|$ is calculated. If the quantity $|y_i - \overline{y}|$ is less than ks, then y_i is within the interval $\overline{y} - ks$ to $\overline{y} + ks$. However, if the quantity $|y_i - \overline{y}|$ exceeds ks, then y_i is outside the interval $\overline{y} \pm ks$. Let n' denote the number of measurements that fall outside the interval $\overline{y} \pm ks$, so that $(n - n')$ is the number that fall within the interval, and $\frac{n-n'}{n}$ is the fraction falling within the interval $\overline{y} \pm ks$. It is necessary to show that this fraction is greater than or equal to $1 - \left(\frac{1}{k^2}\right)$. Now

$$(*) \qquad s^2 = \frac{\sum_{i=1}^{n}(y_i - \overline{y})^2}{n-1} = \frac{1}{n-1}\sum_{i\in A}(y_i - \overline{y})^2 + \frac{1}{n-1}\sum_{i\in B}(y_i - \overline{y})^2$$

where

$$A = \{i\colon\ |y_i - y| \geq ks\} \text{ and } B = \{i\colon\ |y_i - y| < ks\}$$

Consider

$$\frac{1}{n-1}\sum_{i\in A}(y_i - \overline{y})^2$$

If $i \in A$, then $|y_i - \overline{y}| \geq ks$, so that $(y_i - \overline{y})^2 \geq k^2 s^2$. Further, there are n' elements in the set A, by definition of n' above. Hence the following inequality is obtained.

$$\frac{1}{n-1}\sum_{i\in A}(y_i - \overline{y})^2 \geq \frac{1}{n-1}\sum_{i\in A}k^2 s^2 = \frac{n'}{n-1}(k^2 s^2).$$

Then from (*) we have

$$s^2 \geq \frac{n'}{n-1}(k^2 s^2) + \frac{1}{n-1}\sum_{i\in B}(y_i - \overline{y})^2$$

Since the quantity $\frac{1}{n-1}\sum_{i\in B}(y_i - \overline{y})^2$ will always be non negative, we have

$$s^2 \geq \frac{n'}{n-1}(k^2 s^2)$$

Thus,

$$1 \geq \frac{n'}{n-1}(k^2)$$

or $\quad \dfrac{1}{k^2} \geq \dfrac{n'}{n-1} > \dfrac{n'}{n}.$

so $\quad 1 - \dfrac{1}{k^2} \leq 1 - \dfrac{n'}{n}$

and $\quad \dfrac{n-n'}{n} \geq 1 - \dfrac{1}{k^2}$

which is the desired result.

1.31 Using Tchebysheff's theorem with $k = 2$, at least $1 - \left(\frac{1}{k^2}\right) = 1 - \left(\frac{1}{4}\right) = \frac{3}{4}$ or 75% of all measurements should lie within two standard deviations of the mean. Hence the interval is $\mu \pm 2\sigma = 5.5 \pm 5$, or .5 to 10.5.

1.32 The point $y = 13$ lies $13 - 5.5 = 7.5$ units above the mean, or $\frac{7.5}{2.5} = 3$ standard deviations above the mean. Using Tchebysheff's theorem, at least $1 - \left(\frac{1}{k^2}\right) = 1 - \left(\frac{1}{3}\right)^2$ $= \frac{8}{9}$ of the measurements lie within 3 standard deviations of the mean (in the interval 5.5 ± 7.5, or -2 to 13). Hence at most $\frac{1}{9}$ will be outside this interval. Thus, at most $\frac{1}{9}$ of the values will exceed 13.

1.33 Lead content readings must be non negative. Zero is only .33 standard deviations below the mean, so that the population can only extend .33 standard deviations below the mean. This radically skews the distribution, so that it is not normal.

1.34 Tchebysheff's theorem says at least $1 - 1/4 = .75$ (or 75%) of the observations from this population has to lie within 2 standard deviations of the mean(0,240). Similarly at least $1 - 1/9$ $= .89$ (or 89%) has to lie within 3 standard deviations of the mean (0,193). Finally at least $1 - 1/16 = .94$ (or 94%) has to lie within 4 standard deviations of the mean (0,246). Notice the use of the term "has to" as Tchebysheff's rule applies to all populations regardless of the of the underlying distribution. Also notice the intervals were truncated at 0 as we know there cannot be less than 0 amount of chloroform in the water.

CHAPTER 2 PROBABILITY

2.1 Construct the three sets: $A = \{FF\}$; $B = \{MM\}$; $C = \{MF, FM, MM\}$. Then

$A \cap B = \emptyset A \cap C = \emptyset$ $\quad B \cap C = \{MM\}$ $\quad C \cap \overline{B} = \{MF, FM\}$

$A \cup B{:} = \{FF, MM\}$ $\quad A \cup C = S$ $\quad B \cup C = C$

2.2

a. AB

b. $A \cup B$

c. $\overline{A \cup B} = \overline{A} \cap \overline{B}$

d. $A\overline{B} \cup \overline{A}B$

2.3 To verify $\overline{(A \cup B)} = \overline{A} \cap \overline{B}$, we can draw the following:

$\overline{(A \cup B)}$

$A \cup B \Rightarrow$

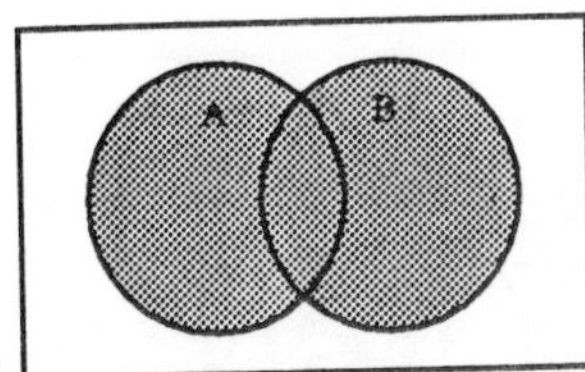

$\overline{A \cup B} \Rightarrow$

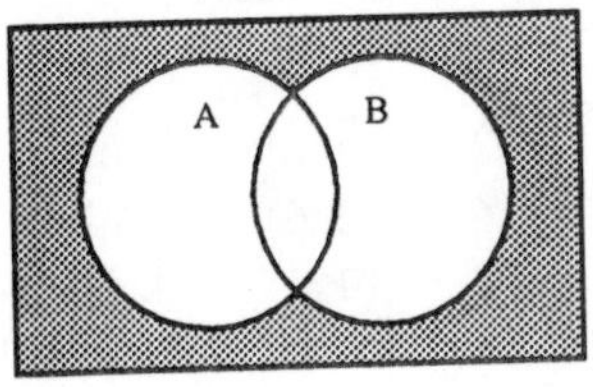

$\overline{A} \cap \overline{B}$

$\overline{A} \Rightarrow$

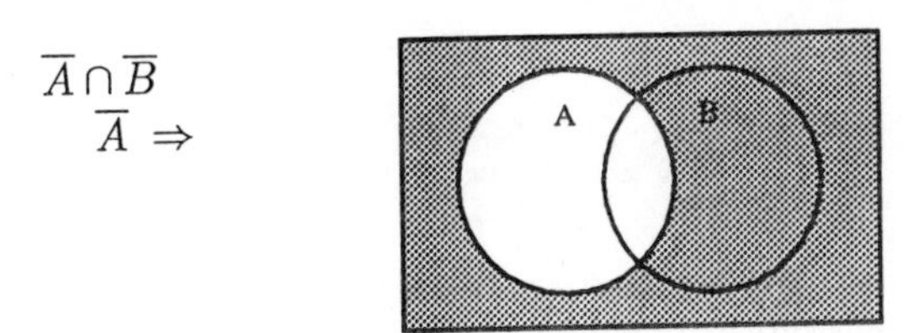

$\overline{B} \Rightarrow$

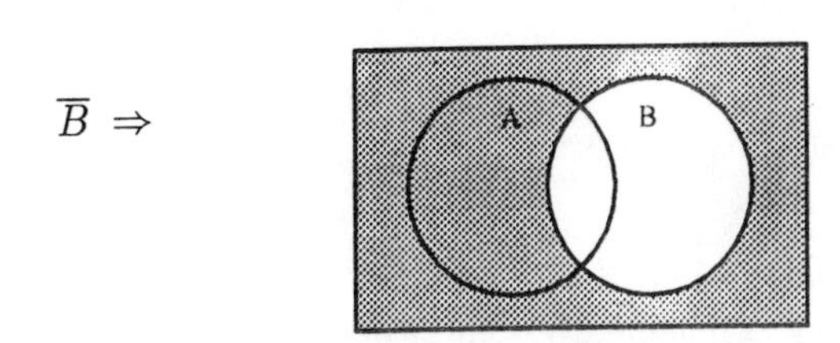

$\overline{A} \cap \overline{B} \Rightarrow$

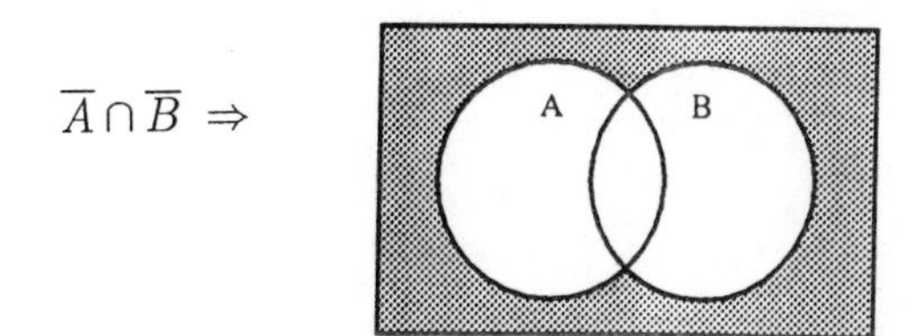

$\overline{A \cap B}$

$A \cap B \Rightarrow$

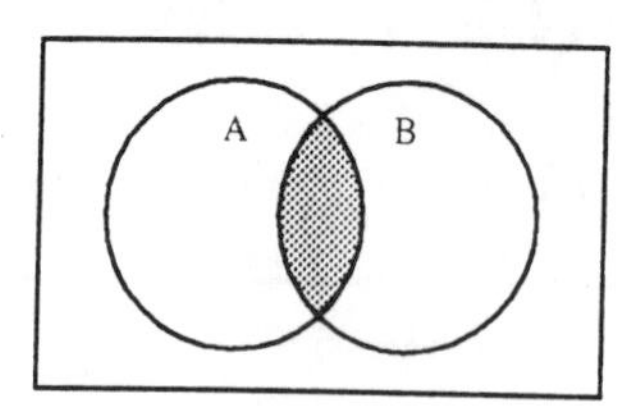

$\overline{A \cap B} \Rightarrow$

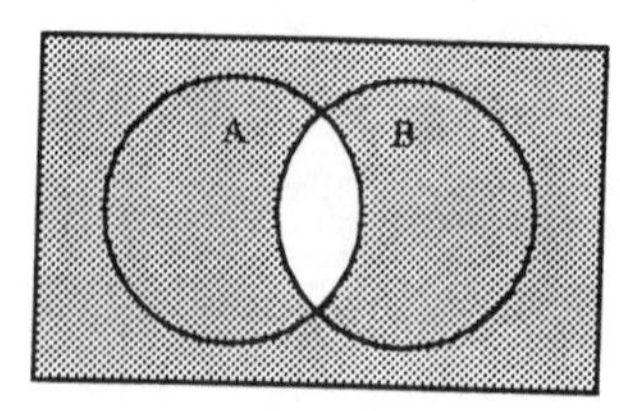

$\overline{A} \cup \overline{B}$

$\overline{A}$, $\overline{B}$ shown above

$\overline{A} \cup \overline{B} \Rightarrow$

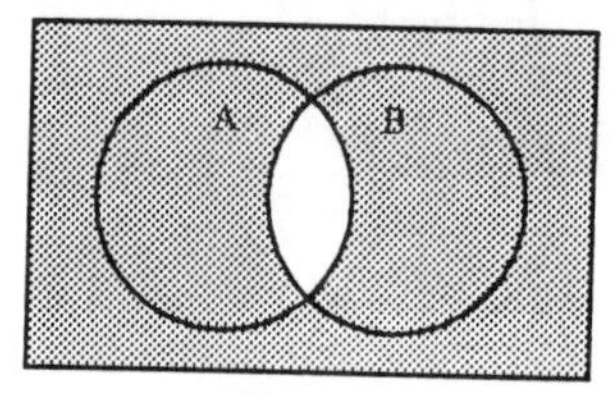

2.4

A:	(1, 2)	(2, 2)	(3, 2)	(4, 2)	(5, 2)	(6, 2)	(1, 4)	(2, 4)	(3, 4)
	(4, 4)	(5, 4)	(6, 4)	(1, 6)	(2, 6)	(3, 6)	(4, 6)	(5, 6)	(6, 6)
$\overline{C}$:	(2, 2)	(2, 4)	(2, 6)	(4, 2)	(4, 4)	(4, 6)	(6, 2)	(6, 4)	(6, 6)
$A \cap B$:	(2, 2)	(4, 2)	(6, 2)	(2, 4)	(4, 4)	(6, 4)	(2, 6)	(4, 6)	(6, 6)
$A \cap \overline{B}$:	(1, 2)	(3, 2)	(5, 2)	(1, 4)	(3, 4)	(5, 4)	(1, 6)	(3, 6)	(5, 6)
$\overline{A} \cup$ B:	all pairs except								
	(1, 2)	(1, 4)	(1,6)	(3, 2)	(3, 4)	(3, 6)	(5, 2)	(5,4)	(5, 6)
$\overline{A} \cap C$:	(1, 1)	(2, 1)	(3, 1)	(4, 1)	(5, 1)	(6, 1)	(1, 3)	(2, 3)	(3, 3)
	(4, 3)	(5, 3)	(6, 3)	(1, 5)	(2, 5)	(3, 5)	(4, 5)	(5, 5)	(6, 5)

Note that $\overline{A} \cap C = \overline{A}$.

2.5

A typical element in the set S is a pair representing the two applicants selected for the job. Then

A ={two males} $= \{M_1M_2,\ M_1M_3,\ M_2M_3\}$

B ={at least one female} $= \{M_1W_1,\ M_2W_1,\ M_3W_1,\ M_1W_2,\ M_2W_2,\ M_3W_2,\ W_1W_2\}$

and hence

$\overline{B}$: = {no females} = {two males} $= A$

$A \cup B$ ={two males or at least one female} $= S$

$A \cap B = \emptyset$ $\qquad A \cap \overline{B} = A \cap A = A.$

2.6

The grid below shows the information given in the exercise in a more convenient form. Note that known quantities are underlined. All unknown counts can be found by subtraction.

	Graduate	Undergraduate	Total
On	18	33	51
Off	6	<u>3</u>	<u>9</u>
Total	24	<u>36</u>	<u>60</u>

a. $36 + 6 = 42$

b. 33

c. 18

2.7 The sample space is all of the possible combinations of blood type and Rhesus factor. That is,
$$S = \{A+, B+, AB+, O+, A-, B-, AB-, O-\}.$$

2.8 **a.** The sample space consists of the four possible blood phenotypes. That is,
$$S = \{A, B, AB, O\}.$$
b. The probabilities that a single caucasian has a given blood type may be assigned as follows.
$$P(\{A\}) = 0.41,\ P(\{B\}) = 0.10,\ P(\{AB\}) = 0.04,\ P(\{O\}) = 0.45$$
c. Since the events $\{A\}$ and $\{B\}$ are mutually exclusive
$$P(A \text{ or } B) = P(A) + P(B) = 0.41 + 0.10 = 0.51$$

2.9 **a.** We know $P(S) = P(E_1 \cup E_2 \cup E_3 \cup E_4 \cup E_5)$
$= P(E_1) + P(E_2) + P(E_3) + P(E_4) + P(E_5)$
$= .15 + .15 + .4 + 2P(E_5) + P(E_5)$
$= .70 + 3P(E_5) = 1.0$
$3P(E_5) = .30$
$P(E_5) = .10$
Then $P(E_4) = .20$

b. Since E_3, E_4, E_5 are all equally probable, let $p = P(E_3) = P(E_4) = P(E_5)$. Then $1 = P(E_1) + P(E_2) + P(E_3) + P(E_4) + P(E_5) = .3 + .1 + 3p$. Thus $p = .2$.

2.10 **a.** Let L = vehicle turns left, R = vehicle turns right, S = vehicle continues straight. Then the sample space would be $\{L, R, S\}$.
b. $P(\text{vehicle turns}) = P(L \cup R) = P(L) + P(R) = \frac{1}{3} + \frac{1}{3} = \frac{2}{3}$.

2.11 **a.** Let VL denote "very likely", SL denote "somewhat likely", U denote "unlikely" and O denote "other". The the simple events are $\{VL\}$, $\{SL\}$, $\{U\}$, $\{O\}$.
b. The simple events are not equally likely. We should assign probabilities based on the given proportions. That is,
$$P(VL) = 0.24,\ P(SL) = 0.24,\ P(U) = 0.4,\ P(O) = 0.12.$$
c. $P(\text{at least } SL) = P(VL) + P(SL) = 0.48$ since $\{VL\}$ and $\{SL\}$ are mutually exclusive.

2.12 **a.** $P(\text{needs glasses}) = 0.44 + 0.14 = 0.58$.
b. $P(\text{needs glasses and does not use them}) = 0.14$.
c. $P(\text{uses glasses}) = 0.44 + 0.02 = 0.46$.

2.13 **a.** We know that $P(S) = P(E_1 \cup E_2 \cup E_3 \cup E_4) = 1$ and, since the four events are pairwise mutually exclusive, $P(S) = P(E_1) + \ldots + P(E_4) = .01 + ? + .09 + .81$. Thus $? = P(E_2) = 1 - .91 = .09$.
b. $P(\text{at least one hit}) = P(E_1 \cup E_2 \cup E_3) = P(E_1) + P(E_2) + P(E_3) = .01 + .09 + .09 = .19$.

2.14 **a.** $\frac{1}{3}$
b. $\frac{1}{3} + \frac{1}{15} = \frac{6}{15}$
c. $\frac{1}{3} + \frac{1}{16} = \frac{19}{48}$
d. $1 - \left[\frac{6}{15} + \frac{19}{48}\right] = 1 - \left[\frac{191}{240}\right] = \frac{49}{240}$

2.15 Let B = assembly has bushing defect and S = shaft defect.
a. $P(B) = .06 + .02 = .08$.
b. $P(B \cup S) = .06 + .08 + .02 = .16$.
c. $P(B\overline{S} \cup S\overline{B}) = .06 + .08 = .14$.
d. $P(\overline{B \cup S}) = 1 - .16 = .84$.

2.16 **a.** The sample points are $\{HH, TH, HT, TT\}$.
b. Assuming the coins are balanced, each sample point is equally probable. The common probability is $\frac{1}{4}$.
c. $A = \{HT, TH\}$ $B = \{HT, TH, HH\}$

d. Using parts **b** and **c**,

$P(A) = \frac{1}{4} + \frac{1}{4} = \frac{1}{2}$ $\quad P(B) = \frac{3}{4}$ $\quad P(A \cap B) = P(A) = \frac{1}{2}$

$P(A \cup B) = P(B) = \frac{3}{4}$ $\quad P(\overline{A} \cup B) = P(S) = 1$

2.17 **a.** Notice that order is important since the pair (V_2, V_3) is different from the pair (V_3, V_2) due to the day on which the order is received. The sample points are:

(V_1, V_1) (V_1, V_2) (V_1, V_3) (V_2, V_1) (V_2, V_2)

(V_2, V_3) (V_3, V_1) (V_3, V_2) (V_3, V_3)

b. The points are equally probable with a common probability of $\frac{1}{9}$.

c. The following sets are of interest:

A: {same vendor gets both} $= \{(V_1, V_1), (V_2, V_2), (V_3, V_3)\}$

B: {V_2 gets at least one} $= \{(V_1, V_2), (V_2, V_1), (V_2, V_3), (V_3, V_2), (V_2, V_2)\}$

Then

$P(A) = \frac{3}{9} = \frac{1}{3}$ $\quad P(B) = \frac{5}{9}$ $\quad P(A \cup B) = \frac{7}{9}$ $\quad P(A \cap B) = P(V_2, V_2) = \frac{1}{9}$

2.18 **a.** Let N_1 and N_2 be the empty cans and let W_1 and W_2 be the cans filled with water. Then the pair $(N_1 W_2)$ will be the simple event that the expert chooses cans N_1 and W_2. The simple events in S are listed below.

$E_1 = (N_1 N_2)$ $\quad E_3 = (N_1 W_2)$ $\quad E_5 = (N_2 W_2)$

$E_2 = (N_1 W_1)$ $\quad E_4 = (N_2 W_1)$ $\quad E_6 = (W_1 W_2)$

b. If the rod is worthless the expert is merely guessing so each simple event is equally likely. Hence

$P(E_i) = \frac{1}{6}, \ i = 1, 2, \ldots, 6$

and the probability that the expert picks the two cans containing water is

$P(E_6) = \frac{1}{6}$.

2.19 **a.** $\{SS, SR, SL, RS, RR, RL, LS, LR, LL\}$

b. $P(SL) + P(RL) + P(LS) + P(LR) + P(LL) = \frac{5}{9}$

c. $P(SS) + P(SR) + P(SL) + P(RS) + P(LS) = \frac{5}{9}$

2.20 Denote the four candidates as A_1, A_2, A_3, and M, where M is the member of a minority group.

a. & b. Order is unimportant in the pair chosen. Hence there are six possible outcomes, each with probability $\frac{1}{6}$.

E_1: $(A_1 A_2)$ $\quad E_2$: $(A_1 A_3)$ $\quad E_3$: $(A_1 M)$

E_4: $(A_2 A_3)$ $\quad E_5$: $(A_2 M)$ $\quad E_6$: $(A_3 M)$

c. $P(\text{minority hired}) = P(E_3) + P(E_5) + P(E_6) = \frac{3}{6} = \frac{1}{2}$.

2.21 **a.** The experiment consists of randomly selecting two jurors from a group of two women and four men.

b. If we let $w1$ and $w2$ denote the women and $m1, m2, m3, m4$ denote the men in the group of potential jurors then the sample space is given by

$w1, w2$	$w2, m1$	$m1, m2$	$m2, m3$	$m3, m4$
$w1, m1$	$w2, m2$	$m1, m3$	$m2, m4$	
$w1, m2$	$w2, m3$	$m1, m4$		
$w1, m3$	$w2, m4$			
$w1, m4$				

c. $P(w1, w2) = 1/15$

2.22 **a.** Let $w1$ denote the first wine, $w2$ the second, and $w3$ the third. Then one sample point would be an ordered triple indicating the rank of each wine. For example, $(w1,\ w2,\ w3)$ would indicate that $w1$ is superior to $w2$and $w3$ while $w2$ is superior to just $w3$.

b. The sample space is given by all of the possible ordered triples. That is,

$(w1,\ w2,\ w3),\ (w1,\ w3,\ w2),\ (w2,\ w1,\ w3),$

$(w2,\ w3,\ w1),\ (w3,\ w1,\ w2),\ (w3,\ w2,\ w1)$

c. Suppose $w1$ is superior to $w2$ and $w3$. Then the probability that the "expert" ranks $w1$ as first or second is
$P[(w1, w2, w3) \text{ or } (w1, w3, w2) \text{ or } (w2, w1, w3) \text{ or } (w3, w1, w2)] = 4/6 = 2/3$.

2.23 a. Two systems are selected from six, two of which are defective. Denote the six systems as $G_1, G_2, G_3, G_4, D_1, D_2$, according to whether they are defective or nondefective. Each sample point will represent a particular pair of systems chosen for testing. The sample space, consisting of 15 pairs, is shown below.

G_1G_2 G_1G_3 G_1G_4 G_1D_1 G_1D_2 G_2G_3 G_2G_4 G_2D_1
G_2D_2 G_3G_4 G_3D_1 G_3D_2 G_4D_1 G_4D_2 D_1D_2

Note that the two systems are drawn simultaneously and that order is unimportant in identifying a sample point. Hence the pairs G_1G_2 and G_2G_1 are not considered to represent two different sample points. Then

$P(\text{at least one defective}) = \frac{9}{15} = \frac{3}{5}$ and $P(\text{both defective}) = P(D_1D_2) = \frac{1}{15}$.

b. If four of the six systems are defective, the 15 sample points are

G_1G_2 G_1D_1 G_1D_2 G_1D_3 G_1D_4 G_2D_1 G_2D_2 G_2D_3
G_2D_4 D_1D_2 D_1D_3 D_1D_4 D_2D_3 D_2D_4 D_3D_4

Then

$P(\text{at least one defective}) = \frac{14}{15}$ and $P(\text{both defective}) = \frac{6}{15} = \frac{2}{5}$.

2.24 a. Let the number 1 represent a customer seeking to purchase style 1 and the number 2 represent a customer seeking to purchase style 2. The sample space then consists of the following 16 four-tuples:

(1, 1, 1, 1) (1, 1, 1, 2) (1, 1, 2, 1) (1, 2, 1, 1) (2, 1, 1, 1) (1, 1, 2, 2)
(1, 2, 1, 2) (2, 1, 1, 2) (1, 2, 2, 1) (2, 1, 2, 1) (2, 2, 1, 1) (2, 2, 2, 1)
(2, 2, 1, 2) (2, 1, 2, 2) (1, 2, 2, 2) (2, 2, 2, 2)

b. Since the styles are in equal demand, each sample point is equally likely. The common probability is $\frac{1}{16}$.

c. $P(A) = P(1111) + P(2222) = \frac{2}{16} = \frac{1}{8}$.

2.25 a. Define the events: E: family's income exceeds $35,353
N: family's income does not exceed $35,353

Then the sample points are

E_1: $(EEEE)$ E_2: $(EEEN)$ E_3: $(EENE)$ E_4: $(ENEE)$
E_5: $(NEEE)$ E_6: $(EENN)$ E_7: $(ENEN)$ E_8: $(NEEN)$
E_9 : $(ENNE)$ E_{10} : $(NENE)$ E_{11} : $(NNEE)$ E_{12} : $(ENNN)$
E_{13} : $(NENN)$ E_{14} : $(NNEN)$ E_{15} : $(NNNE)$ E_{16} : $(NNNN)$

b. $A = \{E_1, E_2, \ldots, E_{11}\}$
$B = \{E_6, E_7, \ldots, E_{11}\}$
$C = \{E_2, E_3, E_4, E_5\}$

c. By the definition of median $P(E) = P(N) = 0.5$. Therefore, each simple event is equally likely and $P(E_i) = \frac{1}{16}$. Then

$P(A) = \frac{11}{16}$ $P(B) = \frac{3}{8}$ $P(C) = \frac{1}{4}$

2.26 a. Three patients enter the hospital and randomly choose stations 1, 2, or 3 for service. Let (1, 2, 1) represent the simple event that patients 1, 2, and 3 choose stations 1, 2, and 1, respectively. Then the sample space, S, will contain the following 27 three-tuples:

(1, 1, 1) (1, 1, 2) (1, 1, 3) (1, 2, 1) (1, 2, 2) (1, 2, 3) (1, 3, 1) (1, 3, 2)
(1, 3, 3) (2, 1, 1) (2, 1, 2) (2, 1, 3) (2, 2, 1) (2, 2, 2) (2, 2, 3) (2, 3, 1)
(2, 3, 2) (2, 3, 3) (3, 1, 1) (3, 1, 2) (3, 1, 3) (3, 2, 1) (3, 2, 2) (3, 2, 3)
(3, 3, 1) (3, 3, 2) (3, 3, 3)

b. The event A, "each station receives a patient," is represented by the six sample points shown below:

(1, 2, 3) (1, 3, 2) (2, 1, 3) (2, 3, 1) (3, 1, 2) (3, 2, 1)

c. Since the patients select stations at random, each sample point is equally likely and is assigned probability $\frac{1}{27}$. Thus, $P(A) = \frac{6}{27} = \frac{2}{9}$.

2.27 The mn rule is used. The flight from New York to California can be chosen in any one of 6 ways, the flight from California to Hawaii in any one of 7 ways. Thus, the total number of flights will be $(6)(7) = 42$.

2.28 Three operations (which may be arranged in any sequence) are required to assemble a piece of equipment. There is only one combination of these three operations, since we must pick all three; however, there are $P_3^3 = \frac{3!}{0!} = 6$ possible arrangements of the operations. Order is important and hence permutations are used.

2.29 **a.** There are $6! = 720$ possible itineraries.

b. If Denver is the first city visited then there are 5 ways for San Francisco to follow and 4! ways to arrange the other 4 cities. If Denver is the second then there are 4 ways for San Francisco to follow and 4! ways to arrange the other cities and so on. Thus, there are $4!(15)$ ways for Denver to precede San Francisco. Hence

$$P(\text{Denver before San Francisco}) = 360/720 = 0.5.$$

2.30 $(4)(3)(4)(5) = 240$ by an application of the mn rule.

2.31 **a.** Use the mn rule. Since the first die can result in one of 6 possible outcomes, and the second can result in one of 6 outcomes, a total of $(6)(6) = 36$ pairs can be formed, each containing one element from the first group and one element from the second.

b. Define the event A, "observe a sum of 7 on the two dice." This event will occur if any one of the following simple events occurs:

$$(1,6),\quad (2,5),\quad (3,4),\quad (4,3),\quad (5,2),\quad (6,1)$$

Then $P(A) = P(\text{observe a sum of } 7) = \frac{6}{36} = \frac{1}{6}$.

2.32 **a.** Use the mn rule. In order to include one of each style-engine-transmission combination, the dealer must stock $(5)(4)(2) = 40$ autos.

b. Further, if he would like to have each of these combinations in every one of the eight colors, he must stock $(40)(8) = 320$ cars.

2.33 Again, an extension of the mn rule is used, since we are concerned with the number of possible combinations from 7 different groups. Notice that the first digit of the 7-digit number can be chosen in 9 ways; that is, we can choose any of the 9 integers, 1, 2, 3, 4, 5, 6, 7, 8, 9. However, each of the remaining digits can be chosen from any one of the 10 integers, 0, 1, 2, 3, 4, 5, 6, 7, 8, 9. Thus, the total number of arrangements will be

$$9(10)(10)(10)(10)(10)(10) = 9(10)^6$$

2.34 Let $(7,3,2)$ represent the simple event that engineers 7, 3, and 2 fill jobs 1, 2, and 3, respectively. This is apparently a different event than $(3,7,2)$ since the same engineers are filling different positions. Hence the order of selection is important as well as the different combinations of engineers that may be selected. The number of ways in which the positions can be filled is

$$P_3^{10} = \frac{10!}{7!} = 10 \cdot 9 \cdot 8 = 720$$

2.35 $\binom{9}{3}\binom{6}{5} = \frac{9!}{3!\,5!\,1!} = 504$ ways

2.36 **a.** Number of ways taxi needing repair can be sent to airport C is

$$\binom{8}{5}\binom{5}{5} = \frac{8!}{3!\,5!\,0!} = 56;\ \frac{56}{504} = \frac{1}{9}$$

b. $3\binom{6}{2}\binom{4}{4} = 45$ so the probability that every airport receives one of the taxis requiring repair is $45/504$.

2.37 $\binom{17}{2\ 5\ 10} = 408{,}408$

2.38 $\binom{8}{5\,3} = \binom{8}{5} = 56 \qquad \binom{8}{3\,5} = \binom{8}{3} = 56$

2.39 **a.** $\binom{130}{2} = 8385$

b. $26 \cdot 26 = 676$ two-letter codes
$26 \cdot 26 \cdot 26 = 17{,}576$ three-letter codes
$= 18{,}252$ total major codes available

c. $8385 + 130 = 8515$ required

d. Yes.

2.40 Two numbers, 4 and 6, are possible for each of the three digits. An extension of the mn rule gives $2 \cdot 2 \cdot 2 = 8$ potential winning three-digit numbers.

2.41 There are $\binom{50}{3} = 19{,}600$ ways to choose the three winners. Since the choice is random, each of the 19,600 sample points is equally likely.

a. There are $\binom{4}{3} = 4$ ways for the organizers to win all of the prizes. Hence, the desired probability is $\frac{4}{19{,}600}$.

b. The organizers can win exactly 2 of the prizes if 1 of the other 46 people wins 1 prize. Using the mn rule, there are $\binom{4}{2}\binom{46}{1} = 276$ ways for this to occur. Hence, the desired probability is $\frac{276}{19{,}600}$.

c. $\binom{4}{1}\binom{46}{2} = 4140$. The probability is $\frac{4140}{19{,}600}$.

d. $\binom{46}{3} = 15{,}180$. The desired probability is $\frac{15{,}180}{19{,}600}$.

2.42 The mn rule is used. The temperature level can be chosen in any one of 3 ways, the pressure level in 3 ways, and the catalyst level in 2 ways. Thus, the total number of experiments will be $(3)(3)(2) = 18$.

2.43 **a.** Define a sample point as a triplet, the first element being the firm chosen for contract 1, the second element the firm chosen for contract 2, and so on. In choosing 3 of the 5 firms, order is important, and the number of sample points is $P_3^5 = \frac{5!}{2!} = 60$.

b. Assuming that F_3 is awarded a contract, the other two contracts can be awarded in $P_2^4 = \frac{4!}{2!} = 12$ ways. However, since F_3 can be awarded one of three contracts, the total number of ways to award F_3 a contract is $3P_2^4 = 36$ and $P(F_3$ is awarded a contract$) = \frac{36}{60} = .6$.

2.44 The total number of ways to choose 4 students from 8 is $\binom{8}{4} = \frac{8!}{4!\,4!} = 70$. Since the choice is random, each of the 70 sample points is equally likely, and it remains only to determine how many sample points result in exactly 2 of the 3 undergraduates and 2 of the 5 graduates. Using the mn rule, this number is $(C_2^3)(C_2^5) = 3(10) = 30$ and the desired probability is $\frac{30}{70} = \frac{3}{7}$.

2.45 **a.** $\binom{90}{10} = \frac{90!}{10!80!}$

b. $\binom{20}{4}\binom{70}{6}/\binom{90}{10}$=0.111

2.46 $P(\text{she can solve all 5 on the exam}) = \frac{\binom{6}{5}}{\binom{10}{5}} = 0.0238$

2.47 In this exercise two cards are drawn from a deck of 52. Define the event A, "the two cards are an ace and a face card." The order of draw is unimportant in considering the event A; hence there are a total of $\binom{52}{2} = \frac{52 \cdot 51}{2} = 1326$ different ways of drawing two cards from the deck. Now in order that the event A occurs, one of the cards must be an ace (of which there are 4) and one must be a face card (of which there are 12). The number of ways of choosing one of the 4 aces and one of the 12 face cards will be, using the rule, $\binom{4}{1}\binom{12}{1} = 4(12) = 48$. Hence the desired probability is $P(A) = \frac{48}{1326}$.

2.48 Refer to Theorem 2.3. The total number of ways to divide the 9 motors into 3 groups of size 3 is $\frac{9!}{3!\,3!\,3!}$. If both of the motors from a particular supplier are assigned to the first line, there are only 7 motors to be assigned, one to line 1 and 3 each to lines 2 and 3. This can be done in $\frac{7!}{1!\,3!\,3!}$ ways. Hence the probability of interest is

$$\frac{\left(\frac{7!}{1!\,3!\,3!}\right)}{\left(\frac{9!}{3!\,3!\,3!}\right)} = \frac{7!\,3!}{9!} = \frac{1}{12}.$$

2.49 There are

$$\binom{8}{5} = C_5^8 = \frac{8!}{5!\,3!} = \frac{8(7)(6)}{3(2)(1)} = 56$$

sample points in the experiment, only one of which results in choosing 5 women. Hence $P(\text{five women}) = \frac{1}{56}$.

2.50 $6!\left(\frac{1}{6}\right)^6 = \frac{5}{324}$

2.51 $5!\left(\frac{2}{6}\right)\left(\frac{1}{6}\right)^4 = \frac{5}{162}$

2.52 **a.** After assigning an ethnic group member to each type of job, there are 16 laborers remaining to assign to the remaining jobs. Let n_a be the number of ways that one ethnic group member can be assigned to each type of job. Then

$$n_a = \binom{4}{1\,1\,1\,1}\binom{16}{5\,3\,4\,4}$$

and hence the probability is $\frac{n_a}{N} = .1238$

b. Using the above logic, there are many ways that the assignments can be made such that no ethnic group member gets assigned to a type 4 job. The total number of ways is the sum

$$\binom{4}{1\,1\,2\,0}\binom{16}{5\,3\,3\,5} + \binom{4}{1\,2\,1\,0}\binom{16}{5\,2\,4\,5} + \binom{4}{2\,1\,1\,0}\binom{16}{4\,3\,4\,5} +$$
$$\binom{4}{2\,2\,0\,0}\binom{16}{4\,2\,5\,5} + \binom{4}{2\,0\,2\,0}\binom{16}{4\,4\,3\,5} + \text{etc.}$$

However, this long and tedious sum can be simplified by rethinking the problem. It doesn't matter how the ethnic group members are assigned to jobs of type 1, 2, and 3. Let n_a be the number of ways that no ethnic member gets assigned to a type 4 job. Then

$$n_a = \binom{4}{0}\binom{16}{5}$$

and the desired probability is

$$\frac{\binom{4}{0}\binom{16}{5}}{\binom{20}{5}} = .2817.$$

2.53 As shown in Example 2.13, $P(E_i) = \frac{1}{M^n} = \frac{1}{10^7}$.

a. Let A denote the event that all of the orders go to different distributors. This will happen if 7 of the 10 distributors are given orders. Thus, A contains

$$n_a = 10 \times 9 \times 8 \times 7 \times 6 \times 5 \times 4 = 604{,}800$$

sample points. The probability is, then,

$$P(A) = \frac{604800}{10^7} = .0605$$

b. Let A be the event that Distributor I gets exactly 2 orders and Distributor II gets exactly 3 orders. The 2 orders assigned to Distributor I can be chosen from the 7 in $\binom{7}{2} = 21$ ways. The 3 orders assigned to Distributor II can then be chosen from the remaining 5 in $\binom{5}{3} = 10$ ways. The final 2 orders can be assigned to any of the other 8 distributors in $8^2 = 64$ ways. Therefore, A contains

$$n_a = \binom{7}{2}\binom{5}{3}8^2 = 13{,}440$$

sample points. Then

$$P(A) = \frac{13{,}440}{10^7} = .001344$$

c. Let A be the event that Distributors I, II, and III get exactly 2, 3, and 1 order(s), respectively. Distributor I can get exactly 2 orders in $\binom{7}{2} = 21$ ways. Distributor II can get exactly 3 orders in $\binom{5}{3} = 10$ ways. Distributor III can get 1 order in $\binom{2}{1} = 2$ ways. The 1 remaining order can be assigned to any of the 7 remaining distributors in 7 ways. Thus, A contains

$$n_a = \binom{7}{2}\binom{5}{3}\binom{2}{1} \cdot 7 = 2940.$$

Therefore, $P(A) = \frac{2940}{10^7} = .00029$

2.54 **a.** For any $n \geq 1$, $\binom{n}{n} = \frac{n!}{n!\,(n-n)!} = 1$.
There is only one way to choose all of the items.

b. For any $n \geq 1$, $\binom{n}{0} = \frac{n!}{0!\,(n-0)!} = 1$.
There is only one way to choose none of the items.

c. For any $n \geq 1$, $\binom{n}{r} = \frac{n!}{r!\,(n-r)!} = \frac{n!}{(n-r)!\,(n-(n-r))!} = \binom{n}{n-r}$.
There are as many ways to choose r out of n objects as to choose $n - r$ out of n objects.

d. $(x+y)^n = \sum_{i=0}^{n} \binom{n}{i} x^{n-i} y^i$.
With $x = y = 1$, this is

$$(1+1)^n = 2^n = \sum_{i=0}^{n} \binom{n}{i} 1^{n-i} 1^i = \sum_{i=0}^{n} \binom{n}{i}$$

The total number of subsets of sizes 0 through n that can be selected is 2^n.

2.55

$$\begin{aligned}\binom{n}{k} + \binom{n}{k-1} &= \frac{n!}{k!\,(n-k)!} + \frac{n!}{(k-1)!\,(n-k+1)!} \\ &= \frac{n!\,(n-k+1)}{k!\,(n-k+1)!} + \frac{n!\,k}{k!\,(n-k+1)!} \\ &= \frac{n!\,(n+1)}{k!\,(n-k+1)!} = \frac{(n+1)!}{k!\,(n+1-k)!} \\ &= \binom{n+1}{k}\end{aligned}$$

2.56 From Theorem 2.3, the expansion of the multinomial term to the nth power is

$$(y_1 + y_2 + y_3 + \ldots + y_k)^n = \sum \binom{n}{n_1\, n_2\, n_3 \cdots n_k} y_1^{n_1} y_2^{n_2} y_3^{n_3} \cdots y_k^{n_k}.$$

Let $y_1 = y_2 = y_3 = \cdots = y_k = 1$. Then

$$k^n = \sum \binom{n}{n_1\, n_2\, n_3 \cdots n_k}.$$

Thus, the total number of distinct partitions equals k^n.

2.57 **a.** $P(A|B) = \frac{P(A \cap B)}{P(B)} = \frac{.1}{.3} = \frac{1}{3}$

b. $P(B|A) = \frac{P(A \cap B)}{P(A)} = \frac{.1}{.5} = \frac{1}{5}$

c. $P(A|A \cup B) = \frac{P(A)}{P(A \cup B)} = \frac{.5}{.5+.3-.1} = \frac{5}{7}$
Notice that we are using the additive law of probability, given in Section 2.8.

d. $P(A|A \cap B) = \frac{P(A \cap B)}{P(A \cap B)} = 1$

e. $P(A \cap B|A \cup B) = \frac{P(A \cap B)}{P(A \cup B)} = \frac{.1}{.5+.3-.1} = \frac{1}{7}$

2.58 The necessary probabilities can be obtained directly from the table. To determine whether or not A and M are independent, look at

$$P(A) = .6 \quad \text{and} \quad P(A|M) = \frac{P(A \cap M)}{P(M)} = \frac{.24}{.4} = .6.$$

Hence A and M are independent. For $\overline{A}$ and F, look at

$$P(\overline{A}) = .4 \quad \text{and} \quad P(\overline{A}|F) = \frac{P(\overline{A} \cap F)}{P(F)} = \frac{.24}{.60} = .4.$$

Hence $\overline{A}$ and F are independent.

2.59 Let r denote recessive white and R denote dominant red. Then

a. $P(\text{at least one } R) = P(\text{Red}) = 3/4$

b. $P(\text{at least one } r) = P(rr \text{ or } rR \text{ or } Rr) = 3/4$

c. $P(\text{one } r \mid \text{Red}) = \frac{0.5}{0.75} = 2/3$

2.60 Define the following events:

U: job is unsatisfactory and A: plumber A does the job

It is given that $P(A) = .4$, $P(U) = .1$, $P(A|U) = .5$.

a. The probability of interest is

$$P(U|A) = \frac{P(AU)}{P(A)} = \frac{P(A|U)\,P(U)}{P(A)} = \frac{.5(.1)}{.4} = .125.$$

b. $P(\overline{U}|A) = 1 - P(U|A) = 1 - .125 = .875.$

2.61 Many of the probabilities can be found directly from the table; others require calculation.

a. $P(A) = .40$

b. $P(B) = .37$

c. $P(AB) = .10$

d. $P(A \cup B) = .4 + .37 - .1 = .67$

e. $P(\overline{A}) = P(9 \text{ years or less of education}) = .60$

f. $P(\overline{A \cup B}) = 1 - P(A \cup B) = 1 - .67 = .33$

g. $P(\overline{AB}) = 1 - P(AB) = 1 - .10 = .90$

h. $P(A|B) = \frac{P(AB)}{P(B)} = \frac{.10}{.37} = .27$

i. $P(B|A) = \frac{P(AB)}{P(A)} = \frac{.10}{.40} = .25$

2.62
1. Assume $P(A|B) = P(A)$. Then
   $P(A \cap B) = P(A|B)P(B) = P(A)P(B)$
   $P(B|A) = \frac{P(B \cap A)}{P(A)} = \frac{P(A)P(B)}{P(A)} = P(B)$
2. Assume $P(B|A) = P(B)$.
   $P(A \cap B) = P(B|A)P(A) = P(B)P(A)$
   $P(A|B) = \frac{P(A \cap B)}{P(B)} = \frac{P(A)P(B)}{P(B)} = P(A)$
3. Assume $P(A \cap B) = P(A)P(B)$.
   The results follow from above.

2.63 We assume that A and B are mutually exclusive, and that $P(A) > 0$ and $P(B) > 0$. In order to show that A and B are not independent, we use the technique of proof by contradiction. Assume that A and B are independent. Then by definition, $P(A \cap B) = P(A)P(B)$. But since A and B are mutually exclusive, we know that $P(A \cap B) = 0$, which implies $P(A)P(B) = 0$. It is given that $P(A) > 0$ and $P(B) > 0$, which implies $P(A)P(B) > 0$, and we have a contradiction. Hence A and B must not be independent.

2.64 Since $A \subset B$, $P(A \cap B) = P(A)$ and $P(B|A) = \frac{P(A \cap B)}{P(A)} = \frac{P(A)}{P(A)} = 1$. Therefore, A and B are not independent unless $P(B) = 1$.

2.65 Given that $P(A) < P(A|B)$

$P(A) < \frac{P(AB)}{P(B)}$	by definition of conditional probability
$P(A) < \frac{P(A)P(B\|A)}{P(B)}$	multiplicative law of probability
$P(A)P(B) < P(A)P(B\|A)$	given $P(B) > 0$
$P(B) < P(B\|A)$	given $P(A) > 0$

2.66 $P(A_1 \cup A_2 \cup A_3) = P(A_1) + P(A_2) + P(A_3) - P(A_1A_2) - P(A_1A_3) - P(A_2A_3) + P(A_1A_2A_3)$

$= P(A_1) + P(A_2) + P(A_3) - P(A_1A_2) - P(A_1A_2) - 0 + 0$

since $P(A_1A_2) = P(A_1A_3)$ and $P(A_2A_3) = 0$ imply $P(A_1A_2A_3) = 0$

$= P(A_1) + P(A_2) + P(A_3) - 2P(A_1A_2)$

2.67 In order to show that A and $\overline{B}$ are independent, it is necessary to show that $P(A \cap \overline{B}) = P(A)P(\overline{B})$. The following relationships are known:

(1) $B \cup \overline{B} = S$, where B and $\overline{B}$ are complementary and hence mutually exclusive events.

(2) $A \cap S = A$.

(3) $P(\overline{B}) = 1 - P(B)$.

Hence we can write

$$P(A) = P(A \cap S) = P\left[A \cap (B \cup \overline{B})\right].$$

Now, using the distributive law given at the end of Section 2.3,

$$P\left[A \cap (B \cup \overline{B})\right] = P\left[(A \cap B) \cup (A \cap \overline{B})\right].$$

Using the additive law (from Section 2.8) and noting that the events $A \cap B$ and $A \cap \overline{B}$ are mutually exclusive, we have

$$P\left[(A \cap B) \cup (A \cap \overline{B})\right] = P(A \cap B) + P(A \cap \overline{B}).$$

Since A and B are independent, $P(A \cap B) = P(A)P(B)$, and the entire equality implies that

$$P(A) = P(A)P(B) + P(A \cap \overline{B}).$$

So,

$$P(A \cap \overline{B}) = P(A) - P(A)P(B) = P(A)[1 - P(B)] = P(A)P(\overline{B}).$$

Hence A and $\overline{B}$ are independent.

In order to show that $\overline{A}$ and $\overline{B}$ are independent, the same reasoning is used.

$$P(\overline{B}) = P(\overline{B} \cap A) + P(\overline{B} \cap \overline{A}) = P(\overline{B})P(A) + P(\overline{B} \cap \overline{A})$$

since A and $\overline{B}$ are independent. Hence

$$P(\overline{A} \cap \overline{B}) = P(\overline{B})[1 - P(A)] = P(\overline{B})P(\overline{A})$$

and the result is proven.

2.68

a. The three tests are independent. Thus, the probability in question is $(.05)^3 = .000125$.

b. The probability of at least one mistake equals 1 minus the probability of no mistakes. The probability of no mistakes is $(.95)^3$. Thus, the probability of at least one mistake is $1 - (.95)^3 = 1 - .857 = .143$.

2.69

Let W demote the event that she wins and assume that she begins with her right hand. Note that there are exactly 3 ways to win once the choice of hand has been made. If H denotes a hit and M denotes a miss these are HHH, HHM, and MHH. Then

$$\begin{aligned} P(W|\text{R}) &= P(HHH|\text{R}) + P(HHM|\text{R}) + P(MHH|\text{R}) \\ &= (.7)(.4)(.7) + (.7)(.4)(.3) + (.3)(.4)(.7) = 0.364 \end{aligned}$$

2.70

Define the events

A: device A detects smoke

B: device B detects smoke

a. $P(A \cup B) = P(A) + P(B) - P(A \cap B) = .95 + .90 - .88 = .97$

b. $P(\text{smoke undetected}) = 1 - P(A \cup B) = 1 - .97 = .03$

2.71

a. $P(A \cup B) = P(A) + P(B) - P(A \cap B)$. Hence $P(A \cap B) = P(A) + P(B) - P(A \cup B) = .2 + .3 - .4 = .1$.

b. From Exercise 2.3, $P(\overline{A} \cup \overline{B}) = P(\overline{A \cap B}) = 1 - P(A \cap B) = 1 - .1 = .9$.

c. $P(\overline{A} \cap \overline{B}) = P(\overline{A \cup B}) = 1 - P(A \cup B) = 1 - .4 = .6$.

d. $P(\overline{A}|B) = \frac{P(\overline{A} \cap B)}{P(B)}$. To find $P(\overline{A} \cap B)$, note that $P(B) = P(A \cap B) + P(\overline{A} \cap B)$, so that $P(\overline{A} \cap B) = P(B) - P(A \cap B) = .3 - .1 = .2$. Thus, $P(\overline{A}|B) = \frac{.2}{.3} = \frac{2}{3}$.

2.72

This is similar to Exercise 2.59. Since A and B are independent

a. $P(A \cup B) = P(A) + P(B) - P(A)P(B) = .5 + .2 - .1 = .6$.

b. $P(\overline{A} \cap \overline{B}) = 1 - P(A \cup B) = 1 - .6 = .4$

c. $P(\overline{A} \cup \overline{B}) = 1 - P(A \cap B) = 1 - .1 = .9$

2.73

a. $P(\text{current flows}) = 1 - P(\text{current is not flowing})$
$= 1 - P(\text{all three relays are open}) = 1 - (.1)^3 = .999$

b. Let A be the event that current flows and B be the event that the relay 1 closed properly

$$P(B|A) = \frac{P(BA)}{P(A)} = \frac{P(B)}{P(A)}; \text{ since } B \subset A$$
$$= \frac{.9}{.999} = .9009$$

2.74

Let A be the event that relay 1 closes properly and B be the event that relay 2 closes properly.

$P(\text{current flows in system arranged in series}) = P(\text{both relays closed})$
$= P(AB) = P(A)P(B) = (.9)^2 = .81$

$P(\text{current flows in system arranged in parallel}) = P(A \cup B)$
$= P(A) + P(B) - P(AB) = .9 + .9 - .81 = .99$

2.75 Given $P(\overline{A \cup B}) = a$; $P(B) = b$; A and B are independent.
Thus, $P(A \cup B) = 1 - a$; $P(AB) = P(A)P(B) = P(A)[b]$.
Consider $P(A \cup B) = P(A) + P(B) - P(AB)$

$$\begin{aligned} 1 - a &= P(A) + b - P(A)(b) \\ \Rightarrow 1 - a &= b + P(A)[1 - b] \\ \Rightarrow 1 - a - b &= P(A)[1 - b] \\ \Rightarrow \frac{1 - b - a}{1 - b} &= P(A) \end{aligned}$$

2.76 The proof follows by using the definition of conditional probability and the additive law for the union of two events.

$$\begin{aligned} P(A \cup B|C) &= \frac{P[(A \cup B) \cap C]}{P(C)} = \frac{P[(A \cap C) \cup (B \cap C)]}{P(C)} \\ &= \frac{P(A \cap C) + P(B \cap C) - P(A \cap B \cap C)}{P(C)} \\ &= \frac{P(A \cap C)}{P(C)} + \frac{P(B \cap C)}{P(C)} - \frac{P(A \cap B \cap C)}{P(C)} \\ &= P(A|C) + P(B|C) - P[(A \cap B)|C] \end{aligned}$$

2.77 Let A be the event that a defective item gets past first inspector and B be the event that a defective item gets past second inspector. Then

$P(A) = .1 \qquad P(B|A) = .5$

The event that the defective item gets past both inspectors is $A \cap B$.

$P(A \cap B) = P(B|A)P(A) = .05$

2.78 Define the following events:

I: disease I is contracted and II: disease II is contracted.

Then $P(I) = .1$, $P(II) = .15$, and $P(I \cap II) = .03$.

a. $P(I \cup II) = P(I) + P(II) - P(I \cap II) = .1 + .15 - .03 = .22$.

b. $P(I \cap II | I \cup II) = \frac{P(I \cap II)}{P(I \cup II)} = \frac{.03}{.22} = \frac{3}{22}$.

2.79 **a.** We assume that the Connecticut and Pennsylvania lotteries are independent. Thus,
$P(\text{666 in Connecticut}|\text{666 in Pennsylvania}) = P(\text{666 in Connecticut})$
$= \frac{1}{10^3} = .001$

b. $P(\text{666 in Connecticut} \cap \text{666 in Pennsylvania})$
$= P(\text{666 in Connecticut})P(\text{666 in Pennsylvania})$
$= (0.001)(1/8) = .000125$

2.80 $P(AB) = 1 - P(\overline{AB})$
$= 1 - P(\overline{A} \cup \overline{B})$ DeMorgan's Law
Aside: $P(\overline{A} \cup \overline{B}) \leq P(\overline{A}) + P(\overline{B})$ from addition law of probability
therefore, $P(AB) \geq 1 - P(\overline{A}) - P(\overline{B})$

2.81 $P(\text{landing safely on both jumps}) \geq 1 - .05 - .05 \geq .90$

2.82 Observe that $P(A) = P(B)$ implies $P(\overline{A}) = P(\overline{B})$. Use exercise 2.80 and set $P(AB) \geq 1 - 2\,P(\overline{A}) \geq 0.98$ which implies that we require $P(A) \geq 0.99$.

2.83 Recall that $P(A|C) = \frac{P(A \cap C)}{P(C)}$ and $P(B|C) = \frac{P(B \cap C)}{P(C)}$. One way to see the claim is to suppose that $C \subset B$ but $A \cap C = \{\}$ and $P(A) > P(B)$. Then, trivially,

$$0 = P(A|C) < P(B|C) = 1$$

even though $P(A) > P(B)$. As a specific example, consider flipping a coin twice. Then the above relationships are satisfied if we define A to be the event that we observe at least one tail, B to be the event that we observe two heads or two tails, and C to be the event that we observe exactly two heads.

2.84 Exercise 2.80 says that $P(U \cap V) \geq 1 - P(\overline{U}) - P(\overline{V})$. Let $U = A \cap B$ and $V = C$. Then $P(A \cap B \cap C) \geq 1 - P(\overline{A \cap B}) - P(\overline{C})$. Note that $P(\overline{A \cap B}) = 1 - P(A \cap B)$ and apply 2.80 again to obtain $1 - P(\overline{A \cap B}) \geq 1 - P(\overline{A}) - P(\overline{B})$ and therefore $P(A \cap B \cap C) \geq 1 - P(\overline{A}) - P(\overline{B}) - P(\overline{C})$.

2.85 This is similar to 2.82. Apply 2.84. Set $0.95 \leq 1 - P(\overline{A}) - P(\overline{B}) - P(\overline{C}) \leq P(ABC)$. Since $P(A) = P(B) = P(C)$ we have $P(\overline{A}) = P(\overline{B}) = P(\overline{C})$. So that $0.95 \leq 1 - 3P(\overline{A})$ or $P(A) \geq 0.9833$ since $P(A) = 1 - P(\overline{A})$.

2.86 Define the following events:

I: item comes from line I $\quad$ II: item comes from line II $\quad$ D: item is defective

Using the approach given in the solution to Exercise 2.57, write

$$P(\overline{D}) = P\left[\overline{D} \cap (I \cup II)\right] = P(\overline{D} \cap I) + P(\overline{D} \cap II) = P(\overline{D}|I)P(I) + P(\overline{D}|II)P(II)$$
$$= .92(.4) + .90(.6) = .908.$$

2.87 Define the following events:

A: buyer sees magazine ad

B: buyer sees corresponding ad on television

C: buyer purchases the product

The following probabilities are known:

$P(A) = .02 \qquad P(B) = .20 \qquad P(A \cap B) = .01.$

Now $P(A \cup B) = P(A) + P(B) - P(A \cap B) = .02 + .20 - .01 = .21$. Further,

$P(\overline{A} \cap \overline{B}) = 1 - P(A \cup B) = 1 - .21 = .79$

where the event $\overline{A} \cap \overline{B}$ is the event that the buyer does not see the ad either on television or in a magazine. Finally, it is given that $P(C|A \cup B) = \frac{1}{3}$ and $P(C|\overline{A} \cap \overline{B}) = \frac{1}{10}$. It is necessary to find $P(C)$.

$$\begin{aligned}
P(C) &= P(\text{buyer purchases the product}) \\
&= P(\text{buyer sees ad and buys}) + P(\text{buyer doesn't see ad and buys}) \\
&= P\left[C \cap (A \cup B)\right] + P\left[C \cap (\overline{A} \cap \overline{B})\right] \\
&= P(C|A \cup B)P(A \cup B) + P(C|\overline{A} \cap \overline{B})P(\overline{A} \cap \overline{B}) \\
&= \left(\tfrac{1}{3}\right)(.21) + \left(\tfrac{1}{10}\right)(.79) = .07 + .079 = .149.
\end{aligned}$$

2.88 Define F to be the event that a radar set fails to detect a plane. The events of interest are intersections of independent events.

a. $P(\text{aircraft undetected}) = P(\text{all three sets fail to detect}) = P(F)P(F)P(F) = (.02)^3$.

b. $P(\text{all three sets detect it}) = P(\overline{F})P(\overline{F})P(\overline{F}) = (.98)^3$.

2.89 Let F be the event that a radar set fails to detect a plane. Consider an independent sequence of F's and $\overline{F}$'s corresponding to the sets "detection" or "nondetection" of aircraft. The event of interest is $\overline{F}\,\overline{F}\,\overline{F}F$ and has probability

$$P(\overline{F}\,\overline{F}\,\overline{F}F) = \left[P(\overline{F})\right]^3 P(F) = (.98)^3(.02).$$

2.90 Let $A\overline{A} = \{\text{positive reading for truthful person, negative reading for liar}\}$.

Then the sample space is

$$S = \left\{AA, A\overline{A}, \overline{A}A, \overline{A}\,\overline{A}\right\}.$$
$$\begin{aligned}
P(AA) &= .10 \times .95 = .095 \\
P(A\overline{A}) &= .10 \times .05 = .005 \\
P(\overline{A}A) &= \ .9 \times .95 = .855 \\
P(\overline{A}\,\overline{A}) &= \ .9 \times .05 = .045
\end{aligned}$$

a. $P(AA) = .095$

b. $P(\overline{A}A) = .855$

c. $P(A\overline{A}) = .005$

d. $1 - P(\overline{A}\,\overline{A}) = .955$.

2.91 Define W to be the event that the team wins. Then

$$P(\text{four wins}) = P(WWWW) = P(W)P(W)P(W)P(W) = (.75)^4.$$

2.92 Using the complement of the event of interest,

$$P(\text{system works}) = 1 - P(\text{system fails}) = 1 - P(I \text{ and } II \text{ and } III \text{ fail}) = 1 - (.01)^3.$$

2.93 Define the following events:

R: driver is rejected; that is, he chooses inspection team 2
$\overline{R}$: driver passes inspection; that is, he chooses inspection team 1

Then $P(R) = P(\overline{R}) = \frac{1}{2}$.

Using a quadruplet, each component of which represents the fate of one of the four drivers, we are interested in the events

A: three of four drivers are rejected and B: all four drivers pass

The first event is a union of four mutually exclusive subevents. That is,

$$P(A) = P(RRR\overline{R}) + P(RR\overline{R}R) + P(R\overline{R}RR) + P(\overline{R}RRR)$$
$$= 4[P(R)]^3P(\overline{R}) = 4\left(\tfrac{1}{2}\right)^4 = \tfrac{4}{16} = \tfrac{1}{4}.$$

The probability that all four will pass is

$$P(B) = P(\overline{R}\,\overline{R}\,\overline{R}\,\overline{R}) = \left[P(\overline{R})\right]^4 = \tfrac{1}{16}.$$

2.94 If the victim is to be saved, a proper donor must be found within eight minutes, allowing two minutes for transfer of blood. Thus four people can be "typed" and the patient will be saved if a proper donor is found on the first, second, third, or fourth try. Note that $P(A) = .4$, where A is the event that an a type A, Rh-positive donor is found. Then

$$P(\text{saving the patient}) = P(A \text{ on 1st trial, or first } A \text{ on 2nd, or first } A \text{ on 3rd or first } A \text{ on 4th})$$
$$= P(A \cup \overline{A}A \cup \overline{A}\,\overline{A}A \cup \overline{A}\,\overline{A}\,\overline{A}A),$$

where, for example, $\overline{A}\,\overline{A}A$ denotes "not A" on the first and second trials and A on the third.

$$P(\text{saving patient}) = P(A) + P(\overline{A}A) + P(\overline{A}\,\overline{A}A) + P(\overline{A}\,\overline{A}\,\overline{A}A)$$
$$= P(A) + P(\overline{A})P(A) + \left[P(\overline{A})\right]^2P(A) + \left[P(\overline{A})\right]^3P(A)$$
$$= .4 + (.6)(.4) + (.6)^2(.4) + (.6)^3(.4) = .8704$$

2.95 **a.** Define the events

A: Obtain a sum of 3
B: Do not obtain a sum of 3 or 7.

In Example 2.5 it was shown that there are 36 sample points corresponding to the numbers on the upper faces of two dice. Of these pairs, two sum to 3 and six sum to 7, leaving 28 that do not sum to either 3 or 7. Hence,

$P(A) = \frac{2}{36}$ and $P(B) = \frac{28}{36}$.

Now, obtaining a sum of 3 before obtaining a sum of 7 can happen on the first toss as

A,

on the second toss as

BA,

on the third toss as

BBA,

and, in general, on the ith toss as

$$\underbrace{BBB\cdots B}_{i-1}A.$$

Note that because the tosses are independent the corresponding probabilities are

1st toss: $P(A) = \frac{2}{36}$
2nd toss: $P(B)P(A) = \left(\frac{28}{36}\right)\left(\frac{2}{36}\right)$

3rd toss: $P(B)^2P(A) = \left(\frac{28}{36}\right)^2\left(\frac{2}{36}\right)$

$\vdots$

ith toss: $P(B)^{i-1}P(A) = \left(\frac{28}{36}\right)^{i-1}\left(\frac{2}{36}\right)$.

Then, the overall probability of obtaining a 3 before a 7 is

$$\sum_{i=1}^{\infty} P(B)^{i-1}\,P(A) = \frac{P(A)}{P(B)}\sum_{i=1}^{\infty} P(B)^i = \frac{P(A)}{P(B)}\frac{P(B)}{1-P(B)} = \frac{P(A)}{1-P(B)} = \frac{2/36}{1-28/36} = \frac{1}{4}.$$

The second equality is from the sum of a geometric series as shown in Appendix A1.11.

b. Similar to part **a.** except replace A and B with

C: obtain a sum of 4

D: do not obtain a sum of 4 or 7

where $P(C) = \frac{3}{36}$ and $P(D) = \frac{27}{36}$.

Then the overall probability of obtaining a 4 before a 7 is

$\frac{P(C)}{1-P(D)} = \frac{3/36}{1-27/36} = \frac{1}{3}$.

2.96

Let D denote defective and G denote good.

a. The event of interest is the union of the following three mutually exclusive events:

$$DGGD \qquad GDGD \qquad GGDD$$

Note that the last defective must be found on the fourth test, but the other may be found on test 1, 2, or 3.

Consider the first event. A defective must be drawn on the first test. This occurs with probability $\frac{2}{6}$. A nondefective must be drawn on each of the next two tests; the probabilities are $\frac{4}{5}$ and $\frac{3}{4}$, respectively. A defective must be drawn on the fourth test. This happens with probability $\frac{1}{3}$. The probability of this intersection is

$$\left(\frac{2}{6}\right)\left(\frac{4}{5}\right)\left(\frac{3}{4}\right)\left(\frac{1}{3}\right).$$

The probabilities associated with the other two events are identical to that of the first. Hence, applying the additive law of probability, we obtain the desired probability:

$$3 \times \frac{2}{6} \times \frac{4}{5} \times \frac{3}{4} \times \frac{1}{3} = 3 \times \frac{4\cdot3\cdot2\cdot1}{6\cdot5\cdot4\cdot3} = \frac{1}{5}.$$

b. We must locate the second defective refrigerator on the second, third, or fourth test. Call this event A. Consider the following events:

A_1: the second defective is found on the second test

A_2: the second defective is found on the third test

A_3: the second defective is found on the fourth test

Then $A = A_1 \cup A_2 \cup A_3$ is the union of three mutually exclusive events and

$$P(A) = P(A_1) + P(A_2) + P(A_3)$$

$P(A_3) = \frac{1}{5}$ was found in part **a.** The event $A_1 = DD$ is the only way to obtain the second defective on the second test, and

$$P(A_1) = P(DD) = \frac{2}{6} \times \frac{1}{5} = \frac{1}{15}$$

A_2 is the union of the two mutually exclusive events, DGD or GDD, which occur with equal probabilities

$$P(DGD) = P(GDD) = \frac{4}{6} \times \frac{2}{5} \times \frac{1}{4} = \frac{1}{15} \qquad \text{so} \qquad P(A_2) = \frac{2}{15}$$

Thus,

$$P(A) = P(A_1) + P(A_2) + P(A_3) = \frac{1}{15} + \frac{2}{15} + \frac{1}{5} = \frac{2}{5}$$

c. One of the two defectives has been found in the first two tests. Thus there are three nondefectives and one defective remaining. The other defective must be found on the third or fourth test. Call this event B, and express it as the union of two mutually exclusive events defined below:

B_1: the second defective is found on the third test

B_2: the second defective is found on the fourth test

Now the probability of event B_1 is $P(B_1) = \frac{1}{4}$. Also, the probability of event B_2, which is the intersection of a nondefective on the third draw and a defective on the fourth, is $P(B_2) = \frac{3}{4} \times \frac{1}{3} = \frac{1}{4}$. The event of interest is

$$P(B) = P(B_1 \cup B_2) = P(B_1) + P(B_2) = \frac{1}{4} + \frac{1}{4} = \frac{1}{2}.$$

2.97 **a.** $\frac{1}{n}$

b. $\left(\frac{n-1}{n}\right) \times \left(\frac{1}{n-1}\right) = \frac{1}{n}$ second try

$\left(\frac{n-1}{n}\right) \times \left(\frac{n-2}{n-1}\right) \times \left(\frac{1}{n-2}\right) = \frac{1}{n}$ third try

c. $P[\text{gain access}] = P(\text{first try}) + P(\text{second try}) + P(\text{third try})$
$= \frac{1}{7} + \frac{1}{7} + \frac{1}{7} = \frac{3}{7}$

2.98 Define these events:

D: voter is Democrat R: voter is Republican F: voter favors a given issue

Using Bayes's rule,

$$P(D|F) = \frac{P(F|D)P(D)}{P(F|D)P(D)+P(F|R)P(R)} = \frac{(.7)(.6)}{(.7)(.6)+(.3)(.4)} = \frac{.42}{.54} = \frac{7}{9}$$

2.99 Define these events:

D: person has the disease and H: test says person has the disease

Then $P(H|D) = .9$; $P(\overline{H}|\overline{D}) = .9$; $P(D) = .01$; $P(\overline{D}) = .99$. Using Bayes's Rule,

$$P(D|H) = \frac{P(H|D)P(D)}{P(H|D)P(D)+P(H|\overline{D})P(\overline{D})} = \frac{(.9)(.01)}{(.9)(.01)+(.1)(.99)} = \frac{.009}{.108} = \frac{1}{12}.$$

2.100 Let R denote the event that the specimen turns red and N denote the event that the specimen contains nitrates.

a. Use the law of total probability to obtain

$P(R) = P(R|N)P(N) + P(R|\overline{N})P(\overline{N}) = (0.95)(0.3) + (0.10)(0.70) = 0.355$

b. Use Bayes's rule and part **a** to obtain

$$P(N|R) = \frac{P(N\cap R)}{P(R)} = \frac{P(R|N)P(N)}{P(R)} = \frac{(0.95)(0.3)}{0.355} = 0.803$$

2.101 Use Bayes's rule to obtain

$$\begin{aligned} P(I_1|H) &= \frac{P(H|I_1)P(I_1)}{P(H|I_1)P(I_1) + P(H|I_2)P(I_2) + P(H|I_3)P(I_3)} \\ &= \frac{0.90(0.01)}{0.90(0.01) + 0.95(0.005) + 0.75(0.02)} \\ &= 0.313 \end{aligned}$$

2.102 **a.** Let $P(A|B) = P(A|\overline{B}) = p$. By the law of total probability,

$$\begin{aligned} P(A) &= P(B)P(A|B) + P(\overline{B})P(A|\overline{B}) \\ &= P(B)p + P(\overline{B})p \\ &= p\left(P(B) + P(\overline{B})\right) \\ &= p. \end{aligned}$$

Hence $P(A|B) = p = P(A)$ so that A and B are independent.

b.By the law of total probability,

$$\begin{aligned} P(A) &= P(A|C)P(C) + P(A|\overline{C})P(\overline{C}) \\ &> P(B \mid C)P(C) + P(B|\overline{C})P(\overline{C}) \\ &= P(B) \end{aligned}$$

again, by the law of total probability.

2.103 Define these events:

P: positive response M: respondent was male F: respondent was female

Then $P(P|F) = .7$; $P(P|M) = .4$; $P(M) = \frac{1}{4}$. Using Bayes's Rule,

$$P(M|\overline{P}) = \frac{P(\overline{P}|M)P(M)}{P(\overline{P}|M)P(M)+P(\overline{P}|F)P(F)} = \frac{(.6)\left(\frac{1}{4}\right)}{(.6)\left(\frac{1}{4}\right)+(.3)\left(\frac{3}{4}\right)} = \frac{.6}{1.5} = .4$$

2.104 Define these events:

C: contract lung cancer S: worked in a shipyard

Then $P(S|C) = .22$ and $P(S|\overline{C}) = .14$. Also, $P(C) = .0004$. Using Bayes's Rule,

$$P(C|S) = \frac{P(S|C)P(C)}{P(S|C)P(C)+P(S|\overline{C})P(\overline{C})} = \frac{(.22)(.0004)}{(.22)(.0004)+(.14)(.9996)} = \frac{.000088}{.140032}$$
$$= .0006$$

2.105 Let$D$ denote death due to lung cancer ad S denote being a smoker. The law of total probability yields
$P(D) = P(D|S)P(S) + P(D|\overline{S})P(\overline{S}) = 10P(D|\overline{S})(0.2) + P(D|\overline{S})(0.8) = 2.8P(D|\overline{S})$
which implies $P(D|\overline{S}) = \frac{0.006}{2.8}$ since $P(D) = 0.006$. Hence $P(D|S) = 10\frac{0.006}{2.8} = 0.021$

2.106 Let W denote the event that the first ball is white and B denote the event that the second ball is black. Then

$$P(W|B) = \frac{P(B|W)P(W)}{P(B)} = \frac{P(B|W)P(W)}{P(B|W)P(W)+P(B|\overline{W})P(\overline{W})}$$

But $P(B|W)P(W) = \frac{b}{w+b+n}\frac{w}{w+b}$ and

$$P(B|W)P(W) + P(B|\overline{W})P(\overline{W}) = \frac{b}{w+n+b}\frac{w}{w+b} + \frac{b+n}{w+b+n}\frac{b}{w+b}.$$

It follows immediately that

$$P(W|B) = \frac{w}{w+b+n}.$$

2.107 The additive law of probability implies that $P(A \triangle B) = P(A \cap \overline{B}) + P(\overline{A} \cap B)$. Now we can write A and B as a union of disjoint events, i.e.

$$A = (A \cap \overline{B}) \cup (A \cap B)$$

and

$$B = (\overline{A} \cap B) \cup (A \cap B)$$

which implies that

$$P(A \cap \overline{B}) = P(A) - P(A \cap B) \text{ and } P(\overline{A} \cap B) = P(B) - P(A \cap B)$$

so that
$P(A \triangle B) = P(A) - P(A \cap B) + P(B) - P(A \cap B) = P(A) + P(B) - 2P(A \cap B).$

2.108 Let F_i be the event that the plane is found in region i when it is searched and N_i the event that it is not found in region i. Also, let $R_i$be the event that the plane is in region i. Then

$$P(F_i|R_i) = 1 - \alpha_i$$

and further $P(R_i) = \frac{1}{3}$ for all i.

a.
$$\begin{aligned} P(R_1|N_1) &= \frac{P(N_1|R_1)P(R_1)}{P(N_1)} \\ &= \frac{P(N_1|R_1)P(R_1)}{P(N_1|R_1)P(R_1) + P(N_1|R_2)P(R_2) + P(N_1|R_3)P(R_3)} \\ &= \frac{\alpha_1\frac{1}{3}}{\alpha_1\frac{1}{3} + \frac{1}{3} + \frac{1}{3}} = \frac{\alpha_1}{\alpha_1 + 2} \end{aligned}$$

b. Similarly, $P(R_2|N_1) = \frac{\frac{1}{3}}{\alpha_1\frac{1}{3}+2\frac{1}{3}} = \frac{1}{\alpha_1+2}$

c. $P(R_3|N_1) = \frac{\frac{1}{3}}{\alpha_1\frac{1}{3}+2\frac{1}{3}} = \frac{1}{\alpha_1+2}$

2.109 Define these events:
D: item is defective C: item goes through complete inspection
We know $P(D) = .1$, $P(C|D) = .6$, and $P(C|\overline{D}) = .2$. Thus,
$P(D|C) = \frac{P(C|D)P(D)}{P(C|D)P(D)+P(C|\overline{D})P(\overline{D})} = \frac{(.6)(.1)}{(.6)(.1)+(.2)(.9)} = \frac{.06}{.24} = .25$

2.110 Let A = athlete disqualified previously and B = athlete disqualified next term. We know

$$P(B|\overline{A}) = 0.15$$
$$P(B|A) = 0.5$$

and

$$P(A) = 0.30.$$

We want to find $P(B)$. By the law of total probability,

$$P(B) = P(B)P(A)P(B|A) + P(\overline{A})P(B|\overline{A})$$
$$= (.30)(.5) + (.70)(.15)$$
$$= .255$$

2.111 Define these events:

G: student guesses $\qquad$ C: student correctly answers question

We know $P(G) = .2$, $P(C|\overline{G}) = 1$, and $P(C|G) = .25$. Thus,

$$P(\overline{G}|C) = \frac{P(C|\overline{G})P(\overline{G})}{P(C|\overline{G})P(\overline{G})+P(C|GP(G)} = \frac{(1)(.8)}{(1)(.8)+(.25)(.2)} = \frac{.8}{.85} = .9412$$

2.112 Define F as "failure to learn." Then $P(F|A) = .2$; $P(F|B) = .10$; $P(B) = .3$; $P(A) = .7$. Thus,

$$P(A|F) = \frac{P(F|A)P(A)}{P(F|A)P(A)+P(F|B)P(B)} = \frac{(.2)(.7)}{(.2)(.7)+(.1)(.3)} = \frac{.14}{.17} = \frac{14}{17}$$

2.113 Let M = major airline

P = private plane

C = commercial plane

B = travel for business

$P(M) = .6$; $P(P) = .3$; $P(C) = .1$

$P(B|M) = .5$; $P(B|P) = .6$; $P(B|C) = .9$

a. $P(B) = P(MB) + P(PB) + P(CB) = P(M)P(B|M) + P(P)P(B|P) + P(C)P(B|C)$
$= (.6)(.5) + (.3)(.6) + (.1)(.9) = .57$

b. $P(PB) = P(P)P(B|P) = (.3)(.6) = .18$

c. $P(P|B) = \frac{P(PB)}{P(B)} = \frac{.18}{.57} = .3158$

d. $P(B|C) = .9$

2.114 Let A = woman's name selected from list 1 and B = woman's name selected from list 2. Then

$$P(A) = \tfrac{5}{7},\ P(\overline{B}|A) = \tfrac{6}{9},\ P(\overline{B}|\overline{A}) = \tfrac{7}{9}.$$

Now

$$P(A|\overline{B}) = \frac{P(\overline{B}|A)P(A)}{P(\overline{B}|A)P(A)+P(\overline{B}|\overline{A})P(\overline{A})} = \frac{\left(\frac{6}{9}\right)\left(\frac{5}{7}\right)}{\left(\frac{6}{9}\right)\left(\frac{5}{7}\right)+\left(\frac{7}{9}\right)\left(\frac{2}{7}\right)} = \frac{30}{44}.$$

2.115 Let A = both balls white

A_i = both balls selected from bowl i are white

B_i = i^{th} bowl is selected

a. $P(A) = \sum P(B_i \cap A_i) = \left(\frac{1}{5}\right)\left[\sum P(A_i|B_i)\right]$ $\qquad$ where $i = 1, \ldots, 5$
$= \left(\frac{1}{5}\right)\left[0 + \left(\frac{2}{5}\right)\left(\frac{1}{4}\right) + \left(\frac{3}{5}\right)\left(\frac{2}{4}\right) + \left(\frac{4}{5}\right)\left(\frac{3}{4}\right) + 1\right] = \frac{2}{5}$

b. $P(B_3|A) = \frac{P(B_3 \cap A)}{P(A)} = \frac{\left(\frac{3}{50}\right)}{\left(\frac{2}{5}\right)} = \frac{15}{100} = \frac{3}{20}$

2.116 Define the events:

A: player wins

B_i: sum of i on first toss

C_k: obtain a sum of k before obtaining a 7

By the law of total probability,

$$P(A) = \sum_{i=2}^{12} P(A \cap B_i).$$

Now

$P(A \cap B_2) = 0$

$P(A \cap B_3) = 0$

$P(A \cap B_{12}) = 0$

$P(A \cap B_7) = P(B_7) = \frac{6}{36}$

$P(A \cap B_{11}) = P(B_{11}) = \frac{2}{36}$

$P(A \cap B_4) = P(C_4 \cap B_4)$
$= P(C_4)P(B_4)$

(since the tosses are independent)

$= \left(\frac{1}{39}\right)\left(\frac{3}{36}\right)$

(as calculated in Exercise 2.95)

$= \frac{1}{36}$.

Similar to the calculations in Exercise 2.95,

$P(C_5) = P(C_9) = \frac{4}{10}$

$P(C_6) = P(C_8) = \frac{5}{11}$

$P(C_{10}) = \frac{3}{9}$

so that

$P(A \cap B_5) = P(A \cap B_9) = \left(\frac{4}{36}\right)\left(\frac{4}{10}\right) = \frac{4}{90} = \frac{2}{45}$

$P(A \cap B_6) = P(A \cap B_8) = \left(\frac{5}{36}\right)\left(\frac{5}{11}\right) = \frac{25}{396}$

$P(A \cap B_{10}) = \left(\frac{3}{9}\right)\left(\frac{3}{36}\right) = \frac{1}{36}$

Hence,

$P(A) = 0 + 0 + \frac{1}{36} + \frac{2}{45} + \frac{25}{396} + \frac{6}{36} + \frac{25}{396} + \frac{2}{45} + \frac{1}{36} + \frac{2}{36} + 0 = .493$

2.117 In Exercise 2.88 we calculated $P(Y = 0) = (.02)^3$ and $P(Y = 3) = (.98)^3$. The event that exactly one plane is detected consists of the following three mutually exclusive events:

$$F\overline{F}F \qquad \overline{F}FF \qquad FF\overline{F}$$

where F is the event that a radar set fails to detect a plane. Hence $P(Y = 1) = 3(.98)(.02)^2$. Similarly, $P(Y = 2) = 3(.98)^2(.02)$.

2.118 The total number of ways to select 3 from 6 refrigerators is $\binom{6}{3} = 20$, and the number of ways to select y defectives and $3 - y$ nondefectives is $\binom{2}{y}\binom{4}{3-y}$. Hence

$P(Y = 0) = \frac{\binom{2}{0}\binom{4}{3}}{20} = \frac{4}{20} = \frac{1}{5}$ $\qquad$ $P(Y = 1) = \frac{\binom{2}{1}\binom{4}{2}}{20} = \frac{12}{20} = \frac{3}{5}$

$P(Y = 2) = \frac{\binom{2}{2}\binom{4}{1}}{20} = \frac{4}{20} = \frac{1}{5}$

2.119 The events $Y = 2$, $Y = 3$, and $Y = 4$ were found in Exercise 2.80 to have probabilities $\frac{1}{15}$, $\frac{2}{15}$, and $\frac{3}{15}$, respectively. The event $Y = 5$ can occur in four ways:

$DGGGD \qquad GDGGD \qquad GGDGD \qquad GGGDD$

Each of these has probability

$$\frac{2(4)(3)(2)}{6(5)(4)(3)} \times \frac{1}{2} = \frac{1}{15}$$

Hence $P(Y = 5) = 4\left(\frac{1}{15}\right) = \frac{4}{15}$.

Another way to see this is to note that if $Y = 5$, then the last two observations are, in order, DG. One of the first four positives must be filled by a D (see Exercise 2.80 "alternate solution"). Thus, $P(Y = 5) = \frac{\binom{4}{1}}{\binom{6}{2}} = \frac{4}{15}$.

The event $Y = 6$ can occur in five ways:

$DGGGGD \qquad GDGGGD \qquad GGDGGD \qquad GGGDGD \qquad GGGGDD$

Each of these has probability

$$\frac{2(4)(3)(2)(1)}{6(5)(4)(3)(2)} = \frac{1}{15}$$

Hence $P(Y = 6) = 5\left(\frac{1}{15}\right) = \frac{5}{15}$.

Note that $P(Y = 6) = \frac{\binom{5}{1}}{\binom{6}{2}}$.

2.120 Let Y = # of positions the spinner did not land on; $Y = 2, 3$

$P(Y = 2) = \binom{4}{2}\left[\frac{1}{16}\right] = \frac{12}{16} = \frac{3}{4}$

$P(Y = 3) = \binom{4}{3}\left[\frac{1}{16}\right] = \frac{4}{16} = \frac{1}{4}$

2.121 The law of total probability gives

$P(B) = P(B \cap A) + P(B \cap \overline{A})$

$\frac{P(B)}{P(B)} = \frac{P(B \cap A)}{P(B)} + \frac{P(B \cap \overline{A})}{P(B)}$

$1 = P(A|B) + P(\overline{A}|B)$

$P(A|B) = 1 - P(\overline{A}|B)$

2.122 **a.**

$\emptyset$	$\{E_1\}$	$\{E_2\}$	$\{E_3\}$	$\{E_4\}$
$\{E_1, E_2\}$	$\{E_1, E_3\}$	$\{E_1, E_4\}$	$\{E_2, E_3\}$	$\{E_2, E_4\}$
$\{E_3, E_4\}$	$\{E_1, E_2, E_3\}$	$\{E_1, E_2, E_4\}$	$\{E_1, E_3, E_4\}$	$\{E_2, E_3, E_4\}$
$\{E_1, E_2, E_3, E_4\} = \text{S}$				

b. Notice that the events enumerated above were in fact obtained by enumerating the number of combinations of four items (E_1, E_2, E_3, and E_4) taken i at a time, for $i = 0$, 1, 2, 3, and 4. The sum of these combinations gives the total number of subsets in S. Hence, using the result of Exercise 2.59, we obtain the total number of events as

$$\binom{4}{0} = \binom{4}{1} + \binom{4}{2} + \binom{4}{3} + \binom{4}{4} = 2^4 = 16$$

which verifies the enumeration obtained in part **a**.

c. $A \cup B = \{E_1, E_2, E_3, E_4\}$
$A \cap B = \{E_2\}$ (since only E_2 appears in both sets)
$\overline{A} \cap \overline{B} = \{E_4\} \cap \{E_1, E_3\} = \emptyset$
$\overline{A} \cup B = \{E_4\} \cup \{E_2, E_4\} = \{E_2, E_4\}$

2.123 The 18 tests may be performed in any sequence, and hence in order to determine which sequence is most efficient, the efficiency expert must study all possible orderings of 18 tests. Permutations are used, and the total number of sequences is

$$P_{18}^{18} = \frac{18!}{0!} = 18!$$

2.124 Define the event A, "all 5 cards are of the same suit." The total number of ways of drawing 5 cards from the deck is $\binom{52}{5}$. Now it is necessary that all 5 cards be of the same suit, and there are four suits to choose from. Once a particular suit is chosen, there are then $\binom{13}{5}$ ways to choose 5 cards from that suit. Hence the total number of sample points resulting in the event A is

$$n_a = 4\binom{13}{5} \quad \text{and} \quad P(A) = \frac{4\binom{13}{5}}{\binom{52}{5}}.$$

2.125 There are $\binom{13}{2}$ ways of getting a set of three of a kind and a set of two of a kind. There are $\binom{4}{3}$ ways to obtain the cards that are three of a kind and $\binom{4}{2}$ ways to obtain the cards that are two of a kind. Thus,

$$P(\text{full house}) = \frac{\binom{13}{2}\binom{4}{3}\binom{4}{2}}{\binom{52}{5}}.$$

2.126 P(each supplier has at least one component tested)

$$= \frac{\binom{3}{2}\binom{4}{1}\binom{5}{1}}{\binom{12}{4}} + \frac{\binom{3}{1}\binom{4}{2}\binom{5}{1}}{\binom{12}{4}} + \frac{\binom{3}{1}\binom{4}{1}\binom{5}{2}}{\binom{12}{4}} = \frac{60}{495} + \frac{90}{495} + \frac{120}{495} = \frac{270}{495} = .545$$

2.127 Let A be the event that the person has symptom A and define B similarly. The following table displays the probabilities given in the exercise (underlined). Others are obtained by subtraction.

	B	$\overline{B}$	Total
A	.1	.2	.3
$\overline{A}$	.3	.4	.7
Total	.4	.6	

a. $P(\overline{A \cup B}) = P(\overline{A} \cap \overline{B}) = .4$

b. $P(A \cup B) = P(A) + P(B) - P(A \cap B) = .3 + .4 - .1 = .6$

c. $P(A \cap B | B) = \frac{P(A \cap B)}{P(B)} = \frac{.1}{.4} = .25$

2.128 Let Y be the number of symptoms possessed by a person. From Exercise 2.105,

$P(Y = 0) = P(\overline{A} \cap \overline{B}) = .4$ $\qquad$ $P(Y = 1) = P(\overline{A} \cap B) + P(A \cap \overline{B}) = .2 + .3 = .5$
$P(Y = 2) = P(A \cap B) = .1$

2.129 Let A denote the event that team A wins and B be the event that team B wins. Then there are 8 ways the series can go exactly 5 games. These are (with the associated probability)

$AAABA$	$p^4(1-p)$
$AABAA$	$p^4(1-p)$
$ABAAA$	$p^4(1-p)$
$BAAAA$	$p^4(1-p)$
$BBBAB$	$p(1-p)^4$
$BBABB$	$p(1-p)^4$
$BABBB$	$p(1-p)^4$
$ABBBB$	$p(1-p)^4$

Summing these probabilities gives $P(\text{exactly 5 games required}) = 4\left[p^4(1-p) + (1-p)^4 p\right]$.

2.130 Let Y be the number of pairs chosen. Then the possible values for Y are 0,1,2.

a. There are $\binom{10}{4} = 210$ ways to choose 4 socks from 10 and there are $\binom{5}{4}2^4$ ways to pick 4 nonmatching socks. Thus, $P(Y=0) = 80/210 = 0.381$.

b. In this case, there are $\binom{2n}{2r}$ total ways to pick $2r$ socks and $\binom{n}{2r}2^{2r}$ ways to pick $2r$ nonmatching socks. Thus

$$P(Y=0) = \frac{\binom{n}{2r}2^{2r}}{\binom{2n}{2r}}.$$

2.131 **a.** $P(A) = .25 + .10 + .05 + .10 = .50$

b. $P(A \cap B) = .10 + .05 = .15$

c. $P(A \cap B \cap \overline{C}) = .10$

d. Using a result proven in Exercise 2.64,

$$P(\overline{A} \cup \overline{B} | C) = \frac{P\left[(\overline{A} \cup \overline{B}) \cap C\right]}{P(C)} = \frac{P(\overline{A} \cap C) + P(\overline{B} \cap C) - P(\overline{A} \cap \overline{B} \cap C)}{P(C)}$$

$$= \frac{.25 + .25 - .15}{.4} = .875$$

2.132 **a.** (i) The probability of interest is

$$\frac{\text{total accidents for ages 15 and up}}{\text{total accidents}} = \frac{89{,}761}{100{,}761} = .89.$$

(ii) $\frac{47{,}038}{100{,}761} = .47.$

(iii) $P[\text{motor vehicle}|(15\text{–}24)] = \frac{P(\text{motor vehicle and 15–24})}{P(15\text{–}24)} = \frac{\left(\frac{16{,}650}{100{,}761}\right)}{\left(\frac{24{,}316}{100{,}761}\right)} = .68.$

(iv) $P(\text{drowning}|\text{not motor vehicle and 34 or less})$

$$= \frac{P(\text{drowning and not motor vehicle and 34 or less})}{P(\text{not motor vehicle and 34 or less})}$$

$$= \frac{1060 + 2090 + 1050 + 720}{19{,}939} = .25.$$

b. In order to determine the probability that a person randomly drawn from the general population had a motor vehicle accident, one would need the total population size, which is not given.

2.133 This exercise is similar to Exercise 2.132.

a. $P(\text{white}) = \frac{177{,}749}{203{,}212} = .87$

b. $P(\text{central city}) = \frac{63{,}922}{203{,}212} = .31$

c. $P(\text{urban fringe}|\text{white}) = \frac{51{,}405}{177{,}749} = .29$

d. $P(\text{white}|\text{urban fringe}) = \frac{51{,}405}{54{,}525} = .94$

e. $P(\text{outside urban}|\text{nonwhite}) = \frac{3057}{25{,}463} = .12$

f. $P(\text{nonwhite and central city } \underline{\text{or}} \text{ white and outside urban}) = \frac{14{,}375 + 27{,}281}{203{,}212} = .21.$

2.134 There are 10 nondefective and 2 defective tubes that have been drawn from the machine. The number of possible (distinct) arrangements of these 12 items is $\binom{12}{2}$ since the two defective tubes may be placed in any of 12 positions.

a. Assuming that the machine is generating defectives in a random manner, each of the $\binom{12}{2}$ arrangements is equally likely, and the probability of observing the arrangement $(NNNNNNNNNNDD)$ is

$$\frac{1}{\binom{12}{2}} = \frac{(2)(1)}{(12)(11)} = \frac{1}{66}$$

b. There are 2 arrangements that would result in 2 runs. These 2 arrangements are

$NNNNNNNNNNDD$ $DDNNNNNNNNNN$

Hence, denoting the number of runs by R, we have

$P(R=2) = \frac{2}{\binom{12}{2}} = \frac{2}{66} = \frac{1}{33}$

2.135 Refer to Exercise 2.134. It is now necessary to enumerate all possible arrangements of 2 defectives and 10 nondefectives that result in 3 runs. Then

$P(R \leq 3) = P(R=2) + P(R=3)$, since the minimum value of R is $R=2$.

The 3 runs of D's and Ns may occur in one of two ways; 1 run of D's and 2 of N's, or 1 run of N's and 2 of D's. The latter will occur only if the following arrangement occurs: $DNNNNNNNNNND$. If the 2 D's occur on consecutive trials (but not in positions 1 and 2 or 11 and 12), the arrangement will result in 1 of D's and 2 of N's. There are 9 such arrangements:

$NDDNNNNNNNNN$ $NNNDDNNNNNNN$ $NNNNNDDNNNNN$

$NNNNNNNDDNNN$ $NNNNNNNNNDDN$ $NNDDNNNNNNNN$

$NNNNDDNNNNNN$ $NNNNNNDDNNNN$ $NNNNNNNNDDNN$

Note that the two D's must be placed together in one of the $(10-1)=9$ "slots" between the N's. Hence

$$P(R \leq 3) = P(R=2) + P(R=3) = \tfrac{2}{66} + \left(\tfrac{1}{66} + \tfrac{9}{66}\right) = \tfrac{12}{66} = \tfrac{2}{11}.$$

2.136 There are 9! ways for the attendant to park the cars. There are 3! ways to park the expensive sports cars together. Among the 9 parking spaces, there are 7 possible groups of 3 parking spaces considered together. Also, there are 6! ways to arrange the remaining cars. Thus

$$P(\text{3 expensive sports cars are parked adjacent}) = 7\left(\tfrac{3!}{9!}\right)6! = 7!\left(\tfrac{3!}{9!}\right) = \tfrac{1}{12}$$

2.137 Let RO_i = relay i is open; RC_i = relay i is closed, where $i = 1, 2, 3, 4$.

Consider design A:

$$\begin{aligned} P(\text{current flows}) &= 1 - P(\text{current doesn't flow}) \\ &= 1 - P\left[(RO_1 \cap RO_2) \cup (RO_3 \cap RO_4)\right] \\ &= 1 - \left[P(RO_1 \cap RO_2) + P(RO_3 \cap RO_4) - P(RO_1 \cap RO_2 \cap RO_3 \cap RO_4)\right] \\ &= 1 - \left[(.1)^2 + (.1)^2 - (.1)^4\right] = 1 - .0199 = .9801 \end{aligned}$$

Consider design B:

$$\begin{aligned} P(\text{current flows}) &= P\left[(RC_1 \cap RC_3) \cup (RC_2 \cap RC_4)\right] \\ &= P(RC_1 \cap RC_3) + P(RC_2 \cap RC_4) - P(RC_1 \cap RC_2 \cap RC_3 \cap RC_4) \\ &= (.9)^2 + (.9)^2 - (.9)^4 = .9639 \end{aligned}$$

Thus, design A yields a higher probability that the current will flow when the relays are activated.

2.138 There are 8 tires, of which the customer selects 4 at random. Order is unimportant, and the total number of possible choices is $\binom{8}{4} = 70$. Now it is necessary to determine how many of the 70 choices result in the event of interest, that the best tire chosen by the customer holds rank #3. In order that this event occur, the customer must choose the tire ranked #3, and then choose 3 of the 5 remaining tires with inferior rankings, 4, 5, 6, 7, or 8. Given that #3 is one of the 4 tires chosen, there are $\binom{5}{3} = 10$ combinations of tires that can be chosen to go with it.

Hence 10 combinations will result in the event of interest, and the desired probability is $\frac{10}{70} = \frac{1}{7}$.

2.139 Y can take on the values 1, 2, 3, 4, and 5. There are always 70 ways that 4 tires can be chosen at random.

$Y = 1$ if the customer chooses the tire ranked #1. There are then $\binom{7}{3} = 35$ ways to choose the other 3 tires with inferior rankings. $P(Y=1) = \frac{35}{70} = \frac{1}{2}$.

$Y = 2$; given that the customer chooses the tire ranked #2, there are $\binom{6}{3} = 20$ ways to choose the other 3 tires with inferior rankings. $P(Y=2) = \frac{20}{70}$.

$Y = 4$; given that the customer chooses the tire ranked #4, there are $\binom{4}{3} = 4$ ways to choose the 3 tires with inferior rankings. $P(Y=4) = \frac{4}{70}$.

$Y = 5$; given that the customer chooses the tire ranked #5, there is $\binom{3}{3} = 1$ way to choose the other 3 tires with inferior rankings. $P(Y = 5) = \frac{1}{70}$.

y	1	2	3	4	5
$P(y)$	$\frac{35}{70}$	$\frac{20}{70}$	$\frac{10}{70}$	$\frac{4}{70}$	$\frac{1}{70}$

Note: the discussion of $Y = 3$ is in Problem 2.138.

2.140 **a.** There are $\binom{8}{4} = 70$ ways to pick 4 tires. The customer must choose those ranked 3 and 7. This leaves those ranked 4, 5, 6 to choose the remaining 2 from. There are $\binom{3}{2} = 3$ ways to do this. Thus $P(\text{best is ranked 3 and the worst is ranked 7}) = 3/70$.

b. There are 4 ways that the range can be 4. Specifically, if the best selected is ranked 1 then the worst must be ranked 5. The remaining ways are 2,6 or 3,7 or 4,8. The probability of each is the same as the probability found in part **a**. So $P(R = 4) = 4(3/70) = 6/35$.

c. The possible values for R are 3,4,5,6,7. calculations that are similar to parts **a** and **b** yield

R	$P(R)$
3	$5/70$
4	$12/70$
5	$18/70$
6	$20/70$
7	$15/70$

2.141 **a.** For each beer drinker there are 24 ways to rank the four beers. So that there are $24^3 = 13824$ total sample points.

b. In order to achieve a combined score of 4 or less the given beer may receive at most one score of 2 with the rest being 1. Consider brand A. If a beer drinker assigns A a one then there are still 3! ways this beer drinker can assign ranks to the other 3 brands. Thus there are $(3!)^3$ ways for brand A to be assigned all ones. Similarly, brand A can be assigned two ones and one two in $3(3!)^3$ ways. Thus, some beer may earn a total rank less than or equal to 4 in $4((3!)^3 + 3(3!)^3) = 3456$ ways. Hence,

$$P(\text{combined score of 4 or less}) = 3456/13824 = 0.25.$$

2.142 The total number of ways to select 3 names from 7 is $\binom{7}{3} = 35$. If the first name on the list is to be included, the other 2 names can be picked in $\binom{6}{2} = 15$ ways. Hence the probability of interest is $\frac{15}{35} = \frac{3}{7}$.

2.143 The probability that Skylab will hit someone is unconditionally $\frac{1}{150}$, regardless of where the person lives. If one wants to know the probability condition on living in a certain area, it is not possible to determine. You can say that in an area containing 4 billion inhabitants, the expected number of casualties is (4 billion) times $\frac{1}{150}$.

2.144 Refer to a table such as the one shown below and rewrite the conditional probabilities according to their definition to see that only $P(A|B) + P(\overline{A}|B) = 1$.

	A	$\overline{A}$	
B	$P(AB)$	$P(\overline{A}B)$	$P(B)$
$\overline{B}$	$P(A\overline{B})$	$P(\overline{A}\,\overline{B})$	$P(\overline{B})$
Total	$P(A)$	$P(\overline{A})$	1

2.145 Denote the four possible simple events as HH, HT, TH, TT. Then the events A, B, C, $A \cap B$, $B \cap C$, $A \cap B \cap C$ can be written as sets of simple events:

$A = \{HH, HT\}$ $\quad B = \{HH, TH\}$ $\quad C = \{HH, TT\}$

$A \cap B = \{HH\}$ $\quad B \cap C = \{HH\}$ $\quad A \cap C = \{HH\}$ $\quad A \cap B \cap C = \{HH\}$

Notice that

$P(A) = P(B) = P(C) = \frac{1}{2}$ $\quad P(A \cap B) = P(A \cap C) = P(B \cap C) = \frac{1}{4}$

$P(A \cap B \cap C) = \frac{1}{4}$

Since $P(A \cap B \cap C) \neq P(A)P(B)P(C)$, the three events are not independent.

2.146 Define the following events:

E: person is exposed to the flu $\qquad$ F: person contracts the flu

Consider two employees, one of whom is inoculated and one not. The probability of interest is the probability that at least one contracts the flu. It is convenient to consider the complement of this even. Then

$P(\text{at least one contracts the flu}) = 1 - P(\text{inoculated does not contract flu and noninoculated does not contract flu}).$

Consider the inoculated employee. For this case,

$$P(\overline{F}) = P(\overline{F}E) + P(\overline{F}\,\overline{E}) = P(\overline{F}|E)P(E) + P(\overline{F}|\overline{E})P(\overline{E})$$
$$= (.8)(.6) + (1)(.4) = .48 + .4 = .88$$

Now consider the noninoculated employee. Again,

$$P(\overline{F}) = P(\overline{F}|E)P(E) + P(\overline{F}|\overline{E})P(\overline{E})$$

However, $P(\overline{F}|E) = .1$ instead of .8. Hence

$$P(\overline{F}) = (.1)(.6) + (1)(.4) = .46$$

Thus,

$$P(\text{neither contracts flu}) = (.88)(.46)$$
$$P(\text{at least one contracts flu}) = 1 - (.88)(.46) = 1 - .4048 = .5952$$

2.147 The game that is being played can be represented by a series of W's and L's, W representing a win for gambler Jones (and hence an increase of \$1 for his bank) and L representing a loss (and hence a decrease of \$1). The tosses are independent and $P(W) = P(L) = \frac{1}{2}$.

a. For this part of the exercise, suppose that the coin has been tossed 6 times, producing a total of 6 outcomes, each of which may be either a W or an L. Each sequence of outcomes is equally likely, and has probability $\left(\frac{1}{2}\right)^6$.
It is necessary only to count the number of such sequences for which the two gamblers break even. If the gamblers are to break even after 6 trials, then each must win 3 times and lose 3 times. If this is the case, the total gain for either gambler will be \$3 – \$3 = \$0, and he will be left with a total of \$6 after 6 trials. Notice that it is irrelevant to consider the order in which the wins and losses occur, since if the gambler is only allowed to lose or win 3 times, he cannot exhaust his bank of \$6 before trial 6 and hence end the game.
Consequently, the total number of outcomes consisting of 3 wins and 3 losses can be found by counting the number of ways to choose 3 trials (out of 6) in which to place the 3 wins. That is,

$n_a = \binom{6}{3} = 20$ $\qquad$ and the desired probability is $\binom{6}{3} \times \left(\frac{1}{2}\right)^6$.

b. For this part of the exercise, suppose that the coin has been tossed 10 times. There are 2^{10} possible sequences of W's and L's, each with probability $\left(\frac{1}{2}\right)^{10}$.
It is necessary to enumerate n_a, the number of sequences that result in gambler Jones winning the game at exactly the tenth toss.
In order for the event of interest to occur, gambler Jones must have \$11 at trial 9 and must win on trial 10. That is, in 9 trials (during which \$9 will change hands), Jones must win 7 times and lose only twice in order that his total gain be \$7 – \$2 = \$5 and that he be left with a total of \$11 after 9 trials.
If the two losses were placed randomly among the 9 trials, which can be done in $\binom{9}{2} = 36$ ways, it is possible that the game might end before the appointed 10 trials. These arrangements must be eliminated from the 36 possibilities. Notice that gambler Jones can only win on an even-numbered trial, since he must gain \$6. That is, the difference between the number of wins and losses is 6, so that the sum of the number of wins and losses must be an even number.
The number of arrangements of 2 L's and 7 W's for which Jones would win on trial 6 are

$WWWWWWWLL \qquad WWWWWWWLLW \qquad WWWWWWWLWL$

The number of arrangements for which Jones would win on trial 8 is found by using the same argument as for the event "win on trial 10" above. The arrangements are

$LWWWWWWWL$ $\quad$ $WLWWWWWWL$ $\quad$ $WWLWWWWWL$
$WWWLWWWWL$ $\quad$ $WWWWLWWWL$ $\quad$ $WWWWWLWWL$

These 9 arrangements are eliminated from consideration, so that $n_a = \binom{9}{2} - 9 = 27$, and the probability of interest is $27\left(\frac{1}{2}\right)^{10}$.

2.148a. The patrolman starts at the center of the 16×16 square grid and walks in any one of 4 directions at each intersection. Hence, in his first 8 blocks there is a total of 4^8 possible paths he can take. Only 4 of these will result in his reaching the boundary of his patrol area, since, in order to get there in 8 "steps," he must walk either 8 blocks north, 8 blocks south, 8 blocks east, or 8 blocks west. Hence the probability of interest is $\frac{4}{4^8} = \left(\frac{1}{4}\right)^7$.

b. It is now necessary to enumerate which of the 4^4 paths (that may be taken in a 4-block excursion) result in a return to the starting point. Consider all paths beginning with a block's walk north that result in a return to the starting point in 4 blocks. There are 9 such paths. If we represent his directional choices by N, W, E, S, he may walk

(1) $NNSS$ $\quad$ (2) $NSNS$ $\quad$ (3) $NSSN$ $\quad$ (4) $NESW$ $\quad$ (5) $NWSE$
(6) $NWES$ $\quad$ (7) $NEWS$ $\quad$ (8) $NSEW$ $\quad$ (9) $NSWE$

and he will have returned to the starting point in 4 blocks. There are, by symmetry, 9 paths each for walks beginning with a block's walk south, east, or west. Hence the total number of paths resulting in the event of interest is 36 and the desired probability is $36\left(\frac{1}{4}\right)^4$.

2.149 Consider using the device of representing the n balls as 0's and creating N boxes by arbitrarily placing bars between the 0's. The space between 2 adjacent bars is a box and the number of 0's between any 2 adjacent bars is the number of balls in the box. Such a device forces us to start and end the sequence with a bar, and in order to create a set of N boxes, a total of $N+1$ bars is needed. Eliminating the 2 that are forced to be at the beginning and end, we have a total of $N-1$ bars to be arranged:

||00|||000|00|... |0||

We have not yet placed any restrictions upon the arrangement of the $N-1$ bars and n 0's, since more than 1 ball is allowed per box and boxes are allowed to remain empty. Thus picking $N-1$ of the $N-1+n$ positions in which to place the bars, we have a total of $\binom{N+n-1}{N-1}$ arrangements.

It is not necessary to determine how many of these result in no box being empty. If no 2 bars are to be placed adjacent to one another, the $N-1$ bars must be placed in the $n-1$ spaces between the 0's. Some of the $n-1$ spaces may be left unfilled, since $n \geq N$, but none may contain more than 1 bar:

|000|00|... |00|0|...0|

The total number of ways to do this is $\binom{n-1}{N-1}$. Thus the probability that no box remains empty is given by $\frac{\binom{n-1}{N-1}}{\binom{N+n-1}{N-1}}$.

CHAPTER 3 DISCRETE RANDOM VARIABLES AND THEIR PROBABILITY DISTRIBUTIONS

3.1 Let A and B denote the event the well has impurities A and B respectively. Notice $P(\overline{A} \cap \overline{B}) = .2$, therefore $P(A \cup B) = .8$. Thus
$P(A \cap B) = P(A) + P(B) - P(A \cup B) = .5 + .4 - .8 = .1$. Then we have:

$$P(Y = 0) = P(\overline{A} \cap \overline{B}) = .2$$
$$P(Y = 2) = P(A \cap B) = .1$$

therefore by the law of total probability: $P(Y = 1) = 1 - .2 - .1 = .7$.

3.2 The simple events and corresponding Y values are

E_i	Y
HH	2
HT	-1
TH	-1
TT	1

Since $P(E_i) = \frac{1}{4}$ for each i, the probability distribution for Y is

y	-1	1	2
$p(y)$	$\frac{1}{2}$	$\frac{1}{4}$	$\frac{1}{4}$

3.3 Similar to Exercise 2.119. The event $Y = 2$ occurs if the first and second components tested are both defective.

$$p(2) = P(DD) = \tfrac{2}{4}\left(\tfrac{1}{3}\right) = \tfrac{1}{6}$$
$$p(3) = P(DGD) + P(GDD) = 2\left(\tfrac{2}{4}\right)\left(\tfrac{2}{3}\right)\left(\tfrac{1}{2}\right) = \tfrac{2}{6}$$
$$p(4) = P(GGDD) + P(DGGD) + P(GDGD) = 3\left(\tfrac{2}{4}\right)\left(\tfrac{1}{3}\right)\left(\tfrac{2}{2}\right) = \tfrac{1}{2}$$

Since there are only four components, $Y = 2$, 3, and 4 are the only possible values for the random variable Y.

3.4 Define the following events:

A: valve 1 fails $\qquad$ B: valve 2 fails $\qquad$ C: valve 3 fails

Notice :

$$P(Y = 2) = P(\overline{A} \cap \overline{B} \cap \overline{C}) = .8^3 = .512,$$
$$P(Y = 0) = P(A \cap (B \cup C)) = P(A)P(B \cup C) = .2(.2 + .2 - (.2)^2) = .072,$$

(by the law of total probability) $P(Y = 1) = 1 - .512 - .072 = .416$.

3.5 Assume that the correct ordering of the animal words is ABC. If the child is guessing, there are 6 possible equally likely permutations he could choose, with Y = number of matches associated with each permutation.

A	B	C	Y = No. of Matches
A	B	C	3
A	C	B	1
B	A	C	1
B	C	A	0
C	A	B	0
C	B	A	1

Then

$$p(0) = \tfrac{2}{6} = \tfrac{1}{3}$$
$$p(1) = \tfrac{3}{6} = \tfrac{1}{2}$$
$$p(2) = 0$$
$$p(3) = \tfrac{1}{6}$$

3.6 There are $\binom{5}{2} = 10$ sample points, all equally likely:

$(1,2)\quad(1,3)\quad(1,4)\quad(1,5)\quad(2,3)\quad(2,4)\quad(2,5)\quad(3,4)\quad(3,5)\quad(4,5)$

a. Let Y = largest of the two sampled numbers. Then

$p(2) = .1 \qquad p(3) = .2 \qquad p(4) = .3 \qquad p(5) = .4$

b. Let Y = sum of the two numbers. Then

$p(3) = .1 \quad p(4) = .1 \quad p(5) = .2 \quad p(6) = .2 \quad p(7) = .2 \quad p(8) = .1 \quad p(9) = .1$

3.7 The random variable Y takes on the values 0, 1, 2, and 3. We can assume that the three entries are independent.

a. Let E denote an error on a single entry; let N denote that there is no error. There are $2^3 = 8$ sample points:

$EEE \quad EEN \quad ENE \quad NEE \quad ENN \quad NEN \quad NNE \quad NNN$

Thus,

$$P(Y=3) = P(EEE) = (.05)^3 = .000125$$
$$P(Y=2) = P(EEN) + P(ENE) + P(NEE)$$
$$= 3(.05)^2(.95) = .007125$$
$$P(Y=1) = P(ENN) + P(NEN) + P(NNE)$$
$$= 3(.05)(.95)^2 = .135375$$
$$P(Y=0) = P(NNN) = (.95)^3 = .857375$$

b. Use the above probabilities to construct the histogram.

c. $P(Y > 1) = P(Y=2) + P(Y=3) = .00725$

3.8 Let R denote the event that a rental occurs on a given day; let N denote no rental. Then $P(R) = \frac{1}{5} = .2$. The variable in question is related to a sequence $RNN\ldots NR$, where the number of N's equals the value of the random variable Y.
Consider the position immediately following the first R. This position is filled by an R with probability .2 and by an N with probability .8. The same probabilities hold for the other positions as well. Note that each sequence under consideration must begin with an R. Thus,

$$P(Y=0) = P(RR) = .2$$
$$P(Y=1) = P(RNR) = (.8)(.2)$$
$$P(Y=2) = P(RNNR) = (.8)^2(.2)$$

In general,

$$P(Y=y) = (.8)^y(.2)$$

3.9 Let O^+ denote the event that a person has type O^+.
Then the sample space consist of the eight events:

$$\{(O^+,O^+,O^+),(O^+,O^+,\overline{O^+}),(O^+,\overline{O^+},O^+),(\overline{O^+},O^+,O^+),$$
$$(\overline{O^+},\overline{O^+},O^+),(\overline{O^+},O^+,\overline{O^+}),(O^+,\overline{O^+},\overline{O^+}),(\overline{O^+},\overline{O^+},\overline{O^+})\}$$

With probabilities $\{\frac{1}{27}, \frac{2}{27}, \frac{2}{27}, \frac{2}{27}, \frac{4}{27}, \frac{4}{27}, \frac{4}{27}, \frac{8}{27}\}$ respectively.
Therefore we have:

x	0	1	2	3
$p(x)$	$\frac{8}{27}$	$\frac{12}{27}$	$\frac{6}{27}$	$\frac{1}{27}$

It will be shown later on that x's probability is governed by the equation:

$$p(x) = \binom{3}{x}\left(\tfrac{1}{3}\right)^x\left(\tfrac{2}{3}\right)^{3-x};\ x = 0, 1, 2, 3.$$

Repeating the same calculation for y we have:

y	0	1	2	3
$p(y)$	$\frac{2744}{3375}$	$\frac{196}{3375}$	$\frac{14}{3375}$	$\frac{1}{3375}$

Considering the distribution of z $= x + y$. Then z = # of people with type O blood,

$z = 0, 1, 2, 3$. Notice the probability that a person has type O blood:

$= $ (Probability a person has O^+ blood) $+$ (Probability a person has O^- blood)

$= \frac{1}{3} + \frac{1}{15} = \frac{6}{15}$.

Let O = person has O blood. Then the sample space is:

$\{(O,O,O),(\overline{O},O,O),(O,\overline{O},O),(O,O,\overline{O}),(\overline{O},\overline{O},O),(\overline{O},O,\overline{O}),(O,\overline{O},\overline{O}),(O,O,O),\}$

with respective probabilities:

$$\left\{\left(\tfrac{6}{15}\right)^3, \left(\tfrac{6}{15}\right)^2\left(\tfrac{9}{15}\right), \left(\tfrac{6}{15}\right)^2\left(\tfrac{9}{15}\right), \left(\tfrac{6}{15}\right)^2\left(\tfrac{9}{15}\right), \left(\tfrac{6}{15}\right)\left(\tfrac{9}{15}\right)^2, \left(\tfrac{6}{15}\right)\left(\tfrac{9}{15}\right)^2, \left(\tfrac{6}{15}\right)\left(\tfrac{9}{15}\right)^2, \left(\tfrac{9}{15}\right)^3\right\}.$$

Then we have

z	0	1	2	3
p(z)	$\frac{729}{3375}$	$\frac{1458}{3375}$	$\frac{972}{3375}$	$\frac{216}{3375}$

3.10 $E(Y) = \Sigma y p(y) = 1(.4) + 2(.3) + 3(.2) + 4(.1) = 2.0$

$E\left(\frac{1}{Y}\right) = \Sigma \frac{1}{y} p(y) = 1(.4) + \frac{1}{2}(.3) + \frac{1}{3}(.2) + \frac{1}{4}(.1) = .6417$

$E(Y^2 - 1) = E(Y^2) - 1 = [1(.4) + 4(.3) + 9(.2) + 16(.1)] - 1 = 5 - 1 = 4$

Using Theorem 3.6,

$$V(Y) = E(Y^2) - [E(Y)]^2 = 5 - (2)^2 = 1$$

3.11 $E(Y) = (-1)\left(\frac{1}{2}\right) + (1)\left(\frac{1}{4}\right) + (2)\left(\frac{1}{4}\right) = \frac{1}{4}$

$E(Y^2) = (-1)^2\left(\frac{1}{2}\right) + (1)^2\left(\frac{1}{4}\right) + (2)^2\left(\frac{1}{4}\right) = \frac{7}{4}$

$V(Y) = E(Y^2) - E(Y)^2 = \frac{7}{4} - \left(\frac{1}{4}\right)^2 = \frac{27}{16}$

Cost of play = C

Net winnings = $Y - C$

$$E(Y - C) = E(Y) - C = 0 \Rightarrow C = \frac{1}{4}$$

3.12 As shown in exercise 2.97 $P(Y = y) = 1/n$ for $y = 1, \ldots, n$. Therefore

$$E[Y] = \sum_{y=1}^{n} \frac{y}{n} = \frac{1}{n}\sum_{y=1}^{n} y = \frac{1}{n}\frac{n(n+1)}{2} = \frac{(n+1)}{2},$$

$$E[Y^2] = \sum_{y=1}^{n} \frac{y^2}{n} = \frac{1}{n}\sum_{y=1}^{n} y^2 = \frac{1}{n}\frac{n(n+1)(2n+1)}{2} = \frac{(n+1)(2n+1)}{2},$$

$$V(Y^2) = E[Y^2] - E[Y]^2 = \frac{(n+1)(2n+1)}{2} - \frac{(n+1)^2}{4} = \frac{(n+1)(3n+1)}{4}.$$

3.13 Let P be a random variable representing the company's profit. Then $P = C - 15$ with probability 98/100 (when A does not occur) and $P = C - 15 - 1000$ with probability 2/100 (when A occurs). Then $E(P) = (C - 15)\frac{98}{100} + (C - 15 - 1000)\frac{2}{100} = 50$. Simplifying we have $C - 15 - 20 = 50$. Thus $C = \$85$.

3.14 With probability .3 the volume is $8 \times 10 \times 30 = 2400$. With probability .7 the volume is $8 \times 10 \times 40 = 3200$. The mean is $(.3)(2400) + (.7)(3200) = 2960$.

3.15 The expected value of N is

$$E(N) = E(8\pi r^2) = 8\pi \mathrm{E}(r^2)$$

Now,

$$E(r^2) = \Sigma r^2 p(r) = 21^2(.05) + 22^2(.20) + 23^2(.30) + 24^2(.25) + 25^2(.15) + 26^2(.05) = 549.1$$

Thus, $E(N) = 8\pi(549.1) = 13{,}800.388$

3.16 Since the die is fair, the probability distribution for Y is

$p(y) = \frac{1}{6} \qquad y = 1, 2, 3, 4, 5, 6$

Then

$$E(Y) = \Sigma y p(y) = \tfrac{1}{6}(1 + 2 + \ldots + 6) = \tfrac{21}{6} = 3.5$$

$$E(Y^2) = \Sigma y^2 p(y) = \tfrac{1}{6}(1 + 4 + 9 + \ldots + 36) = \tfrac{91}{6} = 15.1667$$

$$V(Y) = E(Y^2) - [E(Y)]^2 = 15.1667 - (3.5)^2 = 2.9167$$

3.17 Define G to be the gain to a person in drawing one card. G can take on only three values, \$15, \$5, or \$−4, with probabilities as shown in the accompanying table.

G	$p(G)$
15	$\frac{2}{13}$
5	$\frac{2}{13}$
−4	$\frac{9}{13}$

Then $E(G) = \Sigma G p(G) = 15\left(\frac{2}{13}\right) + 5\left(\frac{2}{13}\right) - 4\left(\frac{9}{13}\right) = \frac{4}{13} = .31$
The expected gain is \$.31.

3.18 Consider first the probability distribution for y:

y	$p(y)$
0	.81
1	.18
2	.01

Then

$$\mu = E(Y) = \Sigma y p(y) = 0(.81) + 1(.18) + 2(.01) = .20$$

and

$$\sigma^2 = E\left(Y^2\right) - \mu^2 = \left[\Sigma\, y^2 p(y)\right] - \mu^2 = [0(.81) + 1(.18) + 4(.01)] - (.2)^2 = .22 - .04 = .18.$$

3.19 Let $X_1 =$ # of contracts assigned to firm 1; $X_2 =$ # of contracts assigned to firm 2.
The sample space would be {(I,I),(I,II),(I,III),(II,I),(II,II),(II,III),(III,I),(III,II),(III,III)} each with equal probability of $\frac{1}{9}$. With this we may easily calculate the probability distributions for X_1 and X_2.

x_1	$p(x_1)$		x_2	$p(x_2)$
0	$\frac{4}{9}$		0	$\frac{4}{9}$
1	$\frac{4}{9}$		1	$\frac{4}{9}$
2	$\frac{1}{9}$		2	$\frac{1}{9}$

The $E(X_1) = \Sigma x_1 p(x_1) = 0\left(\frac{4}{9}\right) + 1\left(\frac{4}{9}\right) + 2\left(\frac{1}{9}\right) = \frac{6}{9} = \frac{2}{3}$
Let $Y =$ profit for firm 1 $= 90{,}000X_1$.
The E[profit for firm 1] $= E(Y) = E[90{,}000X_1] = 90{,}000E(X_1) = 90{,}000\left(\frac{2}{3}\right) = 60{,}000$.
Let $W =$ profit for both firm 1 and 2 $= 90{,}000(X_1 + X_2)$.
Since X_2 follows the same distribution as X_1, we know that $E(X_2) = \frac{2}{3}$.
Thus, the $E(W) = E[90{,}000(X_1 + X_2)] = 90{,}000E(X_1 + X_2) = 90{,}000[E(X_1) + E(X_2)]$
$= 90{,}000\left[\left(\frac{2}{3}\right) + \left(\frac{2}{3}\right)\right] = 120{,}000$

3.20 Let Y be daily sales. Then Y can take on three possible values, $Y = 0$, 50,000, or 100,000. The value $Y = 0$ will occur if the salesperson contacts either one or two customers and fails to make a sale. Then

$$\begin{aligned} P(Y=0) &= P(\text{contact one, fail to sell}) + P(\text{contact two, fail to sell}) \\ &= P(\text{contact one}) \cdot P(\text{fail to sell}) \\ &\quad + P(\text{contact two}) \cdot P(\text{fail with 1st}) \cdot P(\text{fail with 2nd}) \\ &= \left(\tfrac{1}{3}\right)\left(\tfrac{9}{10}\right) + \left(\tfrac{2}{3}\right)\left(\tfrac{9}{10}\right)\left(\tfrac{9}{10}\right) = \tfrac{9}{30} + \tfrac{162}{300} = \tfrac{252}{300} \end{aligned}$$

Similarly,

$$\begin{aligned} P(Y=50{,}000) &= P(\text{contact one, sell}) + P(\text{contact two, sell to one}) \\ &= P(\text{contact one, sell}) + P(\text{contact two, sell to 1st only}) \\ &\quad + P(\text{contact two, sell to 2nd only}) \\ &= \left(\tfrac{1}{3}\right)\left(\tfrac{1}{10}\right) + \left(\tfrac{2}{3}\right)\left(\tfrac{1}{10}\right)\left(\tfrac{9}{10}\right) = \left(\tfrac{2}{3}\right)\left(\tfrac{9}{10}\right)\left(\tfrac{1}{10}\right) = \tfrac{46}{300} \end{aligned}$$

Finally,

$$P(Y=100{,}000) = P(\text{contact two, sell to both}) = \left(\tfrac{2}{3}\right)\left(\tfrac{1}{10}\right)\left(\tfrac{1}{10}\right) = \tfrac{2}{300}$$

Then

$$E(Y) = \sum_y y p(y) = 0\left(\tfrac{252}{300}\right) + 50{,}000\left(\tfrac{46}{300}\right) + 100{,}000\left(\tfrac{2}{300}\right) = \tfrac{25{,}000}{3} = 8333$$

Thus, the expected value of daily sales is \$8333.

The variance of Y is

$$(Y) = \sum_y y^2 p(y) - \mu^2 = (50{,}000)^2\left(\tfrac{46}{300}\right) + (100{,}000)^2\left(\tfrac{2}{300}\right) - (8333)^2 = 380{,}561{,}111$$

$$\sigma = \sqrt{V(Y)} = 19{,}507.98$$

3.21 Let Y be a random variable representing the payout on an individual policy. $P(Y = 85,000) = P(\text{total loss}) = .001$, $P(Y = 45,000) = P(50\%\text{ loss}) = .01$, and $P(Y = 0) = 1 - .001 - .01 = .989$. Let C represent the premium the insurance company charges. Then the company's net gain or loss for this policy is given by $C - y$. To yield a long term average loss of 0 the company should choose C such that: $E(C - Y) = 0$, or $C = E(Y)$. Then we have:

$$\begin{aligned} E(Y) &= \textstyle\sum_y yp(y) \\ &= (85{,}000)(.001) + (42{,}500)(.01) + (0)(.989) \\ &= 510 = C. \end{aligned}$$

3.22 The probability distribution for Y was found in Exercise 3.3 to be as shown in the accompanying table.

y	$p(y)$
2	$\frac{1}{6}$
3	$\frac{2}{6}$
4	$\frac{3}{6}$

Then

$$E(Y) = 2\left(\tfrac{1}{6}\right) + 3\left(\tfrac{2}{6}\right) + 4\left(\tfrac{3}{6}\right) = 3.33$$

The cost for testing and repairing is defined as $C = 2Y + 8$. Hence

$$E(C) = 2E(Y) + 8 = 2(3.33) + 8 = \$14.67$$

3.23 **a.** We let $g_1(Y) = aY$ and $g_2(Y) = b$.
Then, by Theorem 3.5,

$$\begin{aligned} E(aY + b) &= E[g_1(Y) + g_2(Y)] \\ &= E[g_1(Y)] + E[g_2(Y)] = E[aY] + E[b]. \end{aligned}$$

We now use Theorems 3.4 and 3.3 to get $E(aY + b) = aE(Y) + E(b) = a\mu + b$.

b. By Definition 3.5 and exercise **a.**,

$$\begin{aligned} V(aY + b) &= E[aY + b - (a\mu + b)]^2 \\ &= E[aY - a\mu + b - b]^2 = E[a(Y - \mu)]^2 \\ &= E\left[a^2(Y - \mu)^2\right]. \end{aligned}$$

Using Theorem 3.4, this equals $a^2E(Y - \mu)^2$ which equals $a^2V(Y) = a^2\sigma^2$, by Definition 3.5.

3.24 The mean cost is $E(\$10Y) = \$10E(Y)$. Now,

$$E(Y) = 0(.1) + 1(.5) + 2(.4) = 1.3$$

so the mean cost is \$13.
The variance of the cost is $V(10Y) = 100V(Y)$ (see problem 3.23 above). To find $V(Y)$ we find $E(Y^2)$ and use Theorem 3.6.

$$E(Y^2) = 0^2(.1) + 1^2(.5) + 2^2(.4) = 2.1.$$

$$V(Y) = E(Y^2) - \mu^2 = 2.1 - (1.3)^2 = .41.$$

Thus, the variance of the cost is $100(.41) = 41$.

3.25 $B = SS \cup FS$

$$\begin{aligned} P(B) &= P(SS) + P(FS) \qquad \text{SS and SF are mutually exclusive} \\ &= \tfrac{2000}{5000} \times \tfrac{1999}{4999} + \tfrac{3000}{5000} \times \tfrac{2000}{4999} \\ &= \tfrac{2000}{5000}\left(\tfrac{1999}{4999} + \tfrac{3000}{4999}\right) = \tfrac{2000}{5000} = 0.4. \end{aligned}$$

$$P(B|\text{first trial success}) = \tfrac{1999}{4999} = 0.3999$$

which is not markedly different from 0.4.

3.26 **a.** The probability that a given judge will choose formula B will be $\frac{1}{3}$, since the two formulas are equally attractive; the probability that he will choose formula A is thus $\frac{2}{3}$. The random variable Y can take on the values 0, 1, 2, 3, or 4 and the associated probabilities are given below.

$p(0) = \left(\frac{2}{3}\right)^4 = \frac{16}{81}$ $\qquad$ $p(3) = \binom{4}{3}\left(\frac{1}{3}\right)^3\left(\frac{2}{3}\right) = \frac{8}{81}$

$p(1) = \binom{4}{1}\left(\frac{1}{3}\right)\left(\frac{2}{3}\right)^3 = \frac{32}{81}$ $\qquad$ $p(4) = \binom{4}{4}\left(\frac{1}{3}\right)^4 = \frac{1}{81}$

$p(2) = \binom{4}{2}\left(\frac{1}{3}\right)^2\left(\frac{2}{3}\right)^2 = \frac{24}{81}$

b. $P(Y \geq 3) = p(3) + p(4) = \frac{9}{81}$.

c. $E(Y) = \sum_y yp(y) = 0\left(\frac{16}{81}\right) + 1\left(\frac{32}{81}\right) + 2\left(\frac{24}{81}\right) + 3\left(\frac{8}{81}\right) + 4\left(\frac{1}{81}\right) = \frac{4}{3}$.

d. $V(Y) = E(Y^2) - \left(\frac{4}{3}\right)^2 = 0\left(\frac{16}{81}\right) + 1\left(\frac{32}{81}\right) + 4\left(\frac{24}{81}\right) + 9\left(\frac{8}{81}\right) + 16\left(\frac{1}{81}\right) - \frac{16}{9} = \frac{8}{9}$.

(Shortcut formulas for $E(Y)$ and V(Y) are used in later problems for this section.)

3.27 Define Y to be the number of components failing in less than 1000 hours. Then

$p = P(\text{component fails in less than 1000 hours}) = .2$.

There are $n = 4$ independent components.

a. $P(Y = 2) = p(2) = \binom{4}{2}p^2q^2 = \binom{4}{2}(.2)^2(.8)^2 = .1536$.

b. Since the subsystem will operate if 2, 3, or 4 of the components last longer than 1000 hours, it will operate if 2, 1, or none of the components fail in less than 1000 hours. Calculate

$p(0) = \binom{4}{0}(.2)^0(.8)^4 = .4096$ $\qquad$ and $\qquad$ $p(1) = \binom{4}{1}(.2)^1(.8)^3 = .4096$.

Then

$P(\text{subsystem operates}) = p(0) + p(1) + p(2) = .4096 + .4096 + .1536 = .9728$.

3.28 Let Y be the number of patients who survive. Then $p = P(\text{patient survives}) = .2$ and $n = 20$.

a. Since $n = 20$ is one of the sample sizes given in Table 1, Appendix III, it is convenient to use these binomial tables in this situation. Index $n = 20$ and the appropriate value of p (in this case $p = .8$). The tables are cumulative, that is, if one indexes $a = r$, the tabled value is

$$\sum_{y=0}^{r}\binom{n}{y}p^y q^{n-y}$$

Hence, to find $p(r)$ one obtains the values corresponding to $a = r$ and $a = r - 1$ from the table and the two values are subtracted to obtain

$$\sum_{y=0}^{r}\binom{n}{y}p^y q^{n-y} - \sum_{y=0}^{r-1}\binom{n}{y}p^y q^{n-y} = \binom{n}{r}p^r q^{n-r} = p(r)$$

In this case,

$p(14) = P(Y \leq 14) - P(Y \leq 13) = .196 - .087 = .109$.

a. $P(Y \geq 10) = 1 - P(Y \leq 9) = 1 - .001 = .999$.

b. $P(14 \leq Y \leq 18) = P(Y \leq 18) - P(Y \leq 13) = .931 - .087 = .844$.

c. $P(Y \leq 16) = .589$.

3.29 Let Y be the number answered correctly. Then $p = P(\text{correct answer}) = \frac{1}{5}$ and $n = 15$.

$P(Y \geq 10) = 1 - P(Y \leq 9) = 1 - 1.000 = .000$ $\qquad$ (to three decimal places)

using Table 1, Appendix III.

3.30 Let Y be the number of falsified application forms. Then Y has a binomial distribution with $p = .35$ and $n = 5$. Though we could easily use table 1 (appendix III) we perform the calculations exactly.

$P(Y \geq 1) = 1 - P(Y = 0) = 1 - \binom{5}{0}(.35)^0(.65)^5$

$= 1 - .116 = .884$

$P(Y \geq 2) = 1 - P(Y = 0) - P(Y = 1)$

$= 1 - .116 - \binom{5}{1}(.35)^1(.65)^1$

$= .884 - .312 = .572$

3.31 Let Y be the number of qualifying subscribers. Then Y has a binomial distribution with $p = .7$ and $n = 5$. Though we could easily use table 1 (appendix III) we perform the calculations exactly.

a. $P(Y = 5) = \binom{5}{5}(.7)^5 = .1681$

b. $P(Y \geq 4) = P(Y = 4) + P(Y = 5)$
$= \binom{5}{4}(.7)^4(.3) + \binom{5}{5}(.7)^5$
$= .3601 + .1681 = .5282$

3.32 Let Y be the number of successful operations, with $n = 5$.

a. Use Table 1 with $p = .8$.
$p(5) = P(Y \leq 5) - P(Y \leq 4) = 1 - .672 = .328.$

b. For $p = .6$, $p(4) = P(Y \leq 4) - P(Y \leq 3) = .922 - .663 = .259.$

c. For $p = .3$, $P(Y < 2) = P(Y \leq 1) = .528.$

3.33 For $n = 3$ and $p = .8$

$P(Y = 0) = (.2)^3 = .008$ $\qquad P(Y = 2) = \binom{3}{2}(.8)^2(.2) = .384$

$P(Y = 1) = \binom{3}{1}(.8)(.2)^2 = .096$ $\qquad P(Y = 3) = (.8)^3 = .512$

The alarm will function if $Y = 1, 2$, or 3. Hence
$P(\text{alarm functions}) = p(1) + p(2) + p(3) = .096 + .384 + .512 = .992.$

3.34 The binomial probability distribution for $n = 5$ is given as

$$p(y) = \binom{5}{y} p^y q^{5-y} \qquad y = 0, 1, 2, 3, 4, 5$$

Substituting the values $p = .1$, $p = .5$, and $p = .9$ in the above formula, the three probability distributions are obtained.

	$p = .1$	$p = .5$	$p = .9$
y	$p(y)$	$p(y)$	$p(y)$
0	.59049	.03125	.00001
1	.32805	.15625	.00045
2	.07290	.31250	.00810
3	.00810	.31250	.07290
4	.00045	.15625	.32805
5	.00001	.03125	.59049

Note that when $p = .5$, the distribution is symmetric; that is, $p(y) = p(5 - y)$ for $y = 0, 1, 2, 3, 4, 5$. When $p > .5$, the distribution is skewed to the right, and when $p < .5$, the distribution is skewed to the left. The probability distributions for $p = p_0$ and $p = 1 - p_0$ are "mirror images" of each other.
Note that Table 1, Appendix III, could have been used in this exercise.

3.35 Refer to Table 1, Appendix III, indexing $n = 20$ and $p = \frac{1}{2}$. The values of $p(y)$ for $y = 0, 1, 2, \ldots, 20$ are obtained by calculating the differences between tabled values for $a = r$ and $a = r - 1$ for $r = 1, 2, \ldots, 20$. Note that $p(0)$ is obtained directly from the table. The binomial probability distribution and its probability histogram are shown below.

y	$p(y)$	y	$p(y)$
0	.000	11	.160
1	.000	12	.120
2	.000	13	.074
3	.001	14	.037
4	.005	15	.015
5	.015	16	.005
6	.037	17	.001
7	.074	18	.000
8	.120	19	.000
9	.160	20	.000

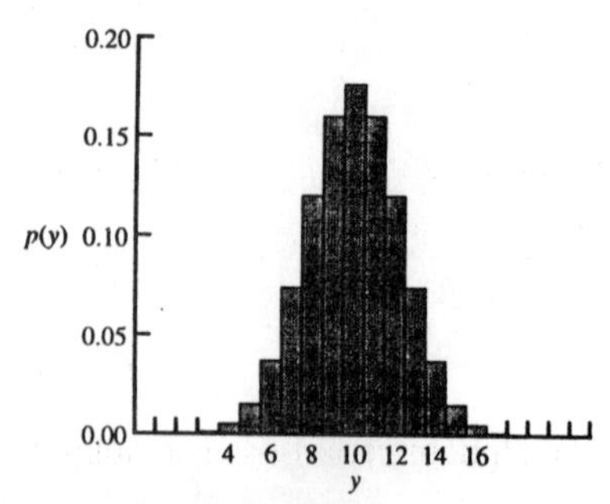

Figure 3.1

3.36 a. Let Y be the number of sets out of 5 that detect the missile. Then

$$P(Y=4) = \binom{5}{4}(.9)^4(.1) = .32805$$

and

$$P(Y \geq 1) = 1 - P(Y=0) = 1 - \binom{5}{0}(.1)^5 = .99999.$$

(notice we could have also used table 1 (appendix III) for these calculations).

b. With n radar sets, the probability of at least one detection is $1-(.1)^n$. We required n such that $1-(.1)^n = .999$. Thus,

$$(.1)^n = 1 - .999 = .001, \text{ so } n = 3.$$

3.37 Let Y be the number of housewives preferring brand A. Under the assumption that there is no difference between brands, $p = P(\text{prefer brand } A) = .5$ and $n = 15$.

a. Using Table 1, Appendix III,

$$P(Y \geq 10) = 1 - P(Y \leq 9) = 1 - .849 = .151$$

b. $P(10 \text{ or more prefer } A \text{ or } B) = P(Y \leq 5 \text{ or } Y \geq 10) = P(Y \leq 5) + [1 - P(Y \leq 9)]$ $= .151 + (1 - .849) = .302$, since 10 or more preferring B is equivalent to 5 or less preferring A.

3.38 The only way team A can win in exactly 5 games is to win 3 in the first 4 games and win the 5th game. Let Y be the number of games team A wins in the first 4 games. Y is distributed binomially with success probability p and 4 trials. Then the probability we want would be

$$\begin{aligned} P(\text{Team A wins in 5}) &= P(Y=3)P(\text{Team A wins game 5}) \\ &= \binom{4}{3}p^3(1-p)\,p \\ &= 4p^4(1-p). \end{aligned}$$

3.39 First we will find $E[Y(Y-1)(Y-2)]$.

$$\begin{aligned} E[Y(Y-1)(Y-2)] &= \sum_{y=0}^{n} \frac{y(y-1)(y-2)\,n!}{y!(n-y)!} p^y(1-p)^{n-y} \\ &= \sum_{y=3}^{n} \frac{n(n-1)(n-2)(n-3)!}{(y-3)!(n-3-(y-3))!} p^3 p^{y-3}(1-p)^{n-3-(y-3)} \\ &= n(n-1)(n-2)p^3 \sum_{z=0}^{n-3}\binom{n-3}{z} p^z(1-p)^{n-3-z} \\ &= n(n-1)(n-2)p^3. \end{aligned}$$

Now we use the fact that $E[Y(Y-1)(Y-2)] = E(Y^3) - 3E(Y^2) + 2E(Y) = E(Y^3) - 3(n(n-1)p^2 + np) + 2np$. Combining our results we have

$$n(n-1)(n-2)p^3 = E(Y^3) - 3(n(n-1)p^2 + np) + 2np.$$

Hence $E(Y^3) = 3(n(n-1)p^2 + np) - n(n-1)(n-2)p^3 - 2np$ or

$$E(Y^3) = 3n(n-1)p^2 - n(n-1)(n-2)p^3 + np.$$

3.40 The random variable of interest is Y, the number of successful explorations in $n = 10$ explorations. Then Y has a binomial distribution with $p = .1$. Hence

$$E(Y) = np = 10(.1) = 1 \qquad V(Y) = npq = 10(.1)(.9) = .9$$

3.41 Refer to Exercise 3.40. The number of successful explorations is Y and the number of unsuccessful explorations is $(10 - Y)$. Hence the cost is

$$C = 20{,}000 + 30{,}000Y + 15{,}000(10 - Y)$$

and

$$E(C) = 20{,}000 + 30{,}000(1) + 15{,}000(10-1) = \$185{,}000$$

3.42 The random variable Y is binomial with $n = 4$, $p = .1$. Hence

$$E(Y) = np = .4$$

and

$$E(Y^2) = V(Y) + [E(Y)]^2 = npq + n^2p^2 = 4(.1)(.9) + (.4)^2 = .52.$$

Then $E(C) = 3E(Y^2) + E(Y) + 2 = 3(.52) + .4 + 2 = 3.96.$

3.43 Let Y be the number of defective motors out of 10. Then Y is binomial with $p = .08$ and $n = 10$. By Theorem 3.7,

$$E(Y) = (.08)(10) = .8.$$

The seller gains \$1000 on the sale of 10 motors and loses \$200 for each defective. The seller's expected net gain is

$$\$1000 - \$200E(Y) = \$1000 - \$200(.8) = \$1000 - \$160 = \$840.$$

3.44 Let Y = # of fish that survive. Y is binomial with $n = 20$ and $p = 0.8$.

a. $P(Y = 14) = P(Y \leq 14) - P(Y \leq 13) = .196 - .087 = .109$

b. $P(Y \geq 10) = 1 - P(Y \leq 9) = 1 - .001 = .999$

c. $P(Y \leq 16) = .589$

d. $\mu = np = 20(.8) = 16;\ \sigma^2 = npq = 20(.8)(.2) = 3.2$

3.45 Let Y = # who have Rh$^+$ blood. Y is binomial with $n = 5$ and $p = .8$.

a. $P(\text{at least one does not have Rh}^+) = P(Y \leq 4) = .672$

b. $P(Y \leq 4) = .672$

c. Y is now binomial with unknown n and $p = .8$. $P(Y \geq 5) = 1 - (Y \leq 4) > .9$. Therefore, $P(Y \leq 4) < .1$. Using Table 1, we find that required sample size is between $n = 5$ and $n = 10$. By calculating $P(Y \leq 4)$ for all values of n between 5 and 10 we find that $n = 8$ is the smallest such that $P(Y \leq 4) < .1$.

3.46 **a.** Independence of the three inspection events.

b. Let Y = # of planes with wing cracks that are detected. Y has a binomial distribution with $n = 3$ and $p = (.9)(.8)(.5) = .36$. $P(Y \geq 1) = 1 - P(Y = 0)$ $= 1 - \binom{3}{0}(.36)^0(.64)^3 = .737856$

3.47 **a.** For any $y = 1, 2, 3, \ldots, n - 1$

$$p(y) = \binom{n}{y} p_y q^{n-y}$$

and

$$p(y-1) = \binom{n}{y-1} p^{y-1} q^{n-y+1},$$

Then

$$\frac{p(y)}{p(y-1)} = \frac{\binom{n}{y} p^y q^{n-y}}{\binom{n}{y-1} p^{y-1} q^{n-y+1}} = \frac{n!}{y!\,(n-y)!}\ \frac{(y-1)!\,(n-y+1)!}{n!}\ \frac{p}{q} = \frac{(n-y+1)p}{yq}.$$

b. We want $\dfrac{(n-y+1)p}{yq} > 1$

or $(n+1)p - yp > yq$

or $(n+1)p > yq + yp$

or $(n+1)p > y.$

We want $\dfrac{(n-y+1)p}{yq} < 1$

or $(n+1)p - yp < yq$

or $(n+1)p < yq + yp$

or $(n+1)p < y.$

Also, we want

$$\frac{(n-y+1)p}{yq} = 1$$

or $\quad (n+1)p - yp = yq$

or $\quad (n+1)p = yq + yp$

or $\quad (n+1)p = y$

which is only possible if $(n+1)p$ is an integer, as y only takes integer values.

c. From the results above,
if $y \leq (n+1)p$ then $p(y) \geq p(y-1) > p(y-2) > \ldots$
and if $y \geq (n+1)p$, then $p(y) \geq p(y+1) > p(y+2) > \ldots$.
Therefore, $p(y)$ is maximized when y is as close to $(n+1)p$ as possible.

3.48 Recall that $P(Y = y_0) = \binom{n}{y_0} p^{y_0}(1-p)^{n-y_0} = f(p)$. Our goal is to maximize $f(p)$.

A standard trick in these problems is to maximize $\ln(f(p))$ instead of $f(p)$ (since $\ln()$ is a strictly increasing function it will not change the maximum) . Notice

$$\ln(f(p)) = \ln\binom{n}{y_0} + y_0\ln(p) + (n-y_0)\ln(1-p).$$

Making

$$\tfrac{d}{dp}\ln(f(p)) = \tfrac{y_0}{p} - \tfrac{(n-y_0)}{(1-p)} = 0.$$

Setting this equal to 0 and solving yields $p = y_0/n$. Notice that

$$\tfrac{d^2}{dp^2}\ln(f(p)) = -\tfrac{y_0}{p^2} - \tfrac{(n-y_0)}{(1-p)^2} \leq 0 \text{ for all } p$$

ensuring that $p = y_0/n$ is a maximum by the second derivative test.

3.49 **a.** $E(Y/n) = E(Y)/n = np/n = p$. That is *on average* our estimator of p will be equal to p.

b. $V(Y/n) = V(Y)/n^2 = np(1-p)/n^2 = p(1-p)/n$. Notice as n gets large the variance of $V(Y/n)$ goes to 0. That is as we collect more data our estimate of p gets better (less variable). The results of **a.** and **b.** suggest Y/n is a good estimator of p.

3.50 **a.** The geometric probability mass function is given by:

$$p(y)q^{y-1}p \qquad y = 1, 2, 3, \ldots$$

To prove this mass function sums to 1 we need to show $\sum_{y=1}^{\infty} q^{y-1}p = 1$, or

equivalently $\sum_{y=1}^{\infty} q^{y-1} = \frac{1}{p} = \frac{1}{1-q}$. This is the same as showing:

$$\sum_{z=0}^{\infty} q^z = \tfrac{1}{1-q} \quad (1).$$

Equation (1) is a commonly known mathematical fact which can be proved as follows. Let $a > 1$ and define $S(a) = \sum_{z=0}^{a} q^z$. We must show that $\lim_{a\to\infty} S(a) = \frac{1}{1-q}$ when $0 < q < 1$. Notice:

$$S(a) = 1 + q + q^2 + \ldots + q^{a-1},$$

and

$$qS(a) = q + q^2 + \ldots + q^{a-1} + q^a.$$

Subtracting the two equations, we have $(1-q)S(a) = 1 - q^a$. Hence, the partial sum is $S(a) = \frac{1-q^a}{1-q}$. Notice $\lim_{a\to\infty} S(a) = \frac{1}{1-q}$ when $0 < q < 1$, proving the result.

b. For any $y = 2, 3, 4, \ldots$

$$\tfrac{p(y)}{p(y-1)} = \tfrac{q^{y-1}p}{q^{y-2}p} = q.$$

$Y = 1$ has the highest probability.

3.51 Let Y be the number of the interview on which the first applicant having advanced training is found. Then Y has a geometric distribution. Thus,

$$P(Y = 5) = (.7)^{5-1}(.3) = .072$$

3.52 $E(Y) = \frac{1}{p} = \frac{1}{.30} = 3.34$

3.53 Since Y = the number of calls until the first person is found who is satisfied with the state of the nation, a success occurs when a person is found who is satisfied with the state of the nation. Assuming the Gallup poll is correct then Y has the geometric distribution with $P = .27$.

3.54 Let Y be the number of holes drilled until the first procedure well is found. Then Y has a geometric distribution with $p = .2$.

a. $p(3) = (.8)^2(.2) = .128$

b. $P(Y > 10) = P(\text{first 10 holes are nonproductive}) = (.8)^{10} = .107$

3.55 **a.** $P(Y > a) = \sum_{y=a+1}^{\infty} q^{y-1}p = q^a \sum_{y=a+1}^{\infty} q^{y-a-1}p = q^a \sum_{z=1}^{\infty} q^{z-1}p = q^a.$

(Notice $\sum_{z=1}^{\infty} q^{z-1}p = 1$ by problem 3.44)

b. Using the result of part **a**,

$$P(Y > a + b|Y > a) = \frac{P(Y>a+b, Y>a)}{P(Y>a)} = \frac{q^{a+b}}{q^a} = q^b = P(Y > b)$$

Let Y represent the time (in years) until failure of an electrical component. Then **b.** suggest that the probability the component last b or more years is q^b regardless of how long the component has already lasted. That is, the life of the component has no memory of the past.

3.56 Refer to Exercise 3.55. Let $a = 10$ and $b = 1$, since "at least two more" is the same as "more than one more." Then

$$P(Y > 11|Y > 10) = P(Y > 1) = \left(\tfrac{1}{2}\right)^1 = \tfrac{1}{2}$$

3.57 Define Y to be the number of the first account containing substantial errors. Then Y has a geometric distribution with $p = .9$.

a. $P(Y = 3) = (.1)^2(.9) = .009$

b. $P(Y \geq 3) = 1 - P(Y \leq 2) = 1 - P(Y = 1) - P(Y = 2) = 1 - .9 - (.1)(.9) = .01$

3.58 $\mu = \frac{1}{0.9} = 1.11; \sigma = \sqrt{\frac{(1-.9)}{(.9)^2}} = \sqrt{.123} = .35$

3.59 Let Y be the number of one – second intervals until the first arrival, so that $p = P(\text{arrival}) = .1$.

a. $P(Y = 3) = q^{3-1}p = q^2p = (.9)^2(.1) = .081.$

b. $P(Y \geq 3) = 1 - P(Y \leq 2) = 1 - (q^{1-1}p + q^{2-1}p) = 1 - (.1) - (.9)(.1) = .81.$

Notice an alternative approach is to use the result of exercise 3.49 **a.**

$P(Y \geq 3) = P(Y > 2) = q^2 = .9^2 = .81.$

3.60 The random variable is Y, the number of consumers interviewed before a success occurs, where a success is defined to be the encountering of a customer who prefers brand A. The random variable Y follows the geometric distribution with $p = P(\text{success}) = .60$.

a. $P(Y = 5) = (.4)^{5-1}(.6) = (.4)^4(.6) = .01536.$

b. $P(Y \geq 5) = 1 - P(Y \leq 4) = 1 - (.6) - (.4)(.6) - (.4)^2(.6) - (.4)^3(.6) = .0256.$

Notice an alternative approach is to use the result of exercise 3.55 **a.**

$P(Y \geq 5) = P(Y > 4) = q^4 = .4^4 = .0256.$

3.61 Define Y to be the number of people questioned before a "yes" answer is given. Then

$$p = P(\text{yes}) = P(\text{smoker} \cap \text{yes}) + P(\text{nonsmoker} \cap \text{yes}) = P(\text{yes}|\text{smoker})P(\text{smoker}) + 0 = .3(.2) = .06.$$

Thus,

$$p(y) = pq^{y-1} = .06(.94)^{y-1}, \qquad y = 1, 2, 3, \ldots$$

3.62 Let Y = # of toss on which first 6 appears. Then Y is geometric with $p = \frac{1}{6}$.

$$\begin{aligned} P(B \text{ tosses first } 6) &= P(Y = 2, 4, 6, \dots) \\ &= p(2) + p(4) + p(6) + \dots \\ &= \sum_{i=1}^{\infty} p(2i) = \sum_{i=1}^{\infty} q^{2i-1} p = \frac{p}{q} \sum_{i=1}^{\infty} q^{2i} \\ &= \frac{p}{q}\left(\frac{1}{1-q^2} - 1\right) = \frac{pq}{1-q^2}. \end{aligned}$$

Since $p = \frac{1}{6}$, $P(B \text{ tosses first } 6) = \frac{5}{11}$.

Then $P(Y = 4 | B \text{ tosses first } 6) = \frac{P(Y=4)}{P(B \text{ tosses first } 6)} = \frac{\left(\frac{5}{6}\right)^3 \left(\frac{1}{6}\right)}{\frac{5}{11}} = \frac{275}{1296}$.

3.63 The number of tosses until the first head appears is a geometric random variable with $p = \frac{1}{2}$. Using Theorem 3.8, $E(Y) = \frac{1}{p} = \frac{1}{\left(\frac{1}{2}\right)} = 2$.

3.64 Let Y be the number of the first productive well. Then Y has a geometric distribution with $p = .2$. From Theorem 3.8, $E(Y) = \frac{1}{.2} = 5$. If the prospector is successful one time in five, then we expect the first success on trial number 5.

3.65 Let Y be the number of trials until the secretary picks the correct password. Then in this setting Y is a geometric random variable with success probability $1/n$. The probability that the secretary picks the correct password on the 6th try is then $(\frac{1}{n})(\frac{n-1}{n})^5$.

3.66 As discussed in exercise 3.65 Y is geometric with success probability $1/n$. Then $E(Y) = n$ and $V(Y) = (1 - \frac{1}{n})n^2 = n(n-1)$.

3.67 Note first that $\frac{d^2}{dq^2} q^y = y(y-1)q^{y-2}$. Hence

$$\frac{d^2}{dq^2} \sum_{y=2}^{\infty} q^y = \sum_{y=2}^{\infty} y(y-1)q^{y-2}$$

(The interchange of derivative and sum can be justified by standard results from calculus). Then

$$\begin{aligned} E(Y(Y-1)) &= \sum_{y=1}^{\infty} y(y-1)pq^{y-1} = pq \sum_{y=2}^{\infty} y(y-1)q^{y-2} = pq \frac{d^2}{dq^2} \sum_{y=2}^{\infty} q^y \\ &= pq \frac{d^2}{dq^2}\left\{\frac{1}{1-q} - 1 - q\right\} = pq \frac{d}{dq}\left\{\frac{1}{(1-q)^2} - 1\right\} = \frac{2pq}{(1-q)^3} = \frac{2q}{p^2} \end{aligned}$$

The variance of Y is then

$$V(Y) = E(Y(Y-1)) + E(Y) - [E(Y)]^2 = \frac{2q}{p^2} + \frac{1}{p} - \frac{1}{p^2} = \frac{2(1-p)+p-1}{p^2} = \frac{q}{p^2}$$

3.68 Note that $P(Y = y_0) = (1-p)^{y_0 - 1} p = f(p)$. A standard trick in this setting is to maximize $\ln(f(p))$ (since ln is a strictly increasing function on (0,1) it will not effect the maximum). Note

$$\ln(f(p)) = (y_0 - 1)\ln(1-p) + \ln(p).$$

Making,

$$\frac{d}{dp}\ln(f(p)) = -\frac{y_0 - 1}{1-p} + \frac{1}{p} = 0.$$

Setting this equal to 0 and solving for p yields $p = 1/y_0$. Secondly notice

$$\frac{d^2}{dp^2}\ln(f(p)) = -\frac{y_0-1}{(1-p)^2} - \frac{1}{p^2} \le 0 \text{ for all } p$$

ensuring that $p = 1/y_0$ is a maximum by the second derivative test.

3.69 $E(1/Y) = \sum_{y=1}^{\infty} \frac{1}{y}(1-p)^{y-1}p = \frac{p}{(1-p)} \sum_{y=1}^{\infty} \frac{(1-p)^y}{y} = \frac{p}{(1-p)}(-\ln(p)) = -\frac{p\ln(p)}{(1-p)}$.

Where the sum is computed using the hint with $r = 1 - p$.

3.70 $P(Y^* = y) = P(Y = y+1) = q^{(y+1)-1}p = q^y p$. Note the range of values Y^* can take is $0, 1, \dots$ as $Y^* = Y - 1$ and Y takes values 1,2,...

3.71 **a.** $E(Y^*) = E(Y) - 1 = \frac{1}{p} - 1 = \frac{q}{p}$. $V(Y^*) = V(Y-1) = V(Y) = \frac{1-p}{p^2}$.

b. Repeating the proof of theorem 3.8 we have

$$\begin{aligned}
E(Y^*) &= \sum_{y^*=0}^{\infty} y^* q^{y^*} p = pq \sum_{y^*=1}^{\infty} y^* q^{y^*-1} \\
&= pq \sum_{y^*=1}^{\infty} \frac{d}{dq} q^{y^*} = pq \frac{d}{dq}\left(\sum_{y^*=1}^{\infty} q^{y^*}\right) \\
&= pq \frac{d}{dq}\frac{q}{(1-q)} = pq \frac{1}{(1-q)^2} \\
&= pq \frac{1}{p^2} \\
&= \frac{q}{p}.
\end{aligned}$$

Now $V(Y^*)$. First we will find $E(Y^{*2})$ by finding $E(Y^*(Y^*-1))$.

$$\begin{aligned}
E(Y^*(Y^*-1)) &= \sum_{y^*=0}^{\infty} y^*(y^*-1) q^{y^*} p = pq^2 \sum_{y^*=2}^{\infty} y^*(y^*-1)\, q^{y^*-2} \\
&= pq^2 \sum_{y^*=2}^{\infty} \frac{d^2}{dq^2} q^{y^*} = pq^2 \frac{d^2}{dq^2}\left(\sum_{y^*=2}^{\infty} q^{y^*}\right) \\
&= pq^2 \frac{d^2}{dq^2}\left(\frac{q}{(1-q)} - q\right) = pq^2 \frac{d}{dq}\left(\frac{1}{(1-q)^2} - 1\right) \\
&= pq^2 \frac{2}{(1-q)^3} \\
&= 2\frac{q^2}{p^2}.
\end{aligned}$$

(Note $\sum_{y^*=2}^{\infty} q^{y^*}$ can be calculated easily as it is $\sum_{y^*=1}^{\infty} q^{y^*} - q = \frac{q}{(1-q)} - q$, recall $\sum_{y^*=1}^{\infty} q^{y^*}$ is given in appendix 1). Now we have

$$E(Y^*(Y^*-1)) = E(Y^{*2}) - E(Y^*) = E(Y^{*2}) - \frac{q}{p} = 2\frac{q^2}{p^2}.$$

This implies

$$E(Y^{*2}) = \frac{q}{p} + 2\frac{q^2}{p^2}.$$

Thus

$$V(Y^*) = \frac{q}{p} + 2\frac{q^2}{p^2} - \frac{q^2}{p^2} = \frac{q(p+q)}{p^2} = \frac{1-p}{p^2}.$$

3.72 The probability of finding an employee with positive indications of asbestos will remain relatively constant from trial to trial provided the number of employees is reasonably large. It then makes sense to define Y as the number of the trial on which the third positive indication of asbestos is observed and model Y as having a negative binomial distribution. Then

$P(10 \text{ employees must be tested in order to find 3 positives}) = P(Y = 10)$
$= \binom{9}{2}(.4)^3(.6)^7 = .06.$

3.73 The total cost of conducting the tests to locate 3 positives is $20X$ dollars. Refer to Theorem 3.9. The expected value of the total cost is

$$E(20X) = 20E(X) = 20\,\frac{r}{p} = \frac{20(3)}{.4} = 150 \text{ dollars}$$

and the variance of the total cost is

$$V(20X) = 400V(X) = 400\,\frac{r(1-p)}{p^2} = \frac{400(3)(.6)}{(.4)^2} = 4500.$$

3.74 Let Y = # of trials until the first non defective engine is found. Y is geometric with $p = .9$.

$$P(Y = 2) = (.9)(.1) = .09.$$

3.75 Let Y = # of trials until the rth non defective engine is found. Y is negative binomial with $r = 3$ and $p = .9$.

a. $P(Y = 5) = \binom{4}{2}(.9)^3(.1)^2 = .04374$

b. $P(Y \le 5) = P(y = 3) + P(Y = 4) + P(Y = 5$
$= \binom{2}{2}(.9)^3(.1)^2 + \binom{3}{2}(.9)^3(.1) + \binom{4}{2}(.9)^3(.1)^2$
$= .729 + .2187 + .04374 = .99144$

3.76 a. $\mu = \frac{1}{p} = \frac{1}{.9} = 1.11; \sigma^2 = \frac{1-.9}{(.9)^2} = .123$

b. $\mu = \frac{r}{p} = \frac{3}{.9} = 3.33; \sigma^2 = \frac{r(1-p)}{p^2} = \frac{.3}{.81} = .3704$

3.77 Recall that Y is geometric with $p = .9$, we use the memory less property (HW 3.55 **b**).

$$P(Y \ge 4|Y > 2) = P(Y > 3|Y > 2) = P(Y > 1) = q^1 = .1.$$

3.78 a. Let Y = # of attempts until you complete your call. Y is geometric with $p = .4$.

P(complete the call on the first try) $= P(Y = 1) = .4$
P(complete the call on the second try) $= P(Y = 2) = (.4)(.6) = .24$
P(complete the call on the third try) $= P(Y = 3) = (.4)(.6)^2 = .144$

b. Let Y = # of attempts until both calls are completed. Y is negative binomial with $r = 2, p = .4$.

$$P(Y = 4) = \binom{3}{1}(.4)^2(.6)^2 = .1728$$

3.79 a. Let Y = # of wells drilled until the first strike of oil. Y is geometric with $p = .2$.

$$P(Y = 3) = (.2)(.8)^2 = .128.$$

b. Let Y = # of wells drilled until the third strike of oil. Y is negative binomial with $r = 3, p = .2$.

$$P(Y = 7) = \binom{6}{2}(.2)^3(.8)^4 = .049$$

c. (1) One of two possible outcomes
(2) The probability of success, *p*, remains constant from oil well to oil well
(3) The trials are independent

d. Y = # of wells drilled until three producing wells are found. Y is negative binomial with $r = 3$ and $p = .2$.

$$\mu = \frac{r}{p} = \frac{3}{.2} = 15; \sigma^2 = \frac{r(1-p)}{p^2} = \frac{2.4}{.04} = 60$$

3.80 a. $p(y) = \binom{y-1}{n-1} p^r q^{y-r} \qquad y = r, r+1, r+2, \ldots; 0 \le p \le 1.$

Then, for $y \ge r + 1$

$$\frac{p(y)}{p(y-1)} = \frac{\frac{(y-1)!}{(r-1)!(y-r)!} p^r q^{y-r}}{\frac{(y-2)!}{(r-1)!(y-r-1)!} p^r q^{y-1-r}} = \frac{(y-1)}{(y-r)} q.$$

b. We want $\frac{y-1}{y-r} q > 1$

or $yq - q > y - r$

or $r - q > y - yq$

or $\frac{r-q}{1-q} > y.$

The second result is similar.

c. If $r = 7$, $p = .5$, then

$$\frac{r-q}{1-q} = \frac{7-.5}{1-.5} = 13.$$

From parts **a** and **b**, $\frac{p(y)}{p(y-1)} > 1$ or $p(y) > p(y-1)$ if y is an integer less than 13.

3.81 There are y trials before the rth success, if the rth success occurs on the $(y+1)^{st}$ trial. Let X = the number of the trial on which the rth success occurs, and X is negative binomial. Then $X = Y + 1$ and

$$P(Y = y) = P(x = y + 1) = \binom{y+1-1}{r-1} p^r q^{y+1-r}$$
$$= \binom{y}{r-1} p^r q^{y+1-r} \qquad y = r-1, r, r+1, \ldots .$$

3.82 **a.** $P(Y^* = y) = P(Y = y + r) = \binom{y+r-1}{r-1} p^r q^{y+r-r} = \binom{y+r-1}{r-1} p^r q^y.$
Now note as Y takes on values $r, r+1, \ldots,$ $Y^* = Y - r$ takes on values $0, 1, \ldots$

b. $E(Y^*) = E(Y) - r = \frac{r}{p} - r = rq/p. V(Y^*) = V(Y - r) = V(Y) = \frac{rq}{p^2}.$

3.83 **a.** Let $f(p) = P(Y = 11) = \binom{10}{4} p^5 (1-p)^6$. A standard trick for these problems is to maximize $\ln(f(p))$ (since ln is a strictly increasing function it will not affect the maximum).

$$\ln(f(p)) = \ln\binom{10}{4} + 5\ln(p) + 6\ln(1-p)$$

Now notice

$$\frac{d}{dp}\ln(f(p)) = \frac{5}{p} - \frac{6}{(1-p)} = 0.$$

Solving for p yields 5/11. It is easy to show that $\frac{d^2}{dp^2}\ln(f(p)) \le 0$ for all p hence $p = 5/11$ is a maximum.

b. Note that $f(p) = P(Y = y_0) = \binom{y_0 - 1}{r-1} p^r (1-p)^{y_0 - r}$. Thus we have

$$\ln(f(p)) = \ln\binom{y_0-1}{r-1} + r\ln(p) + (y_0 - r)\ln(1-p)$$

and hence

$$\frac{d}{dp}\ln(f(p)) = \frac{r}{p} - \frac{y_0 - r}{(1-p)} = 0.$$

Solving for p yields $p = r/y_0$. Again calculating the second derivative shows that this value is indeed a maximum.

3.84 Let Y be the number of green marbles chosen in three draws. Then Y follows the hypergeometric distribution with

$$p(y) = \frac{\binom{5}{y}\binom{2+3}{3-y}}{\binom{10}{3}}$$

For $y = 3$,

$$p(3) = \frac{\binom{5}{3}\binom{5}{0}}{\binom{10}{3}} = \frac{10}{120} = \frac{1}{12}.$$

3.85 Use the hypergeometric probability distribution with $N = 10, r = 6,$ and $n = 5$.

$$P(5 \text{ non defectives}) = \frac{\binom{6}{5}\binom{4}{0}}{\binom{10}{5}} = \frac{6}{252} = \frac{1}{42}$$

3.86 Let C = Total repair cost = $50Y$, where Y is hypergeometric with $N = 10, r = 4,$ and $n = 5$. Using theorem 3.10 we have:

$$E(C) = E(50Y) = 50E(Y) = 50\left[(5)\left(\tfrac{4}{10}\right)\right] = 100$$

and

$$V(C) = V(50Y) = (50)^2 V(Y) = 2500(5)\left(\tfrac{4}{10}\right)\left(\tfrac{6}{10}\right)\left(\tfrac{5}{9}\right) = \tfrac{5000}{3} = 1666.67.$$

3.87 Think of a bowl with 6 balls, 2 red (for programs 1 and 2) and 4 black (for programs 3–6). Let $N = 6, n = 2,$ and $r = 4$. Then Y follows the hypergeometric distribution with

$$P(y) = \frac{\binom{6}{y}\binom{2}{2-y}}{\binom{6}{2}}, \qquad y = 0, 1, 2.$$

3.88 **a.** $P(Y = 1) = \frac{\binom{4}{2}\binom{2}{1}}{\binom{6}{3}} = \frac{(6)(2)}{20} = \frac{3}{5}$

b. $P(Y \ge 1) = p(1) + p(2) = \frac{3}{5} + \frac{\binom{4}{1}\binom{2}{2}}{\binom{6}{3}} = \frac{3}{5} + \frac{1}{5} = \frac{4}{5}$

c. $P(Y \le 1) = p(0) + p(1) = \frac{\binom{4}{3}\binom{2}{0}}{\binom{6}{3}} + \frac{3}{5} = \frac{1}{5} + \frac{3}{5} = \frac{4}{5}$

3.89 **a.** The probability distribution of Y is $\frac{\binom{2}{y}\binom{8}{3-y}}{\binom{10}{3}}$ for $y = 0, 1, 2$.

Thus the distribution (in tabular form) for Y would be

y	0	1	2
$P(y)$	14/30	14/30	2/30

b. The probability distribution of Y is $\frac{\binom{4}{y}\binom{6}{3-y}}{\binom{10}{3}}$ for $y = 0, 1, 2, 3$.

Thus the distribution (in tabular form) for Y would be

y	0	1	2	3
$P(y)$	5/30	15/30	9/30	1/30

3.90 **a.** Let Y denote the number of malfunctioning copiers selected. The distribution of Y is then hypergeometric with probability

$$P(Y = y) = \frac{\binom{3}{y}\binom{5}{4-y}}{\binom{8}{4}} \text{ for } y = 0, 1, 2, 3.$$

Then we have $P(Y = 0) = \frac{\binom{3}{0}\binom{5}{4}}{\binom{8}{4}} = \frac{5}{70} = \frac{1}{14}$.

b. Using part **a.** we have $P(Y \geq 1) = 1 - P(Y = 0) = 1 - \frac{1}{14} = \frac{13}{14}$.

3.91 The probability of an event as rare or rarer than the one observed can be calculated by using the hypergeometric distribution.

$$P(\text{one or fewer black members}) = \frac{\binom{8}{1}\binom{12}{5}}{\binom{20}{6}} + \frac{\binom{8}{0}\binom{12}{6}}{\binom{20}{6}} = \frac{8(792)}{38{,}760} + \frac{924}{38{,}760} = .187$$

This is not a very unlikely event, since it has probability close to $\frac{1}{5}$. It could very well have happened by chance. There is little reason to doubt the randomness of the selection.

3.92 $\mu = \frac{nr}{N} = \frac{6(8)}{20} = 2.4$

$\sigma^2 = n\left(\frac{r}{N}\right)\left(\frac{N-r}{N}\right)\left(\frac{N-n}{N-1}\right) = 6\left(\frac{8}{20}\right)\left(\frac{12}{20}\right)\left(\frac{14}{19}\right)$

$= \frac{8064}{7600} = 1.061$

3.93 The random variable Y follows the hypergeometric distribution with

$$p(y) = \frac{\binom{2}{y}\binom{4}{3-y}}{\binom{6}{3}} \qquad y = 0, 1, 2$$

The probability distribution for Y and the probability histogram are shown below.

y	$p(y)$
0	$\frac{1}{5}$
1	$\frac{3}{5}$
2	$\frac{1}{5}$

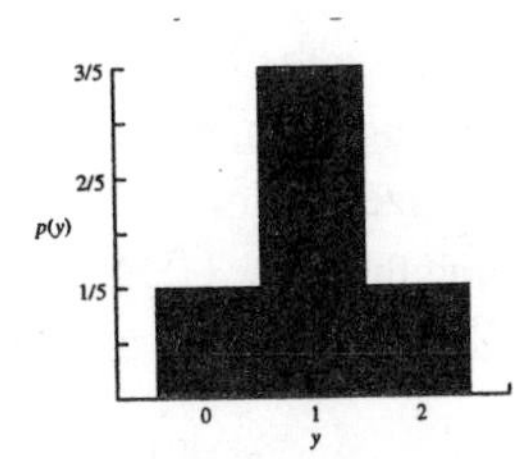

Figure 3.2

3.94 The student is asked to simulate the experiment described in Exercise 3.93. By doing so, he will obtain a probability distribution similar to the theoretical distribution shown above. The resulting data will consist of a sequence of numbers $y_1, y_2, y_3, \ldots, y_{100}$, each of which will assume the values 0, 1, or 2. Notice that in this chapter only theoretical distributions are examined, that is, distributions that describe the total population generated by repeating the experiment an infinite number of times. Thus by

performing the experiment only 100 times, the student is essentially taking a sample of 100 measurements from the population. Hence, the relative frequency of occurrence of each value of the random variable Y will be close to the theoretical probability of occurrence but will not necessarily be exactly the same.
The values that the random variable Y may assume are $Y = 0, 1, 2$. Let f_y denote the number of times that a particular outcome y is observed and $\frac{f_y}{n}$ its corresponding observed relative frequency. A possible outcome of the student's experiment might be as follows.

y	f_y	$\frac{f_y}{n}$
0	21	$\frac{21}{100}$
1	56	$\frac{56}{100}$
2	23	$\frac{23}{100}$

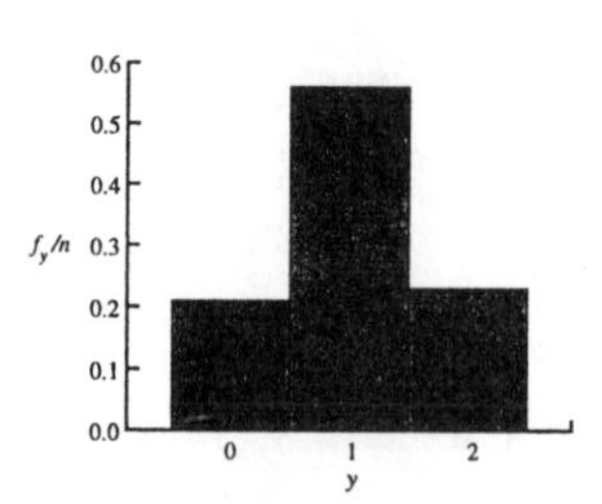

Figure 3.3

3.95 $N = 20$, $n = 5$, $r = 2$. Let Y = the number of improperly drilled gearboxes in the sample of 5. Then Y follows a hypergeometric distribution.

a. $P(Y = 0) = \frac{\binom{2}{0}\binom{18}{5}}{\binom{20}{5}} = .553$

b. The total time, T, that it takes to install the boxes (in minutes) is

$$T = 10Y + (5 - Y)$$
$$= 9Y + 5.$$

First,

$$E(Y) = \frac{nr}{N} = 5\left(\frac{2}{20}\right) = .5$$
$$V(Y) = \text{n}\left(\frac{r}{N}\right)\left(\frac{N-r}{N}\right)\left(\frac{N-n}{N-1}\right)$$
$$= 5(.1)(1 - .1)\left(\frac{20-5}{20-1}\right)$$
$$= .355.$$

It follows that

$$E(T) = 9E(Y) + 5 = 9.5$$

and

$$V(T) = 9^2V(Y) = 81(.355) = 28.755.$$

Thus, installation time should average 9.5 minutes, with a standard deviation of $\sqrt{28.755} = 5.362$ minutes.

3.96 There are N animals in the total population. After taking a sample of k animals, marking and releasing them, there are $N - k$ unmarked animals. We then choose a second sample of size 3 from the N animals. There exist $\binom{N}{3}$ ways of choosing this second sample and there are $\binom{N-k}{2}\binom{k}{1}$ ways of finding exactly one of the originally marked animals. For $k = 4$, the probability of finding just one marked animal is

$$\frac{\binom{N-4}{2}\binom{4}{1}}{\binom{N}{3}} = \frac{\frac{(N-4)(N-5)(4)}{2}}{\frac{N(N-1)(N-2)}{6}} = \frac{12(N-4)(N-5)}{N(N-1)(N-2)}$$

Calculating this probability for various values of N, we find that its value is maximized for $N = 11$ or $N = 12$.

N	4	5	6	7	8	9	10	11	12	13
Prob.	.000	.000	.200	.343	.429	.476	.500	.509	.509	.503

3.97 **a.** $P(Y = 4) = p(4) = \frac{2^4}{4!}e^{-2} = .090$
or, using Table 3, Appendix III,

$P(Y=4) = P(Y \leq 4) - P(Y \leq 3) = .947 - .857 = .090$

b. $P(Y \geq 4) = 1 - P(Y \leq 3) = 1 - \sum_{y=0}^{3} \frac{2^y}{y!} e^{-2} = 1 - .857 = .143$

Note that $P(Y \leq 3) = .857$ can also be obtained from Table 3, Appendix III.

c. $P(Y < 4) = P(Y \leq 3) = .857$

d. Recall from part **b** that $P(Y \geq 4) = .143$
and note that $P(Y \geq 2) = 1 - P(Y \leq 1) = 1 - .406 = .594$
Hence,

$$P(Y \geq 4 | Y \geq 2) = \frac{P(Y \geq 4 \text{ and } Y \geq 2)}{P(Y \geq 2)} = \frac{P(Y \geq 4)}{P(Y \geq 2)} = \frac{.143}{.594} = .241$$

3.98 Let Y be the number of customers arriving. Then Y follows a Poisson distribution with $\lambda = 7$. We perform the calculations exactly, however one could just as easily used table 3 appendix III.

a. $P(Y \leq 3) = p(0) + p(1) + p(2) + p(3) = \frac{7^0 e^{-7}}{0!} + \frac{7^1 e^{-7}}{1!} + \frac{7^2 e^{-7}}{2!} + \frac{7^3 e^{-7}}{3!} = .0818.$

b. $P(Y \geq 2) = 1 - P(Y \leq 1) = 1 - \frac{7^0 e^{-7}}{0!} - \frac{7^1 e^{-7}}{1!} = 1 - 8e^{-7} = .9927.$

c. $P(Y = 5) = \frac{7^5 e^{-7}}{5!} = .1277.$

3.99 Let S = total service time = $10Y$. From 3.98 we know that $Y \sim$ Poisson(7).
Therefore,

$E(S) = 10E(Y) = 10(\lambda) = 10(7) = 70.$
$V(S) = (10)^2 V(Y) = 100\lambda = 100(7) = 700.$

$$P(S > 150) = P(10Y > 150) = P(Y > 15) = 1 - P(Y \leq 15) = 1 - 0.998 = 0.002$$

where Table 3 was used to find $P(Y \leq 15)$. So, we infer that it is unlikely that the total service time will exceed 2.5 hours.

3.100

a. Let Y = number of customers that arrive in a given two-hour period of time. Then Y has a Poisson distribution with $\lambda = 2(7) = 14$ and

$$P(Y = 2) = \frac{14^2}{2!} e^{-14} = 8.45\text{x}10^{-6}.$$

b. The two one-hour time periods are non overlapping, and therefore Y = total number of customers that arrive in the given two-hour time period has a Poisson distribution with $\lambda = 2(7) = 14$, and, as for part **a**, $P(Y = 2) = 8.45\text{x}10^{-6}$. Consistent with this answer, note the following. Let Y_1 = number of customers that arrive between 1:00 p.m. and 2:00 p.m. and Y_2 = number of customers that arrive between 3:00 p.m. and 4:00 p.m. Then Y_1 and Y_2 are each distributed as Poisson with $\lambda = 8$ and

$$\begin{aligned} P(Y_1 + Y_2 = 2) &= P(Y_1 = 0, Y_2 = 2) + P(Y_1 = 1, Y_2 = 1) + P(Y_1 = 2, Y_2 = 0) \\ &= 2 \cdot p(0)p(2) + [p(1)]^2 \\ &= 2\left(\tfrac{7^0}{0!} e^{-7}\right)\left(\tfrac{7^2}{2!} e^{-7}\right) + \left(\tfrac{7^1}{1!} e^{-7}\right)^2 \\ &= 8.45\text{x}10^{-6}, \text{ the same answer as in part } \mathbf{a}.! \end{aligned}$$

3.101 Let Y be the number of typing errors per page. Then Y has a Poisson distribution with $\lambda = 4$.

$$P(Y \leq 4) = \sum_{y=0}^{4} \frac{4^y e^{-4}}{y!} = e^{-4} + 4e^{-4} + \frac{16}{2} e^{-4} + \frac{4^3}{6} e^{-4} + \frac{4^4}{24} e^{-4} = e^{-4}(34.333) = .6288$$

Or equivalently we could have looked this value up from table 3.

3.102 The probabilities of 0, 1, 2, or 3 cars arriving at a particular entrance are shown in the table below.

y	Entrance I ($\lambda = 3$)	Entrance II ($\lambda = 4$)
0	.0497871	.0183156
1	.1493612	.0732626
2	.2240418	.1465251
3	.2240418	.1953668

Each is calculated by using a Poisson distribution with mean $\lambda = 3$ or $\lambda = 4$. Then

$$\begin{aligned} P(3 \text{ cars}) &= P(0 \text{ through I, } 3 \text{ through II}) + P(1 \text{ through I, } 2 \text{ through II}) \\ &\quad + \ldots + P(3 \text{ through I, } 0 \text{ through II}) \\ &= .0497871(.1953668) + \ldots + (.2240418)(.0183156) = .0521 \end{aligned}$$

3.103 Let Y be the number of knots in the wood. Then Y has a Poisson distribution with $\lambda = 1.5$.

$$P(Y \leq 1) = \frac{(1.5)^0 e^{-1.5}}{0!} + \frac{(1.5)^1 e^{-1.5}}{1!} = 2.5e^{-1.5} = .5578$$

Or equivalently we could have looked this value up from table 3.

3.104 Let Y be the number of cars entering the tunnel in a two-minute period. Assume Y has a Poisson distribution with $\lambda = 1$. Then

$$\begin{aligned} P(Y > 3) &= 1 - P(Y \leq 3) = 1 - \sum_{y=0}^{3} \frac{e^{-1} 1^y}{y!} = 1 - \frac{e^{-1}}{0!} - \frac{e^{-1}}{1!} - \frac{e^{-1}}{2!} - \frac{e^{-1}}{3!} \\ &= 1 - \tfrac{8}{3} e^{-1} = .01899 \end{aligned}$$

Or equivalently we could have looked $P(Y \leq 3)$ up in table 3.

3.105 Define a random variable X, the number of times $Y > 3$ in 10 trials. This random variable will have a binomial distribution if we define a success to be the event "$Y > 3$" on a single trial and

$$p = P(Y > 3) = 1 - \tfrac{8}{3} e^{-1}$$

from Exercise 3.104. There are $n = 10$ trials and the probability of interest is

$$P(X \geq 1) = 1 - P(X = 0) = 1 - p^0 q^{10} = 1 - \left(\tfrac{8}{3} e^{-1}\right)^{10} = .1745$$

The Poisson model places a positive probability on any finite number of cars entering the tunnel in a one second interval. Physically only a certain number of cars can enter the tunnel in a one second interval suggesting a possible flaw in the Poisson model (this does not mean the Poisson model is useless for this situation).

3.106 The binomial probabilities for $n = 20$ and $p = .05$ are obtained by using the table as in previous exercises. However, for large n and small p such that $\lambda = np$ is less than 7, the following approximation can be used:

$$P(Y = \text{r}) \approx \frac{e^{-\lambda} \lambda^r}{r!} \qquad \text{where } \lambda = np \text{ and } r = 0, 1, 2, \ldots, n$$

The exact binomial probabilities and their Poisson approximations are shown in the accompanying table. In this case, $\lambda = np = 20(.05) = 1$.

y	$p(y)$ (Exact Binomial)	$p(y)$ (Poisson Approximation)
0	.358	.368
1	.378	.368
2	.189	.184
3	.059	.061
4	.013	.015

Notice that the approximation is not too bad, even though n is fairly small.

3.107 This exercise is similar to previous exercises. The random variable is Y, the number of sales in 100 contacts, and possesses a binomial distribution with $p = P(\text{sale}) = .03$. The probability of interest is

$$P(Y \geq 1) = 1 - P(Y = 0) = 1 - \binom{100}{0} (.03)^0 (.97)^{100} = 1 - (.97)^{100} = .9524$$

If one chooses to use the Poisson approximation with $\lambda = np = 3$, then one obtains

$$P(Y \geq 1) \approx 1 - \frac{3^0 e^{-3}}{0!} = 1 - e^{-3} = .9502$$

3.108 Let Y denote the number of deaths in 200 fires. Then Y has an approximate Poisson distribution with $\lambda = 3$. We use Table 3, Appendix III.

a. $P(Y > 8) = 1 - P(Y \le 8) = 1 - .996 = .004$.

b. Yes. If the region's rate is equal to the national average of 3 then a very rare event has occurred (see part **a**). We suspect that the region's rate is higher than 3 per 200 fires.

3.109 Use the Poisson approximation to the binomial with $\lambda = np = 30(.2) = .6$. Then

$$p(Y \le 3) = \sum_{y=0}^{3} \frac{6^y e^{-6}}{y!} = e^{-6} + 6e^{-6} + \frac{36}{2}e^{-6} + \frac{216}{6}e^{-6} = 61e^{-6} = .1512$$

3.110 Let $P(Y) = \frac{\lambda^y e^{-\lambda}}{y!}$. Then

$$E(Y(Y-1)) = \sum_{y=0}^{\infty} \frac{y(y-1)\lambda^y e^{-\lambda}}{y!} = \lambda^2 \sum_{y=2}^{\infty} \frac{\lambda^{y-2} e^{-\lambda}}{(y-2)!}$$

Using the substitution $z = y - 2$,

$$E(Y(Y-1)) = \lambda^2 \sum_{z=0}^{\infty} \frac{\lambda^z e^{-\lambda}}{z!} = \lambda^2$$

since the summation is that of a Poisson probability distribution over all possible values of z and therefore sums to 1. Now

$V(Y) = E(Y^2) - [E(Y)]^2$ and $E(Y(Y-1)) = E(Y^2) - E(Y)$

Hence $V(Y) = E(Y(Y-1)) - [E(Y)]^2 + E(Y) = \lambda^2 - \lambda^2 + \lambda = \lambda$.

3.111 For a Poisson random variable with $\lambda = 2$,

$$E(Y) = \lambda = 2$$

and

$$E(Y^2) = V(Y) + [E(Y)]^2 = \lambda + \lambda^2 = 2 + 4 = 6.$$

Then $E(X) = 50 - 2E(Y) - E(Y^2) = 50 - 2(2) - 6 = 40$.

3.112 The probability distribution for Y is Poisson with $\lambda = 2$. With $C = 100\left(\frac{1}{2}\right)^Y$,

$$E(C) = \sum_{y=0}^{\infty} \frac{100\left(\frac{1}{2}\right)^y 2^y e^{-2}}{y!} = 100 \sum_{y=0}^{\infty} \frac{1^y e^{-1} e^{-1}}{y!} = 100e^{-1} \sum_{y=0}^{\infty} \frac{1^y e^{-1}}{y!} = 100e^{-1}$$

3.113 Let Y = # of breakdowns per day. Y is Poisson with $\lambda = 2$.

$$\mu_y = E(Y) = 2;\ \text{Var}(Y) = E(Y^2) - \mu_y^2 = 2,$$

thus

$$E(Y^2) = 2 + 4 = 6.$$

Now, $E(R) = E(1600 - 50Y^2) = E(1600) - 50E(Y^2) = 1600 - 50(6) = \$1300$.

3.114

a. $p(y) = \frac{\lambda^y e^{-\lambda}}{y!}$ $\quad \lambda = 0, 1, 2, \ldots$

Then

$$\frac{p(y)}{p(y-1)} = \frac{\frac{\lambda^y e^{-\lambda}}{y!}}{\frac{\lambda^{y-1} e^{-\lambda}}{(y-1)!}} = \frac{\lambda}{y} \qquad y = 1, 2, 3, \ldots$$

b. $p(y) > p(y-1)$ or $\frac{p(y)}{p(y-1)} > 1$ if $\frac{\lambda}{y} > 1$ or $\lambda > y$.

Thus, $p(y) > p(y-1)$ for $1 \le y < \lambda$. Note, also, $p(y) < p(y-1)$ for $y > \lambda$ and $p(y) = p(y-1)$ for $y = \lambda$, if λ is an integer.

c. For λ not an integer, then part **b** implies that for $y - 1 < y < \lambda$, $p(y-1) < p(y) > p(y+1)$. Hence, $p(y)$ is maximized for y = largest integer less than λ. If λ is an integer, then $p(y)$ is maximized at both values $\lambda - 1$ and λ.

3.115 Using the binomial theorem, $(a+b)^n = \sum_{y=0}^{n} \binom{n}{y} a^y b^{n-y}$, we have:

$$m(t) = E(e^{ty}) = \sum_{y=0}^{n} \binom{n}{y} (pe^t)^y q^{n-y} = (pe^t + q)^n.$$

3.116 Refer to Exercise 3.115.

$$E(Y) = \frac{d}{dt} m(t)\Big|_{t=0} = (pe^t + q)^{n-1} npe^t\Big|_{t=0} = n(p+q)^{n-1}p = np.$$

$$\begin{aligned} E\left(Y^2\right) &= \frac{d^2}{dt^2} m(t)\Big|_{t=0} \\ &= npe^t(n-1)pe^t\left(pe^t+q\right)^{n-2} + n\left(pe^t+q\right)^{n-1}pe^t\Big|_{t=0} \\ &= np^2(n-1) + np \end{aligned}$$

$$V(Y) = E\left(Y^2\right) - n^2p^2 = n^2p^2 - np^2 + np - n^2p^2 = np(1-p) = npq$$

3.117 Recall that $\sum_{y=1}^{\infty} q^{y-1}p = 1$ or equivalently $\sum_{y=1}^{\infty} q^{y-1} = \frac{1}{p}$ for $0 < q < 1$. Now we have

$$m(t) = E\left(e^{ty}\right) = \sum_{y=1}^{\infty} pe^{ty}q^{y-1} = pe^t \sum_{y=1}^{\infty} e^{t(y-1)}q^{y-1} = pe^t \sum_{y=1}^{\infty} \left(qe^t\right)^{y-1} = \frac{pe^t}{1-qe^t}$$

if $qe^t < 1$ or equivalently, $t < -\ln q$.

3.118 Refer to Exercise 3.117.

$$\begin{aligned} E(Y) &= \frac{d}{dt} m(t)\Big|_{t=0} = \frac{(1-qe^t)(pe^t) - pe^t(-qe^t)}{(1-qe^t)^2}\Big|_{t=0} = \frac{pe^t}{(1-qe^t)^2}\Big|_{t=0} \\ &= \frac{p}{(1-q)^2} = \frac{1}{p} \end{aligned}$$

$$\begin{aligned} E\left(Y^2\right) &= \frac{d}{dt}\frac{pe^t}{(1-qe^t)^2}\Big|_{t=0} = \frac{(1-qe^t)^2 pe^t - 2pe^t(-qe^t)(1-qe^t)}{(1-qe^t)^4}\Big|_{t=0} \\ &= \frac{p^3+2pq^2}{p^4} = \frac{p+2q}{p^2} = \frac{1+q}{p^2} \end{aligned}$$

Finally,

$$V(Y) = \frac{1+q}{p^2} - \frac{1}{p^2} = \frac{q}{p^2}$$

3.119 Recall that the binomial mgf is $(pe^t + (1-p))^n$ (exercise 3.115). Therefore the distribution in question is binomial with success probability $p = .6$ and $n = 3$ trials.

3.120 Recall (exercise 3.117) that the geometric mgf is $\frac{pe^t}{1-qe^t}$. Therefore the distribution in question is geometric with success probability $p = .3$.

3.121 The distributions can be recognized by recalling the moment-generating functions of some common random variables given in this section.

a. Y has a binomial distribution with $n = 5$, $p = \frac{1}{3}$.

b. The form is closest to the geometric except for the "2" in the denominator. In order to comply with the form of this moment-generating function, we can multiply and divide by $\frac{1}{2}$.

$$m(t) = \frac{e^t}{2-e^t} = \frac{\left(\frac{1}{2}\right)e^t}{\left(\frac{1}{2}\right)(2-e^t)} = \frac{\left(\frac{1}{2}\right)e^t}{1-\left(\frac{1}{2}\right)e^t}$$

which is the mgf for a geometric random variable with $p = \frac{1}{2}$.

c. Refer to Example 3.21. Y is a Poisson random variable with $\lambda = 2$.

3.122 In Exercise 3.121 we found that

a. Y has a binomial distribution with $n = 5$, $p = \frac{1}{3}$. Then

$$E(Y) = 5\left(\tfrac{1}{3}\right) = 1.67, \text{ and } V(Y) = 5\left(\tfrac{1}{3}\right)\left(\tfrac{2}{3}\right) = 1.11.$$

b. Y has a geometric distribution with $p = \frac{1}{2}$. Then

$$E(Y) = \frac{1}{\left(\frac{1}{2}\right)} = 2, \text{ and } V(Y) = \frac{1-\frac{1}{2}}{\left(\frac{1}{2}\right)^2} = 2.$$

c. Y has a Poisson distribution with $\lambda = 2$. Then

$$E(Y) = 2, \text{ and } V(Y) = 2.$$

3.123 **a.** Differentiate $m(t)$ to find the necessary moments.

$$E(Y) = \frac{d}{dt} m(t)\Big|_{t=0} = \frac{1}{6}e^t + \frac{4}{6}e^{2t} + \frac{9}{6}e^{3t}\Big|_{t=0} = \frac{14}{6} = \frac{7}{3}$$

b. $E\left(Y^2\right) = \frac{d^2}{dt^2} m(t)\Big|_{t=0} = \frac{1}{6} + \frac{8}{6} + \frac{27}{6} = 6,\ V(Y) = 6 - \left(\frac{7}{3}\right)^2 = \frac{5}{9}.$

c. Since $m(t) = E\left(e^{ty}\right)$, Y must take only the values $Y = 1, 2,$ and 3, with probabilities $\frac{1}{6}$, $\frac{2}{6}$, and $\frac{3}{6}$, respectively.

3.124 $m_W(t) = E\left(e^{tW}\right) E\left(e^{t(aY+b)}\right) = E\left(e^{atY}e^{tb}\right) = e^{tb}E\left(e^{(at)Y}\right) = e^{tb}m_Y(at)$.

3.125 $m'_W(t) = be^{bt}m_Y(at) + e^{bt}m'_Y(at)a$

$m''_W(t) = b^2e^{bt}m_Y(at) + abe^{bt}m'_Y(at) + abm'_Y(at) + a^2e^{bt}m''_Y(at)$

Note that $m_Y(0) = E\left(e^{0Y}\right) = E(1) = 1$.

$$\begin{aligned} m'_W(t)\big|_{t=0} &= bm_Y(0) + am'_Y(0) \\ &= aE(Y) + b \\ &= E(W). \end{aligned}$$

$$\begin{aligned} m''_W(t)\big|_{t=0} &= b^2 + 2abE(Y) + a^2E\left(Y^2\right) \\ &= E\left(W^2\right). \end{aligned}$$

$$\begin{aligned} V(W) &= E\left(W^2\right) - (E(W))^2 \\ &= a^2E\left(Y^2\right) + 2abE(Y) + b^2 - \left(a^2\left[E(Y)\right]^2 + 2abE(Y) + b^2\right) \\ &= a^2\left[E\left(Y^2\right) - (E(Y))^2\right] \\ &= a^2V(Y). \end{aligned}$$

3.126 $r^{(1)}(t) = \frac{m^{(1)}(t)}{m(t)}$, $r^{(2)}(t) = \frac{m(t)m^{(2)}(t) - \left(m^{(1)}(t)\right)^2}{(m(t))^2}$.

$r^{(1)}(0) = \frac{m^{(1)}(0)}{m(0)} = \frac{E(Y)}{2}$.

$r^{(2)}(0) = \frac{E\left(Y^2\right) - (E(Y))^2}{1^2} = V(Y)$.

3.127 Since $m(t) = e^{5\left(e^t - 1\right)}$, $r(t) = 5\left(e^t - 1\right)$. Then

$E(Y) = r^{(1)}(0) = 5e^t\big|_{t=0} = 5$ $\qquad$ $V(Y) = r^{(2)}(0) = 5e^t\big|_{t=0} = 5$

3.128 $P(t) = E\left(t^Y\right) = \sum_{y=0}^{n} \binom{n}{y} p^y q^{n-y} t^y = (q + pt)^n$.

Differentiating with respect to t,

$$E(Y) = \tfrac{d}{dt}P(t)\big|_{t=1} = np(q+pt)^{n-1}\big|_{t=1} = np$$

3.129 $P(t) = \sum_{y=0}^{\infty} \frac{\lambda^y e^{-\lambda} t^y}{y!} = \frac{e^{-\lambda}}{e^{-\lambda t}} \sum_{y=0}^{\infty} \frac{(\lambda t)^y e^{-\lambda t}}{y!} = e^{-\lambda + \lambda t} = e^{\lambda(t-1)}$.

Differentiating with respect to t,

$$E(Y) = \tfrac{d}{dt}P(t)\big|_{t=1} = \lambda e^{\lambda(t-1)}\big|_{t=1} = \lambda,$$

$$E(Y(Y-1)) = \tfrac{d^2}{dt^2}P(t)\Big|_{t=1} = \lambda^2 e^{\lambda(t-1)}\big|_{t=1} = \lambda^2.$$

Thus $V(Y) = E(Y(Y-1)) + E(Y) - [E(Y)]^2 = \lambda^2 + \lambda - \lambda^2 = \lambda$.

3.130 $E(Y(Y-1)(Y-2)) = \frac{d^3}{dt^3}P(t)\Big|_{t=1} = \lambda^3 e^{\lambda(t-1)}\big|_{t=1} = \lambda^3$.

Now $E(Y(Y-1)(Y-2)) = E\left(Y^3\right) - 3E\left(Y^2\right) + 2E(Y)$, so that

$$E\left(Y^3\right) = \lambda^3 + 3E\left(Y^2\right) - 2E(Y) = \lambda^3 + 3\left(\lambda^2 + \lambda\right) - 2\lambda = \lambda^3 + 3\lambda^2 + \lambda$$

3.131

a. The point $Y = 6$ lies $\frac{11-6}{3} = \frac{5}{3}$ standard deviations below the mean. Similarly, the point $Y = 16$ lies $\frac{16-11}{3} = \frac{5}{3}$ standard deviations above the mean. According to Tchebysheff's theorem with $k = \frac{5}{3}$, at least $1 - \left(\frac{1}{k^2}\right) = 1 - \frac{9}{25} = .64$ of the measurements will be in the interval 6 to 16.

b. The second statement of Tchebysheff's theorem states that $P\left(|Y - \mu| > k\sigma\right) \le \frac{1}{k^2}$. To find C, let $.09 = \frac{1}{k^2}$. Then $k^2 = \frac{1}{.09} = \frac{100}{9}$ and $k = \frac{10}{3}$. Since $\sigma = 3$, $k\sigma = 3\left(\frac{10}{3}\right) = 10 = C$.

3.132 The random variable Y has a binomial distribution with $p = .35$ and $n = 2300$.

a. $E(Y) = np = (2300)(.35) = 805$

b. $V(Y) = npq = (2300)(.35)(.65) = 523.25 = \sigma^2$

$\sigma = \sqrt{523.25} = 22.875$

c. The interval is $(805 - 2(22.875), 805 + 2(22.875)) = (759.25, 850.75)$

d. The observation $Y = 249$ is 24.3 standard deviations below the mean value 805 $\left(\frac{805-249}{22.875} = 24.3\right)$. This value $(Y = 249)$ is not consistent with a rate of 35%

as Tchebyscheff's theorem tells us $P(|Y - 805| > (24.3)(22.875)) \leq \frac{1}{24.3^2} \approx .002$.

3.133 **a.** $E(Y) = (-1)p(-1) + (0)p(0) + (1)p(1)$
$= -1\left(\frac{1}{18}\right) + 0\left(\frac{16}{18}\right) + 1\left(\frac{1}{18}\right)$
$= 0.$

$E\left(Y^2\right) = (-1)^2 p(-1) + (0)^2 p(0) + (1)^2 p(1)$
$= (-1)^2\left(\frac{1}{18}\right) + (0)^2\left(\frac{16}{18}\right) + (1)^2\left(\frac{1}{18}\right)$
$= \frac{1}{9}$
$V(Y) = E\left(Y^2\right) - (E(Y))^2$
$= \frac{1}{9} - 0 = \frac{1}{9}$

b. $\sigma = \sqrt{V(Y)} = \sqrt{\frac{1}{9}} = \frac{1}{3}$.
By Tchebysheff's theorem,
$$P\left(|y - \mu| \geq 3\sigma\right) \leq \frac{1}{3^2} = \frac{1}{9}.$$
According to the probability distribution of Y,
$$P\left(|y - \mu| \geq 3\sigma\right) = P\left(|y| \geq 1\right) = p(-1) + p(1) = \frac{1}{18} + \frac{1}{18} = \frac{1}{9}$$
so that the bound is attained when $k = 3$.

c. Let x have the probability distribution
$$p(-1) = \frac{1}{8} \qquad p(0) = \frac{6}{8} \qquad \text{and} \qquad p(1) = \frac{1}{8}$$
so that $E[x] = 0$ and $V(x) = E\left[x^2\right] = \frac{1}{4}$.
It follows that
$$P\left(|X - \mu_x| \geq 2\sigma_x\right) = P\left(|x| \geq 1\right) = p(-1) + p(1) = \frac{1}{4},$$
as desired.

d. Letting all the probability mass be on values -1, 0, 1, $E(W) = 0$ if
$$p(-1) = p \qquad p(0) = 1 - 2p \qquad \text{and} \qquad p(1) = p$$
for some probability p. We want $k\sigma_W = 1$ so that $\sigma_W = \frac{1}{k}$ and $\sigma_W^2 = \frac{1}{k^2}$.
With $E(W) = 0$, the $V(W) = E\left(W^2\right) = 2p$. Setting $2p = \frac{1}{k^2}$ gives $p = \frac{1}{2k^2}$.
Therefore for any specified $k > 1$, $P\left(|W - \mu_W| \geq k\sigma_W\right) = \frac{1}{k^2}$ if
$$p(-1) = \frac{1}{2k^2} \qquad p(0) = 1 - \frac{1}{k^2} \qquad \text{and} \qquad p(1) = \frac{1}{2k^2}.$$
Alternatively, we can show the same result using complements. We want, with $k\sigma_W = 1$,
$$P(W = 0) = P\left(|W - \mu_W| < k\sigma_W\right) = 1 - \frac{1}{k^2}$$
Then, in order for $E(W) = 0$, we must have the same distribution as above.

3.134 Similar to Exercise 3.131 **a.** The interval .48 to .52 is the interval $|Y - .50| \leq 2\sigma$. Hence the lower bound is $1 - \left(\frac{1}{k^2}\right) = 1 - \left(\frac{1}{4}\right) = \frac{3}{4}$. The expected number of coins is then at least $\left(\frac{3}{4}\right)(400) = 300$.

3.135 Using Tchebysheff's theorem in its first form, we find that the lower bound, $\frac{5}{9}$, must equal $1 - \left(\frac{1}{k^2}\right)$. That is,
$$k^2 = \frac{9}{4} \qquad \text{and} \qquad k = \frac{3}{2}$$
The interval of interest is $\mu \pm \left(\frac{3}{2}\right)\sigma = 100 \pm \left(\frac{3}{2}\right)(10) = 100 \pm 15$, or 85 to 115.

3.136 Refer to Exercise 3.93. The expected value and variance of Y are
$$\mu = E(Y) = 0(.2) + 1(.6) + 2(.2) = 1$$
and
$$\sigma^2 = V(Y) = E\left(Y^2\right) - \mu^2 = 0^2(.2) + 1^2(.6) + 2^2(.2) - 1^2 = .6 + .8 - 1 = .4$$
respectively. Thus
$$\sigma = \sqrt{.4} = .63$$
The interval of interest is $\mu \pm 2\sigma = 1 \pm 2(.63) = 1 \pm 1.26$, or $-.26$ to 2.26. Note that Y can only take the values 0, 1, or 2, so that 100% of the measurements will fall within

two standard deviations of the mean. This is consistent with Tchebysheff's theorem.

3.137 The random variable Y is defined to be the number of heads observed when a coin is flipped three times. Then $p = P(\text{head}) = \frac{1}{2}$.

a. The binomial probabilities are as follows:

$$P(Y=0) = p(0) = \binom{3}{0}\left(\frac{1}{2}\right)^0\left(\frac{1}{2}\right)^3 = \frac{1}{8}$$
$$P(Y=2) = p(2) = \binom{3}{2}\left(\frac{1}{2}\right)^2\left(\frac{1}{2}\right)^1 = \frac{3}{8}$$
$$P(Y=1) = p(1) = \binom{3}{1}\left(\frac{1}{2}\right)^1\left(\frac{1}{2}\right)^2 = \frac{3}{8}$$
$$P(Y=3) = p(3) = \binom{3}{3}\left(\frac{1}{2}\right)^3\left(\frac{1}{2}\right)^0 = \frac{1}{8}$$

b. The associated probability distribution is shown in Figure 3.4.

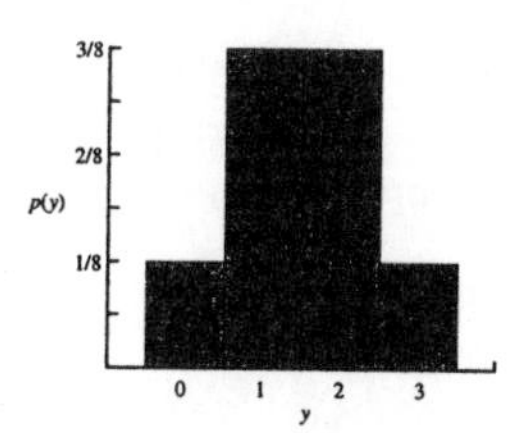

Figure 3.4

c. $\mu = E(Y) = np = 3\left(\frac{1}{2}\right) = 1.5$

$\sigma = \sqrt{V(Y)} = \sqrt{npq} = \sqrt{3\left(\frac{1}{2}\right)\left(\frac{1}{2}\right)} = .866$

d. The desired intervals are

$\mu \pm \sigma = 1.5 + .866$ or $.634$ to 2.366

and $\mu \pm 2\sigma = 1.5 + 1.732$ or $-.232$ to 3.232

The values of the random variable Y that fall within the first interval are the values 1 and 2. Thus the fraction of measurements within this interval will be $\frac{3}{8} + \frac{3}{8} = \frac{3}{4}$. The second interval encloses all four values of Y, and thus the fraction of measurements within two standard deviations of the mean will be 1, or 100% of the measurements. These results are consistent with both Tchebysheff's theorem and the empirical rule.

3.138 Refer to Exercise 3.137. In this case $p = .1$ instead of $p = .5$. The probabilities associated with the random variable Y are as follows:

a. $p(0) = \binom{3}{0}(.1)^0(.9)^3 = .729$ $\quad p(2) = \binom{3}{2}(.1)^2(.9)^1 = .027$

$p(1) = \binom{3}{1}(.1)^1(.9)^2 = .243$ $\quad p(3) = \binom{3}{3}(.1)^3(.9)^0 = .001$

b. Notice that the probability distribution, shown in Figure 3.5, is no longer symmetrical; that is, since the probability of observing a head is so small, the probability of observing a small number of heads on three flips is increased.

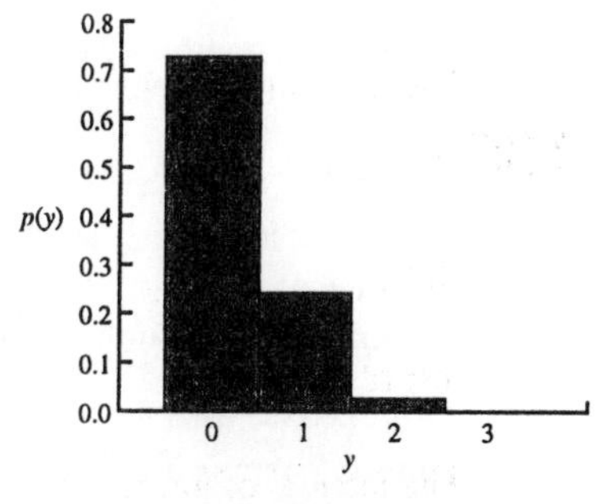

Figure 3.5

c. The mean and standard deviation are

$\mu = E(Y) = 3(.1) = .3$ and $\sigma = \sqrt{V(Y)} = \sqrt{3(.1)(.9)} = .520$

d. The desired intervals are

$\mu \pm \sigma = .3 \pm .520$ or $-.220$ to $.820$

$\mu \pm 2\sigma = .3 \pm 1.04$ or $-.740$ to 1.34

The only value of Y enclosed in the first interval is the value $Y = 0$, so that .729 of the measurements will be within one standard deviation of the mean. The values $Y = 0$ and $Y = 1$ are enclosed by the second interval, so that $.729 + .243$ or .972 of the measurements will be within two standard deviations, consistent with Tchebysheff's theorem and the empirical rule.

3.139 Let Y be the number of fatalities. Then Y is binomial with $p = .0006$ and $n = 40{,}000$.

a. $E(Y) = np = 40{,}000(.0006) = 24$

b. $V(Y) = npq = 24(.9994) = 23.9856 = \sigma^2$

$\sigma = \sqrt{23.9856} = 4.898$

c. No. The value 40 is $\frac{40-24}{4.898} = 3.26$ standard deviations above the mean, and both Tchebyscheff's theorem and the empirical rule suggest this is unlikely. (Tchebyscheff's theorem is probably more relevant in this situation, see 3.120).

3.140 We assume that the 1502 sample observations are independent of each other, so that the number of "successes" is binomial with $n = 1502$. If people have no preference then $p = .5$, so that $E(Y) = 1502(.5) = 751$ and $\sigma = 19.38$. We have observed $Y = (.71)(1502) = 1066$. This value is $\frac{1066-751}{19.38} = 16.25$ standard deviations above the mean!

3.141 The mean of C is $E(C) = \$50 + \$3\,E(Y) = \$50 + \$3(10) = \$80$. The variance is $V(C) = V(50 + 3Y) = 9V(Y) = 9(10) = 90$, so that $\sigma = \sqrt{90} = 9.487$. Using Tchebysheff's theorem with $k = 2$, we have $P(|Y - 80| < 2(9.487)) \geq .75$ so that the required interval is $(80 - 2(9.487), 80 + 2(9.487))$ or $(61.03, 98.97)$.

3.142 The random variable Y is defined to be the number of original tubes that must be replaced. Thus Y is the number of tubes that burn out before their guarantee has expired. Then $p = .1$ and $n = 1000$; $E(Y) = np = 100$; $V(Y) = npq = 90$. One would expect that at least $\frac{3}{4}$ of the 100 values of Y would be within two standard deviations of the mean, according to Tchebysheff's theorem. Using this, one can construct the interval $\mu \pm 2\sigma = 100 \pm 2\sqrt{90} = 100 \pm 19$, or 81 to 119.

3.143 Using Tchebysheff's theorem, Notice:

$$P(Y \geq \mu + k\sigma) = P(Y - \mu \geq k\sigma) \leq P(|Y - \mu| \geq k\sigma) \leq \tfrac{1}{k^2}.$$

Hence to find

$$P(Y \geq 350) \leq \tfrac{1}{k^2}$$

we evaluate $150 + k(67.081) = 350$. Which gives $k = 2.98$. Thus,

$$P(Y \geq 350) \leq \tfrac{1}{(2.98)^2} = .1126.$$

No, this is not highly unlikely.

3.144 Number of combinations $= 26 \cdot 26 \cdot 10 \cdot 10 \cdot 10 \cdot 10 = 6{,}760{,}000$.

$$E(\text{winnings}) = \$100{,}000\left(\tfrac{1}{6{,}760{,}000}\right) + \$50{,}000\left(\tfrac{2}{6{,}760{,}000}\right) + \$1000\left(\tfrac{10}{6{,}760{,}000}\right)$$
$$= \$0.031065.$$

Therefore, it appears that the expected value of the coupon is considerably less than the price of a stamp. However, one might also consider what the probability of winning would be if the coupon isn't mailed back.

3.145 This exercise asks for the probability of accepting a lot of items when the following sampling plan is used: draw a sample of five items and accept the lot if no defectives are observed. Thus,

$$P(\text{acceptance}) = P(\text{observe no defectives}) = \binom{5}{0} p^0 q^5$$

where p is the probability of observing a defective. By substituting the five specific values for p in the above formula, the various probabilities of acceptance are obtained.

p = Fraction Defective	P(Acceptance)
.0	1.0000
.1	.5905
.3	.1681
.5	.0312
1.0	0.0000

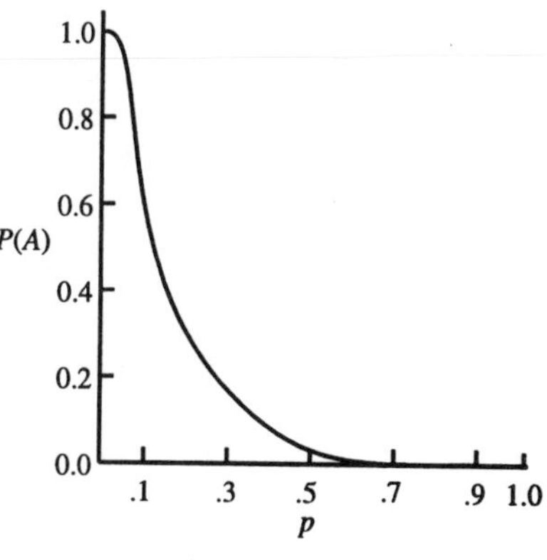

Figure 3.6

Notice that when the fraction defective is 0 (that is, there are no defectives in the lot), the lot will always be accepted and P(acceptance) = 1. The operating characteristic curve for this plan is shown in Figure 3.6.

3.146 **a.** Similar to Exercise 3.145. In this case,

$$P(\text{acceptance}) = \binom{10}{0} p^0 q^{10}$$

Notice that when $p = 0$ and $p = 1$, the probabilities of acceptance are 1 and 0, respectively, regardless of the sampling plan.

p	.00	.05	.10	.30	.50	1.00
P(acceptance)	1.000	.599	.349	.028	.001	.000

b. Similar to part **a**, with $P(\text{acceptance}) = \binom{10}{0} p^0 q^{10} + \binom{10}{1} p^2 q^9$.

p	.00	.05	.10	.30	.50	1.00
P(acceptance)	1.000	.914	.736	.149	.011	.000

c. Here, $P(\text{acceptance}) = \binom{10}{0} p^0 q^{10} + \binom{10}{1} p^2 q^9 + \binom{10}{2} p^2 q^8$.

p	P(acceptance)
.00	1.000
.05	.988
.10	.930
.30	.383
.50	.055
1.00	.000

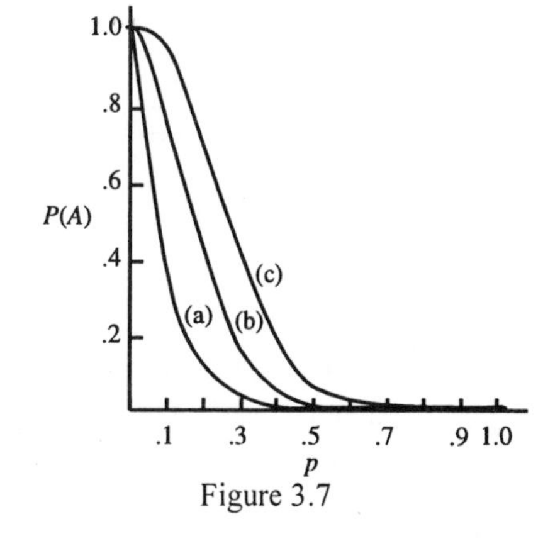

Figure 3.7

The three operating characteristic curves are shown in Figure 3.7. Notice that if n is held constant while a is increased, the probability of acceptance will be increased. For our sample plans, with $a = 0$,

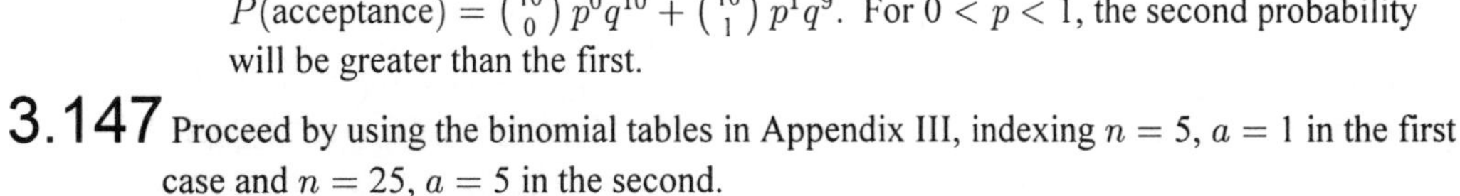

$P(\text{acceptance}) = \binom{10}{0} p^0 q^{10}$, while when $a = 1$, $P(\text{acceptance}) = \binom{10}{0} p^0 q^{10} + \binom{10}{1} p^1 q^9$. For $0 < p < 1$, the second probability will be greater than the first.

3.147 Proceed by using the binomial tables in Appendix III, indexing $n = 5$, $a = 1$ in the first case and $n = 25$, $a = 5$ in the second.

a. If the fraction defective in the lot ranges from $p = 0$ to $p = .10$, the seller would want the probability of accepting in this interval to be as high as possible. Hence he would choose the second plan.

b. If the buyer wishes to be protected against accepting lots with fraction defective greater than .3, he would want the probability of acceptance when p is greater than .3 to be as small as possible. Thus he would also choose the second plan.

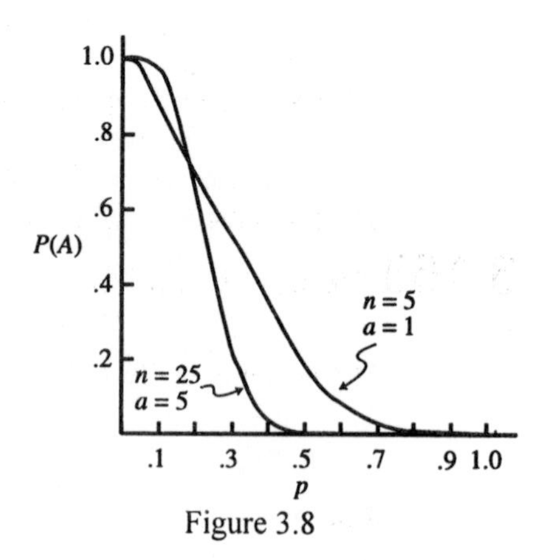

Figure 3.8

3.148 The total cost incurred is $W = 30Y$.

$$E(W) = 30E(Y) = 30\left(\frac{1}{.3}\right) = 100,$$
$$V(W) = 30^2V(Y) = 30^2\left(\frac{.7}{.3^2}\right) = 7000.$$

According to the empirical rule we would expect the interviewer costs to fall within

$$\mu \pm 3\sigma \Rightarrow 100 \pm 3(83.67) \Rightarrow 100 \pm 251, \text{ or } 151 \text{ to } 351.$$

3.149 Let Y = the number of rolls the player throws until the player stops. Then Y is a distributed geometric with $p = \frac{5}{6}$.

a. $P(Y = 3) = q^2p = \left(\frac{1}{6}\right)^2\left(\frac{5}{6}\right) = .023$

b. $E(Y) = \frac{1}{p} = \frac{6}{5} = 1.2$

c. Let X = amount paid to player. Then

$X = 2^{Y-1}$ and $E(X) = E\left(2^{Y-1}\right)$

$$E(X) = \sum_{y=1}^{\infty} 2^{y-1}q^{y-1}p$$
$$= p\sum_{y=1}^{\infty}(2q)^{y-1} \text{ (see exercise problem 3.44 a)}$$
$$= p\frac{1}{1-2q}$$

as the sum of a geometric series, since $2q < 1$

$$= \frac{5}{6}\left(\frac{1}{1-\frac{2}{6}}\right) = \frac{5}{4}$$
$$= \$1.25.$$

3.150
$$P(Y > 1|Y \geq 1) = \frac{P(Y>1 \cap Y\geq 1)}{P(Y\geq 1)}$$
$$= \frac{P(Y>1)}{P(Y\geq 1)}$$
$$= \frac{1-[P(Y=0)+P(Y=1)]}{1-P[Y=0]}$$
$$= \frac{1-(1-p)^n - np(1-p)^{n-1}}{1-(1-p)^n}$$

3.151 The random variable is Y, the number of failures in 10,000 starts, and it possesses a binomial distribution with $p = P(\text{failure}) = .00001$. The probability of interest is

$$P(Y \geq 1) = 1 - P(Y = 0) = 1 - \binom{10{,}000}{0}(.00001)^0(.99999)^{10{,}000}$$
$$= 1 - (.99999)^{10{,}000} = .095$$

3.152 The expected value of Y for the experiment in Exercise 3.93 is calculated to be

$$E(Y) = 0\left(\frac{1}{5}\right) + 1\left(\frac{3}{5}\right) + 2\left(\frac{1}{5}\right) = 1$$

Intuitively, then, one would expect the sample mean generated by the 100 measurements obtained in Exercise 3.78 to be close to 1. To find the sample mean, one can apply the formula $\overline{y} = \sum_{j=1}^{n} \frac{y_j}{n}$, where the y_j's are the elements of the sequence defined in Exercise 3.94. Using the sample data from this exercise, we find that the value $Y = 0$ has been observed 21 times, the value $Y = 1$ observed 56 times, and the value $Y = 2$ observed 23 times. Hence the sample mean is

$$\overline{y} = \frac{\sum_{j=1}^{100} y_j}{100} = \frac{21(0)+56(1)+23(2)}{100} = \frac{102}{100} = 1.02$$

which is a good estimate for the true population mean μ.

3.153 The population variance for Exercise 3.93 is calculated as usual and is found to be

$$V(Y) = \Sigma y^2 p(y) - 1^2 = 0\left(\tfrac{1}{5}\right) + 1\left(\tfrac{3}{5}\right) + 4\left(\tfrac{1}{5}\right) - 1 = \tfrac{2}{5} = .4$$

The sample variance is calculated as

$$s^2 = \frac{\sum_{j=1}^{n}(y_j - \overline{y})^2}{n-1} = \frac{\sum_{j=1}^{n} y_j^2 - \frac{\left(\sum_{j=1}^{n} y_j\right)^2}{n}}{n-1}$$

Refer to Exercise 3.94 and note the number of times that each value of Y occurred. Thus $y = 0$ must be squared 21 times, $Y = 1$ squared 56 times, and so on. The value of s^2 then is

$$s^2 = \frac{[0^2(21)+1^2(56)+2^2(23)] - \frac{[0(21)+1(56)+2(23)]^2}{100}}{99} = \frac{148-104.04}{99}$$

$$= \frac{43.96}{99} = .444$$

which is a good estimate for the population variance σ^2.

3.154 The random variable is Y, the number of dots observed on the upper face. There are 6 sample points in the experiment that, being equally likely, each have probability $\frac{1}{6}$. The probability distribution and histogram are shown below.

y	$p(y)$
1	.167
2	.167
3	.167
4	.167
5	.167
6	.167

Figure 3.9

The mean and variance were found in Exercise 3.16 to be $E(Y) = 3.5$ and $V(Y) = 2.9167$. The desired interval is thus $\mu \pm 2\sigma = 3.5 \pm 3.42$, or .08 to 6.92. Note that Y can take on only the values 1, 2, 3, 4, 5, and 6, so that 100% of the measurements will fall within two standard deviations of the mean.

3.155 Let Y_1 = number of defectives out of the 5 from line I,
and Y_2 = number of defectives out of the 5 from line II.
Then $Y_1 \sim$ binomial $(5, p)$ and $Y_2 \sim$ binomial $(5, p)$, where p is the common probability of a defective regulator. Also, $Y_1 + Y_2 \sim$ binomial $(10, p)$.

$$P(Y_1 = 2 | Y_1 + Y_2 = 4) = \frac{P(Y_1 = 2 \cap Y_1 + Y_2 = 4)}{P(Y_1 + Y_2 = 4)} = \frac{P(Y_1 = 2)P(Y_2 = 2)}{P(Y_1 + Y_2 = 4)}$$

$$= \frac{\binom{5}{2} p^2 q^3 \binom{5}{2} p^2 q^3}{\binom{10}{4} p^4 q^6} = \frac{\binom{5}{2}\binom{5}{2}}{\binom{10}{4}} = .476$$

Notice the end probability does not depend on the value of p!

3.156 If the gambler loses three times in a row, he will have lost his initial capital. To win exactly once he must therefore win either on the first, second, or third trial and then lose continually from that point on, until all of his capital is gone. The three possible sequences of wins and losses, then, that result in the event of interest are

$$WLLLLLLLLLL \qquad LWLLLLLLLLL \qquad LLWLLLLLLLL$$

and the desired probability is

$$(.1)(.9)^{10} + (.9)(.1)(.9)^9 + (.9)^2(.1)(.9)^8 = 3(.1)(.9)^{10} = .104.$$

3.157 a. Define X to be the number of imperfections in one square yard of weave. Then

P(one-square-yard sample will contain at least one imperfection)

is equal to

$$P(X \geq 1) = 1 - P(X = 0) = 1 - \tfrac{4^0}{0!}e^{-4} = 1 - .018 = .982.$$

b. We present two solutions. Let X_1, X_2, and X_3 be the number of imperfections in the first, second, and third one square yards of weave, respectively. Then X_1, X_2, and X_3 are independent, each having a Poisson distribution with $\lambda = 4$. So

P(three-square-yard sample will contain at least one imperfection)

$$= P(X_1 + X_2 + X_3 \geq 1)$$
$$= 1 - P(X_1 + X_2 + X_3 = 0)$$
$$= 1 - P(X_1 = 0)P(X_2 = 0)P(X_3 = 0)$$
$$= 1 - (.018)^3$$

Notice this is the same answer we would get if we had repeated the calculation in **a** with a λ of 3x4=12.

3.158 Let Y be the number of imperfections in an 8-square-yard bolt of the textile. Then Y has a Poisson distribution with mean 32 and variance 32.
Let $X = 10Y$ be the cost of repairing the weave. Then

$$E(X) = 10E(Y) = \$320$$

and

$$V(X) = 10^2V(Y) = \$3200.$$

3.159 a. Let X be the number of bacteria colonies in a one-cubic-centimeter sample. Then X has a Poisson distribution with $\lambda = 2$. Now let Y be the number of the four one-cubic-centimeter samples that have one or more bacterial colonies. Note that Y has a binomial distribution with $n = 4$ and

$$p = P(\text{a sample contains one or more bacteria colonies})$$
$$= P(X \geq 1) = 1 - P(X = 0) = 1 - .135 = .865$$

Thus,

P(at least one sample will contain one or more bacteria colonies)

$$= P(Y \geq 1) = 1 - P(Y = 0) = 1 - \binom{4}{0}(.865)^0(.135)^4 = .9997$$

b. We need to find the number of samples, n, such that approximately

$$P(Y \geq 1) = 1 - P(Y = 0) = 1 - \binom{n}{0}(.865)^0(.135)^n = .95$$

or $\quad (.135)^n = .05$

which implies that $\ln(.135)^n = n\ln(.135) = \ln(.05)$

so $\quad n = \frac{\ln(.05)}{\ln(.135)} = 1.496$

So, being conservative, we take $n = 2$.

3.160 Write

$$(q + pe^t)^n\left[q + p\left(1 + t + \tfrac{t^2}{2!} + \tfrac{t^3}{3!} + \ldots\right)\right]^n = \left(q + p + pt + p\tfrac{t^2}{2!} + p\tfrac{t^3}{3!} + \ldots\right)^n$$
$$= \left(1 + pt + p\tfrac{t^2}{2!} + p\tfrac{t^3}{3!} + \ldots\right)^n$$

The terms that are of interest in this exercise are only those terms that contain either t or t^2, since we are interested in obtaining only μ_1' and μ_2', the coefficients of t and $\frac{t^2}{2!}$. Hence we need only to expand the above multinomial to show the first few terms. Then

$$(q + pe^t)^n = \left[1^n + n(pt)(1)^{n-1} + n\left(p\tfrac{t^2}{2!}\right)(1)^{n-1} + \tfrac{n(n-1)}{2}(pt)^2(1)^{n-2}\right.$$
$$\left. + \text{(terms involving } t^3 \text{ and higher powers)}\right].$$

Recall that the multinomial coefficient was given in Chapter 2 as

$$\frac{n!}{n_1!\,n_2!\cdots n_k!}$$

where n_i represents the exponent given in the ith member of the multinomial sum in the particular term we wish to evaluate. For example, the fourth term in the above expansion is actually

$$\frac{n!}{2!\,(n-2)!\,0!\,0!\cdots}(1)^{n-2}(pt)^2\left(p\,\frac{t^2}{2!}\right)^0\left(p\,\frac{t^3}{3!}\right)^0\cdots=\frac{n(n-1)}{2}\,p^2t^2$$

Thus the coefficient of t is $np(1)^{n-1}=np$. The term involving t^2 is

$$np(1)^{n-1}\,\frac{t^2}{2}+n(n-1)p^2\,\frac{t^2}{2}$$

so that the coefficient of $\frac{t^2}{2}$ is $np+n(n-1)p^2$, which agrees with the results of Exercise 3.100.

3.161

Let Y be the number of defective machines. Then Y follows the hypergeometric distribution with

$$P(y)=\frac{\binom{4}{y}\binom{6}{5-y}}{\binom{10}{5}}$$

The mean of the repair cost is

$$E(50Y)=50E(Y)=50n\left(\tfrac{K}{N}\right)=50(5)\left(\tfrac{4}{10}\right)=100$$

and the variance of the repair cost is

$$V(50Y)=2500V(Y)=2500n\left(\tfrac{K}{N}\right)\left(1-\tfrac{K}{N}\right)\left(\tfrac{N-n}{N-1}\right)$$
$$=2500(5)\left(\tfrac{4}{10}\right)\left(1-\tfrac{4}{10}\right)\left(\tfrac{10-5}{10-1}\right)=1666.67$$

According to Tchebysheff's theorem, with probability of at least .75, $50Y$ will be within two standard deviations of the mean. Using this, one can construct the interval

$$\mu\pm2\sigma=100\pm2\sqrt{1666.67},\text{ or }18.35\text{ to }181.65$$

3.162

Let W = # of drivers who wish to park, and W' = # of cars, which is Poisson with mean λ. We consider

$$\begin{aligned}
P(W=k) &= \sum_{n=k}^{\infty}P(W=k\cap W'=n)=\sum_{n=k}^{\infty}P(W=k|W'=n)\,P(W'=n)\\
&= \sum_{n=k}^{\infty}\left\{\left[\frac{n!}{k!\,(n-k)!}\right]p^k(1-p)^{n-k}\right\}\left[e^{-\lambda}\left(\frac{\lambda^n}{n!}\right)\right]\\
&= \lambda^k e^{-\lambda}\left(\frac{p^k}{k!}\right)\sum_{n=k}^{\infty}\left[\frac{(1-p)^{n-k}}{(n-k)!}\right]\lambda^{n-k}\\
&= \left(\frac{\lambda^k p^k e^{-\lambda}}{k!}\right)\sum_{j=0}^{\infty}\frac{[(1-p)\lambda]^j}{j!}=\left[\frac{(\lambda p)^k}{k!}\right]e^{-\lambda}e^{(1-p)\lambda}\\
&= \left[\frac{(\lambda p)^k}{k!}\right]e^{-\lambda p}.
\end{aligned}$$

Where $j=n-k$, in line 4 of the previous equation.

a. If W = # of drivers who wish to park, then the probability that a space will still be available when you reach the lot = $P(W=0)$. Using the information that was derived above,

$$P(W=0)=\left[\frac{(\lambda p)^0}{0!}\right]e^{-\lambda p}=e^{-\lambda p}.$$

b. Using the information derived above, we see that the probability distribution for W is Poisson with mean λp.

3.163

Note that $Y(t)$ has a negative binomial distribution with parameter $r=k$, $p=e^{-\lambda t}$.

a. $E[Y(t)]=\frac{r}{p}=ke^{\lambda t}$, $\quad V[Y(t)]=\frac{rq}{p^2}=\frac{k\,(1-e^{-\lambda t})}{e^{-2\lambda t}}=k\left(e^{2\lambda t}-e^{\lambda t}\right)$

b. We are given that $k=2$, $\lambda=0.1$, and $t=5$. Therefore, $E[Y(t)]=ke^{\lambda t}=2e^{(5)(0.1)}$ $=2e^{1/2}=3.2974$, $V[Y(t)]=2\left[e-e^{.5}\right]=2.139$

3.164

Let Y = number of left-turning vehicles out of n vehicles arriving while the light is red. Then Y has a binomial distribution with parameters $n=5$, $p=0.2$, and using Table 1, Appendix III, we have $P(Y\le3)=0.993$. This may be computed directly as

$$\begin{aligned} P(Y \le 3) &= 1 - P(Y \ge 4) \\ &= 1 - P(Y = 4) - P(Y = 5) \\ &= 1 - \binom{5}{4}(0.2)^4(0.8)^1 - \binom{5}{5}(0.2)^5(0.8)^0 = 0.993 \end{aligned}$$

3.165 P(toss a die ten times before observing four 6's and a 6 occurs on the ninth and tenth tosses)

$= P$(that two 6's occur in the first eight tosses of the die) $\times \left(\frac{1}{6}\right)^2$

$= \left[P(Y = 2) \text{ where } Y \text{ is binomial with } n = 8 \text{ and } p = \frac{1}{6}\right] \times \left(\frac{1}{6}\right)^2.$

Now, $P(Y = 2) = \binom{8}{2}\left(\frac{1}{6}\right)^2\left(\frac{5}{6}\right)^6 = .26$.

Our desired probability is given by $.26 \times \left(\frac{1}{6}\right)^2 = .00722$. A second solution is obtained by noting the probability that the third success occurs on the 9th trial is the negative binomial probability $\binom{8}{2}\left(\frac{1}{6}\right)\left(\frac{5}{6}\right)^6$. Multiplying this by the probability of a 6 on the 10 trial, $\frac{1}{6}$, gives the same answer as before.

3.166 Let Y represent the gain to the insurance company for a particular insured driver and let P be the premium charged to the driver. With probability .85, no accident occurs and the company gains P dollars. With probability .15, an accident does occur and the company gains either $P - 2400$, $P - 7200$, or $P - 12{,}000$ dollars, with probabilities .80, .12, and .08, respectively. The probability distribution for Y is given in the accompanying table.

y	$p(y)$
P	.85
$P - 2400$	$.15(.80) = .12$
$P - 7200$	$.15(.12) = .018$
$P - 12{,}000$	$.15(.08) = .012$

In order for the company to break even,

$$\begin{aligned} E(Y) &= .85P + .12(P - 2400) + .018(P - 7200) + .012(P - 12{,}000) \\ &= 0 \end{aligned}$$

Solving for P, we have

$$P - 288 - 129.6 - 144 = 0 \quad \text{or} \quad P = \$561.60$$

3.167 Use the Poisson distribution with $\lambda = 5$.

a. $p(2) = \frac{5^2 e^{-5}}{2!} = 12.5e^{-5} = .084$

$$P(Y \le 2) = p(0) + p(1) + p(2) = \frac{5^0 e^{-5}}{0!} + \frac{5^1 e^{-5}}{1!} + \frac{5^2 e^{-5}}{2!} = 18.5e^{-5} = .125$$

b. $P(Y > 10) = 1 - P(Y \le 10) = 1 - .986 = .014$ (using Table 3, Appendix III). Yes, it is unusual that Y will exceed 10.

3.168 If the public is split 50-50, then Y, the number of people in favor of the proposition, is binomial with $n = 2000$ and $p = .5$. In this case,

$$E(Y) = 2000(.5) = 1000$$

and

$$V(Y) = 2000(.5)(.5) = 500 = \sigma^2$$

so that $\sigma = \sqrt{500} = 22.36$.

The value 1373 is $\frac{1373-1000}{22.36} = 16.68$ standard deviations above the mean. Such a value is unlikely.

3.169 We are interested in Y, the number of contacts necessary to obtain the third sale, which is an example of the negative binomial random variable described in Section 3.6. In this case, $r = 3$, $p = .3$, and we have

$$\begin{aligned} P(Y < 5) &= p(3) + p(4) = \binom{2}{2}(.3)^3(.7)^{3-3} + \binom{3}{2}(.3)^3(.7)^1 = (.3)^3 + 3(.3)^3(.7) \\ &= .0837 \end{aligned}$$

3.170 Refer to Example 3.20 with $\mu = \lambda = 3$ and $\sigma^2 = \lambda = 3$. The probability of interest is

$$P\left(|Y - 3| \le 2\sqrt{3}\right) = P(|Y - 3| \le 3.464) = P(-.464 \le Y \le 6.464)$$

Since Y is a Poisson random variable taking only the values $Y = 0, 1, 2, \ldots$, we have

$$P\left(|Y-3| \le 2\sqrt{3}\right) = \sum_{y=0}^{6} \frac{e^{-3}3^y}{y!} = e^{-3}\left(1+3+\frac{9}{2}+\frac{27}{6}+\frac{81}{24}+\frac{243}{120}+\frac{729}{720}\right)$$
$$= (.049787)(19.4125) = .96649$$

This is consistent with the empirical rule, which says that this fraction should be approximately .95.

3.171

There are three possible schemes and with each scheme we can create a probability distribution for X = net profit. Then the merchant's expected net profit will be $E(X)$.

(1) Suppose first of all that she stocks 2 items. Since she knows that she will have a demand for at least 2 items, she will sell both with probability 1. That is, her profit will inevitably be \$.40. The profit table and calculation of $E(X)$ are trivial.

x	\$.40
$p(x)$	1

$$E(X) = \sum_x xp(x) = .40(1) = \$.40$$

(2) Suppose that the merchant stocks 3 items. She has spent \$3 and will realize either \$3.60 or \$2.40 depending on the number of sales. The profit table with associated probabilities is shown below. Notice that she will sell 3 items ($X = .60$) if either 3 or 4 items are demanded. Hence $P(X = .60) = .4 + .5 = .9$.

x	$p(x)$
.60	.9
−.60	.1

$$E(X) = \sum_x xp(x) = .60(.9) + (-.60)(.1) = .54 - .06 = \$.48.$$

(3) Finally, if she stocks 4 items, she has spent \$4.00 and will realize either \$2.40, \$3.60, or \$4.80, depending on demand. The net profits with their associated probabilities are as follows:

x	$p(x)$
.80	.5
−.40	.4
−1.60	.1

$$E(X) = \sum_x xp(x) = .40 - .16 - .16 = \$.08.$$

In order to maximize her expected net profit, she should stock 3 items.

3.172

It is necessary to show that

$$\lim_{N\to\infty} \frac{\binom{r}{y}\binom{N-r}{n-y}}{\binom{N}{n}} = \binom{n}{y} p^y q^{n-y}$$

where $p = \frac{r}{N}$. Consider

$$\frac{\binom{r}{y}\binom{N-r}{n-y}}{\binom{N}{n}} = \frac{r(r-1)\cdots(r-y+1)(N-r)(N-r-1)\cdots(N-r-n+y+1)n(n-1)\cdots(2)(1)}{y(y-1)\cdots(2)(1)(n-y)(n-y-1)\cdots(2)(1)N(N-1)\cdots(N-n+1)}$$
$$= \frac{n(n-1)(n-2)\cdots(2)(1)}{y(y-1)\cdots(2)(1)(n-y)(n-y-1)\cdots(2)(1)} \times \frac{r(r-1)\cdots(r-y+1)N(N-r)(N-r+1)\cdots(N-r-n+y+1)}{N(N-1)(N-2)\cdots(N-n+1)}$$

The left-hand fraction is by definition $\binom{n}{y}$. Counting the number of terms in the right-hand fraction, we see that the numerator $y + (n - y) = n$ terms, as does the denominator. Hence the hypergeometric probability can be written as

$$\binom{n}{y}\left(\frac{r}{N}\right)\left(\frac{r-1}{N-1}\right)\left(\frac{r-2}{N-2}\right)\cdots\left(\frac{r-y+1}{N-y+1}\right)\left(\frac{N-r}{N-y}\right)\left(\frac{N-r-1}{N-y-1}\right)\cdots\left(\frac{N-r-n+y+1}{N-n+1}\right) \quad (*)$$

Consider the limit of the first y fractions. Each is of the form $\frac{r-j}{N-j}$.

$$\lim_{N\to\infty} \frac{r-j}{N-j} = \lim_{N\to\infty}\left(\frac{r}{N-j} - \frac{j}{N-j}\right) = \lim_{N\to\infty}\left(\frac{\frac{r}{N}}{1-\frac{j}{N}} - \frac{j}{N-j}\right) = \frac{r}{N} = p$$

(Note that $\frac{r}{N}$ is held constant.)

Each of the last $n - y$ fractions is of the form $\frac{N-r-k}{N-j}$.

$$\lim_{N\to\infty}\frac{N-r-k}{N-j} = \lim_{N\to\infty}\left(\frac{N}{N-j}-\frac{r}{N-j}\right)$$
$$= \lim_{N\to\infty}\left(\frac{1}{1-\frac{j}{N}}-\frac{\frac{r}{N}}{1-\frac{j}{N}}-\frac{k}{N-j}\right)$$
$$= 1-\mathrm{p}=\mathrm{q}$$

Hence taking the limit in (*), we have

$$\lim_{N\to\infty}\frac{\binom{r}{y}\binom{N-r}{n-y}}{\binom{N}{n}} = \binom{n}{y}p^y q^{n-y}$$

3.173 **a.** The probability of interest is

$$p(10) = \frac{\binom{40}{10}\binom{60}{10}}{\binom{100}{20}}$$
$$= \frac{40(39)(38)(37)(36)(35)(34)(33)(32)(31)(60)(59)(58)(57)(56)(55)(54)(53)(52)}{10(9)(8)(7)(6)(5)(4)(3)(2)(1)(100)(99)(98)}$$
$$\times\frac{(51)(20)(19)(18)(17)(16)(15)(14)(13)(12)(11)(10)(9)(8)(7)(6)(5)(4)(3)(2)}{(97)(96)(95)(94)(93)(92)(91)(90)(89)(88)(87)(86)(85)(84)(83)(82)(81)}$$

The student can simplify this expression first by direct cancellation and then by using a calculator. However, care must be taken to perform operations in such a way that the capacity of the machine is not exceeded. The resulting numerical value will be $p(10) = .119$.

b. Using the binomial approximation with $p = \frac{r}{N} = \frac{40}{100} = .4$ and $n = 20$ yields $p(10) = \binom{20}{10}(.4)^{10}(.6)^{10} = P(Y \le 10) - P(Y \le 9) = .872 - .755 = .117$
This result was obtained by using the binomial tables in the back of the text. Note the accuracy of the approximation and also the comparative ease of calculation.

3.174 Let A = accident next year,
B = accident this year, and
C = safe driver.

We know $P(C) = .7$, $P(A|C) = .1 = P(B|C)$, and $P(A|\overline{C}) = P(B|\overline{C}) = .5$. From Bayes's Rule,

$$P(C|B) = \frac{P(B|C)P(C)}{P(B|C)P(C)+P(B|\overline{C})P(\overline{C})} = \frac{(.1)(.7)}{(.1)(.7)+(.5)(.3)} = \frac{7}{22}.$$

Now, we need

$$P(A|B) = P(A\cap(C\cup\overline{C})|B)$$
$$= P(A\cap C|B) + P(A\cap\overline{C}|B)$$
$$= P(C|B)\mathrm{P}(A|C\cap B) + P(\overline{C}|B)P(A|\overline{C}\cap B)$$
$$= \tfrac{7}{22}(.1) + \tfrac{15}{22}(.5) = .3737.$$

Then the premium should be

$$400(.3727) = \$149.09.$$

3.175 **a.** If method (1) is used, N tests are required, regardless of the number of people having the disease. However, for method (2) there are two possible values for n', the number of tests required for a group of k people. If all k people are healthy, then only one test is required. The probability that $n' = 1$ is

$$P(n'=1) = P(k \text{ people healthy}) = (.95)^k$$

If at least one of the k people has the disease, then the test is positive and k more tests (making a total of $k+1$) are required. Note that

$$P(n'=k+1) = P(\text{at least one diseased person}) = 1 - P(\text{all healthy}) = 1-(.95)^k$$

Hence

$$E(n') = \sum_{n'} n'p(n') = 1(.95)^k + (k+1)(1-.95^k) = .95^k + k + 1 - k(.95)^k - .95^k$$
$$= 1 + k(1-.95^k)$$

This expectation holds for each group, so that for n groups the expected number of tests is $n\left[1+k\left(1-.95^k\right)\right]$.

b. It is necessary to choose k so that the expected number of tests is minimized. Write

$$g(k) = \frac{N}{k}\left[1 + k\left(1 - .95^k\right)\right] = \frac{N}{k} + N\left(1 - .95k^k\right)$$

where $n = \frac{N}{k}$. This quantity must be minimized. Differentiating with respect to k and setting $\frac{d[g(k)]}{dk} = 0$, we obtain

$h(k) = \frac{d[g(k)]}{dk} = \frac{-N}{k^2} + N\left(-.95^k\right)\ln(.95) = -\frac{N}{k^2} - N\left(.95^k\right)\ln(.95) = 0$

which implies

$$h(k) = \frac{1}{k^2} + \left(.95^k\right)\ln(.95) = 0$$

This function $h(k)$ is a strictly decreasing function of k. Hence, for various integer values of k, we can find two values of k for which the function $h(k)$ is closest to zero. These are shown in the accompanying table.

k	$h(k)$
2	.2037
3	.0671
4	.0207
5	.00031
6	$-.0097$

The minimum value is between $k = 5$ and $k = 6$. However, since k can be only integer-valued, we could evaluate the function $g(k)$ at $k = 5$ and $k = 6$ to determine which value of $g(k)$ is smaller. Evaluating, we have

$$g(5) = \frac{N}{5}\left[1 + 5\left(1 - .95^5\right)\right] = N(.4262)$$
$$g(6) = \frac{N}{6}\left[1 + 6\left(1 - .95^6\right)\right] = N(.4316)$$

so that $g(k)$ will be minimized using $k = 5$.

c. The expected number of tests is $.4262N$, compared to N tests if method (1) is used. The number of tests saved is then $N - .4262N = .5738N$.

3.176 a. $P(Y = n) = \frac{\binom{r}{n}\binom{N-r}{0}}{\binom{N}{n}} = \frac{r!}{N!} \times \frac{(N-n)!}{(r-n)!} = \frac{r(r-1)\cdots(r-n+1)}{N(N-1)\cdots(N-n+1)}$.

b. Notice that $\frac{\binom{a}{b}}{\binom{a}{b+1}} = \frac{\frac{a!}{b!(a-b)(a-b-1)!}}{\frac{a!}{(b+1)b!(a-b-1)!}}$

Applying this fact to $\frac{\binom{r_1}{y}}{\binom{r_1}{y+1}}$ and $\frac{\binom{N-r_1}{n-y}}{\binom{N-r_1}{n-y-1}}$, we have

$$\frac{\binom{r_1}{y}\binom{N-r_1}{n-y}}{\binom{r_1}{y+1}\binom{N-r_1}{n-y-1}} = \frac{y+1}{r_1-y} \cdot \frac{N-r_1-n+y+1}{n-y}$$

Thus, $\frac{p(y|r_1)}{p(y+1|r_1)} = \frac{\binom{r_1}{y}\binom{N-r_1}{n-y}}{\binom{r_1}{y+1}\binom{N-r_1}{n-y-1}} = \frac{y+1}{r_1-y} \cdot \frac{N-r_1-n+y+1}{n-y}$

A similar argument gives

$$\frac{p(y|r_2)}{p(y+1|r_2)} = \frac{y+1}{r_2-y} \cdot \frac{N-r_2-n+y+1}{n-y}$$

Now, $r_1 < r_2$ implies $N - r_1 - n + y + 1 > N - r_2 + n + y + 1$ and $r_1 - y < r_2 - y$.

Thus, $\frac{N-r_1-n+y+1}{r_1-y} > \frac{N-r_2-n+y+1}{r_2-y}$

which implies

$$\frac{p(y|r_1)}{p(y+1|r_1)} > \frac{p(y|r_2)}{p(y+1|r_2)}$$

Hence

$$\frac{p(y|r_1)}{p(y|r_2)} > \frac{p(y+1|r_1)}{p(y+1|r_2)}$$

c. Write

$$(1+a)^k = \binom{k}{0} + \binom{k}{1}a + \binom{k}{2}a^2 + \ldots + \binom{k}{k}a^k$$

for the equation given here,

$$\left[\binom{N_1}{0} + \binom{N_1}{1}a + \binom{N_1}{2}a^2 + \ldots + \binom{N_1}{N_1}a^{N_1}\right] \cdot \left[\binom{N_2}{0} + \binom{N_2}{1}a + \ldots + \binom{N_2}{N_2}a^{N_2}\right]$$
$$= \left[\binom{N_1+N_2}{0} + \binom{N_1+N_2}{1}a^1 + \binom{N_1+N_2}{N_1+N_2}a^{N_1+N_2}\right]$$

On the right-hand side of the equation, the coefficient of a^n is $\binom{N_1+N_2}{n}$. On the left-hand side, the coefficient of a^n is the sum of products of pairs of coefficients such that the exponents on a sum to n. Hence the coefficient is

$$\binom{N_1}{0}\binom{N_2}{n} + \binom{N_1}{1}\binom{N_2}{n-1} + \ldots + \binom{N_1}{n}\binom{N_2}{0}.$$

These two coefficients of a^n must be equal, and the result is proved.

From part **c** we know that $\frac{\binom{N_1}{0}\binom{N_2}{n}}{\binom{N_1+N_2}{n}} + \frac{\binom{N_1}{1}\binom{N_2}{n-1}}{\binom{N_1+N_2}{n}} + \ldots + \frac{\binom{N_1}{n}\binom{N_2}{0}}{\binom{N_1+N_2}{n}} = 1.$

Let $r = N_1$ and $N - r = N_2$. Then

$$\sum_{y=0}^{n} p(y) = \sum_{y=0}^{n} \frac{\binom{r}{y}\binom{N-r}{n-y}}{\binom{N}{n}} = 1$$

3.177 $E(Y) = \sum_{y=0}^{n} \frac{y\binom{r}{y}\binom{N-r}{n-y}}{\binom{N}{n}} = r\sum_{y=1}^{n}\left[\frac{(r-1)!}{(y-1)!(r-y)!}\right]\left[\frac{\binom{N-r}{n-y}}{\binom{N}{n}}\right]$

$= r\sum_{y=1}^{n} \frac{\binom{r-1}{y-1}\binom{N-r}{n-y}}{\binom{N}{n}}$ let $x = y - 1$

$= r\sum_{x=0}^{n-1} \frac{\binom{r-1}{x}\binom{N-r}{n-1-x}}{\binom{N}{n}}$

$= r\sum_{x=0}^{n-1} \frac{\binom{r-1}{x}\binom{N-r}{n-1-x}}{\left(\frac{N}{n}\right)\binom{N-1}{n-1}}$

$= \left(\frac{rn}{N}\right)\sum_{x=0}^{n-1} \frac{\binom{r-1}{x}\binom{N-r}{n-1-x}}{\binom{N-1}{n-1}}$

$= \frac{nr}{N}$, since the summation sums to 1.

3.178 $E[Y(Y-1)] = \sum_{y=0}^{n} [y(y-1)] \frac{\binom{r}{y}\binom{N-r}{n-y}}{\binom{N}{n}}$

$= \sum_{y=0}^{n} [y(y-1)]\left[\frac{r(r-1)(r-2)!}{y(y-1)(y-2)!(r-y)!}\right]\left[\frac{\binom{N-r}{n-y}}{\binom{N}{n}}\right]$

$= r(r-1)\sum_{y=2}^{n} \frac{\binom{r-2}{y-2}\binom{N-r}{n-y}}{\binom{N}{n}}$ Let $x = y - 2$

$= \frac{r(r-1)n(n-1)}{N(N-1)}\sum_{x=0}^{n-2} \frac{\binom{r-2}{x}\binom{N-r}{n-2-x}}{\binom{N-2}{n-2}}$

$= [r(r-1)n(n-1)][N(N-1)]$,

since summation sums to 1. Notice the results of 3.177 and 3.178 give the necessary computations to verify the variance the hypergeometric distribution.

CHAPTER 4 CONTINUOUS RANDOM VARIABLES AND THEIR PROBABILITY DISTRIBUTIONS

4.1 By definition, $F(y) = P(Y \leq y)$ for $y = 1, 2, 3, \ldots$

Then

$$P(Y = y) = P(Y \leq y) - P(Y \leq y - 1)$$
$$= F(y) - F(y-1) \qquad y = 2, 3, \ldots .$$

Also, $P(Y = 1) = P(Y \leq 1)$
$= F(1)$.

4.2 **a.** $F(i) = P(Y \leq i) = \sum_{k=1}^{i} q^{k-1}p \qquad i = 0, 1, 2, \ldots$

$$= p\sum_{k=0}^{i-1} q^k = p\left(\frac{1-q^i}{1-q}\right) = \frac{p(1-q^i)}{p}$$
$$= 1 - q^i.$$

Because Y is a discrete random variable, the only changes in $F(y)$ are at the positive integers. The result follows.

b. 1. $F(y) = 0$ for $y < 0$. Hence,
$\lim_{y \to -\infty} F(y) = 0.$

2. $F(y) = 1 - q^i \qquad i \leq y < i + 1$ where $i = 0, 1, 2, 1\ldots$.
Then $\lim_{y \to \infty} F(y) = 1 - \lim_{i \to \infty} q^i \qquad$ for i an integer and $0 < q < 1$
$= 1$

3. Suppose $i \leq y_1 < y_2 < i + 1 \qquad$ for $i = 0, 1, 2, \ldots$.
Then $F(y_1) = 1 - q^i = F(y_2)$.
On the other hand, suppose
$i - 1 \leq y_1 < i \leq y_2 < i + 1$. Then
$F(y_1) = 1 - q^{i-1} < 1 - q^i = F(y_2)$.

4.3 **a.** $P(2 \leq Y \leq 5) = P(Y \leq 5) - P(Y \leq 1) = 0.994 - 0.376 = 0.618$
$P(2 < Y \leq 5) = P(Y \leq 5) - P(Y \leq 2) = 0.994 - 0.678 = 0.316$

The probabilities are not equal because Y is discrete and $P(Y = 2) > 0$.

b. $P(2 \leq Y < 5) = P(Y \leq 4) - P(Y \leq 1) = 0.967 - 0.376 = 0.591$
$P(2 < Y < 5) = P(Y \leq 4) - P(Y \leq 2) = 0.967 - 0.678 = 0.289$

c. $P(a \leq Y < b) = P(a < Y < b) + P(Y = a)$. When Y is continuous $P(Y = a) = 0$ but if Y is discrete it may be that $P(Y = a) > 0$. Thus, the result in part **a.** does not contradict the claim.

4.4 **a.** We need to find k such that

$$1 = \int_0^1 ky(1-y)\,dy = \tfrac{k}{6} \text{ so that } k = 6.$$

b. $P(0.4 \leq Y < 1) = \int_{0.4}^{1} 6y(1-y)\,dy = 0.648.$

c. Continuity implies $P(0.4 \leq Y < 1) = P(0.4 \leq Y \leq 1) = 0.94133.$

d. Note that, by definition, $P(Y \leq 0.4 | Y \leq 0.8) = \frac{P(Y \leq .04)}{P(Y \leq 0.8)}$

and

$$P(Y \leq 0.4) = \int_0^{0.4} 6y(1-y)\,dy = 0.352$$
$$P(Y \leq 0.8) = \int_0^{0.8} 6y(1-y)\,dy = 0.896.$$

Thus,

$$P(Y \leq 0.4 | Y \leq 0.8) = \frac{0.352}{0.896} = 0.393.$$

e. $P(Y < 0.4|Y < 0.8) = \frac{P(Y < 0.4)}{P(Y < 0.8)} = \frac{P(Y \le 0.4)}{P(Y \le 0.8)} = 0.393.$

4.5

a. We must find the value of c such that

$$F(\infty) = \int_{-\infty}^{\infty} f(y)\, dy = 1.$$

That is,

$$F(\infty) = \int_0^2 cy\, dy = c\left[\frac{y^2}{2}\right]_0^2 = c\left(\frac{4}{2}\right) = 1.$$

Hence $c = \frac{2}{4} = \frac{1}{2}$. The density function for Y is

$$f(y) = \begin{cases} \frac{y}{2}, & 0 \le y \le 2 \\ 0, & \text{elsewhere} \end{cases}$$

b. $F(y) = \int_{-\infty}^{y} f(t)\, dt = \int_0^y \frac{t}{2}\, dt = \left.\frac{t^2}{4}\right]_0^y = \frac{y^2}{4}$ for $0 \le y \le 2$

Note that $F(y) = 0$ for $y < 0$ and $F(y) = 1$ for $y > 2$.

c. The graphs of $f(y)$ and $F(y)$ are shown in Figures 4.1 and 4.2.

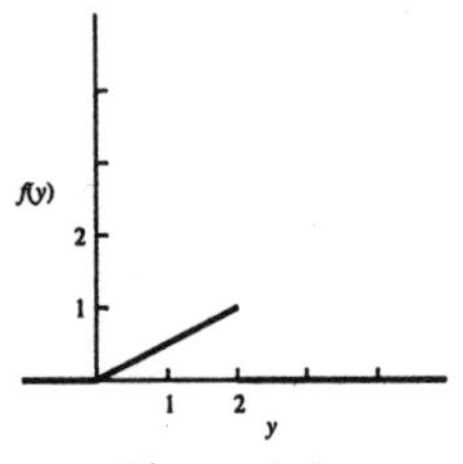

Figure 4.1

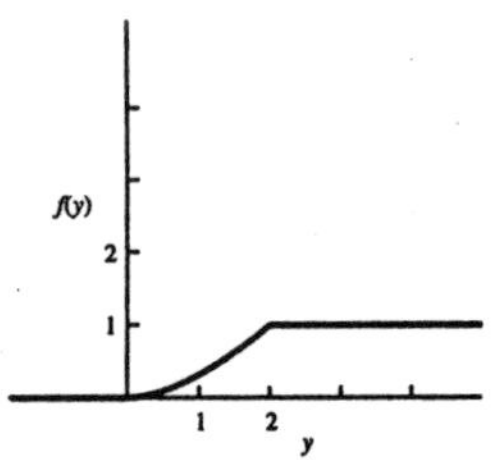

Figure 4.2

d. From part **b** we have

$$P(1 \le Y \le 2) = P(Y \le 2) - P(Y \le 1) + \mathrm{P}(Y = 1) = F(2) - F(1) + 0$$
$$= \frac{2^2}{4} - \frac{1^2}{4} = 1 - \frac{1}{4} = \frac{3}{4}$$

e. Refer to Figure 4.1. It is necessary to calculate the area under the density function $f(y)$ from 1 to 2. Note that the total area under $f(y)$ is 1 and that the area from 0 to 1 is the area of a triangle with base 1 and height $\frac{1}{2}$. Hence the area from 1 to 2 is

$$1 - \left(\frac{1}{2}\right)(1)\left(\frac{1}{2}\right) = 1 - \frac{1}{4} = \frac{3}{4}$$

4.6

a. The properties of a distribution function are satisfied since:

(1) $F(-\infty) = 0$.

(2) $F(\infty) = 1 - e^{-\infty} = 1 - 0 = 1$.

(3) $F(y_1) - F(y_2) = e^{-y_2^2} - e^{-y_1^2}$, which is positive if $y_1 > y_2$.

b. By Definition 4.3,

$$f(y) = F'(y) = \begin{cases} 2ye^{-y^2} & \text{for } y > 0 \\ 0 & \text{for } y \le 0 \end{cases}$$

c. $P(Y \ge 2) = 1 - P(Y < 2) = 1 - P(Y \le 2)$, since the probability at any particular point is 0. Thus,

$$P(Y \ge 2) = 1 - F(2) = 1 - (1 - e^{-4}) = e^{-4}.$$

d. $P(Y > 1|Y \le 2) = \frac{P(1 < Y \le 2)}{P(Y \le 2)}$

So we need

$$P(1 < Y \le 2) = F(2) - F(1) = (1 - e^{-4}) - (1 - e^{-1}) = e^{-1} - e^{-4}.$$

Next we get $P(Y \le 2) = 1 - e^{-4}$ (using part **b.**). Thus

$$P(Y > 1|Y \le 2) = \frac{(e^{-1} - e^{-4})}{1 - e^{-4}}.$$

4.7 **a.** $F(y) = \int_{-\infty}^{y} f(t)\,dt$. For $y < 0$, $F(y) = P(Y \le y) = 0$

For $0 \le y \le 1$, $F(y) = \int_{0}^{y} t\,dt = \frac{y^2}{2}$

For $1 \le y \le 1.5$, $F(y) = P(Y \le y) = P(0 \le Y \le 1) + P(1 < Y \le y)$

$$= \tfrac{1}{2} + \int_{1}^{y} dt = \tfrac{1}{2} + (y-1) = y - \tfrac{1}{2}$$

For $y > 1.5$, $F(y) = P(Y \le y) = P(Y \le 1.5) + P(1.5 < Y \le y) = F(1.5) + 0 = 1$

Hence

$$F(y) = \begin{cases} 0, & y < 0 \\ \frac{y^2}{2}, & 0 \le y \le 1 \\ y - \left(\frac{1}{2}\right), & 1 \le y \le 1.5 \\ 1, & 1.5 < y \end{cases}$$

b. $P(0 \le Y \le .5) = F(.5) = \frac{1}{8} = .125$

c. $P(.5 \le Y \le 1.2) = F(1.2) - F(.5) = .7 - .125 = .575$

4.8 **a.** $f(y)$

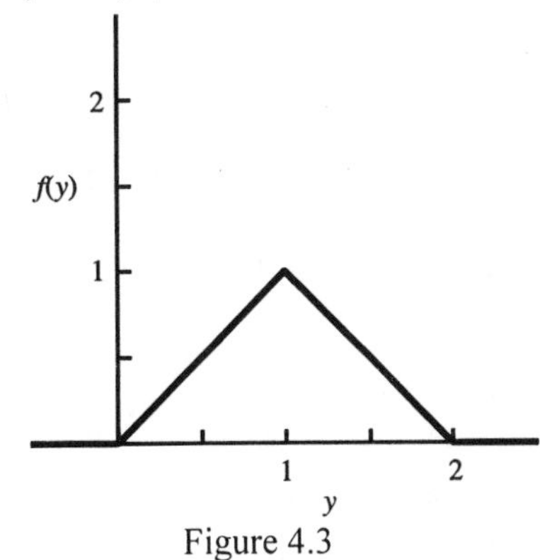

Figure 4.3

b. For $y < 0$, $F(y) = 0$.
For $y > 2$, $F(y) = 1$.
For $0 \le y \le 1$,

$$F(y) = \int_{0}^{y} t\,dt = \frac{y^2}{2}$$

For $1 \le y \le 2$,

$$F(y) = \int_{0}^{1} t\,dt + \int_{1}^{y} (2-t)\,dt = \tfrac{1}{2} + \left[2t - \tfrac{t^2}{2}\right]_{1}^{y} = 2y - \tfrac{y^2}{2} - 1$$

c. $P(.8 \le Y \le 1.2) = F(1.2) - F(.8) = (2.4 - .72 - 1) - .32 = .36$

d. $P(Y > 1.5 \mid Y > 1) = \frac{P(Y>1.5)}{P(Y>1)} = \frac{1-(3-1.125-1)}{\frac{1}{2}} = \frac{.125}{.5} = .25$

4.9 **a.** For $b \ge 0$ and for any value of y, $f(y) \ge 0$. Moreover,

$$\int_{-\infty}^{\infty} f(y)\,dy = \int_{b}^{\infty} \frac{b}{y^2}\,dy = -\frac{b}{y}\Big]_{b}^{\infty} = \frac{b}{b} = 1$$

Since b is defined as a traversal time, it will always be positive. Thus $f(y) \ge 0$ and $f(y)$ has the properties of a density function.

b. $F(y) = \int_{b}^{y} \frac{b}{t^2}\,dt = -\frac{b}{t}\Big]_{b}^{y} = 1 - \frac{b}{y}$ for $y \ge b$; $F(y) = 0$ for $y < b$.

c. $P(Y > b + c) = 1 - F(b + c) = 1 - \left(1 - \frac{b}{b+c}\right) = \frac{b}{b+c}$

d. Apply part **c** to obtain

$$P(Y > b + d \mid Y > b + c) = \frac{P(Y>b+d)}{P(Y>b+c)} = \frac{b+c}{b+d}.$$

4.10 **a.** $F(\infty) = \int_{0}^{2} c(2-y)\,dy = c\left[2y - \frac{y^2}{2}\right]_{0}^{2} = c(4-2) = 2c$

In order for $F(y)$ to be a cumulative distribution function, we need $F(\infty) = 1$.

Hence $c = \frac{1}{2}$.

b. For $0 \le y \le 2$, $F(y) = \int_{-\infty}^{y} f(t)\,dt = \int_{0}^{y} \left(1 - \frac{t}{2}\right) dt = \left[t - \frac{t^2}{4}\right]_0^y = y - \frac{y^2}{4}$

$F(y) = 0$ for $y < 0$ and $F(y) = 1$ for $y > 2$.

c. The graphs of $f(y)$ and $F(y)$ are shown in Figures 4.4 and 4.5.

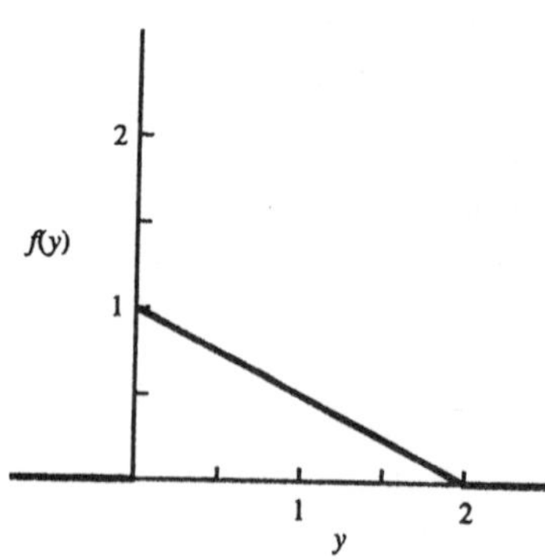

Figure 4.4

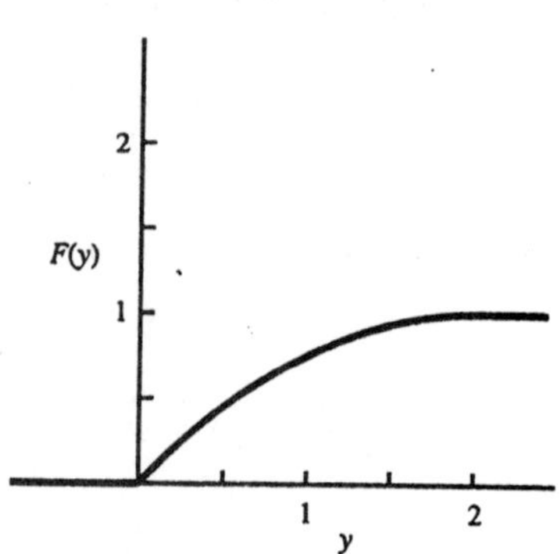

Figure 4.5

d. $P(1 \le Y \le 2) = F(2) - F(1) = \left(2 - \frac{2^2}{4}\right) - \left(1 - \frac{1^2}{4}\right) = \frac{1}{4}$

e. The area under the density function, $f(y)$, from 1 to 2 is the area of a triangle with base 1 and height $\frac{1}{2}$. Hence

$$P(1 \le Y \le 2) = \tfrac{1}{2}\,bh = \tfrac{1}{2}\,(1)\left(\tfrac{1}{2}\right) = \tfrac{1}{4}$$

4.11 a. $F(\infty) = \int_{-\infty}^{\infty} f(y)\,dy = \int_{0}^{1} (cy^2 + y)\,dy = c\left[\frac{y^3}{3}\right]_0^1 + \left[\frac{y^2}{2}\right]_0^1 = \frac{c}{3} + \frac{1}{2} = 1$

Hence $\frac{c}{3} = \frac{1}{2}$ and $c = \frac{3}{2}$.

b. $F(y) = \int_{-\infty}^{y} f(t)\,dt = \int_{0}^{y} \left(\frac{3}{2}t^2 + t\right) dt = \frac{t^3}{2}\Big]_0^y + \frac{t^2}{2}\Big]_0^y = \frac{y^3}{2} + \frac{y^2}{2}$ for $0 \le y \le 1$

and $F(y) = 0$ for $y < 0$, $F(y) = 1$ for $y > 1$.

c. The graphs of $F(y)$ and $f(y)$ are shown in Figures 4.6 and 4.7.

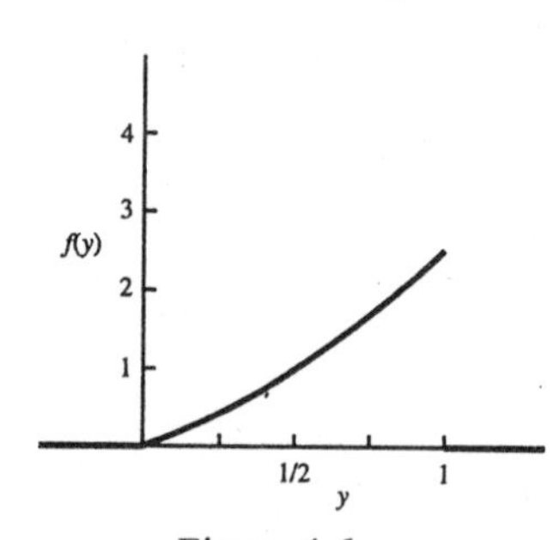

Figure 4.6

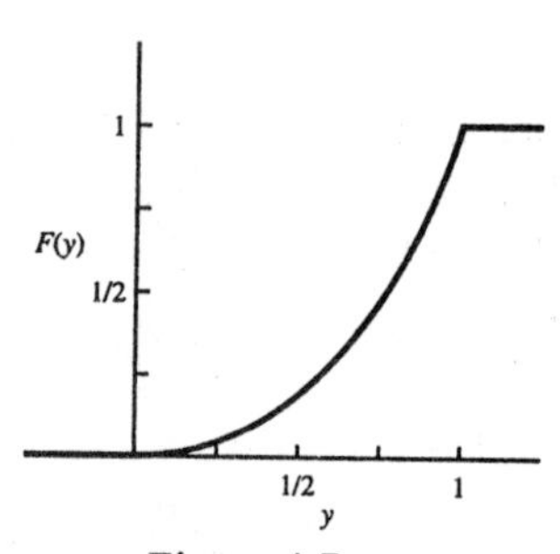

Figure 4.7

d. $F(-1) = 0$ since $y < 0$; $F(0) = 0$; $F(1) = \frac{1}{2} + \frac{1}{2} = 1$

e. $P(0 \le Y \le .5) = F(.5) - F(0) = \frac{(.5)^3}{2} + \frac{(.5)^2}{2} - 0 = \frac{1}{16} + \frac{1}{8} = \frac{3}{16}$

f. $P\left(Y > \frac{1}{2} \middle| Y > \frac{1}{4}\right) = \frac{P\left(Y > \frac{1}{2}\right)}{P\left(Y > \frac{1}{4}\right)} = \frac{1 - \left(\frac{3}{16}\right)}{1 - \left(\frac{1}{128} + \frac{1}{32}\right)} = \frac{\frac{13}{16}}{\frac{123}{128}} = \frac{104}{123}$

4.12 a. $F(\infty) = \int_{-\infty}^{\infty} f(y)\,dy = \int_{-1}^{0} .2\,dy + \int_{0}^{1} (.2 + cy)\,dy = .2y]_{-1}^{0} + \left[.2y + \frac{cy^2}{2}\right]_0^1$

$= .2 + .2 + \frac{c}{2} = 1$ so that $c = 1.2$ and the density function is

$$f(y) = \begin{cases} .2, & -1 < y \le 0 \\ .2 + 1.2y, & 0 < y < 1 \\ 0, & \text{elsewhere} \end{cases}$$

b. $F(y) = 0$ for $y < -1$.

$$F(y) = \int_{-\infty}^{y} f(t)\,dy = \int_{-1}^{y} .2\,dt = .2t]_{-1}^{y} = .2y + .2 \qquad \text{for } -1 \le y \le 0$$

$$F(y) = \int_{-\infty}^{y} f(t)\,dy = \int_{-1}^{0} .2\,dt + \int_{0}^{y} (.2 + 1.2t)\,dt = .2 + [.2t + .6t^2]_0^y$$

$= .2 + .2y + .6y^2$ for $0 \le y \le 1$. $F(y) = 1$ for $y > 1$.

Collecting results, we have

$$F(y) = \begin{cases} 0, & y < -1 \\ .2(y+1), & -1 \le y \le 0 \\ .2\,(1 + y + 3\,y^2), & 0 < y \le 1 \\ 1, & y > 1 \end{cases}$$

c. The graphs of $f(y)$ and $F(y)$ are shown in Figures 4.8 and 4.9.

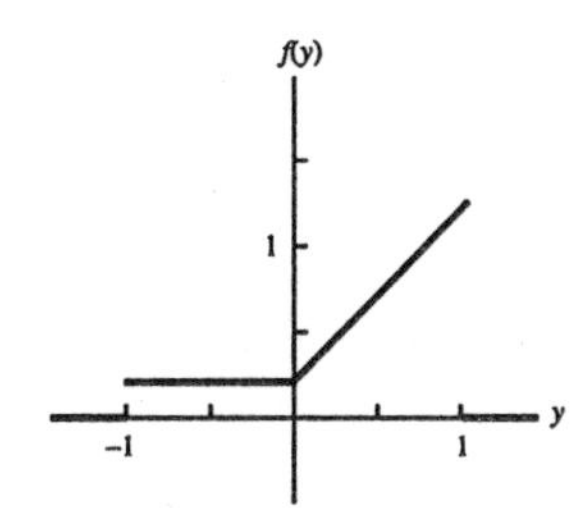

Figure 4.8

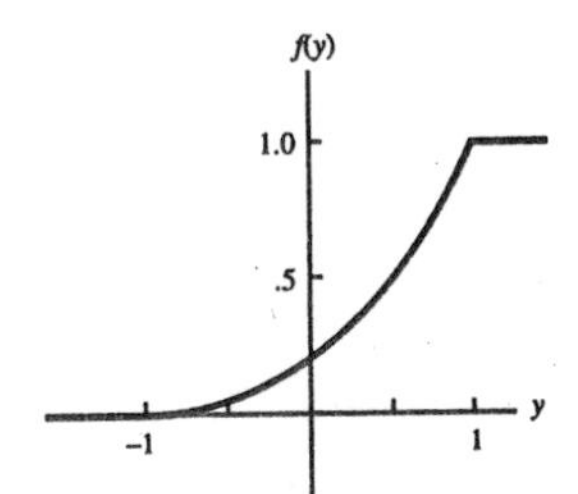

Figure 4.9

d. $F(-1) = .2(-1+1) = 0$, $F(0) = .2(0+1) = .2$, $F(1) = .2(1+1+3) = .2(5) = 1$

e. $P(0 \le Y \le .5) = F(.5) - F(0) = .2[1 + .5 + 3(.25)] - .2 = .2(2.25) - .2 = .25$

f. $P(Y > .5|Y > .1) = \frac{P(Y>.5)}{P(Y>.1)} = \frac{1-.45}{1-.2(1+.1+.03)} = \frac{.55}{.774} = .71$

4.13

a. Differentiating $F(y)$ with respect to y, we have

$$f(y) = \begin{cases} 0, \; y \le 0 \\ \frac{1}{8}, \; 0 < y < 2 \\ \frac{y}{8}, \; 2 \le y < 4 \\ 0, \; y \ge 4 \end{cases}$$

Notice that $F(y)$ is not differentiable at $y = 0$, 2, and 4.

b. $P(1 \le Y \le 3) = F(3) - F(1) = \frac{9}{16} - \frac{2}{16} = \frac{7}{16}$

c. $P(Y \ge 1.5) = 1 - F(1.5) = 1 - \frac{1.5}{8} = \frac{13}{16}$

d. $P(Y \ge 1|Y \le 3) = \frac{P(1 \le Y \le 3)}{P(Y \le 3)} = \frac{\frac{7}{16}}{\frac{9}{16}} = \frac{7}{9}$

4.14

Refer to Exercise 4.10.

$$E(Y) = \tfrac{1}{2}\int_0^2 y(2-y)\,dy = \left[\tfrac{y^2}{2} - \tfrac{y^3}{6}\right]_0^2 = \tfrac{2}{3}$$

$$E(Y^2) = \tfrac{1}{2}\int_0^2 y^2(2-y)\,dy = \left[\tfrac{y^3}{3} - \tfrac{y^4}{8}\right]_0^2 = \tfrac{2}{3}$$

and

$$V(Y) = E(Y^2) - [E(Y)]^2 = \left(\tfrac{2}{3}\right) - \left(\tfrac{4}{9}\right) = \tfrac{2}{9}.$$

4.15

Refer to Exercise 4.11.

$$E(Y) = \int_0^1 \left(\tfrac{3}{2}y^3 + y^2\right)dy = \left[\tfrac{3}{8}y^4 + \tfrac{y^3}{3}\right]_0^1 = \tfrac{3}{8} + \tfrac{1}{3} = \tfrac{9+8}{24} = \tfrac{17}{24} = .708$$

$$E(Y^2) = \int_0^1 \left(\tfrac{3}{2}y^4 + y^3\right)dy = \left[\tfrac{3}{10}y^5 + \tfrac{1}{4}y^4\right]_0^1 = \tfrac{3}{10} + \tfrac{1}{4} = .55$$

so that $V(Y) = E(Y^2) - (EY)^2 = .55 - (.708)^2 = .0487$.

4.16 Refer to Exercise 4.12.

$$E(Y) = \int_{-1}^{0} .2y\,dy + \int_{0}^{1} (.2y + 1.2y^2)\,dy = \frac{.2y^2}{2}\Big]_{-1}^{0} + \left[\frac{.2y^2}{2} + \frac{1.2y^3}{3}\right]_{0}^{1}$$
$$= -.1 + .1 + \frac{1.2}{3} = .4$$
$$E(Y^2) = \int_{-1}^{0} .2y^2\,dy + \int_{0}^{1} (.2y^2 + 1.2y^3)\,dy = \frac{.2y^3}{3}\Big]_{-1}^{0} + \left[\frac{.2y^3}{3} + \frac{1.2y^4}{4}\right]_{0}^{1}$$
$$= \frac{.2}{3} + \frac{.2}{3} + \frac{1.2}{4} = \frac{1.3}{3}$$

so that $V(Y) = \left(\frac{1.3}{3}\right) - (.4)^2 = .2733.$

4.17 1. $E(c) = \int_{-\infty}^{\infty} cf(y)\,dy = c\int_{-\infty}^{\infty} f(y)\,dy = c$ since $\int_{-\infty}^{\infty} f(y)\,dy = 1.$

2. $E[cg(y)] = \int_{-\infty}^{\infty} cg(y)f(y)\,dy = c\int_{-\infty}^{\infty} g(y)f(y)\,dy = cE[g(y)].$

3. $E[g_1(y) + g_2(y) + \ldots + g_k(y)]$
$$= \int_{-\infty}^{\infty} [g_1(y) + g_2(y) + \ldots + g_k(y)]f(y)\,dy$$
$$= \int_{-\infty}^{\infty} g_1(y)f(y)\,dy + \int_{-\infty}^{\infty} g_2(y)f(y)\,dy + \ldots + \int_{-\infty}^{\infty} g_k(y)f(y)\,dy$$
$$= E[g_1(y)] + E[g_2(y)] + \ldots + E[g_k(y)].$$

4.18 $V(Y) = E(Y - EY)^2 = E(Y^2 - 2YEY + (EY)^2) = EY^2 - 2(EY)^2 + (EY)^2$
$= EY^2 - (EY)^2.$

4.19 Recall that we found the density function, $f(y)$, in exercise 4.13. Then

$$E(Y) = \int_{-\infty}^{\infty} yf(y)\,dy = \int_{0}^{2} y/8\,dy + \int_{2}^{4} y^2/8\,dy = \left[\frac{y^2}{16}\right]_0^2 + \left[\frac{y^3}{24}\right]_2^4 = 31/12.$$

To find the variance we apply the result of exercise 4.18. That is, we need to calculate

$$E(Y^2) = \int_{-\infty}^{\infty} y^2 f(y)\,dy = \int_{0}^{2} y^2/8\,dy + \int_{2}^{4} y^3/8\,dy = \left[\frac{y^3}{24}\right]_0^2 + \left[\frac{y^4}{32}\right]_2^4 = 47/6.$$

Thus $V(Y) = E(Y^2) - (EY)^2 = 47/6 - (31/12)^2 \approx 7.83 - 6.67 = 1.16.$

4.20 The proofs of these facts for a continuous random variable are identical to the proofs for a discrete random variable. See Exercise 3.23.

Let $g_1(Y) = aY$ and let $g_2(Y) = b$. Then

$E(aY + b) = E(g_1(Y) + g_2(Y)] = E[g_1(Y)] + E[g_2(Y)] = aE(Y) + b = a\mu + b.$
$V(aY + b) = E[aY + b - (a\mu + b)]^2 = E[a^2(Y - \mu)^2] = a^2E[(Y - \mu)^2] = a^2\sigma^2.$

4.21 First calculate

$$E(Y) = \int_{0}^{1} \left(\tfrac{3}{2}y^3 + y^2\right)dy = \left[\tfrac{3}{8}y^4 + \tfrac{y^3}{3}\right]_0^1 = \tfrac{3}{8} + \tfrac{1}{3} = \tfrac{17}{24} = .708$$

Then $E(Y) = 5 - .5E(Y) = 5 - .5\left(\frac{17}{24}\right) = \$4.65$. Now

$$E(Y^2) = \int_{0}^{1} \left(\tfrac{3}{2}y^4 + y^3\right)dy = \left[\tfrac{3}{10}y^5 + \tfrac{y^4}{4}\right]_0^1 = \tfrac{3}{10} + \tfrac{1}{4} = .55$$

Thus,

$$V(W) = V(5 - .5Y) = (.5)^2V(Y) = (.5)^2[E(Y^2) - [E(Y)]^2] = (.5)^2[.55 - (.708)^2] = .012$$

4.22 **a.** To solve for c, we need to consider $\int_{0}^{1} cy^2(1 - y)^4\,dy = 1.$

Integrating, we have $c\left(\frac{y^3}{3} - y^4 + \frac{6y^5}{5} - \frac{4y^6}{6} + \frac{y^7}{7}\right)\Big]_0^1 = 1.$

$c\left[\left(\frac{1}{3}\right) - 1 + \left(\frac{6}{5}\right) - \left(\frac{4}{6}\right) + \left(\frac{1}{7}\right)\right] = 1.$

$c = (1)\left(\frac{630}{6}\right) = 105.$

b. $E(Y) = 105\int_{0}^{1} yy^2(1 - y)^4\,dy = 105\left[\left(\frac{y^4}{4}\right) + \left(-\frac{4y^5}{5}\right) + y^6 + \left(-\frac{4y^7}{7} + \frac{y^8}{8}\right)\right]_0^1 = \frac{3}{8}.$

4.23 $E(Y) = \int_{59}^{61} y\left(\frac{1}{2}\right) dy = \left(\frac{1}{4}\right) y^2\Big]_{59}^{61} = 60$

$V(Y) = E(Y^2) - [E(Y)]^2 = \left[\int_{59}^{61} y^2\left(\frac{1}{2}\right) dy\right] - (60)^2 = \left(\frac{1}{6}\right) y^3\Big]_{59}^{61} - 3600 = \frac{1}{3}$

4.24 **a.** $E(Y) = \int_0^1 y(2y)\, dy = \frac{2y^3}{3}\Big|_0^1 = \frac{2}{3}$

$V(Y) = E(Y^2) - [E(Y)]^2 = \left[\int_0^1 y^2(2y)\, dy\right] - \left(\frac{2}{3}\right)^2 = \left(\frac{1}{2}\right) y^4\Big]_0^1 - \left(\frac{4}{9}\right) = \frac{1}{18}$

b. $E(X) = E(200Y - 60) = 200E(Y) - 60 = 200\left(\frac{2}{3}\right) - 60 = \frac{220}{3}$
$V(X) = V(200Y - 60) = V(200Y) = (200)^2 V(Y) = 40{,}000\left(\frac{1}{18}\right) = \frac{20{,}000}{9}$

c. Recall Tchebysheff's theorem from exercise 1.24.
$P(\mu_x \pm k\sigma_x) \geq 1 - \left(\frac{1}{k^2}\right)$ where $1 - \left(\frac{1}{k^2}\right) = \frac{3}{4}$. Solving, $k = 2$.
The desired interval is $\left(\left[\frac{220}{3}\right] \pm 2\sqrt{\frac{20{,}000}{9}}\right) = (-20.948,\ 167.614)$

4.25 $E(Y) = \int_2^6 y\left(\frac{3}{32}\right)(y-2)(6-y)\, dy = \left(\frac{3}{32}\right)\left[\left(-\frac{y^4}{4}\right) + \left(\frac{8y^3}{3}\right) - 6y^2\right]_2^6 = \left(\frac{3}{32}\right)(42.67) = 4$

4.26 **a.** $E(Y) = \left(\frac{3}{64}\right)\int_0^4 y^3(4-y)\, dy = \left(\frac{3}{64}\right)\left[y^4 - \left(\frac{y^5}{5}\right)\right]_0^4 = \left(\frac{3}{64}\right)(256 - 204.8) = 2.4$

$V(Y) = E(Y^2) - [E(Y)]^2 = \left[\left(\frac{3}{64}\right)\int_0^4 y^4(4-y)\, dy\right] - (2.4)^2$

$= \left(\frac{3}{64}\right)\left[\left(\frac{4}{5}\right)y^5 - \left(\frac{y^6}{6}\right)\right]_0^4 - 5.76 = \left(\frac{3}{64}\right)(819.2 - 682.66) - 5.76 = .64$

b. The weekly cost for CPU time is $X = 200Y$ where Y is the amount of CPU time. Then
$E(X) = E(200Y) = 200E(Y) = 200(2.4) = 480$
$V(X) = V(200Y) = (200)^2 V(Y) = 40{,}000(0.64) = 25{,}600$

c. $P(X > 600) = P(Y > 3) = \left(\frac{3}{64}\right)\int_3^4 y^2(4-y)\, dy = \left(\frac{3}{64}\right)\left[\left(\frac{4}{3}\right)y^3 - \left(\frac{1}{4}\right)y^4\right]_3^4$
$= \left(\frac{3}{64}\right)(5.58) = .26$
We expect the weekly cost to exceed \$600 about 26% of the time.

4.27 **a.** $E(Y) = \left(\frac{3}{8}\right)\int_5^7 y(7-y)^2\, dy = \left(\frac{3}{8}\right)\left[\left(\frac{49}{2}\right)y^2 - \left(\frac{14}{3}\right)y^3 + \frac{y^4}{4}\right]_5^7 = \left(\frac{3}{8}\right)(14.66) = 5.5$

$V(Y) = E(Y^2) - [E(Y)]^2 = \left[\left(\frac{3}{8}\right)\int_5^7 y^2(7-y)^2\, dy\right] - (5.5)^2$

$= \left(\frac{3}{8}\right)\left[\left(\frac{49}{3}\right)y^3 - \left(\frac{14}{4}\right)y^4 + \left(\frac{1}{5}\right)y^5\right]_5^7 - 30.25 = \left(\frac{3}{8}\right)(81.07) - 30.25 = .15$

b. Using Tchebysheff's inequality with $k = 2$, we evaluate
$\left(\mu_x \pm 2\sqrt{V(Y)}\right) = \left(5.5 \pm 2\sqrt{.15}\right) = (5.5 \pm .775) = (4.725,\ 6.275) = (5,\ 6.275)$

c. $P(Y < 5.5) = \left(\frac{3}{8}\right)\int_5^{5.5} (7-y)^2\, dy = \left(\frac{3}{8}\right)\left[49y - 7y^2 + \frac{y^3}{3}\right]_5^{5.5} = \left(\frac{3}{8}\right)(1.54) = .58$
We would expect to see a pH measurement below 5.5 about 58% of the time.

4.28 It is given that $f(y) = 1$ for $0 \leq y \leq 1$.

a. $F(y) = \int_0^y dt = y$ for $0 \leq y \leq 1$; $F(y) = 0$ for $y < 0$; $F(y) = 1$ for $y > 1$.

b. Using part **a**, $P(a \leq Y \leq a + b) = F(a + b) - F(a) = a + b - a = b$.

4.29 Since the parachutist is landing at a random point in the interval (A, B), the point of landing is a continuous random variable Y, with a uniform distribution over (A, B). Hence

$$f(y) = \frac{1}{B-A} \qquad A \leq y \leq B$$

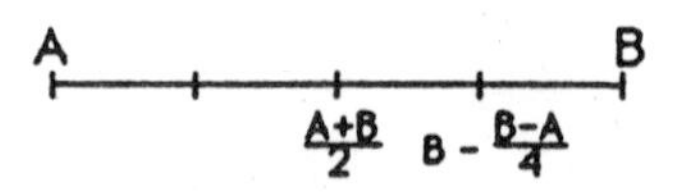

Figure 4.10

a. Refer to Figure 4.10. If he lands closer to A than to B, he has landed in the interval $\left(A, \frac{A+B}{2}\right)$. The probability is

$$\int_A^{(B+A)/2} \frac{1}{B-A}\,dy = \frac{\left(\frac{A+B}{2}\right)-A}{B-A} = \frac{1}{2}$$

b. The point at which the distance to A is exactly 3 times the distance to B is the point $B - \left(\frac{1}{4}\right)(B-A) = \frac{3B+A}{4}$. Then

$P(\text{distance to } A \text{ is more than 3 times distance to } B) = P\left(\frac{3B+A}{4} \le Y \le B\right)$

$$= \frac{B-\left(\frac{3B+A}{4}\right)}{B-A} = \frac{\left(\frac{B-A}{4}\right)}{B-A} = \frac{1}{4}$$

4.30 Let X = # of parachutists that land past the midpoint.

p = probability that one lands past the midpoint = $\int_{(A+B)/2}^{B} \frac{1}{B-A}\,dy$

$= \left(\frac{1}{B-A}\right) y\Big]_{(A+B)/2}^{B} = \frac{1}{2}$

X is binomial with $n = 3$, $p = \frac{1}{2}$.

$$P(X=1) = 3\left(\frac{1}{2}\right)\left(\frac{1}{2}\right)^2 = \frac{3}{8}$$

4.31 Recall Theorem 4.6.

$$V(Y) = E\left(Y^2\right) - (E(Y))^2 = \left[\int_{\theta_1}^{\theta_2} y^2\left(\frac{1}{\theta_2-\theta_1}\,dy\right)\right] - \left(\frac{\theta_2+\theta_1}{2}\right)^2$$

$$= \left[\frac{1}{3(\theta_2-\theta_1)}\right] y^3\Big]_{\theta_1}^{\theta_2} - \left(\frac{1}{4}\right)(\theta_2-\theta_1)^2$$

$$= \left[\frac{1}{3(\theta_2-\theta_1)}\right](\theta_2-\theta_1)\left(\theta_2^2+\theta_1\theta_2+\theta_1^2\right) - \left(\frac{1}{4}\right)\left(\theta_2^2+2\theta_1\theta_2+\theta_1^2\right)$$

$$= \left(\frac{4}{12}\right)\left(\theta_2^2+\theta_1\theta_2+\theta_1^2\right) - \left(\frac{3}{12}\right)\left(\theta_2^2+2\theta_1\theta_2+\theta_1^2\right)$$

$$= \frac{\theta_2^2-2\theta_1\theta_2+\theta_1^2}{12} = \frac{(\theta_2-\theta_1)^2}{12}$$

4.32 a. $\int_{-\infty}^{\infty} f(y)\,dy = \int_{-2}^{2} k\,dy = 4k$. In order for $f(y)$ to be a density function, $4k = 1$, or $k = \frac{1}{4}$.

b. $F(y) = 0$ for $y < -2$. For $-2 \le y \le 2$,

$$F(y) = \int_{-2}^{y} \frac{1}{4}\,dt = \frac{y}{4} + \frac{2}{4} = \frac{y}{4} + \frac{1}{2}$$

and $F(y) = 1$ for $y > 2$.

4.33 Observe that the density is $f(y) = 1/5$ for $20 \le y \le 25$ and 0 otherwise.

a. $P(Y < 22) = \int_{20}^{22} 1/5\,dy = 0.4$

b. $P(Y > 24) = \int_{24}^{25} 1/5\,dy = 0.2$

4.34 Apply Theorem 4.6 to obtain

$$E(Y) = \frac{25+20}{2} = 22.5.$$

4.35 The density of the delivery time Y is $f(y) = 1/4$ for $1 \le y \le 5$ and 0 otherwise.

a. $P(Y > 2) = \int_2^5 1/4\, dy = 3/4$

b.
$$\begin{aligned} E(C) &= E(c_0 + c_1 Y^2) = c_0 + c_1 E(Y^2) = c_0 + c_1(V(Y) - (E(Y))^2) \\ &= c_0 + c_1(16/12 + 36/4) = c_0 + \frac{31}{3}c_1 \end{aligned}$$

4.36 Let the interval be $(0, 500)$ so that the location Y has density $f(y) = 1/500$ for $0 < y < 500$ and 0 otherwise.

a. $P(475 < Y < 500) = \int_{475}^{500} 1/500\, dy = 1/20$

b. $P(0 < Y < 25) = \int_0^{25} 1/500\, dy = 1/20$

c. $P(0 < Y < 250) = \int_0^{250} 1/500\, dy = 1/2$

4.37 The arrival time Y of the call is uniformly distributed over the one minute interval. Thus $f(y) = 1$ for $0 < y < 1$ and 0 otherwise. Then the probability of interest is

$$P(1/4 < Y < 1) = \int_{1/4}^{1} dy = 3/4.$$

4.38 The time Y that the call comes in is a uniformly distributed random variable on the interval 0(midnight) to 5am. Thus $f(y) = 1/5$ for $0 < y < 5$ and 0 otherwise. Then the probability of interest is

$$P(0 < Y < 1) + P(3 < Y < 4) = \int_0^1 1/5\, dy + \int_3^4 1/5\, dy = 0.4.$$

4.39 Let Y = cycle time. Then

$$f(y) = \frac{1}{70-50} = \frac{1}{20} \text{ for } 50 \le y \le 70$$

and

$$F(y) = \int_{50}^{y} \frac{1}{20}\, dt = \frac{y-50}{20} \text{ for } 50 \le y \le 70;$$
$$F(y) = 0 \text{ for } y < 50;$$
$$F(y) = 1 \text{ for } y > 70.$$

Thus

$$P(Y > 65 | Y > 55) = \frac{P(Y>65)}{P(Y>55)} = \frac{1-\left(\frac{65-50}{20}\right)}{1-\left(\frac{55-50}{20}\right)} = \frac{20-15}{20-5} = \frac{1}{3}$$

4.40 $E(Y) = \frac{\theta_1+\theta_2}{2} = \frac{50+70}{2} = 60$

$V(Y) = \frac{(\theta_2-\theta_1)^2}{12} = \frac{(70-50)^2}{12} = \frac{400}{12}$

4.41 Let Y = time the defective board is detected.

a. $P(0 < Y < 1) = \int_0^1 \left(\frac{1}{8}\right) dx = \frac{1}{8}$

b. $P(7 < y < 8) = \int_7^8 \left(\frac{1}{8}\right) dx = \frac{1}{8}$

c. $P(4 < Y < 5 | Y > 4) = \frac{\int_4^5 \left(\frac{1}{8}\right) dx}{\int_4^8 \left(\frac{1}{8}\right) dx} = \frac{\left(\frac{1}{8}\right)}{\left(\frac{1}{2}\right)} = \frac{1}{4}$

4.42 Let Y = amount of measurement error. Y is $U(-.05, .05)$.

a. $P(-.01 < Y < .01) = \int_{-.01}^{.01} \left(\frac{1}{.1}\right) dx = .2$

b. $E(Y) = E(Y) = \frac{-.05+.05}{2} = 0$
$V(Y) = \frac{(.05+.05)^2}{12} = \frac{(.1)^2}{12} = .00083$

4.43 The measurement errors have density $f(y) = 1/0.07$ for $-0.02 \le y \le 0.05$ and 0 otherwise.

a. $P(-0.01 \le Y \le 0.01) = \int_{-0.01}^{0.01} 1/0.07\, dy = 2/7$

b. Apply Theorem 4.6 to obtain

$$E(Y) = \frac{0.03}{2} \text{ and } V(Y) = \frac{(0.07)^2}{12} = 0.00041.$$

4.44 Recall that the arrival time follows a uniform distribution over the interval $(0, 30)$. We want

$$P(25 \le Y \le 30 | Y > 10) = \frac{P(25 \le Y \le 30)}{P(Y > 10)}.$$

Now

$$P(25 \le Y \le 30) = \int_{25}^{30} 1/30\, dy = 1/6$$

$$P(Y > 10) = \int_{10}^{30} 1/30\, dy = 2/3$$

Thus

$$P(25 \le Y \le 30 | Y > 10) = \frac{1/6}{2/3} = 1/4.$$

4.45 Let r = radius and $d = 2r$ be the diameter of sphere. d is $U(.01, .05)$.

$$V = \left(\tfrac{4}{3}\right)\pi r^3 = \left(\tfrac{4}{3}\right)\pi\left(\tfrac{d}{2}\right)^3 = \left(\tfrac{\pi}{6}\right)d^3$$

$$E(Y) = E(Y) = E\left[\left(\tfrac{\pi}{6}\right)d^3\right] = \left(\tfrac{\pi}{6}\right)\int_{.01}^{.05}\left(\tfrac{d^3}{.04}\right)dd = \left(\tfrac{\pi}{6}\right)\left(\tfrac{d^4}{.16}\right)\Big]_{.01}^{.05}$$

$$= \left(\tfrac{25\pi}{24}\right)\left[(.05)^4 - (.01)^4\right] = (6.5 \times 10^{-6})\,\pi$$

$$V(Y) = V\left[\left(\tfrac{4\pi}{3}\right)\left(\tfrac{d}{2}\right)^3\right] = \left(\tfrac{\pi^2}{36}\right)V(d^3) = \left(\tfrac{\pi^2}{36}\right)\left[\mathrm{E}(d^3)^2 - (E(d^3))^2\right]$$

$$= \left(\tfrac{\pi^2}{36}\right)\left[E(d^6) - (E(d^3))^2\right]$$

$$= \left(\tfrac{\pi^2}{36}\right)\left[\int_{.01}^{.05}\left(\tfrac{d^6}{.04}\right)dd - (E(d^3))^2\right]$$

$$= \left(\tfrac{\pi^2}{36}\right)\left\{\left[\tfrac{d^7}{.28}\right]_{.01}^{.05} - \left[\tfrac{(.05)^4-(.01)^4}{.16}\right]^2\right\}$$

$$= \left(\tfrac{\pi^2}{36}\right)\left[2.780 \times 10^{-9}\right]$$

$$= \left(\tfrac{\pi^2}{36}\right)\left[2.790 \times 10^{-9} - 1.521 \times 10^{-9}\right] = \left[3.525 \times 10^{-4}\right]\pi^2$$

4.46 The next few exercises are designed to provide practice for the student in evaluating areas under the normal curve. The following notes may be of some assistance.

(1) Table 4 tabulates the area under the standard normal curve to the right of a specified value z_0. See Figure 4.11. Denote the area obtained by indexing $z = z_0$ in Table 4 by $A(z_0)$ and the desired area by A.

(2) Because of the symmetry of the normal distribution, and since the total area under the curve is 1, the total area lying on one side of 0 will be .5. Thus in order to calculate the area between 0 and z_0 (when $z_0 > 0$) we index z_0, which gives us $A(z_0)$. We then subtract $A(z_0)$ from .5. That is, $A = .5 - A(z_0)$.

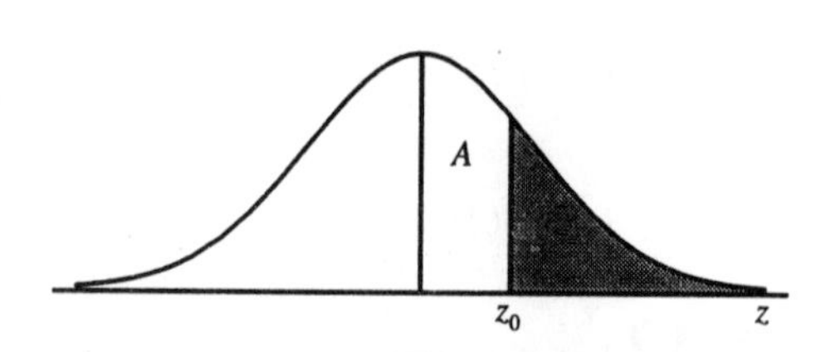

Figure 4.11

(3) Notice that Z is actually a random variable that may take on an infinite number of values, both positive and negative. However, since the standardized normal curve is symmetric about 0, a left-hand area (i.e., an area corresponding to a negative value of z) may be evaluated by indexing the corresponding positive value in Table 4.

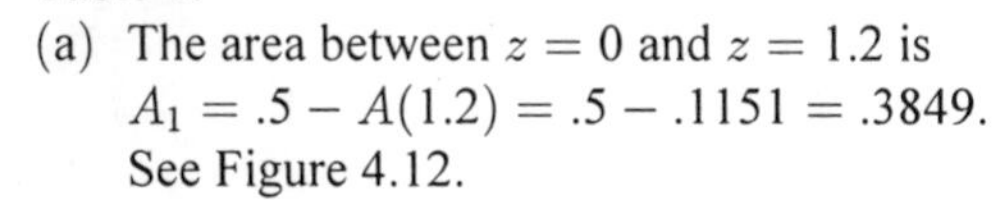

(a) The area between $z = 0$ and $z = 1.2$ is $A_1 = .5 - A(1.2) = .5 - .1151 = .3849$. See Figure 4.12.

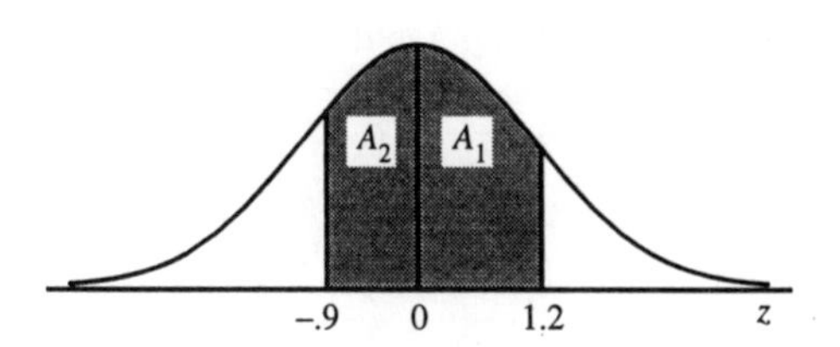

Figure 4.12

(b) The area between $z = 0$ and $z = -.9$ is $A_2 = .5 - A(-.9) = .5 - A(.9)$ $.5 - .1841 = .3159$.

(c) The desired area is A_1, as shown in Figure 4.13. Note that $A(.3) = .3821$ and $A(1.56) = .0594$.
$A_1 = A(.3) - A(1.56) = .3821 - .0594$ $= .3227$.

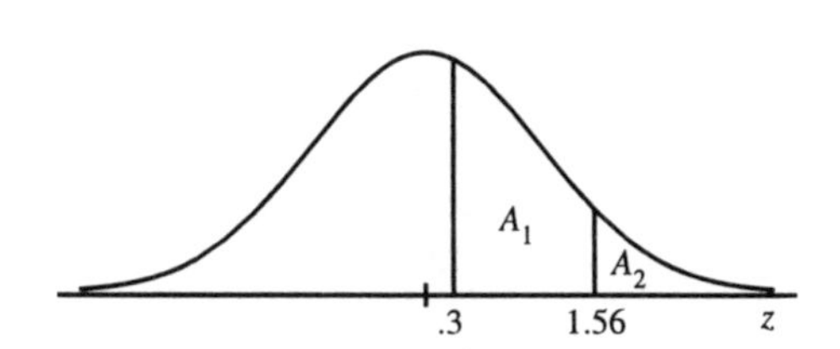

Figure 4.13

(d) The desired area is $A_1 + A_2$
$= .5 - A(-.2) + .5 - A(.2)$
$= 1 - 2(.4207) = .1586$. See Figure 4.14.

(e) The desired area is $A(-.2) - A(-1.56)$
$= .4207 - .0594 = .3613$.

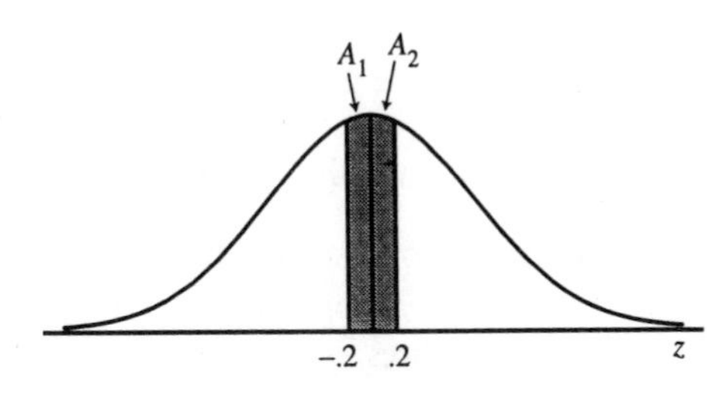

Figure 4.14

4.47 a. The procedure is reversed now, because the area under the curve is known. The objective is to determine the particular values, z_0, that will yield the given probability. In this exercise it is necessary to find a z_0 such that $P(Z > z_0) = .5000$. By the symmetry of the normal distribution, half of the area falls on each side of the mean. Thus, $P(Z > 0) = .5000$ and the desired value of z_0 is 0.

b. A value of z_0 is desired such that $P(Z < z_0) = .8643$. Thus the probability, .8643, will be the entire area under the curve to the left of the value $z = z_0$. Notice that the probability is greater than .5, so that z_0 must be in the right-hand half of the curve (i.e., $z_0 > 0$). See Figure 4.15. Then

$P(Z < z_0) = 1 - A(z_0) = .8643$
$A(z_0) = .1357$

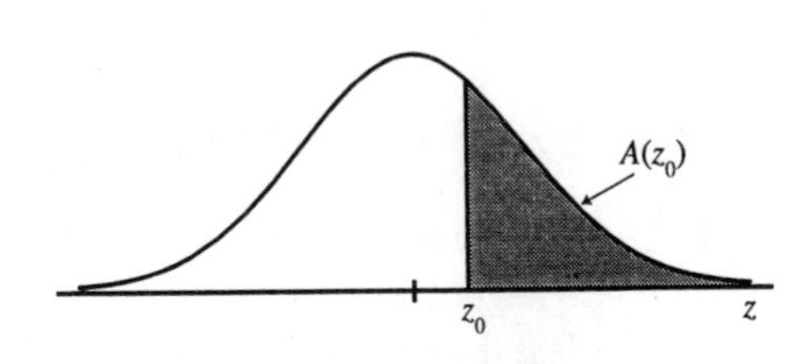

Figure 4.15

From Table 4 we get $z_0 = 1.10$. If the exact probability cannot be found in the table, we may choose to search for the probability closest to the one desired and perform an interpolation that will determine the exact value of z_0.

c. It is given that $P(-z_0 < Z < z_0) = .9000$.
See Figure 4.16. That is,

$A_1 + A_2 = .9000 = 1 - 2A(z_0)$
$2A(z_0) = .1$
$A(z_0) = .05$

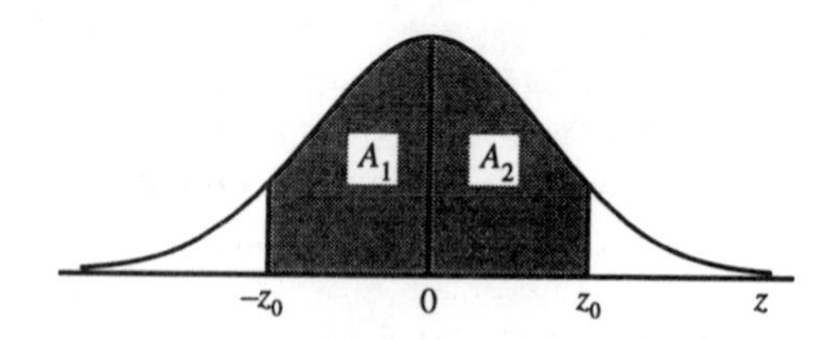

Figure 4.16

The desired value is not tabulated in Table 4 but falls between two tabulated values, .0505 and .0495. Hence z_0 will lie between 1.64 and 1.65, which are the z values associated with the above probabilities. Intuitively, one can see that the value .05 is halfway between the two tabulated values, and thus the desired value of z_0 will be halfway between 1.64 and 1.65, or $z_0 = 1.645$. This method of evaluation is called "linear interpolation."

d. Similar to part **c**. In this case,

$P(-z_0 < Z < z_0) = 1 - 2A(z_0) = .9900$

so that $A(z_0) = .0050$. Linear interpolation may now be used to determine the value of z_0, which will be between $z_1 = 2.57$ and $z_2 = 2.58$. Hence

$$z = 2.57 + \frac{.0050-.0049}{.0051-.0049}(2.58 - 2.57) = 2.57 + \frac{.0001}{.0002}(.01) = 2.57 + .005 = 2.575$$

4.48 Assume that $f(y)$ is indeed a density function. That is, $f(y) > 0$ for all y. But since

$$e^{-(y-\mu)^2/2\sigma^2} > 0$$

for all y, it follows that $\frac{1}{\sqrt{2\pi}\sigma} > 0$ implying that $\sigma > 0$.

4.49 In this case, the z-score is $z = \frac{17-16}{1} = 1$. Thus, we want $P(z > 1) = 0.1587$.

4.50 This normal distribution has $\mu = 400$ and $\sigma = 20$. Probabilities associated with any normal random variable Y can be obtained by converting the necessary values of y to their corresponding z values. This conversion is made by using the formula $z = \frac{y-\mu}{\sigma}$. Note that z is the distance from the mean, $y - \mu$, measured in units of σ. In this case, the desired probability is $P(Y > y_1) = 450$. The z value corresponding to $y_1 = 450$ is $z_1 = \frac{450-400}{20} = 2.5$. Then

$$P(Y > 450) = P(Z > 2.5) = A(2.5) = .0062$$

4.51 Refer to Figure 4.17. It is necessary to find a value for y such that $P(Y > y_0) = .1$. In terms of the corresponding z value,

$$z = \frac{y_0-\mu}{\sigma} = \frac{y_0-400}{20}$$

It is necessary to have

$P(Z > z) = .10$ or $A(z) = .1$

Using Table 4, the necessary value for z is $z = 1.28$ and

$$z = \frac{y_0-400}{20} = 1.28$$

Solving for y_0, $y_0 = 400 + 1.28(20) = \$425.60$.

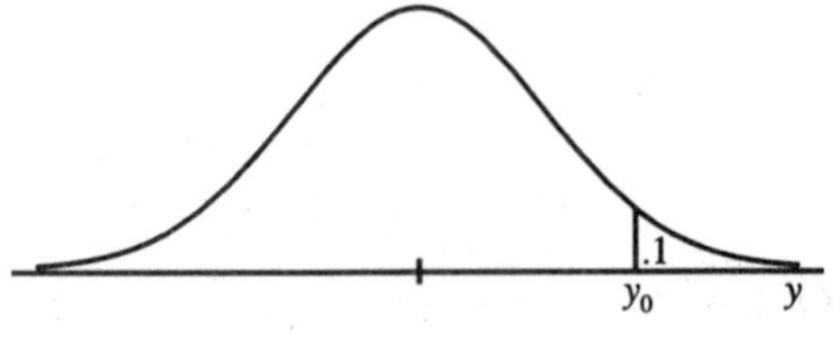

Figure 4.17

4.52 The random variable Y (bearing diameter) is normally distributed with $\mu = 3.0005$ and $\sigma = .0010$. Since "scrap" bearings are those whose diameters are greater than 3.002 or less than 2.998, the necessary area is $P(Y > 3.002) = P(Y < 2.998)$. Corresponding z values are

$$z_1 = \frac{3.002-3.0005}{.0010} = 1.5 \quad \text{and} \quad z_2 = \frac{2.998-3.0005}{.0010} = -2.5$$

and the necessary fraction is

$P(Y > 3.002) + P(Y < 2.998) = P(Z > 1.5) + P(Z < -2.5) = A(1.5) + A(-2.5)$
$= .0668 + .0062 = .0730.$

4.53 By inspection of the normal curve tail areas, one can see that if one marks off a fixed distance (in this case, 2(.001) $= .004$) along the y axis and considers the fraction outside this interval, the minimum area is obtained if the interval is centered about the mean. See Figure 4.18. Hence if this normal distribution is centered on the midpoint of the interval $3.000 \pm .002$, the fraction scrapped will be minimum. The mean diameter should be $\mu = 3$.

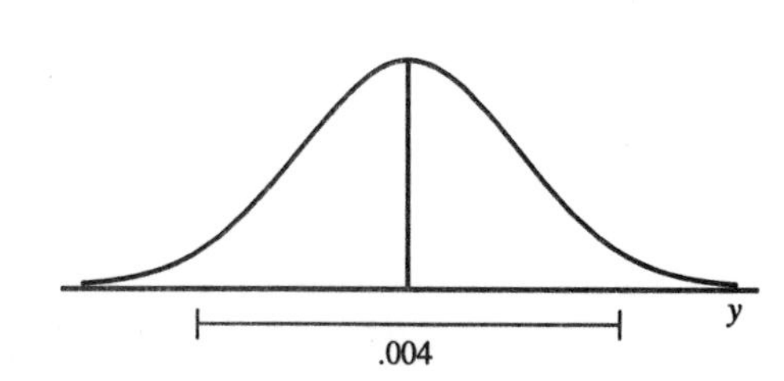

Figure 4.18

4.54 The fraction of students with grade point averages greater than 3.0 is given by $A_1 = P(Y > 3.0)$ (shown in Figure 4.19). Then the z value corresponding to the point $y = 3.0$ is

$$z = \frac{y-\mu}{\sigma} = \frac{3.0-2.4}{.8} = .75$$

Hence

$$A_1 = P(Y > 3.0) = P(Z > .75) = A(.75) = .2266$$

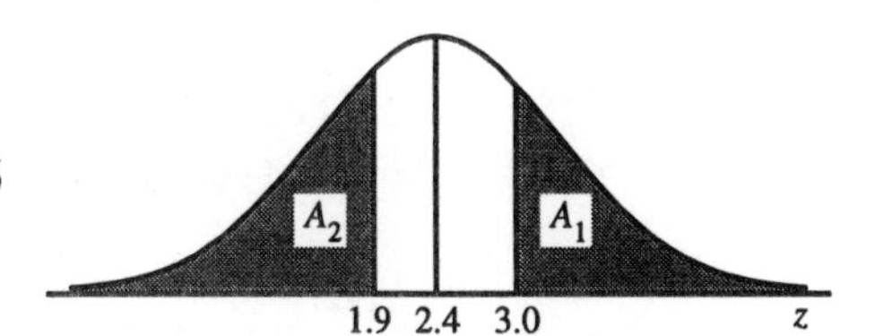

Figure 4.19

4.55 The probability of interest is $P(Y < 1.9)$ with corresponding z value

$$z = \frac{y-\mu}{\sigma} = \frac{1.9-2.4}{.8} = -.625$$

(Recall that a negative value of z implies a value to the left of the mean.) Then

$$A_2 = P(Y < 1.9) = P(Z < -.625) = P(Z > .625) = A(.625) = .2660$$

(after interpolating).

4.56 Let X be the number of students with a grade point average in excess of 3.0 when 3 students are randomly selected. Then X has a binomial distribution with $n = 3$ and $p = P$(student's GPA exceeds 3.0) $= .2266$, from Exercise 4.54. The probability of interest is

$$P(X = 3) = \binom{3}{3} p^3 q^0 = (.2266)^3 = .0116$$

4.57 Let Y be the measured resistance of a randomly selected wire.

a. The required probability is

$P(.12 \leq Y \leq .14) = P\left(\frac{.12-.13}{.005} \leq \frac{Y-\mu}{\sigma} \leq \frac{.14-.13}{.005}\right)$
$= P(-2 \leq Z \leq 2) = 1 - 2P(Z > 2) = 1 - 2A(2)$
$= 1 - 2(.0228) = .9544$

b. Define X to be the number of wires that meet the specifications. Then X has a Binomial distribution with $n = 4$ and $p = .9544$. Thus,

P(all four will meet the specifications) $= P(X = 4) = \binom{4}{4}(.9544)^4(.0456)^0 = .83$

4.58 Let Y denote the analyst's forecast. Then Y has a normal distribution with mean .14 and standard deviation .026.

a. $P(Y > .11) = P\left(Z > \frac{.11-.07}{.026}\right) = A(1.54) = .0618$

b. $P(Y < .09) = P\left(Z < \frac{.09-.07}{.026}\right) = P(Z < .77) = 1 - P(Z > .77) = 1 - A(.77)$
$= .7794$

4.59 **a.** The z values corresponding to $y_1 = 947$ and $y_2 = 958$ are

$$z_1 = \frac{947-950}{10} = -.3 \qquad \text{and} \qquad z_2 = \frac{958-950}{10} = .8$$

Then

$$P(947 \le Y \le 958) = P(-.3 \le Z \le .8) = 1 - P(Z < -.3) - P(Z > .8)$$
$$= 1 - .3821 - .2119 = .406$$

b. It is necessary that $P(Y \le C) = .8531$. Apparently, then, C must be to the right of the mean, $\mu = 950$, and must be associated with a z value such that

$$A(z) = A\left(\frac{C-950}{10}\right) = .1469$$

This value of z is $z = 1.05$ (from Table 4). Hence

$$\frac{C-950}{10} = 1.05 \qquad \text{and} \qquad C = 960.5$$

4.60

Let Y denote the variable "exam score."

a. $P(Y > 72) = P\left(Z > \frac{72-78}{6}\right) = P(Z > -1) = 1 - P(Z < -1) = 1 - P(Z > 1)$
$= 1 - A(1) = .8413$

b. We seek c such that $P(Y > c) = .1$. Now

$$.1 = P(Y > c) = P\left(Z > \frac{c-78}{6}\right) = A\left(\frac{c-78}{6}\right)$$

The value of z_0 such that $A(z_0) = .1$ is $z_0 = 1.28$. So it must be that

$$1.28 = \frac{c-78}{6}$$

and

$$c = 6(1.28) + 78 = 85.68$$

c. We seek c such that $P(Y > c) = .281$. So

$$.281 = P(Y > c) = P\left(Z > \frac{c-78}{6}\right) = A\left(\frac{c-78}{6}\right)$$

which implies that $\frac{c-78}{6} = .58$ and $c = 81.48$.

d. The score that cuts off the lowest 25% is the score c such that

$$P\left(Z < \frac{c-78}{6}\right) = .25$$

Now, $P(Z > .67) = .25$ so

$$P(Z < -.67) = .25.$$

Hence

$$\frac{c-78}{6} = -.67$$

and

$$c = (-.67)(6) + 78 = 73.98$$

We must now find $P(Y > 73.95 + 5)$. This probability is

$$P(Y > 78.95) = P\left(Z > \frac{78.95-78}{6}\right)$$
$$= P(Z > .16) = .4364$$

e. $P(Y > 84 | Y > 72) = \frac{P(Y>84)}{P(Y>72)} = \frac{P\left(Z>\frac{84-78}{6}\right)}{P\left(Z>\frac{72-78}{6}\right)} = \frac{P(Z>1)}{P(Z>-1)} = \frac{P(Z>1)}{1-P(Z>1)}$
$= \frac{.1587}{.8413} = .1886$

4.61

It is given that the random variable Y (ounces of fill) is normally distributed with mean μ and standard deviation $\sigma = .3$. The objective is to find a value of μ so that $P(Y > 8) = .01$. That is, an 8-ounce cup will overflow when $Y > 8$, and this should happen only 1% of the time. The z value corresponding to $Y = 8$ is

$$z = \frac{y-\mu}{\sigma} = \frac{8-\mu}{.3}$$

Thus,

$$P(Y > 8) = P\left(Z > \frac{8-\mu}{.3}\right) = .01 \qquad \text{or} \qquad A\left(\frac{8-\mu}{.3}\right) = .01$$

Consider $z_0 = \frac{8-\mu}{.3}$ and determine the value of z_0 that satisfies the equality shown above. This value is 2.33. Hence the value for μ can be obtained as

$$\frac{8-\mu}{.3} = 2.33 \qquad \text{or} \qquad \mu = 7.301$$

4.62

$0.95 = P(|x - \mu| < 1) = P(|z| < 1/\sigma)$ implies $1/\sigma = 1.96$ so that $\sigma = 1/1.96$.

4.63

a. $z = \frac{550-480}{100} = \frac{70}{100} = 0.7$

$P(Z \ge z) = 0.2420$ so that $1 - 0.2420 = 0.758$ or about 76% will score below 550.

b. $z = \frac{x-18}{6}$ and $P(Z \ge z) = 0.2420$ so set $z = \frac{x-18}{6} = 0.7$ so that $x = 18 + 6(0.7) = 22.2$. That is, a comparable score on the ACT math test would be 22.2.

4.64 $f(y) = \left(\frac{1}{\sigma\sqrt{2\pi}}\right) e^{-(y-\mu)^2/2\sigma^2} = \left(\frac{1}{\sigma\sqrt{2\pi}}\right) e^{(-y^2+2\mu y-\mu^2)/2\sigma^2}$

$f'(y) = \left(\frac{1}{\sigma\sqrt{2\pi}}\right) e^{-(y-\mu)^2/2\sigma^2} \left[\left(\frac{1}{2\sigma^2}\right)(-2y+2\mu)\right] = 0$

When $y = \mu$, $f'(y) = \left(\frac{1}{\sigma\sqrt{2\pi}}\right) e^0 \left[\left(\frac{1}{2\sigma^2}\right)(0)\right] = 0.$

Evaluating $f(\mu) = \left(\frac{1}{\sigma\sqrt{2\pi}}\right) e^0 = \frac{1}{\sigma\sqrt{2\pi}}$.

Recall the first derivative test. When $y < \mu$ then $f'(y) > 0$. Similarly, when $y > \mu$ then $f'(y) < 0$. Hence a maximum occurs at $y = \mu$.

4.65 $f'(y) = \left(\frac{-y+\mu}{\sigma^3\sqrt{2\pi}}\right) e^{-(y-\mu)^2/2\sigma^2} = \left(\frac{1}{\sigma^3\sqrt{2\pi}}\right)\left[-ye^{-(y-\mu)^2/2\sigma^2} + \mu e^{-(y-\mu)^2/2\sigma^2}\right]$

$$f''(y) = \left(\frac{1}{\sigma^3\sqrt{2\pi}}\right)\left\{e^{-(y-\mu)^2/2\sigma^2} - ye^{-(y-\mu)^2/2\sigma^2}\left[\left(\frac{1}{2\sigma^2}\right)(-2y+2\mu)\right] + \mu e^{-(y-\mu)^2/2\sigma^2}\left[\left(\frac{1}{2\sigma^2}\right)(-2y+2\mu)\right]\right\}$$

$$= \left(\frac{1}{\sigma^3\sqrt{2\pi}}\right) e^{-(y-\mu)^2/2\sigma^2}\left[1 - \frac{(\mu-y)^2}{\sigma^2}\right] = 0$$

We evaluate

$1 - \frac{(\mu-y)^2}{\sigma^2} = 0$

$(\mu - y)^2 = \sigma^2$ or $\mu - y = \pm\sigma$ so that

$y = \mu - \sigma$ or $y = \mu + \sigma$

4.66 $A = L \times W = |Y| \times 3|Y| = 3Y^2$

$$E(A) = E(3Y^2) = 3E(Y^2) = 3(V(Y) + \mu^2) = 3(\sigma^2 + \mu^2)$$

4.67 **a.** $\Gamma(1) = \int_0^\infty e^{-y}\,dy = 1 - e^{-\infty} = 1.$

b. Suppose we have two variables v and w. Integrating by parts, we may write $\int w\,dv = wv - \int v\,dw$, where w and v are suitably chosen to simplify the integration. The integral that must be evaluated is, in this case,

$$\Gamma(\mathrm{u}) = \int_0^\infty y^{u-1}e^{-y}\,dy$$

Let $w = y^{u-1}$ and $dv = e^{-y}\,dy$. Then

$dw = (u-1)y^{u-2}\,dy$ and $v = \int e^{-y}\,dy = -e^{-y}$

Integrating by parts, we have

$$\Gamma(u) = -e^{-y}y^{u-1}\big]_0^\infty + \int_0^\infty (u-1)y^{u-2}e^{-y}\,dy = 0 + (u-1)\int_0^\infty y^{u-2}e^{-y}\,dy$$

$$= (u-1)\Gamma(u-1)$$

Note that the quantity $e^{-y}y^{u-1}$ when evaluated at $y = \infty$ is of the indeterminate form $(0)(\infty)$, and L'Hôpital's rule is used to evaluate it.

4.68 Note that $\Gamma(1) = 1 = (1-1)! = 0!$ and generally, i.e. for any $n > 1$,

$$\Gamma(n-1) = (n-1)\Gamma(n-1) = (n-1)(n-2)\Gamma(n-2) = \cdots = (n-1)!$$

$\Gamma(2) = 1! = 1$, $\Gamma(4) = 3! = 6$, $\Gamma(7) = 6! = 730.$

4.69 Let Y = magnitude of earthquake. Y is exponential with $\beta = 2.4$.

a. $P(Y > 3) = \int_3^\infty \left(\frac{1}{2.4}\right) e^{-y/2.4}\,dy = -e^{-y/2.4}\big]_3^\infty = e^{-3/2.4} = .2865$

b. $P(2 < Y < 3) = \int_2^3 \left(\frac{1}{2.4}\right) e^{-y/2.4}\,dy = -e^{-y/2.4}\big]_2^3 = .1481$

4.70 Let X = # of earthquakes to exceed 5.0 on the Richter scale.

Y = magnitude of the earthquake. Y is Exponential with $\beta = 2.4$ and X is $B(10, p = P(Y > 5))$ where $p = \int_5^\infty \left(\frac{1}{2.4}\right) e^{-y/2.4}\,dy = e^{-y/2.4}\big]_5^\infty = .1245$

$$P(X \geq 1) = 1 - P(X = 0) = 1 - (.8755)^{10} = 1 - .2646 = .7354.$$

4.71 **a.** Let Y = demand for water. Y is exponential with $\beta = 100$.

$$P(Y > 200) = \int_{200}^{\infty} \left(\tfrac{1}{100}\right) e^{-y/100}\, dy = -e^{-y/100}\Big]_{200}^{\infty} = .1353$$

b. Let C = capacity. $P(Y > c) = \int_{c}^{\infty} \left(\tfrac{1}{100}\right) e^{-y/100}\, dy = -e^{-y/100}\Big]_{c}^{\infty} = e^{-c/100} = .01.$

so that $c = -100\, ln\,(0.01) = 460.52$ cfs.

4.72 Since Y is exponential with mean 10 apply Theorem 4.1.10 to obtain

$$V(Y) = \beta^2 = (10)^2 = 100.$$

We will need $E(y^2)$, $E(y^3)$ and $E(y^4)$.

$$E(y^2) = V(Y) + (E(Y))^2 = 100 + 100 = 200$$

$$E(y^3) = \int_0^{\infty} \tfrac{y^3}{10} e^{-y/10}\, dy = \Gamma(4)10^3 \int_0^{\infty} \tfrac{y^3 e^{-y/10}}{\Gamma(4)10^4}\, dy = \Gamma(4)(1000) = 3!(1000)$$
$$= 6000$$

$$E(y^4) = \int_0^{\infty} \tfrac{y^4}{10} e^{-y/10}\, dy = \Gamma(5)10^4 \int_0^{\infty} \tfrac{y^4 e^{-y/10}}{\Gamma(5)10^5}\, dy = \Gamma(5)10^4 = \; = 4!\,(10^4) =$$
$$= 240{,}000$$

Thus,

$$E(C) = 100 + 40E(y) + 3E(Y^2)$$
$$= 100 + 40(10) + 3(200) = 1100$$

$$E(C^2) = E\left[(100 + 40y + 3y^2)^2\right]$$
$$= E(10{,}000 + 1600y^2 + 9y^4 + 8000y + 600y^2 + 240y^3)$$
$$= 10{,}000 + 8000E(y) + 2200E(y^2) + 240E(y^3) + 9E(y^4)$$
$$= 10{,}000 + 8000(10) + 2200(200) + 240(6000) + 9(240{,}000)$$
$$= 4{,}130{,}000$$

Finally,

$$V(C) = E(C^2) - [E(C)]^2 = 4{,}130{,}000 - 1{,}210{,}000 = 2{,}920{,}000$$

4.73 **a.** $P(Y \le 31) = \int_0^{31} \left(\tfrac{1}{44}\right) e^{-y/44}\, dy = 1 - e^{-31/44} = .5057$

b. $V(Y) = \beta^2 = (44)^2 = 1936$

4.74 **a.** $P(Y > 9) = \int_9^{\infty} \left(\tfrac{1}{3.6}\right) e^{-y/3.6}\, dy = -e^{-y/3.6}\Big]_9^{\infty} = e^{-9/3.6} = .0821.$

b. $P(Y > 9) = \int_9^{\infty} \left(\tfrac{1}{2.5}\right) e^{-y/2.5}\, dy = -e^{-y/2.5}\Big]_9^{\infty} = e^{-9/2.5} = .0273.$

4.75 **a.** For any $k = 1, 2, 3, \ldots$

$$P(X = k) = P(k - 1 \le Y < k)$$
$$= P(Y < k) - P(Y \le k)$$
$$= \left(1 - e^{-k/\beta}\right) - \left(1 - e^{-(k-1)/\beta}\right)$$
$$= e^{-(k-1)/\beta} - e^{-k/\beta}$$

b. For any $k = 1, 2, 3, \ldots$

$$P(X = k) = e^{-(k-1)/\beta} - e^{-k/\beta}$$
$$= e^{-(k-1)/\beta} - e^{-(k-1)/\beta} e^{-1/\beta}$$
$$= e^{-(k-1)/\beta}\left(1 - e^{-1/\beta}\right)$$

$$= \left(e^{-1/\beta}\right)^{k-1}\left(1 - e^{-1/\beta}\right).$$

Thus, X has a geometric distribution with $p = 1 - e^{-1/\beta}$.

4.76 Notice that the variable part of $f(y)$ is that of a gamma-type random variable with $\alpha = 4$ and $\beta = 2$. In order for the density to integrate to 1, k must be the constant that accompanies this density. Hence

$$k = \frac{1}{\beta^{\alpha}\Gamma(\alpha)} = \frac{1}{\Gamma(4)2^4} = \frac{1}{6(16)} = \frac{1}{96}$$

4.77 The density function for Y is $f(y) = \left(\frac{1}{4}\right) e^{-y/4}$ for $y \geq 0$. Then

$$P(Y > 4) = \int_4^\infty \tfrac{1}{4} e^{-y/4}\, dy = -e^{-y/4}\Big]_4^\infty = e^{-1} = .368$$

4.78 It is necessary to find a value C so that $P(Y > C) = .05$. That is,

$$\int_C^\infty \tfrac{1}{4} e^{-y/4}\, dy = .05 = -e^{-y/4}\Big]_C^\infty = e^{-C/4} = .05$$

Implying,

$$-\tfrac{C}{4} = \ln .05$$
$$C = -4 \ln .05 = 11.98$$

4.79 $P(Y > \lambda) = \int_\lambda^\infty \frac{y^{\alpha-1}e^{-y}}{\Gamma(\alpha)}\, dy = \sum_{y=0}^{\alpha-1} \frac{\lambda^y e^{-\lambda}}{y!}$

For $\lambda = 1$ and $\alpha = 2$, this becomes $P(Y > 1) = \sum_{y=0}^{1} \frac{e^{\ 1}}{y!} = e^{-1} + e^{-1} = 2e^{-1} = .736$.

4.80 **a.** $P(X_1 = 0) = \frac{\lambda_1 e^{-\lambda_1}}{0!} = e^{-\lambda_1}$ and similarly $P(X_2 = 0) = e^{-\lambda_2}$. Since $\lambda_2 > \lambda_1$ we have

$$P(X_2 = 0) = e^{-\lambda_2} < e^{-\lambda_1} = P(X_1 = 0).$$

b. Using the result of 4.79 and that $Y \sim \text{Gamma}(\alpha = k+1, \beta = 1)$ obtain

$$P(Y > \lambda_1) = \sum_{x=0}^{k} \frac{\lambda_1 e^{-\lambda_1}}{x!} = P(X_1 \leq k)$$

when $X_1$is Poisson with mean λ_1. Similarly, $P(X_2 \leq k) = P(Y > \lambda_2)$.

c. Let $Y \sim \text{Gamma}(k+1, 1)$ then because the cdf is nondecreasing it follows from part **b** that

$$P(X_1 \leq k) = P(Y > \lambda_1) > P(Y > \lambda_2) = P(X_2 \leq k).$$

d. This means that X_2 is stochastically greater than X_1, i.e. in a probabilistic sense $X_2 > X_1$.

4.81 Let R denote the radius of a crater. Then $R \sim \text{Exponential}(10)$. Now thw area is $A = \pi R^2$so that

$$E(A) = \pi E(R^2) = \pi\left[V(R) + [E(R)]^2\right] = \pi(100 + 10^2) = 200\pi$$

and

$$V(A) = \pi^2 V(R^2) = \pi^2\left[E(R^4) - [E(R^2)]^2\right] = \pi^2\left[E(R^4) - 200^2\right].\ \text{Next,}$$
$$E(R^4) = (1/10)\int_0^\infty r^4 e^{-y/10}\, dy = (1/10)\, 10^5\, \Gamma(5) = 24(10^4)$$

since the integrand is the kernel of a Gamma(5,100) random variable.
Therefore, $V(A) = \pi^2(24(10^4) - 200^2) = 200000\pi^2$.

4.82 Let Y be the number of the 3 components operating more than 200 hours. Then since the components operate independently, Y has a binomial distribution with $n = 3$ and

$$\backslash p = \int_{200}^\infty \tfrac{1}{100} e^{-x/100}\, dx = \left[-e^{-x/100}\right]_{200}^\infty = e^{-2}$$

Then

$$P(\text{equipment operates at least 200 hours}) = P(Y \geq 2) = \binom{3}{2} p^2 q + \binom{3}{3} p^3$$
$$= 3\left(e^{-2}\right)^2\left(1 - e^{-2}\right) + \left(e^{-2}\right)^3 = .05$$

4.83 Let Y be the rainfall totals. Then

$$E(Y) = \alpha\beta = (1.6)(2) = 3.2$$

and

$$V(Y) = \alpha\beta^2 = (1.6)(4) = 6.4.$$

4.84 Let Y = response time, the mean of Y would be $\mu = \alpha\beta = 4$. Further the variance of Y would be $\sigma^2 = \alpha\beta^2 = (\alpha\beta)\beta = 8$. Thus $4\beta = 8$, implying $\beta = 2, \alpha = 2$. Then the density for Y would be

$$f(y) = \left[\frac{1}{\Gamma(2)2^2}\right] ye^{-y/2}, y > 0$$

or

$$\left(\frac{y}{4}\right) e^{-y/2}, y > 0.$$

4.85 Using Tchebycheff's theorem, with $k = 2$, consider the interval $(\mu \pm k\sigma) = \left(4 \pm 2\left(\sqrt{8}\right)\right)$ $= (4 \pm 5.6569) = (-1.6569, 9.6569)$. Since the lower bound for y is 0, our interval is (0, 9.6569).

4.86 Since Y has a gamma distribution with $\alpha = 1000$ and $\beta = 20$,

$$E(Y) = \alpha\beta = 20{,}000 \qquad \text{and} \qquad V(Y) = \alpha\beta^2 = 400{,}000$$

The standard deviation is $\sigma = \sqrt{400{,}000} = 632.46$. Hence the point $y = 40{,}000$ lies $\frac{40{,}000-20{,}000}{632.46} = 31.62$ standard deviations above the mean. It is highly unlikely that such a value would be observed.

4.87 From Theorem 4.8, a gamma-type random variable with $\alpha = 3$ and $\beta = 2$ has $E(Y) = \alpha\beta = 6$ and $E(Y^2) = V(Y) + (E(Y))^2 = \alpha(\alpha+1)\beta^2 = 3(4)(4) = 48$. Hence

$$E(L) = 30E(Y) + 2E(Y^2) = 30(6) + 2(48) = 276$$

Since

$V(L) = E(L^2) - [E(L)]^2 = E(L^2) - 76{,}176 = E(900Y^2 + 120Y^3 + 4Y^4) - 76{,}176$

we need the third and fourth moments about the origin.

$$E(Y^3) = \int_0^\infty \frac{y^5 e^{-y/2}}{\Gamma(3)2^3}\, dy = \frac{\Gamma(6)2^3}{\Gamma(3)} \int_0^\infty \frac{y^5 e^{-y/2}}{\Gamma(6)2^6}\, dy = \frac{5!2^3}{2!} = 480$$

$$E(Y^4) = \int_0^\infty \frac{y^6 e^{-y/2}}{\Gamma(3)2^3}\, dy = \frac{\Gamma(7)2^4}{\Gamma(3)} \int_0^\infty \frac{y^6 e^{-y/2}}{\Gamma(7)2^7}\, dy = \frac{6!\,2^4}{2!} = 5760$$

Then $V(L) = 900(48) + 120(480) + 4(5760) - 76{,}176 = 47{,}664$.

4.88 Since Y has a gamma distribution with $\alpha = 3$, $\beta = \frac{1}{2}$,

$$E(Y) = \alpha\beta = \frac{3}{2} \qquad \text{and} \qquad V(Y) = \alpha\beta^2 = 3\left(\frac{1}{4}\right) = \frac{3}{4}$$

4.89 **a.** $E(Y^a) = \int_0^\infty \frac{y^a y^{\alpha-1}}{\beta^\alpha\Gamma(\alpha)} e^{-y/\beta}\, dy = \int_0^\infty \frac{y^{\alpha+a-1}e^{-y/\beta}}{\beta^\alpha\Gamma(\alpha)}\, dy$

$$= \frac{\beta^a\Gamma(\alpha+a)}{\Gamma(\alpha)} \int_0^\infty \frac{y^{\alpha+a-1}e^{-y/\beta}}{\Gamma(\alpha+a)\beta^{\alpha+a}}\, dy = \frac{\beta^a\Gamma(\alpha+a)}{\Gamma(\alpha)} \qquad \text{if } \alpha + a > 0.$$

b. Because $\int_0^\infty \frac{y^{\alpha+a-1}e^{-y/\beta}}{\Gamma(\alpha+a)\beta^{\alpha+a}}\, dy = 1$ requires $\alpha + a > 0$ and $\beta > 0$.

c. With $a = 1$

$$E(Y^1) = \frac{\beta^1\Gamma(\alpha+1)}{\Gamma(\alpha)} = \beta\alpha\frac{\Gamma(\alpha)}{\Gamma(\alpha)} = \alpha\beta.$$

d. $E\left(\sqrt{Y}\right) = E\left(Y^{1/2}\right) = \frac{\beta^{1/2}\Gamma\left(\alpha+\frac{1}{2}\right)}{\Gamma(\alpha)}$.

We assume $\alpha > 0$.

e. $E\left(\frac{1}{Y}\right) = E\left(Y^{-1}\right) = \frac{\beta^{-1}\Gamma(\alpha-1)}{\Gamma(\alpha)} = \frac{1}{\beta(\alpha-1)}$.

We assume $\alpha > 1$.

$$E\left(\frac{1}{\sqrt{Y}}\right) = E\left(Y^{-1/2}\right) = \frac{\beta^{-1/2}\Gamma\left(\alpha-\frac{1}{2}\right)}{\Gamma(\alpha)} = \frac{\Gamma\left(\alpha-\frac{1}{2}\right)}{\sqrt{\beta}\Gamma(\alpha)}.$$

We assume $\alpha > \frac{1}{2}$.

$$E\left(\frac{1}{Y^2}\right) = E\left(Y^{-2}\right) = \frac{\beta^{-2}\Gamma(\alpha-2)}{\Gamma(\alpha)} = \frac{1}{\beta^2(\alpha-1)(\alpha-2)}.$$

We assume $\alpha > 2$.

4.90 **a.** The chi-square distribution with ν degrees of freedom is a gamma distribution with $\alpha = \frac{\nu}{2}$ and $\beta = 2$. If $\nu > -2a$, then

$$E(Y^a) = \frac{2^a\Gamma\left(\frac{\nu}{2}+a\right)}{\Gamma\left(\frac{\nu}{2}\right)}.$$

b. Because when $\alpha = \frac{\nu}{2}$, $\alpha + a > 0$ if and only if

$$\frac{\nu}{2} + a > 0$$
$$\frac{\nu}{2} > -a$$
$$\nu > -2a.$$

c. $E\left(\sqrt{Y}\right) = E\left(Y^{1/2}\right) = \frac{2^{1/2}\Gamma\left(\frac{\nu}{2}+\frac{1}{2}\right)}{\Gamma\left(\frac{\nu}{2}\right)}$.

We assume $\nu > -2a = -1$ or $\nu > 0$.

d. $\mathrm{E}\left(\frac{1}{Y}\right) = E\left(Y^{-1}\right) = \frac{2^{-1}\Gamma\left(\frac{\nu}{2}-1\right)}{\Gamma\left(\frac{\nu}{2}\right)} = \frac{\Gamma\left(\frac{\nu}{2}-1\right)}{2\left(\frac{\nu}{2}-1\right)\Gamma\left(\frac{\nu}{2}-1\right)} = \frac{1}{\nu-2}$.

We assume $\nu > 2$.

$$E\left(\frac{1}{\sqrt{Y}}\right) = E\left(Y^{-1/2}\right) = \frac{2^{-1/2}\Gamma\left(\frac{\nu}{2}-\frac{1}{2}\right)}{\Gamma\left(\frac{\nu}{2}\right)}.$$

We assume $\nu > 1$.

$$E\left(\frac{1}{Y^2}\right) = E\left(Y^{-2}\right) = \frac{2^{-2}\Gamma\left(\frac{\nu}{2}-2\right)}{\Gamma\left(\frac{\nu}{2}\right)} = \frac{2^{-2}\Gamma\left(\frac{\nu}{2}-2\right)}{\left(\frac{\nu}{2}-1\right)\left(\frac{\nu}{2}-2\right)\Gamma\left(\frac{\nu}{2}-2\right)} = \frac{1}{(\nu-2)(\nu-4)}.$$

We assume $\nu > 4$.

4.91 Similar to Exercise 4.76. The value of k is the constant part of the beta density with $\alpha = 4$ and $\beta = 3$. Hence

$$k = \frac{\Gamma(\alpha+\beta)}{\Gamma(\alpha)\Gamma(\beta)} = \frac{\Gamma(7)}{\Gamma(4)\Gamma(3)} = \frac{6!}{3!\,2!} = 60$$

4.92 $P(Y \geq .4) = \int_{.4}^{1} (12y^2 - 12y^3)\, dy = \left[4y^3 - 3y^4\right]_{.4}^{1} = 4 - 3 - (.256 - .0768) = .8208$

4.93 Since Y has a beta distribution with $\alpha = 3$, $\beta = 2$,

$$E(Y) = \frac{\alpha}{\alpha+\beta} = \frac{3}{5} \qquad \text{and} \qquad V(Y) = \frac{\alpha\beta}{(\alpha+\beta)^2(\alpha+\beta+1)} = \frac{6}{25(6)} = \frac{1}{25}$$

4.94 **a.** $F(y) = \int_0^y (6t - 6t^2)\, dt = \left[3t^2 - 2t^3\right]_0^y = 3y^2 - 2y^3$ for $0 \leq y \leq 1$. $F(y) = 0$ for $y < 0$; $F(y) = 1$ for $y > 1$.

b. The graphs for $F(y)$ and $f(y)$ are shown in Figures 4.20 and 4.21.

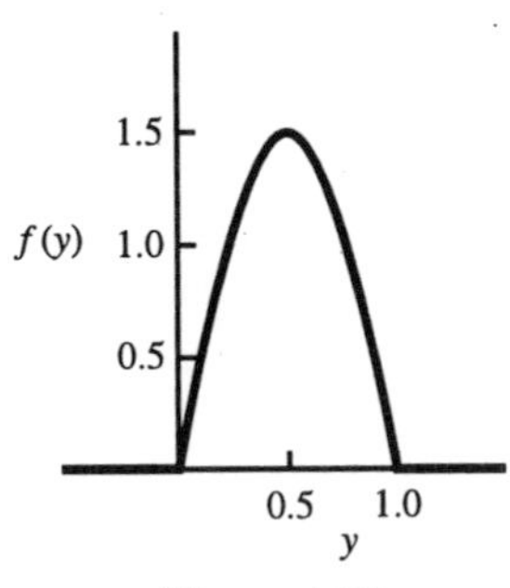

Figure 4.20

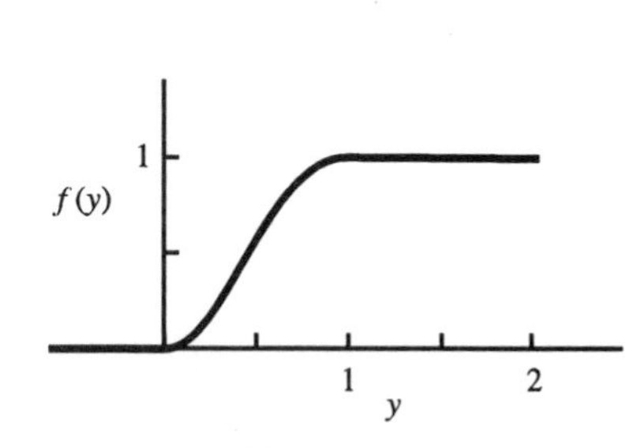

Figure 4.21

c. $P(.5 \leq Y \leq .8) = F(.8) - F(.5) = 1.92 - 1.024 - .75 + .25 = .396$

4.95 For $\alpha = \beta = 1$, $f(y) = \frac{\Gamma(2)}{\Gamma(1)\Gamma(1)} y^{1-1}(1-y)^{1-1} = 1$ for $0 \leq y \leq 1$, which is the uniform distribution.

4.96 Let x be the budgeted cost. We must have $P(Y > x) > .10$. Thus,

$$\int_x^1 3(1-y)^2\, dy = .10$$

We make the substitution $t = 1 - y$ and $dt = -\,dy$ to get

$$-\int_{1-x}^{0} 3t^2\, dt = \int_0^{1-x} 3t^2 dt \;= t^3\big]_0^{1-x} = (1-x)^3$$

Hence x must satisfy

$$(1-x)^3 = .10$$
$$1 - x = (.10)^{1/3}$$
$$x = 1 - (.10)^{1/3} = .5358$$

so that the budgeted cost should be $100(.5358) = \$53.58$.

4.97 $E(C) = 10 + 20E(X) + 4E\left(X^2\right)$

$$= 10 + 20E(X) + 4(V(X) + (E(X)^2))$$
$$= 10 + 20\left(\tfrac{\alpha}{\alpha+\beta}\right) + 4\left[\tfrac{\alpha(\alpha+1)}{(\alpha+\beta)(\alpha+\beta+1)}\right]$$
$$= 10 + 20\left(\tfrac{1}{3}\right) + 4\left(\tfrac{2}{12}\right) = \tfrac{52}{3}$$

To compute $V(C)$ we will need $E\left(X^3\right)$ and $E\left(X^4\right)$.

$$E\left(X^3\right) = \int_0^1 \left(2x^3 - 2x^4\right) dx = \left[\tfrac{1}{2}x^4 - \tfrac{2}{5}x^5\right]_0^1 = \tfrac{1}{2} - \tfrac{2}{5} = \tfrac{1}{10}$$

$$E\left(X^4\right) = \int_0^1 \left(2x^4 - 2x^5\right) dx = \left[\tfrac{2}{5}x^5 - \tfrac{1}{3}x^6\right]_0^1 = \tfrac{2}{5} - \tfrac{1}{3} = \tfrac{1}{15}$$

$$V(C) = V(10 + 20X + 4X^2) = V(20X + 4X^2)$$
$$= E\left[\left(20X + 4X^2\right)^2\right] - \left[E\left(20X + 4X^2\right)\right]^2$$
$$= 400E\left(X^2\right) + 160E\left(X^3\right) + 16E\left(X^4\right) - \left(\tfrac{22}{3}\right)^2$$
$$= 400\left(\tfrac{1}{6}\right) + 160\left(\tfrac{1}{10}\right) + 16\left(\tfrac{1}{15}\right) - \left(\tfrac{22}{3}\right)^2 = 29.96$$

4.98 The density function for Y is $f(y) = \frac{\Gamma(\alpha+\beta)}{\Gamma(\alpha)\Gamma(\beta)} y^{\alpha-1}(1-y)^{\beta-1}$ for $0 \le y \le 1$. Thus,

$$E(Y) = \frac{\Gamma(\alpha+\beta)}{\Gamma(\alpha)\Gamma(\beta)} \int_0^1 y^{\alpha}(1-y)^{\beta-1}\, dy$$

The quantity $y^{\alpha}(1-y)^{\beta-1}$ is the variable factor of a beta density function with parameters $\alpha + 1$ and β. Hence

$$E(Y) = \frac{\Gamma(\alpha+\beta)\Gamma(\alpha+1)\Gamma(\beta)}{\Gamma(\alpha)\Gamma(\beta)\Gamma(\alpha+\beta+1)} \int_0^1 \frac{y^{\alpha}(1-y)^{\beta-1}}{B(\alpha+1,\beta)}\, dy = \frac{\Gamma(\alpha+\beta)\Gamma(\alpha+1)}{\Gamma(\alpha)\Gamma(\alpha+\beta+1)} = \frac{\alpha}{\alpha+\beta}$$

since the integral of a complete density function is 1. Similarly,

$$E\left(Y^2\right) = \frac{\Gamma(\alpha+\beta)}{\Gamma(\alpha)\Gamma(\beta)} \int_0^1 y^{\alpha+1}(1-y)^{\beta-1}\, dy = \frac{\Gamma(\alpha+\beta)\Gamma(\alpha+2)\Gamma(\beta)}{\Gamma(\alpha)\Gamma(\beta)\Gamma(\alpha+\beta+2)}$$
$$= \frac{(\alpha+1)\alpha}{(\alpha+\beta)(\alpha+\beta+1)}$$

and

$$V(Y) = \frac{(\alpha+1)\alpha}{(\alpha+\beta)(\alpha+\beta+1)} - \frac{\alpha^2}{(\alpha+\beta)^2} = \frac{(\alpha+1)\alpha(\alpha+\beta) - \alpha^2(\alpha+\beta+1)}{(\alpha+\beta)^2(\alpha+\beta+1)}$$
$$= \frac{\alpha\beta}{(\alpha+\beta)^2(\alpha+\beta+1)}$$

4.99 Let Y = measurement error.

a. $P(Y < .5) = \int_0^{.5} \left[\frac{\Gamma(1+2)}{\Gamma(1)\Gamma(2)}\right] y^{1-1}(1-y)^{2-1}\, dy$

$$= \int_0^{.5} 2(1-y)\, dy = 2y - y^2\Big]_0^{.5} = 1 - .25 = .75$$

b. $\mu = E(Y) = \frac{\alpha}{\alpha+\beta} = \frac{1}{3}$.

$$V(Y) = \frac{\alpha\beta}{(\alpha+\beta)^2(\alpha+\beta+1)} = \frac{2}{3^2(4)} = \frac{1}{18}\,;\ \sigma = \frac{1}{\sqrt{18}} = .2357$$

4.100 Let Y = proportion of weight contributed by the fine powders.

Recall that $E(Y) = \frac{\alpha}{\alpha+\beta}$ and $V(Y) = \frac{\alpha\beta}{(\alpha+\beta)^2(\alpha+\beta+1)}$

a. $E(Y) = \frac{3}{6} = \frac{1}{2};\ V(Y) = \frac{9}{(6^2)(7)} = \frac{1}{28}$

b. $E(Y) = \frac{2}{4} = \frac{1}{2};\ V(Y) = \frac{4}{(4^2)(5)} = \frac{1}{20}$

c. $E(Y) = \frac{1}{2};\ V(Y) = \frac{1}{(2^2)(3)} = \frac{1}{12}$

d. case (a) since it has the smallest variance

4.101 **a.** The variable factor is that of a beta density with $\alpha = 3$ and $\beta = 5$. Hence

$$c = \frac{\Gamma(\alpha+\beta)}{\Gamma(\alpha)\Gamma(\beta)} = \frac{\Gamma(8)}{\Gamma(3)\Gamma(5)} = \frac{7!}{2!\,4!} = 105$$

b. For a beta random variable, $E(Y) = \frac{\alpha}{\alpha+\beta} = \frac{3}{8}$.

Note that this agrees with the solution to Exercise 4.22.

4.102 **a.** If $\alpha = 4$ and $\beta = 7$ then we want to find

$$F(0.7) = \sum_{i=4}^{10} \binom{10}{i}(0.7)^i(0.3)^{10-i}.$$

That is, we want $P(4 \le X \le 10)$ where $X \sim$ Binomial(10,0.7). Using Table 1 we find that

$$P(4 \le X \le 10) = 1 - P(X \le 3) = 1 - 0.011 = 0.989.$$

b. This time we want $P(12 \le X \le 25)$ where $X \sim$ Binomial(25,0.6). Table 1 yields

$$P(12 \le X \le 25) = 1 - P(X \le 11) = 1 - 0.078 = 0.922.$$

4.103 **a.** $P(Y_1 = 0) = \binom{n}{0}p_1^0(1-p_1)^n = (1-p_1)^n$ and similarly $P(Y_2 = 0) = (1-p_2)^n$ so that $p_1 < p_2$ implies $P(Y_1 = 0) > P(Y_2 = 0)$.

b.

$$\begin{aligned}
P(Y_1 \le k) &= \sum_{i=0}^{k}\binom{n}{i}p_1^i(1-p_1)^{n-i} = 1 - \sum_{i=k+1}^{n}\binom{n}{i}p_1^i(1-p_1)^{n-i} \\
&= 1 - \int_0^{p_1}\frac{t^k(1-t)^{n-k-1}}{B(k+1,n-k)}dt = 1 - P(X \le p_1) \ \{X \sim \text{Beta}(k+1, n-k)\} \\
&= P(X > p_1) = \int_{p_1}^{1}\frac{t^k(1-t)^{n-k-1}}{B(k+1,n-k)}dt
\end{aligned}$$

c. Note that, by using the result of part **c.**, it is easy to see that the integrands for $P(Y_1 \le k)$ and $P(Y_2 \le k)$ will be identical but since $p_1 < p_2$ the regions of integration will be different. This trivially yields the result that Y_2 is stochastically greater than Y_1.

4.104 **a.** Refer to Example 4.13 with $\beta = \theta$, $\alpha = 1$. The mgf for y is $m(t) = \frac{1}{(1-\beta t)^\alpha}$ $= \frac{1}{(1-\theta t)}$.

b. Differentiating with respect to t,

$$E(Y) = m'(0) = \left.\frac{\theta}{(1-\theta t)^2}\right|_{t=0} = \theta \quad \text{and} \quad E(Y^2) = m''(0) = \left.\frac{2\theta^2}{(1-\theta t)^3}\right|_{t=0} = 2\theta^2$$

so that $V(Y) = 2\theta^2 - \theta^2 = \theta^2$.

4.105 Let $m_Y(t)$ be the moment-generating function of Y. Then

$$\begin{aligned}
m_u(t) &= E(e^{tu}) = E(e^{bt+atY}) = e^{bt}E(e^{atY}) \\
&= e^{bt}m_Y(at).
\end{aligned}$$

Differentiating $m_u(t)$, we obtain

$$\begin{aligned}
E(U) &= m_u'(t)|_{t=0} = ae^{bt}m_Y'(at) + bm_Y(at)e^{bt}|_{t=0} \\
&= am_Y'(0) + bE(e^{aY0}) = am_Y'(0) + bE(1) \\
&= b + aE(Y). \\
E(U^2) &= m_u''(t)|_{t=0} = a^2e^{bt}m_Y''(at) + m_Y'(at)bae^{bt} + b^2m_Y(at)e^{bt} + e^{bt}bam_Y'(at)|_{t=0} \\
&= a^2m_Y''(0) + 2m_Y'(0)ab + m_Y(0)b^2 \\
&= a^2E(Y^2) + 2abE(Y) + b^2.
\end{aligned}$$

Hence,

$$\begin{aligned}
V(U) &= a^2E(Y^2) + 2abE(Y) + b^2 - [b + aE(Y)]^2 \\
&= a^2[E(Y^2) - E(Y)^2] = a^2V(Y).
\end{aligned}$$

4.106 **a.** From Example 4.16, $m_{Y-\mu}(t) = E\left[e^{t(Y-\mu)}\right] = e^{t^2\sigma^2/2}$. Since

$$E\left[e^{t(Y-\mu)}\right] = E(e^{tY_e - \mu t}) = e^{-\mu t}E(e^{tY})$$

we have $e^{t^2\sigma^2/2} = e^{-\mu t}E(e^{tY})$, so the mgf for Y must be $m_Y(t) = E(e^{tY})$ $= e^{\mu t + (t^2\sigma^2/2)}$.

b. Differentiating with respect to t,

$$E(Y) = m'(0) = (\mu + t\sigma^2)\, e^{\mu t + (t^2\sigma^2/2)}\Big|_{t=0} = \mu$$

Then $V(Y) = \mu^2 + \sigma^2 - \mu^2 = \sigma^2$.

4.107 Using exercise 4.105 with $a = -3$ and $b = 4$ it is easy to see that

$$m_x(t) = e^{4t} m(-3t) = e^{(4-3\mu)t + t^2(9\sigma^2)/2}$$

Thus $X \sim \text{Normal}(4 - 3\mu, 9\sigma^2)$ by the uniqueness of moment generating functions.

4.108 **a.** Gamma(2,4)

b. Exponential(3.2)

c. Normal($-$ 5,12)

4.109 $m_y(t) = E(e^{ty}) = \int_{\theta_1}^{\theta_2} e^{ty} \frac{1}{\theta_2 - \theta_1}\, dy = \frac{e^{t\theta_2} - e^{t\theta_1}}{t(\theta_2 - \theta_1)}$

4.110 **a.** Set $\theta_1 = 0$ and $\theta_2 = 1$ to obtain $m_y(t) = \frac{e^t - 1}{t}$.

b. Apply 4.105 with $b = 0$ to get

$$m_w(t) = \frac{e^{at} - 1}{at}$$

which implies that $W \sim \text{Uniform}(0, a)$ by the uniqueness of mgfs.

c. In this case,

$$m_x(t) = \frac{e^{-at} - 1}{-at}$$

so that $X \sim \text{Uniform}(-a, 0)$ by the uniqueness of mgfs.

d. $m_v(t) = e^{bt}\left(\frac{e^{at} - 1}{at}\right) = \frac{e^{(b+a)t} - e^{bt}}{at}$

so that $V \sim \text{Uniform}(b, b + a)$.

4.111 Differentiating $m(t)$ with respect to t, where $m(t) = \frac{1}{(1-\beta t)^\alpha} = (1 - \beta t)^{-\alpha}$

$$E(Y) = m'(0) = \frac{\alpha\beta}{(1-\beta t)^{\alpha+1}}\Big|_{t=0} = \alpha\beta$$

$$E(Y^2) = m''(0) = \frac{d}{dt}\, \alpha\beta(1 - \beta t)^{-\alpha-1}\big|_{t=0} = \frac{\alpha\beta^2(\alpha+1)}{(1-\beta t)^{\alpha+2}}\Big|_{t=0} = \alpha\beta^2(\alpha + 1)$$

and $V(Y) = \alpha\beta^2(\alpha + 1) - \alpha^2\beta^2 = \alpha\beta^2$.

4.112 **a.** The density shown is that of a normal random variable with $\mu = 0$ and $\sigma^2 = 1$, which has constant

$$k = \frac{1}{\sqrt{2\pi}\,\sigma} = \frac{1}{\sqrt{2\pi}}$$

b. Substituting $\mu = 0$ and $\sigma^2 = 1$ in the general formula derived in Exercise 4.106, we obtain

$$m(t) = e^{t^2/2}.$$

c. Again using the result of Exercise 4.106, $E(Y) = \mu = 0$ and $V(Y) = \sigma^2 = 1$.

4.113 **a.** $E\left(e^{3Y/2}\right) = \int_{\infty}^{0} e^{(3y/2)+y}\, dy = \frac{2}{5}\, e^{5y/2}\big]_{-\infty}^{0} = \frac{2}{5}$

b. $m(t) = E\left(e^{tY}\right) = \int_{-\infty}^{0} e^{(t+1)y}\, dy = \frac{e^{(t+1)y}}{t+1}\Big]_{-\infty}^{0} = \frac{1}{t+1}$

c. Differentiating with respect to t, we obtain

$E(Y) = m'(t)|_{t=0} = \frac{-1}{(t+1)^2}\Big|_{t=0} = -1 \qquad E(Y^2) = m''(0) = \frac{2}{(t+1)^3}\Big|_{t=0} = 2$

$V(Y) = 2 - (-1)^2 = 1$

4.114 The interval must include 90% of all mileage on tires he sells. Using Tchebysheff's theorem, we must have

$$P(|Y - \mu| \leq k\sigma) \geq .90 = 1 - \frac{1}{k^2}.$$

Then $k = \sqrt{\frac{1}{.10}} = \sqrt{10} = 3.1622$. The necessary interval is then

$$|Y - 25{,}000| \leq 3.1622(4000) \qquad \text{or} \qquad 12{,}351 \leq Y \leq 37{,}649$$

4.115 It is necessary to have $P(|Y - \mu| \leq 1) \geq .75$. Hence, $1 - \frac{1}{k^2} = .75$ and $k = 2$. According to Tchebysheff's inequality, then, $1 = k\sigma$ and $\sigma = \frac{1}{k} = \frac{1}{2}$.

4.116 From Exercise 4.14 we have $\mu = \frac{2}{3}$ and $\sigma^2 = \frac{2}{9}$. Hence $2\sigma = 2\sqrt{\frac{2}{9}} = .943$. Then

$$P(|Y - \mu| \le 2\sigma) = P(\mu - 2\sigma \le Y \le \mu + 2\sigma) = P(-.276 \le Y \le 1.610) = F(1.61)$$
$$= 1.61 - \frac{(1.61)^2}{4} = .962 \qquad \text{from exercise 4.10}$$

Since Y has a positive density function only for $0 \le y \le 2$, $P(-.276 \le Y \le 0) = 0$. Notice that, according to Tchebysheff's theorem, this probability should be at least $1 = \frac{1}{k^2} = 1 - \left(\frac{1}{4}\right) = \frac{3}{4} = .75$, which it is. According to the empirical rule, it should be approximately .95. Notice that the approximation is not too bad, even though the probability distribution (see Figure 4.4) is far from mound-shaped.

4.117 From Table A2.2, Appendix II, we find that when Y is uniform over (θ_1, θ_2)

$$E(Y) = \frac{\theta_1+\theta_2}{2} \qquad V(Y) = \frac{(\theta_2-\theta_1)^2}{12}$$

Thus,

$$2\sigma = 2\sqrt{V(Y)} = \frac{2(\theta_2-\theta_1)}{\sqrt{12}} = \frac{\theta_2-\theta_1}{\sqrt{3}}$$

The probability of interest is $P(|Y - \mu| \le 2\sigma) = P(\mu - 2\sigma \le Y \le \mu + 2\sigma)$. Now

$$\theta_2 - \frac{\theta_1+\theta_2}{2} = \frac{\theta_2-\theta_1}{2} < \frac{\theta_2-\theta_1}{\sqrt{3}} \quad \text{and hence} \quad \theta_2 < \frac{\theta_1+\theta_2}{2} + \frac{\theta_2-\theta_1}{\sqrt{3}} = \mu + 2\sigma$$

Similarly,

$$\frac{\theta_1+\theta_1}{2} - \theta_1 = \frac{\theta_2-\theta_1}{2} < \frac{\theta_2-\theta_1}{\sqrt{3}} \quad \text{so that} \quad \theta_1 > \frac{\theta_1+\theta_2}{2} - \frac{\theta_2-\theta_1}{\sqrt{3}} = \mu - 2\sigma$$

But θ_1 and θ_2 are the upper and lower limits on Y. Hence

$$P(\mu - 2\sigma \le Y \le \mu + 2\sigma) = P\left[\left(\frac{\theta_1+\theta_2}{2} - \frac{\theta_2-\theta_1}{\sqrt{3}}\right) \le Y \le \left(\frac{\theta_1+\theta_2}{2} + \frac{\theta_2-\theta_1}{\sqrt{3}}\right)\right]$$
$$= P(\theta_1 \le Y \le \theta_2) = 1$$

Tchebysheff's theorem is satisfied, but the approximation suggested by the empirical rule is inaccurate. This is because the probability distribution for the uniform random variable is far from mound-shaped.

4.118 The density function for the exponential random variable Y is

$$f(y) = \frac{e^{-y/\beta}}{\beta} \qquad \beta > 0 \qquad y \ge 0$$

which is gamma with $\alpha = 1$. Hence

$$\mu = E(Y) = \alpha\beta = \beta \qquad V(Y) = \alpha\beta^2 = \beta^2$$

The probability of interest is

$$P(|Y - \mu| \le 2\sigma) = P(\mu - 2\sigma \le Y \le \mu + 2\sigma) = P(\beta - 2\beta \le Y \le \beta + 2\beta)$$
$$= P(0 \le Y \le 3\beta)$$

since the density function will be positive only for $y \ge 0$. Thus,

$$P(|Y - \mu| \le 2\sigma) = \int_0^{3\beta} \frac{e^{-y/\beta}}{\beta}\,dy = \left.\frac{-\beta e^{-y/\beta}}{\beta}\right]_0^{3\beta} = -e^{-3\beta/\beta} + e^0 = 1 - e^{-3}$$
$$= 1 - .0498 = .9502$$

The empirical rule and Tchebysheff's theorem are both accurate.

4.119 From Exercise 4.52, $E(C) = 1100$ and $V(C) = 2{,}920{,}000 = \sigma^2$ so that $\sigma = 1708.8$. The value 2000 is $\frac{2000-1100}{1708.8} = .53$ standard deviations above the mean. Thus we would expect C to exceed 2000 fairly often.

4.120 We need $P(|L - \mu| < k\sigma) \ge .89$. Using Tchebysheff's theorem, we have $1 - \frac{1}{k^2} = .89$, which implies that $k = \sqrt{17.11} = 3.015$. From Exercise 4.87, $\mu = 276$ and $\sigma = \sqrt{47{,}664} = 218.32$. The interval is

$$|L - 276| < 3.015(218.32) \qquad \text{or} \qquad (-382.23, 934.23)$$

Since we know that L cannot be negative, we can use the interval $(0, 934.23)$.

4.121 We need $P(|C - \mu| < k\sigma) \ge .75$. Hence, $1 - \frac{1}{k^2} = .75$ and $k = 2$. From Exercise 4.97, $\mu = \frac{52}{3}$ and $\sigma = \sqrt{29.96} = 5.474$. The necessary interval is

$$|C - 17.33| < 10.95 \qquad \text{or} \qquad (6.38, 28.28)$$

4.122 **a.** $E(Y) = \nu = 7$ and $V(Y) = 2\nu = 14$.

b. $\sigma = \sqrt{V(Y)} = \sqrt{14} = 3.742$

23 is $\frac{23-7}{3.742} = 4.276$ standard deviations above the mean. It is not likely that Y takes on a value of 23 or more.

4.123 The random variable Y has a uniform distribution over the interval (1, 4). That is, $f(y) = \frac{1}{3}$ for $1 \le y \le 4$. The cost of delay is given as

$$c(y) = \begin{cases} 100, & \text{for } 1 \le y \le 2 \\ 100 + 20(y-2), & \text{for } 2 \le y \le 4 \end{cases}$$

Then

$$E(c(y)) = \int_{-\infty}^{\infty} c(y)f(y)\,dy = \int_1^2 \tfrac{100}{3}\,dy + \int_2^4 \tfrac{60+20y}{3}\,dy = \tfrac{100}{3} + \left[20y + \tfrac{20}{6}y^2\right]_2^4$$
$$= \tfrac{100}{3} + (80-40) + \tfrac{320-80}{6} = \tfrac{100}{3} + (80-40) + 40 = \$113.33$$

4.124 Y is a mixture of two random variables, X_1 and X_2, with $c_1 = \frac{3}{10}$ (i.e. c_1 is the accumulated probability of the discrete points). See Definition 4.14. Here X_1 takes two values, 3 and 6, with probabilities $\frac{2}{3}$ and $\frac{1}{3}$. See the accompanying table. Similarly,

X_1	$P(X_1 \mid Y = 3 \text{ or t})$
3	$\frac{2}{3}$
6	$\frac{1}{3}$

$c_2 = 1 - \left(\frac{3}{10}\right) = \frac{7}{10}$ and X_2 has the density given in the text. Now,
$E(X_1) = 3\left(\frac{2}{3}\right) + 6\left(\frac{1}{3}\right) = 4$. Since X_2 has a gamma distribution with $\alpha = 2, \beta = 2$, $E(X_2) = 4$, hence using Definition 4.14,

$$E(Y) = c_1E(X_1) + c_2E(X_2) = (.3)(4) + (.7)(4) = 4.0$$

4.125 **a.**
$$F(x) = \begin{cases} 0, & x < 0 \\ \int_0^x \left(\frac{1}{100}\right)e^{-y/100}\,dy = 1 - e^{-x/100}, & 0 \le x < 200 \\ 1, & x \ge 200 \end{cases}$$

b. $E(X) = \int_0^{200} \left(\frac{1}{100}\right) xe^{-x/100}\,dx + (.1353)(200)$ where $.1353 = P(x \ge 200)$.

Using integration by parts, we find that $\int_0^{200} \left(\frac{1}{100}\right) xe^{-x/100}\,dx = 59.4$.

Thus, $E(X) = 59.4 + (.1353)(200) = 59.4 + 27.06 = 86.47$.

4.126 There is only one discrete point $v = 0$ and this point has probability 0.02. Hence $c_1 = 0.02$ and $c_2 = 0.98$. Apply definition 4.19 and use the fact that $V \sim \text{Gamma}(4, 500)$ to find that

$$E(K) = 0.98\tfrac{m}{2}E(V^2) = 0.98\tfrac{m}{2}\{V(V) + (E(V))^2\}$$
$$= 0.98\,\tfrac{m}{2}\{4(500)^2 + 2000^2\} = 2450000m$$

4.127 **a.**
$$F_1(y) = \begin{cases} 0 & y < 0 \\ 0.4 & 0 \le y < 0.5 \\ 1 & y \ge 0.5 \end{cases}$$

$$F_2(y) = \begin{cases} 0 & y < 0 \\ 4y^2/3 & 0 \le y < 0.5 \\ (4y-1)/3 & 0.5 \le y < 1 \\ 1 & y \ge 1 \end{cases}$$

b. $F(y) = 0.25F_1(y) + 0.75F_2(y)$

c. First, note that

$$f_2(y) = \frac{dF_2(y)}{dy} = \begin{cases} 8y/3 & 0 \le y < 0.5 \\ 4/3 & 0.5 \le y < 1 \end{cases}$$

so that

$$E(Y) = 0.25(1 - 0.1/0.25)(0.5) + \int_0^{0.5} 8y^2/3\,dy + \int_{0.5}^1 4y/3\,dy = .5333$$

and similarly, $E(Y^2) = 0.3604$ so that $V(Y) = 0.076$.

4.128 **a.** $F(y) = \int_{-1}^{y} \frac{2}{\pi(1=t^2)}\, dt = \frac{2}{\pi}\left[\tan^{-1}(t)\right]_{-1}^{y} = \frac{2}{\pi}\left[\tan^{-1}(y) - \tan^{-1}(-1)\right]$

$= \frac{2}{\pi}\left[\tan^{-1}(y) - \left(-\frac{\pi}{4}\right)\right] = \frac{2}{\pi}\tan^{-1}(y) + \frac{1}{2}$ if $-1 \le y \le 1$;

$F(y) = 0$ if $y < -1$.

$F(y) = 1$ if $y > 1$

b. $E(Y) = \int_{-1}^{1} \frac{2y}{\pi(1+y^2)}\, dy = \frac{1}{\pi}\left[\ln(1+y^2)\right]_{-1}^{1} = \frac{1}{\pi}(\ln 2 - \ln 2) = 0$

4.129 For this exercise $\mu = 70$ and $\sigma = 12$. The objective is to determine a particular value, y_0, for the random variable Y so that $P(Y < y_0) = .90$ (i.e., 90% of the students will finish the examination before the set time limit). Referring to Figure 4.22, we obtain $A_1 = P(Y \ge y_0) = .10$. Corresponding to the value $Y = y_0$ is the z value

$z = \frac{y-\mu}{\sigma} = \frac{y_0 - 70}{12}$

Hence

$A\left[\frac{y_0-70}{12}\right] = .1$ or $\frac{y_0-70}{12} = 1.28$.

Thus $y_0 = 85.36$

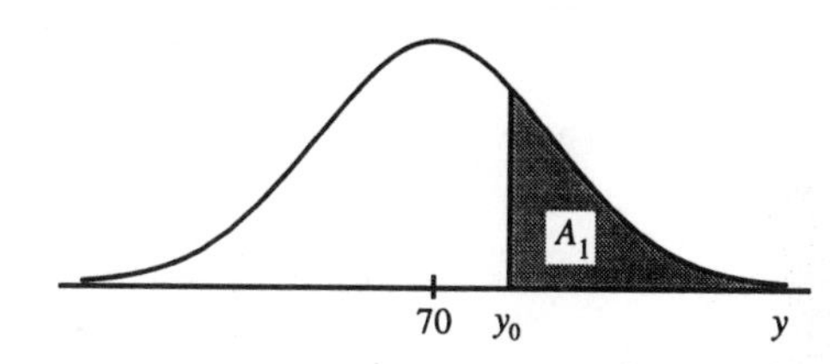

Figure 4.22

4.130 The 3000 light bulbs utilized by the manufacturing plant comprise the entire population (i.e., this is not a sample from the population) whose length of life is normally distributed with mean $\mu = 500$, and standard deviation $\sigma = 50$. The objective is to find a particular value, y_0, so that $P(Y \le y_0) = .01$. That is, only 1% of the bulbs will burn out before they are replaced at time y_0. Then

$P(Y \le y_0) = P(Z \le z_0) = .01$ when $z_0 = \frac{y_0 - 500}{50}$

From Table 4, Appendix III, the value of z corresponding to an area (in the left tail of the distribution) of .01 is $z_0 = -2.327$ (see Figure 4.23). Solving for y_0 corresponding to $z_0 = -2.327$, we obtain

$-2.327 = \frac{y_0 - 500}{50}$

$-116.35 = y_0 - 500$ or $y_0 = 383.65$

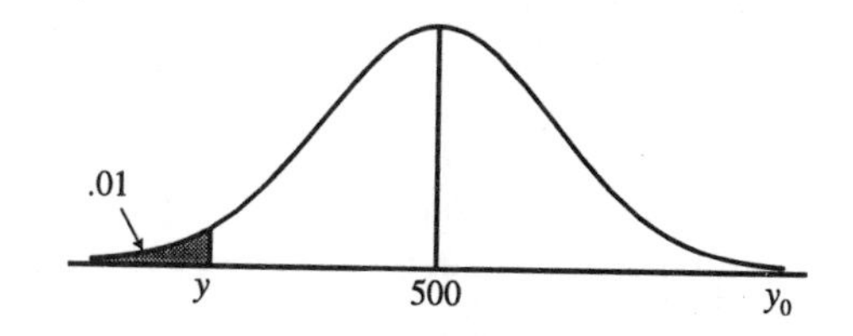

Figure 4.23

4.131 Refer to Exercise 4.52, and let X be the number of defective bearings. Then X will have a binomial distribution with $n = 5$, $p = P(\text{defective}) = .073$, and

$P(X > 1) = 1 - P(X = 0) = 1 - \binom{5}{0} p^0 (1-p)^5 = 1 - (.927)^5 = .3155$

4.132 Let Y denote the length of life of a drill bit. Then Y has a normal distribution with $\mu = 75$ and $\sigma = 12$.

a. $P(Y < 60) = P\left(Z < \frac{60-75}{12}\right) = P(Z < -1.25) = .1056$

b. $P(Y \ge 60) = 1 - P(Y < 60) = 1 - .1056 = .8944$

c. $P(Y > 90) = P\left(Z > \frac{90-75}{12}\right) = P(Z > 1.25) = .1056$

4.133 **a.** The variable factor of $f(y)$ is that of a gamma density with $\alpha = 2$ and $\beta = \frac{1}{2}$. Hence

$$c = \frac{1}{\Gamma(\alpha)\beta^{\alpha}} = \frac{1}{\Gamma(2)\left(\frac{1}{2}\right)^2} = \frac{4}{1!} = 4$$

b. Since Y has a gamma distribution with $\alpha = 2$, $\beta = \frac{1}{2}$,

$E(Y) = \alpha\beta = 1 \qquad V(Y) = \alpha\beta^2 = 2\left(\frac{1}{4}\right) = \frac{1}{2}$

c. Recall that the moment-generating function of a gamma random variable is

$m(t) = \frac{1}{(1-\beta t)^{\alpha}}$ and in this case $m(t) = \frac{1}{\left[1-\left(\frac{t}{2}\right)\right]^2} = \left(1 - \frac{t}{2}\right)^{-2}$

4.134

In Example 4.16 it is shown that $m(t) = e^{t^2\sigma^2/2}$ is the moment-generating function of the variable $X = Y - \mu$, where Y is normally distributed with mean μ and variance σ^2. That is,

$$m(t) = E\left(e^{tX}\right) = E\left[e^{t(Y-\mu)}\right] = E\left[1 + t(Y-\mu) + \frac{t^2(Y-\mu)^2}{2!} + \cdots\right]$$

Hence the coefficient of $\frac{t^r}{r!}$ in the series expansion of $m(t)$ will yield $E\left(X^r\right) = E\left[(Y-\mu)^r\right] = \mu_r$, where μ_r is the rth central moment of the random variable Y. Expanding $m(t)$, we obtain

$$m(t) = e^{t^2\sigma^2/2}$$
$$= 1 + \left(\frac{t^2\sigma^2}{2}\right) + \left(\frac{t^2\sigma^2}{2}\right)^2 \frac{1}{2!} + \left(\frac{t^2\sigma^2}{2}\right)^3 \frac{1}{3!} + \cdots = 1 + \frac{t^2\sigma^2}{2} + \frac{t^4\sigma^4}{8} + \cdots$$

Then

$\mu_1 = \text{coefficient of } t = 0 \qquad \mu_2 = \text{coefficient of } \frac{t^2}{2} = \sigma^2$

$\mu_3 = \text{coefficient of } \frac{t^3}{6} = 0 \qquad \mu_4 = \text{coefficient of } \frac{t^4}{24} = 3\sigma^4$

4.135

For the beta random variable given in Section 4.7,

$$\mu_k' = E\left(Y^k\right) = \int_{-\infty}^{\infty} y^k f(y)\,dy = \int_0^1 y^k \frac{\Gamma(\alpha+\beta)}{\Gamma(\alpha)\Gamma(\beta)} y^{\alpha-1}(1-y)^{\beta-1}\,dy$$
$$= \frac{\Gamma(\alpha+\beta)}{\Gamma(\alpha)\Gamma(\beta)} \int_0^1 y^{\alpha+k-1}(1-y)^{\beta-1}\,dy$$

The quantity $y^{\alpha+k-1}(1-y)^{\beta-1}$ is the variable factor of a beta density function with parameters $\alpha + k$ and β. Hence

$$\mu_k' = \frac{\Gamma(\alpha+\beta)}{\Gamma(\alpha)\Gamma(\beta)} \times \frac{\Gamma(\alpha+k)\Gamma(\beta)}{\Gamma(\alpha+k+\beta)} \int_0^1 \frac{y^{\alpha+k-1}(1-y)^{\beta-1}}{B(\alpha+k,\beta)}\,dy = \frac{\Gamma(\alpha+\beta)\Gamma(\alpha+k)}{\Gamma(\alpha)\Gamma(\alpha+k+\beta)}$$

since the integral of a complete density function is 1.

4.136

The probability distribution for n, the number of arrivals in time $(0, t)$, is Poisson with mean λt, so that

$$P[n \text{ arrivals in } (0,t)] = \frac{(\lambda t)^n e^{-\lambda t}}{n!}$$

Let T be the length of time until the first arrival, and consider the distribution function for T.

$$F(t) = P(T \le t) = 1 - P(T > t) = 1 - \text{P}[n = 0 \text{ in } (0,t)] = 1 - \frac{(\lambda t)^0 e^{-\lambda t}}{0!} = 1 - e^{-\lambda t}$$

Since T is a continuous random variable and $F(t)$ is its distribution function, the density function for T may be found, using Definition 4.3, to be

$$f(t) = \frac{d}{dt}F(t) = \lambda e^{-\lambda t} \qquad \text{for} \qquad t > 0$$

Note that the first derivative exists and is everywhere continuous, as is $F(t)$ itself. Hence Definitions 4.2 and 4.3 are satisfied. Moreover, $f(t)$ is the density of an Exponential random variable with mean $1/\lambda$.

4.137

Let Y be the time between the arrival of two calls, measured in hours. We require $P\left(Y > \frac{1}{4}\right)$. Since $\lambda t = 10$ and $t = 1$ (hour), $\frac{1}{\lambda} = \frac{1}{10}$, and

$$f(y) = \frac{1}{.1} e^{-y/.1} = 10e^{-10y}$$

and

$$P\left(Y > \frac{1}{4}\right) = \int_{1/4}^{\infty} 10e^{-10y}\,dy = -e^{-10y}\Big]_{1/4}^{\infty} = e^{-2.5} = .082$$

4.138a. Similar to Exercise 4.136, the second arrival will occur after time t if either one arrival has occurred in $(0, t)$ or no arrivals have occurred in $(0, t)$.

$$P(U > t) = P[1 \text{ arrival in } (0, t)] + P[0 \text{ arrivals in } (0, t)]$$

$$= \frac{(\lambda t)^1 e^{-\lambda t}}{1!} + \frac{(\lambda t)^0 e^{-\lambda t}}{0!} = (\lambda t + 1)e^{-\lambda t}.$$

Then,

$$F(t) = 1 - P(U > t) = 1 - (\lambda t + 1)e^{-\lambda t}$$

$$f(t) = \tfrac{d}{dt}F(t) = -\left[(\lambda t + 1)(-\lambda)e^{-\lambda t} + (e^{-\lambda t})\lambda\right]$$

$$= (\lambda^2 t + \lambda - \lambda)\, e^{-\lambda t}$$

$$= \lambda^2 t e^{-\lambda t}.$$

This is the gamma density function with $\alpha = 2$ and $\beta = \frac{1}{\lambda}$.

b. Similar to part **a**. Let X = time until the kth arrival.

$$P(X > t) = \sum_{n=0}^{k-1} \frac{(\lambda t)^n e^{-\lambda t}}{n!}$$

$$F(t) = 1 - P(x > t) = 1 - \sum_{n=0}^{k-1} \frac{(\lambda t)^n e^{-\lambda t}}{n!}$$

$$f(t) = \tfrac{d}{dt}F(t) = -\left[(-\lambda)e^{-\lambda t}\left(\sum_{n=0}^{k-1} \frac{(\lambda t)^n}{n!}\right) + e^{-\lambda t}\sum_{n=1}^{k-1} \frac{\lambda^n t^{n-1}}{(n-1)!}\right]$$

$$= \lambda e^{-\lambda t}\left[\sum_{n=0}^{k-1} \frac{(\lambda t)^n}{n!} - \sum_{n=1}^{k-1} \frac{(\lambda t)^{n-1}}{(n-1)!}\right]$$

$$= \lambda e^{-\lambda t}\frac{(\lambda t)^{k-1}}{(k-1)!} = \frac{\lambda^k t^{k-1} e^{-\lambda t}}{(k-1)!}$$

which is the density of a gamma random variable with $\alpha = k$ and $\beta = \frac{1}{\lambda}$.

4.139 Recall exercise 4.137. Assume the arrivals follow a Poisson distribution with mean 2. Then the inter arrival times follow an exponential distribution with mean 1/2. Let W denote the waiting times.

a. $E(W) = 1/2$ and $V(W) = 1/4$.

b. $\int_0^3 2e^{-2y}\,dy = 1 - e^{-6}$

4.140 $P(T > 1/5) = \int_{1/5}^{\infty} 4e^{-4t}dt = e^{-4/5}$

where T is the time (in minutes) until the next call.

4.141 Let R be the distance to the nearest neighbor. Then

$$P(R > r) = P(\text{no plants in a circle of radius } r) = P(\text{no plants in an area of } \pi r^2)$$

Since the number of plants in an area of 1 unit has a Poisson distribution with mean λ, the number of plants in an area of πr^2 units has a Poisson distribution with mean $\lambda\pi r^2$. Hence

$$P(R > r) = \frac{(\lambda\pi r^2)^0 e^{-\lambda\pi r^2}}{0!} = e^{-\lambda\pi r^2}$$

Then

$$F(r) = P(R \le r) = 1 - e^{-\lambda\pi r^2} \quad \text{and} \quad f(r) = F'(r) = 2\lambda\pi r e^{-\lambda\pi r^2} \quad r > 0$$

That is, R is a Gamma random variable with $\alpha = 2$ and $\beta = 1/\lambda\pi$.

4.142 Let Y be the time it takes to interview a job applicant, with $f(y) = 2e^{-2y}$, $y > 0$. The second applicant will have to wait only if the time to interview the first applicant exceeds 15 minutes. Hence

$$P(\text{wait}) = P\left(Y > \tfrac{1}{4}\right) = \int_{1/4}^{\infty} 2e^{-2y}\,dy = -e^{-2y}\Big]_{1/4}^{\infty} = e^{-1/2} = 0.61$$

4.143 From Exercise 4.5, we have the following:

$$F(y) = \begin{cases} 0, & y < 0 \\ \frac{y^2}{4}, & 0 \le y \le 2 \\ 1, & y > 2 \end{cases}$$

The median value of Y will be that value of y such that

$$F(y) = .5 \qquad \text{or} \qquad \frac{y^2}{4} = .5 \qquad \text{or} \qquad y^2 = 2$$

Since the range of y for which $0 < F(y) < 1$ is $0 \le y \le 2$, the median must be $y = +\sqrt{2} = 1.414$.

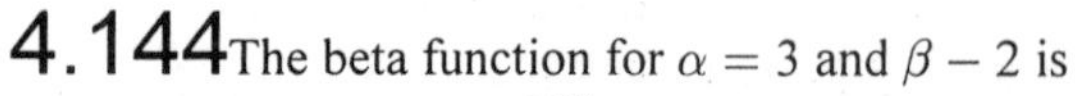

4.144 The beta function for $\alpha = 3$ and $\beta - 2$ is

$$f(y) = \frac{\Gamma(5)}{\Gamma(3)\Gamma(2)} y^2(1-y) = 12y^2(1-y)$$

The graph is shown in Figure 4.24. If Y has this density, then

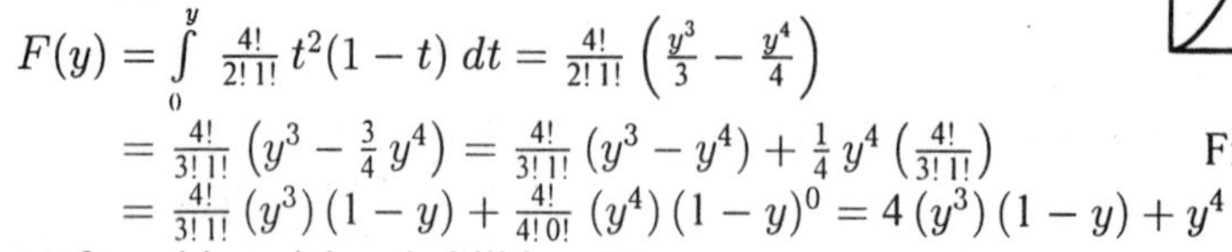

$$F(y) = \int_0^y \frac{4!}{2!\,1!} t^2(1-t)\,dt = \frac{4!}{2!\,1!}\left(\frac{y^3}{3} - \frac{y^4}{4}\right)$$

$$= \frac{4!}{3!\,1!}\left(y^3 - \frac{3}{4}y^4\right) = \frac{4!}{3!\,1!}\left(y^3 - y^4\right) + \frac{1}{4}y^4\left(\frac{4!}{3!\,1!}\right)$$

$$= \frac{4!}{3!\,1!}\left(y^3\right)(1-y) + \frac{4!}{4!\,0!}\left(y^4\right)(1-y)^0 = 4\left(y^3\right)(1-y) + y^4$$

Figure 4.24

a sum of two binomial probabilities. Now

$$P(.1 \le Y \le .2) = F(.2) - F(.1) = 4(.2)^3(.8) + (.2)^4 - 4(.1)^3(.9) - (.1)^4$$

$$= .0256 + .0016 - .0036 - .0001 = .0235$$

4.145 Let X be the grocer's profit. In general, her profit (in cents) on an order of $100k$ pounds of food will be $X = 1000Y - 600k$, as long as $Y < k$, that is, as long as the demand does not exceed the amount ordered. If $Y > k$, however, the grocer's profit will be $X = 1000k - 600k = 400k$. Define Y' as Y for $0 \le Y \le k$ and $Y' = k$ for $Y \ge k$. Then we can write $g(Y') = X = 1000Y' - 600k$ for all values of Y'. Y' has a mixed distribution with one discrete point, namely, $Y' = k$. Note that

$$c_1 = P(Y' = k) = P(Y \ge k) = \int_k^1 3y^2\,dy = 1 - k^3$$

so that $c_2 = 1 - c_1 = k^3$. We can write Y' as a mixture of Y_1' and Y_2', where

$$F_1(y) = \begin{cases} 0, & y < k \\ 1, & y \ge k \end{cases}$$

$$F_2(y) = P\left(Y_2' \le y | 0 \le Y' < k\right) = \frac{\int_0^y f(t)\,dt}{k^3} = \frac{\int_0^y 3t^2\,dt}{k^3} = \frac{y^3}{k^3} \qquad 0 < y < k$$

Note that

$$f_2(y) = \begin{cases} \frac{3y^2}{k^3}, & 0 \le y \le k \\ 0, & \text{elsewhere} \end{cases}$$

Now using Definition 4.14,

$$E(X) = E\left[g(Y')\right] = c_1 E\left[g(Y_1')\right] + c_2 E\left[g(Y_2')\right]$$

$$= \left(1 - k^3\right)400k + k^3 \int_0^k (1000y - 600k)\,\frac{3y^2}{k^3} = 400k - 400k^4 + 1000\left[\frac{3y^4}{4}\right]_0^k$$

$$= 400k - 400k^4 + 750k^4 - 600k^4 = 400k - 250k^4$$

To maximize expected profit, then, set $\frac{d}{dk}\left[E(X)\right] = 400 - 1000k^3$ equal to 0. The desired value of k is $\left(\frac{400}{1000}\right)^{1/3} = (.4)^{1/3}$. (Note that the second derivative is negative, so that we do have a maximum.)

4.146 Using the result of Exercise 4.79 with $\alpha = 3$ and $\beta = 1$,

$$P(Y \le 4) = 1 - \frac{1}{\Gamma(3)} \int_4^\infty y^2 e^{-y}\,dy = 1 - \sum_{y=0}^{2} \frac{4^y e^{-4}}{y!} = 1 - \left(e^{-4} + 4e^{-4} + \frac{16}{2}e^{-4}\right)$$

$$= 1 - 13e^{-4} = .7619$$

4.147 From Example 4.16 we know that the moment-generating function of $U = Y - \mu$ is

$$m_U = e^{t^2\sigma^2/2}$$

Then

$$m_Z(t) = m_{U/\sigma}(t) = m_U\left(\tfrac{t}{\sigma}\right) = e^{(t/\sigma)^2(\sigma^2/2)} = e^{t^2/2}$$
$$E(Y) = m'_Z(t)\big|_{t=0} = te^{t^2/2}\big|_{t=0} = 0$$
$$E(Y^2) = m''_Z(t)\big|_{t=0} = t^2e^{t^2/2} + e^{t^2/2}\big|_{t=0} = 1$$
$$V(Y) = 1 - 0 = 1$$

4.148 **a.** $P(Y \le 4) = P(X \le \ln 4) = P\left(Z \le \frac{\ln 4 - 4}{1}\right) = P(Z \le -2.61) = .0045.$

b. $P(Y > 8) = P(X > \ln 8) = P\left(Z > \frac{\ln 8 - 4}{1}\right) = P(Z > -1.92) = 1 - .0274 = .9726.$

4.149 **a.** $E(Y) = e^{\mu+\sigma^2/2} = e^{3+16/2} = e^{11}\,(10^{-2}g) = 598.74g$

$$V(Y) = e^{2\mu+\sigma^2}\left(e^{\sigma^2} - 1\right) = e^{6+16}\left(e^{16} - 1\right) = e^{22}\left(e^{16} - 1\right)$$
$$= \left(e^{38} - e^{22}\right)\left(10^{-4}g^2\right) = 3.1856 \times 10^{12}$$

b. With $k = 2$, consider the interval $\left(E[Y] \pm 2\sqrt{V[Y]}\right)$

$(598.74 \pm 3{,}569{,}637.4) = (-3{,}569{,}038.7,\ 3{,}570{,}236.1)$

Since weights are positive, give as the final interval $(0,\ 3{,}570{,}236.1)$.

c. $P(Y < 598.74) = P(\ln(Y) < 6.3948) = P\left[Z < \frac{6.3948-3}{4}\right] = P(Z < .8487) = .8023.$

4.150 $M(t) = E\left(e^{ty}\right) = \left(\tfrac{1}{2}\right)\int_{-\infty}^{0} e^{ty}e^{y}\,dy + \left(\tfrac{1}{2}\right)\int_{0}^{\infty} e^{ty}e^{-y}\,dy$

$$= \left(\tfrac{1}{2}\right)\int_{-\infty}^{0} e^{y(1+t)}\,dy + \left(\tfrac{1}{2}\right)\int_{0}^{\infty} e^{-y(1-t)}\,dy$$
$$= \frac{1}{2(1+t)}\,e^{u}\Big]_{-\infty}^{0} - \frac{1}{2(1-t)}\,e^{-u}\Big]_{0}^{\infty}$$
$$= \frac{1}{2(1+t)} + \frac{1}{2(1-t)} = \frac{1-t+1+t}{2\left(1-t^2\right)}$$
$$= \frac{1}{1-t^2}\,.$$

To find $E(Y)$, consider $M'(t)$ evaluated at $t = 0$.

$$M'(t) = \frac{2t}{\left(1-t^2\right)^2} \qquad \text{and} \qquad E(Y) = M'(0) = 0.$$

4.151 **a.** Note that $f_i(y) \ge 0$ and $\int_{-\infty}^{\infty} f_i(y)\,dy = 1 \qquad i = 1, 2$

Since $0 \le a \le 1$, it immediately follows that

$f(y) = af_1(y) + (1-a)f_2(y) \ge 0$. Also,

$$\int_{-\infty}^{\infty} f(y)\,dy = \int_{-\infty}^{\infty} [af_1(y0 + (1-a)f_2(y)\,dy]$$
$$= e\int_{-\infty}^{\infty} f_1(y)\,dy + (1-a)\int_{-\infty}^{\infty} f_2(y)\,dy$$
$$= a + 1 - a = 1.$$

b. (i) $E(Y) = \int_{-\infty}^{\infty} yf(y)\,dy = \int_{-\infty}^{\infty} y[af_1(y) + (1-a)f_2(y)]\,dy$

$$= a\int_{-\infty}^{\infty} yf_1(y)\,dy + (1-a)\int_{-\infty}^{\infty} yf_2(y)\,dy$$
$$= a\mu_1 + (1-a)\mu_2.$$

(ii) $\sigma^2 = E\left(Y^2\right) - [E(Y)]^2$

Proceeding as in part **b** (i), $E\left(Y^2\right) = \int_{-\infty}^{\infty} y^2 f(y)\,dy$

$$= a\int_{-\infty}^{\infty} y^2 f_1(y)\,dy + (1-a)\int_{-\infty}^{\infty} y^2 f_2(y)\,dy$$

Using the hint, $E\left(Y^2\right) = a\left[\mu_1^2 + \sigma_1^2\right] + (1-a)\left[\mu_2^2 + \sigma_2^2\right]$

Thus,

$$\text{Var}(Y) = a\left[\mu_1^2 + \sigma_1^2\right] + (1-a)\left[\mu_2^2 + \sigma_2^2\right] - [a\mu_1 + (1-a)\mu_2]^2$$
$$= a\sigma_1^2 + (1-a)\sigma_2^2 + \left(a - a^2\right)\mu_1^2 - 2a(1-a)\mu_1\mu_2 +$$
$$\left[1 - a - (1-a)^2\right]\mu_2^2$$
$$= a\sigma_1^2 + (1-a)\sigma_2^2 + a(1-a)[\mu_1 - \mu_2]^2.$$

4.152 For $m = 2$,

$$E(Y) = \int_0^\infty \frac{2y^2 e^{-y^2/\alpha}}{\alpha}\, dy$$

Let $z = y^2$. Then $dz = 2y\, dy$ and

$$E(Y) = \int_0^\infty \frac{\sqrt{z}\, e^{-z/\alpha}}{\alpha}\, dz = \int_0^\infty \frac{z^{1/2} e^{-z/\alpha}}{\alpha}\, dz$$

which, when the proper constant is added, will be the integral of the density function of a gamma random variable with parameters $\frac{3}{2}$ and α. Then

$$E(Y) = \frac{\alpha^{3/2}\Gamma\left(\frac{3}{2}\right)}{\alpha} \int_0^\infty \frac{z^{1/2} e^{-z/\alpha}}{\Gamma\left(\frac{3}{2}\right)\alpha^{3/2}}\, dz = \frac{\alpha^{3/2}\Gamma\left(\frac{3}{2}\right)}{\alpha} = \alpha^{1/2}\Gamma\left(\tfrac{3}{2}\right) = \alpha^{1/2}\left(\tfrac{1}{2}\right)\Gamma\left(\tfrac{1}{2}\right)$$
$$= \frac{(\alpha\pi)^{1/2}}{2}$$

where $\Gamma\left(\frac{1}{2}\right) = \sqrt{\pi}$ is shown in exercise 4.162.
Again using the transformation $z = y^2$, we find

$$E\left(Y^2\right) = \int_0^\infty \frac{2y^3 e^{-y^2/\alpha}}{\alpha}\, dy = \int_0^\infty \frac{z e^{-z/\alpha}}{\alpha}\, dz = \frac{\Gamma(2)\alpha^2}{\alpha} \int_0^\infty \frac{z e^{-z/\alpha}}{\Gamma(2)\alpha^2}\, dz = \alpha$$

so that $V(Y) = \alpha - \left[\alpha^{1/2}\Gamma\left(\frac{3}{2}\right)\right]^2 = \alpha\left\{1 - \left[\Gamma\left(\frac{3}{2}\right)\right]^2\right\} = \alpha\left[1 - \frac{\pi}{4}\right]$

See Exercise 4.162.

4.153 The density for Y, the life length of a resistor in thousands of hours, is

$$f(y) = \frac{2ye^{-y^2/10}}{10}, \qquad 0 \le y < \infty$$

a. $P(Y > 5) = 1 - P(Y \le 5) = 1 - \int_0^5 \frac{2ye^{-y^2/10}}{10}\, dy = 1 - \left[-e^{-y^2/10}\right]_0^5$
$= 1 - \left[-e^{-2.5} + 1\right] = e^{-2.5}$

b. Let X be the number of resistors that burn out prior to 5000 hours. Then X is a binomial random variable with $n = 3$ and $p = 1 - e^{-2.5}$. Thus,

$$P(X = 1) = \binom{3}{1}\left(1 - e^{-2.5}\right)\left(e^{-2.5}\right)^2 = .0186$$

4.154 **a** When $m = 1$

$$f(y) = \frac{1}{\alpha} e^{-y/\alpha} \text{ for } 0 \le y < \infty,\ \alpha > 0$$

which is the density for an Exponential(α) random variable.

b.
$$\begin{aligned}
E(Y) &= \int_0^\infty \frac{m}{\alpha} y^m e^{-y^m/\alpha}\, dy \\
&= \int_0^\infty \frac{1}{\alpha} u^{1/m} e^{-u/\alpha}\, du \text{ (after making the substitution } u = y^m) \\
&= \alpha^{1/m}\Gamma(1 + 1/m)
\end{aligned}$$

$$\begin{aligned}
E(Y^2) &= \int_0^\infty \frac{m}{\alpha} y^{m+1} e^{-y^m/\alpha}\, dy \\
&= \int_0^\infty \frac{1}{\alpha} u^{2/m} e^{-u/\alpha}\, du \text{ (after making the substitution } u = y^m) \\
&= \alpha^{2/m}\Gamma(1 + 2/m)
\end{aligned}$$

Therefore,

$$V(Y) = \alpha^{2/m}\{\Gamma(1 + 2/m) - \Gamma^2(1 + 1/m)\}.$$

4.155 $$E(Y) = \int_{-1}^{1} \frac{1}{B(1/2,(n-2)/2)} y(1-y^2)^{(n-4)/2}\,dy$$
$$= \frac{1}{B(1/2,(n-2)/2)}\left\{\int_1^0 (u+1)^{\frac{1}{2}} u^{(n-4)/2} \frac{-1}{2\sqrt{u+1}}\,du + \int_0^1 -\sqrt{u+1}\,u^{(n-4)/2}\frac{1}{2\sqrt{u+1}}\,du\right\} = 0$$

where we used the substitution $u = 1 - y^2$. Thus

$$V(Y) = E(Y^2) = \frac{1}{B(1/2,(n-2)/2)}\int_{-1}^{1} y^2(1-y^2)^{(n-4)/2}\,dy$$
$$= \frac{2}{B(1/2,(n-2)/2)}\int_0^1 u(1-u)^{(n-4)/2}\,du$$
$$= \frac{B(3/2,(n-2)/2)}{B(1/2,(n-2)/2)}$$
$$= \frac{1}{n-1}$$

Note that the second equality follows after making the substitution $u = y^2$.

4.156 **a.** Let Y have an exponential distribution with parameter θ. Then

$$f(t) = \tfrac{1}{\theta}e^{-t/\theta} \quad \text{and} \quad 1 - F(t) = \int_t^\infty \tfrac{1}{\theta}e^{-y/\theta}\,dy = -e^{-y/\theta}\Big]_t^\infty = e^{-t/\theta}$$

forming the ratio $r(t) = \frac{f(t)}{1-F(t)} = \frac{1}{\theta}$, which is constant.

b. Refer to Exercise 4.152, for $f(t)$ given there,

$$1 - F(t) = \int_t^\infty \frac{my^{m-1}e^{-y^m/\alpha}}{\alpha}\,dy$$

Let $z = y^m$ so that $dz = my^{m-1}\,dy$. Then

$$1 - F(t) = \int_{t^m}^\infty \frac{e^{-z/\alpha}}{\alpha}\,dz = -e^{-z/\alpha}\Big]_{t^m}^\infty = e^{-t^m/\alpha}$$

forming the ratio

$$r(t) = \frac{f(t)}{1-F(t)} = \frac{mt^{m-1}}{\alpha}$$

which is increasing in t for $m > 1$.

4.157 **a.** $P(Y \le y|Y \ge c) = \frac{P(c \le Y \le y)}{P(Y \ge c)} = \frac{F(y)-F(c)}{1-F(c)}$

b. Refer to the properties of distribution functions given in Section 4.2. Define $P(Y \le y|Y \ge c) = G(y)$. Then, since we are given $Y \ge c$,

$$G(-\infty) = G(c) = \frac{F(c)-F(c)}{1-F(c)} = 0 \qquad G(\infty) = \frac{F(\infty)-F(c)}{1-F(c)} = \frac{1-F(c)}{1-F(c)} = 1$$

since F is a distribution function. Finally, for $b \ge a$,

$$G(b) - G(a) = \frac{F(b)-F(a)}{1-F(c)} \ge 0$$

since F is a distribution function. Hence G must also be a distribution function.

c. It is given that $f(y) = \frac{2ye^{-y^2/3}}{3}$. From Exercise 4.156, $1 - F(y) = e^{-y^2/3}$, so that

$$F(y) = 1 - e^{-y^2/3}.$$

Using the result of part **a**,

$$P(Y \le 4|Y \ge 2) = \frac{1-e^{-4^2/3} - \left(1-e^{-2^2/3}\right)}{e^{-2^2/3}} = \frac{e^{-4/3}-e^{-16/3}}{e^{-4/3}} = 1 - e^{-4}$$

4.158 **a.** $E(V) = 4\pi\left(\frac{m}{2\pi KT}\right)^{3/2}\int_0^\infty v^3 e^{-v^2(m/2KT)}\,dv$

To evaluate this integral, let us make the transformation $u = v^2$. Then $du = 2v\,dv$ and

$$E(V) = 4\pi\left(\frac{m}{2\pi KT}\right)^{3/2}\int_0^\infty \tfrac{1}{2}ue^{-u(m/2KT)}\,du$$
$$= \frac{2}{\sqrt{\pi}}\left(\frac{m}{2Kt}\right)^{-1/2}\int_0^\infty \frac{ue^{-u(m/2KT)}}{\left(\frac{2KT}{m}\right)^2}\,du$$
$$= \frac{2}{\sqrt{\pi}}\left(\frac{m}{2Kt}\right)^{-1/2}$$

since the function inside the last integral is a gamma density function with $\alpha = 2$ and $\beta = \frac{2KT}{m}$ and so integrates to 1.

b. $E\left(\frac{mV^2}{2}\right) = 2\pi m\left(\frac{m}{2\pi KT}\right)^{3/2}\int_0^\infty v^4 e^{-v^2(m/2KT)}\,dv$

To evaluate this integral let us make the transformation $y = v^2$. Then $dy = 2v\,dv$ and

$$E\left(\frac{mV^2}{2}\right) = 2\pi m\left(\frac{m}{2\pi KT}\right)^{3/2}\int_0^\infty \frac{1}{2}y^{3/2}e^{-y(m/2KT)}\,dy$$

$$= \frac{m}{\sqrt{\pi}}\left(\frac{m}{2KT}\right)^{-1}\Gamma\left(\frac{5}{2}\right)\int_0^\infty \frac{y^{3/2}e^{-y(m/2KT)}}{\Gamma\left(\frac{5}{2}\right)\left(\frac{2KT}{m}\right)^{5/2}}\,dy$$

$$= \frac{m}{\sqrt{\pi}}\left(\frac{2KT}{m}\right)\Gamma\left(\frac{5}{2}\right) = \frac{2KT}{\sqrt{\pi}}\Gamma\left(\frac{5}{2}\right) = \frac{2KT}{\sqrt{\pi}}\left(\frac{3}{2}\right)\left(\frac{1}{2}\right)\Gamma\left(\frac{1}{2}\right) = \frac{3}{2}KT$$

since the function inside the last integral is a gamma density function with $\alpha = \frac{5}{2}$ and $\beta = \frac{2KT}{m}$ and so integrates to 1. $\Gamma\left(\frac{1}{2}\right) = \sqrt{\pi}$ is shown in Exercise 4.162.

4.159 It is given that $f(y) = \left(\frac{1}{100}\right)e^{-y/100}$. Then $1 - F(y) = e^{-y/100}$. Using the given result, we have

$$E(Y|Y \geq 50) = \frac{1}{e^{-1/2}}\int_{50}^\infty \frac{ye^{-y/100}}{100}\,dy$$

Integrating by parts with $u = y$, $du = dy$, $dv = \left(\frac{e^{-y/100}}{100}\right)dy$, and $v = \int\left(\frac{e^{-y/100}}{100}\right)dy = -e^{-y/100}$, we obtain

$$E(Y|Y \geq 50) = \frac{1}{e^{-1/2}}\left(\left[-ye^{-y/100}\right]_{50}^{\infty} - \int_{50}^\infty -e^{-y/100}\,dy\right)$$

$$= \frac{1}{e^{-1/2}}\left[50e^{-1/2} + 100e^{-1/2}\right] = 150$$

Note that the quantity ye^{-y} when evaluated at $y = \infty$ is of the indeterminate form $(0)(\infty)$, and L'Hôpital's rule is used to evaluate it.

4.160 In order to transform to polar coordinates, we use the transformation $x = r\cos\theta$ and $y = r\sin\theta$. The Jacobian of the transformation is

$$|J| = \begin{vmatrix} \frac{dx}{dr} & \frac{dx}{d\theta} \\ \frac{dy}{dr} & \frac{dy}{d\theta} \end{vmatrix} = \begin{vmatrix} \cos\theta & -r\sin\theta \\ \sin\theta & r\cos\theta \end{vmatrix}$$

$$= r\cos^2\theta + r\sin^2\theta = r(\cos^2\theta + \sin^2\theta) = r$$

That is, $dx\,dy = r\,dr\,d\theta$. Then the double integral becomes

$$\frac{1}{2\pi}\int_{-\infty}^\infty\int_{-\infty}^\infty e^{-(1/2)u(x^2+y^2)}\,dx\,dy = \frac{1}{2\pi}\int_0^{2\pi}\int_0^\infty re^{-(1/2)ur^2}\,dr\,d\theta$$

$$= \frac{1}{2\pi}\int_0^{2\pi}\int_0^\infty \frac{-ur}{-u}e^{-(1/2)ur^2}\,dr\,d\theta = \frac{1}{2\pi}\int_0^{2\pi} -\frac{1}{u}\left[e^{-(1/2)ur^2}\right]_0^\infty d\theta = \frac{1}{2\pi}\int_0^{2\pi}\frac{1}{u}\,d\theta = \frac{1}{u}$$

4.161 **a.** First note that $W = (Z^2 + 3Z)^2 = Z^4 + 6Z^3 + 9Z^2$. Now, odd moments of a standard normal random variable are 0 so that $E(Z^3) = 0$. Also, $V(Z) = E(Z^2) = 1$ since Z is standard normal. Finally, $Z^2 \sim \chi_1^2$ so that if $Y = Z^2$then

$$E(Z^4) = E(Y^2) = V(Y) + E(Y^2) = 2 + 1 = 3.$$

Therefore, $E(W) = 3 + 6(0) + 9(1) = 12$.

b. Applying exercise 4.164 and the result of part a obtain

$$P(W \leq w) \geq 1 - \frac{E(W)}{w}$$

Now set the right hand side equal to 0.90 and solve for w to get $w = 120$.

4.162 Write $y = \left(\frac{1}{2}\right)x^2$ so that $dy = x\,dx$. Using this transformation, we obtain

$$\Gamma\left(\frac{1}{2}\right) = \int_0^\infty y^{-(1/2)}e^{-y}\,dy = \int_0^\infty \frac{\sqrt{2}}{x}e^{-(1/2)x^2}x\,dx = \int_0^\infty \sqrt{2}\,e^{-(1/2)x^2}\,dx$$

$$= \frac{1}{2}\int_{-\infty}^\infty \sqrt{2}\,e^{-(1/2)x^2}\,dx = \frac{\sqrt{2}\left(\sqrt{2\pi}\right)}{2}\int_{-\infty}^\infty e^{-(1/2)x^2}\,dx = \sqrt{\pi}$$

Note that the function $g(x) = e^{-(1/2)x^2}$ is an even function of x; that is, $g(x) = g(-x)$, and hence

$$\int_{-\infty}^{\infty} g(x)\,dx = 2\int_0^{\infty} g(x)\,dx$$

4.163 a. Let $y = \sin^2\theta$, so that $dy = 2\sin\theta\cos\theta\,d\theta$. Now

$$B(\alpha,\beta) = \int_0^1 y^{\alpha-1}(1-y)^{\beta-1}\,dy = 2\int_0^{\pi/2} \sin^{2\alpha-2}\theta\,(1-\sin^2\theta)^{\beta-1}\sin\theta\cos\theta\,d\theta$$

$$= 2\int_0^{\pi/2} \sin^{2\alpha-2}\theta\cos^{2\beta-2}\theta\sin\theta\cos\theta\,d\theta = 2\int_0^{\pi/2}\sin^{2\alpha-1}\theta\cos^{2\beta-1}\theta\,d\theta$$

b. Following the instructions given in the text, we consider

$$\Gamma(\alpha)\Gamma(\beta) = \int_0^{\infty} y^{\alpha-1}e^{-y}\,dy\int_0^{\infty} z^{\beta-1}e^{-z}\,dz = \int_0^{\infty}\int_0^{\infty} y^{\alpha-1}z^{\beta-1}e^{-(y+z)}\,dy\,dz$$

and transform to polar coordinates using the transformation $y = r^2\cos^2\theta$ and $z = r^2\sin^2\theta$. Then the Jacobian is

$$|J| = \begin{vmatrix} \frac{dy}{dr} & \frac{dy}{d\theta} \\ \frac{dz}{dr} & \frac{dz}{d\theta}\end{vmatrix} = \begin{vmatrix} 2r\cos^2\theta & -2r^2\cos\theta\sin\theta \\ 2r\sin^2\theta & 2r^2\cos\theta\sin\theta\end{vmatrix}$$

$$= 4r^3\cos^3\theta\sin\theta + 4r^3\sin^3\theta\cos\theta = 4r^3\cos\theta\sin\theta$$

Then

$$\Gamma(\alpha)\Gamma(\beta) = \int_0^{\infty}\int_0^{\infty} y^{\alpha-1}z^{\beta-1}e^{-(y+z)}\,dy\,dz$$

$$= \int_0^{\pi/2}\int_0^{\infty} r^{2\alpha-2}\cos^{2\alpha-2}\theta\,r^{2\beta-2}\sin^{2\beta-2}\theta\,e^{-r^2}4r^3\cos\theta\sin\theta\,dr\,d\theta$$

$$= 2\int_0^{\pi/2}\cos^{2\alpha-1}\theta\sin^{2\beta-1}\theta\,d\theta\int_0^{\infty} 2r^{2\alpha+2\beta-1}e^{-r^2}\,dr$$

$$= B(\alpha,\beta)\int_0^{\infty} r^{2(\alpha+\beta-1)}e^{-r^2}(2r)\,dr$$

Using the transformation $x = r^2$ so that $dx = 2r\,dr$, we have

$$\Gamma(\alpha)\Gamma(\beta) = B(\alpha,\beta)\int_0^{\infty} x^{\alpha+\beta-1}e^{-x}\,dx = B(\alpha,\beta)\Gamma(\alpha+\beta)$$

and the result is proven.

4.164 Write

$$E\,|g(y)| = \int_{-\infty}^{\infty} |g(y)|\,f(y)\,dy = \int_{|g(y)|\le k} |g(y)|\,f(y)\,dy + \int_{|g(y)|>k} |g(y)|\,f(y)\,dy.$$

Since $|g(y)| > 0$, the first integral is nonnative and may be eliminated to form the inequality

$$E\,|g(y)| \ge \int_{|g(y)|>k} |g(y)|\,f(y)\,dy$$

But

$$\int_{|g(y)|>k} |g(y)|\,f(y)\,dy > \int_{|g(y)|>k} kf(y)\,dy = kP\left(|g(y)| > k\right)$$

since $|g(y)| > k$ for this integral. The inequality then becomes

$$E\,|g(y)| \ge kP\left(|g(y)| > k\right) = k\left[1 - P\left(|g(y)| \le k\right)\right],$$

which can be rewritten as

$$P\left(|g(y)| \le k\right) \ge 1 - \frac{E\,|g(y)|}{k}$$

and the result is proven.

CHAPTER 5 MULTIVARIATE PROBABILITY DISTRIBUTIONS

5.1 Denote a sample space in terms of the firm that received the first and second contracts:

S	(y_1, y_2)
AA	(2, 0)
AB	(1, 1)
AC	(1, 0)
BA	(1, 1)
BB	(0, 2)
BC	(0, 1)
CA	(1, 0)
CB	(0, 1)
CC	(0, 0)

Each sample point is equally likely with probability $\frac{1}{9}$. Setting up a table for the joint function for Y_1 and Y_2,

		y_1		
		0	1	2
	0	$\frac{1}{9}$	$\frac{2}{9}$	$\frac{1}{9}$
y_2	1	$\frac{2}{9}$	$\frac{2}{9}$	0
	2	$\frac{1}{9}$	0	0

b. $F(1, 0) = P(Y_1 \le 1, Y_2 \le 0) = p(0, 0) + p(1, 0) = \frac{1}{9} + \frac{2}{9} = \frac{1}{3}$

5.2 The sample space for the toss of three balanced coins, the values for Y_1 and Y_2 at each outcome, and the probability of each outcome are given below:

OUTCOMES	(y_1, y_2)	PROBABILITY
HHH	(3, 1)	$\frac{1}{8}$
HHT	(3, 1)	$\frac{1}{8}$
HTH	(2, 1)	$\frac{1}{8}$
HTT	(1, 1)	$\frac{1}{8}$
THH	(2, 2)	$\frac{1}{8}$
THT	(1, 2)	$\frac{1}{8}$
TTH	(1, 3)	$\frac{1}{8}$
TTT	(0, −1)	$\frac{1}{8}$

		y_1			
		0	1	2	3
	−1	$\frac{1}{8}$	0	0	0
y_2	1	0	$\frac{1}{8}$	$\frac{2}{8}$	$\frac{1}{8}$
	2	0	$\frac{1}{8}$	$\frac{1}{8}$	0
	3	0	$\frac{1}{8}$	0	0

b. $F(2, 1) = P(Y_1 \le 2, Y_2 \le 1) = p(0, -1) + p(1, 1) + p(2, 1) = \frac{1}{8} + \frac{1}{8} + \frac{2}{8} = \frac{1}{2}$

5.3 In this exercise Y_1 and Y_2 are both discrete random variables, and the joint distribution for Y_1 and Y_2 is given by

$$P(Y_1 = y_1, Y_2, y_2) = p(y_1, y_2)$$

We must calculate $p(y_1, y_2)$ for $y_1 = 0, 1, 2, 3$ and $y_2 = 0, 1, 2, 3$. The total number of ways of choosing 3 persons for the committee is $\binom{9}{3} = 84$. Now,

$$P(Y_1 = 0, Y_2 = 0) = P(3 \text{ divorced}) = 0$$

since there are only 2 divorced executives available. However,

$$P(Y_1 = 1, Y_2 = 0) = P(1 \text{ married, 0 never married, 2 divorced}) = \frac{\binom{4}{1}\binom{3}{0}\binom{2}{2}}{\binom{9}{3}} = \frac{4}{84}.$$

Similar calculations, using an extension of the hypergeometric probability distribution discussed in Chapter 3, will allow one to obtain all 16 probabilities, and the joint probability distribution of Y_1 and Y_2 may be written in the form of a table.

$p(2,0) = \frac{\binom{4}{2}\binom{3}{0}\binom{2}{1}}{84} = \frac{12}{84}$ $\qquad p(0,2) = \frac{\binom{4}{0}\binom{3}{2}\binom{2}{1}}{84} = \frac{6}{84}$

$p(1,1) = \frac{\binom{4}{1}\binom{3}{1}\binom{2}{1}}{84} = \frac{24}{84}$

$p(3,0) = \frac{\binom{4}{3}}{84} = \frac{4}{84}$ $\qquad p(1,2) = \frac{\binom{4}{1}\binom{3}{2}}{84} = \frac{12}{84}$

$p(2,1) = \frac{\binom{4}{2}\binom{3}{1}\binom{2}{0}}{84} = \frac{18}{84}$

$p(0,3) = \frac{\binom{3}{3}}{84} = \frac{1}{84}$ $\qquad p(0,1) = \frac{\binom{4}{0}\binom{3}{1}\binom{2}{2}}{84} = \frac{3}{84}$

$p(3,1) = p(2,2) = p(3,2) = p(3,3) = p(1,3) = p(2,3) = 0$

Note that $\sum_{y_1=0}^{3}\sum_{y_2=0}^{3} p(y_1, y_2) = 1.$

		y_2			
		0	1	2	3
	0	0	$\frac{3}{84}$	$\frac{6}{84}$	$\frac{1}{84}$
y_1	1	$\frac{4}{84}$	$\frac{24}{84}$	$\frac{12}{84}$	0
	2	$\frac{12}{84}$	$\frac{18}{84}$	0	0
	3	$\frac{4}{84}$	0	0	0

5.4

a. Notice that all of the probabilities are at least 0 and sum to 1

b. Note $F(1,2) = P(Y_1 \le 1, Y_2 \le 2) = 1$. The interpretation of this value is that every child in the experiment either survived or didn't and used either 0, 1 or 2 seatbelts.

5.5

a.

$$\begin{aligned} P(Y_1 \le \frac{1}{2}, Y_2 \le \frac{1}{3}) &= \int_0^{1/3}\int_{y_2}^{1/2} 3y_1\,dy_1\,dy_2 = \int_0^{1/3}\left[\frac{3}{2}y_1^2\right]_{y_2}^{1/2} dy_2 \\ &= \int_0^{1/3}\left(\frac{3}{8} - \frac{3}{2}y_2^2\right)dy_2 = \left[\frac{3}{8}y_2 - \frac{1}{2}y_2^3\right]_0^{1/3} \\ &= \frac{1}{8} - \frac{1}{54} \approx .11. \end{aligned}$$

Note that performing the integration in this order prevented splitting the integral into two parts.

b. $P(Y_2 \le Y_1/2) = \int_0^1\int_0^{y_1/2} (3y_1)\,dy_2\,dy_1 = \int_0^1 (3y_1)\,\frac{y_1}{2}\,dy_1 = \left[\frac{1}{2}y_1^3\right]_0^1 = \frac{1}{2}.$

5.6

a. We must have

$$F(\infty,\infty) = \int_0^1\int_0^1 Ky_1y_2\,dy_1\,dy_2 = 1.$$

Then

$\int_0^1\int_0^1 Ky_1y_2\,dy_1\,dy_2 = K\int_0^1 (y_2)\left[\frac{y_1^2}{2}\right]_0^1 dy_2 = \frac{K}{2}\int_0^1 y_2\,dy_2 = \frac{K}{2}\left[\frac{y_2^2}{2}\right]_0^1 = \frac{K}{4} = 1$

so that $K = 4$.

b. $F(y_1,y_2) = \int_0^{y_2}\int_0^{y_1} 4t_1t_2\,dt_1\,dt_2 = \int_0^{y_2}\left[\frac{4t_1^2}{2}\right]_0^{y_1} dt_2 = \int_0^{y_2} 2y_1^2t_2\,dt_2 = y_1^2y_2^2$

for $0 \le y_1 \le 1$ and $0 \le y_2 \le 1$. Recall that

$$F(y_1,y_2) = \begin{cases} 0, & \text{for } y_1 \le 0 \text{ or } y_2 \le 0 \\ 1, & \text{for } y_1 \ge 1 \text{ and } y_2 \ge 1. \end{cases}$$

c. $P\left(Y_1 \le \frac{1}{2}, Y_2 \le \frac{3}{4}\right) = F\left(\frac{1}{2},\frac{3}{4}\right) = \left(\frac{1}{2}\right)^2\left(\frac{3}{4}\right)^2 = \frac{9}{64}$

5.7

a. Integrating the joint density function over the region indicated under the restriction that $y_1 \le y_2$, we have

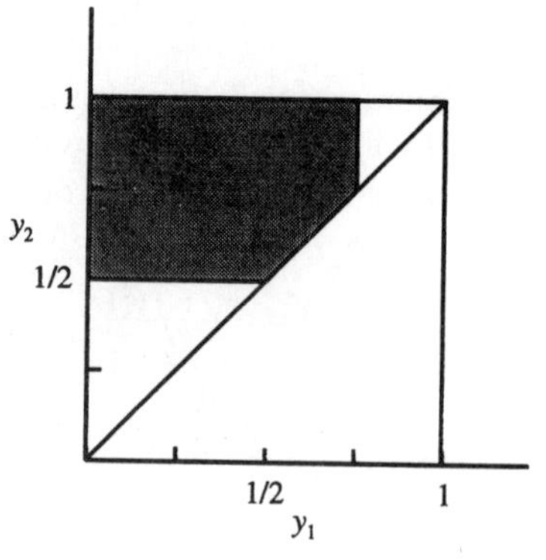

Figure 5.1

$$\int_0^1 \int_0^{y_2} K(1-y_2)\,dy_1\,dy_2$$
$$= \int_0^1 K\left(y_2 - y_2^2\right) dy_2 = \left(\tfrac{1}{2} - \tfrac{1}{3}\right) K$$

and hence $K = 6$.

b. The region of integration is given in Figure 5.1.

$$\begin{aligned} P\left(Y_1 \le \frac{3}{4}, Y_2 \ge \frac{1}{2}\right) &= \int_{1/2}^{3/4}\int_0^{y_2} 6(1-y_2)\,dy_1\,dy_2 + \int_{3/4}^{1}\int_0^{3/4} 6(1-y_2)\,dy_1\,dy_2 \\ &= \int_{1/2}^{3/4} 6\left(y_2 - y_2^2\right) dy_2 + \int_{3/4}^{1} \frac{9}{2}(1-y_2)\,dy_2 \\ &= \left[3y^2 - 2y^3\right]_{1/2}^{3/4} + \frac{9}{2}\left[y - \frac{y^2}{2}\right]_{3/4}^{1} \\ &= \frac{22}{64} + \frac{9}{64} = \frac{31}{64} \end{aligned}$$

5.8 **a.** The region over which Y_1 and Y_2 are uniformly distributed is the larger triangle shown in Figure 5.2. To find K, integrate

$$\int_0^1 \int_{2y_2}^{2} K\,dy_1\,dy_2 = \int_0^1 K(2-2y_2)\,dy_2$$
$$= K\left[2y_2 - y_2^2\right]_0^1 = K$$

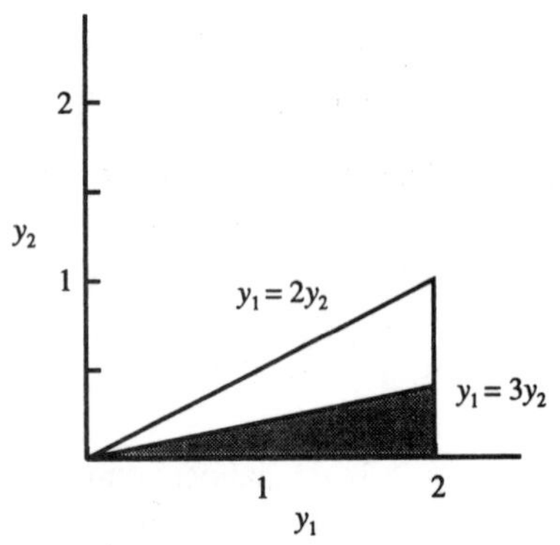

Figure 5.2

Hence $K = 1$. Alternatively we may have avoided integration by calculating the volume under the joint density geometrically. That is we know $(\frac{1}{2})(2)(1)(K) = 1$, implying $K = 1$.

b. The region within which $3Y_2 \le Y_1$ is the shaded triangular region in Figure 5.2. Hence

$$P(Y_1 \ge 3Y_2) = \int_0^{2/3}\int_{3y_2}^{2} dy_1\,dy_2 = \int_0^{2/3} (2-3y_2)\,dy_2 = \left[2y_2 - \tfrac{3}{2}y_2^2\right]_0^{2/3} = \tfrac{2}{3}$$

Again we may have alternatively calculated the volume over the shaded region without integration. That is, $(\frac{1}{2})(2)(\frac{2}{3})(1) = \frac{2}{3}$.

5.9 The two lines that define the region shaded in the text's diagram are the lines $y_2 - y_1 = 1$ and $y_1 + y_2 = 1$. The triangle has area $\left(\frac{1}{2}\right)(1)(2) = 1$. Hence the joint density can be written as

$$f(y_1, y_2) = 1 \qquad \begin{array}{lll} y_2 - y_1 \le 1 & \text{and} & -1 \le y_1 \le 0 \\ y_1 + y_2 \le 1 & \text{and} & 0 \le y_1 \le 1 \end{array}$$

a. The region of interest is the shaded area in Figure 5.3.

$$P\left(Y_1 \le \tfrac{3}{4}, Y_2 \le \tfrac{3}{4}\right)$$
$$= \int_0^{3/4}\int_{y_2-1}^{1/4} dy_1\, dy_2 + \int_{1/4}^{3/4}\int_0^{1-y_1} dy_2\, dy_1$$
$$= \int_0^{3/4}\left(\tfrac{5}{4} - y_2\right) dy_2 + \int_{1/4}^{3/4}(1-y_1)\, dy_1$$
$$= \left[\tfrac{5}{4}y_2 - \tfrac{y_2^2}{2}\right]_0^{3/4} + \left[y_1 - \tfrac{y_1^2}{2}\right]_{1/4}^{3/4} = \tfrac{29}{32}$$

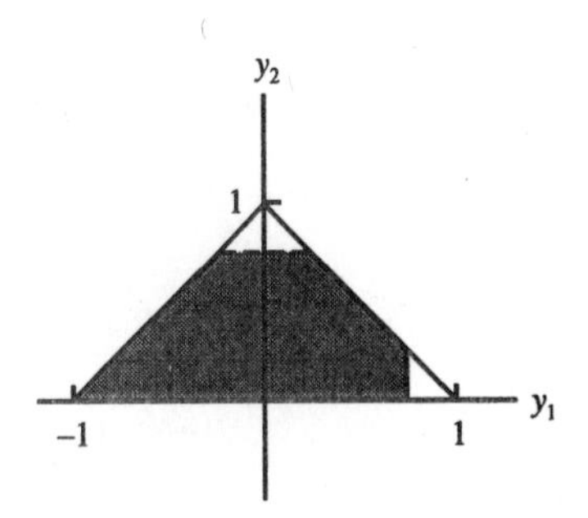

Figure 5.3

b. Refer to Figure 5.4. The region of interest, which is the region on which $Y_1 \ge Y_2$, is the shaded area. Then

$$P(Y_1 - Y_2 \ge 0)$$
$$= P(Y_1 \ge Y_2) = \int_0^{1/2}\int_{y_2}^{1-y_2} dy_1\, dy_2$$
$$= \int_0^{1/2}(1-2y_2)\, dy_2 = \left[y_2 - y_2^2\right]_0^{1/2}$$
$$= \tfrac{1}{2} - \tfrac{1}{4} = \tfrac{1}{4}$$

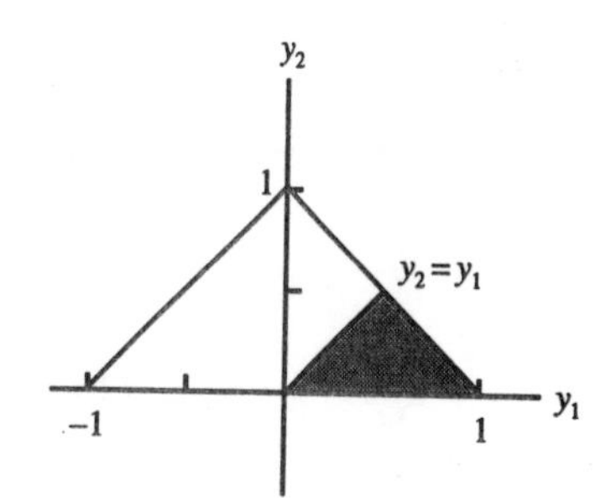

Figure 5.4

This can be verified by calculating the area corresponding to the shaded region in Figure 5.4.

5.10 The region over which the joint density function is positive is the triangular region shown in Figure 5.5. The shaded area is the region in which $Y_1 \le \frac{3}{4}$ and $Y_2 \le \frac{3}{4}$.

a. $P\left(Y_1 \le \frac{3}{4}, Y_2 \le \frac{3}{4}\right)$

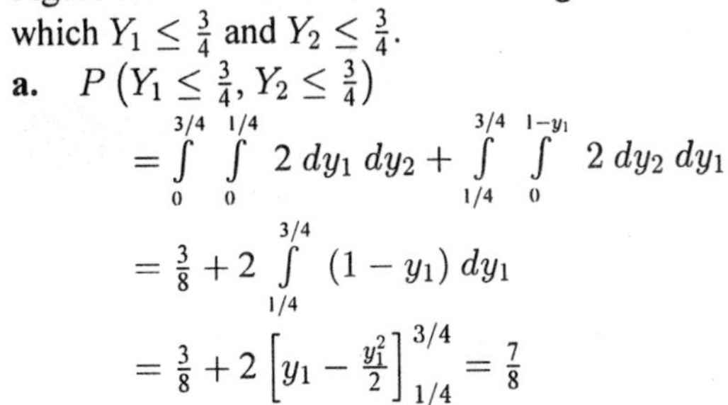

$$= \int_0^{3/4}\int_0^{1/4} 2\, dy_1\, dy_2 + \int_{1/4}^{3/4}\int_0^{1-y_1} 2\, dy_2\, dy_1$$
$$= \tfrac{3}{8} + 2\int_{1/4}^{3/4}(1-y_1)\, dy_1$$
$$= \tfrac{3}{8} + 2\left[y_1 - \tfrac{y_1^2}{2}\right]_{1/4}^{3/4} = \tfrac{7}{8}$$

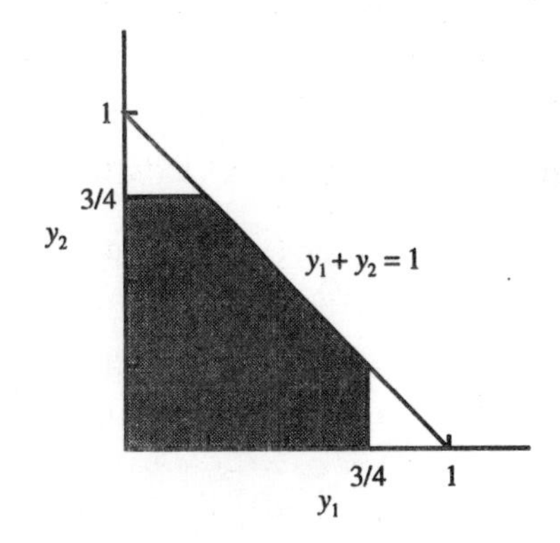

Figure 5.5

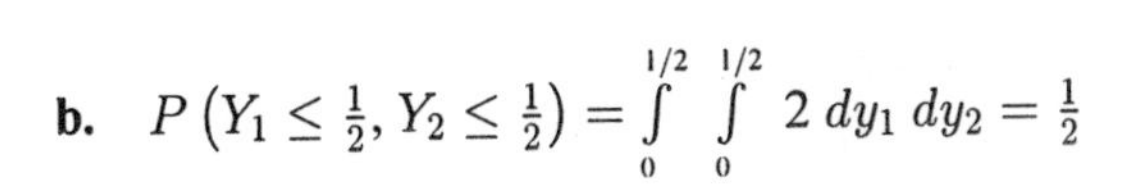

b. $P\left(Y_1 \le \frac{1}{2}, Y_2 \le \frac{1}{2}\right) = \int_0^{1/2}\int_0^{1/2} 2\, dy_1\, dy_2 = \frac{1}{2}$

5.11 a.

$$F\left(\frac{1}{2},\frac{1}{2}\right) = P\left(Y_1 \le \frac{1}{2}, Y_2 \le \frac{1}{2}\right) = \int_0^{1/2}\int_{y_1-1}^{1/2} 30y_1y_2^2\,dy_2\,dy_1$$

$$= \int_0^{1/2} 10y_1\left[y_2^3\right]_{y_1-1}^{1/2} = 10\int_0^{1/2} y_1\left(\frac{1}{8} - (y_1-1)^3\right)dy_1$$

$$= 10\int_0^{1/2}\left(\frac{1}{8}y_1 + y_1 - 3y_1^2 + 3\,y_1^3 - y_1^4\right)dy_1$$

$$= 10\left(\frac{1}{16}y_1^2 + \frac{1}{2}y_1^2 - y_1^3 + \frac{3}{4}\,y_1^4 - \frac{1}{5}y_1^5\right)\Bigg|_0^{1/2}$$

$$= 10\left(\frac{1}{64} + \frac{1}{8} - \frac{1}{8} + \frac{3}{64} - \frac{1}{5(32)}\right) = \frac{9}{16}.$$

b.

$$F\left(\frac{1}{2},2\right) = F\left(\frac{1}{2},1\right) = P\left(Y_1 \le \frac{1}{2}, Y_2 \le 1\right)$$

$$= P\left(Y_1 \le \frac{1}{2}, Y_2 \le \frac{1}{2}\right) + P\left(Y_1 \le \frac{1}{2}, Y_2 \ge \frac{1}{2}\right)$$

$$= F\left(\frac{1}{2},\frac{1}{2}\right) + P\left(Y_1 \le \frac{1}{2}, Y_2 \ge \frac{1}{2}\right)$$

$$= \frac{9}{16} + \int_{1/2}^{1}\int_0^{1-y_2} 30y_1y_2^2\,dy_1\,dy_2$$

$$= \frac{9}{16} + \int_{1/2}^{1} 15\,y_2^2\,(1-y_2)^2 dy_2$$

$$= \frac{9}{16} + 15\int_{1/2}^{1}\left(y_2^2 - 2y_2^3 + y_2^4\right)dy_2$$

$$= \frac{9}{16} + 15\left[\frac{1}{3}y_2^3 - \frac{2}{4}y_2^4 + \frac{1}{5}y_2^5\right]_{1/2}^{1}$$

$$= \frac{9}{16} + 15\left(\frac{1}{3} - \frac{2}{4} + \frac{1}{5} - \frac{1}{24} + \frac{1}{32} - \frac{1}{5(32)}\right)$$

$$= \frac{9}{16} + \frac{4}{16} = 13/16.$$

c.

$$P(Y_1 > Y_2) = 1 - P(Y_1 \le Y_2) = 1 - \int_0^{1/2}\int_{y_1}^{1-y_1} 30y_1y_2^2\,dy_2\,dy_1$$

$$= 1 - \int_0^{1/2} 10\,y_1\left((1-y_1)^3 - y_1^3\right)dy_1$$

$$= 1 - \int_0^{1/2} 10\left(y_1 - 3y_1^2 + 3y_1^3 - 2y_1^5\right)dy_1$$

$$= 1 - 10\left(\frac{1}{8} - \frac{1}{8} + \frac{3}{64} - \frac{1}{5(16)}\right) = 1 - \frac{11}{32} = 21/32 = .65625.$$

5.12 a. Clearly $f(y_1, y_2) \geq 0$. Then note $\int_{-\infty}^{\infty}\int_{-\infty}^{\infty} f(y_1, y_2)dy_2\, dy_1$

$$= \int_0^1 \int_{y_1}^{2-y_1} 6y_1y_2^2\, dy_2\, dy_1 = \int_0^1 3y_1((2-y_1)^2 - y_1^2)dy_1$$

$$= \int_0^1 (12y_1^2 - 12y_1^3)dy_1 = 4y_1^3 - 3y_1^4\Big|_0^1 = 4 - 3 = 1.$$

b. $P(Y_1 + Y_2 < 1) = \int_0^{1/2}\int_{y_1}^{1-y_1} 6y_1y_2^2\, dy_2\, dy_1 = \int_0^{1/2} 3y_1((1-y_1)^2 - y_1^2)dy_1$

$$= \int_0^{1/2} (3y_1^2 - 6y_1^3)dy_1 = y_1^3 - \tfrac{6}{4}y_1^4\Big|_0^{1/2} = \tfrac{1}{8} - \tfrac{6}{64} = 1/16.$$

Note by performing the integration in this order eliminated the need to split the integral into two parts in both parts **a.** and **b.**.

5.13 The region over which the density is positive is the area in the first quadrant ($y_1 \geq 0$, $y_2 \geq 0$) below the line $y_1 = y_2$. (See Figure 5.6).

a. $P(Y_1 < 2, Y_2 > 1) = \int_1^2 \int_{y_2}^2 e^{-y_1}\, dy_1\, dy_2$

$$= \int_1^2 -e^{-y_1}\big]_{y_2}^2\, dy_2 = \int_1^2 (e^{-y_2} - e^{-2})\, dy_2$$

$$= \left[-e^{-y_2} - y_2e^{-2}\right]_1^2$$

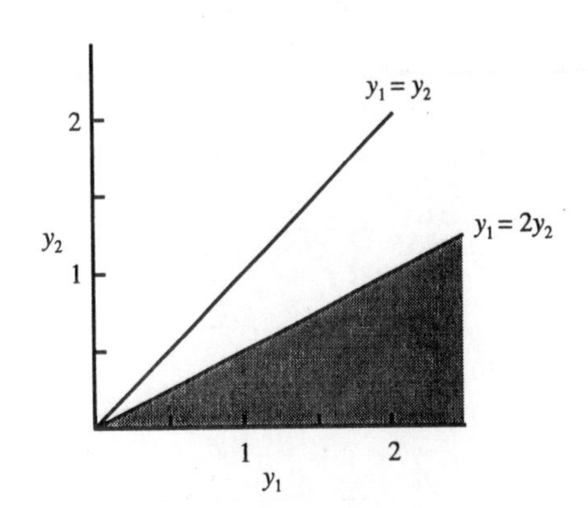

Figure 5.6

b. Refer to Figure 5.6. Integrating over the shaded region, we obtain

$$P(Y_1 \geq 2Y_2) = \int_0^{\infty}\int_{2y_2}^{\infty} e^{-y_1}\, dy_1\, dy_2 = \int_0^{\infty} e^{-2y_2}\, dy_2 = -\tfrac{1}{2}e^{-2y_2}\big]_0^{\infty} = \tfrac{1}{2}.$$

c. Refer to Figure 5.7. Integrating over the shaded region, we obtain

$$P(Y_1 - Y_2 \geq 1) = \int_0^{\infty}\int_{1+y_2}^{\infty} e^{-y_1}\, dy_1\, dy_2$$

$$= \int_0^{\infty} e^{-(1+y_2)}\, dy_2$$

$$= -e^{-(1+y_2)}\big]_0^{\infty} = e^{-1}.$$

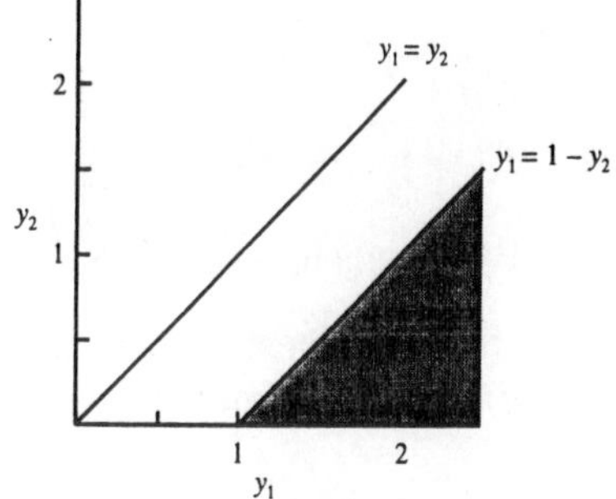

Figure 5.7

5.14 a. $P\left(Y_1 < \tfrac{1}{2}, Y_2 > \tfrac{1}{4}\right) = \int_{1/4}^{1}\int_0^{1/2} (y_1 + y_2)\, dy_1\, dy_2 = \int_{1/4}^{1} \left(\tfrac{1}{8} + \tfrac{y_2}{2}\right) dy_2 = \tfrac{21}{64}$

b. Refer to Figure 5.8. Integrating over the shaded region, we obtain

$$P(Y_1 + Y_2 \leq 1) = \int_0^1\int_0^{1-y_2} (y_1 + y_2)\, dy_1\, dy_2$$

$$= \int_0^1 \left(\tfrac{1}{2} - \tfrac{y_2^2}{2}\right) dy_2 = \tfrac{1}{2} - \tfrac{1}{6} = \tfrac{1}{3}.$$

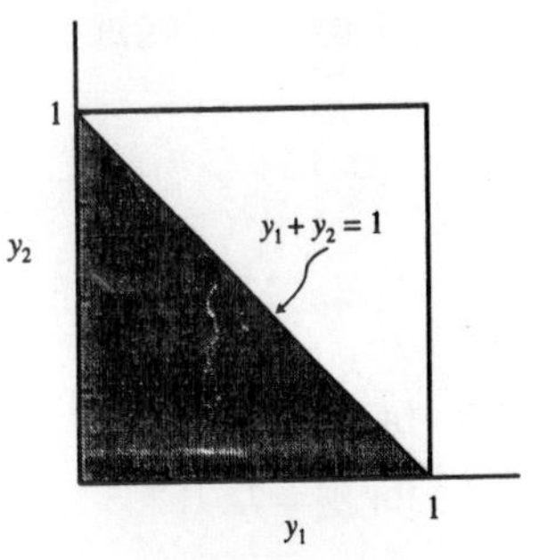

Figure 5.8

5.15 The region on which $Y_1 \le Y_2$ is the shaded region shown in Figure 5.9. The angle θ is $\theta = 45° = \frac{\pi}{4}$. Transforming to polar coordinates, we have $y_1 = r\cos\theta$ and $y_2 = r\sin\theta$. Also, $dy_1 dy_2 = r\,dr\,d\theta$ (see Exercise 4.128). Then

$$P(Y_1 \le Y_2) = \int\int_{y_1 \le y_2} f(y_1, y_2)\,dy_1\,dy_2$$
$$= \int_{\pi/4}^{5\pi/4}\int_0^1 \frac{r}{\pi}\,dr\,d\theta = \int_{\pi/4}^{5\pi/4} \frac{r^2}{2\pi}\Big]_0^1 d\theta$$
$$= \frac{\theta}{2\pi}\Big]_{\pi/4}^{5\pi/4} = \frac{\frac{4\pi}{4}}{2\pi} = \frac{1}{2}$$

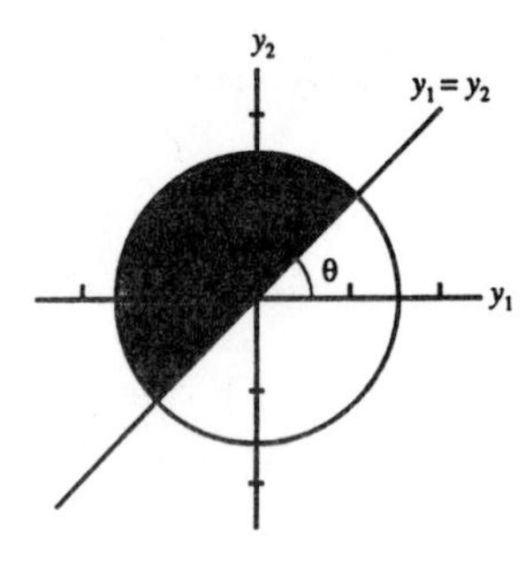

Figure 5.9

This answer can be verified by calculating the area corresponding to the shaded region given in Figure 5.9.

5.16 $P(Y_1 > 1, Y_2 > 1) = \int_1^\infty\int_1^\infty \frac{1}{8} y_1 e^{-y_1/2} e^{-y_2/2}\,dy_2\,dy_1 = \frac{-e^{-y_2/2}}{4}\Big]_1^\infty \int_1^\infty y_1 e^{-y_1/2}\,dy_1$

$$= \frac{1}{4} e^{-1/2} \int_1^\infty y_1 e^{-y_1/2}\,dy_1$$

Integrating by parts with $w = y_1$ and $dv = e^{-y_1/2}\,dy_1$, we can evaluate

$$\int_1^\infty y_1 e^{-y_1/2}\,dy_1 = -2y_1 e^{-y_1/2}\Big]_1^\infty + \int_1^\infty 2e^{-y_1/2}\,dy_1 = 0 + 2e^{-1/2} - \left[4e^{-y_1/2}\right]_1^\infty$$
$$= 6e^{-1/2}$$

so that

$$P(Y_1 > 1, Y_2 > 1) = \left(\tfrac{1}{4} e^{-1/2}\right)\left(6e^{-1/2}\right) = \tfrac{6}{4} e^{-1} = \tfrac{3}{2} e^{-1}.$$

5.17 **a.**

y_1	0	1	2
$p(y_1)$	$\frac{4}{9}$	$\frac{4}{9}$	$\frac{1}{9}$

b. No. Evaluating $f(y) = \binom{2}{y}\left(\frac{1}{3}\right)^y\left(\frac{2}{3}\right)^{2-y}$ for each value of Y_1 will result in the same probabilities as those given in part **a**.

5.18 **a.**

y_2	−1	1	2	3
$p(y_2)$	$\frac{1}{8}$	$\frac{4}{8}$	$\frac{2}{8}$	$\frac{1}{8}$

b. $P(Y_1 = 3 | Y_2 = 1) = \frac{P(Y_1=3, Y_2=1)}{P(Y_2=1)} = \frac{\left(\frac{1}{8}\right)}{\left(\frac{4}{8}\right)} = \frac{1}{4}$

5.19 **a.** By definition, $p_1(y_1) = \sum_{y_2=0}^{3} p(y_1, y_2)$ for $y_1 = 0, 1, 2, 3$. These probabilities may be obtained by summing across the rows in the table given in the solution to Exercise 5.3. However, we may calculate the marginal distribution of Y_1 directly as follows:

$$p_1(y_1) = \sum_{y_2=0}^{3-y_1} \frac{\binom{4}{y_1}\binom{3}{y_2}\binom{2}{3-y_1-y_2}}{\binom{9}{3}} = \frac{\binom{4}{y_1}\binom{5}{3-y_1}}{\binom{9}{3}} \sum_{y_2=0}^{3-y_1} \frac{\binom{3}{y_2}\binom{2}{3-y_1-y_2}}{\binom{5}{3-y_1}} = \frac{\binom{4}{y_1}\binom{5}{3-y_1}}{\binom{9}{3}}.$$

Where we recognize the sum as that of a hypergeometric random variable. Notice also that the marginal distribution of Y_1 is hypergeometric with $N = 9$, $n = 3$, and $r = 4$. Plugging in for y_1 gives the following values for the marginal distribution of Y_1

y_1	$p_1(y_1)$
0	10/84
1	40/84
2	30/84
3	4/84

In an exactly similar fashion we could calculate the marginal distribution for Y_2, $\frac{\binom{3}{y_2}\binom{6}{3-y_2}}{\binom{9}{3}}$, for $y_2 = 0, 1, 2, 3$ (this distribution will be needed for parts **b** and **c**).

b. Using the work from part **a.** we have:

$$P(Y_1 = 1|Y_2 = 2) = \frac{P(Y_1=1, Y_2=2)}{P(Y_2=2)} = \frac{\frac{\binom{4}{1}\binom{3}{2}\binom{2}{0}}{\binom{9}{3}}}{\frac{\binom{3}{2}\binom{6}{1}}{\binom{9}{3}}} = \frac{12}{18} = \frac{2}{3}.$$

c. Again, use the hypergeometric and the extended hypergeometric distribution.

$$P(Y_3 = 1|Y_2 = 1) = P(Y_1 = 1|Y_2 = 1) = \frac{P(Y_1=1, Y_2=1)}{P(Y_2=1)} = \frac{\frac{\binom{3}{1}\binom{2}{1}\binom{4}{1}}{\binom{9}{3}}}{\frac{\binom{3}{1}\binom{6}{2}}{\binom{9}{3}}} = \frac{24}{45} = \frac{8}{15}$$

d. As we have shown, the probabilities are identical.

5.20 a. The marginal distributions for Y_1 and Y_2 are given in the margins of the table. That is, the marginal distribution for Y_1 is $P(Y_1 = 0) = .76$ and $P(Y_1 = 1) = .24$ and the marginal distribution for Y_2 is given by $P(Y_2 = 0) = .55$, $P(Y_2 = 1) = .16$ and $P(Y_2 = 2) = .29$.

b. $P(Y_2 = 0|Y_1 = 0) = \frac{P(Y_2=0,Y_1=0)}{P(Y_1=0)} = \frac{.38}{.76} = .5$, $P(Y_2 = 1|Y_1 = 0) = \frac{.14}{.76} = .18$
$P(Y_2 = 2|Y_1 = 0) = \frac{.24}{.76} = .32$.

c. The desired probability $P(Y_1 = 0|Y_2 = 2) = \frac{.38}{.55} = .69$.

5.21 a. $f(y_2) = \int_{y_2}^{1} 3y_1 dy_1 = \frac{3}{2}y_1^2\Big|_{y_2}^{1} = \frac{3}{2} - \frac{3}{2}y_2^2$ for $0 \le y_2 \le 1$.

b. $f(y_1|y_2)$ is defined for $0 \le y_2 < 1$.

c. Note that $f(y_1) = \int_0^{y_1} 3y_1 dy_2 = 3y_1^2$. Then $f(y_2|y_1) = f(y_1,y_2)/\, f(y_1) = 1/y_1$ for $0 \le y_2 \le y_1$. Specific to this problem we consider $f(y_2|y_1 = \frac{3}{4}) = \frac{4}{3}$. The probability in question is $\int_{1/2}^{3/4} \frac{4}{3} dy_2 = 1 - \frac{4}{6} = 1/3$.

5.22 a. By definition,

$$f_1(y_1) = \int_{-\infty}^{\infty} f(y_1, y_2)\, dy_2 = \int_0^1 4y_1y_2\, dy_2 = (4y_1)\left(\frac{y_2^2}{2}\right)\Big]_0^1 = 2y_1 \quad \text{for } 0 \le y_1 \le 1$$

and

$$f_2(y_2) = \int_0^1 4y_1y_2\, dy_1 = (4y_2)\left(\frac{y_1^2}{2}\right)\Big]_0^1 = 2y_2 \qquad \text{for } 0 \le y_2 \le 1$$

b. By the definition of conditional probability,

$$P\left(Y_1 \le \tfrac{1}{2}|Y_2 > \tfrac{3}{4}\right) = \frac{P\left(Y_1 \le \frac{1}{2}, Y_2 > \frac{3}{4}\right)}{P\left(Y_2 > \frac{3}{4}\right)}.$$

Now

$$P\left(Y_1 \le \tfrac{1}{2}, Y_2 > \tfrac{3}{4}\right) = \int_0^{1/2}\int_{3/4}^{1} 4y_1y_2\, dy_2\, dy_1 = \int_0^{1/2} 2y_1\, [y_2^2]_{3/4}^{1}\, dy_1 = \tfrac{7}{16}\, y_1^2\Big|_0^{1/2}$$
$$= \tfrac{7}{64}$$

and

$$P\left(Y_2 > \tfrac{3}{4}\right) = \int_{3/4}^{1} f_2(y_2)\, dy_2 = \int_{3/4}^{1} 2y_2\, dy_2 = y_2^2\Big]_{3/4}^{1} = \tfrac{7}{16}.$$

Hence

$$P\left(Y_1 \le \tfrac{1}{2}|Y_2 > \tfrac{3}{4}\right) = \frac{\left(\frac{7}{64}\right)}{\left(\frac{7}{16}\right)} = \frac{1}{4}.$$

Notice this the same probability as $P(Y_1 \le \frac{1}{2})$.

c. By Definition 5.7, if $0 < y_2 \le 1$

$$f(y_1|y_2) = \frac{f(y_1, y_2)}{f_2(y_2)} = \frac{4y_1y_2}{2y_2} = 2y_1, \qquad 0 \le y_1 \le 1.$$

Notice this is the same as $f(y_1)$.

d. If $0 < y_1 \le 1$,

$$f(y_2|y_1) = \frac{f(y_1, y_2)}{f_1(y_1)} = \frac{4y_1y_2}{2y_1} = 2y_2, \qquad 0 \le y_2 \le 1.$$

Notice this is the same as $f(y_2)$

e. $P\left(Y_1 \le \frac{3}{4} | Y_2 = \frac{1}{2}\right) = \int_0^{3/4} f\left(y_1 | y_2 = \frac{1}{2}\right) dy_2 = \int_0^{3/4} 2y_1\, dy_1 = y_1^2\Big]_0^{3/4} = \frac{9}{16}$

5.23 **a.** For this joint density function, $0 \le y_1 \le y_2 \le 1$. Integrating over y_2, we have

$$f_1(y_1) = \int_{y_1}^{1} 6(1 - y_2)\, dy_2 = 6\left(y_2 - \frac{y_2^2}{2}\right)\Big]_{y_1}^{1} = 3(1 - y_1)^2 \qquad \text{for } 0 \le y_1 \le 1.$$

Similarly,

$$f_2(y_2) = \int_0^{y_2} 6(1 - y_2)\, dy_1 = 6y_2(1 - y_2) \qquad \text{for } 0 \le y_2 \le 1.$$

b. Using part **a**, calculate

$$P\left(Y_1 \le \frac{3}{4}\right) = \int_0^{3/4} 3(1 - y_1)^2\, dy_1$$

$$= 3\left[y_1 - y_1^2 + \frac{y_1^3}{3}\right]_0^{3/4} = \frac{63}{64}.$$

Refer to Figure 5.10 and integrate the joint density over the shaded region to obtain

$$P\left(Y_2 \le \frac{1}{2}, Y_1 \le \frac{3}{4}\right) = \int_0^{1/2}\int_0^{y_2} 6(1 - y_2)\, dy_1\, dy_2$$

$$= \int_0^{1/2} (6y_2 - 6y_2^2)\, dy_2 = \frac{3}{4} - \frac{2}{8} = \frac{1}{2}$$

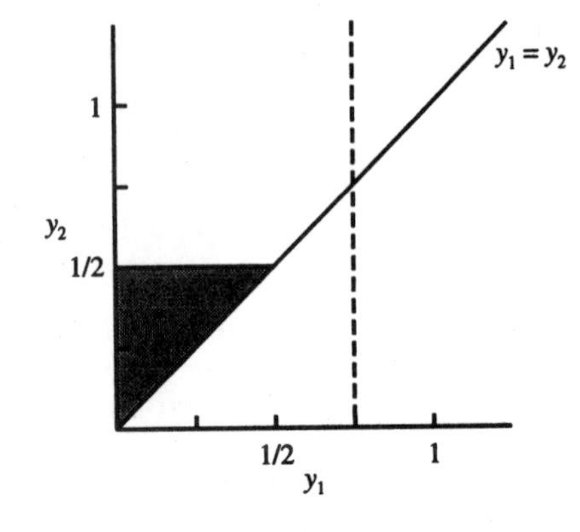

Figure 5.10

Then

$$P\left(Y_1 \le \frac{1}{2} | Y_2 \ge \frac{3}{4}\right) = \frac{\left(\frac{1}{2}\right)}{\left(\frac{63}{64}\right)} = \frac{32}{63}$$

c. If $0 < y_2 < 1$,

$$f(y_1|y_2) = \frac{f(y_1, y_2)}{f_2(y_2)} = \frac{6(1-y_2)}{6y_2(1-y_2)} = \frac{1}{y_2} \qquad \text{for } 0 \le y_1 \le y_2.$$

d. If $0 < y_1 < 1$,

$$f(y_2|y_1) = \frac{f(y_1, y_2)}{f_1(y_1)} = \frac{6(1-y_2)}{3y_2(1-y_1)^2} = \frac{2(1-y_2)}{(1-y_1)^2} \qquad \text{for } y_1 \le y_2 \le 1.$$

e. Refer to part **d.** Since $f\left(y_2|y_1 = \frac{1}{2}\right) = \frac{2(1-y_2)}{\left(\frac{1}{2}\right)^2} = 8(1 - y_2)$ for $\frac{1}{2} \le y_2 \le 1$,

$$P\left(Y_2 \ge \frac{3}{4} | Y_1 = \frac{1}{2}\right) = \int_{3/4}^{1} 8(1 - y_2)\, dy_2 = [8y_2 - 4y_2^2]_{3/4}^{1} = \frac{1}{4}$$

5.24 Refer to Exercise 5.1.

a. Calculate

$$f_2(y_2) = \int_{2y_2}^{2} dy_1 = 2(1 - y_2) \qquad \text{for } 0 \le y_2 \le 1.$$

Thus,

$$P(Y_2 \ge .5) = \int_{.5}^{1} (2 - 2y_2)\, dy_2 = [2y_2 - y_2^2]_{.5}^{1} = \frac{1}{4}.$$

b. Calculate the conditional density of Y_1 given Y_2 as follows: If $0 \le y_2 < 1$

$$f(y_1|y_2) = \frac{1}{2(1-y_2)} \qquad \text{for } 2y_2 \le y_1 \le 2$$

Since it is given that $y_2 = \frac{1}{2}$, then $f\left(y_1|y_2 = \frac{1}{2}\right) = 1$ for $1 \le y_1 \le 2$. Hence

$$P\left(Y_1 \ge 1.5 | Y_2 = \frac{1}{2}\right) = \int_{1.5}^{2} 1\, dy_1 = 2 - 1.5 = .5$$

5.25 **a.** Refer to Exercise 5.9.

$$f_2(y_2) = \int_{y_2-1}^{1-y_2} 1\, dy_1 = (1 - y_2) - (y_2 - 1) = 2(1 - y_2) \qquad 0 \le y_2 \le 1$$

In order to find $f_1(y_1)$, notice that the limits of integration are different for $0 \le y_1 \le 1$ and $-1 \le y_1 \le 0$. For the first case,

$$f_1(y_1) = \int_0^{1-y_1} dy_2 = 1 - y_1 \qquad \text{for } 0 \le y_1 \le 1.$$

Then, for $-1 \le y_1 \le 0$,

$$f_1(y_1) = \int_0^{1+y_1} dy_2 = 1 + y_1$$

This can be written as $f_1(y_1) = 1 - |y_1|$ for $-1 \le y_1 \le 1$.

b. The conditional distribution of Y_2 given Y_1 is, for $-1 \le y_1 < 1$

$$f(y_2|y_1) = \frac{1}{1-|y_1|} \qquad \text{for } 0 \le y_2 \le 1 - |y_1|$$

Since $Y_1 = \frac{1}{4}$,

$$f\left(y_2|y_1 = \tfrac{1}{2}\right) = \frac{1}{\left(\frac{3}{4}\right)} = \frac{4}{3} \qquad \text{for } 0 \le y_2 \le \tfrac{3}{4}$$

Then

$$P\left(Y_2 > \tfrac{1}{2}|Y_1 = \tfrac{1}{4}\right) = \int_{1/2}^{3/4} \tfrac{4}{3}\, dy_1 = \tfrac{4}{3}\left(\tfrac{3}{4} - \tfrac{1}{2}\right) = \tfrac{1}{3}$$

5.26 a. $P\left(Y_1 \ge \frac{1}{2}, Y_2 \le \frac{1}{4}\right) = \int_0^{1/4} \int_{1/2}^{1-y_2} 2\, dy_1\, dy_2 = \int_0^{1/4} 2\left(1 - y_2 - \frac{1}{2}\right) dy_2 + \left[y_2 - y_2^2\right]_9^{1/4}$

$$= \frac{3}{16}$$

and

$$P\left(Y_2 \le \tfrac{1}{4}\right) = \int_0^{1/4} 2(1 - y_2)\, dy_2 = \tfrac{1}{2} - \tfrac{1}{16} = \tfrac{7}{16}$$

Hence

$$P\left(Y_1 \ge \tfrac{1}{2}|Y_2 \le \tfrac{1}{4}\right) = \frac{\left(\frac{3}{16}\right)}{\left(\frac{7}{16}\right)} = \tfrac{3}{7}.$$

Notice we could have this probability without integration as the joint density is constant.

b. If $0 \le y_2 < 1$, the conditional distribution of Y_1 given Y_2 is:

$$\frac{f(y_1, y_2)}{f_2(y_2)} = \frac{2}{2(1-y_2)},$$

or

$$f(y_1|y_2) = \frac{1}{1-y_2} \qquad 0 \le y_1 \le 1 - y_2.$$

If $Y_2 = \frac{1}{4}$ then,

$$f\left(y_1|y_2 = \tfrac{1}{4}\right) = \frac{1}{\left(\frac{3}{4}\right)} = \frac{4}{3} \qquad \text{for } 0 \le y_1 \le \tfrac{3}{4}$$

and hence,

$$P\left(Y_1 \le \tfrac{1}{2}|Y_2 = \tfrac{1}{4}\right) = \int_{1/2}^{3/4} \tfrac{4}{3}\, dy_1 = \tfrac{4}{3}\left(\tfrac{1}{4}\right) = \tfrac{1}{3}.$$

Again notice we could have this probability without integration as the conditional density is constant.

5.27 a.

$$\begin{aligned} f(y_1) &= \int_{y_1-1}^{1-y_1} 30 y_1 y_2^2 dy_2 = 10 y_1 \left[y_2^3\right]_{y_1-1}^{1-y_1} \\ &= 10y_1((1-y_1)^3 - (y_1-1)^3) = 20y_1(1-y_1)^3. \end{aligned}$$

b. Note that the integral must be split into two parts, thus the marginal density of Y_2 has two components. Note that for $0 \le y_2 \le 1$ we have $f(y_2) = \int_0^{1-y_2} 30 y_1 y_2^2 dy_1$ $= 15y_2^2(1-y_2)^2$. Notice this is half of a beta(3,3) density. For $0 \ge y_2 \ge -1$ we have $f(y_2) = \int_0^{1+y_2} 30 y_1 y_2^2 dy_1 = 15y_2^2(1+y_2)^2$. (One may notice that this is half of the density of the negative of a beta(3,3) random variable). Therefore we have

$$f(y_2) = \begin{cases} 5y_2^2(1-y_2)^2 & \text{for } 0 \le y_2 \le 1 \\ 15y_2^2(1+y_2)^2 & \text{for } -1 \le y_2 \le 0 \end{cases}.$$

c. $f(y_2|Y_1 = y_1) = f(y_1, y_2)/f(y_1) = 30y_1y_2^2/20y_1(1-y_1)^3 = \frac{3}{2}y_2^2(1-y_1)^{-3}$ for $y_1 - 1 \le y_2 \le 1 - y_1$.

d. The probability in question is $P(Y_2 > 0|Y_1 = .75)$. This has to be $\frac{1}{2}$ as the conditional distribution of Y_2 given Y_1 is symmetric about 0 (from part **c.**). More tedious calculations yield $f(y_2|Y_1 = .75) = \frac{3}{2}y_2^2(.25)^{-3}$ for $-.25 \le y_2 \le .25$, hence

$$P(Y_2 > 0|Y_1 = .75) = \int_0^{.25} \tfrac{3}{2}y_2^2(.25)^{-3}dy_2 = \tfrac{1}{2}(.25)^3(.25)^{-3} = 1/2.$$

5.28 a.

$$f(y_1) = \int_{y_1}^{2-y_1} 6y_1^2y_2dy_2 = 3y_1^2\left[y_2^2\right]_{y_1}^{2-y_1}$$
$$= 3y_1^2((2-y_1)^2 - y_1^2) = 3y_1^2(4-4y_1)$$
$$= 12y_1^2(1-y_1)$$

for $0 \le y_1 \le 1$. This is a beta(3,2) density.

b. Note that the integral must be split into two parts, thus the marginal density of Y_2 has two components. Note that for $1 \le y_2 \le 2$ we have

$$f(y_2) = \int_0^{2-y_2} 6y_1^2y_2dy_1 = 2y_2[y_1^3]_0^{2-y_2} = 2y_2(2-y_2)^3.$$

For $0 \le y_2 \le 1$ we have

$$f(y_2) = \int_0^{y_2} 6y_1^2y_2dy_1 = 2y_2[y_1^3]_0^{y_2} = 2y_2^4.$$

c. $f(y_2|Y_1 = y_1) = f(y_1, y_2)/f(y_1) = 6y_1^2y_2/12y_1^2(1-y_1) = \frac{1}{2}y_2/(1-y_1)$ for $y_1 \le y_2 \le 2 - y_1$.

d. Using the density found in part **c.** we have $P(Y_2 < 1.1\,|Y_1 = .6) = \frac{1}{2}\int_{.6}^{1.1} \frac{y_2}{.4}dy_2$

$$\tfrac{10}{16}y_2^2\Big|_{.6}^{1.1} = \tfrac{10}{16}(1.1^2 - .6^2) = .53.$$

5.29 Refer to Exercise 5.13. The probability of interest is $P(Y_2 < 1|Y_1 = 2)$. Calculate $f_1(y_1)$ as

$$f_1(y_1) = \int_0^{y_1} e^{-y_1}\,dy_2 = y_1e^{-y_1} \qquad 0 \le y_1 \le \infty$$

Then if $y_1 > 0$, $f(y_2|y_1) = \frac{1}{y_1}$ for $0 \le y_2 \le y_1$, which for $y_1 = 2$ is

$$f(y_2|Y_1 = 2) = \tfrac{1}{2} \qquad 0 \le y_2 \le 2$$

Finally,

$$P(Y_2 < 1|Y_1 = 2) = \int_0^1 \tfrac{1}{2}\,dy_2 = \tfrac{1}{2}$$

5.30 a.

$$f_1(y_1) = \int_0^1 (y_1 + y_2)\,dy_2 = y_1 + \tfrac{1}{2} \qquad 0 \le y_1 \le 1$$
$$f_2(y_2) = \int_0^1 (y_1 + y_2)\,dy_1 = y_2 + \tfrac{1}{2} \qquad 0 \le y_2 \le 1$$

b. Calculate

$$P\left(Y_2 \ge \tfrac{1}{2}\right) = \int_{1/2}^1 \left(y_2 + \tfrac{1}{2}\right)dy_2 = \left[\tfrac{1}{2}\,y_2 + \tfrac{y_2^2}{2}\right]_{1/2}^1 = \tfrac{5}{8}$$
$$P\left(Y_1 \ge \tfrac{1}{2}, Y_2 \ge \tfrac{1}{2}\right) = \int_{1/2}^1\int_{1/2}^1 (y_1 + y_2)\,dy_1\,dy_2 = \int_{1/2}^1 \left(\tfrac{3}{8} + \tfrac{y_2}{2}\right)dy_2 = \tfrac{3}{8}$$

Hence

$$P\left(Y_1 \ge \tfrac{1}{2}|Y_2 \ge \tfrac{1}{2}\right) = \frac{\left(\frac{6}{16}\right)}{\left(\frac{5}{8}\right)} = \tfrac{3}{5}$$

c. First consider $f(y_1|y_2) = \frac{f(y_1, y_2)}{f(y_2)}$. If $0 \le y_2 \le 1$ we have

$$f(y_1|y_2) = \frac{y_1+y_2}{y_2+\frac{1}{2}} \qquad 0 \le y_1 \le 1$$

Then

$$P(Y_1 > .75|Y_2 = .5) = \int_{.75}^{1} \frac{y_1 + \frac{1}{2}}{\frac{1}{2} + \frac{1}{2}}\, dy_1$$
$$= \left(\frac{1}{2}\right) y_1^2 + \left(\frac{1}{2}\right) y_1 \Big]_{.75}^{1}$$
$$= \left(\frac{1}{2}\right) + \left(\frac{1}{2}\right) - .28125 - .375$$
$$= .34375$$

5.31 Calculate

$$f_2(y_2) = \int_0^\infty \frac{y_1}{8} e^{-(y_1+y_2)/2}\, dy_1 = \frac{e^{-y^2/2}}{8} \int_0^\infty \frac{y_1 e^{-y_1/2}}{4}\, dy_1 = \frac{e^{-y^2/2}}{2} \qquad y_2 > 0.$$

Notice that the preceding integral is that of a complete gamma density with $\alpha = 2$ and $\beta = 2$. Then

$$P(Y_2 > 2) = \int_2^\infty \frac{e^{\ y^2/2}}{2}\, dy_2 = -e^{-y^2/2}\Big]_2^\infty = e^{-1}$$

5.32 It is given that $f_1(y_1) = 1$ for $0 \le y_1 \le 1$, where Y_1 = amount stocked. Further, for a fixed value of Y_1, say $Y_1 = y_1$, $f(y_2|y_1) = \frac{1}{y_1}$ for $0 \le y_2 \le y_1$, where Y_2 = amount sold.

a. By definition,

$$f(y_1, y_2) = f_1(y_1) f(y_2|y_1) = \begin{cases} \frac{1}{y_1}, & \text{for } 0 \le y_2 \le y_1 \le 1 \\ 0, & \text{elsewhere} \end{cases}$$

b. Given that $Y_1 = \frac{1}{2}$, it is necessary to find $P\left(Y_2 > \frac{1}{4}|Y_1 = \frac{1}{2}\right)$. Using the conditional density of Y_2 given $Y_1 = \frac{1}{2}$, which is

$$f\left(y_2|y_1 = \tfrac{1}{2}\right) = \frac{1}{\left(\frac{1}{2}\right)} = \begin{cases} 2, & 0 \le y_2 \le \frac{1}{2} \\ 0, & \text{elsewhere} \end{cases}$$

we have

$$P\left(Y_2 > \tfrac{1}{4}|Y_1 = \tfrac{1}{2}\right) = \int_{1/4}^{1/2} 2\, dy_2 = 2y_2\Big]_{1/4}^{1/2} = \tfrac{1}{2}$$

c. The probability of interest is $P\left(Y_1 \ge \frac{1}{2}|Y_2 = \frac{1}{4}\right)$. Hence it is necessary to calculate $f(y_1|y_2)$. Note that

$$f_2(y_2) = \int_{y_2}^{1} \frac{1}{y_1}\, dy_1 = \ln y_1\Big]_{y_2}^{1} = -\ln y_2 \qquad 0 \le y_2 \le 1.$$

Then if $0 \le y_2 \le 1$

$$f(y_1|y_2) = \frac{f(y_1, y_2)}{f_2(y_2)} = \frac{1}{y_1(-\ln y_2)} \qquad y_2 \le y_1 \le 1$$

or

$$f\left(y_1|y_2 = \tfrac{1}{4}\right) = \frac{1}{y_1 \ln 4} \qquad \tfrac{1}{4} \le y_1 \le 1.$$

Finally,

$$P\left(Y_1 \ge \tfrac{1}{2}|Y_2 = \tfrac{1}{4}\right) = \int_{1/2}^{1} \frac{1}{y_1 \ln 4}\, dy_1 = \frac{1}{\ln 4} \ln y_1\Big]_{1/2}^{1} = \frac{\ln 2}{\ln 4} = \frac{1}{2}$$

5.33 By Definition 5.5, if $w = 0, 1, 2, \ldots$

$$P(Y_1 = y_1|W = w) = \frac{P(Y_1 = y_1, W = w)}{P(W = w)} = \frac{P(Y_1 = y_1, Y_1 + Y_2 = w)}{P(W = w)}$$
$$= \frac{P(Y_1 = y_1, Y_2 = w - y_1)}{P(W = w)} = \frac{P(Y_1 = y_1)P(Y_2 = w - y_1)}{P(W = w)}$$

(since Y_1 and Y_2 are independent) If $y_1 = 0, 1, \ldots, w$,

$$= \frac{\left(\frac{\lambda_2^{y_1} e^{-\lambda_2}}{y_1!}\right)\left(\frac{\lambda_2^{w-y_1} e^{-\lambda_2}}{(w-y_1)!}\right)}{\left(\frac{(\lambda_1+\lambda_2)^w e^{-(\lambda_1+\lambda_2)}}{w!}\right)}$$
$$= \frac{w!}{y_1!(w-y_1)!} \frac{\lambda_1^{y_1}}{(\lambda_1+\lambda_2)^{y_1}} \frac{\lambda_2^{w-y_1}}{(\lambda_1+\lambda_2)^{w-y_1}}$$
$$= \binom{w}{y_1}\left(\frac{\lambda_1}{\lambda_1+\lambda_2}\right)^{y_1}\left(1 - \frac{\lambda_1}{\lambda_1+\lambda_2}\right)^{w-y_1} \qquad y_1 = 0, 1, \ldots, w$$

which is the probability mass function of a binomial random variable with $n = w$ and $p = \frac{\lambda_1}{\lambda_1+\lambda_1}$.

5.34 $P(Y_1 = y_1|W = w) = \frac{P(Y_1=y_1,\, Y_1+Y_2=w)}{P(W=w)} = \frac{P(Y_1=y_1,\, Y_2=w-y_1)}{P(W=w)}$

$= \frac{P(Y_1=y_1)P(Y_2=w-y_1)}{P(W=w)}$

(since Y_1 and Y_2 are independent)

If $y_1 = 0, 1, \ldots, w \leq n_1$,

$$= \frac{\left[\binom{n_1}{y_1} p^{y_2}(1-p)^{n_1-y_1}\right]\left[\binom{n_2}{w-y_1} p^{w-y_1}(1-p)^{n_2-(w-y_1)}\right]}{\binom{n_1+n_2}{w} p^w(1-p)^{n_1+n_2-w}}$$

$$= \frac{\binom{n_1}{y_1}\binom{n_2}{w-y_1}}{\binom{n_1+n_2}{w}} \frac{p^{y_1+w-y_1}(1-p)^{n_1-y_1+n_2-w+y_1}}{p^w(1-p)^{n_1+n_2-w}}$$

$$= \frac{\binom{n_1}{y_1}\binom{n_2}{w-y_1}}{\binom{n_1+n_2}{w}} \qquad \text{if } y_1 \text{ is an integer, } 0 \leq y_1 \leq n_1 \text{ and } 0 \leq w - y_1 \leq n_2$$

which is the probability mass function of a hypergeometric random variable with $N = n_1 + n_2$, $n = w$, and $r = n_1$.

5.35 Let Y be the number of defectives in a random selection of three items. Conditional on p, fixed, the probability distribution of Y is

$$P(Y = y|p) = \binom{3}{y} p^y q^{3-y} \qquad y = 0, 1, 2, 3$$

It is given that P is distributed uniformly on (0, 1). That is, $f(p) = 1$ for $0 \leq p \leq 1$. We are interested in the unconditional marginal probability

$$P(Y = 2) = \int_0^1 P(Y = 2,\, p)\, dp = \int_0^1 P(Y = 2|p)f(p) = \int_0^1 \binom{3}{2} p^2(1-p)^1\, dp$$

$$= 3\int_0^1 (p^2 - p^3)\, dp = 3\left(\frac{1}{3} - \frac{1}{4}\right) = \frac{3}{12} = \frac{1}{4}.$$

5.36 Similar to Exercise 5.35. We have, if $\lambda > 0$,

$$P(Y = y|\lambda) = \frac{\lambda^y e^{-\lambda}}{y!} \qquad y = 0, 1, 2, \ldots$$

and

$$f(\lambda) = \begin{cases} e^{-\lambda}, & 0 \leq \lambda \leq \infty \\ 0, & \text{elsewhere.} \end{cases}$$

Then

$$P(y) = \int_0^\infty P(Y = y,\, \lambda)\, d\lambda = \int_0^\infty P(Y = y|\lambda)f(\lambda)\, d\lambda = \int_0^\infty \frac{\lambda^y e^{-\lambda}}{y!} e^{-\lambda}\, d\lambda$$

$$= \frac{1}{y!}\int_0^\infty \lambda^y e^{-2\lambda}\, d\lambda$$

$$= \frac{\Gamma(y+1)\left(\frac{1}{2}\right)^{y+1}}{y!}\int_0^\infty \frac{\lambda^y e^{-2\lambda}}{\Gamma(y+1)\left(\frac{1}{2}\right)^{y+1}}\, d\lambda = \frac{\Gamma(y+1)\left(\frac{1}{2}\right)^{y+1}}{\Gamma(y+1)}$$

$$= \left(\frac{1}{2}\right)^{y+1} \qquad \text{for } y = 0, 1, 2, \ldots$$

5.37 Assume $f(y_1|y_2) = f_1(y_1)$, then $f(y_1, y_2) = f(y_1|y_2)f_2(y_2) = f_1(y_1)f_2(y_2)$ implying Y_1 and Y_2 are independent. Conversely assume Y_1 and Y_2 are independent. Then $f(y_1, y_2) = g(y_1)h(y_2)$ implying $1 = \int\int f(y_1, y_2)dy_1dy_2 = \int g(y_1)dy_1 \int h(y_2)dy_2$. Hence $f(y_1, y_2) = \frac{g(y_1)h(y_2)}{\int g(y_1)dy_1 \int h(y_2)dy_2}$, then notice

$$f_1(y_1) = \int \frac{g(y_1)h(y_2)}{\int g(y_1)dy_1 \int h(y_2)dy_2} dy_2 = \frac{g(y_1)}{\int g(y_1)dy_1}$$

and similarly

$$f_2(y_2) = \frac{h(y_2)}{\int h(y_2)dy_2}.$$

That is $f(y_1, y_2) = f_1(y_1)\, f_2(y_2)$ (that is if Y_1 and Y_2 are independent then the joint density factors into the marginals not just arbitrary functions g and h). Now notice $f(y_1|y_2) = f(y_1, y_2)/\, f_2(y_2) = f_1(y_1)\, f_2(y_2)\, /\, f_2(y_2) = f_1(y_1)$. Notice how the last step requires $f_2(y_2) > 0$ as it appears in the denominator.

5.38 The argument follow exactly as 5.37 with the integrals replaced by sums and the densities replaced by probability mass functions.

5.39 No. For example, consider $P(Y_1 = 0, Y_2 = 0)$ and $p(Y_1 = 0)p(Y_2 = 0)$

$$p(0,0) = \tfrac{1}{9} \neq \left(\tfrac{4}{9}\right)\left(\tfrac{4}{9}\right) = p_1(0)p_2(0).$$

Thus, $Y_1$and Y_2 are not independent.

5.40 No. Considering $P(Y_1 = 3, Y_2 = 1)$ and $p(Y_1 = 3)p(Y_2 = 1)$

$$p(3,1) = \tfrac{1}{8} \neq \left(\tfrac{1}{8}\right)\left(\tfrac{4}{8}\right) = p_1(3)p_2(1)$$

Thus, $Y_1$and Y_2 are not independent.

5.41 Dependent, for example $P(Y_1 = 1, Y_2 = 2) \neq P(Y_1 = 1)P(Y_2 = 2)$.

5.42 Dependent, for example $P(Y_1 = 0, Y_2 = 0) \neq P(Y_1 = 0)P(Y_2 = 0)$.

5.43 Dependent as the range of y_1 values on which $f(y_1, y_2)$ is defined depends on y_2. More rigorously recall from exercise 5.21 $f(y_2|y_1) = 1/y_1$ for $0 \leq y_2 \leq y_1$and $f(y_2) = \frac{3}{2} - \frac{3}{2}y_2$ for $0 \leq y_2 \leq 1$. Thus $f(y_2|y_1) \neq f(y_2)$ for all values for which $f(y_2) > 0$, hence by problem 5.37 $Y_1$and Y_2 are dependent.

5.44 Independent as $f(y_1, y_2)$ can be factored (Theorem 5.5).

5.45 Dependent as the range of y_1 values on which $f(y_1, y_2)$ is defined depends on y_2 (and vice versa).

5.46 Dependent as the range of y_1 values on which $f(y_1, y_2)$ is defined depends on y_2

5.47 Dependent as the range of y_1 values on which $f(y_1, y_2)$ is defined depends on y_2.

5.48 Dependent as the range of y_1 values on which $f(y_1, y_2)$ is defined depends on y_2.

5.49 Dependent as the range of y_1 values on which $f(y_1, y_2)$ is defined depends on y_2. More rigorously one could verify from the solution to exercise 5.27 that $f(y_2|Y_1 = y_1) \neq f(y_2)$.

5.50 Dependent as the range of y_1 values on which $f(y_1, y_2)$ is defined depends on y_2. More rigorously one could verify from the solution to exercise 5.28 that $f(y_2|Y_1 = y_1) \neq f(y_2)$.

5.51 Dependent as the range of y_1 values on which $f(y_1, y_2)$ is defined depends on y_2.

5.52 Dependent as $f(y_1, y_2)$ cannot be factored.

5.53 Independent as $f(y_1, y_2)$ can be factored.

5.54 Let X, Y be the number on which person A, B flips a head on the coin respectively. Then X and Y are independent geometric random variables. The probability that they stop on the same number toss would be

$$\begin{aligned}
&P(X = 1, Y = 1) + P(X = 2, Y = 2) + \ldots \\
&= P(X = 1)P(Y = 1) + P(X = 2)P(Y = 2) + \ldots \\
&= \sum_{i=1}^{\infty} P(X = i)P(Y = i) = \sum_{i=1}^{\infty} p(1-p)^{i-1}p(1-p)^{i-1} \\
&= p^2 \sum_{i=1}^{\infty} (1-p)^{2(i-1)} = p^2 \sum_{k=0}^{\infty} ((1-p)^2)^k \\
&= p^2/(1-(1-p)^2).
\end{aligned}$$

5.55 Note that both

$$P(Y_1 > Y_2, Y_1 < 2Y_2) = \int_0^\infty \int_{y_1/2}^{y_1} e^{-y_1-y_2} dy_2 dy_1 = \int_0^\infty e^{-y_1}\left(-e^{y_2}\Big|_{y_1/2}^{y_1}\right) dy_1$$
$$= \int_0^\infty e^{-y_1}\left(-e^{-y_1} + e^{-y_1/2}\right) dy_1 = -\int_0^\infty e^{-2y_1} dy_1 + \int_0^\infty e^{-\frac{3}{2}y_1} dy_1$$
$$= -\tfrac{1}{2} + \tfrac{2}{3} = \tfrac{1}{6}$$

and

$$P(Y_1 < 2Y_2) = \int_0^\infty \int_{y_1/2}^{\infty} e^{-y_1-y_2} dy_2 dy_1 = \int_0^\infty e^{-y_1}\left(-e^{y_2}\Big|_{y_1/2}^{\infty}\right) dy_1$$
$$= \int_0^\infty e^{-\frac{3}{2}y_1} dy_1 = \tfrac{2}{3} = \tfrac{4}{6}.$$

Thus $P(Y_1 > Y_2 | Y_1 < 2Y_2) = P(Y_1 > Y_2, Y_1 < 2Y_2)\,/P(Y_1 < 2Y_2) = 1/4$.

5.56 Note that both

$$P(Y_1 > Y_2, Y_1 < 2Y_2) = \int_0^1 \int_{y_1/2}^{y_1} dy_2 dy_1 = 1/4$$

and

$$P(Y_1 < 2Y_2) = 1 - P(Y_1 > 2Y_2) = 1 - \int_0^1 \int_0^{y_1/2} dy_2 dy_1 = 1 - 1/4 = 3/4.$$

Thus $P(Y_1 > Y_2 | Y_1 < 2Y_2) = P(Y_1 > Y_2, Y_1 < 2Y_2)\,/P(Y_1 < 2Y_2) = 1/3$.
Notice also for this problem the joint density is simply a cube with height 1. Therefore, calculating the necessary probabilities for this problem can be accomplished by geometric arguments alone.

5.57 **a.** $\int_0^\infty f(y_1, y_2)\, dy_2 = \int_0^\infty (1 - \alpha(1 - 2e^{-y_1})(1 - 2e^{-y_2}))e^{-y_1-y_2} dy_2$

$$= e^{-y_1}\left[\int_0^\infty e^{-y_2} dy_2 - \alpha(1 - 2e^{-y_1})\int_0^\infty (1 - 2e^{-y_2})e^{-y_2} dy_2\right]$$
$$= e^{-y_1}(1 - \alpha(1 - 2e^{-y_1})(1 - 1)) = e^{-y_1} \text{ for } 0 < y_1.$$

b. Notice this density is symmetric as a function of y_1 and y_2. Therefore the marginal distribution of y_2 is also exponential with mean 1.

c. Clearly if $\alpha = 0$ then $y_1$and $y_2$are independent. Conversely suppose $y_1$and $y_2$are independent.
Then $E(y_1 y_2) = E(y_1)E(y_2) = 1$. Then note

$$E(y_1 y_2) = \int_0^\infty \int_0^\infty y_1 y_2 (1 - \alpha(1 - 2e^{-y_1})(1 - 2e^{-y_2}))e^{-y_1-y_2} dy_1 dy_2$$

$$= \int_0^\infty \int_0^\infty y_1 y_2 e^{-y_1-y_2} dy_1 dy_2$$
$$- \alpha\left(\int_0^\infty y_1(1 - 2e^{-y_1})e^{-y_1} dy_1\right)\left(\int_0^\infty y_2(1 - 2e^{-y_2})e^{-y_1} dy_2\right)$$

$$= \left(\int_0^\infty y_1 e^{-y_1} dy_1\right)\left(\int_0^\infty y_2 e^{-y_2} dy_2\right)$$
$$- \alpha\left(\int_0^\infty y_1 e^{-y_1} dy_1 - \int_0^\infty 2y_1 e^{-2y_1} dy_1\right)\left(\int_0^\infty y_2 e^{-y_2} dy_2 - \int_0^\infty 2y_2 e^{-2y_2} dy_2\right)$$
$$= 1 - \alpha(1 - \tfrac{1}{2})(1 - \tfrac{1}{2}) = 1 - \tfrac{\alpha}{4}. \text{ Which will only equal 1 when } \alpha = 0.$$

5.58 It is given that

$$p_1(y_1) = \binom{2}{y_1}(.2)^{y_1}(.8)^{2-y_1} \quad y_2 = 0, 1, 2; \quad p_2(y_2) = \binom{1}{y_2}(.3)^{y_2}(.7)^{1-y_2} \quad y_2 = 0, 1.$$

a. $p(y_1, y_2) = p_1(y_1)p_2(y_2) = \binom{2}{y_1}(.2)^{y_1}(.3)^{y_2}(.8)^{2-y_1}(.7)^{1-y_2}$
for $y_1 = 0, 1, 2$ and $y_2 = 0, 1$.

b. Since Y_i is the number of customers in line i, $i = 1, 2$, purchasing more than \$50 in groceries, the probability of interest is

$$P(Y_1 + Y_2 \leq 1) = P(Y_1 = 0, Y_2 = 0) + P(Y_1 = 1, Y_2 = 0) + P(Y_1 = 0, Y_2 = 1)$$
$$= (.8)^2(.7) + 2(.2)(.8)(.7) + (.3)(.8)^2 = .864$$

5.59 a. Because of the independence of Y_1 and Y_2,

$$f(y_1, y_2) = f(y_1)f(y_2) = \tfrac{1}{9}\, e^{-(y_1+y_2)/3}$$

for $y_1 > 0, y_2 > 0$.

b. The probability of interest is the shaded area in Figure 5.11. Hence

$$P(Y_1 + Y_2 \leq\ ) = \int_0^1 \int_0^{1-y_2} f(y_1, y_2)\, dy_1\, dy_2$$
$$= \int_0^1 \left[1 - e^{-(1-y_2)/3}\right] \tfrac{1}{3}\, e^{-y_2/3}\, dy_2$$

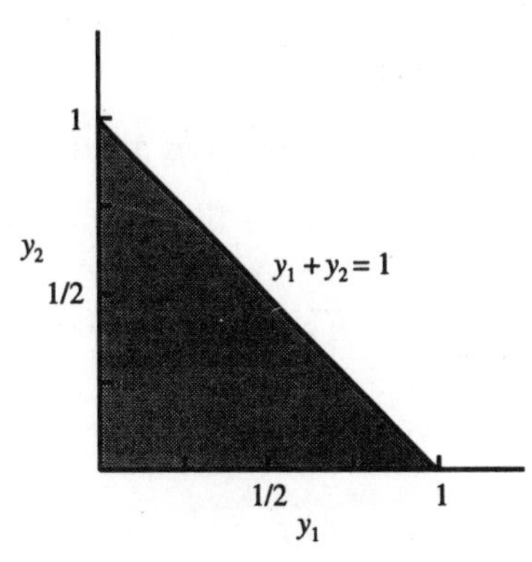

Figure 5.11

$$= \int_0^1 \left(\tfrac{1}{3}\, e^{-y_2/3} - \tfrac{1}{3}\, e^{-1/3}\right) dy_2 - e^{-y_2/3}\Big]_0^1 - \tfrac{1}{3}\, e^{-1/3} = 1 - \tfrac{4}{3}\, e^{-1/3}$$

5.60 $f(y_1, y_2) = f(y_1)f(y_2) = \begin{cases} 1, & 0 \leq y_1 \leq 1, 0 \leq y_2 \leq 1 \\ 0, & \text{otherwise} \end{cases}$

$$P\left(Y_2 \leq Y_1 \leq Y_2 + \tfrac{1}{4}\right) = \int_0^{1/4} \int_0^{y_1} 1\, dy_2\, dy_1 + \int_{1/4}^{1} \int_{y_1-(1/4)}^{y_1} 1\, dy_2\, dy_1$$
$$= \int_0^{1/4} y_1\, dy_1 + \int_{1/4}^{1} \left(\tfrac{1}{4}\right) dy_1$$
$$= \left(\tfrac{1}{2}\right) y_1^2\Big]_0^{1/4} + \left(\tfrac{1}{4}\right) y_1\Big]_{1/4}^{1} = \left(\tfrac{1}{32}\right) + \left(\tfrac{1}{4}\right) - \left(\tfrac{1}{16}\right) = \tfrac{7}{32}$$

5.61 Let Y_1 = calling time to the switchboard of the first call, then

$$f(y_1) = 1; \qquad 0 \leq y_1 \leq 1$$

Y_2 = calling time to the switchboard of the second call, then

$$f(y_2) = 1; \qquad 0 \leq y_2 \leq 1$$

Then we have $f(y_1, y_2) = 1$.

a. $P\left(Y_1 \leq \tfrac{1}{2}, Y_2 \leq \tfrac{1}{2}\right) = \left(\int_0^{1/2} 1\, dy_1\right)\left(\int_0^{1/2} 1\, dy_2\right) = \left(\tfrac{1}{2}\right)\left(\tfrac{1}{2}\right) = \tfrac{1}{4}$

(since Y_1 and Y_2 are independent).

b. Note that 5 minutes = $\tfrac{1}{12}$ of 1 hour.

$$P\left(|Y_1 = Y_2| < \tfrac{1}{12}\right) = \int_0^{1/12} \int_0^{y_1+(1/12)} dy^2\, dy_1 + \int_{1/12}^{11/12} \int_{y_1-(1/12)}^{y_1+(1/12)} dy_2\, dy_1$$
$$+ \int_{11/12}^{1} \left[\left(\tfrac{13}{12}\right) - y_1\right] dy_1$$
$$= \left(\tfrac{y_1^2}{2}\right) + \tfrac{y_1}{12}\Big]_0^{1/12} + \tfrac{2y_1}{12}\Big]_{1/12}^{11/12} + \left(\tfrac{13y_1}{12}\right) - \tfrac{y_1^2}{2}\Big]_{11/12}^{1} = \tfrac{46}{288} = \tfrac{23}{144}.$$

5.62 a. $E(Y_1) = np = 2\left(\tfrac{1}{3}\right) = \tfrac{2}{3}$.

b. $V(Y_1) = np(1-p) = 2\left(\tfrac{1}{3}\right)\left(\tfrac{2}{3}\right) = \tfrac{4}{9}$.

c. $E(Y_1 - Y_2) = E(Y_1)\_ - E(Y_2) = \left(\tfrac{2}{3}\right) - \left(\tfrac{2}{3}\right) = 0$.

5.63 Refer to the solution for Exercise 5.19, where the marginal distribution for Y_1 is shown to be hypergeometric with $r = 4$, $N = 9$, and, $n = 3$. Therefore:

$$E(Y_1) = \tfrac{(3)(4)}{9} = \tfrac{4}{3}.$$

5.64 Refer to Exercises 5.6 and 5.22. Recall $f_1(y_1) = 2y_1$ for $0 \le y_1 \le 1$.

a. $E(Y_1) = \int_0^1 2y_1 y_1\, dy_1 = \int_0^1 2y_1^2\, dy_1 = \frac{2}{3}$

b. $E(Y_1^2) = \int_0^1 2y_1^3\, dy_1 = \frac{1}{2}$ so that $V(Y_1) = \frac{1}{2} - \frac{4}{9} = \frac{1}{18}$.

c. Since $E(Y_2) = \int_0^1 2y_2^2\, dy_2 = \frac{2}{3}$, $E(Y_1 - Y_2) = 0$.

5.65 Refer to Exercises 5.7 and 5.13.

a. $E(Y_1) = \int_0^1 3y_1(1-y_1)^2\, dy_1 = \int_0^1 (3y_1 - 6y_1^2 + 3y_1^3)\, dy_1 = \frac{1}{4}$

$E(Y_2) = \int_0^1 (6y_2^2 - 6y_2^3)\, dy_2 = \frac{1}{2}$

b. $E(Y_1^2) = \int_0^1 (3y_1^2 - 6y_1^3 + 3y_1^4)\, dy_1 = \frac{1}{10}$ $\qquad V(Y_1) = \frac{1}{10} - \frac{1}{16} = \frac{3}{80}$

$E(Y_2^2) = \int_0^1 (6y_2^3 - 6y_2^4)\, dy_2 = \frac{6}{4} - \frac{6}{5} = \frac{3}{10}$ $\qquad V(Y_2) = \frac{3}{10} - \frac{1}{4} = \frac{1}{20}$

c. $E(Y_1 - 3Y_2) = \frac{1}{4} - 3\left(\frac{1}{2}\right) = -\frac{5}{4}$

5.66 **a.** The marginal distributions for Y_1 and Y_2 are

$f_1(y_1) = \frac{y_1}{2} \quad 0 \le y_1 \le 2$ $\qquad$ and $\qquad f_2(y_2) = 2 - 2y_2 \quad 0 \le y_2 \le 1$

[See Exercise 5.18 **a.**.] Then

$$E(Y_1) = \int_0^2 \frac{y_1^2}{2}\, dy_1 = \frac{4}{3}$$

$$E(Y_2) = \int_0^1 (2y_2 - 2y_2^2)\, dy_2 = \frac{1}{3}$$

b. $E(Y_1^2) = \int_0^2 \frac{y_1^3}{2}\, dy_1 = 2$ $\qquad E(Y_2^2) = \int_0^1 (2y_2^2 - 2y_2^3)\, dy_2 = \frac{2}{3} - \frac{1}{2} = \frac{1}{6}$

$V(Y_1) = 2 - \frac{16}{9} = \frac{2}{9}$ $\qquad V(Y_2) = \frac{1}{6} - \frac{1}{9} = \frac{1}{18}$

c. $E(Y_1 - Y_2) = \frac{4}{3} - \frac{1}{3} = 1$

d. $V(Y_1 - Y_2) = E[(Y_1 - Y_2)^2] - [E(Y_1 - Y_2)]^2$

$= E[Y_1^2 - 2Y_1Y_2 + Y_2^2] - 1$

$= E[Y_1^2] - 2E[Y_1Y_2] + E[Y_2^2] - 1$

Aside, $E[Y_1Y_2] = \int_0^1 \int_{2y_2}^2 y_1 y_2\, dy_1\, dy_2 = \left(\frac{1}{2}\right) \int_0^1 (4y_2 - 4y_2^3)\, dy_2$

$= \left(\frac{1}{2}\right)(2y_2^2 - y_2^4)\big]_0^1 = \left(\frac{1}{2}\right)(1) = \frac{1}{2}$

Using results from part **b.**, we find

$$V(Y_1 - Y_2) = 2 - 2\left(\frac{1}{2}\right) + \frac{1}{6} - 1 = \frac{1}{6}.$$

Using Tchebysheff's theorem, with $k = 2$,

$P(\mu - 2\sigma \le Y_1 - Y_2 \le \mu + 2\sigma) \ge \frac{3}{4}$

$= (1 - 2(.41) \le Y_1 - Y_2 \le 1 + 2(.41))$

$= (1 - .81 \le Y_1 - Y_2 \le 1 + .81) = (.19, 1.81)$.

5.67 Refer to Exercise 5.9. Integrating $f(y_1, y_2)$ over the two regions of integration, we have

$$E(Y_1Y_2) = \int_{-1}^0 \int_0^{1+y_1} y_1 y_2\, dy_2 dy_1 + \int_0^1 \int_0^{1-y_1} y_1 y_2\, dy_2\, dy_1$$

$$= \int_{-1}^0 \frac{y_1(1+y_1)^2}{2}\, dy_1 + \int_0^1 \frac{y_1(1-y_1)^2}{2}\, dy_1$$

$$= \left[\frac{y_1^2}{4} + \frac{2y_1^3}{6} + \frac{y_1^4}{8}\right]_{-1}^0 + \left[\frac{y_1^2}{4} - \frac{2y_1^3}{6} + \frac{y_1^4}{8}\right]_0^1 = 0$$

5.68 Refer to Exercise 5.14.

$$E(Y_1) = \int_{-\infty}^{\infty}\int_{-\infty}^{\infty} y_1 f(y_1, y_2)\,dy_1\,dy_2 = \int_0^1\int_0^1 y_1(y_1+y_2)\,dy_1\,dy_2$$

$$= \int_0^1 \left[\frac{y_1^3}{3} + \frac{y_1^2 y_2}{2}\right]_0^1 dy_2 = \left[\frac{1}{3}y_2 + \frac{y_2^2}{4}\right]_0^1 = \frac{7}{12}$$

By symmetry, $E(Y_2) = \frac{7}{12}$ and $E(30Y_1 + 25Y_2) = (30+25)\left(\frac{7}{12}\right) = 32.08$.

5.69 Since Y_1 and Y_2 are independent, with $f_1(y_1) = \frac{1}{4}y_1e^{-y_1/2}$ and $f_2(y_2) = \frac{1}{2}e^{-y_2/2}$,

$$E\left(\frac{Y_2}{Y_1}\right) = E\left(\frac{1}{Y_1}\right)E(Y_2) = \frac{1}{8}\int_0^{\infty} e^{-y_1/2}\,dy_1 \int_0^{\infty} y_2e^{-y_2/2}\,dy_2$$

$$= \frac{1}{8}\left[-2e^{-y_1/2}\right]_0^{\infty}(4) = \frac{1}{4}(4) = 1$$

since the second integral is the variable factor of a gamma distribution with $\alpha = 2$, $\beta = 2$ and integrates to $\Gamma(2)2^2 = 4$.

5.70 The marginal distribution of Y_1 is $f_1(y_1) = 1$ for $0 \le y_1 \le 1$, so that $E(Y_1) = \int_0^1 y_1\,dy_1$ $= \frac{1}{2}$. Using the joint distribution of Y_1 and Y_2, we obtain

$$E(Y_2) = \int_0^1\int_0^{y_1} \frac{y_2}{y_1}\,dy_2\,dy_1 = \int_0^1 \frac{y_1^2}{2y_1}\,dy_1 = \frac{y_1^2}{4}\Big]_0^1 = \frac{1}{4}$$

Thus, $E(Y_1 - Y_2) = \frac{1}{2} - \frac{1}{4} = \frac{1}{4}$.

5.71 Using $p(y)$ as derived in Exercise 5.36, we obtain

$$E(Y) = \sum_{y=0}^{\infty} y\left(\tfrac{1}{2}\right)^{y+1} = \sum_{y=1}^{\infty} y\left(\tfrac{1}{2}\right)^{y+1} = \tfrac{1}{2}\sum_{y=1}^{\infty} y\tfrac{1}{2}\left(\tfrac{1}{2}\right)^{y-1}.$$

We recognize the sum as the expected value of a geometric random variable with $p = \frac{1}{2}$(which happens to be 2). Therefore we have: $E(Y) = \frac{1}{2}(2) = 1$.

5.72

a. As we know Y_1 and Y_2 are geometric(p) random variables, $E(Y_1) = E(Y_2) = 1/p$. Then $E(Y_1 - Y_2) = 0$.

b. We know $V(Y_1) = \frac{1-p}{p^2} = E(Y_1^2) - E^2(Y_1) = E(Y_1^2) - \frac{1}{p^2}$. Therefore, $E(Y_1^2) = E(Y_2^2) = \frac{2-p}{p^2}$. Note then $E(Y_1Y_2) = E(Y_1)E(Y_2) = \frac{1}{p^2}$.

c. $E((Y_1 - Y_2)^2) = E(Y_1^2) - 2E(Y_1Y_2) + E(Y_2^2) = 2\left(\frac{2-p}{p^2} - \frac{1}{p^2}\right) = 2\left(\frac{1-p}{p^2}\right)$.

$V(Y_1 - Y_2) = E((Y_1 - Y_2)^2) = 2\left(\frac{1-p}{p^2}\right)$. Recall the following general fact. If $Y_1$and $Y_2$are independent and identically distributed then $V(Y_1 - Y_2) = 2V(Y_1) = 2V(Y_2)$.

d. Tcheyscheff's theorem says (Theorem 3.14)

$$P(|Y_1 - Y_2 - 0| < 3\sigma) \ge 8/9.$$

Therefore $(-3\sigma, 3\sigma)$ where $\sigma = \sqrt{2\left(\frac{1-p}{p^2}\right)}$ will contain $Y_1 - Y_2$ with probability at least $\frac{8}{9}$.

5.73

a. $E(Y_1) = E(Y_2) = 1$ as they are both exponential 1.

b. $V(Y_1) = V(Y_1) = 1$ as they are both exponential 1.

c. $E(Y_1 - Y_2) = 0$.

d. From exercise 5.57 we have $E(Y_1Y_2) = 1 - \alpha/4$. Note then $\text{Cov}(Y_1, Y_2) = -\alpha/4$.

e. $V(Y_1 - Y_2) = V(Y_1) + V(Y_2) - 2\text{Cov}(Y_1, Y_2) = 2 + 2\alpha/4 = 2 + \alpha/2$. We would expect by Tchebyscheff's theorem $Y_1 - Y_2$ to fall within 2 standard deviations from 0, or in the interval $(-2\sqrt{2+\alpha/2}, 2\sqrt{2+\alpha/2})$.

5.74

a. Using the hint and Theorem 5.9,

$$E(W) = E\left(Z\frac{1}{\sqrt{Y_1}}\right) = E(Z)E\left(Y_1^{-1/2}\right) = 0E\left(Y_1^{-1/2}\right) = 0 \quad \text{if } \nu_1 > 1.$$

Now,

$$V(W) = E(W^2) - [E(W)]^2 = E(W^2) - 0^2 = E(W^2).$$

But, using Theorem 5.9 again,

$$E(W^2) = E\left(Z^2\left(\frac{1}{Y_1}\right)\right) = E(Z^2)E(Y_1^{-1})$$

$= E\left(Y_2^{-1}\right)$ since $E(Z) = 0$ and $E\left(Z^2\right) = 1$.

Using Exercise 4.70(d), if $\nu_1 > 2$

$$E\left(Y_1^{-1}\right) = \frac{\Gamma\left(\frac{\nu_2}{2}-1\right)}{2\Gamma\left(\frac{\nu_2}{2}\right)} = \frac{1}{2\left(\frac{\nu_2}{2}-1\right)} = \frac{1}{\nu_1-2}.$$

Thus, $V(W) = \frac{1}{\nu_1-2}$, if $\nu_1 > 2$.

b. Using the results of Exercise 4.70(d), we obtain

$$E(Y_1) = \nu_1;\ E\left(Y_1^2\right) = \frac{4\Gamma\left(\frac{\nu_1}{2}+2\right)}{\Gamma\left(\frac{\nu_1}{2}\right)} = \nu_1(\nu_1+2)$$

and

$$E\left(\frac{1}{Y_2}\right) = \frac{1}{\nu_2-2} \qquad \text{if } \nu_2 > 2,$$

$$E\left(\frac{1}{Y_2^2}\right) = \frac{1}{(\nu_2-2)(\nu_2-4)} \qquad \text{if } \nu_2 > 4.$$

By Theorem 5.9, since Y_1 and Y_2 are independent,

$$E(u) = E(Y_1)E\left(\frac{1}{Y_2}\right) = \frac{\nu_1}{\nu_2-2} \qquad \text{if } \nu_2 > 2,$$

$$E\left(u^2\right) = E\left(Y_1^2\right)E\left(\frac{1}{Y_2^2}\right) = \frac{\nu_1(\nu_1+2)}{(\nu_2-2)(\nu_2-4)} \qquad \text{if } \nu_2 > 4.$$

Thus, if $\nu_2 > 4$

$$V(u) = E\left(u^2\right) - [E(u)]^2 = \frac{\nu_1(\nu_1+2)}{(\nu_2-2)(\nu_2-4)} - \left(\frac{\nu_1}{\nu_2-2}\right)^2 = \frac{2\nu_1(\nu_1+\nu_2-2)}{(\nu_2-2)^2(\nu_2-4)}.$$

5.75 $\text{Cov}(Y_1, Y_2) = E(Y_1Y_2) - E(Y_1)E(Y_2)$.

$$E(Y_1Y_2) = \sum_{y_1}\sum_{y_2} y_1y_2p(y_1, y_2) = (0)(0)\left(\tfrac{1}{9}\right) + (1)(0)\left(\tfrac{2}{9}\right) + (2)(0)\left(\tfrac{1}{9}\right) + (0)(1)\left(\tfrac{2}{9}\right)$$
$$+ (1)(1)\left(\tfrac{2}{9}\right) + (0)(2)\left(\tfrac{1}{9}\right) = \tfrac{2}{9}.$$

Since Y_1 and Y_2 are both binomial with $n = 2$ and $p = \frac{1}{3}$,

$$E(Y_1) = E(Y_2) = 2\left(\tfrac{1}{3}\right) = \tfrac{2}{3}.$$

Thus, $\text{Cov}(Y_1, Y_2) = \left(\frac{2}{9}\right) - \left(\frac{2}{3}\right)\left(\frac{2}{3}\right) = -\frac{2}{9}$.

No, as value of Y_1 increases, value of Y_2 tends to decrease.

5.76 Refer to the joint distribution of Y_1 and Y_2 given in the solution to Exercise 5.3.

$$E(Y_1) = \tfrac{4}{3} \qquad \text{(from Exercise 5.45)}$$

$$E(Y_2) = 1\left(\tfrac{45}{84}\right) + 2\left(\tfrac{18}{84}\right) + \tfrac{3}{84} = \tfrac{84}{84} = 1$$

$$E(Y_1Y_2) = 1(1)\left(\tfrac{24}{84}\right) + 2(1)\left(\tfrac{12}{84}\right) + 1(2)\left(\tfrac{18}{84}\right) = \tfrac{84}{84} = 1$$

since all other products involve a zero term and hence add zero to the expectation.

Thus $\text{Cov}(Y_1, Y_2) = 1 - \left(\frac{4}{3}\right) = -\frac{1}{3}$.

5.77 From Exercise 5.46; $E(Y_1) = E(Y_2) = \frac{2}{3}$. Then

$$E(Y_1Y_2) = \int_0^1\int_0^1 4y_1^2y_2^2\,dy_1\,dy_2 = \int_0^1 \tfrac{4}{3}y_2^2\,dy_2 = \tfrac{4}{9}$$

$$\text{Cov}(Y_1, Y_2) = \tfrac{4}{9} - \tfrac{4}{9} = 0.$$

No, this is not surprising since Y_1 and Y_2 are independent.

5.78 From Exercise 5.65, $E(Y_1) = \frac{1}{4}$ and $E(Y_2) = \frac{1}{2}$.

$$E(Y_1Y_2) = \int_0^1\int_0^{y_2} 6y_1y_2(1-y_2)\,dy_1\,dy_2 = \int_0^1 3\left(y_2^3 - y_2^4\right)dy_2 = \tfrac{3}{4} - \tfrac{3}{5} = \tfrac{3}{20}$$

$$\text{Cov}(Y_1, Y_2) = \tfrac{3}{20} - \tfrac{1}{8} = \tfrac{1}{40}$$

Since $\text{Cov}(Y_1, Y_2) \neq 0$, Y_1 and Y_2 are not independent.

5.79 From Exercise 5.67, $E(Y_1Y_2) = 0$. Calculate

$$E(Y_1) = \int_{-1}^{0}\int_0^{1+y_1} y_1\,dy_2\,dy_1 = \int_0^1\int_0^{1-y_1} y_1\,dy_2\,dy_1$$

$$= \int_{-1}^{0} y_1(1+y_1)\,dy_1 + \int_0^1 y_1(1-y_1)\,dy_1 = -\tfrac{1}{2} + \tfrac{1}{3} + \tfrac{1}{2} - \tfrac{1}{3} = 0$$

Hence

$$\text{Cov}(Y_1, Y_2) = E(Y_1Y_2) - E(Y_1)E(Y_2) = 0 - (0)E(Y_2) = 0$$

and $\rho = \frac{\text{Cov}(Y_1, Y_2)}{\sigma_1\sigma_2} = 0$. Note that $\text{Cov}(Y_1, Y_2) = 0$, even though Y_1 and Y_2 are dependent (c.f. Exercise 5.35).

5.80
$$\begin{aligned}\text{Cov}(U_1, U_2) &= E\{(Y_1 + Y_2)(Y_1 - Y_2 - [E(Y_1) + E(Y_2)][E(Y_1) - E(Y_2)]\} \\ &= E(Y_1Y_2) + E(Y_1^2) - E(Y_1Y_2) - E(Y_2^2) - [E(Y_1)]^2 - E(Y_1)E(Y_2) \\ &\quad + E(Y_1)E(Y_2) + [E(Y_2)]^2 \\ &= \sigma_1^2 - \sigma_2^2\end{aligned}$$

Now
$$\begin{aligned}V(U_1) &= E[U_1^2] - [E(U_1)]^2 \\ &= E(Y_1^2 + 2Y_1Y_2 + Y_2^2) - [(EY_1)^2 + 2(EY_1)(EY_2) + (EY_2)^2] \\ &= V(Y_1) + V(Y_2) + 2[E(Y_1Y_2) - (EY_1)EY_2)] \\ &= \sigma_1^2 + \sigma_2^2 + 2\text{Cov}(Y_1, Y_2) \\ &= \sigma_1^2 + \sigma_2^2\end{aligned}$$

since Y_1 and Y_2 are uncorrelated. A similar calculation yields $V(U_2) = \sigma_1^2 + \sigma_2^2$. Hence
$$\rho = \frac{\sigma_1^2 - \sigma_2^2}{\sqrt{(\sigma_1^2+\sigma_2^2)(\sigma_1^2+\sigma_2^2)}} = \frac{\sigma_1^2 - \sigma_2^2}{\sigma_1^2+\sigma_2^2}$$

5.81 The marginal distributions for Y_1 and Y_2 are shown in the accompanying tables.

y_1	$p_1(y_1)$
-1	$\frac{1}{3}$
0	$\frac{1}{3}$
1	$\frac{1}{3}$

y_2	$p_2(y_2)$
0	$\frac{2}{3}$
1	$\frac{1}{3}$

Since, for example, $p(-1, 0) \neq p(-1)p(0)$, Y_1 and Y_2 are not independent. However,
$$E(Y_1) = -1\left(\tfrac{1}{3}\right) + 0\left(\tfrac{1}{3}\right) + 1\left(\tfrac{1}{3}\right) = 0$$
$$E(Y_1Y_2) = (-1)(0)\left(\tfrac{1}{3}\right) + (0)(1)\left(\tfrac{1}{3}\right) + (1)(0)\left(\tfrac{1}{3}\right) = 0$$
so that $\text{Cov}(Y_1, Y_2) = 0$.

5.82 **a.** $\text{Cov}(Y_1, Y_2) = E((Y_1 - \mu_1)(Y_2 - \mu_2)) = E((Y_2 - \mu_2)(Y_1 - \mu_1)) = \text{Cov}(Y_1, Y_2)$.
b. $\text{Cov}(Y_1, Y_1) = E((Y_1 - \mu_1)(Y_1 - \mu_1)) = V(Y_1)$.

5.83 **a.** $\text{Cov}(Y_1, Y_1) = V(Y_1) = 2$.
b. Notice that $\rho = 7/\sqrt{(2)(8)} = 7/4 > 1$ which cannot happen.
c. If $\rho = 1$ then $\text{Cov}(Y_1, Y_2) = \sigma_1\sigma_2 = 4$. This implies a perfect positive relationship.
d. If $\rho = -1$ then $\text{Cov}(Y_1, Y_2) = \sigma_1\sigma_2 = -4$. This implies a perfect negative relationship.

5.84 **a.** $E(Y_1) = E(Z) = 0$. $E(Y_2) = E(Z^2) = V(Z) = 1$.
b. $E(Y_1Y_2) = E(Z^3)$. To find the the expected value of the cube of a standard normal, we will use the method of moment generating functions. Recall the the mgf of a standard normal is $e^{t^2/2}$. Note then
$$\frac{\partial^3}{\partial t^3}e^{t^2/2} = \frac{\partial^2}{\partial t^2}\left(te^{t^2/2}\right) = \frac{\partial}{\partial t}\left(e^{t^2/2} + t^2e^{t^2/2}\right) = te^{t^2/2} + 2te^{t^2/2} + t^3e^{t^2/2}$$
which is 0 when $t = 0$.
c. $\text{Cov}(Y_1, Y_2) = E(Y_1Y_2) - E(Y_1)E(Y_2) = 0 - 0 = 0$.
d. Y_1 and Y_2 are not independent as $P(Y_2 > 1|Y_1 > 1) = 1 \neq P(Y_2 > 1)$. This is an example where 0 covariance does not imply independence. Do not confuse this with the result that two normally distributed random variables are independent if and only if their covariance is 0. Specifically Y_2 is not normally distributed rather it is a function of a normally distributed random variable (in specific $Y_2$is distributed as a chi-squared with 1degree of freedom).

5.85 **a.** Recall from exercise 5.57 that $E(Y_1Y_2) = 1 - \alpha/4$ and $E(Y_1) = E(Y_2) = 1$.
Therefore $\text{Cov}(Y_1,Y_2) = -\alpha/4$.
b. This is clear from **a.**
c. As discussed in 5.57 it is clear that if $\alpha = 0$ then $Y_1$and $Y_2$are independent exponential 1's. Conversely if $Y_1$and $Y_2$are independent then $\text{Cov}(Y_1Y_2) = 0$, implying $\alpha = 0$. Thus $Y_1$and $Y_2$are independent if and only if $\text{Cov}(Y_1Y_2) = 0$. Note for this problem $V(Y_1) = V(Y_2) = 1$, thus $\text{Cov}(Y_1Y_2) = \rho$. Thus $Y_2$are independent if and only if $\rho = 0$.

5.86 Let X = dollar amount spent per week = $3Y_1 + 5Y_2$.

$E(X) = E[3Y_1 + EY_2] = 3E(Y_1) + 5E(Y_2) = 3(40) + 5(65) = 445.$

$V(X) = V[3Y_1 + 5Y_2] = 9V(Y_1) + 25V(Y_2)$ since Y_1 and Y_2 are independent
$= 9(4) + 25(8) = 236.$

5.87 Refer to Theorem 5.12.

$$E(3Y_1 + 4Y_2 - 6Y_3) = 3(2) + 4(-1) - 6(4) = -22$$
$$V(3Y_1 + 4Y_2 - 6Y_3) = 9(4) + 16(6) + 36(8) + (2)(3)(4)(1) + (2)(3)(-6)(1) + 2(4)(-6)(0) = 480$$

5.88 **a.** The probability distribution of $X = Y_1 + Y_2$ can be found from the table given in the solution to Exercise 5.3 as follows.

x	1	2	3
$p(x)$	$\frac{7}{84}$	$\frac{42}{84}$	$\frac{35}{84}$

Then

$$E(X) = \tfrac{7}{84} + \tfrac{84}{84} + \tfrac{105}{84} = \tfrac{196}{84} = \tfrac{7}{3}$$
$$E\left(X^2\right) = \tfrac{7}{84} + \tfrac{168}{84} + \tfrac{315}{84} = \tfrac{490}{84}$$
$$V(X) = \tfrac{490}{84} - \tfrac{49}{9} = .3889$$

b. From previous exercises, $E(Y_1) = \frac{4}{3}$, $E(Y_2) = 1$, $\text{Cov}(Y_1, Y_2) = -\frac{1}{3}$. Calculate

$$V(Y_1) = 1\left(\tfrac{40}{84}\right) + 4\left(\tfrac{30}{84}\right) + 9\left(\tfrac{4}{84}\right) - \tfrac{16}{9} = \tfrac{10}{18}$$

and

$$V(Y_2) = 1\left(\tfrac{45}{84}\right) + 4\left(\tfrac{18}{84}\right) + 9\left(\tfrac{1}{84}\right) - 1 = \tfrac{42}{84}.$$

Using Theorem 5.12, we obtain

$$E(Y_1 + Y_2) = \tfrac{7}{3}$$
$$V(Y_1 + Y_2) = \tfrac{10}{18} + \tfrac{42}{84} + 2\left(-\tfrac{1}{3}\right) = \tfrac{7}{18} = .3889$$

5.89 $V(Y_1 - Y_2) = \frac{1}{18} + \frac{1}{18} - 2(0) = \frac{1}{9}$

(See Exercise 5.64 for $V(Y_1)$. Also, $V(Y_2) = V(Y_1)$ by symmetry.)

5.90 $V(Y_1 - 3Y_2) = \frac{3}{80} + 9\left(\frac{1}{20}\right) - 6\left(\frac{1}{40}\right) = \frac{27}{80} = .3375$

(See Exercise 5.65 for $V(Y_1)$ and $V(Y_2)$.)

5.91 Calculate

$$E(Y_1) = \int_0^1 \int_0^{1-y_2} 2y_1\, dy_1\, dy_2 = \int_0^1 (1 - y_2)^2\, dy_2 = \tfrac{1}{3}$$
$$E\left(Y_1^2\right) = \int_0^1 \int_0^{1-y_2} 2y_1^2\, dy_1\, dy_2 = \int_0^1 \tfrac{2}{3}(1 - y_2)^3\, dy_2 = \tfrac{1}{6}$$
$$V(Y_1) = \tfrac{1}{6} - \tfrac{1}{9} = \tfrac{1}{18}$$

By symmetry, $E(Y_2) = \frac{1}{3}$ and $V(Y_2) = \frac{1}{18}$.

$$E(Y_1Y_2) = \int_0^1 \int_0^{1-y_2} 2y_1y_2\, dy_1\, dy_2 = \int_0^1 (1 - y_2)^2\, dy_2 = \tfrac{1}{12}$$

and

$$\text{Cov}(Y_1, Y_2) = \tfrac{1}{12} - \tfrac{1}{9} = -\tfrac{1}{36}.$$

Then

$E(Y_1 + Y_2) = \frac{2}{3}$ and $V(Y_1 + Y_2) = V(Y_1) + V(Y_2) + 2\text{Cov}(Y_1, Y_2) = \frac{2}{18} - \frac{2}{36} = \frac{1}{18}$.

5.92 From Exercise 5.29, $f_1(y_1) = y_1e^{-y_1}$, which is a gamma distribution with $\alpha = 2$, $\beta = 1$. Hence $E(Y_1) = 2(1) = 2$ and $V(Y_1) = \alpha\beta^2 = 2$.

$$f_2(y_2) = \int_{y_2}^{\infty} e^{-y_1}\, dy_1 = -e^{-y_1}\Big]_{y_2}^{\infty} = e^{-y_2}$$

which has a gamma distribution with $\alpha = \beta = 1$. Hence $E(Y_2) = V(Y_2) = 1$. Finally,

$$E(Y_1Y_2) = \int_0^{\infty} \int_0^{y_1} y_1y_2e^{-y_1}\, dy_2\, dy_1 = \int_0^{\infty} \tfrac{y_1^3}{2} e^{-y_1}\, dy_1 = \tfrac{\Gamma(4)1^4}{2} = 3$$

$$\text{Cov}(Y_1, Y_2) = 3 - (1)(2) = 1 \qquad E(Y_1 - Y_2) = 2 - 1 = 1$$
$$V(Y_1 - Y_2) = 2 + 1 - 2(1) = 1$$

It is unlikely that a customer would spend more than 4 minutes at the service window because this is 3 standard deviations above the mean.

5.93 Several intermediate results will be necessary.

(i) From Exercise 5.50, $E(Y_1) = \frac{7}{12}$ and $E(Y_2) = \frac{7}{12}$.

(ii) $E(Y_1Y_2) = \int_0^1 \int_0^1 (y_1 + y_2)\, y_1 y_2 \, dy_1 \, dy_2 = \int_0^1 \left[\frac{y_1^3 y_2}{3} + \frac{y_1^2 y_2^2}{2}\right]_0^1 dy_2$

$= \int_0^1 \left(\frac{y_2}{3} + \frac{y_2^2}{2}\right) dy_2 = \left[\frac{y_2^2}{6} + \frac{y_2^3}{6}\right]_0^1 = \frac{1}{3}$

(iii) $V(Y_1) = \int_0^1 \int_0^1 (y_1^3 + y_1^2 y_2)\, dy_2 \, dy_1 - [E(Y_1)]^2 = \int_0^1 \left(y_1^3 + \frac{1}{2} y_1^2\right) dy_1 - \frac{49}{144}$

$= \left[\frac{y_1^4}{4} + \frac{y_1^3}{6}\right]_0^1 - \frac{49}{144} = \frac{11}{144}$

and $V(Y_2) = V(Y_1) = \frac{11}{144}$.

(iv) $\text{Cov}(Y_1, Y_2) = E(Y_1Y_2) - E(Y_1)E(Y_2) = \frac{1}{3} - \left(\frac{7}{12}\right)\left(\frac{7}{12}\right) = -\frac{1}{144} = .0069$

Thus

$$E(30Y_1 + 25Y_2) = 30E(Y_1) + 25E(Y_2) = 32.08$$

and

$$V(30Y_1 + 25Y_2) = 900\left(\frac{11}{144}\right) + 625\left(\frac{11}{144}\right) + 2(750)\left(-\frac{1}{144}\right) = 106.08.$$

Then $\sigma = \sqrt{V(30Y_1 + 25Y_2)} = 10.30$.

Using Tchebysheff's theorem with $k = 2$, the necessary interval is $\mu \pm 2\sigma = 32.08 \pm 2(10.30) = 32.08 \pm 20.6$, or 11.48 to 52.68

5.94 Notice $f(y_1, y_2) = \frac{1}{4} y_1 e^{-y_1/2} \frac{1}{2} e^{-y_2/2}$ for $0 \le y_1, y_2 \le \infty$. Therefore Y_1 and Y_2 are independent. Notice also this implies that Y_1 is distributed as a gamma with $\alpha = 2$ and $\beta = 2$, and Y_2 is distributed as an exponential with $\beta = 2$. Hence $E(Y_2) = 2$ and $V(Y_2) = 4$, and $E(Y_1) = 4$, $V(Y_1) = 8$. Since Y_1 and Y_2 are independent $\text{Cov}(Y_1, Y_2) = 0$. Then

$$E(C) = 50 + 2(4) + 4(2) = 66 \qquad V(C) = 4(8) + 16(4) + 0 = 96$$

5.95 It is known that X, the daily gain, is normally distributed with $E(X) = 50$ and $V(X) = 9$. Also, Y, the daily cost, is distributed as a gamma variable with $\alpha = 4$ and $\beta = 2$, so that $E(Y) = \alpha\beta = 8$ and $V(Y) = \alpha\beta^2 = 16$. The net daily gain is then $G = X - Y$. Since X and Y are independent,

$$E(X - Y) = E(X) - E(Y) = 50 - 8 = 42$$

and

$$V(X - Y) = V(X) + V(Y) = 9 + 16 = 25$$

Note that a gain of $G = 70$ lies 5.6 standard deviations away from the mean, $E(G) = 42$. We would not expect her gain to be higher than 70 since, even using the conservative Tchebysheff's theorem, at most $\left(\frac{1}{k^2}\right) = \left(\frac{1}{5.6^2}\right) = .032$ of the measurements will fall beyond 5.6 standard deviations.

5.96 Since Y_1 has a gamma distribution with $\alpha = 4$, $\beta = 1$, then $E(Y_1) = 4$ and $V(Y_1) = 4$. Similarly, Y_2 has a gamma distribution with $\alpha = 1$, $\beta = 2$, so that $E(Y_2) = 2$ and $V(Y_2) = 4$. $\text{Cov}(Y_1, Y_2) = 0$ since Y_1 and Y_2 are independent.

a. $E(U) = 4 - 2 = 2$

b. $V(U) = 4 + 4 = 8$

c. The point $U = 0$ lies two units or $\frac{2}{\sqrt{8}} = .707$ standard deviations below the mean. Hence, it is quite likely that the daily profit will drop below zero.

5.97 Refer to Example 5.29 in the text. The situation here is analogous to drawing n balls from an urn containing N balls, r_1 of which are red, r_2 of which are black, and $N - r_1 - r_2$ of which are of another color. Using the argument given there, we can deduce that

$$E(Y_1) = np_1 \qquad V(Y_1) = np_1(1 - p_1)\left(\frac{N-n}{N-1}\right) \qquad \text{where} \qquad p_1 = \frac{r_1}{N}$$

$$E(Y_2) = np_2 \qquad V(Y_2) = np_2(1 - p_2)\left(\frac{N-n}{N-1}\right) \qquad \text{where} \qquad p_2 = \frac{r_2}{N}$$

Define two new random variables for $i = 1, 2, \ldots, n$:

$$U_i = \begin{cases} 1, & \text{if alligator } i \text{ is male} \\ 0, & \text{if not} \end{cases} \qquad V_i = \begin{cases} 1, & \text{if alligator } i \text{ is female} \\ 0, & \text{if not} \end{cases}$$

Then $Y_1 = \sum_{i=1}^{n} U_i$ and $Y_2 = \sum_{i=1}^{n} V_i$. In order to find the variance of a linear form involving Y_1 and Y_2, it is necessary to find $\text{Cov}(Y_1, Y_2)$, which involves finding $E(Y_1 Y_2)$.

$$E(Y_1Y_2) = E\left[(\Sigma U_i)(\Sigma V_i)\right] = E\left[\sum_{i=1}^{n} U_iV_i + \sum_{i \neq j}\sum U_iV_j\right]$$
$$= \sum_{i=1}^{n} E(U_iV_i) + \sum_{i \neq j} E(U_iV_j)$$

The only situation in which U_iV_i would be nonzero is when $U_i = 1$ and $V_i = 1$, that is, when the ith alligator is both male and female. Since this event is impossible, $E(U_iV_i) = 0$ for all i. For U_iV_j to be nonzero, we need $U_i = 1$ and $V_j = 1$, which happens with probability

$$P(U_i = 1, V_j = 1) = P(U_i = 1)P(V_j = 1|U = 1) = \frac{r_1}{N} \times \frac{r_2}{N-1} = p_1 \times \frac{Np_2}{N-1}$$
$$= \frac{N}{N-1} \times p_1p_2$$

Since there are $n(n-1)$ terms in $\sum_{i \neq j} E(U_iV_j)$

$$E(Y_1Y_2) = 0 + [n(n-1)] \times \frac{N}{N-1} \times p_1p_2$$
$$\text{Cov}(Y_1, Y_2) = n(n-1) \times \frac{N}{N-1} \times p_1p_2 - n^2p_1p_2 = \frac{n^2p_1p_2}{N-1} - \frac{nN}{M-1} \times p_1p_2$$
$$= \frac{n(n-N)}{N-1} \times p_1p_2$$

Then

$$E\left[\frac{Y_1}{n} - \frac{Y_2}{n}\right] = \frac{1}{n}(np_1 - np_2) = p_1 - p_1$$
$$V\left(\frac{Y_1}{n} - \frac{Y_2}{n}\right) = \frac{1}{n^2}V(Y_1 - Y_2) = \frac{1}{n^2}\left(V(Y_1) + V(Y_2) - 2\text{Cov}(Y_1Y_2)\right)$$
$$= \frac{1}{n^2}\left(\frac{n(N-n)}{N-1}p_1(1-p_1) + \frac{n(N-n)}{N-1}p_2(1-p_2) - \frac{2n(n-N)}{N-1}p_1p_2\right)$$
$$= \frac{N-n}{n(N-1)}\left[p_1(1-p_1) + p_2(1-p_2) + 2p_1p_2\right]$$
$$= \frac{N-n}{n(N-1)}\left(p_1 + p_2 - (p_1 - p_2)^2\right)$$

It is also possible to solve this problem by directly working with the joint distribution of Y_1 and Y_2.

5.98 Let $Y = X_1 + X_2$, the total sustained load on the footing.

a. Since X_1 and X_2 have gamma distributions, $E(X_1) = \alpha_1\beta_1 = 100$ and $E(X_2) = \alpha_2\beta_2 = 40$. Also, $V(X_1) = \alpha_1\beta_1^2 = 200$ and $V(X_2) = \alpha_2\beta_2^2 = 80$. Thus,
$$E(Y) = E(X_1 + X_2) = 100 + 40 = 140.$$
Since X_1 and X_2 are independent,
$$V(Y) = V(X_1 + X_2) = V(X_1) + V(X_2) = 200 + 80 = 280.$$

b. Consider Tchebysheff's theorem with $k = 4$, $P(|Y - \mu| \geq 4\sigma) \leq \frac{1}{16}$. The corresponding interval is $\left(140 - 4\sqrt{280}, 140 + 4\sqrt{280}\right)$ or $(73.07, 206.93)$. Thus, the sustained load will exceed 206.93 with a probability less than $\frac{1}{16}$.

5.99 **a.** Using the multinomial distribution with $p_1 = p_2 = p_3 = \frac{1}{3}$,
$$P(Y_1 = 3, Y_2 = 1, Y_3 = 2) = \frac{6!}{3!\,1!\,2!} = \left(\frac{1}{3}\right)^3\left(\frac{1}{3}\right)^1\left(\frac{1}{3}\right)^2 = 60\left(\frac{1}{3}\right)^6 = .0823$$

b. Refer to Theorem 5.13 in the text.
$$E(Y_1) = np_1 = \frac{n}{3} \qquad \text{and} \qquad V(Y_1) = np_1q_1 = \frac{2n}{9}$$

c. Refer to Theorem 5.13 in the text.
$$\text{Cov}(Y_1, Y_2) = -np_2p_3 = -n\left(\frac{1}{3}\right)\left(\frac{1}{3}\right) = -\frac{n}{9}$$

d. $E(Y_2 - Y_3) = \frac{n}{3} - \frac{n}{3} = 0$ and
$$V(Y_2 - Y_3) = \frac{2n}{9} + \frac{2n}{9} - 2\left(-\frac{n}{9}\right) = \frac{6n}{9} = \frac{2n}{3}$$

5.100 $E(C) = E(Y_1) + 3E(Y_2) = np_1 + 3np_2$
$$V(C) = V(Y_1) + 9V(Y_2) + 6\text{Cov}(Y_1, Y_2) = np_1q_1 + 9np_2q_2 - 6np_1p_2$$

5.101 **a.** If N is large, the multinomial distribution is appropriate and

$$P(Y_1 = 2, Y_2 = 1) = \frac{5!}{2!\,1!\,2!} \times (.3)^2(.1)^1(.6)^2 = .0972$$

b. Using the multinomial means, variances, and covariances, we have

$$E\left[\frac{Y_1}{n} - \frac{Y_2}{n}\right] = p_1 - p_2 = .2$$

$$V\left[\frac{Y_1}{n} - \frac{Y_2}{n}\right] = \frac{p_1q_2}{n} + \frac{p_2q_2}{n} + 2\,\frac{p_1p_2}{n} = \frac{.3(.7)}{5} + \frac{.1(.9)}{5} + \frac{2(.3)(.1)}{5} = .072$$

5.102 **a.** For this multinomial situation, let Y_1 be the number of mice weighing between 80 and 100 grams, and let Y_2 be the number weighing over 100 grams. Then

$$p_1 = P(80 \le Z \le 100) = P(-1 \le z \le 0) = .3413$$

$$p_2 = P(X \ge 100) = P(z > 0) = .5$$

The probability of interest is

$$\frac{4!}{2!\,1!\,1!} \times (.3415)^2(.5)^1(.1587)^1 = .1109$$

b. $P(Y_2 = 4) = \frac{4!}{0!\,4!\,0!} \times (.5)^4 = .0625$

5.103 Let Y_1 = # of family home fires, Y_2 = # of apartment fires, Y_3 = # of fires in other types. The joint probability function for Y_1, Y_2, Y_3 is multinomial with $n = 4$, $p_1 = .73, p_2 = .2, p_3 = .07$.

$$P(Y_1 = 2, Y_2 = 1, Y_3 = 1) = \left(\frac{4!}{2!\,1!\,1!}\right)(.73)^2(.2)(.07) = .08953.$$

5.104 Let C = total cost $= 20{,}000Y_1 + 10{,}000Y_2 + 2000Y_3$.

a.
$$\begin{aligned} E(C) &= 20{,}000E(Y_1) + 0{,}000E(Y_2) + 2000E(Y_3) \\ &= 20{,}000[4(.73)] + 10{,}000[4(.2)] + 2000[4(.07)] \\ &= 58{,}400 + 8000 + 560 = 66{,}960. \end{aligned}$$

b.
$$\begin{aligned} V(C) &= (20{,}000)^2V(Y_1) + (10{,}000)^2V(Y_2) + (2000)^2V(Y_3) + 2\,\Sigma\Sigma\, a_ia_j\,\mathrm{Cov}(Y_i, Y_j) \\ &= [(20{,}000)^2(4)(.73)(.27) + (10{,}000)^2(4)(.2)(.8) + (2000)^2(4)(.07)(.93)] \\ &\quad + 2[(20{,}000)(10{,}000)(-4)(.73)(.2) + (20{,}000)(2000)(-4)(.73)(.07) \\ &\quad + (10{,}000)(2000)(-4)(.2)(.07)] \\ &= 380{,}401{,}600 - 252{,}192{,}000 = 128{,}209{,}600. \end{aligned}$$

5.105 Let Y_1 = # of planes with no wing cracks, Y_2 = # of planes with detectable wing cracks, Y_3 = # of planes with critical wing cracks. The joint probability function for Y_1, Y_2, Y_3 is multinomial with $n = 5$, $p_1 = .7$, $p_2 = .25$, $p_3 = .05$.

a. $P(Y_1 = 2, Y_2 = 2, Y_3 = 1) = \left(\frac{5!}{2!\,2!\,1!}\right)(.7)^2(.25)^2(.05) = .046.$

b. The distribution of Y_3 is binomial with $n = 5$, $p = .05$. Thus,

$$P(Y_3 \ge 1) = 1 - P(Y_3 = 0) = 1 - (.95)^5 = 1 - .7738 = .2262.$$

5.106 $E(Y_1) = 10(.01) = .1;\ V(Y_1) = 10(.1)(.9) = .9.$

$E(Y_2) = 10(.05) = .5;\ V(Y_2) = 10(.05)(.95) = .475.$

$\mathrm{Cov}(Y_1, Y_2) = -10(.1)(.05) = -.05.$

$E[Y_1 + 3Y_2] = E(Y_1) + 3E(Y_2) = 1 + 3(.5) = 2.5.$

$V[Y_1 + 3Y_2] = V(Y_1) + 3^2V(Y_2) + 2(3)\mathrm{Cov}(Y_1, Y_2) = .9 + 9(.475) + 6(-.05) = 4.875.$

5.107 Let Y = # of items with at least one defect. Y is binomial with $n = 10$, $p = .10 + .05 = .15$.

a. $P(Y = 2) = \binom{10}{2}(.15)^2(.85)^8 = .2759$

b. $P(Y \ge 1) = 1 - P(Y = 0) = 1 - (.85)^{10} = .8031.$

5.108 The bivariate normal density function is

$$f(y_1, y_2) = \frac{e^{-Q/2}}{2\pi\sigma_1\sigma_2\sqrt{1-p^2}}, \qquad -\infty < y_1 < \infty, \quad -\infty < y_2 < \infty$$

where

$$Q = \frac{1}{1-p^2}\left[\frac{(y_2-\mu_2)^2}{\sigma_1^2} - \frac{2p(y_2-\mu_1)(y_2-\mu_2)}{\sigma_1\sigma_2} + \frac{(y_2-\mu_2)^2}{\sigma_2^2}\right].$$

Note that

$$Q - \frac{(y_2-\mu_1)^2}{\sigma_1^2} = \frac{1}{\sigma_2^2(1-p^2)}\left[(y_2-\mu_2)^2 - 2(y_2-\mu_2)\frac{p\sigma_2}{\sigma_2}(y_1-\mu_1) + \frac{p^2\sigma_2^2}{\sigma_2^2}(y_1-\mu_1)^2\right]$$
$$= \frac{\left[y_2-\mu_2-\left(\frac{p\sigma_2}{\sigma_1}\right)(y_1-\mu_1)\right]^2}{\sigma_2^2(1-p^2)}.$$

The joint density can then be written as

$$f(y_1, y_2) = \frac{\exp\left\{-\frac{1}{2}\left[\frac{(y_1-\mu_1)^2}{\sigma_1^2}\right]\right\}}{\sqrt{2\pi\,\sigma_1^2}}\;\frac{\exp\left\{-\frac{1}{2}\left[\frac{(y_2-\theta)^2}{\tau^2}\right]\right\}}{\sqrt{2\pi\tau^2}}$$

where

$$\theta = \mu_2 + p\sigma_1^{-1}\sigma_2(y_1-\mu_1) \qquad \text{and} \qquad \tau^2 = \sigma_2^2\,(1-p_2)\,.$$

Consequently,

$$f(y_1) = \int_{-\infty}^{\infty} f(y_1, y_2)\,dy_2 = \frac{\exp\left\{-\frac{1}{2}\left[\frac{(y_1-\mu_1)^2}{\sigma_1^2}\right]\right\}}{\sqrt{2\pi\,\sigma_1^2}}\int_{-\infty}^{\infty}\frac{\exp\left\{-\frac{1}{2}\left[\frac{(y_2-\theta)^2}{\tau^2}\right]\right\}}{\sqrt{2\pi\tau^2}}\,dy_2$$
$$= \frac{\exp\left\{-\frac{1}{2}\left[\frac{(y_1-\mu_1)^2}{T_1^2}\right]\right\}}{\sqrt{2\pi}\,\sigma_1},$$

because the above integrand is recognized as being the density function of a normal distribution. Very similar steps will show that the marginal distribution of Y_2 is normal with mean μ_2 and variance σ_2^2.

5.109 Similar to Exercise 5.108. It can be shown that

$$Q - \frac{(y_2-\mu_2)^2}{\sigma_2^2} = \frac{\left[y_1-\mu_1-\left(\frac{p\sigma_1}{\sigma_2}\right)(y_2-\mu_2)\right]^2}{\sigma_1^2(1-p^2)}$$

Then, using the results in Exercise 5.84,

$$f(y_1|y_2) = \frac{f(y_1, y_2)}{f(y_2)} = \frac{\left(\frac{e^{-Q/2}}{2\pi\sigma_1\sigma_2\sqrt{1-p^2}}\right)}{\left(\frac{e^{(-1/2)(y^2-\mu_2)^2/\sigma_2^2}}{\sqrt{2\pi\sigma_2^2}}\right)} = \frac{\exp\left[-\frac{1}{2}\left(Q-\frac{(y_2-\mu_2)^2}{\sigma_2^2}\right)\right]}{\sqrt{2\pi\sigma_1^2(1-p^2)}}$$

$$= \frac{1}{\sqrt{2\pi\sigma_1^2(1-p^2)}}\exp\left\{-\frac{1}{2}\,\frac{\left[y_1-\left(\mu_1+\left(\frac{p\sigma_1}{\sigma_2}\right)\right)(y_2-\mu_2)\right]^2}{\sigma_1^2(1-p^2)}\right\}$$

which is the density function of a normal distribution with mean $\mu_1 + p\left(\frac{\sigma_2}{\sigma_2}\right)(y_2-\mu_2)$ and variance $\sigma_1^2(1-p^2)$.

5.110

a. If U_1 and U_2 are orthogonal, then $\text{Cov}(U_1, U_2) = 0$. But

$$\text{Cov}(U_1, U_2) = \sum_{i=1}^{n}\sum_{j=1}^{n} a_i b_j\,\text{Cov}(Y_i, Y_j) = \sum_{i=1}^{n} a_i b_i V(Y_i)$$

(since the Y_i's are independent)

$$= \sigma^2\sum_{i=1}^{n} a_i b_i$$

Hence $\Sigma\, a_i b_i = 0$, since $\sigma^2 > 0$. Conversely, if $\Sigma\, a_i b_i = 0$, then $\text{Cov}(U_1, U_2) = 0$ above, so that U_1 and U_2 must be orthogonal.

b. Recall it is always true that if U_1 and U_2 are independent then $\text{Cov}(U_1, U_2) = 0$. Conversely suppose $Y_1, Y_2, \dots, Y_n$ are normally distributed and that U_1 and U_2 are orthogonal. Then $\text{Cov}(U_1, U_2) = 0$ so that $\rho_{U_1 \cdot U_2} = 0$. Now if $Y_1, Y_2, \dots, Y_n$ are normal, then U_1 and U_2 are bivariate normal (this will be proved in Exercise 6.33), with (from Theorem 5.12)

$$E(U_1) = \mu\sum_{i=1}^{n} a_i \qquad E(U_2) = \mu\sum_{i=1}^{n} b_i \qquad V(U_1) = \sigma^2\,\Sigma\, a_i^2 \qquad V(U_2) = \sigma^2\,\Sigma\, b_i^2$$

and $\text{Cov}(U_1, U_2) = 0$. As explained in Section 5.10, this implies that U_1 and U_2 are independent.

5.111 **a.** Note $f(y_1, y_2) = (2\pi\sigma^2)^{-2/2}\exp\left(-\frac{1}{2\sigma^2}((y_1-\mu_1)^2 + (y_2-\mu_2)^2)\right)$ which we recognize as the bivariate normal density with $\sigma_1 = \sigma_2 = \sigma$ and $\rho = 0$.

b. We simply apply exercise 5.10 with $a_1 = a_2 = b_1 = 1$, and $b_2 = -1$. Note that $\Sigma\, a_i b_i = 0$. Hence $U_1$and U_2 follow a bivariate normal distribution with $\rho = 0$. This implies $\text{Cov}(U_1, U_2) = 0$ which, in turn, implies $U_1$and U_2 are independent.

5.112 As discussed in exercise 5.111, $U_1$and U_2 follow independent normal distributions. Note that $E(U_1) = \mu_1 + \mu_2$and $\text{Var}(U_1) = 2\sigma^2$. Therefore $U_1 \sim \text{N}(\mu_1 + \mu_2, 2\sigma^2)$. Similar arguments show that U_2 follows the same distribution.

5.113 Refer to Exercise 5.23.

a. The conditional distribution of Y_1 given Y_2 is $f(y_1|y_2) = \frac{1}{y_2}$ for $0 \le y_1 \le y_2$. Thus,

$$E(Y_1|Y_2 = y_2) = \int_0^{y_2} y_1 \left(\frac{1}{y_2}\right) dy_1 = \left[\frac{y_1^2}{2y_2}\right]_0^{y_2} = \frac{y_2}{2}$$

b. $E(Y_1) = E[E(Y_1|Y_2 = y_2)] = \int_0^1 \frac{y_2}{2}\,(6y_2 - 6y_2^2)\, dy_2 = \left[\frac{3y_2^3}{3} - \frac{3}{4}\,y_2^4\right]_0^1 = \frac{1}{4}$

(same as answer to Exercise 5.47). A more statistical solution to this problem would have noted that distribution of Y_1 given $Y_2 = y_2$ is uniform on the interval 0 to $y_2$and that Y_2 is distributed as a Beta with $\alpha = 2$ and $\beta = 2$. Using the properties of these distributions we could have avoided all integration.

5.114 $z = \frac{6-1.25}{\sqrt{1.5625}} = 3.8.$

Because 6 is 3.8 standard deviations, it is not likely that Y will exceed 6.

5.115 Refer to Exercise 5.35.

a. $E(Y|p) = 3p$ as given p, Y is a binomial random variable.
Then, since $f(p) = 1, 0 \le p \le 1$,
$E(Y) = E[E(Y|p)] = E[3p] = 3E[p] = \frac{3}{2}$ (recall p is uniform(0,1))

b. $V(Y|p) = 3p(1-p)$ and from part **a.**, $E(Y|p) = 3p$.
$V(Y) = E(3p(1-p)) + V(3p)$
Since p is uniformly distributed on the interval (0, 1), $E(p) = \frac{1}{2}$, $V(p) = \frac{1}{12}$,
and $E(p_2) = V(p) + [E(p)]^2 = \frac{1}{12} + \frac{1}{4} = \frac{1}{3}$. Then
$V(Y) = 3\,[E(p) - E(p^2)] + 9V(p) = 3\left(\frac{1}{2} - \frac{1}{3}\right) + 9\left(\frac{1}{12}\right) = 1.25$

5.116 **a.** From Theorem 5.14, we have $E(Y) = E[E(Y|\lambda)]$. For a given value of λ, Y has a Poisson distribution with parameter λ, so that $E(Y|\lambda) = \lambda$. Notice also that λ is distributed (marginally) as an exponential random variable with $\beta = 1$. Then

$$E(Y) = E(\lambda) = 1$$

b. $V(Y|\lambda) = \lambda$ and from part (a), $E(Y|\lambda) = \lambda$. Now

$$V(Y) = E[V(Y|\lambda)] + V[E(Y|\lambda)] = E(\lambda) + V(\lambda).$$

Since λ is a distributed exponential with parameter 1, $E(\lambda) = 1$ (as shown in part (a)) and $V(\lambda) = 1$. Then

$$V(Y) = 2.$$

c. $9 = E(Y) + 5.657\sigma_Y$ so it is unlikely that Y exceeds 9.

5.117 Refer to Exercise 5.32. There we obtained $f(y_2|y_1) = \frac{1}{y_1}$ for $0 \le y_2 \le y_1$. That is, given y_1, Y_2 is distributed uniformly on the interval $[0, y_1]$. Therefore $E[Y_2|Y_1 = y_1] = \frac{y_1}{2}$. Plugging in $y_1 = \frac{3}{4}$ gives the answer $E[Y_2|Y_1 = \frac{3}{4}] = \frac{3}{8}$.

5.118 **a.** Since λ is random, the Poisson assumption applies to the conditional distribution of Y for fixed λ. Thus,

$$p(y|\lambda) = \frac{\lambda^y e^{-\lambda}}{y!}, \qquad y = 0, 1, 2, \ldots$$

and

$$E(Y|\lambda) = \lambda.$$

Since λ has a gamma distribution with parameters α and β, the expected number of bacteria per cubic centimeter is given by $\alpha\beta$.

b. From part (a), $E(Y|\lambda) = \lambda$. Also, $V(Y|\lambda) = \lambda$. Since λ is distributed gamma with parameters α and β, where α is a positive integer,

$$E(\lambda) = \alpha\beta \qquad \text{and} \qquad V(\lambda) = \alpha\beta^2.$$

Then

$$V(Y) = E[V(Y|\lambda)] + V[E(Y|\lambda)] = E(\lambda) + V(\lambda) = \alpha\beta + \alpha\beta^2$$

and

$$\sigma = \sqrt{\alpha\beta(1+\beta)}.$$

5.119 Consider the random variable $y_1 Y_2$ for such a fixed value of Y_1. Then $y_1 Y_2$ has a normal distribution with mean 0 and variance y_1^2. Hence

$$E\left(e^{ty_1Y_2}|Y_1 = y_1\right) = e^{t^2y_1^2/2}$$

Using Theorem 5.14, we have

$$m_U(t) = E(e^{tU}) = E\left(e^{tY_1Y_2}\right) = E\Big[E\left(e^{ty_1Y_2}|Y_1 = y_1\right)\Big]$$

$$= E\left(e^{t^2Y_1^2/2}\right) = \int_{-\infty}^{\infty} \frac{1}{\sqrt{2\pi}} e^{t^2y_1^2/2} e^{-y_1^2/2}\, dy_1$$

$$= \int_{-\infty}^{\infty} \frac{1}{\sqrt{2\pi}} e^{(-y_1^2/2)(1-t^2)}\, dy_1 = (1-t^2)^{-1/2}$$

since the variable factor is that of a normal random variable with mean 0 and variance $\frac{1}{(1-t^2)}$. Now

$$m'(0) = E(U) = t\,(1-t^2)^{-3/2}\Big|_{t=0} = 0$$

$$m''(0) = E\left(U^2\right) = 3t^2\,(1-t^2)^{-5/2} + (1-t^2)^{-3/2}\Big]_{t=0} = 1$$

Evaluating $E(U)$ and $V(U)$ directly and using the independence of Y_1 and Y_2, we have

$$E(U) = E(Y_1Y_2) = E(Y_1)E(Y_2) = 0$$

$$E\left(U^2\right) = E\left(Y_1^2Y_2^2\right) = E\left(Y_1^2\right)E\left(Y_2^2\right) = 1(1) = 1$$

$$V(U) = 1$$

5.120 Let (Y_1, Y_2) represent the coordinates of the landing point of the bomb. Since the radius $= 1$, $0 \le y_1^2 + y_2^2 \le 1$.

The probability that the target is destroyed
= the probability that the bomb destroys everything within $\frac{1}{2}$ mile of the landing point

$$= P\left(Y_1^2 + Y_2^2 \le \left(\tfrac{1}{2}\right)^2\right)$$

Graphically, this is the same as the

$$P(\text{bomb falls within the shaded circle}) = \frac{\text{Area of shaded circle}}{\text{Area of outside circle}}$$

$$= \frac{\pi\left(\frac{1}{2}\right)^2}{\pi(1)^2} = \frac{1}{4}.$$

Figure 5.12

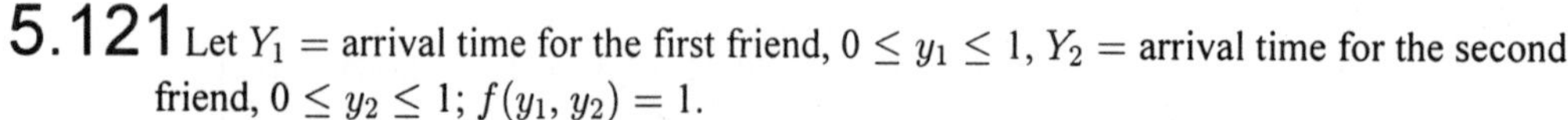

5.121 Let $Y_1 =$ arrival time for the first friend, $0 \le y_1 \le 1$, $Y_2 =$ arrival time for the second friend, $0 \le y_2 \le 1$; $f(y_1, y_2) = 1$.
If friend 2 arrives $\frac{1}{6}$ hour (10 min.) before or $\frac{1}{6}$ hour after friend 1, they will meet. We can represent this as $|Y_1 - Y_2| < \frac{1}{3}$. We consider

$$P\left[|Y_1 - Y_2| < \tfrac{1}{3}\right] = \int_0^{1/6}\int_0^{y_1+(1/6)} dy_2\,dy_1 + \int_{1/6}^{5/6}\int_{y_1-(1/6)}^{y_1+(1/6)} dy_2\,dy_1 + \int_{5/6}^{1}\int_{y_1-(1/6)}^{1} dy_2\,dy_1$$

$$= \int_0^{1/6}\left(y_1 + \tfrac{1}{6}\right) dy_1 + \int_{1/6}^{5/6}\left(\tfrac{1}{3}\right) dy_1 + \int_{5/6}^{1}\left[\left(\tfrac{7}{6}\right) - y_1\right] dy_1$$

$$= \left(\frac{y_1^2}{2} + \frac{y_1}{6}\right)\Big]_0^{1/6} + \left(\frac{y_1}{3}\right)\Big]_{1/6}^{5/6} + \left(7y_1 - \frac{y_1^2}{2}\right)\Big]_{5/6}^{1} = \frac{22}{72} = \frac{11}{36}.$$

5.122 **a.** $f(y_1, y_2) = \dfrac{\binom{4}{y_1}\binom{3}{y_2}\binom{2}{3-y_1-y_2}}{\binom{9}{3}}$ $\quad y_1 = 0, 1, 2, 3$
$y_2 = 0, 1, 2, 3$
$0 \le y_1 + y_2 \le 3$

b. $f(y_1) = \dfrac{\binom{4}{y_1}\binom{5}{3-y_1}}{\binom{9}{3}}$ $\quad y_1 = 0, 1, 2, 3$ $\qquad f(y_2) = \dfrac{\binom{3}{y_2}\binom{6}{3-y_2}}{\binom{9}{3}}$ $\quad y_2 = 0, 1, 2, 3$

c. $P(Y_1 = 1|Y_2 \ge 1) = \dfrac{p(1,1)+p(1,2)}{1-P(Y_2=0)} = \dfrac{\left(\frac{24}{84}+\frac{12}{84}\right)}{\left(1-\frac{20}{84}\right)} = \dfrac{36}{64} = \dfrac{9}{16}$

5.123 **a.** $f(y_1) = \int_0^{y_1} 3y_1\,dy_2 = 3y_1y_2]_0^{y_1} = 3y_1^2;\ 0 \le y_1 \le 1$

$f(y_2) = \int_{y_2}^{1} 3y_1\,dy_1 = \frac{3y_1^2}{2}\Big]_{y_2}^{1} = \left(\frac{3}{2}\right)(1 - y_2^2)\,;\ 0 \le y_2 \le 1$

b. $P\left(Y_1 \le \frac{3}{4} | Y_2 \le \frac{1}{2}\right) = \frac{P\left(Y_1 \le \frac{3}{4},\, Y_2 \le \frac{1}{2}\right)}{P\left(Y_2 \le \frac{1}{2}\right)}$

$$= \frac{\int_0^{1/2}\int_0^{y_1} 3y_1\,dy_2\,dy_1 + \int_{1/2}^{3/4}\int_0^{1/2} 3y_1\,dy_2\,dy_1}{\int_0^{1/2} \left(\frac{3}{2}\right)(1 - y_2^2)\,dy_2}$$

$$= \frac{\frac{1}{8} + \frac{15}{64}}{\frac{11}{16}} = \frac{23}{44}$$

c. If $0 \le y_2 < 1$, $f(y_1|y_2) = \frac{f(y_1, y_2)}{f(y_2)} = \frac{3y_1}{\left(\frac{3}{2}\right)(1 - y_2^2)} = \frac{2y_1}{(1 - y_2^2)} \quad y_2 \le y_1 \le 1$

d. $P\left(Y_1 \le \frac{3}{4} | Y_2 = \frac{1}{2}\right) = \int_{1/4}^{3/4} f\left(y_1|y_2 = \frac{1}{2}\right)dy_1 = \int_{1/4}^{3/4} \frac{2y_1}{\left(1 - \frac{1}{4}\right)}\,dy_1 = \frac{8y_1^2}{6}\Big]_{1/4}^{3/4} = \frac{2}{3}$

5.124

a. $E(Y_2|Y_1 = y_1) = \int_0^{y_1} y_2 f(y_2|y_1)\,dy_2$

Now, if $0 \le y_1 \le 1$, $f(y_2|y_1) = \frac{f(y_1, y_2)}{f(y_1)} = \frac{3y_1}{3y_1^2} = \frac{1}{y_1} \quad 0 \le y_2 \le y_1$

Thus, $\int_0^{y_1} y_2\left(\frac{1}{y_1}\right)dy_2 = \frac{y_2^2}{2y_1}\Big]_0^{y_1} = \frac{y_1}{2}$.

b. $E(Y_2) = E[E(Y_2|Y_1)] = E\left(\frac{Y_1}{2}\right) = \int_0^1 \left(\frac{y_1}{2}\right)3y_1^2\,dy_1 = \frac{3y_1^4}{8}\Big]_0^1 = \frac{3}{8}$.

c. $E(Y_2) = \int_0^1 y_2\left(\frac{3}{2}\right)(1 - y_2^2)\,dy_2 = \int_0^1 \left(\frac{3y_2}{2} - \frac{3y_2^3}{2}\right)dy_2 = \left(\frac{3y_2^2}{4} - \frac{3y_2^4}{8}\right)\Big]_0^1 = \frac{3}{8}$.

5.125

a. Since Y_1 and Y_2 are independent,

$f(y_1, y_2) = f(y_1)f(y_2) = \left(\frac{e^{-y_1/\beta}}{\beta}\right)\left(\frac{e^{-y_2/\beta}}{\beta}\right)$

$= \left(\frac{1}{\beta^2}\right)e^{-(y_1+y_2)/\beta};\ y_1 > 0,\ y_2 > 0.$

Note the similarity with Exercise 5.59.

b. $P(Y_1 + Y_2 \le a) = \int_0^{a}\int_0^{a-y_2} \left(\frac{1}{\beta^2}\right)e^{-(y_1+y_2)/\beta}\,dy_1\,dy_2$

$= \left(\frac{1}{\beta}\right)\int_0^a \left[1 - e^{-(a-y_2)/\beta}\right]e^{-y_2/\beta}\,dy_2$

$= -e^{-y_2/\beta}\Big]_0^a - \left(\frac{a}{\beta}\right)e^{-a/\beta}$

$= 1 - \left[1 + \left(\frac{a}{\beta}\right)\right]e^{-a/\beta} \quad a > 0$

5.126 Refer to Figure 5.13. The region in which $Y_1Y_2 \le .5$ is the shaded area of the figure. For ease of integration, calculate the area of <u>unshaded</u> portion and subtract from the total area, which is 1. Then

$P(Y_1Y_2 > .5) = 18\int_{.5}^{1}\int_{.5/y_2}^{1} (y_1 - y_1^2)\,y_2^2\,dy_1\,dy_2$

$= 18\int_{1/2}^{1}\left(\frac{1}{6}y_2^2 - \frac{1}{8} + \frac{1}{24y_2}\right)dy_2$

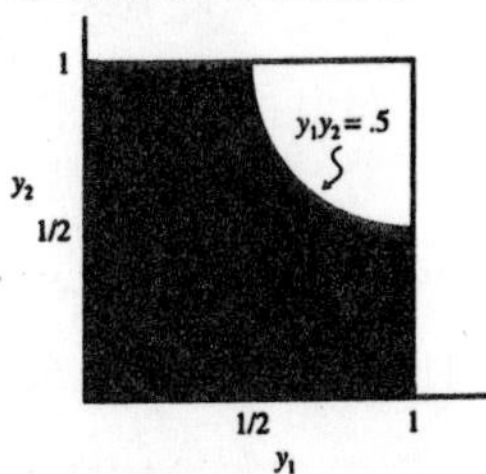

Figure 5.13

$= y_2^3 - \frac{9}{4}y_2 + \frac{3}{4}\ln y\Big]_{1/2}^{1} = -\frac{1}{4} + \frac{3}{4}\ln 2$

Then $P(Y_1Y_2 \le .5) = 1 + \frac{1}{4} - \frac{3}{4}\ln 2 = \frac{5}{4} - \frac{3}{4}\ln 2$.

5.127 a. Let N be the number of eggs laid by an insect, and let Y be the number of eggs hatched. Given that $N = n$ eggs were laid, Y has a binomial distribution with success probability p and n trials. Hence $E(Y|N = n) = pn$. Since N has a Poisson distribution with parameter λ, $E(Y) = E[E(Y|N)] = E(pN) = pE(N) = p\lambda$.

b. $V(Y|N = n) = np(1-p)$ and, from part **a.**, $E(Y|N = n) = np$.
$V(Y) = E[V[Y|N] + V[E(Y \mid N)] = E[Np(1-p)] + V[Np]$.
$E(N) = \lambda$ and $V(N) = \lambda$. Then
$V(Y) = p(1-p)E(N) + p^2V(N) = p\lambda - p^2\lambda + p^2\lambda = p\lambda$.

5.128 The conditional distribution of Y given $P = p$ is binomial with parameter p, and the marginal distribution of p is given.

a. $f(y) = \int_0^1 f(y, p) = \int_0^1 f(y|p)\, f(p)\, dp = \int_0^1 \binom{n}{y} p^y(1-p)^{n-y} 12p^2(1-p)\, dp$

if $y = 0, 1, 2, \ldots, n$

$= \binom{n}{y} 12 \int_0^1 p^{2+y}(1-p)^{n-y+1}\, dp$

The variable factor is that of a beta random variable with $\alpha = 3 + y$ and $\beta = n - y + 2$. Hence

$$p(y) = \frac{12\binom{n}{y}\Gamma(n-y+2)\Gamma(3+y)}{\Gamma(n+5)} \qquad y = 0, 1, \ldots, n$$

b. Since $p(y)$ is intractable, use Theorem 5.14. For $n = 2$, $E(Y|p) = 2p$. Hence

$E(Y) = E[E(Y|P)] = E(2P) = 2E(P) = 2(12) \int_0^1 p^3(1-p)dp = \frac{24\Gamma(4)\Gamma(2)}{\Gamma(6)} = \frac{24}{20}$

5.129 $f(y) = \int_0^\infty p(y|\lambda)f(\lambda)\, d\lambda = \int_0^\infty \frac{\lambda^y e^{-\lambda}}{y!} \times \frac{1}{\Gamma(\alpha)\beta^\alpha} e^{-\lambda/\beta}\lambda^{\alpha-1}\, d\lambda$

$= \int_0^\infty \frac{\lambda^{y+\alpha-1}e^{-\lambda[(\beta+1)/\beta]}}{\Gamma(y+1)\Gamma(\alpha)\beta^\alpha}\, d\lambda \quad y = 0, 1, 2, \ldots$

$= \frac{\Gamma(y+\alpha)\left(\frac{\beta}{\beta+1}\right)^{y+\alpha}}{\beta^\alpha\Gamma(y+1)\Gamma(\alpha)} = \binom{y+\alpha-1}{y}\left(\frac{\beta}{\beta+1}\right)^y\left(\frac{\beta}{\beta+1}\right)^\alpha\left(\frac{1}{\beta}\right)^\alpha$

$= \binom{y+\alpha-1}{y}\left(\frac{\beta}{\beta+1}\right)^y\left(\frac{1}{\beta+1}\right)^\alpha = \frac{\Gamma(y+\alpha)}{\Gamma(y+1)\Gamma(\alpha)}\left(\frac{\beta}{\beta+1}\right)^y\left(\frac{1}{\beta+1}\right)^\alpha,$

$y = 0, 1, 2, \ldots$

5.130 We have

$$X_1 = \begin{cases} 1, & \text{if the toss results in a head} \\ 0, & \text{if the toss results in a tail} \end{cases}$$

Hence the probability distribution for X_1 is as shown in the accompanying table. Then

x_i	$p(x_i)$
1	p
0	q

$E(X_i) = 1 \cdot p + 0 \cdot q = p$
$V(X_i) = 1^2 \cdot p + 0^2 \cdot q - p^2 = p - p^2 = pq$

Since X_i and X_j are independent, $\text{Cov}(X_i, X_j) = 0$ for all pairs $i \neq j$, and

$$E(Y) = E\left(\sum_{i=1}^n X_i\right) = \sum_{i=1}^n E(X_i) = np$$

$$V(Y) = V\left(\sum_{i=1}^n X_i\right) = \sum_{i=1}^n V(X_i) = npq$$

(which are exactly what we would expect to get).

5.131 The mean and variance of the geometric random variable W_i are given in Table 3.2 as

$$E(W_i) = \frac{1}{p} \qquad \text{and} \qquad V(W_i) = \frac{q}{p^2}$$

Since W_i and W_j are independent, $\text{Cov}(W_i, W_j) = 0$ for $i \neq j$, and

$$E(Y) = E\left(\sum_{i=1}^r W_i\right) = \sum_{i=1}^r E(W_i) = \frac{r}{p}$$

$$V(Y) = E\left(\sum_{i=1}^{r} W_i\right) = \sum_{i=1}^{r} V(W_i) = \frac{rq}{p^2}$$

5.132 The joint and marginal probabilities may be written down directly.

$P(X_1 = 1) = P(\text{ball 2 or 1}) = \frac{1}{2}$ $\quad P(X_1 = 0) = \frac{1}{2}$

$P(X_2 = 1) = \frac{1}{2}$ $\quad P(X_2 = 0) = \frac{1}{2}$

$P(X_3 = 1) = P(X_3 = 0) = \frac{1}{2}$

$P(X_1 = 1, X_2 = 1) = P(\text{ball 1}) = \frac{1}{4}$ $\quad P(X_1 = 1, X_2 = 0) = P(\text{ball 2}) = \frac{1}{4}$

$P(X_1 = 0, X_2 = 1) = P(\text{ball 3}) = \frac{1}{4}$ $\quad P(X_1 = 0, X_2 = 0) = P(\text{ball 4}) = \frac{1}{4}$

The joint distributions, $p_{13}(x_1, x_3)$ and $p_{23}(x_2, x_3)$, are identical. Note that X_i and X_j are pairwise independent since $p_{ij}(x_i, x_j) = p_i(x_i)p_j(x_j)$ for all $i \neq j$. However,

$P(X_1 = 1, X_2 = 1, X_3 = 1) = P(\text{ball 1}) = \frac{1}{4}$

$P(X_1 = 1, X_2 = 1, X_3 = 1) \neq P(X_1 = 1) \cdot P(X_2 = 1) \cdot P9X_3 = 1)$

Thus X_1, X_2, and X_3 are not mutually independent.

5.133 $E\left(\overline{Y} - \overline{X}\right) = E\left(\frac{\Sigma Y_i}{n} - \frac{\Sigma X_i}{m}\right) = \mu_1 - \mu_2$

$V\left(\overline{Y} - \overline{X}\right) = V\left(\overline{Y}\right) + V\left(\overline{X}\right) = \frac{n\sigma_1^2}{n^2} + \frac{m\sigma_2^2}{m^2} = \frac{\sigma_1^2}{n} + \frac{\sigma_2^2}{m}$ since $\overline{Y}$ and $\overline{X}$ are independent.

5.134 The joint moment-generating function of X_1, X_2, and X_3 is

$m(t_1, t_2, t_3) = E\left(e^{t_1X_1 + t_2X_2 + t_3X_3}\right)$.

a. If $t_1 = t_2 = t_3 = t$, $m(t, t, t) = E\left[e^{t(X_1+X_2+X_3)}\right]$, which is, by definition, the moment-generating function of the random variable $X_1 + X_2 + X_3$.

b. Similarly, $m(t, t, 0) = E\left[e^{t(X_1+X_2)}\right]$, which is the moment-generating function of $X_1 + X_2$.

c. Let X_1, X_2, and X_3 be continuous random variables with joint density function $f(x_1, x_2, x_3)$. Then

$$m(t_1, t_2, t_3) = \int_{-\infty}^{\infty}\int_{-\infty}^{\infty}\int_{-\infty}^{\infty} e^{t_1x_1+t_2x_2+t_3x_3} f(x_1, x_2, x_3)\, dx_1\, dx_2\, dx_3$$

Differentiating k_1 times with respect to t_1, we have

$$\frac{\partial^{k_1} m(t_1, t_2, t_3)}{\partial^{k_1} t_1} = \int\int\int_{-\infty}^{\infty} x_1^{k_1} e^{t_1x_1+t_2x_2+t_3x_3} f(x_1, x_2, x_3)\, dx_1\, dx_2\, dx_3$$

(Interchanging the derivative and the integral can be justified.)

Similarly, differentiating with respect to t_2 and t_3, we have

$$\left.\frac{\partial^{k_1+k_2+k_3} m(t_1, t_2, t_3)}{\partial^{k_1} t_1 \partial^{k_2} t_2 \partial^{k_3} t_3}\right|_{t_1=t_2=t_3=0} = \int\int\int_{-\infty}^{\infty} x_1^{k_1} x_2^{k_2} x_3^{k_3} f(x_1, x_2, x_3)\, dx_1\, dx_2\, dx_3$$

$$= E\left[X_1^{k_1} X_2^{k_2} X_3^{k_3}\right]$$

The proof is similar for the discrete case.

5.135 **a.** $m(t_1, t_2, t_3) = \sum_{x_1}\sum_{x_2}\sum_{x_3} \frac{n!}{x_1!\,x_2!\,x_3!} e^{t_1x_1+t_2x_2+t_3x_3} p_1^{x_1} p_2^{x_2} p_3^{x_3}$

$= \sum_{x_1}\sum_{x_2}\sum_{x_3} \frac{n!}{x_1!\,x_2!\,x_3!} \left(p_1e^{t_1}\right)^{x_1} \left(p_2e^{t_2}\right)^{x_2} \left(p_3e^{t_3}\right)^{x_3}$

$= \left(p_1e^{t_1} + p_2e^{t_2} + p_3e^{t_3}\right)^n$

$(\Sigma x_i = n)$, using the multinomial theorem.

b. The moment-generating function of X_1 is given by $m(t, 0, 0) = \left(p_1e^t + p_2 + p_3\right)^n$, which is the moment-generating function of a binomial random variable with $p = p_1$ and $q = p_2 + p_3$. Hence, by Theorem 6.1, X_1 must be binomial with parameter p_1.

c. $E(X_1) = \left.\frac{\partial m(t_1, t_2, t_3)}{\partial t_1}\right|_{t_1=t_2=t_3=0} = n\left(p_1e^{t_1} + p_2e^{t_2} + p_3e^{t_3}\right)^{n-1}\left.\left(p_1e^{t_1}\right)\right|_{t_i=0}$

$= np_1$

Similarly,

$$E(X_2) = \left.\frac{\partial m(t_1, t_2, t_3)}{\partial t_2}\right|_{t_1=0} = n\left(\sum_{i=1}^{3} p_ie^{t_i}\right)^{n-1} \left.\left(p_2e^{t_2}\right)\right|_{t_i=0} = np_2$$

Also,

$$E(X_1X_2) = \left.\frac{\partial^2 m(t_1, t_2, t_3)}{\partial t_1 \partial t_2}\right|_{t_1=0} = n(n-1)\left(\sum_{i=1}^{3} p_ie^{t_i}\right)^{n-2} \left. p_1e^{t_1}p_2e^{t_2}\right|_{t_i=0}$$

$= n(n-1)p_1p_2$

so that $\mathrm{Cov}(X_1, X_2) = -np_1p_2$.

5.136 The joint distribution of Y_1, Y_2, and Y_3 is an extension of the hypergeometric distribution of Chapter 3. As in Exercise 5.3, the joint distribution of Y_1 and Y_2 is given as

$$P(Y_1 = y_1, Y_2 = y_2) = \frac{\binom{N_1}{y_1}\binom{N_2}{y_2}\binom{N_3}{y_3}}{\binom{N}{n}} = \frac{\binom{Np_1}{y_1}\binom{Np_2}{y_2}\binom{Np_3}{y_3}}{\binom{N}{n}}$$

with $y_1 + y_2 + y_3 = n$, $N_1 + N_2 + N_3 = N$, and $p_i = \frac{N_i}{N}$ for $i = 1, 2, 3$. The marginal distributions of Y_1 and Y_2 are hypergeometric of the form

$$P(Y_1 = y_1) = \frac{\binom{Np_1}{y_1}\binom{N-Np_1}{n-y_1}}{\binom{N}{n}}$$

$$P(Y_2 = y_2) = \frac{\binom{Np_2}{y_2}\binom{N-Np_2}{n-y_2}}{\binom{N}{n}}$$

Hence, from Example 5.29 in the text, we obtain

$$E(Y_1) = n\left(\frac{N_1}{N}\right) = np_1$$

$$E(Y_2) = n\left(\frac{N_2}{N}\right) = np_2$$

$$V(Y_1) = n\left(\frac{N_1}{N}\right)\left(1 - \frac{N_1}{N}\right)\left(\frac{N-n}{N-1}\right) = np_1(1-p_1)\left(\frac{N-n}{N-1}\right)$$

$$V(Y_2) = np_2(1-p_2)\left(\frac{N-n}{N-1}\right)$$

Then

$$E(Y_1Y_2) = \sum_{y_1=0}^{Np_1}\sum_{y_2=0}^{Np_2} \frac{y_1y_2\binom{Np_1}{y_1}\binom{Np_2}{y_2}\binom{Np_3}{y_3}}{\binom{N}{n}}$$

$$= \frac{n(n-1)N^2p_1p_2}{N(N-1)}\sum_{y_1=1}^{Np_1}\sum_{y_2=1}^{Np_2}\frac{\binom{Np_1-1}{y_1-1}\binom{Np_2-1}{y_2-1}\binom{Np_3}{y_3}}{\binom{N-2}{n-2}}$$

$$= \frac{n(n-1)N^2p_1p_2}{N(N-1)}\sum_{z_1=0}^{Np_1-1}\sum_{z_2=0}^{Np_2-1}\frac{\binom{Np_1-1}{z_1}\binom{Np_2-1}{z_2}\binom{Np_3}{y_3}}{\binom{N-2}{n-2}}$$

$$= \frac{nNp_1p_2(n-1)}{N-1}$$

The double sum equals one, being that of an extended hypergeometric variable with $z_1 + z_2 + y_3 = n - 2$ and $(Np_1 - 1) + (Np_2 - 1) + (Np_3) = N - 2$. Now

$$\mathrm{Cov}(Y_1, Y_2) = \frac{nNp_1p_2(n-1) - n^2p_1p_2(N-1)}{N-1} = -p_1p_2n\left(\frac{N-n}{N-1}\right)$$

and

$$\rho = \frac{\mathrm{Cov}(Y_1, Y_2)}{\sqrt{V(Y_1)V(Y_2)}} = \frac{-p_1p_2n\left(\frac{N-n}{N-1}\right)}{n\sqrt{p_1p_2(1-p_1)(1-p_2)}} = -\sqrt{\frac{p_1p_2}{(1-p_1)(1-p_2)}}$$

5.137 **a.** For this exercise a quadratic form of interest is

$$At^2 + Bt + C = E\left(Y_1^2\right)t^2 + [-2E(Y_1Y_2)]t + E(Y_2)^2$$

Since $E\left[(ty_1 - Y_2)^2\right]$ is the integral of a nonnegative quantity, it must be nonnegative, so that we must have $At^2 + Bt + C \geq 0$. In order to satisfy this inequality, the quadratic form must not dip below the horizontal axis in Figure 5.14. That is, the two roots of the equation $At^2 + Bt + C$ must either be imaginary (a) or equal (b). In terms of the discriminant, then,

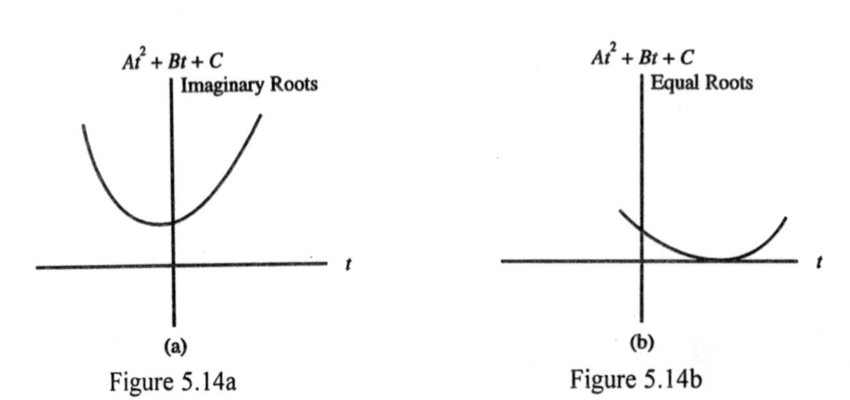

Figure 5.14a

Figure 5.14b

$$B^2 - 4AC \leq 0$$
$$[-2E(Y_1Y_2)]^2 - 4E\left(Y_1^2\right)E\left(Y_2^2\right) \leq 0$$

and thus,

$$[E(Y_1Y_2)]^2 \leq E\left(Y_1^2\right)E\left(Y_2^2\right)$$

b. Consider

$$\rho^2 = \frac{[E(Y_1-\mu_1)(Y_2-\mu_2)]^2}{E(Y_1-\mu_1)^2E(Y_2-\mu_2)^2} = \frac{[E(Z_1Z_2)]^2}{E\left(Z_1^2\right)E\left(Z_2^2\right)} \quad \text{where } Z_i = Y_i - \mu_i \quad i = 1, 2$$

Using part (a), we know

$$[E(Z_1Z_2)]^1 \leq E\left(Z_1^2\right)E\left(Z_2^2\right) \qquad \text{or} \qquad \rho^2 = \frac{[E(Z_1Z_2)]^2}{E\left(Z_1^2\right)E\left(Z_2^2\right)} \leq 1$$

CHAPTER 6 FUNCTIONS OF RANDOM VARIABLES

6.1 **a.** Using the distribution function approach, we write

$$F_{U_1}(u) = P(U_1 \le u) = P(2Y - 1 \le u) = P\left(Y \le \tfrac{u+1}{2}\right) = F_Y\left(\tfrac{u+1}{2}\right)$$

Now

$$F_Y(y) = \int_0^y (2 - 2t)\,dt = 2y - y^2, \qquad \text{for } 0 \le y \le 1$$
$$F_Y(y) = 0, \qquad \text{for } y < 0$$
$$F_Y(y) = 1, \qquad \text{for } y > 1$$

Since $F_{U_1}(u) = F_Y\left(\frac{u+1}{2}\right)$, the distribution function can be written as

$$F_{U_1}(u) = \begin{cases} 0, & \frac{u+1}{2} < 0 \\ 2\left(\frac{u+1}{2}\right) - \left(\frac{u+1}{2}\right)^2, & 0 \le \frac{u+1}{2} \le 1 \\ 1, & \frac{u+1}{2} > 1 \end{cases}$$

The density function can be obtained by differentiating $F_{U_1}(u)$ with respect to u. Thus,

$$f_{U_1}(u) = -\tfrac{u}{2} + \tfrac{1}{2} = \tfrac{1-u}{2}, \qquad -1 \le u \le 1$$
$$= 0, \qquad \text{elsewhere}$$

b. Similar to part **a.**

$$F_{U_2}(u) = P(1 - 2Y \le u) = P\left(Y \ge \tfrac{1-u}{2}\right) = 1 - F_Y\left(\tfrac{1-u}{2}\right)$$

$F_Y(y)$ was given in part **a.**, so that

$$F_{U_2}(u) = \begin{cases} 1 - 0, & \frac{1-u}{2} \le 0 \\ 1 - 2\left(\frac{1-u}{2}\right) + \left(\frac{1-u}{2}\right)^2, & 0 \le \frac{1-u}{2} \le 1 \\ 1 - 1, & \frac{1-u}{2} > 1 \end{cases}$$

or

$$F_{U_2}(u) = \begin{cases} 0, & u < -1 \\ \frac{1+2u+u^2}{4}, & -1 \le u \le 1 \\ 1, & u > 1 \end{cases}$$

By differentiating $F_{U_2}(u)$, we obtain

$$f_{U_2}(u) = \tfrac{1}{2} + \tfrac{u}{2} = \tfrac{u+1}{2}, \qquad -1 \le u \le 1$$
$$= 0, \qquad \text{elsewhere}$$

c. Write

$$F_{U_3}(u) = P(U \le u) = P(Y^2 \le u) = P\left(Y \le \sqrt{u}\right) = F_Y\left(\sqrt{u}\right)$$

since Y is defined only on the range $0 \le y \le 1$. Hence

$$F_{U_3}(u) = \begin{cases} 0, & u < 0 \\ 2\sqrt{u} - u, & 0 \le u \le 1 \\ 1, & u > 1 \end{cases}$$

and

$$f_{U_3}(u) = \tfrac{1}{\sqrt{u}} - 1 = \tfrac{1-\sqrt{u}}{\sqrt{u}}, \qquad 0 \le u \le 1$$
$$= 0, \qquad \text{elsewhere}$$

d. Using the derived density functions, we have

$$E(U_1) = \int_{-\infty}^{\infty} u f_{U_1}(u)\,du = \int_{-1}^{1} u\left(\tfrac{1-u}{2}\right)du = \left[\tfrac{1}{4}u^2 - \tfrac{1}{6}u^3\right]_{-1}^{1} = \tfrac{1}{4} - \tfrac{1}{6} - \tfrac{1}{4} - \tfrac{1}{6} = -\tfrac{1}{3}$$

$$E(U_2) = \int_{-\infty}^{\infty} u f_{U_2}(u)\,du = \int_{-1}^{1}\left(\tfrac{u^2}{2} + \tfrac{1}{2}u\right)du = \left[\tfrac{u^3}{6} + \tfrac{1}{4}u^2\right]_{-1}^{1} = \tfrac{1}{6} + \tfrac{1}{4} + \tfrac{1}{6} - \tfrac{1}{4} = \tfrac{1}{3}$$

$$E(U_3) = \int_{-\infty}^{\infty} u f_{U_3}(u)\,du = \int_0^1\left(\tfrac{u - u\sqrt{u}}{\sqrt{u}}\right)du = \int_0^1\left(\sqrt{u} - u\right)du = \left[\tfrac{2}{3}u^{3/2} - \tfrac{1}{2}u^2\right]_0^1 = \tfrac{1}{6}.$$

e. Calculate

$$E(Y) = \int_0^1 (2 - 2y)y\,dy = \left[y^2 - \tfrac{2}{3}y^3\right]_0^1 = \tfrac{1}{3}$$

$$E\left(Y^2\right) = \int_0^1 (2 - 2y)y^2\,dy = \left[\tfrac{2}{3}y^3 - \tfrac{2}{4}y^4\right]_0^1 = \tfrac{2}{12} = \tfrac{1}{6}.$$

Then

$$E(U_1) = 2E(Y) - 1 = 2\left(\tfrac{1}{3}\right) - 1 = -\tfrac{1}{3}$$
$$E(U_2) = 1 - 2E(Y) = 1 - 2\left(\tfrac{1}{3}\right) = \tfrac{1}{3}$$
$$E(U_3) = E\left(Y^2\right) = \tfrac{1}{6}.$$

6.2 Calculate

$$F_Y(y) = \int_{-1}^{y} \tfrac{3}{2} t^2\, dt = \tfrac{1}{2}\left(y^3 + 1\right)$$

for $-1 \le y \le 1$

a. $F_{U_1}(u) = P(3Y \le u) = P\left(Y \le \frac{u}{3}\right) = F_Y\left(\frac{u}{3}\right) = \frac{1}{2}\left(\frac{u^3}{27} + 1\right)$ for $-3 \le u \le 3$

Differentiating with respect to u, we have

$$f_{U_1}(u) = \begin{cases} \frac{u^2}{18}, & -3 \le u \le 3 \\ 0, & \text{elsewhere} \end{cases}$$

b. $F_{U_2}(u) = P(3 - Y \le u) = P(Y \ge 3 - u) = 1 - F_Y(3-u) = \frac{1}{2}\left[1 - (3-u)^3\right]$ for $-1 \le (3-u) \le 1$ or $2 \le u \le 4$. Differentiating with respect to u, we have

$$f_{U_2}(u) = \begin{cases} \left(\frac{3}{2}\right)(3-u)^2, & 2 \le u \le 4 \\ 0, & \text{elsewhere} \end{cases}$$

c. $F_{U_3}(u) = P\left(Y^2 \le u\right) = P\left(-\sqrt{u} \le Y \le \sqrt{u}\right) = F_Y\left(\sqrt{u}\right) - F_Y\left(-\sqrt{u}\right)$
$= \frac{1}{2} u^{3/2} + \frac{1}{2} u^{3/2} = u^{3/2}$ for $0 \le y^2 \le 1$ or $0 \le u \le 1$.

Hence

$$f_{U_3}(u) = \tfrac{3}{2} u^{1/2} = \tfrac{3}{2}\sqrt{u}$$

for $0 \le u \le 1$

6.3 Similar to Exercise 6.2. Calculate

$$F_Y(y) = \begin{cases} \frac{y^2}{2}, & 0 \le y \le 1 \\ \left(\frac{1}{2}\right) + (y-1), & 1 \le y \le 1.5 \\ 1, & y \ge 1.5 \end{cases}.$$

a. $F_U(u) = P(10Y - 4 \le u) = P\left(Y \le \frac{u+4}{10}\right) = F_Y\left(\frac{u+4}{10}\right)$

Hence

$$F_U(u) = \begin{cases} \frac{(u+4)^2}{200}, & -4 \le u \le 6 \\ \left(\frac{1}{2}\right) + \frac{u-6}{10}, & 6 \le u \le 11 \\ 1, & u > 11 \end{cases}.$$

Differentiating with respect to u, we have

$$f_U(u) = \begin{cases} \frac{u+4}{100}, & -4 \le u \le 6 \\ \frac{1}{10}, & 6 \le u \le 11 \\ 0, & \text{elsewhere} \end{cases}.$$

b. $E(U) = \int_{-4}^{6} \frac{u^2+4u}{100}\, du + \int_{6}^{11} \frac{u}{10}\, du = \left[\frac{u^3}{300} + \frac{u^2}{50}\right]_{-4}^{6} + \left[\frac{u^2}{20}\right]_{6}^{11} = 5.583$

c. Using an alternative procedure, calculate

$$E(Y) = \int_0^1 y^2\, dy + \int_1^{1.5} y\, dy = \tfrac{1}{3} + \left[\tfrac{y^2}{2}\right]_1^{1.5} = \tfrac{23}{24}.$$

Then $E(U) = 10\left(\frac{23}{24}\right) - 4 = 5.58$.

6.4 It is given that $f(y) = \left(\frac{1}{4}\right) e^{-y/4}$ for $y \ge 0$. Then

$$F_Y(y) = \int_0^y \tfrac{1}{4} e^{-t/4}\, dt = 1 - e^{-y/4}, \qquad y > 0$$

a. $F_U(u) = P(3Y + 1 \le u) = P\left(Y \le \frac{u-1}{3}\right) = 1 - e^{-(u-1)/12}$ for $u \ge 1$

Finally,

$$f_U(u) = \tfrac{d}{du} F_U(u) = \begin{cases} \left(\frac{1}{12}\right) e^{-(u-1)/12}, & u \ge 1 \\ 0, & \text{elsewhere} \end{cases}.$$

b. $E(U) = \int_1^\infty \frac{u}{12} e^{-(u-1)/12}\, du = e^{12} \int_1^\infty \frac{u}{12} e^{-u/12} = 12e^{12} \int_{1/12}^\infty ze^{-z}\, dz = 13$

where $z = u/12$ and we used integration by parts to evaluate the integral.

6.5

It is given that $f(y) = \frac{1}{4}$ for $1 \le y \le 5$. Then

$F_Y(y) = \frac{y}{4}$ for $1 \le y \le 5$

$F_U(u) = P\left(2Y^2 + 3 \le u\right) = P\left[Y \le \sqrt{\frac{u-3}{2}}\right]$ for $1 \le \sqrt{\frac{u-3}{2}} \le 5$

$= \frac{1}{4}\sqrt{\frac{u-3}{2}}$ for $5 \le u \le 53$.

Differentiating, we have

$$f_U(u) = \begin{cases} \frac{1}{16}\left(\frac{u-3}{2}\right)^{-1/2} = \frac{1}{8\sqrt{2(u-3)}}, & 5 \le u \le 53 \\ 0, & \text{elsewhere} \end{cases}$$

6.6

a. Consider $U = Y_1 - Y_2$, where Y_1 and Y_2 lie in the large triangle shown in Figure 6.1. $F_U(u)$ can be found directly by integrating the joint density over various regions, which will change depending on the value taken by u. Note that the line $Y_1 - Y_2 = u$ or $Y_1 = Y_2 + u$ is a set of parallel lines having intercept u on the Y_1 axis.

Figure 6.1

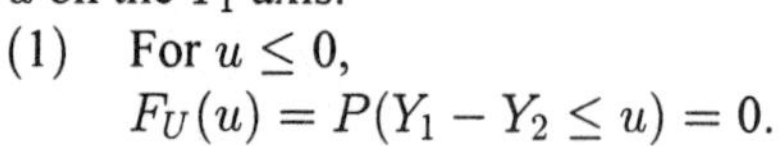

(1) For $u \le 0$,
$F_U(u) = P(Y_1 - Y_2 \le u) = 0$.

(2) For $0 \le u \le 1$,

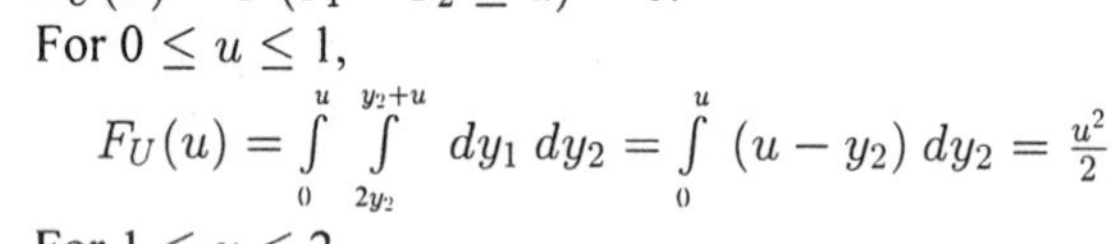

$$F_U(u) = \int_0^u \int_{2y_2}^{y_2+u} dy_1\, dy_2 = \int_0^u (u - y_2)\, dy_2 = \frac{u^2}{2}$$

(3) For $1 \le u \le 2$,

$$F_U(u) = 1 - \int_0^{2-u} \int_{y_2+u}^{2} dy_1\, dy_2 = 1 - \int_0^{2-u} (2 - y_2 - u)\, dy_2 = 1 - \frac{(2-u)^2}{2}$$

(4) For $u > 2$, $F_U(u) = 1$.

Differentiating with respect to u, we have

$$f_U(u) = \begin{cases} u, & 0 \le u \le 1 \\ 2 - u, & 1 \le u \le 2 \\ 0, & \text{elsewhere} \end{cases}$$

b. $E(U) = \int_0^1 u^2\, du + \int_1^2 (2u - u^2)\, du = \frac{1}{3} + \left[u^2 - \frac{u^3}{3}\right]_1^2 = 1$

6.7

The region on which the density is positive is shown in Figure 6.2. This triangle has area $\frac{1}{2}$, so that the uniform density will be

$f(y_1, y_2) = 2 \qquad 0 \le y_1 \le 1, 0 \le y_2 \le 1, 0 \le y_1 + y_2 \ge 1$

a. Let $Y_1 + Y_2 = u$, where $0 \le u \le 1$.

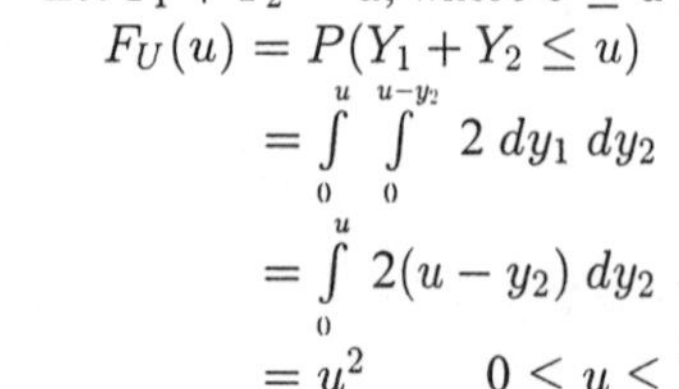

$$F_U(u) = P(Y_1 + Y_2 \le u) = \int_0^u \int_0^{u-y_2} 2\, dy_1\, dy_2 = \int_0^u 2(u - y_2)\, dy_2 = u^2 \qquad 0 \le u \le 1$$

Differentiating, we have

$$f_U(u) = \begin{cases} 2u, & 0 \le u \le 1 \\ 0, & \text{elsewhere} \end{cases}$$

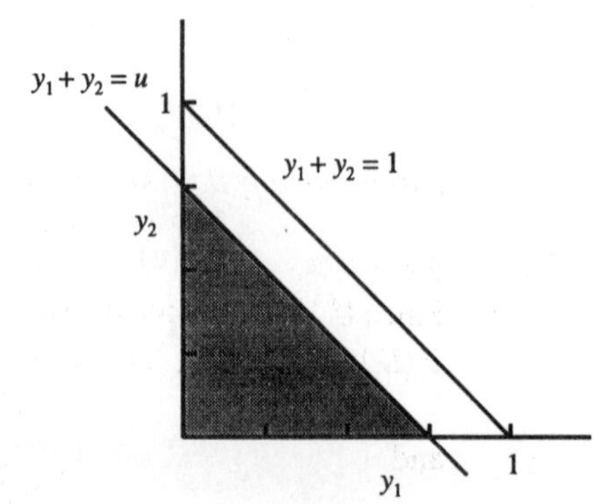

Figure 6.2

b. $E(U) = \int_0^1 2u^2\,du = = \frac{2}{3}u^3\Big]_0^1 = \frac{2}{3}$

c. We must first obtain the marginal densities of Y_1 and Y_2. Integrating over the ranges shown in Figure 6.2, we have

$$f_1(y_1) = \int_0^{1-y_1} 2\,dy_2 = 2(1-y_1) \qquad 0 \le y_1 \le 1$$

$$f_2(y_2) = \int_0^{1-y_2} 2\,dy_1 = 2(1-y_2) \qquad 0 \le y_2 \le 1$$

The marginal densities are identical. Then

$$E(Y_i) = \int_0^1 2y_i(1-y_i)\,dy_i = \left[y_i^2 - \tfrac{2}{3}y_i^3\right]_0^1 = \tfrac{1}{3},$$

and $E(Y_1 + Y_2) = E(Y_1) + E(Y_2) = \frac{2}{3}$.

6.8 The region for which $Y_1 - Y_2 \le u$ is shown in Figure 6.3 over the region for which Y_1 and Y_2 are positive and $y_2 \le y_1$.

a. $F_U(u) = P(Y_1 - Y_2 \le u)$

$$= \int_0^{\infty}\int_{y_2}^{y_2+u} e^{-y_1}\,dy_1\,dy_2$$

for $u \ge 0$. Then $f_U(u) = e^{-u}$ for $u \ge 0$, which is a gamma density with $\alpha = \beta = 1$.

b. $E(U) = \alpha\beta = 1$; $V(U) = \alpha\beta^2 = 1$.

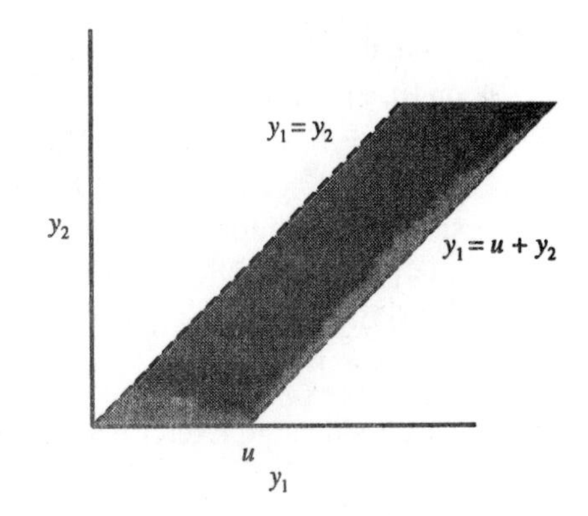

Figure 6.3

6.9 It is given that $f(y_i) = e^{-y_i}$ for $i = 1, 2$. Since Y_1 and Y_2 are independent, $f(y_1, y_2) = e^{-(y_1+y_2)}$ for $y_1 \ge 0$, $y_2 \ge 0$.

a. Let $U = \frac{Y_1+Y_2}{2}$. Then (see Figure 6.4)

$$\begin{aligned} F_U(u) &= P\left(\tfrac{Y_1+Y_2}{2} \le u\right) \\ &= P(Y_1 + Y_2 \le 2u) \\ &= \int_0^{2u}\int_0^{2u-y_2} e^{-(y_1+y_2)}\,dy_1\,dy_2 \\ &= \int_0^{2u} (e^{-y_2} - e^{-2u})\,dy_2 \\ &= 1 - e^{-2u} - 2ue^{-2u} \qquad \text{for } u \ge 0 \end{aligned}$$

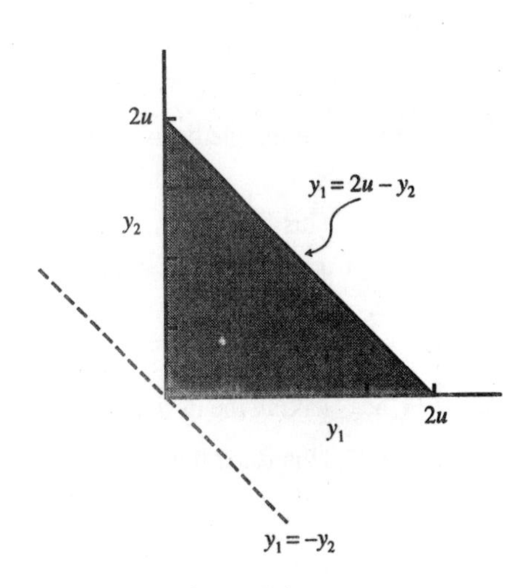

Figure 6.4

Differentiating with respect to u, we have

$$f_U(u) = 2e^{-2u} + 4ue^{-2u} - 2e^{-2u} = 4ue^{-2u}$$

for $\mu \ge 0$.

b. Since U has a gamma distribution with $\alpha = 2$, $\beta = \frac{1}{2}$, then $E(U) = \alpha\beta = 1$ and $V(U) = \alpha\beta^2 = \frac{1}{2}$. Using Theorem 5.12 with $E(Y_i) = V(Y_i) = 1$, we obtain

$$E(U) = \tfrac{1}{2}(1) + \tfrac{1}{2}(1) = 1$$

and

$$V(U) = \tfrac{1}{4}(1+1) = \tfrac{1}{2}.$$

6.10 If Y_1 and Y_2 are independent, then $f(y_1, y_2) = 18\,(y_1 - y_1^2)\,y_2^2$ for $0 \le y_1 \le 1, 0 \le y_2 \le 1$. Using the distribution function approach, we see that $F_U(u)$ is the unshaded area in Figure 6.5. Integrating the shaded area and subtracting from 1, we have

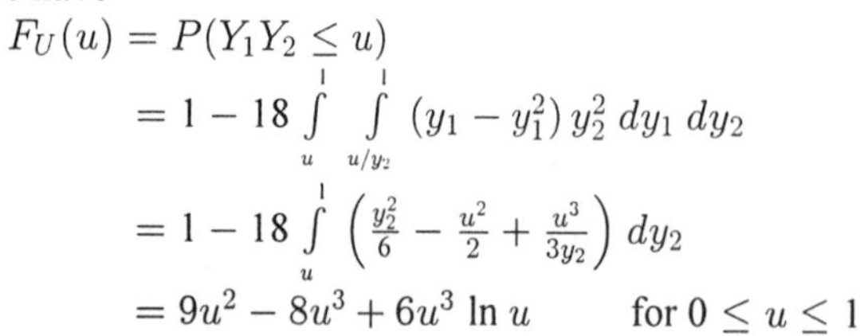

$$F_U(u) = P(Y_1Y_2 \le u)$$
$$= 1 - 18\int_u^1 \int_{u/y_2}^1 (y_1 - y_1^2)\,y_2^2\,dy_1\,dy_2$$
$$= 1 - 18\int_u^1 \left(\frac{y_2^2}{6} - \frac{u^2}{2} + \frac{u^3}{3y_2}\right) dy_2$$
$$= 9u^2 - 8u^3 + 6u^3 \ln u \qquad \text{for } 0 \le u \le 1$$

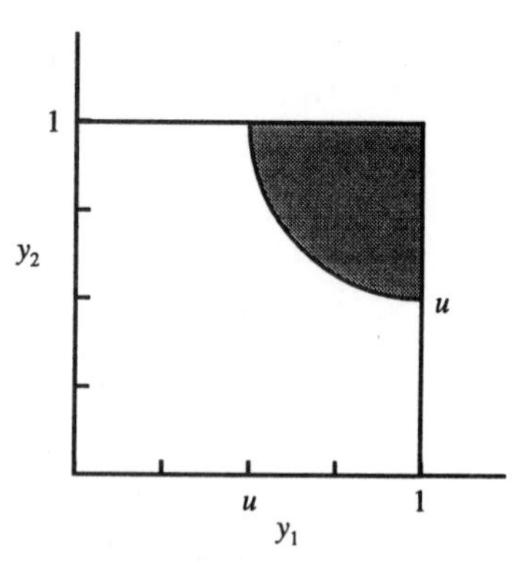

Figure 6.5

Differentiating with respect to u, we have

$$f_U(u) = 18u - 24u^2 + \frac{6u^3}{u} + 18u^2 \ln u = 18u(1 - u + u \ln u) \qquad \text{for } 0 \le u \le 1$$
$$= 0 \qquad \text{elsewhere}$$

6.11 Note that $F_Y(y) = 1 - e^{-y^2}, y \ge 0$. Suppose that U has a uniform distribution on [0, 1], in which case $P(U \le u) = u, 0 < u < 1$. We want to find $Y = G(U)$ such that $F_Y(y) = 1 - e^{-y^2}, w \ge 0$. Now, $G^{-1}(U) = Y$, and so

$$F_Y(y) = P(Y \le y) = P(G(U) \le y) = P\left(U \le G^{-1}(y)\right) = P(U \le u) = u, \quad 0 < u < 1$$

since U is uniform on [0, 1]. Thus, we want $u = 1 - e^{-y^2}$ so $-y^2 = \ln(1 - u)$ and $y = [-\ln(1-u)]^{1/2}$. So $G(U) = [-\ln(1-U)]^{1/2}$.
(NOTE: $F_Y(y) = P\left([-\ln(1-U)]^{1/2} \le y\right) = P\left(1 - U \ge e^{-y^2}\right) = P\left(U \le 1 - e^{-y^2}\right)$ $= 1 - e^{-y^2}, y > 0$.)

6.12 This is similar to Exercise 6.11. From Exercise 4.7, $f(y) = \frac{b}{y^2}, y \ge b$ so that

$$F_Y(y) = b\int_b^y t^{-2}\,dt = 1 - \tfrac{b}{y}, y \ge \text{b}.$$

Let $Y = G(U)$ so that $U = G^{-1}(Y)$

$$F_Y(y) = P(G(U) \le y) = P(U \le u) = u, 0 < u < 1.$$

We need $u = 1 - \frac{b}{y}$ or $y = \frac{b}{1-u}$. Thus, $G(U) = \frac{b}{1-U}$.

6.13
a. Taking the derivative of F, we have

$$f(y) = \frac{\alpha y^{\alpha-1}}{\theta^\alpha}, 0 \le y \le \theta$$

b. This is similar to Exercise 6.11. Let $W = G(U)$. We need $F_Y(y) = u = \left(\frac{y}{\theta}\right)^\alpha$ or $y = \theta u^{1/\alpha}$. So, $G(U) = \theta U^{1/\alpha}$.

c. From part **b**, $y = 4\sqrt{u}$. The values are $4\sqrt{.2700} = 2.0785$, $4\sqrt{.6901} = 3.229$, $4\sqrt{.1413} = 1.5036$, $4\sqrt{.1523} = 1.5610$, $4\sqrt{.3609} = 2.403$.

6.14
a. Taking the derivative of F yields $f(y) = \alpha\beta^\alpha y^{-\alpha-1}, y \ge \beta$.

b. This is similar to Exercise 6.11. We need $F_Y(y) = u = 1 - \left(\frac{\beta}{y}\right)^\alpha$. Thus,

$$\left(\frac{\beta}{y}\right)^\alpha = 1 - u$$

and

$$y = \frac{\beta}{(1-u)^{1/\alpha}}.$$

So, $G(U) = \beta(1-U)^{-1/\alpha}$.

c. From part **b.**, $y = \frac{3}{\sqrt{1-u}}$. The values are $\frac{3}{\sqrt{1-.0058}} = 3.0087$, $\frac{3}{\sqrt{1-.2048}} = 3.3642$, $\frac{3}{\sqrt{1-.7692}} = 6.2446$, $\frac{3}{\sqrt{1-.2475}} = 3.4583$, $\frac{3}{\sqrt{1-.6078}} = 4.7904$.

6.15 From Exercise 6.14, Y is always positive, so that $X = \frac{1}{Y}$ is always positive. Thus, $F_X(x) = 0$ if $x < 0$. Consider $x \geq 0$.

$$\begin{aligned} F_X(x) &= P\left(\tfrac{1}{Y} \leq x\right) = P\left(Y \geq \tfrac{1}{x}\right) = 1 - F_Y\left(\tfrac{1}{x}\right) \\ &= 1 - \{1 - (\beta x)^\alpha\} \qquad \text{if } \tfrac{1}{x} \geq \beta \\ &= 1 \qquad \text{if } \tfrac{1}{x} < \beta \end{aligned}$$

Thus,

$$F_X(x) = \begin{cases} 0, & x < 0 \\ \left(\frac{x}{\beta^{-1}}\right)^\alpha, & 0 \leq x \leq \beta^{-1} \\ 1, & x > \beta^{-1} \end{cases}$$

so the result is shown.

6.16 **a.** $F_W(w) = P(W \leq w) = P(Y^2 \leq w) = P\left(-\sqrt{w} < Y < \sqrt{w}\right) = \int\limits_0^{\sqrt{w}} 1\, dy$

$$= \sqrt{w},\ 0 < w < 1$$

$$f(w) = \left(\tfrac{1}{2}\right) w^{-1/2}, 0 \leq w \leq 1$$

b. $F_W(w) = P(W \leq w) = P\left(\sqrt{Y} \leq w\right) = P\left(Y \leq w^2\right) = \int\limits_0^{w^2} 1\, dy$

$$= w^2,\ 0 < w < 1$$

$$f(w) = 2w, 0 \leq w \leq 1$$

6.17 By definition,

$$\begin{aligned} P(X = i) &= P[F(i-1) < U \leq F(i)] = P[U \leq F(i)] - P[U \leq F(i-1)] \\ &= F(i) - F(i-1), i = 1, 2, 3, \ldots \end{aligned}$$

since $P(U \leq a) = a$ for any $0 \leq a \leq 1$. From Exercise 4.1

$$P(Y = i) = F(i) - F(i-1), i = 1, 2, 3, \ldots .$$

Therefore X has the same distribution as Y.

6.18 Let Y be a geometrically distributed random variable with probability p. Then Y satisfies the conditions of the random variable Y in Exercise 6.17.
Let U and X be defined as in Exercise 6.17. Then

X $= k$ if and only if $F(k-1) < U \leq F(k)$, $k = 1, 2, 3,$

or, by Exercise 4.2(a),

$X = k$ if and only if $1 - q^{k-1} < U \leq 1 - q^k$
if and only if $q^k \leq 1 - U < q^{k-1}$
if and only if $k \log(q) \leq \log(1-u) < (k-1)\log(q)$
if and only if $k \geq \frac{\log(1-u)}{\log(q)} > k - 1$.

6.19 **a.** If $U = 2Y - 1$, then $Y = \frac{U+1}{2}$. Differentiating, we have $\frac{dy}{du} = \frac{1}{2}$ and

$$f_U(u) = \tfrac{1}{2}(2)\left(1 - \tfrac{u+1}{2}\right) = \tfrac{1-u}{2}$$

for $0 \leq \frac{u+1}{2} \leq 1$ or $-1 \leq u \leq 1$.

b. If $U = 1 - 2Y$, then $Y = \frac{1-U}{2}$, and $\frac{dy}{du} = \frac{1}{2}$. Then

$$f_U(u) = \tfrac{1}{2}(2)\left(1 - \tfrac{1-u}{2}\right) = \tfrac{1+u}{2}$$

for $0 \leq \frac{1-u}{2} \leq 1$ or $-1 \leq u \leq 1$.

c. If $U = Y^2$, then $Y = \sqrt{U}$ and $\frac{dy}{du} = \frac{1}{2\sqrt{U}}$. Then (since $u = y^2$ is increasing for $y > 0$)

$$f_U(u) = \tfrac{1}{2\sqrt{u}}(2)\left(1 - \sqrt{u}\right) = \tfrac{1-\sqrt{u}}{\sqrt{u}}$$

for $0 \leq \sqrt{u} \leq 1$ or $0 \leq u \leq 1$.

6.20 If $U = 3Y + 1$, then $Y = \frac{U-1}{3}$ and $\frac{dy}{du} = \frac{1}{3}$. Since $f_Y(y) = \frac{1}{4}e^{y/4}$ obtain

$$f_U(u) = \tfrac{1}{3} \times \tfrac{1}{4} \times e^{-(u-1)/12} = \tfrac{1}{12}\, e^{-(u-1)/12}$$

for $\frac{u-1}{3} \geq 0$ or $u \geq 1$.

6.21 The variable of interest is $U = \frac{Y_1+Y_2}{2}$. Fix $Y_2 = y_2$. Then $Y_1 = 2u - y_2$ and $\frac{dy_1}{du} = 2$. The joint density for U and Y_2 is $g(u, y_2) = 2e^{-2u}$, where $u \geq 0$, $y_2 \geq 0$, and

$2u - y_2 \geq 0$, or $y_2 \leq 2u$. Integrating over all possible values of y_2, we have

$$f_U(u) = \int_0^{2u} 2e^{-2u}\, dy_2 = 4ue^{-2u}$$

for $u \geq 0$.

6.22 **a.** Let $U = Y^m$. Then using the transformation approach, we have

$$Y = U^{1/m}$$

and

$$\frac{dy}{du} = \frac{1}{m}\, u^{(1-m)/m} = \frac{1}{m} u^{-(m-1)/m}$$

so that

$$g_U(u) = \frac{1}{\alpha}\, m\left(u^{1/m}\right)^{m-1} e^{-u/\alpha} \times \frac{1}{m}\, u^{-(m-1)/m} = \frac{1}{\alpha}\, e^{-u/\alpha}$$

for $u > 0$.

b. $E\left(Y^k\right) = E\left(U^{k/m}\right) = \int_0^{\infty} \frac{u^{k/m} e^{-u/\alpha}}{\alpha}\, du = \frac{1}{\alpha}\Gamma\left(\frac{k}{m}+1\right)\alpha^{(k/m)+1} = \Gamma\left(\frac{k}{m}+1\right)\alpha^{k/m}$

Note that the integrand is the density (except for constants) of a gamma variable with parameters $\left(\frac{k}{m}\right) + 1$ and α, so that integration can be done by choosing the necessary constants.

6.23 $f_Y = \left(\frac{1}{\beta}\right) e^{-y/\beta}$

a. $F_w(w) = P(W \leq w) = P\left(Y \leq w^2\right)$

$f_w(w) = f_Y\left(w^2\right)(2w) = \left(\frac{2w}{\beta}\right) e^{-w^2/\beta}$ for $w > 0$

which is Weibull with $\alpha = \beta$ and $m = 2$.

b. In Exercise 6.22, part **b**, it was shown that

$$E\left(Y^k\right) = \Gamma\left(1+\frac{k}{m}\right)\alpha^{k/m}.$$

Thus,

$$E\left(Y^{k/2}\right) = E\left(W^k\right) = \Gamma\left(1+\frac{k}{2}\right)\beta^{k/2}.$$

6.24 $f(y) = 1$ for $0 \leq y \leq 1$

$U = -2 \ln Y$; solving for y,

$\ln y = -\frac{u}{2}$ which gives $y = e^{-u/2}$.

$\frac{dy}{du} = \left(-\frac{1}{2}\right) e^{-u/2}$.

Thus,

$f_u(u) = f(y)\left|\frac{dy}{du}\right| = 1\left|\left(-\frac{1}{2}\right) e^{-u/2}\right| = \left(\frac{1}{2}\right) e^{-u/2}$ for $u > 0$

which is exponential with $\beta = 2$.

6.25 **a.** If $W = \frac{mV^2}{2}$, then $V = \sqrt{\frac{2W}{m}}$ and $\left|\frac{dv}{dw}\right| = \left(\frac{1}{2}\right)\sqrt{\frac{2}{m}}\left(w^{-1/2}\right) = \frac{1}{\sqrt{2mw}}$. Then

$$f_W(w) = \frac{a\left(\frac{2w}{m}\right)}{\sqrt{2mw}} \times e^{-2bw/m} = \frac{a\sqrt{2w}}{m^{3/2}} \times e^{-w/kT} = \frac{a\sqrt{2}}{m^{3/2}} \times w^{1/2} \times e^{-w/kT}$$

Since the above density must integrate to 1 and since the variable part of the density is that of a gamma variable with $\alpha = \frac{3}{2}$ and $\beta = kT$, a must be chosen so that

$$\frac{a\sqrt{2}}{m^{3/2}} = \frac{1}{\Gamma\left(\frac{3}{2}\right)(kT)^{3/2}}$$

and the density is

$$f_W(w) = \frac{1}{\Gamma\left(\frac{3}{2}\right)(kT)^{3/2}} \times w^{1/2} \times e^{-w/kT}$$

for $w \geq 0$.

b. For a gamma-type variable, $E(W) = \alpha\beta = \left(\frac{3}{2}\right) kT$.

6.26 It is given that $f(i) = \frac{1}{2}$, for $9 \leq i \leq 11$. If $P = 2i^2$, then $i = \sqrt{\frac{P}{2}}$ and

$$\left|\frac{di}{dp}\right| = \left(\frac{1}{4}\right)\left(\frac{p}{2}\right)^{-1/2} = \left(\frac{1}{2}\right)^{3/2} p^{-1/2}.$$

Then

$$f_p(p) = \frac{1}{2^{5/2} p^{1/2}} = \frac{1}{4\sqrt{2p}}$$

for $9 \le \sqrt{\frac{p}{2}} \le 11$ or in other words $162 \le p \le 242$.

6.27 Similar to Exercise 6.21. Fix $Y_1 = y_1$. Then if $U = \frac{Y_2}{y_1}$, $Y_2 = y_1 U$ and $\left|\frac{dy_2}{du}\right| = y_1$.
The joint density of Y_1 and U is

$$f(y_1, u) = \tfrac{1}{8}\, y_1^2\, e^{-(y_1+uy_1)/2} = \tfrac{1}{8}\, y_1^2\, e^{-y_1(1+u)/2} \qquad y_1 \ge 0,\ u \ge 0.$$

Integrating over all possible values of y_1, we have

$$f_U(u) = \tfrac{1}{8}\int_0^\infty y_1^2\, e^{-y_1(1+u)/2}\, dy_1 = \frac{\Gamma(3)\left[\frac{2}{(1+u)}\right]^3}{8} = \frac{2}{(1+u)^3} \qquad u \ge 0$$

since the variable part of the integrand is that of a gamma variable with $\alpha = 3$, $\beta = \frac{2}{1+u}$.

6.28 If $U = 2Y^2 + 3$, then $Y = \left(\frac{U-3}{2}\right)^{1/2}$ and $\left|\frac{dy}{du}\right| = \left(\frac{1}{4}\right)\left(\frac{\sqrt{2}}{\sqrt{u-3}}\right)$. Then

$$f_U(u) = \frac{\sqrt{2}}{16\sqrt{u-3}} = \frac{1}{8\sqrt{2(u-3)}} \qquad \text{for} \quad 1 \le \left(\frac{u-3}{2}\right)^{1/2} \le 5 \quad \text{or} \quad 5 \le u \le 53$$

$f_U(u) = 0$ elsewhere

6.29 If $U = 5 - \left(\frac{Y}{2}\right)$, then $Y = 2(5-U)$ and $\left|\frac{dy}{du}\right| = 2$. Then

$$f_U(u) = 2\left[\left(\tfrac{3}{2}\right)(4)(5-u)^2 + 2(5-u)\right] = 4\left(80 - 31u + 3u^2\right)$$

for $0 \le 2(5-u) \le 1$ or $4.5 \le u \le 5$ (and 0 elsewhere).

6.30 **a.** If $U = Y^2$, then $Y = \sqrt{U}$ and $\left|\frac{dy}{du}\right| = \frac{1}{2\sqrt{u}}$. Then

$$f_U(u) = \frac{2\sqrt{u}}{\theta} \times e^{-u/\theta} \times \frac{1}{2\sqrt{u}} = \frac{1}{\theta}\, e^{-u/\theta} \qquad u \ge 0$$

which is an exponential density with mean θ.

b. From part **a.**, $E\left(Y^2\right) = E(U) = \theta$. Further, integrating directly, we have

$$E(Y) = E\left(U^{1/2}\right) = \int_0^\infty \frac{u^{1/2} e^{-u/\theta}}{\theta}\, du = \frac{\Gamma\left(\frac{3}{2}\right)\theta^{3/2}}{\theta} = \tfrac{1}{2}\Gamma\left(\tfrac{1}{2}\right)\theta^{1/2} = \frac{\sqrt{\pi\theta}}{2}$$

Then $V(Y) = E\left(Y^2\right) - [E(Y)]^2 = \theta - \left(\frac{\pi\theta}{4}\right) = \theta\left[1 - \left(\frac{\pi}{4}\right)\right]$.

6.31 By independence, $f(y_1, y_2) = f(y_1)f(y_2) = 1$ for $0 < y_1 < 1$, $0 < y_2 < 1$ and 0 elsewhere.
Let $U = Y_1 Y_2$. Now fix a value of Y_1 at y_1, say, with $0 < y_1 < 1$. Then $y_2 = u/y_1 = h^{-1}(u)$ so that

$$\frac{dh^{-1}(u)}{du} = 1/y_1$$

and hence

$$g(y_1, u) = \begin{cases} 1/y_1 & 0 < y_1 < 1,\ 0 < u \le y_1 \\ 0 & \text{otherwise} \end{cases}$$

Now

$$f_U(u) = \int_u^1 \frac{dy_1}{y_1} = -\log(u), \text{ for } 0 < u < 1.$$

6.32 By independence we know that

$$f(y_1, y_2) = \frac{4y_1y_2}{\theta^2} e^{-(y_1^2+y_2^2)/\theta} \text{ when } y_1 > 0 \text{ and } y_2 > 0.$$

Let $U = Y_1^2 + Y_2^2$. Now fix a value of Y_1at y_1, say. Then set $U = y_1^2 + Y_2^2$ so that $y_2 = \sqrt{u - y_1^2}$. Note that

$$\frac{dh^{-1}(u)}{du} = \frac{1}{2\sqrt{u-y_1^2}}.$$

Then we have

$$g(y_1, u) = \frac{4y_1\sqrt{u-y_1^2}}{\theta^2} e^{-u/\theta} \frac{1}{2\sqrt{u-y_1^2}} = \frac{2e^{-u/\theta}}{\theta^2} y_1$$

for $0 < y_1 < \sqrt{u}$. Then

$$f_U(u) = \int_0^{\sqrt{u}} \frac{2e^{-u/\theta}}{\theta^2} y_1 dy_1 = \frac{1}{\theta^2} u e^{-u/\theta}.$$

for $u > 0$. That is, $U \sim \text{Gamma}(2, \theta)$.

6.33 Since Y_i has an exponential distribution, $m_{Y_i}(t) = \frac{1}{1-t}$, for $i = 1, 2$. Using Theorem 6.2, we have for $U = \frac{1}{2}(Y_1 + Y_2)$

$$m_U(t) = m_{Y_1}\left(\tfrac{t}{2}\right) m_{Y_2}\left(\tfrac{t}{2}\right) = \frac{1}{\left[1-\left(\frac{t}{2}\right)\right]^2}.$$

which is the moment-generating function for a gamma random variable with $\alpha = 2$, $\beta = \frac{1}{2}$. Hence $f_U(u) = 4ue^{-2u}$ for $u \geq 0$.

6.34 Since Y_1 and Y_2 are independent standard normal random variables, the moment-generating functions for Y_1^2 and Y_2^2 can be written, from Example 6.11, as

$$m_{Y_1^2}(t) = \frac{1}{(1-2t)^{1/2}}$$
$$m_{Y_2^2}(t) = \frac{1}{(1-2t)^{1/2}}.$$

Now using Theorem 6.2, we have

$$m_U(t) = m_{Y_1^2}(t)\, m_{Y_2^2}(t) = \frac{1}{(1-2t)^1},$$

which is the moment-generating function of a gamma random variable with $\alpha = 1$ and $\beta = 2$. Hence by Theorem 6.1, U has a gamma distribution with $\alpha = 1$ and $\beta = 2$. Equivalently, U has a χ^2 distribution with 2 degrees of freedom.

6.35 The moment-generating function of any of the random variables Y_i, $i = 1, 2, \dots, n$, is given in Appendix II as

$$m_{Y_i}(t) = e^{\mu t + (t^2\sigma^2/2)}$$

Then

$$m_{a_iY_i}(t) = m_{Y_i}(a_i t) = e^{a_i\mu t + (a_i^2 t^2\sigma^2/2)}$$

Using Theorem 6.2, we have

$$m_U(t) = \prod_{i=1}^{n} m_{a_iY_i}(t) = \prod_{i=1}^{n} e^{a_i\mu t + (a_i^2t^2\sigma^2/2)} = e^{\mu t\Sigma a_i + (t^2\sigma^2/2)\,\Sigma a_i^2}$$

which is the moment-generating function of a normal random variable with mean $\mu\Sigma\, a_i$ and variance $\sigma^2\Sigma\, a_i^2$. Hence U is distributed normally with

$$E(U) = \mu \sum_{i=1}^{n} a_i$$

and

$$V(U) = \sigma^2 \sum_{i=1}^{n} a_i^2.$$

6.36 The probability of interest is $P(Y_2 > Y_1) = P(Y_2 - Y_1 > 0)$. Refer to Theorem 6.3. The distribution of $U = Y_2 - Y_1$ will be normal with mean $E(U) = E(Y_2) - E(Y_1)$ $= 4000 - 5000 = -1000$ and variance $V(U) = V(Y_2) + V(Y_1) = 400^2 + 300^2 = 250{,}000$. Then $Z = \frac{U - E(U)}{\sqrt{V(U)}}$ has a standard normal distribution. Hence

$$P(Y_2 - Y_1 > 0) = P\left(Z > \frac{0-(-1000)}{500}\right) = P(Z > +2) = .0228$$

6.37 **a.** Applying the results of exercise 6.35 we find that $\overline{Y} \sim N(\mu, \sigma^2/n)$.

b. $P(|\overline{Y} - \mu| \leq 1) = P(-1 \leq \overline{Y} - \mu \leq 1) = P(\frac{-\sqrt{n}}{\sigma} \leq Z \leq \frac{\sqrt{n}}{\sigma})$

$= P(-1.25 \leq Z \leq 1.25) = 1 - 2P(Z \geq 1.25) = 0.7888$

c. Each of the required calculations is similar to part **b.**

$n = 36$ $\quad \frac{\sqrt{n}}{\sigma} = 1.5$ $\quad P(|\overline{Y} - \mu| \leq 1) = 0.8664$

$n = 64$ $\quad \frac{\sqrt{n}}{\sigma} = 2$ $\quad P(|\overline{Y} - \mu| \leq 1) = 0.9544$

$n = 81$ $\quad \frac{\sqrt{n}}{\sigma} = 2.25$ $\quad P(|\overline{Y} - \mu| \leq 1) = 0.9756$

Thus, as n increases so does the probability that the sample mean is within one unit of the population mean.

6.38 By Theorem 6.3 we have $U = \sum_{1}^{n} Y_i \sim N(n\mu, n\sigma^2)$. Note that U is the total weight of watermelons in the packing container. Then

$$0.05 = P(U > 140) = P(Z > \frac{140 - n\mu}{n\sigma^2}).$$

Looking at the standard normal table shows that

$$\frac{140 - n\mu}{n\sigma^2} = 1.645$$

Thus, we can solve for n to obtain

$$n = \frac{140}{\mu + 1.645\sigma^2}.$$

Now, if we use $\mu = 15$ and $\sigma^2 = 4$ we see that $n = 6.5$. Therefore, the maximum number of melons that should be put in the container is 6.

6.39 Assuming that Y_1 and Y_2 are independent, Theorem 6.3 can be used to find the distribution of U, which will be normal with $E(U) = 100 + 7(10) + 3(4) = 182$ and $V(U) = 49(.5)^2 + 9(.2)^2 = 12.61$. It is necessary to find a value c such that $P(U > c) = .01$. Now $P(U > c) = P\left[Z < \frac{c-182}{\sqrt{12.61}}\right]$. The value of the standard normal random variable that satisfies the requirement is $z = 2.33$. Hence

$$\frac{c-182}{\sqrt{12.61}} = 2.33 \quad \text{and} \quad c = \$190.27$$

6.40 $m_W(t) = E\left(e^{tW}\right) = E\left(e^{t2Y/\beta}\right) = m_Y\left(\frac{2t}{\beta}\right) = E\left(e^{(2t/\beta)Y}\right)$

$$= \frac{1}{\left[1-\beta\left(\frac{2t}{\beta}\right)\right]^{\alpha}} = \frac{1}{(1-2t)^{n/2}}.$$

Thus W has a gamma distribution with $\alpha = \frac{n}{2}$ and $\beta = 2$, i.e., a χ^2 distribution with n degrees of freedom.

6.41 Looking in the table under $n = 2\alpha = 7$ degrees of freedom, we find

$$P(Y > 33.627) = P\left(\frac{2Y}{\beta} > 16.0128\right) = .025.$$

6.42 It is known from Exercise 6.34 that $V = U^2 = Y_1^2 + Y_2^2$ has a gamma distribution with $\alpha = 1, \beta = 2$ (equivalently, a chi-square distribution with 2 degrees of freedom). Then $f_V(v) = \left(\frac{1}{2}\right) e^{-v/2}$, $v \geq 0$. Using the transformation approach, we have $U = V^{1/2}$, $V = U^2$, and $\left|\frac{dv}{du}\right| = 2u$. Then

$$f_U(u) = \frac{2u}{2} \times e^{-u^2/2} = ue^{-u^2/2}$$

for $u \geq 0$.

6.43 Using the moment-generating function approach, we have

$$m_{Y_1}(t) = (pe^t + q)^{n_1}$$

and

$$m_{Y_2}(t) = (pe^t + q)^{n_2}.$$

Further as Y_1 and Y_2 are independent, we have

$$m_{Y_1+Y_2}(t) = m_{Y_1}(t)\, m_{y_2}(t) = (pe^t + q)^{n_1+n_2}$$

which is the moment-generating function of the binomial random variable with parameters $n_1 + n_2$ and p. By Theorem 6.1,

$$P(Y_1 + Y_2 = k) = \binom{n_1+n_2}{k} p^k q^{n_1+n_2-k}$$

for $k = 0, 1, 2, \ldots, n_1 + n_2$.

6.44 **a.** Because $Y_1$and Y_2 are Poisson random variables,

$$m_{Y_1}(t) = e^{\lambda_1(e^t-1)}$$

and

$$m_{Y_2}(t) = e^{\lambda_2(e^t-1)}.$$

So that $m_{Y_1+Y_2}(t) = \exp\left[-(\lambda_1 + \lambda_2)\left(1 - e^t\right)\right]$; which is the moment-generating function of a Poisson random variable with mean $\lambda_1 + \lambda_2$. (Recall that exp( ) is simply a convenient way to write $e^{(\ )}$). By Theorem 6.1, then,

$$P(Y_1 + Y_2 = k) = \frac{e^{-(\lambda_1+\lambda_2)}(\lambda_1+\lambda_2)^k}{k!}$$

for $k = 0, 1, 2, \ldots$

b. By definition,

$$P(Y_1 = k|Y_1 + Y_2 = m) = \frac{P(Y_1=k, Y_1+Y_2=m)}{P(Y_1+Y_2=m)} = \frac{P(Y_1=k, Y_2=m-k)}{P(Y_1+Y_2=m)}$$

$$= \frac{\left[\frac{e^{-\lambda_1}\lambda_1^k}{k!}\right]\left[\frac{e^{-\lambda_2}\lambda_2^{m-k}}{(m-k)!}\right]}{\left[e^{-(\lambda_1+\lambda_2)}\left(\frac{(\lambda_1+\lambda_2)^m}{m!}\right)\right]} = \binom{m}{k}\left(\frac{\lambda_1}{\lambda_1+\lambda_2}\right)^k\left(\frac{\lambda_2}{\lambda_1+\lambda_2}\right)^{m-k} \quad k = 0, 1, 2, \ldots, m$$

which is the probability distribution function for a binomial random variable with parameters m and $\frac{\lambda_1}{\lambda_1+\lambda_2}$.

6.45 Recall that the moment generating function for a binomial random variable Y is $m_Y(t) = \{pe^t + (1-p)\}^n$. Now let $U = \sum Y_i$. Then by independence we have

$$m_U(t) = \prod m_{Y_i}(t) = \prod \{p_i e^t + (1-p_i)\}^{n_i}$$

a. Let $p_i = p$ and $n_i = m$ for $i = 1, ..., n$. Then

$$m_U(t) = \{pe^t + (1-p)\}^{nm}$$

Thus, $U \sim \text{binomial}(nm, p)$.

b. In this case,

$$m_U(t) = \{pe^t + (1-p)\}^{\sum n_i}$$

So that $U \sim \text{binomial}(\sum n_i, p)$.

c. By definition,

$$\begin{aligned}
P(Y_1 = k | \textstyle\sum Y_i = m) &= \frac{P(Y_1 = k, \sum Y_i = m)}{P(\sum Y_i = m)} \\
&= \frac{P(Y_1 = k, \sum_2^n Y_i = m - k)}{P(\sum Y_i = m)} \\
&= \frac{P(Y_1 = k) P(\sum_2^n Y_i = m - k)}{P(\sum Y_i = m)} \\
&= \frac{\binom{n_1}{k} p^k (1-p)^{n_1 - k} \binom{\sum_2^n n_i}{m-k} p^{m-k} (1-p)^{\sum_2^n n_i - m + k}}{\binom{\sum_1^n n_i}{m} p^m (1-p)^{\sum_1^n n_i - m}} \\
&= \frac{\binom{n_1}{k} \binom{\sum_2^n n_i}{m-k}}{\binom{\sum_1^n n_i}{m}}
\end{aligned}$$

so that the distribution of $Y_1 = k | \sum Y_i = m$ is hypergeometric($r = n_i$, $N = \sum_1^n n_i$).

d. This is similar to part **c**. By definition,

$$P(Y_1 + Y_2 = k|\sum Y_i = m) = \frac{P(Y_1 + Y_2 = k, \sum Y_i = m)}{P(\sum Y_i = m)}$$

$$= \frac{P(Y_1 + Y_2 = k, \sum_{3}^{n} Y_i = m - k)}{P(\sum Y_i = m)}$$

$$= \frac{P(Y_1 + Y_2 = k)P(\sum_{3}^{n} Y_i = m - k)}{P(\sum Y_i = m)}$$

$$= \frac{\binom{n_1+n_2}{k}\binom{\sum_{3}^{n} n_i}{m-k}}{\binom{\sum_{1}^{n} n_i}{m}}$$

So that the distribution of $Y_1 + Y_2 = k|\sum Y_i = m$ is hypergeometric.

d. No, as the product, $\prod\{p_i e^t + (1 - p_i)\}^{n_i}$, does not combine into a recognizable form.

6.46 a. Let $U = \sum_{1}^{n} Y_i$. Then

$$m_U(t) = \prod_{1}^{n} m_{Y_i}(t) = e^{(e^t-1)\sum_{1}^{n}\lambda_i}$$

Therefore, $U \sim$ Poisson($\sum_{1}^{n}\lambda_i$).

b. This is similar to parts **c.** and **d.** of Exercise 6.45.

$$P(Y_1 = k|\sum Y_i = m) = \frac{P(Y_1 = k, \sum Y_i = m)}{P(\sum Y_i = m)}$$

$$= \frac{P(Y_1 = k, \sum_{2}^{n} Y_i = m - k)}{P(\sum Y_i = m)}$$

$$= \frac{P(Y_1 = k)P(\sum_{2}^{n} Y_i = m - k)}{P(\sum Y_i = m)}$$

$$= \frac{\frac{\lambda_1^k e^{-\lambda_1}}{k!}\frac{\left(\sum_{2}^{n}\lambda_i\right)^{m-k} e^{-\sum_{2}^{n}\lambda_i}}{(m-k)!}}{\frac{\left(\sum_{1}^{n}\lambda_i\right)^{m} e^{-\sum_{1}^{n}\lambda_i}}{m!}}$$

$$= \binom{m}{k}\left(\frac{\lambda_1}{\sum_{1}^{n}\lambda_i}\right)^k\left(\frac{\sum_{2}^{n}\lambda_i}{\sum_{1}^{n}\lambda_i}\right)^{m-k}$$

So that the distribution of $Y_1 = k|\sum Y_i = m$ is binomial with m trials and success probability

$$p = \left(\frac{\lambda_1}{\sum_{1}^{n}\lambda_i}\right)^k.$$

c. Once again this is similar to parts **a.** and **b.** of Exercise 6.45. Note that $Z = Y_1 + Y_2$ is Poisson with mean $\lambda_1 + \lambda_2$ indep of $Y_3 \ldots Y_n$. Then the question is identical to part **b.** where we want the distribution of Z given that $Z + \sum_3^n Y_i = m$. Thus we know this distribution must be binomial with m trials and success probability $\left(\frac{\lambda_1+\lambda_2}{\sum_1^n \lambda_i}\right)$. Rather, we could do the problem directly by

$$P(Y_1 + Y_2 = k | \sum Y_i = m) = \frac{P(Y_1+Y_2=k, \sum Y_i = m)}{P(\sum Y_i = m)} = \frac{P(Y_1+Y_2=k, \sum_3^n Y_i = m-k)}{P(\sum Y_i = m)}$$

$$= \frac{P(Y_1+Y_2=k)P(\sum_3^n Y_i = m-k)}{P(\sum Y_i = m)} = \binom{m}{k}\left(\frac{\lambda_1+\lambda_2}{\sum_1^n \lambda_i}\right)^k \left(\frac{\sum_3^n \lambda_i}{\sum_1^n \lambda_i}\right)^{m-k}$$

which is, of course, a binomial probability mass function.

6.47 Let $Y = Y_1 + Y_2$with Y_1, Y_2 the number that arrive in each hour. Using the result in Exercise 6.44, Y is Poisson with $\mu = 14$.

$$P(Y > 20) = 1 - P(Y \leq 19) = 1 - .923 = .077.$$

6.48 Let U = total service time for two cars. U is gamma with $\alpha = 2$, $\beta = .5$.

$$P(U > 1.5) = 1 - \int_0^{1.5} \frac{1}{\Gamma(2)(.5)^2} ue^{-u/.5}\, du$$

$$= 1 - 4\left[.5ue^{-u/.5} - .25e^{-u/.5}\right]_0^{1.5}$$

$$= 1 - 4[-.03734 - .01245 + .25] = 1 - .8 = .20.$$

6.49 The moment generating function of each Y_i is $m_{Y_i}(t) = (1 - \beta t)^{-\alpha_i}$. Then the moment generating function of the sum would be

$$\begin{aligned} m_Y(t) &= m_{Y_1}(t)m_{Y_2}(t)\cdots m_{Y_n}(t) \\ &= (1 - \beta t)^{-\alpha_1}(1 - \beta t)^{-\alpha_2}\cdots(1 - \beta t)^{-\alpha_n} \\ &= (1 - \beta t)^{-(\alpha_1+\alpha_2+\ldots+\alpha_n)} \end{aligned}$$

which is a gamma distribution with parameters $(\alpha_1 + \ldots + \alpha_n)$ and β.

6.50 **a.** Recall that $m_{W_i}(t) = \frac{pe^t}{(1-qe^t)}$. Using Theorem 6.2, we have

$$m_Y(t) = \prod_{i=1}^r m_{W_i}(t) = \left(\frac{pe^t}{1-qe^t}\right)^r$$

b. Differentiating with respect to t, we have

$$E(Y) = m'(0) = r\left(\frac{pe^t}{1-qe^t}\right)^{r-1} \times \left.\frac{pe^t}{(1-qe^t)^2}\right|_{t=0} = \frac{r}{p}.$$

Further,

$$E\left(Y^2\right) = m''(0) = \frac{d}{dt}\left.\frac{r\,(pe^t)^r}{(1-qe^t)^{r+1}}\right|_{t=0}$$

$$= \left.\frac{(1-qe^t)^{r+1}r^2pe^t\,(pe^t)^{r-1} - r\,(pe^t)^r(r+1)(-qe^t)\,(1-qe^t)^r}{(1-qe^t)^{2(r+1)}}\right|_{t=0}$$

$$= \frac{pr^2+r(r+1)q}{p^2}$$

Then

$$V(Y) = \frac{pr^2+r(r+1)q-r^2}{p^2} = \frac{rq}{p^2}.$$

c. This is similar to exercises 6.45 and 6.46.

$$P(W_1 = k|\sum W_i = m) = \frac{P(W_1 = k, \sum W_i = m)}{P(\sum W_i = m)}$$

$$= \frac{P(W_1 = k, \sum_2^r W_i = m - k)}{P(\sum W_i = m)}$$

$$= \frac{P(W_1 = k)P(\sum_2^r W_i = m - k)}{P(\sum W_i = m)}$$

$$= \frac{\binom{m-k-1}{r-2}}{\binom{m-1}{r-1}}.$$

6.51 Y_i is Gamma with $\alpha = \frac{\nu_i}{2}$ and $\beta = 2$,so that $m_{Y_i}(t) = (1-2t)^{\nu_i/2}$. From Theorem 6.2,

$$m_U(t) = m_{Y_1}(t)m_{Y_2}(t) = (1-2t)^{-(\nu_1+\nu_2)/2}$$

which is the moment-generating function of a chi-square random variable with $\nu_1 + \nu_2$ degrees of freedom. The result follows from Theorem 6.1.

6.52 $E\{\exp[t_1(Y_1+Y_2) + t_2(Y_1 - Y_2)]\}$

$= E\{\exp[Y_1(t_1+t_2) + Y_2(t_1-t_2)]\}$

$= E\{\exp[(t_1+t_2)Y_1]\exp[(t_1-t_2)Y_2]\}$

$= m_{Y_1}(t_1+t_2)\, m_{Y_2}(t_1-t_2)$

$= \exp\left[\frac{\sigma^2}{2}(t_1+t_2)^2\right]\exp\left[\frac{\sigma^2}{2}(t_1-t_2)^2\right]$

$= \exp[\sigma^2 t_1^2]\exp[\sigma^2 t_2^2]$

$= m_{U_1}(t_1)\, m_{U_2}(t_2)$.

Since the joint moment-generating function is the product of the marginal moment-generating functions, U_1 and U_2 are independent.

6.53 **a.** Notice that

$$f_{U_1}(u_1) = \int f(u_1,u_2)du_2 = \int_0^\infty \frac{1}{\beta}u_2 e^{-u_2/\beta} I(0<u_1<1)du_2$$

$$= I(0<u_1<1)$$

Thus, $U_1 \sim \text{uniform}(0,1)$.

b. One can note this easily using the method of moment generating functions. Otherwise we could proceed with direct calculations

$$f_{U_2}(u_2) = \int f(u_1,u_2)du_1 = \int_0^1 \frac{1}{\beta}u_2 e^{-u_2/\beta}du = \frac{1}{\beta}u_2 e^{-u_2/\beta}$$

Thus, $U_2 \sim \text{Gamma}(2,\beta)$.

c. $f_{U_1U_2}(u_1,u_2) = f_{U_1}(u_1)f_{U_2}(u_2)$. Thus U_1 and $U_2$are independent.

6.54 **a.** First note that, by independence, we have

$$f_{Y_1Y_2}(y_1,y_2) = f_{Y_1}(y_1)f_{Y_2}(y_2) = \frac{1}{\Gamma(\alpha_1)\Gamma(\alpha_2)\beta^{\alpha_1+\alpha_2}} y_1^{\alpha_1-1}y_2^{\alpha_2-1}e^{-(y_1+y_2)/\beta}.$$

Now let $U = \frac{Y_1}{Y_1+Y_2}$ and $V = Y_1 + Y_2$. Then $y_1 = uv$ and $y_2 = v(1-u)$ so that the Jacobian of the transformation is $J = v$ (see example 6.14). Thus, the joint density of U and V is

$$f_{U,V}(u,v) = \frac{1}{\Gamma(\alpha_1)\Gamma(\alpha_2)\beta^{\alpha_1+\alpha_2}}(uv)^{\alpha_1-1}\{v(1-u)\}^{\alpha_2-1}e^{-v/\beta}v$$

$$= \frac{1}{\Gamma(\alpha_1)\Gamma(\alpha_2)\beta^{\alpha_1+\alpha_2}}u^{\alpha_1-1}(1-u)^{\alpha_2-1}v^{\alpha_1+\alpha_2-1}e^{-v/\beta}$$

when $0<u<1$ and $v>0$.

b. Notice that

$$\begin{aligned} f_U(u) &= \frac{1}{\Gamma(\alpha_1)\Gamma(\alpha_2)\beta^{\alpha_1+\alpha_2}} u^{\alpha_1-1}(1-u)^{\alpha_2-1}\int_0^\infty v^{\alpha_1+\alpha_2-1}e^{-v/\beta}dv \\ &= \frac{1}{\Gamma(\alpha_1)\Gamma(\alpha_2)\beta^{\alpha_1+\alpha_2}} u^{\alpha_1-1}(1-u)^{\alpha_2-1}\Gamma(\alpha_1+\alpha_2)\beta^{\alpha_1+\alpha_2} \\ &= \frac{\Gamma(\alpha_1+\alpha_2)}{\Gamma(\alpha_1)\Gamma(\alpha_2)} u^{\alpha_1-1}(1-u)^{\alpha_2-1} \end{aligned}$$

for $0 < u < 1$. Thus, $U \sim \text{Beta}(\alpha_1, \alpha_2)$.

c. This problem is very straightforward using the method of moment generating functions. Alternatively we could calculate the density of V directly,

$$\begin{aligned} f_V(v) &= \frac{1}{\Gamma(\alpha_1)\Gamma(\alpha_2)\beta^{\alpha_1+\alpha_2}}\int_0^1 u^{\alpha_1-1}(1-u)^{\alpha_2-1}v^{\alpha_1+\alpha_2-1}e^{-v/\beta}du \\ &= \frac{1}{\Gamma(\alpha_1)\Gamma(\alpha_2)\beta^{\alpha_1+\alpha_2}} v^{\alpha_1+\alpha_2-1}e^{-v/\beta}\frac{\Gamma(\alpha_1)\Gamma(\alpha_2)}{\Gamma(\alpha_1+\alpha_2)} \\ &= \frac{1}{\Gamma(\alpha_1+\alpha_2)\beta^{\alpha_1+\alpha_2}} v^{\alpha_1+\alpha_2-1}e^{-v/\beta} \end{aligned}$$

for $v > 0$. Thus, $V \sim \text{Gamma}(\alpha_1+\alpha_2, \beta)$.

d. $f_{U,V}(u,v) = \frac{1}{\Gamma(\alpha_1)\Gamma(\alpha_2)\beta^{\alpha_1+\alpha_2}} u^{\alpha_1-1}(1-u)^{\alpha_2-1}v^{\alpha_1+\alpha_2-1}e^{-v/\beta}$

$= \left(\frac{\Gamma(\alpha_1+\alpha_2)}{\Gamma(\alpha_1)\Gamma(\alpha_2)} u^{\alpha_1-1}(1-u)^{\alpha_2-1}\right)\left(\frac{1}{\Gamma(\alpha_1+\alpha_2)} v^{\alpha_1+\alpha_2-1}e^{-v/\beta}\right)$

$= f_U(u)f_V(v)$ which implies the independence of U and V.

6.55 a. By independence we have

$$f_{Y_1,Y_2}(y_1,y_2) = \frac{1}{y_1^2 y_2^2}$$

for $y_1 > 1$ and $y_2 > 1$.

b. Let $U = \frac{Y_1}{Y_1+Y_2}$ and $V = Y_1 + Y_2$. Then $y_1 = uv$ and $y_2 = v(1-u)$ so that the Jacobian of the transformation is v. Then

$$f_{U,V}(u,v) = \frac{1}{u^2v^3(1-u)^2}$$

for $v > 1/u$ and $v(1-u) > 1$. The conditions on the support may be simplified as follows

$$1/u < v,$$
$$0 < u < 1/2$$

which we may simplify to

$$1/(1-u) < v,$$
$$1/2 \le u \le 1.$$

d. If $0 < u < 1/2$ then

$$f_U(u) = \frac{1}{u^2(1-u)^2}\int_{1/u}^\infty v^{-3}dv = \frac{1}{2(1-u)^2}.$$

If $1/2 \le u \le 1$ then

$$f_U(u) = \frac{1}{u^2(1-u)^2}\int_{1/(u-1)}^\infty v^{-3}dv = \frac{1}{2u^2}.$$

e. No, as their joint density does not factor. Specifically the range of values U can take depends on V and vice versa.

6.56 a. By independence we have

$$f_{Y_1,Y_2}(y_1,y_2) = f_{Y_1}(y_1)f_{Y_2}(y_2) = 1$$

for $0 < y_1 < 1$ $0 < y_2 < 1$. Then the inverse transformations are

$$y_1 = \frac{u+v}{2}$$

and

$$y_2 = \frac{u-v}{2}.$$

Thus the Jacobian is

$$J = \begin{vmatrix} \frac{1}{2} & \frac{1}{2} \\ -\frac{1}{2} & \frac{1}{2} \end{vmatrix} = \frac{1}{2}$$

so that

$$f_{U,V}(u,v) = \frac{1}{2}$$

for $0 < \frac{u+v}{2} < 1$, and $0 < \frac{u-v}{2} < 1$. or, taking on a case by case basis, the support may be expressed as $0 < u < 1, \ -u < v < u$ or $1 \le u \le 2,\ u-2 < v < 2-u$.

c. If $0 < u < 1$ then

$$f_U(u) = \int_{-u}^{u} \frac{1}{2} dv = u$$

but if $1 \le u < 2$ then

$$f_U(u) = \int_{u-2}^{2-u} \frac{1}{2} dv = 2 - u.$$

d. If $-1 < v < 0$ then

$$f_V(v) = \int_{-v}^{2+v} \frac{1}{2} du = 1 + v$$

but if $1 \le u < 2$ then

$$f_V(v) = \int_v^{2-v} \frac{1}{2} du = 1 - u.$$

e. No, as their joint density does not factor. Specifically the range of values U can take depends on V and vice versa.

6.57 a. By independence we have

$$f_{Y_1,Y_2}(y_1,y_2) = f_{Y_1}(y_1)f_{Y_2}(y_2) = \frac{1}{\beta^2}e^{-(y_1+y_2)/\beta}$$

for $y_1 > 0$ and $y_2 > 0$. The inverse transformations are $y_1 = \frac{uv}{1+v}$ and $y_2 = \frac{u}{1+v}$ so that the Jacobian is

$$J = \begin{vmatrix} \frac{v}{1+v} & \frac{u}{(1+v)^2} \\ \frac{1}{1+v} & \frac{-u}{(1+v)^2} \end{vmatrix} = \frac{-u}{(1+v)^2}.$$

Then

$$f_{U,V}(u,v) = \frac{1}{\beta^2}e^{-u/\beta}\ \frac{u}{(1+v)^2} \text{ for } 0 < u \text{ and } 0 < v.$$

b. Yes, as the joint density factors into

$$f_{U,V}(u,v) = f_U(u)f_V(v)$$

where

$$f_U(u) = \frac{1}{\beta^2}ue^{-u/\beta}\ ,\ u > 0$$

and

$$f_V(v) = \frac{1}{(1+v)^2},\ v > 0.$$

6.58 Y_1 and Y_2 are independently and identically distributed with density function $f(y) = 1$ for $0 \le y \le 1$ and cumulative density function

$$F(Y) = \begin{cases} 0, & y < 0 \\ y, & 0 \le y \le 1 \\ 1, & y > 1 \end{cases}$$

a. Referring to Section 6.6, with $n = 2$, we have

$$g_1(u) = 2[1 - F(u)]^{2-1}f(u) = 2(1-u) \qquad 0 \le u \le 1$$

b. $E(u_1) = \int_0^1 u[2(1-u)]\,du = \frac{1}{3}$.

$E(u_1^2) = \int_0^1 u^2[2(1-u)]\,du = \frac{1}{6}$.

Then

$$V(u_1) = E(u_1^2) - [E(u_1)]^2 = \frac{1}{6} - \frac{1}{9} = \frac{1}{18}$$

6.59 a. Similarly, $g_2(u) = 2[F(u)]^{2-1}f(u) = 2u \qquad$ for $0 \le u \le 1$.

b. Notice that

$$E(u_2) = \int_0^1 u[2u]\,du = \frac{2}{3},$$

and

$$E\left(u_2^2\right) = \int_0^1 u^2[2u]\,du = \tfrac{1}{2}.$$

Then

$$V(u_2) = E\left(u_2^2\right) - [E(u_2)]^2 = \tfrac{1}{2} - \tfrac{4}{9} = \tfrac{1}{18}.$$

6.60 $F(y) = \left(\frac{y}{\theta}\right)$

a. Now $[F(y)]^n = \left(\frac{y}{\theta}\right)^n$. Thus,

$$F_{Y_{(n)}}(y) = \begin{cases} 0, & y < 0 \\ \left(\frac{y}{\theta}\right)^n, & 0 \le y \le \theta \\ 1, & y > 0 \end{cases}$$

b. $f_n(y) = \frac{ny^{n-1}}{\theta^n}, \qquad 0 \le y \le \theta$

c. $E(Y_{(n)}) = \int_0^\theta \frac{yny^{n-1}}{\theta^n}\,dy = \left(\frac{n}{n+1}\right)\theta.$

$$E\left(Y_{(n)}^2\right) = \int_0^\theta \frac{y^2ny^{n-1}}{\theta^n}\,dy = \left(\frac{n}{n+2}\right)\theta^2.$$

Then

$$V\left(Y_{(n)}\right) = E\left(Y_{(n)}^2\right) - \left[E\left(Y_{(n)}\right)\right]^2 = \left(\frac{n}{n+2}\right)\theta^2 - \left(\frac{n}{n+1}\theta\right)^2$$
$$= n\theta^2\left(\frac{1}{n+2} - \frac{n}{(n+1)^2}\right) = \frac{n\theta^2}{(n+1)^2(n+2)}$$

6.61 $P(Y_{(5)} < 10) = F_{Y_{(5)}}(10) = \left(\frac{10}{15}\right)^5 = \left(\frac{2}{3}\right)^5.$

6.62 **a.** The density and cdf of the Y_i are

$$f(y) = \begin{cases} \frac{1}{\theta}, & 0 \le y \le \theta \\ 0, & \text{otherwise} \end{cases}$$

and

$$F(y) = \begin{cases} 0, & y < 0 \\ \frac{y}{\theta}, & 0 \le y \le \theta \\ 1, & y > \theta \end{cases}$$

Then, from Theorem 6.5,

$$f_{(k)}(y_k) = \frac{n!}{(k-1)!(n-k)!}\left(\frac{y}{\theta}\right)^{k-1}\left[\frac{\theta-y}{\theta}\right]^{n-k}\left(\frac{1}{\theta}\right)$$
$$= \frac{n!}{(k-1)!(n-k)!}\,\frac{y^{k-1}(\theta-y)^{n-k}}{\theta^n} \qquad 0 \le y_k \le \theta.$$

b. $E\left(Y_{(k)}\right) = \int_0^\theta yf_{(k)}(y)\,dy = \int_0^\theta \frac{n!}{(k-1)!(n-k)!}\,\frac{y^k(\theta-y)^{n-k}}{\theta^n}\,dy$

$$= \frac{k}{n+1}\int_0^\theta \frac{\Gamma(n+2)}{\Gamma(k+1)\Gamma(n-k+1)}\left(\frac{y}{\theta}\right)^k\left(1-\frac{y}{\theta}\right)^{n-k}dy$$

Let $z = \frac{y}{\theta}$ and $dy = \theta\,dz$. Then

$$E\left(Y_{(k)}\right) = \frac{k}{n+1}\theta\int_0^1 \frac{z^k(1-z)^{n-k}}{B(k+1,\,n-k+1)}\,dz$$

Notice that the integral is that y is beta density with $\alpha = k+1$ and $\beta = n-k+1$ and, hence, the integral is 1. Then

$$E\left(Y_{(k)}\right) = \frac{k}{n+1}\theta$$

c. Similar to part **b.**

$$E\left(Y_{(k)}^2\right) = \int_0^\theta y^2\frac{n!}{(k-1)!(n-k)!}\left(\frac{y}{\theta}\right)^{k-1}\left(1-\frac{y}{\theta}\right)^{n-k}\left(\frac{1}{\theta}\right)dy$$

Letting $z = y/\theta$, this becomes

$$\frac{\theta^2k(k+1)}{(n+1)(n+2)}\int_0^1 \frac{\Gamma(n+3)}{\Gamma(k+2)\Gamma(n-k+1)}z^{k+1}(1-z)^{n-k}\,dz = \frac{\theta^2k(k+1)}{(n+1)(n+2)}.$$

Then $V\left(Y_{(k)}\right) = E\left(Y_{(k)}^2\right) - \left[E\left(Y_{(k)}\right)\right]^2$

$$= \frac{\theta^2k(k+1)}{(n+1)(n+2)} - \frac{k^2\theta^2}{(n+1)^2} = \frac{\theta^2k}{n+1}\left[\frac{k+1}{n+2} - \frac{k}{n+1}\right] = \frac{(n-k+1)k\theta^2}{(n+1)^2(n+2)}$$

d. $E\left(Y_{(k)} - Y_{(k-1)}\right) = E\left(Y_{(k)}\right) - E\left(Y_{(k-1)}\right) = \frac{k\theta}{n+1} - \frac{(k-1)\theta}{n+1}$ (by part **b.**)

$$= \frac{\theta}{n+1},$$

which is constant for all k. Thus, the order statistics are, on the average, equally spaced.

6.63 a. $f_{(j)(k)}(y_j, y_k) = \frac{n!}{(j-1)!(k-1-j)!(n-k)!}\left(\frac{y_j}{\theta}\right)^{j-1}$
$\times\left[\frac{y_k}{\theta} - \frac{y_j}{\theta}\right]^{k-1-j}\left[1 - \frac{y_k}{\theta}\right]^{n-k}\left(\frac{1}{\theta}\right)^2 \qquad 0 \le y_j \le y_k \le \theta$

b. $\text{Cov}\left(Y_{(j)}, Y_{(k)}\right) = E\left(Y_{(j)}Y_{(k)}\right) - E\left(Y_{(j)}\right)E\left(Y_{(k)}\right)$.

To simplify calculations, let $u = \frac{Y}{\theta}$. Then $|J| = \theta^2$ so that

$$f(u_j, u_k) = \frac{n!}{(j-1)!(k-1-j)!(n-k)!} u_j^{j-1}[u_k - u_j]^{k-1-j}[1 - u_k]^{n-k}$$
$$0 \le u_j \le u_k \le 1.$$

$$E\left[u_{(j)}u_{(k)}\right] = L\int_0^1\int_0^{u_k} u_j^j[u_k - u_j]^{k-1-j}u_k[1 - u_k]^{n-k}\,du_j\,du_k$$

where $L = \frac{n!}{(j-1)!(k-1-j)!(n-k)!}$.

Let $w = \frac{u_j}{u_k}$, $u_j = wu_k$, and $du_j = u_k\,dw$. The expected value becomes

$$= L\int_0^1\int_0^1 (wu_k)^j u_k^{k-1-j}[1 - w]^{k-1-j}\,u_k[1 - u_k]^{n-k}\,u_k\,dw\,du_k$$
$$= L\int_0^1 u_k^{k+1}[1 - u_k]^{n-k}\int_0^1 w^j(1 - w)^{k-1-j}\,dw\,du_k$$
$$= LB(k + 2, n - k + 1)\,B(j + 1, k - j)$$
$$= \frac{n!}{(j-1)!(k-1-j)!(n-k)!}\,\frac{(k+1)!(n-k)!}{(n+2)!}\,\frac{j!(k-j-1)!}{k!} = \frac{(k+1)j}{(n+2)(n+1)}.$$

From Exercise 6.62, $E\left[u_{(j)}\right] = \frac{j}{n+1}$. Then

$$\text{Cov}\left(U_{(j)}U_{(k)}\right) = \frac{(k+1)j}{(n+2)(n+1)} - \left(\frac{k}{n+1}\right)\left(\frac{j}{n+1}\right) = \frac{(n+1)(nk+j)-(n+2)jk}{(n+2)(n+1)^2}$$
$$= \frac{(n-k+1)j}{(n+1)^2(n+2)}.$$

Finally, $\text{Cov}\left(Y_{(j)}, Y_{(k)}\right) = \frac{(n-k+1)}{(n+1)^2(n+2)}\,\theta^2$.

c. $V\left(Y_{(k)} - Y_{(j)}\right) = V\left(Y_{(k)}\right) + V\left(Y_{(j)}\right) - 2\text{Cov}\left(Y_{(k)}, Y_{(j)}\right)$

$$= \left[\frac{k(n-k+1)}{(n+1)^2(n+2)} + \frac{j(n-j+1)}{(n+1)^2(n+2)} - \frac{2(n-k+1)j}{(n+1)^2(n+2)}\right]\theta^2$$
$$= \left[\frac{(k+j-2j)(n-k+1)+j(k-j)}{(n+1)^2(n+2)}\right]\theta^2$$
$$= \frac{(k-j)(n-k+j+1)}{(n+1)^2(n+2)}\,\theta^2.$$

6.64 $f_Y(y) = \left[\frac{1}{B(2,2)}\right]y(1-y) = \left[\frac{\Gamma(4)}{\Gamma(2)\Gamma(2)}\right]y(1-y) = 6y(1-y) \qquad 0 \le y \le 1.$

$$F_Y(y) = \int_0^y 6y(1-y)\,dy = 3y^2 - 2y^3 \qquad 0 \le y \le 1$$

a. $[F_Y(y)]^n = (3y^2 - 2y^3)^n$

Thus,

$$F_{Y_{(n)}}(y) = \begin{cases} 0, & y < 0 \\ (3y^2 - 2y^3)^n, & 0 \le y \le 1 \\ 1, & y > 1 \end{cases}$$

b. $f_{Y_{(n)}}(y) = n\left(3y^2 - 2y^3\right)^{n-1}\left(6y - 6y^2\right) = 6ny(1-y)\left(3y^2 - 2y^3\right)^{n-1}, \qquad 0 \le y \le 1$

c. For $n = 2$,

$$f_{Y_{(2)}}(y) = 12y(1-y)\left(3y^2 - 2y^3\right) = 36y^3 - 60y^4 + 24y^5, \qquad 0 \le y \le 1.$$

$$E\left(Y_{(2)}\right) = \int_0^1 y\left(36y^3 - 60y^4 + 24y^5\right)dy = \frac{36}{5}y^5\Big|_0^1 - \frac{60}{6}y^6\Big|_0^1 + \frac{24}{7}y^7\Big|_0^1 = .6286$$

6.65 a. $f(y) = \left(\frac{1}{\beta}\right)e^{-y/\beta};\ F(y) = 1 - e^{-y/\beta} \qquad \text{if } y > 0$

$$f_{Y_{(1)}}(y) = n\left[1 - F(y)^{n-1}\right]f(y) = n\left[\left(e^{-y/\beta}\right)^{n-1}\right]\left[\left(\frac{1}{\beta}\right)e^{-y/\beta}\right]$$
$$= \left(\frac{n}{\beta}\right)e^{-ny/\beta}, \text{ which is exponential with } \mu = \frac{\beta}{n}.$$

b. $f_{Y_{(1)}}(y) = 2.5e^{-y/.4} \qquad y > 0$

$$F_{Y_{(1)}}(3.6) = 1 - e^{-(3.6)/(.4)} = 1 - e^{-9}$$

6.66 a. For the exponential random variable, the density function is
$$f(y) = \tfrac{1}{\beta} e^{-y/\beta}$$
for $0 \le y < \infty$ while the cumulative distribution function is given by
$$F(y) = 1 - e^{-y/\beta}$$
for $0 \le y < \infty$. Then, from Theorem 6.5,
$$\begin{aligned} f_{(k)}(y_k) &= \tfrac{n!}{(k-1)!(n-k)!} \left(1 - e^{-y_k/\beta}\right)^{k-1} \left(e^{-y_k/\beta}\right)^{n-k} \left(\tfrac{1}{\beta} e^{-y_k/\beta}\right), \quad 0 < y_k < \infty \\ &= \tfrac{n!}{(k-1)!(n-k)!} \left(\tfrac{1}{\beta}\right) \left(e^{-y_k/\beta}\right)^{n-k+1} \left(1 - e^{-y_k/\beta}\right)^{k-1} \end{aligned}$$

b. $$\begin{aligned} f_{(j)(k)}(y_j, y_k) &= \tfrac{n!}{(j-1)!(k-1-j)!(n-k)!} \times F^{j-1}(y_j)\,[F(y_k) - F(y_j)]^{k-1-j} \\ &\quad \times [1 - F(y_k)]^{n-k} f(y_j) f(y_k) \qquad -\infty < y_i < y_k < \infty \\ &= \tfrac{n!}{(j-1)!(k-1-j)!(n-j)!} \left[1 - e^{-y_j/\beta}\right]^{j-1} \left[e^{-y_j/\beta} - e^{-y_k/\beta}\right]^{k-1-j} \\ &\quad \times \left[e^{-y_k/\beta}\right]^{n-k} \left(\tfrac{1}{\beta} e^{-y_j/\beta}\right) \left(\tfrac{1}{\beta} e^{-y_k/\beta}\right) \\ &= \tfrac{1}{\beta^2} \tfrac{n!}{(j-1)!(k-j-1)!(n-j)!} \left[1 - e^{-y_j/\beta}\right]^{j-1} \left[e^{-y_j/\beta}\right] \\ &\quad \times \left[e^{-y_j/\beta} - e^{-y_k/\beta}\right]^{k-j-1} \left[e^{-y_k/\beta}\right]^{n-k+1} \end{aligned}$$

6.67 a. This is similar to Exercise 6.58. The density function of $Y_{(1)}$ is $g_1(u) = n[1 - F(u)]^{n-1} f(u)$. Now
$$\begin{aligned} 1 - F(u) &= \int_u^\infty \tfrac{1}{2} e^{-(1/2)(t-4)}\,dt = \int_{u-4}^\infty \tfrac{1}{2} e^{-(1/2)s}\,ds = \left[-e^{-(1/2)s}\right]_{u-4}^\infty \\ &= e^{-(1/2(u-4))} \end{aligned}$$
so that
$$g_1(u) = 2\left[e^{-(1/2)(u-4)}\right]^{2-1} \times \tfrac{1}{2} e^{-(1/2)(u-4)} = e^{-(u-4)} \qquad \text{for } u \ge 4$$

b. $E(u) = \int_4^\infty u e^{-(u-4)}\,du$

Let $y = u - 4$, so that $dy = du$ and
$$E(u) = \int_0^\infty (y+4)e^{-y}\,dy = \int_0^\infty y e^{-y}\,dy + \int_0^\infty 4e^{-y}\,dy = 1 + 4 = 5$$

6.68 a. This is similar to Exercise 6.67. Calculate
$$1 - F(y) = \int_y^\infty e^{-(t-\theta)}\,dt = e^{-(y-\theta)}$$
Then
$$g_1(y) = n\left[e^{-(y-\theta)}\right]^{n-1} e^{-(y-\theta)} = ne^{-n(y-\theta)}$$
for $y \ge 0$.

b. $$\begin{aligned} E\left(Y_{(1)}\right) &= \int_0^\infty nye^{-n(y-\theta)}\,dy = n\int_0^\infty (z+\theta)e^{-nz}\,dz \qquad (\text{with } z = y - \theta) \\ &= n\Gamma(2)\left(\tfrac{1}{n}\right)^2 + n\theta\left(\tfrac{1}{n}\right)^1 = \tfrac{1}{n} + \theta \end{aligned}$$
Notice that the integral is calculated in two parts, using the fact that the constants associated with the integrals must be constants of gamma random variables.

6.69 Since $f(y) = 1$ and $F(y) = y$ for $0 \le y \le 1$, Theorem 6.5 gives
$$f_{(1)(n)}(y_1, y_n) = \tfrac{n!}{(n-2)!}[y_n - y_1]^{n-2} \qquad 0 \le y_1 \le y_n \le 1$$
Now use the method of transformations to get the joint distribution of $R = Y_{(n)} - Y_{(1)}$ and $Y_{(1)}$ by letting $Y_{(1)} = y_1$ so that $Y_{(n)} = R + y_1$, $\frac{dy_n}{dR} = 1$, and
$$g(r, y_1) = \tfrac{n!}{(n-2)!} r^{n-2} = n(n-1)\,r^{n-2}$$
The ranges of $\left(Y_{(1)}, Y_{(n)}\right)$ and $(R, Y_{(1)})$ are shown in Figure 6.6.

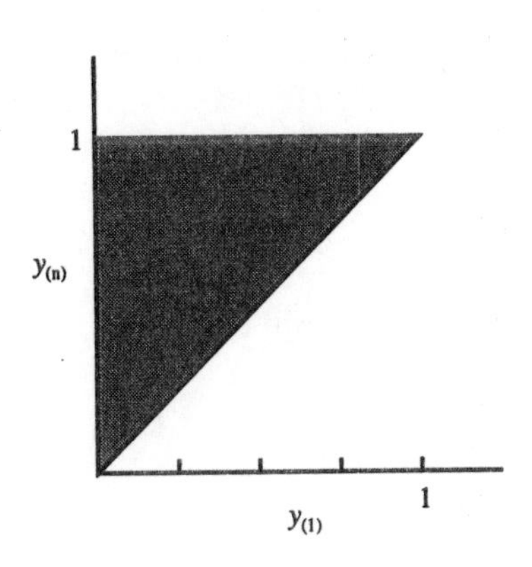

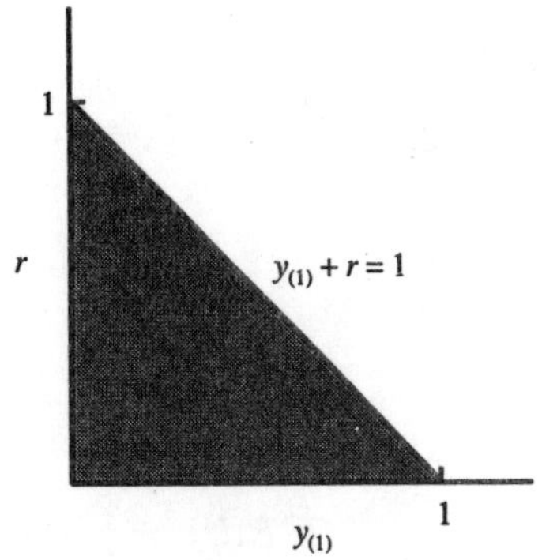

Figure 6.6a

Figure 6.6b

Since $Y_{(1)} \leq Y_{(n)}$, $0 \leq r \leq 1 - y_1$, or $0 \leq y_1 \leq 1 - r$. Hence the marginal density of R is

$$g(r) = \int_0^{1-r} g(r, y_1)\, dy_1 = \int_0^{1-r} n(n-1)\, r^{n-2}\, dy_1 = n(n-1)\, r^{n-2}(1-r) \qquad 0 \leq r \leq 1$$

$$g(r) = 0 \qquad \text{elsewhere}$$

6.70 Both parts of this exercise concern the waiting time until the fourth call, given that only four calls have occurred. This is defined as $W_{(4)}$.

a. Since the points on the interval $(0, t)$ at which the calls occur are uniformly distributed, we have

$$f(w) = \tfrac{1}{t} \qquad 0 \leq w \leq t \qquad\qquad F(w) = \tfrac{w}{t} \qquad 0 \leq w \leq t$$

The density of $W_{(4)}$ can be found to be

$$g_4(w) = 4[F(w)]^3 f(w) = \tfrac{4w^3}{t^4} \qquad 0 \leq w \leq t$$

and, with $t = 2$,

$$P\left(W_{(4)} \leq 1\right) = \int_0^1 \tfrac{4w^3}{16}\, dw = \tfrac{w^4}{16}\Big]_0^1 = \tfrac{1}{16}$$

b. $E\left(W_{(4)}\right) = \int_0^2 \tfrac{w^4}{4}\, dw = \tfrac{w^5}{20}\Big]_0^2 = \tfrac{32}{20} = 1.6$

6.71 a. It is given that $f(y) = \left(\frac{1}{\theta}\right) e^{-y/\theta}$ for $y \geq 0$, $\theta \geq 0$. Then $F(y) = 1 - e^{-y/\theta}$ for $y \geq 0$. In order to solve the exercise, we need the joint distribution of the two order statistics, W_{i-1} and W_j. Using Theorem 6.5, we obtain

$$f(w_j, w_{j-1}) = \tfrac{n!}{(j-2)!(n-j)!}\left(1 - e^{-w_{j-1}/\theta}\right)^{j-2}\left(e^{-w_j/\theta}\right)^{n-j}\left(\tfrac{1}{\theta^2}\right)$$
$$\times \left(e^{-(w_j + w_{j-1})/\theta}\right) \qquad 0 \leq w_{j-1} \leq w_j < \infty$$

Fix $W_{j-1} = w_{j-1}$ and let $T_j = W_j - w_{j-1}$. Then $\frac{dw_j}{dT_j} = 1$ and the joint distribution of T_j and w_{j-1} is

$$f(t_j, w_{j-1}) = \tfrac{n!}{(j-2)!(n-j)!}\left(1 - e^{-w_{j-1}/\theta}\right)^{j-2}\left(e^{-(t_j + w_{j-1}/\theta)}\right)^{n-j}\left(\tfrac{1}{\theta^2}\right)$$
$$\times \left(e^{-(t_j + 2w_{j-1})/\theta}\right) \qquad 0 \leq t_j;\ 0 < w_{j-1}$$

Integrating over all possible values of W_{j-1}, we obtain

$$f(t_j) = \tfrac{n!}{(j-2)!(n-j)!} \exp\left[\tfrac{-t_j(n-j) - t_j}{\theta}\right] \times \tfrac{1}{\theta^2} \int_0^{\infty} \left(1 - e^{-w_{j-1}/\theta}\right)^{j-2}$$
$$\times \left(e^{-w_{j-1}/\theta}\right)^{n-j+2} dw_{j-1}$$

Let $u = e^{-w_{j-1}/\theta}$ so that $w_{j-1} = -\theta \ln u$ and $\frac{dw_{j-1}}{du} = -\frac{\theta}{u}$. Then

$$f(t_j) = \frac{1}{\theta^2} e^{-t_j(n-j+1)/\theta} \int_0^1 \frac{n!}{(j-2)!(n-j)!} (1-u)^{j-2} u^{n-j+2} \left(\frac{\theta}{u}\right) du$$
$$= \frac{n-j+1}{\theta} e^{-t_j(n-j+1)/\theta} \int_0^1 \frac{n!}{(j-2)!(n-j+1)!} (1-u)^{j-2} u^{n-j+1} du$$
$$= \frac{n-j+1}{\theta} e^{-t_j(n-j+1)/\theta} \qquad 0 < t_j$$

since the integrand is that of a complete beta random variable with $\alpha = j-1$, $\beta = n-j+2$. Hence T_j has an exponential distribution with mean $\frac{\theta}{n-j+1}$.

b. Look at

$$\sum_{j=1}^{r} (n-j+1)T_j = nW_1 + (n-1)(W_2 - W_1) + (n-2)(W_3 - W_2)$$
$$+ (n-3)(W_4 - W_3) + \ldots + (n-r+2)(W_{r-1} - W_{r-2})$$
$$+ (n-r+1)(W_r - W_{r-1})$$
$$= W_1 + W_2 + \ldots + W_{r-1} + (n-r+1)W_r$$
$$= \sum_{j=1}^{r} W_j + (n-r)W_r = U_r$$

Hence

$$E(U_r) = \sum_{j=1}^{r} (n-j+1)E(T_j) = \sum_{j=1}^{r} \frac{(n-j+1)\theta}{n-j+1} = r\theta$$

6.72

Using Theorem 6.3, we see that U will have a normal distribution with

$$E(U) = \tfrac{1}{2}(\mu - 3\mu) = -\mu$$

and

$$V(U) = \tfrac{1}{4}\sigma^2 + \tfrac{9}{4}\sigma^2 = \tfrac{5}{2}\sigma^2.$$

6.73

Since I and R are independent,

$$f(i,r) = f(i)f(r) = 2r \qquad \text{for } 0 \le r \le 1, 0 \le i \le 1$$

Fix $R = r$. Then $W = I^2 r$ and $I = \sqrt{\frac{W}{r}}$ so that $\left|\frac{di}{dw}\right| = \left(\frac{1}{2r}\right)\left(\frac{W}{r}\right)^{-1/2}$ for $0 \le w \le r \le 1$. Notice that since $W = I^2 r$ and I is in the interval (0, 1), w will always be less than r. Then $f(w,r) = \sqrt{\frac{r}{w}}$ and

$$f(w) = \int_w^1 \sqrt{\tfrac{r}{w}}\, dr = \tfrac{1}{\sqrt{w}} \left(\tfrac{2}{3}\right) r^{3/2}\Big]_w^1 = \tfrac{2}{3}\left(\tfrac{1}{\sqrt{w}} - w\right) \qquad \text{for } 0 \le w \le 1$$
$$f(w) = 0 \qquad \text{otherwise}$$

6.74

Y_1 and Y_2 are both independently and identically distributed as gamma random variables with parameters $\alpha = 2$ and $\beta = 2$. Hence the moment-generating function for Y_i, $i = 1, 2$, is

$$m_{Y_i}(t) = (1-\beta t)^{-\alpha} = (1-2t)^{-2}$$

Now

$$m_{Y_i/2}(t) = m_{Y_i}\left(\tfrac{t}{2}\right) = (1-t)^{-2}$$

and

$$m_U(t) = m_{Y_1/2}(t) m_{Y_2/2}(t) = (1-t)^{-4}.$$

Evidently U has a gamma distribution with $\alpha = 4$ and $\beta = 1$.

6.75

The joint density of Y_1 and Y_2 can be written as

$$f(y_1, y_2) = f_{Y_1}(y_1) f_{Y_2}(y_2) = 1 \qquad 0 \le y_1 \le 1, 0 \le y_2 \le 1$$

a. On the region defined above, the function $U_1 = Y_1^2$ is an increasing function of Y. Hence

$$f_{U_1}(u) = f_{Y_1}\left(\sqrt{u}\right)\left|\tfrac{dy_1}{du}\right| = 1 \times \tfrac{1}{2\sqrt{u}} = \tfrac{1}{2\sqrt{u}} \qquad 0 \le u \le 1$$
$$f_{U_1}(u) = 0 \qquad \text{elsewhere}$$

Note that we used the marginal density of Y_1, $f_{Y_1}(y_1) = 1$ for $0 \le y_1 \le 1$.

b. Consider the joint distribution of $U = \frac{Y_1}{Y_2}$ and Y_2. Letting Y_2 be fixed at y_2, we can write $U = \frac{Y_1}{y_2}$. Then $Y_1 = y_2 U$ and $\frac{dy_1}{du} = y_2$, so that the joint density of Y_2 and U is $g(y_2, u) = y_2$. The ranges of Y_2 and U are shown in Figure 6.7.

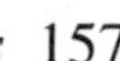

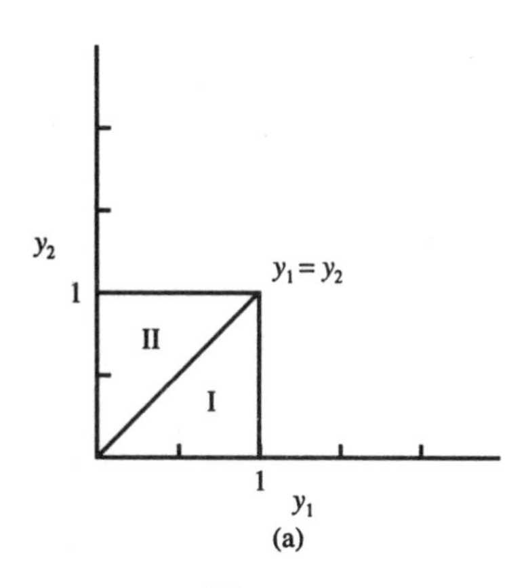

Figure 6.7a Figure 6.7b

In region II of Figure 6.7(a) we have $0 \le y_1 < y_2 \le 1$, so that $U = \frac{Y_1}{Y_2}$ is in the interval $0 \le u \le 1$ and $0 \le y_2 \le 1$. However, region I, where $0 \le y_2 \le y_1 \le 1$, U has minimum value $u = \frac{y_2}{y_2} = 1$ and maximum value $u = \frac{1}{y_2}$, so that

$$1 \le u \le \tfrac{1}{y_2} \qquad 0 \le y_2 \le 1$$

Written in another form, the ranges are $0 \le y_2 \le \frac{1}{u}$, $u > 1$. The marginal density of U is given by

$$f_U(u) = \int_{-\infty}^{\infty} g(y_2, u)\, dy_2 = \int_0^1 y_2\, dy_2 = \frac{y_2^2}{2}\bigg]_0^1 = \frac{1}{2} \qquad \text{if } 0 \le u \le 1$$

and

$$f_U(u) = \int_0^{1/u} y_2\, dy_2 = \frac{y_2^2}{2}\bigg]_0^{1/u} = \frac{1}{2u^2} \qquad \text{if } u > 1$$

c. Consider the joint distribution of $U = -\ln Y_1 Y_2$ and Y_1. Letting $Y_1 = y_1$, we have $U = -\ln y_1 Y_2$, so that

$$Y_2 = \frac{e^{-U}}{y_1} \qquad \text{and} \qquad \frac{dy_2}{du} = \frac{-1}{y_1} e^{-u}$$

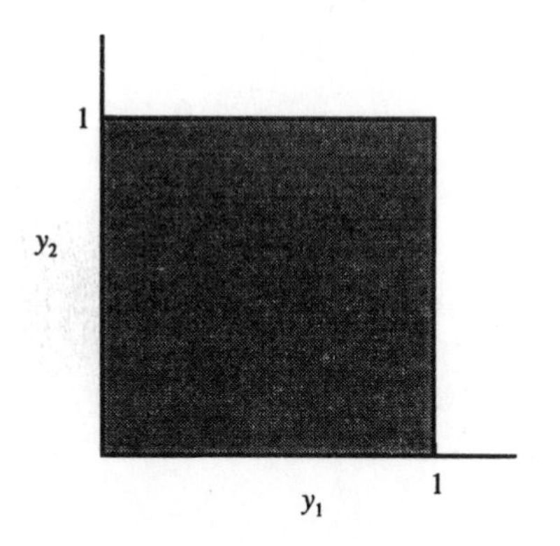

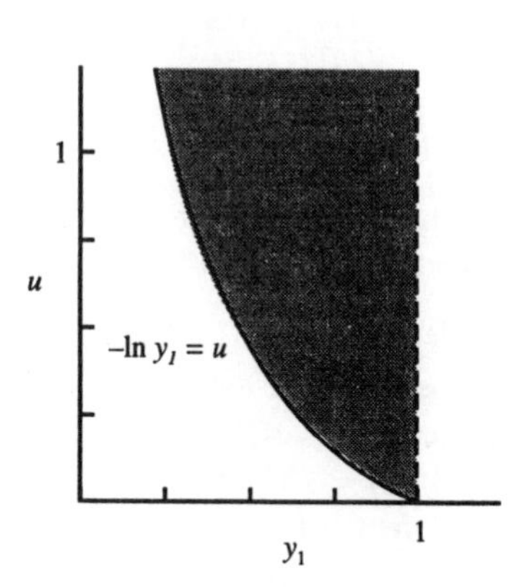

Figure 6.8a Figure 6.8b

The joint density of Y_1 and U is then $g(y_1, u) = \frac{1}{y_1} e^{-u}$. The ranges of Y_1 and U are shown in Figure 6.8. When $0 \le y_2 \le 1$, $U = -\ln Y_1 Y_2$ ranges in the interval $-\ln y_1 \le u \le \infty$ and $0 \le y_1 \le 1$. Written in another form, the ranges are $e^{-u} \le y_1 \le 1, 0 \le u \le \infty$. The marginal density of U is then

$$f_U(u) = \int_{e^{-u}}^{1} \frac{1}{y_1} e^{-u}\, dy_1 = e^{-u}[\ln y_1]_{e^{-u}}^{1} = ue^{-u} \qquad \text{for } 0 \le u \le \infty.$$

d. Consider the joint distribution of $U = Y_1Y_2$ and Y_1. For $Y_1 = y_1$ we have

$$Y_2 = \frac{U}{y_1} \qquad \frac{dy_2}{du} = \frac{1}{y_1}$$

and

$$g(y_1, u) = \frac{1}{y_1}$$

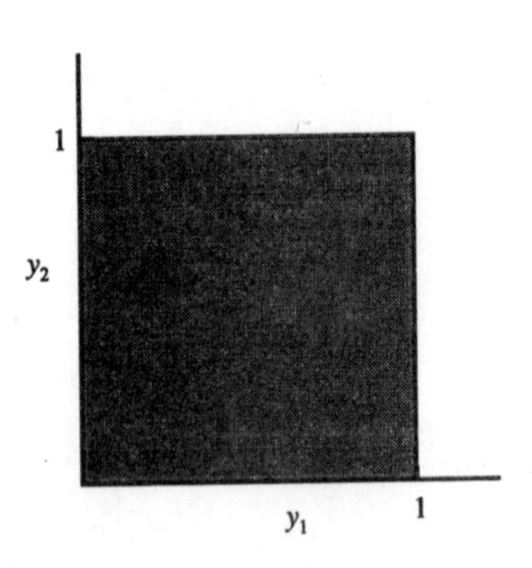

Figure 6.9a Figure 6.9b

The ranges of Y_1 and U are $0 \le y_1 \le 1$ and $0 \le u \le y_1$, or $u \le y_1 \le 1$ and $0 \le u \le 1$, as shown in Figure 6.9. Then

$$f_U(u) = \int_{u}^{1} \frac{1}{y_1}\, dy_1 = \ln y_1]_u^1 = -\ln u \qquad 0 \le u \le 1$$

6.76 The joint distribution of Y_1 and Y_2 is given as

$$f(y_1, y_2) = f_1(y_1)f_2(y_2) = e^{-(y_1+y_2)} \qquad y_1 > 0,\ y_2 > 0$$

The variable of interest is $U = \frac{Y_1}{Y_1+Y_2}$. Now fixing $Y_2 = y_2$, we obtain $Y_1 = \frac{Uy_2}{1-U}$, an increasing function of U in the range $(0, 1)$ and

$$\left|\frac{dy_1}{du}\right| = \left|\frac{y_2(1-u+u)}{(1-u)^2}\right| = \frac{y_2}{(1-u)^2}.$$

Then

$$g(y_2, u) = \frac{y_2}{(1-u)^2} \exp\left[-\left(\frac{y_2 u}{1-u} + y_2\right)\right] = \frac{y_2^2}{(1-u)^2} \exp\left[-y_2\left(\frac{1}{1-u}\right)\right]$$

for $0 \le u \le 1, 0 \le y_2 \le \infty$. Now the marginal density of U will be

$$g_U(u) = \int_0^\infty g(y_2, u)\, dy_2 = \int_0^\infty \frac{y_2}{(1-u)^2} \exp\left[-y_2\left(\frac{1}{1-u}\right)\right] dy_2 = 1 \qquad 0 \le u \le 1$$

since the integrand represents the density function for a gamma random variable with parameters $\alpha = 2$ and $\beta = 1 - u$.

6.77 If we let $(A, B) = (-1, 1)$ and $T = 0$, then the density function for X, the landing point, is $f(x) = \frac{1}{2}$ for $-1 \le x \le 1$. It is necessary to obtain the probability density function for $U = |X|$. Write

$$F_U(u) = P(U \le u) = P(|X| \le u) = P(-u \le X \le u) = F_X(u) - F_X(-u)$$

On differentiating with respect to u, we obtain

$$f_U(u) = f_X(u) + f_X(-u) = \tfrac{1}{2} + \tfrac{1}{2} = 1 \qquad 0 \le u \le 1$$

Note that since $U = |X|$, U has positive density only on the range $0 \le u \le 1$.

6.78 Let Y_1 be the point on the one-mile stretch chosen for sentry 1 and let Y_2 be the point chosen for sentry 2. It is given that $f_1(y_1) = 1$ for $0 \le y_1 \le 1$ and $f_2(y_2) = 1$ for $0 \le y_2 \le 1$. Since Y_1 and Y_2 are independent, their joint distribution is

$$f(y_1, y_2) = f_1(y_1)f_2(y_2) = 1 \qquad 0 \le y_1 \le 1, 0 \le y_2 \le 1$$

The distance between the two sentries, which is the variable of interest, can be represented by $|Y_1 - Y_2|$. Let us consider first the joint distribution of $u = Y_1 - Y_2$ and Y_2. Letting $Y_2 = y_2$, we have $Y_1 = U + y_2$ and $\left|\frac{dy_1}{du}\right| = 1$. Hence $g(y_2, u) = 1$. The region on which the density is positive is shown in Figure 6.10.

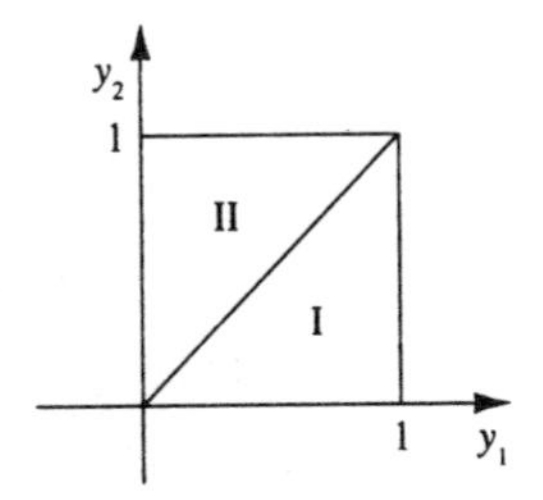

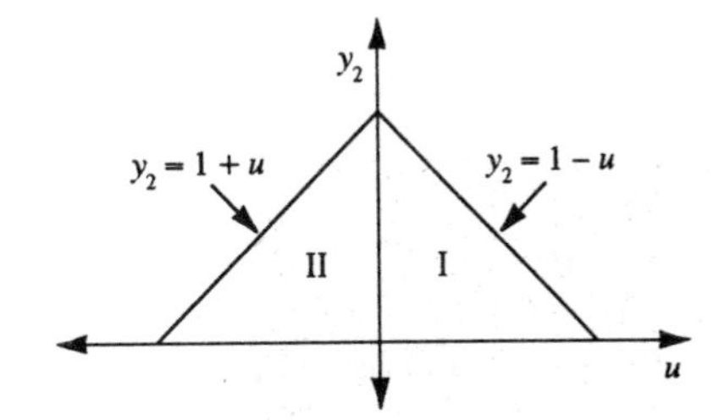

Figure 6.10

$$g(u) = \int_0^{1+u} dy_2 = 1 + u \qquad \text{for} \qquad -1 \le u \le 0$$

$$g(u) = \int_0^{1-u} dy_2 = 1 - u \qquad \text{for} \qquad 0 \le u \le 1.$$

Now the probability of interest is

$$P\left(|Y_1 - Y_2| < \tfrac{1}{2}\right) = P\left(|U| < \tfrac{1}{2}\right) = P\left(-\tfrac{1}{2} \le U \le \tfrac{1}{2}\right)$$
$$= \int_{-1/2}^{0} (1 + u)\, du + \int_0^{1/2} (1 - u)\, du$$
$$= \tfrac{3}{4}$$

6.79 The joint distribution of Y_1 and Y_2 is

$$f(y_1, y_2) = \tfrac{1}{2\pi} \exp\left[-\tfrac{1}{2}\left(y_1^2 + y_2^2\right)\right] \qquad -\infty \le y_1 \le \infty, -\infty \le y_2 \le \infty$$

Consider the joint distribution of $U = \frac{Y_1}{Y_2}$ and Y_2 by letting $Y_2 = y_2$ so that $Y_1 = Uy_2$,

$$\left|\tfrac{dy_1}{du}\right| = |y_2|, \text{ and}$$

$$g(y_2, u) = \tfrac{|y_2|}{2\pi} \exp\left\{-\tfrac{1}{2}\left[y_2^2\left(1 + u^2\right)\right]\right\}$$

Now

$$g(u) = \int_{-\infty}^{\infty} \tfrac{|y_2|}{2\pi} \exp\left\{-\tfrac{1}{2}\left[y_2^2\left(1 + u^2\right)\right]\right\} dy_2 = \tfrac{2}{2\pi} \int_0^{\infty} y_2 \exp\left\{-\tfrac{1}{2}\left[y_2^2\left(1 + u^2\right)\right]\right\} dy_2$$

since the integrand is an even function. Making transformation $y_2^2 = z$, we have

$$g(u) = \left(\tfrac{1}{2\pi}\right) \tfrac{1}{(1+u^2)} \int_0^{\infty} \left(1 + u^2\right) \exp\left[-\tfrac{1}{2} z\left(1 + u^2\right)\right] dz = \tfrac{1}{2\pi\left(1+u^2\right)} \qquad -\infty \le u \le \infty$$

since the integral is that of an exponential variable with $\beta = \frac{2}{(1+u^2)}$.

6.80 Let $U = Y_1 - Y_2$. We want to find $P(U = u) = P(Y_1 - Y_2 = u)$. If $u > 0$ then $y_1 = u + y_2 > 0$ so that we can write

$$
\begin{aligned}
P(U=u) &= P(Y_1 - Y_2 = u) \\
&= \sum_{y_2=1}^{\infty} P(Y_1 = u + y_2) P(Y_2 = y_2) \\
&= p^2(1-p)^u \sum_{y_2=1}^{\infty} (1-p)^{2y_2-1} \\
&= \frac{p^2(1-p)^{u+1}}{(1-(1-p)^2)} \sum_{y_2=1}^{\infty} [1-(1-p)^2]\left[(1-p)^2\right]^{y_2-1} \\
&= \frac{p^2(1-p)^{u+1}}{(1-(1-p)^2)} \\
&= \frac{p(1-p)^{u+1}}{(2-p)}
\end{aligned}
$$

Where the infinite sum is 1 as it is the sum of values of a geometric random variable with success probability $1-(1-p)^2$. On the other hand, if $u < 0$ then $y_2 = y_1 - u > 0$ and we can proceed in a fashion similar to the above approach. Specifically,

$$P(U=u) = P(Y_1 - Y_2 = u) = \sum_{y_1=1}^{\infty} P(Y_2 = y_1 - u) P(Y_1 = y_1)$$

$$= p^2(1-p)^{-u-1} \sum_{y_1=1}^{\infty} (1-p)^{2y_2} = \frac{p^2(1-p)^{-u+1}}{(1-(1-p)^2)} = \frac{p(1-p)^{-u+1}}{(2-p)}.$$

Putting these results together gives us

$$P(u=u) = \frac{p(1-p)^{|u|+1}}{(2-p)}.$$

6.81 Note that the inverse transformation is $y = 1/u - 1$. Then

$$f_U(u) = \frac{1}{B(\alpha,\beta)}\left(\frac{1-u}{u}\right)^{\alpha-1} u^{\alpha+\beta}\frac{1}{u^2} = \frac{1}{B(\alpha,\beta)} u^{\beta-1}(1-u)^{\alpha-1}$$

Thus, $U \sim \text{Beta}(\beta, \alpha)$.

6.82 Write $U = F(Y)$, an increasing function of Y. Then $Y = F^{-1}(u)$ and $\frac{du}{dy} = f(y)$, so that $\frac{dy}{du} = \frac{1}{f(y)}$. Since $f(y)$ is positive, $\left|\frac{dy}{du}\right| = \frac{dy}{du}$ and, using the method of transformations, we have

$$g(u) = f\left(F^{-1}(u)\right)\frac{dy}{du} = f\left(F^{-1}(u)\right)\frac{1}{f\left(F^{-1}(u)\right)} = 1 \qquad \text{for} \qquad 0 \le u \le 1$$

Notice that the range of $U = F(Y)$ reflects the fact that $0 \le F(y) \le 1$.

6.83 If Y is distributed uniformly on the interval $(-1, 3)$, then

$$f(y) = \begin{cases} \frac{1}{4}, & -1 \le y \le 3 \\ 0, & \text{elsewhere} \end{cases}$$

Now, if $U = Y^2$, then

$$g(u) = \frac{1}{2\sqrt{u}}\left[f\left(\sqrt{u}\right) + f\left(-\sqrt{u}\right)\right]$$

as in Example 6.4. If $-1 \le y \le 3$, then $0 \le u \le 9$; however, if $1 \le u \le 9$, $f\left(-\sqrt{u}\right)$ is not positive. That is,

$$g(u) = \frac{1}{2\sqrt{u}}\left(\frac{1}{4} + \frac{1}{4}\right) = \frac{1}{4\sqrt{u}} \qquad \text{if } 0 \le u \le 1$$

but

$$g(u) = \frac{1}{2\sqrt{u}}\left(\frac{1}{4} + 0\right) = \frac{1}{8\sqrt{u}} \qquad \text{if } 1 \le u \le 9$$

6.84 The reliability of a component with length of life Y is defined as $P(Y > y)$. For a system with four independent components, C_1, C_2, C_3, and C_4, define Y_i to be the length of life for component C_i and define X to be the length of life for the entire system. Now the reliability of the system can be written as $P(X > y)$.
Notice that the system will operate as long as there is an unbroken chain of components between A and B. Hence either C_3 or C_4 or both must be in operation (as well as C_1 and C_2). Then

$$P(X > y) = P(Y_1 > y)P(Y_2 > y)P(Y_3 > y)P(Y_4 < y)$$

$$+ P(Y_1 > y)P(Y_2 > y)P(Y_3 < y)P(Y_4 > y) + \prod_{i=1}^{4} P(Y_i y)$$
$$= 2[1 - F(y)]^3 F(y) + [1 - F(y)]^4 = [1 - F(y)]^3[2F(y) + 1 - F(y)]$$
$$= [1 - F(y)]^3[1 + F(y)]$$

6.85 Let C_3 be production cost. Then the profit function (per gallon) is

$$U = \begin{cases} C_1 - C_3, & \frac{1}{3} < y < \frac{2}{3} \\ C_2 - C_3, & 0 \le y \le \frac{1}{3} \text{ or } \frac{2}{3} \le y \le 1 \end{cases}$$

Since profit can only take on two different values, depending on the region into which y falls, the profit distribution is a discrete random variable with

$$P(U = C_1 - C_3) = \int_{1/3}^{2/3} (20y^3 - 20y^4)\, dy = [5y^4 - 4y^5]_{1/3}^{2/3} = .4156$$
$$P(U = C_2 - C_3) = 1 - .4156 = .5844$$

6.86 **a.** Let X denote the next gap. Then

$$P(X \le 60) = \int_0^{60} \tfrac{1}{10} e^{-x/10}\, dx = \left[-e^{-x/10}\right]_0^{60} = 1 - e^{-6}$$

b. If we assume that the next four gap times, X_1, X_2, X_3, and X_4 are independent of one another, then the sum, Y, of the next four gap times is the sum of four independent random variables each having an exponential distribution with $\theta = 10$. The moment-generating function of X_i is $(1 - 10t)^{-1}$. By Theorem 6.2, $m_Y(t) = (1 - 10t)^{-4}$. Thus, Y will have a gamma distribution with $\alpha = 4$ and $\beta = 10$ so

$$f(y) = \frac{y^3 e^{-y/10}}{\Gamma(4)(10)^4} \qquad y > 0.$$

6.87 $P(\text{largest} > \text{median}) = 1 - P(\text{largest} < \text{median}) = 1 - P(\text{all } Y\text{'s} < \text{median})$
$= 1 - \left(\frac{1}{2}\right)^4.$

6.88 Recall that $Y_{(1)}$ is exponential with mean $\frac{\beta}{n}$ [refer to Exercise 6.65].
Thus, $Y_{(1)}$ is exponential with $\mu = \frac{15}{5} = 3$.

a. $P(Y > 9) = \int_9^{\infty} \left(\frac{1}{3}\right) e^{-y/3}\, dy = e^{-y/3}\Big]_9^{\infty} = e^{-3}.$

b. $P(Y < 12) = \int_0^{12} \left(\frac{1}{3}\right) e^{-y/3}\, dy = e^{-y/3}\Big]_0^{12} = 1 - e^{-4}.$

6.89 **a.** Let $U = \ln Y$ so that $\frac{dU}{dy} = \frac{1}{y}$. Since the density function of U is known, $f_Y(y)$ can be derived as

$$f_Y(y) = f_U(\ln y)\left|\frac{du}{dy}\right| = \frac{1}{y\sigma\sqrt{2\pi}} \exp\left[\frac{-(\ln y - \mu)^2}{2\sigma^2}\right] \qquad \text{for } y \ge 0$$

b. $E(Y) = \int_0^{\infty} \frac{1}{\sigma\sqrt{2\pi}} \exp\left[\frac{-(\ln y - \mu)^2}{2\sigma^2}\right] dy.$

Let $z = \ln y$. Then $-\infty \le z \le \infty$ and $dz = \left(\frac{1}{y}\right) dy$. Thus,

$$E(Y) = \int_{-\infty}^{\infty} \frac{e^z}{\sigma\sqrt{2\pi}} \exp\left[\frac{-(z-\mu)^2}{2\sigma^2}\right] dz.$$

Let $x = z - \mu$ so that $z = x + \mu$ and $dz = dx$. Then

$$E(Y) = e^{\mu} \int_{-\infty}^{\infty} \frac{1}{\sigma\sqrt{2\pi}} \exp\left[-\frac{x^2 - 2\sigma^2 x}{2\sigma^2}\right] dx$$
$$= e^{\mu + (\sigma^2/2)} \int_{-\infty}^{\infty} \frac{1}{\sigma\sqrt{2\pi}} \exp\left[\frac{-(x-\sigma^2)^2}{2\sigma^2}\right] dx = e^{\mu + (\sigma^2/2)}$$

Note that the second equality is obtained by completing the square in the exponent.

6.90 The volume of a sphere is given by $V = \frac{4}{3}\pi r^3$ or $r = \left(\frac{3}{4\pi}V\right)^{1/3}$ with $\left|\frac{dr}{dV}\right| = \left(\frac{1}{3}\right)\left(\frac{3}{4\pi}\right)^{1/3} V^{-2/3}$. Making the transformation, we have

$$f(v) = 2\left(\frac{3V}{4\pi}\right)^{1/3} \frac{1}{3}\left(\frac{3}{4\pi}\right)^{1/3} V^{-2/3} = \frac{2}{3}\left(\frac{3}{4\pi}\right)^{2/3} V^{-1/3} \text{ for } 0 \le V \le \frac{4\pi}{3}$$

6.91 a. Let R be the distance from a randomly chosen point to the nearest particle. Consider

$$P(R > r) = P(\text{no particles in the sphere of radius } r)$$
$$= P\left[Y = 0 \text{ in a volume of } \left(\tfrac{4}{3}\right)\pi r^3\right]$$

Since the number of particles in a volume V has a Poisson distribution with mean λV,

$$P(R > r) = \frac{\left[\left(\frac{4}{3}\right)\pi r^3 \lambda\right]^0 e^{-(4/3)\pi r^3 \lambda}}{0!} = e^{-(4/3)\pi r^3 \lambda}$$

But $P(R > r) = 1 - F(r)$, so that $F(r) = 1 - e^{-(4/3)\pi r^3 \lambda}$ and $f(r) = 4\pi r^2 \lambda e^{-(4/3)\pi r^3 \lambda}$ for $r > 0$.

b. Let $U = R^3$ so that $R = U^{1/3}$ and $\frac{dr}{du} = \left(\frac{1}{3}\right) u^{-2/3}$. Then

$$f(u) = 4\pi \lambda u^{2/3} e^{-(4/3)\pi \lambda u} \left(\tfrac{1}{3} u^{-2/3}\right) = \tfrac{4\lambda\pi}{3} e^{-(4\pi\lambda/3)u}$$

which is an exponential distribution with $\beta = \left(\frac{3}{4}\right)\pi\lambda$.

CHAPTER 7 SAMPLING DISTRIBUTIONS AND THE CENTRAL LIMIT THEOREM

7.1 **a.** $P(|\overline{Y}-\mu| \le .3) = P[-.3 \le (\overline{Y}-\mu) \le .3] = P\left(\frac{-.3}{\left(\frac{\sigma}{\sqrt{n}}\right)} \le \frac{Y-\mu}{\left(\frac{\sigma}{\sqrt{n}}\right)} \le \frac{.3}{\left(\frac{\sigma}{\sqrt{n}}\right)}\right)$
$= P(-1.2 \le z \le 1.2) = 1 - 2P(z > 1.2) = 1 - 2(.1151) = .7698.$

b. Repeating the same calculation as in part **a.** we see that

$$P(|\overline{Y}-\mu| \le .3) = P\left(\frac{-.3}{\left(\frac{\sigma}{\sqrt{n}}\right)} \le \frac{Y-\mu}{\left(\frac{\sigma}{\sqrt{n}}\right)} \le \frac{.3}{\left(\frac{\sigma}{\sqrt{n}}\right)}\right) = P(-.3\sqrt{n} \le z \le .3\sqrt{n}).$$
$$= 1 - 2P(z > .3\sqrt{n}).$$

Thus, for $n = 25$ we have $P(|\overline{Y}-\mu| \le .3) = 1 - 2P(z > 1.5) = .8664$, for $n = 36$ we have $1 - 2P(z > 1.8) = .9284$, for $n = 49$ we have $1 - 2P(z > 2.1) = .9642$, and finally for $n = 64$ we have $1 - 2P(z > 2.4) = .9836$.

c. We see that the probabilities increase with n. This is because the variability of $\overline{Y}$ (which is σ^2/n)decreases with n while the mean (μ) stays the same. That is, the density of $\overline{Y}$ gets places more mass around μ as n increases.

d. These results are consistent with example 7.2. That is $P(|\overline{Y}-\mu| \le .3)$ is less than .95 for for the values of n we tried that were lower than 43 and vice-versa for all the values of n that we tried that were larger than 43.

7.2 **a.** $P(|\overline{Y}-\mu| \le .3) = P[-.3 \le (\overline{Y}-\mu) \le .3] = P\left(\frac{-.3}{\left(\frac{\sigma}{\sqrt{n}}\right)} \le \frac{Y-\mu}{\left(\frac{\sigma}{\sqrt{n}}\right)} \le \frac{.3}{\left(\frac{\sigma}{\sqrt{n}}\right)}\right)$
$= P(-.3\sqrt{n}/2 \le z \le .3\sqrt{n}/2) = P(-.15\sqrt{n} \le z \le .15\sqrt{n})$
$= 1 - 2P(z > .15\sqrt{n}) = 1 - 2P(z > .45) = 1 - 2(.3264) = .3472$. This probability is less than that found in example 7.1 (as σ has increased).

b. The probability in question is $1 - 2P(z > .15\sqrt{n})$ which is $1 - 2P(z > .75) = .5468$ for $n = 25$, .6318 for $n = 36$, .7062 for $n = 49$, and .7698 for $n = 64$.

c. The probabilities increase as n increases.

d. As we increase σ the variability of $\overline{Y}$ increases. Therefore the probabilities for exercise 7.2 part **b** should be smaller than those for exercise 7.1 part **b.** (which is exactly what we see).

7.3 Since the distribution of basal areas is normally distributed with mean μ and variance $\sigma^2 = 16$, the sample mean will also be normally distributed, from Theorem 7.1. Then

$$P(|\overline{Y}-\mu| \le 2) = P[-2 \le (\overline{Y}-\mu) \le 2] = P\left(\frac{-2}{\left(\frac{\sigma}{\sqrt{n}}\right)} \le \frac{Y-\mu}{\left(\frac{\sigma}{\sqrt{n}}\right)} \le \frac{2}{\left(\frac{\sigma}{\sqrt{n}}\right)}\right)$$
$$= P(-1.5 \le z \le 1.5)$$
$$= 1 - 2P(Z > 1.5) = 1 - 2(.0668) = .8664$$

7.4 Refer to Exercise 7.3. It is necessary to have

$$P(|\overline{Y}-\mu \le 1|) = .90 \qquad \text{or} \qquad P\left(\frac{-\sqrt{n}}{4} \le z \le \frac{\sqrt{n}}{4}\right) = .90$$

The inequality will be satisfied if we take $\frac{\sqrt{n}}{4} = 1.645$, or $n = 43.30$. Hence 44 trees must be sampled.

7.5 Similar to Exercise 7.3. It is necessary to calculate

$$P(|\overline{Y}-\mu| \le .5) = P\left(\frac{|\overline{Y}-\mu|}{\left(\frac{\sigma}{\sqrt{n}}\right)} \le \frac{.5}{\left(\frac{\sigma}{\sqrt{n}}\right)}\right) = P(|x| \le 2.5) = 1 - 2(.0062) = .9876$$

7.6 Similar to Exercise 7.4. It is necessary to have

$$P(|\overline{Y}-\mu| \le .5) = .95 \qquad \text{or} \qquad P\left(|z| \le \frac{.5\sqrt{n}}{\sqrt{.4}}\right) = .95$$

That is, $\frac{.5\sqrt{n}}{\sqrt{.4}} = 1.96$ or $n = 6.15$. Thus at least 7 tests must be run.

7.7 Use Theorems 6.3 and 7.1 together,

a. $E(\overline{X}-\overline{Y}) = E(\overline{X}) - E(\overline{Y}) = \mu_1 - \mu_2$

b. $V(\overline{X}-\overline{Y}) = V(\overline{X})V + (\overline{Y}) = \frac{\sigma_1^2}{m} + \frac{\sigma_2^2}{n}$

c. It is necessary that

$$P(|(\overline{X}-\overline{Y}) - (\mu_1-\mu_2)| \le 1) = .95$$

or

$$P\left(\frac{|(\overline{X}-\overline{Y})-(\mu_1-\mu_2)|}{\sqrt{\frac{2}{n}+\frac{2.5}{n}}} \le \frac{1}{\sqrt{\frac{2}{n}+\frac{2.5}{n}}}\right) = .95$$

That is, $\frac{1}{\sqrt{\frac{4.5}{n}}} = 1.96$, or $n = 17.29$. Thus two samples of size $m = n = 18$ should be drawn.

7.8 Using the results of Exercise 7.7, one must find

$$P\left[(\overline{X}_A - \overline{X}_B) \ge 1\right] = P\left[\frac{(\overline{X}_A-\overline{X}_B)-(\mu_A-\mu_B)}{\sqrt{\frac{4}{10}+\frac{8}{10}}} \ge \frac{1-0}{\sqrt{\frac{4+.8}{10}}}\right] = P(Z > 2.89)$$
$$= .0019$$

Notice that it is unnecessary to know the values of μ_A and μ_B since their difference is zero.

7.9 We are given that $s^2 = .065$ and $n = 10$. Suppose that $\sigma^2 = .04$. Then from Theorem 7.3, $\frac{9S^2}{.04}$ has a χ^2 distribution with 9 degrees of freedom. Thus,

$$P(S^2 > .065) = P\left(\frac{9S^2}{.04} > \frac{9(.065)}{.04}\right) = P(\chi_9^2 > 14.625) = .10$$

There is some suggestion that $\sigma > .2$.

7.10 a. Since a χ^2 random variable is defined as a gamma random variable with $\alpha = \frac{\nu}{2}$ and $\beta = 2$, the expected value and variance are

$$E(U) = \alpha\beta = \nu \qquad \text{and} \qquad V(U) = \alpha\beta^2 = 2\nu$$

b. Using Theorem 7.3, we see that the quantity $\frac{(n-1)S^2}{\sigma^2}$ has a χ^2 distribution with $\nu = n-1$. Hence, from part **a**,

$$E\left(\frac{(n-1)S^2}{\sigma^2}\right) = n-1 \quad \text{or} \quad \left(\frac{n-1}{\sigma^2}\right)E(\mathrm{S}^2) = n-1 \quad \text{or} \quad E(S^2) = \sigma^2$$

Similarly,

$$V\left(\frac{(n-1)S^2}{\sigma^2}\right) = 2(n-1) \quad \text{or} \quad \left[\frac{(n-1)^2}{\sigma^4}\right]V(\mathrm{S}^2) = 2(n-1) \quad \text{or} \quad \mathrm{V}(S^2) = \frac{2\sigma^4}{n-1}$$

7.11 a. Note $P(S^2 \le b) = P\left(\frac{(n-1)S^2}{\sigma^2} \le \frac{(n-1)b}{\sigma^2}\right) = P\left(\chi^2_{(n-1)} \le \frac{(n-1)b}{\sigma^2}\right)$ where $\chi^2_{(n-1)}$refers to a chi-squared random variable with $n-1$ df. Therefore if we set $\frac{(n-1)b}{\sigma^2} = \chi^2_{.025,(n-1)} = 32.8523$ we get $b = 2.42$.

b. Note $P(a \le S^2) = P\left(\frac{(n-1)a}{\sigma^2} \le \frac{(n-1)S^2}{\sigma^2}\right) = P\left(\frac{(n-1)a}{\sigma^2} \le \chi^2_{(n-1)}\right)$. Then if we set $\frac{(n-1)a}{\sigma^2} = \chi^2_{.975,(n-1)} = 8.96055$ we get $a = .656$.

c. Note that $P(a \le S^2 \le b)$

$= 1 - P(S^2 < a) - P(S^2 > b)$

$= 1 - (1 - P(S^2 \ge a)) - (1 - P(S^2 \le b))$

$= 1 - (1 - .975) - (1 - .975) = 1 - .025 - .025 = .95.$

7.12 Similar to Exercise 7.11. Since, from Definition 7.2,

$$P(g_1 \le (\overline{Y}-\mu) \le g_2) = P\left(\frac{g_1}{\left(\frac{s}{\sqrt{n}}\right)} \le t \le \frac{g_2}{\left(\frac{s}{\sqrt{n}}\right)}\right)$$

the necessary probability will be obtained if we take

$$\frac{g_1}{\left(\frac{s}{\sqrt{n}}\right)} = -t_{.05} \qquad \text{and} \qquad \frac{g_2}{\left(\frac{s}{\sqrt{n}}\right)} = t_{.05}$$

Indexing $n-1 = 8$ degrees of freedom, we have $t_{.05} = 1.86$, $g_1 = \left(\frac{-1.86}{3}\right)S$, and $g_2 = \left(\frac{1.86}{3}\right)S$.

7.13 Since Y has an F distribution, $Y = \frac{\left(\frac{\chi_1^2}{\nu_1}\right)}{\left(\frac{\chi_2^2}{\nu_2}\right)}$.

$$U = \frac{1}{Y} = \frac{\left(\frac{\chi_1^2}{\nu_2}\right)}{\left(\frac{\chi_2^2}{\nu_2}\right)} = \frac{\left(\frac{\chi_2^2}{\nu_2}\right)}{\left(\frac{\chi_1^2}{\nu_1}\right)}$$

which has an F distribution with ν_2 numerator and ν_1 denominator degrees of freedom.

7.14 **a.** $E(Z) = 0$
$E(Z^2) = V(Z) = [E(Z)]^2 = V(Z) = 1$
since, by definition, a standard normal distribution has mean 0 and variance 1.

b. Similar to Exercise 5.108.

$$E(T) = E\left(\frac{Z}{\sqrt{\frac{Y}{\nu}}}\right) = \sqrt{\nu}\,E(Z)\,E\left(\frac{1}{\sqrt{Y}}\right) = \sqrt{\nu}E(Z)E\left(Y^{-1/2}\right)$$
$$= \sqrt{\nu}(0)\,\frac{\Gamma\left(\frac{\nu}{2}-\frac{1}{2}\right)2^{-1/2}}{\Gamma\left(\frac{\nu}{2}\right)} = 0.$$
$$E(T^2) = E\left(\frac{Z^2}{\left(\frac{Y}{\nu}\right)}\right) = \nu E(Z^2)\,E\left(\frac{1}{Y}\right) = \nu(1)\,\frac{\Gamma\left(\frac{\nu}{2}-1\right)}{\Gamma\left(\frac{\nu}{2}\right)2}$$
$$= \frac{\nu\Gamma\left(\frac{\nu}{2}-1\right)}{\left(\frac{\nu}{2}-1\right)\left(\frac{\nu}{2}-1\right)2} = \frac{\nu}{\nu-2}.$$

Then

$$V(T) = E(T^2) - [E(T)]^2 = \frac{\nu}{\nu-2}.$$

7.15 It was shown that $T = \frac{Z}{\sqrt{\frac{Y}{\nu}}}$ has a t distribution with ν degrees of freedom where Z has a standard normal distribution and Y is an independent χ^2 random variable. Then

$$U = T^2 = \frac{\left(\frac{Z^2}{1}\right)}{\left(\frac{Y}{\nu}\right)}.$$

Theorem 7.2 shows that Z^2 has a χ^2 distribution with 1 degree of freedom (and still independent of Y). Finally, by Definition 7.3, U has an F distribution with 1 numerator degree of freedom and ν denominator degrees of freedom.

7.16 **a.** $E(F) = E\frac{\left(\frac{W_1}{\nu_1}\right)}{\left(\frac{W_2}{\nu_2}\right)} = \frac{\nu_2}{\nu_1}\,E\left(\frac{W_1}{W_2}\right) = \frac{\nu_2}{\nu_1}\,E(W_1)\,E\left(\frac{1}{W_2}\right)$

since W_1 and W_2 are independent. We know

$$E(W_1) = \nu_1$$

and, by the result summarized in Exercise 7.14, part **b**,

$$E\left(\frac{1}{W_2}\right) = \frac{\Gamma\left(\frac{\nu_2}{2}-1\right)2^{-1}}{\Gamma\left(\frac{\nu_2}{2}\right)} = \frac{\Gamma\left(\frac{\nu_2}{2}-1\right)}{\Gamma\left(\frac{\nu_2}{2}-1\right)\left(\frac{\nu_2}{2}-1\right)2} = \frac{1}{\nu_2-2}, \qquad \text{for } \nu_2 > 2.$$

Hence,

$$E(F) = \frac{\nu_2}{\nu_1}\,\frac{\nu_2}{(\nu_2-2)} = \frac{\nu_2}{n_2-2}, \qquad \text{for } \nu_2 > 2.$$

b. $V(F) = E(F^2) - [E(F)]^2$.

$$E(F^2) = E\frac{\left(\frac{W_1^2}{\nu_1^2}\right)}{\left(\frac{W_2^2}{\nu_2^2}\right)} = \frac{\nu_2^2}{\nu_1^2}\,E\left(\frac{W_1^2}{W_2^2}\right) = \frac{\nu_2^2}{\nu_1^2}\,E(W_1^2)\,E\left(\frac{1}{W_2^2}\right)$$

since W_1 and W_2 are independent. Using the result summarized in Exercise 7.14, part **b**,

$$E(W_1^2) = \frac{\Gamma\left(\frac{\nu_1}{2}+2\right)4}{\Gamma\left(\frac{\nu_1}{2}\right)} = \nu_1(\nu_1+2)$$

and

$$E\left(\frac{1}{W_2^2}\right) = \frac{\Gamma\left(\frac{\nu_2}{2}-2\right)}{\Gamma\left(\frac{\nu_2}{2}\right)4} = \frac{\Gamma\left(\frac{\nu_2}{2}-2\right)}{\left(\frac{\nu_2}{2}-1\right)\left(\frac{\nu_2}{2}-2\right)\Gamma\left(\frac{\nu_2}{2}-2\right)4} = \frac{1}{(\nu_2-2)(\nu_2-4)},$$

if $\nu_2 > 4$.

Then

$$E(F^2) = \frac{\nu_2^2}{\nu_1^2}\,\frac{\nu_1(\nu_1+2)}{(\nu_2-2)(\nu_2-4)} = \frac{\nu_2^2(\nu_1+2)}{\nu_1(\nu_2-2)(\nu_2-4)}$$

Finally, using the result in part **a**,

$$V(F) = \frac{\nu_2^2(\nu_1+2)}{\nu_1(\nu_2-2)(\nu_2-4)} - \left(\frac{\nu_2}{\nu_2-2}\right)^2 = \frac{2\nu_2^2(\nu_1+\nu_2-2)}{\nu_1(\nu_2-2)^2(\nu_2-4)} \qquad \text{if } \nu_2 > 4.$$

7.17 **a.** Using the result in Exercise 7.16, part **a**

$$E(F) = \frac{\nu_2}{\nu_2 - 2} = \frac{70}{68} = 1.029$$

b. Using the result in Exercise 7.16, part **b**,

$$V(F) = \frac{2\nu_2^2(\nu_1+\nu_2-2)}{\nu_1(\nu_2-2)^2(\nu_2-4)} = \frac{2(70)^2(118)}{50(68)^2(66)} = .076$$

c. $3 = \mu + 7.15\sigma$. It is not likely that F will exceed 3.

7.18 **a.** Assume that $\sigma_1^2 = 2\sigma_2^2$, with $n_1 = 10$ and $n_2 = 8$. Using Definition 7.3 and Theorem 7.3, we know that

$$\frac{\left(\frac{S_1^2}{\sigma_1^2}\right)}{\left(\frac{S_2^2}{\sigma_2^2}\right)} = \frac{\left(\frac{S_1^2}{2\sigma_2^2}\right)}{\left(\frac{S_2^2}{\sigma_2^2}\right)} = \frac{S_1^2}{2S_2^2}$$

has an F distribution with $n_1 - 1 = 9$ and $n_2 - 1 = 7$ degrees of freedom. Therefore $.95 = P\left(\frac{S_1^2}{S_2^2} \le b\right) = P\left(\frac{S_1^2}{2S_2^2} \le \frac{b}{2}\right)$. By setting $\frac{b}{2} = F_{.05} = 3.68$ we get $b = 7.36$.

b. Now we want $.95 = P\left(a \le \frac{S_1^2}{S_2^2}\right) = P\left(\frac{a}{2} \le \frac{S_1^2}{2S_2^2}\right)$. Therefore we set $\frac{a}{2} = F_{.95}$.

We note that $F_{.95}$ satisfies $P\left(F_{.95} \le \frac{S_1^2}{2S_2^2}\right) = .95$, implying $P\left(\frac{S_1^2}{2S_2^2} < F_{.95}\right) = .05$.

Now we use the fact that $\frac{2S_2^2}{S_1^2}$ has an F distribution with $n_2 - 1 = 7$ and $n_1 - 1 = 9$ degrees of freedom, and

$$P\left(\frac{S_1^2}{2S_2^2} < F_{.95}\right) = P\left(\frac{2S_2^2}{S_1^2} > \frac{1}{F_{.95}}\right) = .05$$

Then we have

$$\frac{1}{F_{.95}} = 3.29 \qquad \text{or} \qquad F_{.95} = \frac{1}{3.29} = .304.$$

Therefore $a = .304(2) = .608$.

c.

$$\begin{aligned} P\left(a \le \frac{S_1^2}{S_2^2} \le b\right) &= 1 - P\left(a > \frac{S_1^2}{S_2^2}\right) - P\left(\frac{S_1^2}{S_2^2} > b\right) \\ &= 1 - \left(1 - P\left(a \le \frac{S_1^2}{S_2^2}\right)\right) - \left(1 - P\left(\frac{S_1^2}{S_2^2} \le b\right)\right) \\ &= 1 - (1 - .95) - (1 - .95) = 1 - .05 - .05 = .90. \end{aligned}$$

7.19 **a.** χ^2 with 5 degrees of freedom, by Theorem 7.2

b. χ^2 with 4 degrees of freedom, by Theorem 7.3. Note that $\sigma^2 = 1$.

c. χ^2 with 5 degrees of freedom. Let $R = \sum_{i=1}^{5} (Y_i - \overline{Y})^2 + Y_6^2 = U + Y_6^2$, which is χ^2 with $(4+1)$ degrees of freedom.

7.20 **a.** $\frac{\sqrt{5}Y_6}{\sqrt{W}} = \frac{Y_6}{\sqrt{\frac{W}{5}}}$, which has a t distribution with 5 degrees of freedom where Y_6 is a standard normal random variable and W is chi-square with 5 degrees of freedom.

b. $\frac{2Y_6}{\sqrt{U}} = \frac{Y_6}{\sqrt{\frac{U}{4}}}$, which has a t distribution with 4 degrees of freedom where Y_6 is a standard normal random variable and U is chi-square with 4 degrees of freedom.

c. $\overline{Y}$ is $N\left(0, \frac{1}{\sqrt{5}}\right)$. Thus, $\sqrt{5}\,\overline{Y}$ is $N(0, 1)$.

$\left(\sqrt{5}\,\overline{Y}\right)^2 = 5\overline{Y}^2$ is chi-square with 1 df.

Y_6^2 is $\chi^2(1)$ and U is $\chi^2(4)$.

Now, $\frac{2\left(5\overline{Y}^2 + Y_6^2\right)}{U} = \frac{\left(\frac{5\overline{Y}^2+Y_6^2}{2}\right)}{\left(\frac{U}{4}\right)}$, which has an F distribution with numerator degrees of freedom $= 2$ and denominator degrees of freedom $= 4$.

7.21 **a.** Note that $\overline{X}_i$ for $i = 1, \ldots, k$ are normally distributed with variances $\sigma^2/n_i$and means μ_i respectively (by theorem 7.1). Further we know the $\overline{X}_i$ are independent as each is a function of only one of the independent groups of random variables. Then by theorem 6.3, $\sum_{i=1}^{k} c_i\,\overline{X}_i$

is also normally distributed with mean $\sum_{i=1}^{k} c_i\mu_i$ and variance $\sum_{i=1}^{k}(c_i\sigma)^2/n_i$.

b. Note that $(n_i - 1)S_i^2/\sigma^2$ are all chi squared random variables with $n_i - 1$ degrees of freedom (by theorem 7.3). Further, as argued for the sample means in part **a** , each of the $(n_i - 1)S_i^2/\sigma^2$ are independent. Recall from chapter 6 that the sum of chi squared random variables is chi squared with degrees of equal to the sum of the degrees of freedom o f the summands (this is easily verified using the method of moment generating functions). Therefore, $SSE/\sigma^2 = \sum_{i=1}^{k}(n_i - 1)S_i^2/\sigma^2$ is chi squared with $\sum_{i=1}^{k}(n_i - 1) = n - k$ degrees of freedom (where n is the total sample size, $n = \sum_{i=1}^{k} n_i$).

c. Notice, $t = \frac{\hat{\theta}-\theta}{\sqrt{MSE\sum_{i=1}^{k} c_i/n_i}} = \frac{\hat{\theta}-\theta}{\sqrt{\frac{SSE}{(n-k)\sigma^2}\left(\sum_{i=1}^{k} c_i\sigma^2/n_i\right)}} = \frac{\hat{\theta}-\theta}{\sqrt{\left(\sum_{i=1}^{k} c_i\sigma^2/n_i\right)}} \Big/ \frac{SSE}{(n-k)\sigma^2}$, is a standard normal divided by a chi squared divided by its degrees of freedom. Thus, provided SSE is independent of $\hat{\theta}$, we know that the quantity t is distributed as a t random variable with $n - k$ df. To see that SSE is independent of $\hat{\theta}$ notice that $\overline{X}_i$ is independent of S_i because sample means are independent of sample standard deviations. Further $\overline{X}_i$ is independent of S_j for $i \neq j$ because they are functions of different independent groups of random variables. Therefore, the $\overline{X}_i$ are mutually independent of the S_i^2 implying SSE is independent of $\hat{\theta}$.

7.22 **a.** $P(\overline{X} > 4.5) = P\left(\frac{\overline{X}-4}{2/\sqrt{100}} > \frac{4.5-4}{2/\sqrt{100})}\right) \approx P(Z > 2.5) = .0062.$

b. There are several correct answers to this problem. We choose the answer that produces an interval symmetric about μ. That is we want to find an a so that

$$P(\mu - a < \overline{X} < \mu + a) = P(|\overline{X} - \mu| > a) = .95.$$

Note that

$$P(|\overline{X} - \mu| > a) = P\left(|\tfrac{\overline{X}-\mu}{\sigma/\sqrt{n}}| > \tfrac{a}{\sigma/\sqrt{n}}\right) \approx P\left(|Z| > \tfrac{a}{\sigma/\sqrt{n}}\right).$$

where Z is a standard normal random variable. Then setting $\frac{a}{\sigma/\sqrt{n}} = 1.96$, or $a = 1.96\,\sigma/\sqrt{n}$, will give an approximate 95% interval. Therefore our interval is $\mu \pm 1.96\,\sigma/\sqrt{n}$. For this particular problem we get the interval $14 \pm 1.96(2)/\sqrt{100}$, $(13.608, 14.392)$.

7.23 The Central Limit Theorem (Theorem 7.4) states that $Y_n = \frac{\sqrt{n}(\overline{X}-\mu)}{\sigma}$ converges in distribution to a standard normal random variable, which is denoted by Z. For this exercise, $n = 100$, $\sigma = 2.5$, and the approximation is

$$P\left(|\overline{X} - \mu| \le .5\right) = P\left(-.5 \le \overline{X} - \mu \le .5\right) = P\left[\tfrac{-.5(10)}{2.5} \le Z \le \tfrac{.5(10)}{2.5}\right]$$
$$= P(-2 \le Z \le 2) = 1 - 2(.0228) = .9544$$

7.24 Refer to Exercise 7.23. It is now necessary to choose n such that

$$P\left(|\overline{X} - \mu| \le .4\right) = .95$$

or

$$P\left(-.4 \le \overline{X} - \mu \le .4\right) = P\left(\tfrac{-.4\sqrt{n}}{2.5} \le Z \le \tfrac{.4\sqrt{n}}{2.5}\right) = .95.$$

This probability statement will be satisfied by taking $\frac{.4\sqrt{n}}{2.5} = 1.96$, which implies $\sqrt{n} = \frac{4.9}{.4}$ or $n = 150.0625$. Thus, $n = 151$ men should be chosen.

7.25 Similar to Exercise 7.23. for this exercise, $\mu = 7.00$, $\sigma = .50$, $n = 64$, and the is approximation

$$P(\overline{X} \le 6.90) = P\left[Z \le \tfrac{\sqrt{64}\,(6.90-7.00)}{.5}\right] = P(Z \le -1.6) = .0548$$

7.26 With $n = 40$ and $\sigma = \frac{\text{range}}{4} = \frac{3}{4}$, the approximation is

$$P\left(|\overline{Y} - \mu| \le .2\right) = P\left[|Z| \le \tfrac{\sqrt{40}(.2)}{.75}\right] = P(|Z| \le 1.69) = -2(.0455) = .9090$$

7.27 Refer to Exercise 7.26. It is necessary to have

$$P\left(|\overline{Y}-\mu|\leq .1\right)=P\left[|Z|\leq \frac{\sqrt{n}(.1)}{.75}\right]=.90$$

Hence, take $\frac{\sqrt{n}(.1)}{.75}=1.645$, or $n=152.21$. At least 153 core samples should be taken.

7.28 **a.** The population from which we are randomly sampling $n=35$ measurements is not necessarily normally distributed. However, the sampling distribution of $\overline{X}$ does have an approximate normal distribution, with mean μ and standard deviation $\frac{\sigma}{\sqrt{n}}$.

The probability of interest is

$$P\left[|\overline{X}-\mu|<1\right]=P\left[-1<(\overline{X}-\mu)<1\right].$$

Since

$$Z=\frac{\overline{X}-\mu}{\left(\frac{\sigma}{\sqrt{n}}\right)}$$

has a standard normal distribution, we need only find $\frac{\sigma}{\sqrt{n}}$ to approximate the above probability. Though σ is unknown, it can be approximated by $s=12$ and $\frac{\sigma}{\sqrt{n}}=\frac{12}{\sqrt{35}}=2.028$. Then

$$P\left[|\overline{X}-\mu|<1\right]=P\left[-\tfrac{1}{2.028}<Z<\tfrac{1}{2.028}\right]=P[-.49<Z<.49]=1-2(.3121)$$
$$=.3758.$$

b. No. The population mean μ is only the average of the estimates. It is possible that all of the estimates are too high, for example.

7.29 We are given that $\mu=1.4$ and $\sigma=.7$. Although X, the service time on one automobile, may not have a normal distribution, $\overline{X}$ will. Thus,

$$P(\overline{X}>1.6)=P\left(Z>\frac{\sqrt{50}(1.6-1.4)}{.7}\right)=P(Z>2.02)=.0217$$

7.30 We want $P(|\overline{X}-\mu|<1)\approx P(|Z|<\frac{1}{\sigma/\sqrt{n}})=P(|Z|<1)=.6826.$

7.31 We want $.99=P(|\overline{X}-\mu|<1)\approx P(|Z|<\frac{1}{\sigma/\sqrt{n}})=P(|Z|<\sqrt{n}/10)$. Setting $\frac{\sqrt{n}}{10}=z_{.01/2}=2.576$ yields $n=(25.76)^2=663.57$ which we round up to 664.

7.32 **a.** $P(199<\overline{X}<202)=P\left(\frac{\sqrt{25}(199-200)}{10}<\frac{\sqrt{25}(\overline{X}-200)}{10}<\frac{\sqrt{25}(202-200)}{10}\right)$

$\approx P(-.5<Z<1)=1-.3085-.1587=.5328.$

b. Let X_i be the resistance in the ith resistor Then we want

$$P(\textstyle\sum X_i\leq 5,100)=P(\overline{X}\leq 204).$$

A quick way to solve this problem is to note that 204 is 2 standard deviations above the mean (recall the standard deviation of $\overline{X}$ is $10/\sqrt{25}=2$). An application of he empirical rule indicates the answer should be close to .975. More rigorously

$$P(\overline{X}\leq 204)\approx P\left(Z\leq\frac{\sqrt{25}(204-200)}{10}\right)=P(Z\leq 2)=1-.0228=.9772.$$

7.33 **a.** It is not reasonable to assume that carbon monoxide concentrations are normally distributed with a mean of 12 ppm and a standard deviation of 9 ppm, as it would allow a non-negligible probability of having negative concentrations of carbon monoxide, which is physically impossible.

b. $P(\overline{Y}>14)\approx P\left(Z>\frac{\sqrt{100}(14-12)}{9}\right)=P(Z>2.22)=.0132.$

7.34 $P(\overline{Y}<1.3)\approx P\left(Z<\frac{\sqrt{25}(1.3-1.4)}{.05}\right)=P(Z<10)=1.$

7.35 **a.** i.) We assume that we have a random sample. Also note that the standard deviation is .8 which indicates that the population variance is finite.

ii) The standard deviation of the average daily downtime for 30 days is $\frac{.8^2}{30}=.02$. Then each of the endpoints of the interval (1 hour, 5 hours) are substantially beyond 3 standard deviations from the mean. Therefore the probability is approximately 1.

b. Let Y_i be the downtime on day i. Then we want

$$P\left(\sum_{i=1}^{30} Y_i < 115\right) = P(\overline{Y} < 3.83) \approx P\left(Z < \frac{\sqrt{30}(3.83-4)}{.8}\right)$$
$$= P(Z < -1.14) = .1271.$$

7.36 Let Y_i be the volume of the ith sample. Then the total volume of the composite sample is $\Sigma\, Y_i$, and this total must exceed 200 with probability .95. That is,

$$P\left(\Sigma\, Y_i > 200\right) = P\left(\overline{Y} - \mu > \tfrac{200}{50} - \mu\right) = P\left(Z > \frac{4-\mu}{\sqrt{\frac{4}{50}}}\right) = .95$$

Hence $(4-\mu)\sqrt{\frac{4}{50}} = -1.645$ and $\mu = 4.47$.

7.37 Let X_i be the length of life for the ith heat lamp, $i = 1, 2, \ldots, 25$. It is given that the X_i's are independent, each with mean 50 and standard deviation 4. Then by the Central Limit Theorem, the random variable

$$Y_n = \frac{\sqrt{n}(X-\mu)}{\sigma} = \frac{n(\overline{X}-\mu)}{\left(\frac{\sigma}{\sqrt{n}}\right)} = \frac{\Sigma\, X_i - n\mu}{\left(\frac{\sigma}{\sqrt{n}}\right)}$$

converges in distribution to a standard normal random variable. Hence, since the lifetime of the lamp system is represented by $V = \sum_{i=1}^{25} X_i$, the probability of interest is

$$P\left(\sum_{i=1}^{25} X_1 \geq 1300\right) = P\left(\frac{\Sigma\, X_i - n\mu}{\left(\frac{\sigma}{\sqrt{n}}\right)} \geq \frac{1300-1250}{\sqrt{400}}\right) = P(Z > 2.5) = .0062$$

7.38 It is given that $X_1, X_2, \ldots, X_n$ are independent and identically distributed with $E(X_i) = \mu_1$ and $V(X_i) = \sigma_1^2$. Similarly, $Y_1, Y_2, \ldots, Y_n$ are independent and identically distributed with $E(Y_i) = \mu_2$ and $V(Y_i) = \sigma_2^2$. Consider $d_i = X_i - Y_i$, for $i = 1, 2, \ldots, n$. The d_i's are independent and identically distributed with $E(d_i) = E(X_i) - E(Y_i)$ $= \mu_1 - \mu_2$ and $V(d_i) = V(X_i) + V(Y_i) = \sigma_1^2 + \sigma_2^2 < \infty$. Hence, applying Theorem 7.4 to the set $d_1, d_2, \ldots, d_n$, we have

$$Y_n = \frac{\left[\overline{d} - (\mu_1-\mu_2)\right]\sqrt{n}}{\sqrt{\sigma_1^2+\sigma_2^2}} = \frac{(\overline{X}-\overline{Y})-(\mu_1-\mu_2)}{\sqrt{\frac{\sigma_1^2+\sigma_2^2}{n}}}$$

which converges in distribution to a standard normal random variable.

7.39 Use the results of Exercise 7.38. It is given that $n = 50$, $\sigma_1 = \sigma_2 = 2$, and $\mu_1 = \mu_2$. Let $\overline{X}$ be the mean for operator B and $\overline{Y}$ be the mean for operator A. Then the probability of interest is

$$P(\overline{X} - \overline{Y} > 1) = P\left[\frac{(\overline{X}-\overline{Y})-(\mu_1-\mu_2)}{\sqrt{\frac{\sigma_1^2+\sigma_2^2}{n}}}\right] > \frac{1-0}{\sqrt{\frac{4+4}{50}}} = P(Z > 2.5) = .0062$$

7.40 An extension of the Central Limit Theorem states that for large values of n_1 and n_2, $\overline{X} - \overline{Y}$ is approximately normally distributed with mean $\mu_1 - \mu_2$ and variance $\frac{\sigma_1^2}{n_1} + \frac{\sigma_2^2}{n_2}$. Hence the probability of interest is approximately

$$P\left[\left|(\overline{X}-\overline{Y}) - (\mu_1-\mu_2)\right| \leq .05\right] = P\left(|Z| \leq \frac{.05}{\sqrt{\frac{.01}{50}+\frac{.02}{100}}}\right) = P\left(|Z| \leq 2.5\right)$$
$$= 1 - 2(.0062) = .9876$$

7.41 It is necessary to have

$$P\left[\left|(\overline{X}-\overline{Y}) - (\mu_1-\mu_2)\right| \leq .04\right] = P\left(|Z| \leq \frac{.04}{\sqrt{\frac{.01+.02}{n}}}\right) = .90$$

Hence $1.645 = \frac{.04}{\sqrt{\frac{.03}{n}}}$, or $n = 50.74$. Each sample must be at least of size $n = 51$.

7.42 Let X_i represent the time to process the ith person's order, where $i = 1, 2, \ldots, 100$. X_i has $\mu = 2.5$ minutes and $\sigma = 2$ minutes. Since 4 hours $= 240$ minutes, we consider

$$P\left(\sum_{i=1}^{100} X_i > 240\right) = P\left(\overline{X} > \tfrac{240}{100}\right) = P\left(\overline{X} > 2.4\right)$$
$$= P\left(Z > \tfrac{2.4-2.5}{2}\right) = .6915.$$

7.43 We need to consider $P\left(\sum_{i=1}^{n} X_i < 120\right) = .1$.

$$P\left(\sum_{i=1}^{n} x_i < 120\right) = P\left(\overline{X} < \tfrac{120}{n}\right) = P\left[Z < \left(\frac{\frac{120}{n}-2.5}{\frac{2}{\sqrt{n}}}\right)\right] = P(Z < z_0) = .1.$$

Going to the normal table, we find that $z_0 \cong -1.28$. Thus,

$$\frac{\frac{120}{n}-2.5}{\frac{2}{\sqrt{n}}} = -1.28.$$

Solving for n,

$\sqrt{n}\left(\frac{120}{n} - 2.5\right) = 2(-1.28)$.

$2.5n - 2.56\left(\sqrt{n}\right) - 120 = 0$.

Using the quadratic formula, where $a = 2.5$, $b = -2.56$, $c = -120$,

$$\sqrt{n} = \frac{2.56 \pm \sqrt{(-2.56)^2 - 4(2.5)(-120)}}{2(2.5)} = \frac{2.56 \pm 34.74}{5} = 7.46 \text{ or } -6.44.$$

Taking the positive value, we find $n = 55.65 \cong 56$.

7.44 **a.** $p + 3\sqrt{\frac{pq}{n}} < 1 \Leftrightarrow 3\sqrt{\frac{pq}{n}} < q \Leftrightarrow 9\frac{pq}{n} < q^2 \Leftrightarrow 9\frac{p}{q} < n$.

b. $0 < p - 3\sqrt{\frac{pq}{n}} \Leftrightarrow 3\sqrt{\frac{pq}{n}} < p \Leftrightarrow 9\frac{pq}{n} < p^2 \Leftrightarrow 9\frac{q}{p} < n$.

c. Parts **a** and **b** imply that $n > 9\max(\frac{p}{q}, \frac{q}{p})$. Therefore, we will be done if we show that $\max(\frac{p}{q}, \frac{q}{p}) = \frac{\max(p,q)}{\min(p,q)}$. The implication is clear if $p = q = .5$. If $p > q$ then $\frac{p}{q} > 1 > \frac{q}{p}$ implying $\max(\frac{p}{q}, \frac{q}{p}) = \frac{p}{q} = \frac{\max(p,q)}{\min(p,q)}$. Similar calculations work for $q > p$, proving the result. Note we assuming that $p \neq 1, 0$ and using the fact that p and q are both positive.

7.45 **a.** $n > 9\frac{\max(p,q)}{\min(p,q)} = 9$.

b. $n > 14$, $n > 14$, $n > 36$, $n > 36$, $n > 891$, $n > 8991$.

7.46 Use the approximation given in Example 7.10. The quantity $Y_n = \frac{\frac{Y}{n}-p}{\sqrt{\frac{p(1-p)}{n}}}$ converges in distribution to a standard normal random variable. Hence, with $p = .10$ and $n = 100$, we have

$$P(Y \geq 15) \approx P\left(z \geq \tfrac{14.5-10}{\sqrt{9}}\right) = P(z \geq 1.5) = .0668$$

7.47 The random variable of interest is X, the number of persons not showing up for a given flight. This is a binomial random variable with $n = 160$ and

$$p = P(\text{person does not show up}) = .05.$$

If there is to be a seat available for every person planning to fly, then there must be at least five persons not showing up. Hence, the probability of interest is $P(X \geq 5)$. Calculate

$$\mu = np = 160(.05) = 8,$$

and

$$\sigma = \sqrt{npq} = \sqrt{160(.05)(.95)} = \sqrt{7.6} = 2.76.$$

A correction for continuity is made to include the entire area under the rectangle associated with the value $X = 5$, and the approximation becomes $P(X \geq 4.5)$. The Z value corresponding to $X = 4.5$ is

$$Z = \frac{4.5-8}{2.76} = 1.27,$$

so that

$$P(X \geq 4.5) = P(Z \geq -1.27) = 1 - P(Z \leq 1.27) = 1 - .1020 = .8980.$$

7.48 a. It is given that $n = 1500$ residents are selected, and $p = P(\text{lawyer}) = \frac{1}{410}$. Then if X is the number of lawyers in the sample,

$$\begin{aligned} P(\text{at least one lawyer}) &= 1 - P(\text{no lawyers}) \\ &= 1 - \binom{1500}{0}\left(\frac{1}{410}\right)^0\left(\frac{409}{410}\right)^{1500} \\ &= 1 - .0496 = .9504. \end{aligned}$$

Notice that the exact binomial probability was easier to calculate than an approximation.

b. To use the normal approximation, $n = 1500$, $p = P(\text{lawyer}|\text{Washington}) = \frac{1}{64}$, $\mu = np = 23.4375$, $\sigma = \sqrt{npq} = \sqrt{23.0713} = 4.803$. Then,

$$P(X > 30) = P\left(Z > \tfrac{30.5 - 23.4375}{4.803}\right) = P(Z > 1.47) = .0708.$$

c. The value $x = 30$ lies

$$z = \frac{x-\mu}{\sigma} = \frac{30 - 23.4375}{4.803} = 1.37$$

standard deviations above the mean. Hence, it is not a very unlikely event. There is no reason to believe that p is higher than usual at this corner.

7.49 Similar to previous exercises. With $p = .20$ and $n = 64$, the probability of interest is

$$\begin{aligned} P\left(\left|\tfrac{Y}{n} - p\right| \le .06\right) = P\left(-.06 \le \tfrac{Y}{n} - p \le .06\right) &= P\left(\frac{-.06}{\sqrt{\frac{(.2)(.8)}{64}}} \le Z \le \frac{.06}{\sqrt{\frac{(.2)(.8)}{64}}}\right) \\ = P(-1.2 \le Z \le 1.2) &= 1 - 2(.1151) = .7698 \end{aligned}$$

7.50 a. It is known that $V\left(\frac{Y}{n}\right) = \frac{p(1-p)}{n}$. Consider $V\left(\frac{Y}{n}\right)$ for fixed n and let $f(p) = p(1-p)$. The value of p for which this function is maximized is found by looking at $f'(p) = 0$, or $1 - 2p = 0$. Hence, $p = \frac{1}{2}$. Since $f''(p) = -2$, which is negative for all values of p, this value, $p = \frac{1}{2}$, must be a maximum.

b. It is necessary to have

$$P\left(\left|\tfrac{Y}{n} - p\right| \le .1\right) = .95 \qquad \text{or} \qquad P\left(|Z| \le \frac{.1}{\sqrt{\frac{pq}{n}}}\right) = .95$$

Hence $\frac{.1}{\sqrt{\frac{pq}{n}}} = 1.96$. Since p is unknown, we replace it with the value of p for which pq is maximized $\left(p = \frac{1}{2}\right)$ and solve for n.

$$\sqrt{n} = \frac{1.96(.5)}{.1} = 9.8 \qquad \text{or} \qquad n = 96.04$$

Hence, sample $\mu = 97$ items.

7.51 Similar to Exercise 7.50. It is necessary to solve for n in the following equality: $\frac{.15\sqrt{n}}{\sqrt{pq}} = 2.33$. Using $p = \frac{1}{2}$ to approximate p, we have

$$\sqrt{n} = \frac{2.33(.5)}{.15} \qquad \text{or} \qquad n = 60.32$$

Hence $n = 61$ customers should be sampled.

7.52 Refer to Exercise 7.51. If $p = .9$, then

$$P\left(\left|\tfrac{Y}{n} - p\right| \le .15\right) = P\left(|Z| \le \frac{.15}{\sqrt{\frac{(.9)(.1)}{50}}}\right) = P(|Z| \le 3.54) = 1.00$$

7.53 Similar to Example 7.10 in the text.

a. To use the normal approximation to the binomial, treat Y as having approximately the same distribution as W, where W is normally distributed with mean np and variance $np(1-p)$. Then

$$P(Y \ge 2) = P(W \ge 1.5) = P\left(Z \ge \frac{1.5 - 2.5}{\sqrt{25(.1)(.9)}}\right) = P(Z > -.67) = .7486$$

b. Indexing $n = 25$ and $p = .1$ in Table 1, Appendix III, we have

$$P(Y \ge 2) = 1 - P(Y \le 1) = 1 - .271 = .729$$

which is quite close to the approximate probability.

7.54 Let X be the number of our sample that are younger than 31. As 31 is the median age, X will be binomially distributed with $p = .5$ and $n = 100$. We want

$$P(X \geq 60) = P(X \geq 55.5) \approx P\left(Z \geq \frac{55.5 - 50}{\sqrt{(.5)(.5)(100)}}\right) = P(Z > 1.1) = .1357.$$

7.55 **a.** Let X be the count of non-conforming items of our sample. We want

$$P(X \leq 5) = P(X \leq 5.5) \approx P\left(Z \leq \frac{5.5-(.1)50}{\sqrt{(.1)(.9)50}}\right) = P(Z \leq .24) = 1 - .4052 = .5948.$$

b. For $p = .2$ we have $\approx P\left(Z \leq \frac{5.5-(.2)50}{\sqrt{(.2)(.8)50}}\right) = P(Z \leq -1.59) = .0559.$

For $p = .3$ we have $\approx P\left(Z \leq \frac{5.5-(.3)50}{\sqrt{(.3)(.7)50}}\right) = P(Z \leq -.293) = .0017.$

7.56 As 80% of the disk contain no missing pulses, 20% contain missing pulses. Let X represent the count of our sample that have missing pulses. Then X is binomial with $p = .2$ and $n = 100$. We want

$$P(X \geq 15) = P(X \geq 14.5) \approx P\left(Z \geq \frac{14.5-(.2)100}{\sqrt{(.2)(.8)100}}\right) = P(Z \geq -1.38) = 1 - .0838 = .9162.$$

7.57 **a.** Let X represent the number of vehicles that turn right. Then X is binomial with $p = 1/3$ and $n = 500$. We want

$$P(X \leq 150) = P(X \leq 150.5) \approx P\left(Z \leq \frac{150.5-500(1/3)}{\sqrt{\left(\frac{1}{3}\right)\left(\frac{2}{3}\right)500}}\right)$$
$$= P(Z \leq -1.53) = .0630.$$

b. Let X now represent the number of vehicles that turn. Then X is binomial with $p = 2/3$ and $n = 500$. We want

$$P(X \geq 350) = P(X \geq 349.5) \approx P\left(Z \geq \frac{349.5-500(2/3)}{\sqrt{\left(\frac{2}{3}\right)\left(\frac{1}{3}\right)500}}\right)$$
$$= P(Z \geq 1.53) = .0630.$$

7.58 Use the results of Theorem 5.12.

a.
$$E\left(\frac{Y_1}{n_1} - \frac{Y_2}{n_2}\right) = E\left(\frac{Y_1}{n_1}\right) - E\left(\frac{Y_2}{n_2}\right)$$
$$= \frac{1}{n_1}E(Y_1) - \frac{1}{n_2}E(Y_2)$$
$$= \frac{n_1 p_1}{n_1} - \frac{n_2 p_2}{n_2} = p_1 - p_2$$

b.
$$V\left(\frac{Y_1}{n_1} - \frac{Y_2}{n_2}\right) = V\left(\frac{Y_1}{n_1}\right) + V\left(\frac{Y_2}{n_2}\right)$$
$$= \left(\frac{1}{n_1}\right)^2 V(Y_1) - \left(\frac{1}{n_2}\right)^2 V(Y_2)$$
$$= \frac{n_1 p_1 q_1}{n_1^2} + \frac{n_2 p_2 q_2}{n_2^2} = \frac{p_1 q_1}{n_1} + \frac{p_2 q_2}{n_2}.$$

7.59 It is given that $p_1 = .1$ and $p_2 = .2$. Using the results of Exercise 7.58, we obtain

$$P\left[\left|\left(\frac{Y_1}{n_1} - \frac{Y_2}{n_2}\right) - (p_1 - p_2)\right| \leq .1\right] = P\left(|Z| \leq \frac{.1}{\sqrt{\frac{(.1)(.9)}{50} + \frac{(.2)(.8)}{50}}}\right) = P(|Z| \leq 1.41)$$
$$= 1 - 2(.0793) = .8414$$

7.60 Let X = # of travel vouchers improperly documented. X is binomial with $n = 100$, $p = .20$. Using the normal approximation to the binomial, we consider

$$P(X > 30) = P(W \geq 31) = P(W \geq 30.5)$$
$$\approx P\left(Z \geq \frac{30.5 - 100(.2)}{\sqrt{100(.2)(.8)}}\right) = P(z \geq 2.625) = .0043.$$

We would conclude that the firm's claim of 20% is probably incorrect, since the probability is so small that we would observe 30% if 20% is the correct percentage.

7.61 Let X = waiting time over a 2-day period. X is exponential with $\beta = 10$ minutes. Let Y = # of customers whose waiting time is greater than 10 minutes. Y is binomial with $n = 100$ and success probability

$$p = P(X > 10) = \int_{10}^{\infty} \left(\tfrac{1}{10}\right) e^{-x/10}\, dx = e^{-1} = .3679.$$

Then we have

$$P(Y \geq 50) \approx P\left(Z \geq \frac{49.5-36.79}{\sqrt{100(.3679)(.6321)}}\right) = P(Z \geq 2.636) \approx .0041.$$

7.62 Since the efficiency measurements are normally distributed with common mean 9.5 and $\sigma = .5$, $\overline{X}$ has an exact normal distribution.

$$P\left[\overline{X} > 10\right] = P\left[Z > \frac{10-9.5}{\frac{.5}{\sqrt{8}}}\right] = P[Z > 2.83] = .0023$$

7.63 It is necessary that

$$P\left[\overline{X} > 10\right] = P\left[Z > \frac{10-\mu}{\frac{.5}{\sqrt{8}}}\right] = .8$$

But, $P[Z > -.84] = .8$. Thus, $-.84 = \frac{10-\mu}{\frac{.5}{\sqrt{8}}}$ or $\mu = 10.15$.

7.64 The random variable Y has a binomial distribution with $n = 100$ and $p = .65$. Using the technique from Example 7.10, we have

$$P(Y > 70) \approx P\left(Z \geq \frac{70.5-100(.65)}{\sqrt{100(.65)(.35)}}\right) = P(Z > 1.15) = .1251$$

7.65 Since X, Y, and Z are normally distributed, $\overline{X}$, $\overline{Y}$, and $\overline{Z}$ are normally distributed, and Theorem 6.3 states that

$$E(U) = .4\mu_1 + .2\mu_2 + .4\mu_3$$

$$V(U) = .16\left(\frac{\sigma_1^2}{n_1}\right) + .04\left(\frac{\sigma_2^2}{n_2}\right) + .16\left(\frac{\sigma_3^2}{n_3}\right)$$

7.66 The desired probability is

$$P\left(\left|\overline{Y}_1 - \overline{Y}_2\right| > .6\right) = P\left(|Z| > \frac{.6-0}{\sqrt{\frac{(6.4)^2}{64}+\frac{(7.2)^2}{64}}}\right) = P\left(|Z| > .50\right) = 2(.3085) = .617$$

7.67 Since Y has an exponential distribution with mean θ, $f(y) = \left(\frac{1}{\theta}\right) e^{-y/\theta}$ for $y > 0$. Let $u = \frac{2Y}{\theta}$. Then $Y = \left(\frac{\theta}{2}\right) U$ and $\left|\frac{dy}{du}\right| = \frac{\theta}{2}$.
Finally,

$$f(u) = \frac{1}{\theta}\, e^{-u/2}\, \frac{\theta}{2} = \frac{1}{2}\, e^{-u/2} \qquad \text{for } u > 0$$

Hence U has a gamma distribution with $\alpha = 1$, $\beta = 2$ or, equivalently, a χ^2 distribution with $\alpha = \frac{\nu}{2} = 1$ or $\nu = 2$, or equivalently, an exponential distribution with $\beta = 2$.

7.68 Using the results of Exercise 7.67, the quantity $\frac{2Y_i}{20} = U_i$ has a χ^2 distribution with 2 degrees of freedom. Using the fact that the sum of n independent χ^2 random variables is also χ^2, with degrees of freedom equal to the sum of the individual degrees of freedom, we know that

$$\frac{2\sum_{i=1}^{5} Y_i}{20} = \frac{\sum_{i=1}^{5} Y_i}{10}$$

has a χ^2 distribution with $2(5) = 10$ degrees of freedom. Then

$$P\left(\Sigma Y_i > c\right) = P\left(\Sigma \tfrac{Y_i}{10} > \tfrac{c}{10}\right) = .05.$$

The value of $\frac{c}{10}$ that satisfies this equality is $\frac{c}{10} = \chi^2_{.05} = 18.307$, or $c = 183.07$.

7.69 **a. & b.** Since $\mu = 0$, we have that $T = \frac{\overline{Y}}{\left(\frac{S}{\sqrt{n}}\right)}$ (see the discussion following Definition 7.2) and $T^2 = \frac{n\overline{Y}^2}{S^2}$ has an F distribution with 1 and $(n-1)$ degrees of freedom. By Definition 7.3, $\frac{1}{T^2} = \frac{S^2}{n\overline{Y}^2}$ also has an F distribution with $(n-1)$ and 1 degrees of freedom.

c. Consider the probability statement given at the beginning of the exercise.

$$P\left(-c \leq \frac{S}{\overline{Y}} \leq c\right) = P\left(0 \leq \frac{S^2}{\overline{Y}^2} \leq c^2\right) = P\left(0 \leq \frac{S^2}{10\overline{Y}^2} \leq \frac{c^2}{10}\right) = .95$$

The constant $\frac{c^2}{10}$ is thus the same as $F_{.05}$ for an F distribution with 9 and 1 degrees of freedom. Then $\frac{c^2}{10} = 240.5$, or $c = 49.04$.

7.70 For the given random variable Y,

$$\mu = E(Y) = \int_0^1 3y^3\, dy = \frac{3}{4}$$

and

$$\sigma^2 = V(Y) = \int_0^1 3y^4\, dy - \frac{9}{16} = \frac{3}{5} - \frac{9}{16} = .0375$$

Using the Central Limit Theorem,

$$P(\overline{Y} > .7) = P\left(Z > \frac{.7-.75}{\sqrt{\frac{.0375}{40}}}\right) = P(Z > -1.63) = P(Z < 1.63) = 1 - .0516 = .9484$$

7.71 **a.** Note that $E(X_i) = 1$ and $V(X_i) = 2$. By the Central Limit Theorem,

$$\overline{X} \sim N\left(1, \sqrt{\frac{2}{n}}\right)$$

or

$$\frac{\frac{Y}{n}-1}{\sqrt{\frac{2}{n}}} = \frac{Y-n}{\sqrt{2n}} \sim N(0, 1).$$

b. It is given that Y has a normal distribution with mean $\mu = 6$ and variance $\sigma^2 = .2$. Let C_i be the cost for a single rod, $i = 1, 2, \ldots, 50$. That is, $C_i = 4(Y_i - \mu)^2$. The total cost for the day, then, will be

$$\sum_{i=1}^{50} C_i = 4 \sum_{i=1}^{50} (y_i - \mu)^2$$

where the Y_i are independent and distributed as Y above. Recall from Chapter 6 that

$$\frac{\sum_{i=1}^{50} (Y_i-\mu)^2}{\sigma^2} = \sum_{i=1}^{50} \left(\frac{Y_i-\mu}{\sigma}\right)^2 = \sum_{i=1}^{50} Z_i^2$$

has a chi-square distribution with 50 degrees of freedom, since each Z_i is standard normal, and hence Z_i^2 has a chi-square distribution with 1 degree of freedom, $i = 1, 2, \ldots, 50$. The probability of interest, using the results of part **a**, is

$$P(\Sigma C_i > 48) = P[\Sigma (Y_i - \mu)^2 > 12] = P\left[\frac{\Sigma (Y_i-\mu)^2}{.2} > 60\right] = P(X > 60)$$

where X is a chi-square random variable with $n = 50$ degrees of freedom. Hence the approximation is

$$P(X > 60) = P\left(\frac{X-n}{\sqrt{2n}} > \frac{60-50}{\sqrt{100}}\right) = P(Z > 1) = .1587$$

7.72 **a.** Let $T = \frac{Z}{\sqrt{\frac{W}{\nu}}}$, where W is a standard normal random variable and W is a χ^2 variable with ν degrees of freedom. Using the moment-generating function approach, we can write $m_w(t) = e^{t^2/2}$. Further, for fixed $W = w.\nu$, the moment-generating function of $W = w$, is

$$m_{Z/\sqrt{w/\nu}}(t) = m_Z\left(\frac{t}{\sqrt{\frac{\nu}{w}}}\right) = \exp\left(\frac{t^2\nu}{2w}\right)$$

Evidently, then, the conditional distribution of T given $W = w$ is normal with mean 0 and variance $\frac{\nu}{w}$. That is,

$$f(t|w) = \frac{\sqrt{w}}{\sqrt{2\pi}\sqrt{\nu}} e^{-wt^2/2\nu} \qquad -\infty \le t \le \infty$$

b. Now, since W has a χ^2 distribution with ν degrees of freedom,

$$f(t, w) = f(t|w)f(w) = \frac{\sqrt{w}}{\sqrt{2\pi}\sqrt{\nu}} \times e^{-t^2 w/2\nu} \times \frac{w^{(\nu/2)-1} e^{-w/2}}{\Gamma\left(\frac{\nu}{2}\right) 2^{\nu/2}}$$

$$= \frac{\exp\left[-\frac{w}{2}\left(1+\frac{t^2}{\nu}\right)\right] w^{[(\nu+1)/2]-1}}{\sqrt{\pi}\sqrt{\nu}\, 2^{(\nu+1)/2}\, \Gamma\left(\frac{\nu}{2}\right)}$$

c. Integrating over w, we obtain

$$f(t) = \int_0^\infty f(t, w)\, dw = \int_0^\infty \frac{\exp\left[-\frac{w}{2}\left(1+\frac{t^2}{\nu}\right)\right] w^{[(\nu+1)/2]-1}}{\sqrt{\pi\nu}\,\Gamma\left(\frac{\nu}{2}\right) 2^{(\nu+1)/2}}\, dw$$

$$= \frac{\Gamma\left(\frac{\nu+1}{2}\right)\left(1+\frac{t^2}{\nu}\right)^{-(\nu+1)/2}}{\sqrt{\pi\nu}\,\Gamma\left(\frac{\nu}{2}\right)} \int_0^\infty \frac{\exp\left[-\frac{w}{2}\left(1+\frac{t^2}{\nu}\right)\right] w^{[(\nu+1)/2]-1}}{\Gamma\left(\frac{\nu+1}{2}\right) 2^{(\nu+1)/2}\left(1+\frac{t^2}{\nu}\right)^{-(\nu+1)/2}}\, dw$$

$$= \frac{\Gamma\left(\frac{\nu+1}{2}\right)\left(1+\frac{t^2}{\nu}\right)^{-(\nu+1)/2}}{\sqrt{\pi\nu}\,\Gamma\left(\frac{\nu}{2}\right)}$$

since the integrand is the density of a gamma random variable with parameters $\alpha = \frac{\nu+1}{2}$ and $\beta = \frac{2}{1+\left(\frac{t^2}{\nu}\right)}$ and hence must integrate to 1.

7.73

a. Similar to Exercise 7.72. For fixed W_2, let $F = \frac{W_1}{c}$, where $c = \frac{w_2\nu_1}{\nu_2}$. Since W_1 has a chi-square distribution, we have

$$m_{W_1}(t) = (1-2t)^{-\nu_1/2}$$

implying

$$m_{W_1/c}(t) = m_{W_1}\left(\tfrac{t}{c}\right) = \left(1 - \tfrac{2}{c}\,t\right)^{-\nu_1/2}.$$

Hence the distribution of F, conditional on $W_2 = w_2$, is that of a gamma random variable with parameters $\alpha = \frac{\nu_1}{2}$ and $\beta = \frac{2}{c} = \frac{2\nu_2}{w_2\nu_1}$.

b. Now, since W_2 has a chi-square distribution, we can write

$$g(f, w_2) = g(f|w_2)g(w_2) = \frac{f^{(\nu_1/2)-1} e^{-fw_2\nu_1/2\nu_2} w_2^{(\nu_2/2)-1} e^{-w_2/2}}{\Gamma\left(\frac{\nu_1}{2}\right)\left(\frac{2\nu_2}{w_2\nu_1}\right)^{\nu_1/2}\Gamma\left(\frac{\nu_2}{2}\right) 2^{\nu_2/2}}$$

$$= \frac{f^{(\nu_1/2)-1} e^{-(w_2/2)[(f\nu_1/\nu_2)+1]} w_2^{(\nu_2/2)-1} w_2^{\nu_1/2}}{\Gamma\left(\frac{\nu_1}{2}\right)\left(\frac{2\nu_2}{\nu_1}\right)^{\nu_1/2}\Gamma\left(\frac{\nu_2}{2}\right) 2^{\nu_2/2}} \qquad \text{for } f \ge 0,\ w_2 \ge 0$$

c. Finally,

$$g(f) = \int_0^\infty g(f, w_2)\, dw_2$$

$$= \frac{f^{(\nu_1/2)-1}\left(\frac{\nu_1}{\nu_2}\right)^{\nu_1/2}}{\Gamma\left(\frac{\nu_1}{2}\right)\Gamma\left(\frac{\nu_2}{2}\right) 2^{(\nu_1+\nu_2)/2}} \int_0^\infty w_2^{[(\nu_1+\nu_2)/2]-1} e^{-(w_2/2)[(f\nu_1/\nu_2)+1]}\, dw_2$$

$$= \frac{f^{(\nu_1/2)-1}\left(\frac{\nu_1}{\nu_2}\right)^{\nu_1/2}\Gamma\left(\frac{\nu_1+\nu_2}{2}\right)\left(1+\frac{f\nu_1}{\nu_2}\right)^{-(\nu_1+\nu_2)/2}}{\Gamma\left(\frac{\nu_1}{2}\right)\Gamma\left(\frac{\nu_2}{2}\right)} \qquad f \ge 0$$

7.74

a. The moment-generating function of the Poisson random variable X is $m_X(t) = e^{\lambda(e^t-1)}$ and hence

$$m_Y(t) = m_X\left(\tfrac{t}{\sqrt{\lambda}}\right) e^{-\sqrt{\lambda}t} = e^{-\sqrt{\lambda}t}\exp\left[\lambda\left(e^{t/\sqrt{\lambda}} - 1\right)\right]$$

$$= \exp\left(\lambda e^{t/\sqrt{\lambda}} - \sqrt{\lambda}t - \lambda\right)$$

b. Since $e^{e/\sqrt{\lambda}} = 1 + \left(\frac{t}{\sqrt{\lambda}}\right) + \left(\frac{t^2}{2\lambda}\right) + \ldots$, then

$$m_Y(t) = \exp\left[-\sqrt{\lambda}t + \lambda\left(\tfrac{t}{\sqrt{\lambda}} + \tfrac{t^2}{2\lambda} + \tfrac{t^3}{6\lambda^{3/2}} + \ldots\right)\right] = \exp\left(\tfrac{t^2}{2} + \tfrac{t^3}{6\lambda^{1/2}} + \ldots\right)$$

Taking the limit as $\lambda \to 0$, all terms after the first will tend to zero so that $\lim m_Y(t) = e^{t^2/2}$.

c. Since this is the moment-generating function for a standard normal variable, the result is proven by Theorem 7.5.

7.75

Using the result of Exercise 7.74, we have

$$P(X \le 110) = P\left(\tfrac{X-\lambda}{\sqrt{\lambda}} \le \tfrac{110-\lambda}{\sqrt{\lambda}}\right) = P\left(Z \le \tfrac{110-100}{\sqrt{100}}\right) = P(Z \le 1)$$

$$= 1 - .1587 = .8413$$

7.76

Use the result of Exercise 7.74, that $Y = \frac{X-\lambda}{\sqrt{\lambda}}$ converges in distribution to a standard normal distribution (denoted by Z) as λ tends to infinity, when X has a Poisson probability distribution with mean λ. With $\lambda = 36$, the approximation is

$$P(X \ge 45) = P\left(\tfrac{X-\lambda}{\sqrt{\lambda}} \ge \tfrac{45-36}{6}\right) = P(Z > 1.5) = .0668$$

7.77 From Exercise 7.74, X converges in distribution to a normal distribution with mean λ_1 and variance λ_1, while Y converges in distribution to a normal distribution with mean λ_2 and variance λ_2. It can be shown that, as in Exercise 7.30, the quantity

$$\frac{(X-Y)-(\lambda_1-\lambda_2)}{\sqrt{\lambda_1+\lambda_2}}$$

will also converge in distribution to a standard normal random variable. Using this result, the approximation is

$$P(X-Y>10)=P\left(\frac{X-Y}{\sqrt{\lambda_1+\lambda_2}}>\frac{100}{\sqrt{100}}\right)=P(Z>1)=.1587$$

7.78 The mgf of Y_n is given by

$$\begin{aligned} m_n(t) &= (q+pe^t)^n \qquad \text{where } q = 1-\text{p} \\ &= \left[(1-p)+pe^t\right]^n \\ &= \left[1+p\left(e^t-1\right)\right]^n. \end{aligned}$$

Let $np=\lambda$, thus $p=\frac{\lambda}{n}$. Substituting into $m_n(t)$,

$$m_n(t)=\left[1+\left(\tfrac{\lambda}{n}\right)\left(e^t-1\right)\right]^n.$$

Consider taking $\lim_{n\to\infty} m_n(t)$, $\lim_{n\to\infty}\left[1+\left(\frac{\lambda}{n}\right)\left(e^t-1\right)\right]^n$.

Recall from calculus, $\lim_{n\to\infty}\left(1+\frac{k}{n}\right)^n=e^k$.

If we let $k=\lambda\left(e^t-1\right)$, we have

$$\lim_{n\to\infty} m_n(t)=\exp\left[\lambda\left(e^t-1\right)\right]$$

which we have shown to be the mgf for a Poisson random variable. Therefore, the distribution of Y_n converges to a Poisson distribution with mean λ.

7.79 Let Y = # of people that suffer an adverse reaction. Y is binomial with $n=1000$, $p=.001$. Using the result of Exercise 7.62, we let

$$\lambda=1000(.001)=1.$$

Using the Poisson distribution, we evaluate

$$P(Y\geq 2)=1-P(Y\leq 1)=1-.736=.264.$$

CHAPTER 8 ESTIMATION

8.1 Write

$$\text{MSE}\left(\hat{\theta}\right) = E\left[\left(\hat{\theta} - \theta\right)\right]^2 = E\left[\hat{\theta} - E\left(\hat{\theta}\right) + B\right]^2 = E\left\{[\hat{\theta} - E(\hat{\theta})]^2\right\} + E\left(B^2\right)$$
$$+ 2B\left\{E[\hat{\theta} - E(\hat{\theta})]\right\}$$
$$= V(\hat{\theta}) + B^2 + 2B[E(\hat{\theta}) - E(\hat{\theta})] = V(\hat{\theta}) + B^2$$

8.2 **a.** $E\left[\hat{\theta}_3\right] = E\left[a\hat{\theta}_1 + (1-a)\hat{\theta}_2\right] = aE\left[\hat{\theta}_1\right] + (1-a)E\left[\hat{\theta}_2\right] = a\theta + (1-a)\theta = \theta$

b. It is given that $E(\hat{\theta}_1) = E(\hat{\theta}_2) = \theta$ and $V(\hat{\theta}_1) = \sigma_1^2$, $V(\hat{\theta}_2) = \sigma_2^2$. Assuming that $\hat{\theta}_1$ and $\hat{\theta}_2$ are independent, the variance of the new estimator, $\hat{\theta}_3$, will be

$$V\left(\hat{\theta}_3\right) = V\left[a\hat{\theta}_1 + (1-a)\hat{\theta}_2\right] = a^2 V\left(\hat{\theta}_1\right) + (1-a)^2 V\left(\hat{\theta}_2\right) = a^2\sigma_1^2 + (1-a)^2\sigma_2^2$$

In order to choose a value of a such that $V(\hat{\theta}_3)$ is minimized, look at

$$\tfrac{d}{da} V\left(\hat{\theta}_3\right) = 2a\sigma_1^2 - 2(1-a)\sigma_2^2.$$

Setting the derivative equal to 0, we obtain

$$a\sigma_1^2 - (1-a)\sigma_2^2 = 0$$

or

$$a = \tfrac{\sigma_2^2}{\sigma_1^2 + \sigma_2^2}$$

Notice that $\tfrac{d^2}{da^2} V\left(\hat{\theta}_3\right) = 2\sigma_1^2 + 2\sigma_2^2 > 0$, so that the value is in fact a minimum.

8.3 Given that $E\left(\hat{\theta}_1\right) = E\left(\hat{\theta}_2\right) = \theta$ and $V\left(\hat{\theta}_1\right) = \sigma_1^2$ and $V\left(\hat{\theta}_2\right) = \sigma_2^2$ and $\hat{\theta}_3 = a\hat{\theta}_1 + (1-a)\hat{\theta}_2$ and finally that $\text{cov}(\hat{\theta}_1, \hat{\theta}_2) = c$ we have

$$V\left(\hat{\theta}_3\right) = V\left[a\hat{\theta}_1 + (1-a)\hat{\theta}_2\right]$$
$$= a^2 V\left(\hat{\theta}_1\right) + (1-a)^2 V\left(\hat{\theta}_2\right) + 2a(1-a)\,\text{cov}\left(\hat{\theta}_1, \hat{\theta}_2\right)$$
$$= a^2\sigma_1^2 + (1-a)^2\sigma_2^2 + 2a(1-a)c.$$

To choose a value of a so that $V(\hat{\theta}_3)$ is minimized, consider

$$\tfrac{d}{da} V(\hat{\theta}_3) = 2a\sigma_1^2 - 2(1-a)\sigma_2^2 + 2c(1-2a).$$

Setting this equal to 0 and solving for a,

$$a\sigma_1^2 - (1-a)\sigma_2^2 + c(1-2a) = 0$$
$$\Rightarrow \quad a\sigma_1^2 + a\sigma_2^2 - 2ac = \sigma_2^2 - c$$
$$\Rightarrow \quad a(\sigma_1^2 + \sigma_2^2 - 2c) = \sigma_2^2 - c$$
$$\Rightarrow \quad a = \frac{\sigma_2^2 - c}{\sigma_1^2 + \sigma_2^2 - 2c}.$$

8.4 Recall that if Y_i is Exponential(θ) then $E(Y_i) = \theta$ and $V(Y_i) = \theta^2$. Hence we can use Theorem 5.12 to obtain

$$E(\hat{\theta}_1) = E(\hat{\theta}_2) = E(\hat{\theta}_3) = E(\hat{\theta}_5) = \theta$$
$$V(\hat{\theta}_1) = \theta^2$$
$$V(\hat{\theta}_2) = \tfrac{1}{4}\left(2\theta^2\right) = \tfrac{\theta^2}{2}$$
$$V(\hat{\theta}_3) = \tfrac{1}{9}\left(\theta^2 + 4\theta^2\right) = \tfrac{5\theta^2}{9}$$
$$V(\hat{\theta}_5) = \tfrac{1}{9}\left(3\theta^2\right) = \tfrac{\theta^2}{3}$$

The distribution of $\hat{\theta}_4$ can be obtained by using the methods of Section 6.6 in the text, with $F(y) = 1 - e^{-y/\theta}$. Then

$$g_1(y) = \tfrac{3}{\theta}\, e^{-y/\theta}\left(e^{-y/\theta}\right)^2 = \tfrac{3}{\theta}\, e^{-3y/\theta}$$

which is an exponential distribution with mean $\frac{\theta}{3}$.

$$E(\hat{\theta}_4) = \tfrac{\theta}{3} \qquad\qquad V(\hat{\theta}_4) = \tfrac{\theta^2}{9}$$

a. The unbiased estimators are $\hat{\theta}_1, \hat{\theta}_2, \hat{\theta}_3$, and $\hat{\theta}_5$.

b. Among these four estimators, $\hat{\theta}_5 = \overline{Y}$ has the smallest variance.

8.5

Since Y has an exponential distribution with mean $\theta + 1$, $E(Y) = \theta + 1$ and $E(\overline{Y}) = \theta + 1$. Hence if we use $\hat{\theta} = \overline{Y} - 1$, $E(\hat{\theta}) = \theta$ and we have constructed an unbiased estimator.

8.6

a. For the Poisson distribution, $E(Y_i) = \lambda$ and $E(\overline{Y}) = \lambda$. Hence $\hat{\lambda} = \overline{Y}$ is an unbiased estimator for λ.

b. In order to find $E(Y^2)$, use the fact that $V(Y) = \lambda$ and $E(Y^2) = V(Y) + [E(Y)]^2 = \lambda + \lambda^2$. Then $E(C) = 3E(Y) + E(Y^2) = 4\lambda + \lambda^2$.

c. Since $E(\overline{Y}) = \lambda$, $E(\overline{Y})^2 = V(\overline{Y}) + [E(\overline{Y})]^2 = \frac{\lambda}{n} + \lambda^2$, we construct as an estimator $\hat{\theta} = \overline{Y}^2 + \overline{Y}\left(4 - \frac{1}{n}\right)$. Considering

$$E(\hat{\theta}) = \tfrac{\lambda}{n} + \lambda^2 + 4\lambda - \left(\tfrac{1}{n}\right)\lambda = 4\lambda + \lambda^2.$$

Thus, $\hat{\theta}$ is an unbiased estimator of $E(C)$.

8.7

The third central moment is defined to be

$$\begin{aligned} E\left[(Y-\mu)^3\right] &= E\left[(Y-3)^3\right] \\ &= E\left[Y^3 - 9Y^2 + 27Y - 27\right] \\ &= E\left[Y^3\right] - 9E\left[Y^2\right] + 27E[Y] - 27 \\ &= E\left[Y^3\right] - 9E\left[Y^2\right] + 27(3) - 27 \\ &= E\left[Y^3\right] - 9E\left[Y^2\right] + 54. \end{aligned}$$

Given that $\hat{\theta}_3$ is an unbiased estimator of $E\left[Y^3\right]$ and $\hat{\theta}_2$ is an unbiased estimator of $E\left[Y^2\right]$, we consider $\hat{\theta}_3 - 9\hat{\theta}_2 + 54$ as an estimator. By the definition of unbiasness,

$$E[\hat{\theta}_3 - 9\hat{\theta}_2 + 54] = E[\hat{\theta}_3] - 9E[\hat{\theta}_2] + 54 = E\left[Y^3\right] - 9E\left[Y^2\right] + 54 = E\left[(Y-3)^3\right].$$

8.8

a. For the uniform distribution given here, $E(Y_i) = \theta + \frac{1}{2}$. Hence $E(\overline{Y}) = \theta + \frac{1}{2}$ and the bias is $B = E(\overline{Y}) - \theta = \frac{1}{2}$.

b. An unbiased estimator of θ can be constructed by using $\hat{\theta} = \overline{Y} - \frac{1}{2}$, which has

$$E(\hat{\theta}) = \theta.$$

c. If $\overline{Y}$ is used as an estimator, then

$$V(\overline{Y}) = \frac{V(Y)}{n} = \frac{1}{12n} \qquad \text{and} \qquad \text{MSE} = V(\overline{Y}) + B^2 = \frac{1}{12n} + \frac{1}{4}.$$

8.9

a. For a binomial random variable Y, $E(Y) = np$ and $E(Y^2) = V(Y) + n^2p^2 = npq + n^2p^2$. Hence

$$E\left\{n\left(\frac{Y}{n}\right)\left[1 - \left(\frac{Y}{n}\right)\right]\right\} = E(Y) - \tfrac{1}{n}E(Y^2) = np - pq - np^2 = np(1-p) - pq$$
$$= (n-1)pq$$

b. An unbiased estimator $\hat{\theta}$ has expected value npq. Hence we can use

$$\left(\tfrac{n}{n-1}\right) n \times \tfrac{Y}{n}\left(1 - \tfrac{Y}{n}\right) = \tfrac{n^2}{n-1}\left(\tfrac{Y}{n}\right)\left(1 - \tfrac{Y}{n}\right)$$

8.10

The following information is required to answer the question.

$$E(Y) = \int_0^\theta \left[\tfrac{\alpha y^\alpha}{\theta^\alpha}\right] dy = \left[\tfrac{\alpha y^{\alpha+1}}{(\alpha+1)\theta^\alpha}\right]_0^\theta = \tfrac{\alpha\theta}{\alpha+1}$$

$$E(Y^2) = \int_0^\theta \left[\tfrac{\alpha y^{\alpha+1}}{\theta^\alpha}\right] dy = \left[\tfrac{\alpha y^{\alpha+2}}{(\alpha+2)\theta^\alpha}\right]_0^\theta = \tfrac{\alpha\theta^2}{\alpha+2}$$

$$f(y) = \tfrac{\alpha y^{\alpha-1}}{\theta^\alpha}$$

$$F(y) = \int_0^y \tfrac{\alpha t^{\alpha-1}}{\theta^\alpha}\, dt = \left(\tfrac{y}{\theta}\right)^\alpha$$

$$F_{Y(n)}(y) = \left(\tfrac{y}{\theta}\right)^{n\alpha}, 0 \le y \le \theta$$

$$f_{Y(n)}(y) = \tfrac{n\alpha y^{n\alpha-1}}{\theta^{n\alpha}}\ 0 \le y \le \theta$$

So that $Y_{(n)}$ is also distributed as the power family with parameters $n\alpha$ and θ.

a. $E(Y_{(n)}) = \frac{n\alpha\theta}{n\alpha+1} \neq \theta.$

b. $\left(\frac{n\alpha+1}{n\alpha}\right) Y_{(n)}$ would be unbiased.

c.
$$\begin{aligned}\text{MSE}(Y_{(n)}) &= E\left[(Y_{(n)} - \theta)^2\right] = E\left(Y_{(n)}^2\right) - 2\theta E(Y_{(n)}) + \theta^2 \\ &= \frac{n\alpha\theta^2}{n\alpha+2} - 2\theta\left(\frac{n\alpha\theta}{n\alpha+1}\right) + \theta^2 \\ &= \frac{2\theta^2}{(n\alpha+1)(n\alpha+2)}.\end{aligned}$$

8.11 The following information is required to obtain the answer.

$$f(y) = 3\beta^3 y^{-4}, \qquad \beta \le y$$
$$F(y) = \int_\beta^y 3\beta^3 y^{-4}\, dy = 1 - \left(\frac{\beta}{y}\right)^3$$
$$E(Y) = \tfrac{3}{2}\beta$$
$$E\left(Y^2\right) = 3\beta^2$$
$$F_{Y_{(1)}}(y) = 1 - \left(\frac{\beta}{y}\right)^{3n}$$
$$f_{Y_{(1)}}(y) = -3n\left(\frac{\beta}{y}\right)^{3n-1}(-\beta y^{-2}) = 3n\beta^{3n} y^{-3n-1}$$
$$E(Y_{(1)}) = \frac{3n\beta}{3n-1} = \frac{3n\beta}{3n-1}$$
$$E(Y_{(1)}^2) = \frac{3n\beta^2}{3n-2}$$

So that $Y_{(1)}$ is again a Pareto with parameters $3n$ and β.

a.
$$\begin{aligned}\text{Bias} &= \frac{3n\beta}{3n-1} - \beta \\ &= \frac{3n\beta - 3n\beta + \beta}{3n-1} \\ &= \frac{\beta}{3n-1}\end{aligned}$$

b.
$$\begin{aligned}\text{MSE}(Y_{(1)}) &= E\left(Y_{(1)}^2\right) - 2\beta E(Y_{(1)}) + \beta^2 \\ &= \frac{3n\beta^2}{3n-2} - 2\beta\left(\frac{3n\beta}{3n-1}\right) + \beta^2 \\ &= \frac{2\beta^2}{(3n-1)(3n-2)}\end{aligned}$$

8.12 **a.** Let $S = \sqrt{S^2}$, where $\frac{(n-1)S^2}{\sigma^2}$ has a χ^2 distribution with $n-1$ degrees of freedom. It is necessary to find $E(S)$. Let $X = \frac{(n-1)S^2}{\sigma^2}$. Then

$$f(x) = \frac{x^{[(n-1)/2]} e^{-x/2}}{\Gamma\left(\frac{n-1}{2}\right) 2^{(n-1)/2}} \qquad \text{for } x > 0$$

The density function for $Y = S^2 = \frac{\sigma^2 X}{n-1}$ is obtained by the transformation method.

$$g(y) = f\left[\frac{(n-1)y}{\sigma^2}\right]\left|\frac{dx}{dy}\right| = \frac{\left[\frac{(n-1)}{\sigma^2}\right]^{(n-1)/2} y^{[(n-1)/2]-1} e^{-(n-1)y/2\sigma^2}}{\Gamma\left(\frac{n-1}{2}\right) 2^{(n-1)/2}}$$

Now

$$\begin{aligned}E(S) &= E\left(\sqrt{Y}\right) = \int_0^\infty y^{1/2} g(y)\, dy \\ &= \int_0^\infty \frac{y^{1/2}\left(\frac{n-1}{\sigma^2}\right)^{(n-1)/2} y^{(n-1)/2-1} e^{-(n-1)y/2\sigma^2}}{\Gamma\left(\frac{n-1}{2}\right) 2^{(n-1)/2}}\, dy \\ &= \frac{\left(\frac{n-1}{\sigma^2}\right)^{(n-1)/2} \Gamma\left(\frac{n}{2}\right)(2\sigma^2)^{n/2}}{\Gamma\left(\frac{n-1}{2}\right) 2^{(n-1)/2}(n-1)^{n/2}} \int_0^\infty \frac{y^{(n/2)-1} e^{-(n-1)y/2\sigma^2}}{\Gamma\left(\frac{n}{2}\right)\left(\frac{2\sigma^2}{n-1}\right)^{n/2}}\, dy \\ &= \frac{(n-1)^{-1/2}\Gamma\left(\frac{n}{2}\right)(\sigma^2)^{1/2}}{\Gamma\left(\frac{n-1}{2}\right) 2^{-1/2}} = \frac{\Gamma\left(\frac{n}{2}\right) 2^{1/2}\sigma}{\Gamma\left(\frac{n-1}{2}\right)\sqrt{n-1}}\end{aligned}$$

which is a biased estimator for σ.

b. In order to adjust S so that it is not biased, take

$$\hat{\sigma} = S\left(\frac{\Gamma\left(\frac{n-1}{2}\right)\sqrt{n-1}}{\Gamma\left(\frac{n}{2}\right)\sqrt{2}}\right)$$

c. Since $E(\overline{Y}) = \mu$, an unbiased estimator of $\mu - Z_\alpha\sigma$ is $\hat{\theta} = \overline{Y} - Z_\alpha\hat{\sigma}$.

8.13 Note that

$$E(\hat{p}_1) = E\left(\frac{Y}{n}\right) = \left(\frac{1}{n}\right)(np) = p$$
$$E(\hat{p}_2) = E\left(\frac{Y+1}{n+2}\right) = \frac{1}{(n+2)}(np+1) = \frac{np+1}{n+2}$$

a. $\text{Bias} = \frac{np+1}{n+2} - p = \frac{np+1-np-2p}{n+2} = \frac{1-2p}{n+2}.$

b. $\text{MSE}(\hat{p}_1) = V(\hat{p}_1) + B^2 = V\left(\frac{Y}{n}\right) + 0 = \left(\frac{1}{n^2}\right) np(1-p) = \frac{p(1-p)}{n}$.

$$\text{MSE}(\hat{p}_2) = V(\hat{p}_2) + B^2 = V\left(\frac{Y+1}{n+2}\right) + \left(\frac{1-2p}{n+2}\right)^2$$
$$= \left[\frac{1}{(n+2)^2}\right] V(Y+1) + \frac{(1-2p)^2}{(n+2)^2}$$
$$= \frac{np(1-p)+(1-2p)^2}{(n+2)^2}.$$

c. We need to consider $\text{MSE}(\hat{p}_2) < \text{MSE}(\hat{p}_1)$.

$$\frac{np(1-p)+(1-2p)^2}{(n+2)^2} < \frac{p(1-p)}{n}$$
$$n^2p(1-p) + n(1-2p)^2 - p(1-p)(n+2)^2 < 0$$

This simplifies to

$$(8n+4)p^2 - (8n+4)p + n < 0$$

By the quadratic formula

$$\text{p} = \frac{8n+4 \pm \sqrt{(8n+4)^2 - 4(8n+4)n}}{2(8n+4)} = \frac{1}{2} \pm \sqrt{\frac{n+1}{8n+4}}$$

That is, p will be close to $\frac{1}{2}$.

8.14 $F_{Y_{(1)}}(t) = P(Y_{(1)} \le t) = 1 - P(Y_{(1)} > t) = 1 - [1 - F(t)]^n$

$$f_{Y_{(1)}}(t) = n[1-F(t)]^{n-1} f(t) = n\left[1 - \left(\frac{t}{\theta}\right)\right]^{n-1}\left(\frac{1}{\theta}\right)$$
$$E(Y_{(1)}) = \int_0^\theta nt\left[1 - \left(\frac{t}{\theta}\right)\right]^{n-1}\left(\frac{1}{\theta}\right) dt$$

Letting $w = \frac{t}{\theta}$, $dw = \frac{dt}{\theta}$, $t = \theta w$;

$$= \theta \int_0^1 nw(1-w)^{n-1}\, dw$$
$$= n\theta B(2, n) = \frac{n\theta\Gamma(2)\Gamma(n)}{\Gamma(n+2)}$$
$$= \frac{n\theta(n-1)!}{(n+1)!} = \frac{\theta}{(n+1)}.$$

An unbiased estimator for θ is given by $(n+1)Y_{(1)}$.

8.15 From exercise 6.64 we know

$$E(Y_{(1)}) = \frac{\beta}{n}.$$
$$E(\hat{\theta}) = E[nY_{(1)}] = nE(Y_{(1)}) = n\left(\frac{\beta}{n}\right) = \beta = \theta.$$

Thus, $nY_{(1)}$ is an unbiased estimator for θ.

$$\text{MSE}(\hat{\theta}) = V(\hat{\theta}) + B^2$$
$$= \text{V}(nY_{(1)}) = n^2V(Y_{(1)}) = n^2\left(\frac{\beta}{n}\right)^2 = \beta^2.$$

8.16 **a.** $E\left(\sqrt{Y}\right) = \int_0^\infty \left(\frac{1}{\theta}\right) y^{1/2} e^{-y/\theta}\, dy = \left(\frac{1}{\theta}\right)\left[\Gamma\left(\frac{3}{2}\right)\theta^{3/2}\right] = \left(\frac{1}{2}\right)\Gamma\left(\frac{1}{2}\right)\theta^{1/2} = \frac{\sqrt{\pi\theta}}{2}$

so that by independence

$$E\left(\sqrt{Y_1Y_2}\right) = \left[\left(\frac{1}{2}\right)\Gamma\left(\frac{1}{2}\right)\theta^{1/2}\right]^2 = \frac{\pi\theta}{4}.$$

Therefore, $\left(\frac{4}{\pi}\right) X$ is unbiased for θ.

b. Again by independence

$$E\left(\sqrt{Y_1Y_2Y_3Y_4}\right) = \left(\frac{\sqrt{\pi\theta}}{2}\right)^2 = \frac{\pi^2\theta^2}{16}.$$

Therefore, $\left(\frac{16}{\pi^2}\right) W$ is unbiased for θ^2.

8.17 A point estimate of the population mean, μ, is the sample mean $\overline{x} = 11.5$, and the bound on the error of estimation with $s = 3.5$ and $n = 50$ is

$$2\sigma_{\overline{y}} = 2\frac{\sigma}{\sqrt{n}} \approx 2\frac{s}{\sqrt{n}} = \frac{2(3.5)}{\sqrt{50}} = 0.99.$$

See table 8.1 for common point estimators and their standard errors.

8.18 The point estimate of μ is $\overline{y} = 7.2\%$, and the bound on the error of estimation is $2\sigma_{\overline{y}}$. With $n = 200$ and $s = 5.6\%$, we have

$$2\sigma_{\overline{y}} = 2\frac{\sigma}{\sqrt{n}} \approx 2\frac{s}{\sqrt{n}} = \frac{2(5.6)}{\sqrt{200}} = .79$$

8.19 **a.** The point estimate of μ is $\overline{y} = 11.3$, and the bound on the error of estimation is $2\sigma_{\overline{y}}$. With $n = 467$ and $s = 16.6$, this bound is

$$2\sigma_{\overline{y}} = 2\,\frac{\sigma}{\sqrt{n}} = 2\,\frac{s}{\sqrt{n}} = \frac{2(16.6)}{\sqrt{467}} = 1.54$$

b. The point estimate of $\mu_R - \mu_C$ is $\overline{y}_R - \overline{y}_C = 46.4 - 45.1 = 1.3$. The bound on the error of estimation is

$$2\sqrt{\frac{\sigma_R^2}{n_R} + \frac{\sigma_C^2}{n_C}} = 2\sqrt{\frac{s_R^2}{n_R} + \frac{s_C^2}{n_C}} = 2\sqrt{\frac{(9.8)^2}{191} + \frac{(10.2)^2}{467}} = 1.7$$

c. The point estimate of $p_C - p_R$ is $\hat{p}_C - \hat{p}_R = .78 - .61 = .17$. The bound on the error of estimation is

$$2\sqrt{\frac{\hat{p}_C\hat{q}_C}{n_C} + \frac{\hat{p}_R\hat{q}_R}{n_R}} = 2\sqrt{\frac{(.78)(.23)}{467} + \frac{(.61)(.39)}{191}} = .08$$

8.20 The value .54 is a point estimate of p. A two-standard-deviation bound on the error of estimation is

$$2\sqrt{\frac{pq}{n}} = 2\sqrt{\frac{\hat{p}\hat{q}}{n}} = 2\sqrt{\frac{(.54)(.46)}{1000}} = .03$$

Note that $.54 - .03 = .51$. Thus we an conclude that a majority of individuals in this age group feel that religion is a very important part of their lives.

8.21 We estimate the difference to be

$$\overline{y}_1 - \overline{y}_2 = 2.4 - 3.1 = -.7.$$

A bound for the error of estimation is

$$b = 2\sqrt{\frac{1.44+2.64}{100}} = 2\sqrt{.0408} = .404.$$

8.22 The point estimate for p is $\hat{p} = \frac{2}{3}$. The bound on the error of estimation is

$$2\sqrt{\frac{\hat{p}\hat{q}}{n}} = 2\sqrt{\frac{\left(\frac{2}{3}\right)\left(\frac{1}{3}\right)}{1752}} = .023$$

8.23 a. The bound is $2\sqrt{\frac{\hat{p}(1-\hat{p})}{n}} = 2\sqrt{\frac{(.67)(.33)}{308,007}} = .0017$.

b. The bound is $2\sqrt{\frac{\hat{p}(1-\hat{p})}{n}} = 2\sqrt{\frac{(.71)(.29)}{308,007}} = .0016$.

c. No, ± 2 percentage points is too large for the margin of error. The bound on the margin of error is closer to $\pm .2$ percentage points.

8.24 p_1 = proportion of college graduates; $\hat{p}_1 = \frac{126}{180} = .70$;
p_2 = proportion of non college graduates; $\hat{p}_2 = \frac{54}{100} = .54$;
The point estimate of $p_1 - p_2$ is $\hat{p}_1 - \hat{p}_2 = .70 - .54 = .16$.
The bound on the error of estimation is $2\sqrt{\frac{\hat{p}_1\hat{q}_1}{n_1} + \frac{\hat{p}_2\hat{q}_2}{n_2}} = 2\sqrt{\frac{(.7)(.3)}{180} + \frac{(.54)(.46)}{100}} = .121$.

8.25 a. Let p_1 = proportion of Americans who ate the recommended amount of fibrous foods in 1983 and p_2 = proportion of Americans who ate the recommended amount of fibrous foods in 1992. Then $n_1 = 1250$, $n_2 = 1251$, $\hat{p}_1 = .59$, and $\hat{p}_2 = .53$. The point estimator for the difference in proportions is

$$\hat{p}_1 - \hat{p}_2 = .59 - .53 = .06.$$

The bound on the error of estimation is

$$2\sqrt{\frac{\hat{p}_1(1-\hat{p}_1)}{n_1} + \frac{\hat{p}_2(1-\hat{p}_2)}{n_2}} = 2\sqrt{\frac{(.59)(.41)}{1250} + \frac{(.53)(.47)}{1251}} = .04$$

b. Since $.06 - .04 > 0$, we can conclude that there has been a demonstrable decrease in the proportion of Americans who eat the recommended amount of fibrous foods.

8.26 Avoided Fat: $\hat{p}_1 - \hat{p}_2 = .55 - .51 = .04$

$$b = 2\sqrt{\frac{(.55)(.45)}{1250} + \frac{(.51)(.49)}{1251}} = .04$$

$.04 - .04 = 0$. This is barely a demonstrable decrease.
Avoided Excess Salt: $\hat{p}_1 - \hat{p}_2 = .53 - .46 = .07$

$$b = 2\sqrt{\frac{(.53)(.47)}{1250} + \frac{(.46)(.54)}{1251}} = .04$$

$.07 - .04 > 0$. Yes, there is a demonstrable decrease.
These results indicate that a demonstrable decrease in the proportion of Americans eating fiber and avoiding fat and excess salt does exist. However, the size of the decrease does not appear to be too large.

8.27 Let p = proportion of individuals who regularly used seat belts in 1992. Then the point estimator is $\hat{p} = .70$ and the bound is

$$2\sqrt{\frac{(.7)(.3)}{1251}} = .026$$

It is not likely that the estimate is off by as much as 10%. The bound on the margin of error is only 2.6%.

8.28 The point estimate of the total accounts receivable is $500\overline{y} = 500(197.1) = 98{,}550$. To find a bound on the error of estimation, we need to find s^2:

$$s^2 = \frac{\sum_1^n y_i^2 - n(\overline{y})^2}{n-1} = \frac{933{,}814 - 20(197.1)^2}{19} = 8255.04$$

The variance of $500(\overline{y})$ is $500^2\,\sigma_{\overline{y}}^2 = \frac{500^2\sigma^2}{20}$, which we estimate as $\frac{500^2 s^2}{20}$. A bound on the error of estimation is

$$2\sqrt{\frac{500^2 s^2}{20}} = 20{,}316.3$$

The point estimate of the average accounts receivable, μ, is point $\overline{y} = 197.1$. A bound on the error of estimation is

$$2\left(\frac{\sigma}{\sqrt{n}}\right)$$

which may be estimated by

$$2\left(\frac{s}{\sqrt{n}}\right) = \frac{2(90.857)}{\sqrt{20}} = 40.63.$$

The value 250 is beyond the point estimate plus the bound on the error of estimation. Thus, it is unlikely that the average account receivable exceeds \$250.

8.29 The point estimate is $\hat{p} = .3$. A bound on the error of estimation is

$$2\sqrt{\frac{\hat{p}\hat{q}}{n}} = \sqrt{\frac{(.3)(.7)}{20}} = .205$$

It is fairly likely that the proportion in compliance exceeds .80 since the value .20 is well within the margin of error from the estimation of the proportion not in compliance.

8.30 An unbiased estimator of λ is $\hat{\lambda} = \overline{Y}$. Since $\sigma_{\overline{Y}}^2 = \frac{\lambda}{n}$, an unbiased estimator of $\sigma_{\overline{Y}}^2$ can be obtained by using $\sigma_{\overline{Y}}^2 = \frac{\overline{Y}}{n}$. An estimate of the standard deviation of $\hat{\lambda}$ is $\sqrt{\frac{\overline{Y}}{n}}$.

8.31 We will use the result of Exercise 8.30.

a. The point estimate is $\hat{\lambda}_A = \overline{y} = 20$. A bound on the error of estimation is

$$2\sqrt{\frac{\overline{y}}{n}} = 2\sqrt{\frac{20}{50}} = 1.265$$

b. The point estimate is $\hat{\lambda}_A - \hat{\lambda}_B = \overline{y}_A - \overline{y}_B = 20 - 23 = -3$. A bound on the error of estimation is

$$2\sqrt{\frac{\overline{y}_A}{n_A} + \frac{\overline{y}_B}{n_B}} = \sqrt{\frac{20}{50} + \frac{23}{50}} = 1.855$$

If the two regimes are identical, then $|\lambda_A - \lambda_B| = 0$. The value $|-3|$ is well beyond the bound of the error of estimation. Thus, we would say the regime B tends to produce a larger mean number of nucleation sites.

8.32 The unbiased estimator for θ is $\overline{Y}$, since $E(\overline{Y}) = \theta$, and since $\sigma_{\overline{Y}} = \frac{\theta}{\sqrt{n}}$, an unbiased estimate of $\sigma_{\overline{Y}}$ is $\hat{\sigma}_{\overline{Y}} = \frac{\overline{Y}}{\sqrt{n}}$.

8.33 Refer to Exercise 8.32. An estimate of θ is $\hat{\theta} = \overline{y} = 1020$, and the bound on error is approximately

$$2\hat{\sigma}_{\overline{Y}} = 2\,\frac{\overline{y}}{\sqrt{n}} = 2\left(\frac{1020}{\sqrt{10}}\right) = 645.10$$

8.34 It is necessary to find an unbiased estimator of $V(Y) = \frac{1-p}{p^2} = \left(\frac{1}{p^2}\right) - \left(\frac{1}{p}\right)$. An unbiased estimator of $\frac{1}{p}$ is available in Y, and we must only find an estimate of $\frac{1}{p^2}$. Consider Y^2.

$$E\left(Y^2\right) = V(Y) + [E(Y)]^2 = \frac{1-p}{p^2} + \frac{1}{p^2} = \frac{2}{p^2} - \frac{1}{p}$$

so that

$$E(Y^2+Y) = \frac{2}{p^2} \qquad \text{and} \qquad E\left(\frac{Y^2+Y}{2}\right) = \frac{1}{p^2}$$

The unbiased estimate of $V(Y)$ is then

$$\frac{Y^2+Y}{2} - Y = \frac{Y^2-Y}{2}.$$

8.35 Using Table 6 and indexing $\nu = 4$ degrees of freedom, we can write

$$P(\chi^2_{.95} \le X \le \chi^2_{.05}) = .90$$
$$P\left(.710721 < \frac{2Y}{\beta} < 9.48773\right) = .90$$
$$P\left(\frac{2Y}{9.48773} < \beta < \frac{2Y}{.710721}\right) = .90$$

Hence the interval $\left(\frac{2Y}{9.48773}, \frac{2Y}{.710721}\right)$ forms a 90% confidence interval for β.

8.36 Use the fact that $Z = \frac{Y-\mu}{\sigma} = Y - \mu$ has a standard normal distribution.

a. The 95% confidence interval for μ is $(Y - 1.96, Y + 1.96)$ since

$$P(-1.96 \le Z \le 1.96) = .95$$
$$P(-1.96 \le Y - \mu \le 1.96) = .95$$
$$P(Y - 1.96 \le \mu \le Y + 1.96) = .95$$

b. Since

$$P(Z \le -1.645) = .05$$
$$P(Y - \mu \le -1.645) = .05$$
$$P(\mu \ge Y + 1.645) = .05$$

Hence $Y + 1.645$ is the 95% upper limit for μ.

c. Similarly, $Y - 1.645$ is the 95% lower limit for μ.

8.37 a. $.95 = P\left(\chi^2_{.975} \le \frac{Y^2}{\sigma^2} \le \chi^2_{.025}\right) = P\left(.0009821 \le \frac{Y^2}{\sigma^2} \le 5.02389\right)$

$= P\left(\frac{Y^2}{5.02389} \le \sigma^2 \le \frac{Y^2}{.0009821}\right)$

b. $.95 = P\left(\chi^2_{.95} \le \frac{Y^2}{\sigma^2}\right) = P\left(\sigma^2 \le \frac{Y^2}{.0039321}\right)$

c. $.95 = P\left(\frac{Y^2}{\sigma^2} \le \chi^2_{.05}\right) = P\left(\sigma^2 \ge \frac{Y^2}{3.84146}\right)$

8.38 Using the results of Exercise 8.37 and taking the square root of the interval boundaries, the three intervals are obtained.

a. $\frac{Y}{2.24} \le \sigma \le \frac{Y}{.0313}$

b. $\sigma \le \frac{Y}{.0627}$

c. $\sigma < \frac{Y}{1.96}$

8.39 $f(y) = \frac{1}{\theta}$; $F(y) = \frac{y}{\theta}$

a. $F_u(u) = P(U \le u) = P(Y_{(n)} \le \theta u) = F_{Y(n)}(\theta u)$.

$F_{Y(n)}(y) = n[F(y)]^{n-1} f(y) = \left(\frac{y}{\theta}\right)^n$

$F_{Y(n)}(\theta u) = \left(\frac{\theta u}{\theta}\right)^n = u^n$.

Thus,

$$F(u) = \begin{cases} 0, & u < 0 \\ u^n, & 0 \le u < 1 \\ 1, & u > 1 \end{cases}$$

b. $P\left[\left(\frac{Y_{(n)}}{\theta}\right) < a\right] = .95$

$a^n = .95$

$a = (.95)^{1/n}$.

Thus, a 95% lower confidence bound for θ is

$$\frac{Y_{(n)}}{[(.95)^{1/n}]}.$$

8.40 a. $F_Y(y) = \int_0^y \frac{2(\theta-t)}{\theta^2} dt = \frac{y}{\theta}\left(2 - \frac{y}{\theta}\right)$ for $0 < y < \theta$

Note that $F_Y(\theta) = 1$ and that $f_Y(y) = 0$ for $y > \theta$ so that $F_Y(y) = 1$ for $y \ge \theta$.

b. Let $U = Y/\theta$ so that

$$f_U(u) = \frac{2(\theta - \theta u)}{\theta^2}\theta = 2(1-u) \quad \text{for } 0 < u < 1.$$

Therefore, Y/θ is a pivotal quantity.

c. Set $0.90 = P(U \le a) = \int_0^a 2(1-u)du = 2a(1-a)$. Now solve the quadratic equation for a and obtain the following valid solution: $a = 1 - \sqrt{0.10} = 0.6838$.
Then $0.90 = P(U \le 0.6838) = P(Y/\theta \ge 0.6838) = P(\theta \ge Y/0.6838)$.
So that a 90% lower confidence bound for θ is $Y/0.6838$.

8.41 a. $0.90 = P(U \ge b) = 1 - P(U < b) = 1 - \int_0^b 2(1-u)du = 1 - 2b + b^2$

so that $b = 1 - \sqrt{0.90} = 0.05132$. Then

$$0.90 = P(U \ge 0.05132) = P(Y/\theta \ge 0.05132) = P(\theta \le Y/0.05132).$$

Thus, a 90%upper confidence limit is given by $Y/0.05132$.

b. $P(Y/0.6838 < \theta < Y/0.05132) = P(\theta < Y/0.05132) - P(\theta < Y/0.6838) \approx 0.80$
Thus $(Y/0.6838,\ Y/0.05132)$ is an 80% confidence interval for θ.

8.42 a. $\hat{p} = \frac{268}{500} = .536$. Therefore, an approximate 98% confidence interval for p is

$$\hat{p} \pm z_{.01}\sqrt{\frac{\hat{p}\hat{q}}{n}} = .536 \pm 2.33\sqrt{\frac{(.536)(.464)}{500}} = .536 \pm .052 \text{ or } (.484, .588).$$

b. Since the interval does include $p = .51$, we cannot conclude that there is a difference in the graduation rates before and after Proposition 48.

8.43 $\hat{p} = \frac{1912}{2374} = .805$; $\alpha = .01$, $z_{\alpha/2} = z_{.005} = 2.576$

$$\hat{p} \pm Z_{\alpha/2}\sqrt{\frac{\hat{p}(1-\hat{p})}{n}} = .805 \pm 2.576\sqrt{\frac{(.805)(1-.805)}{2374}} = .805 \pm .021 = (.784, .826)$$

Thus, at the 99% confidence level, the proportion of adults in the continental United States registered to vote is between .784 and .826.

8.44 The parameter to be estimated in this exercise is μ, the average number of days required for treatment of patients. The 95% confidence interval is approximately

$$\bar{y} \pm z_{.025}\left(\frac{s}{\sqrt{n}}\right) \quad \text{or} \quad 5.4 \pm 1.96\left(\frac{3.1}{\sqrt{500}}\right) \quad \text{or} \quad 5.4 \pm .27 \quad \text{or} \quad (5.13\ ,\ 5.67)$$

8.45 a. We are given that $n = 224$, $\hat{p} = \frac{2}{3}$, and $\alpha = .10$. Then $z_{.05} = 1.645$ and hence a 90% confidence interval for the proportion of children aged 9 to 17 who would like to experience space travel is

$$\hat{p} \pm z_{\alpha/2}\sqrt{\frac{\hat{p}(1-\hat{p})}{n}} = \frac{2}{3} \pm 1.645\sqrt{\frac{\left(\frac{2}{3}\right)\left(\frac{1}{3}\right)}{224}} = .667 \pm .052 = (.615, .719)$$

b. Since the entire interval is greater than one-half, it is believable that most of the children in the group think that they would like to experience space travel.

8.46 We are given that $n = 75$, $\bar{y} = 4.2$, $s = 1.5$, and $\alpha = .05$. Then $z_{.025} = 1.96$ and hence a 95% confidence interval for the average biomass for North America's northern forests is

$$\bar{y} \pm z_{\alpha/2}\left(\frac{s}{\sqrt{n}}\right) = 4.2 \pm 1.96\left(\frac{1.5}{\sqrt{75}}\right) = 4.2 \pm .34 = (3.86, 4.54).$$

8.47 For a 95% confidence interval, $z_{.025} = 1.96$ so that the interval is

$$(167.1 - 140.9) \pm (1.96)\sqrt{\frac{(24.3)^2}{30} + \frac{(17.6)^2}{30}} = 26.2 \pm 10.74 = (15.46, 36.94).$$

8.48 The approximate 99% confidence interval is, since $z_{.005} = 2.575$,

$$(\bar{y}_1 - \bar{y}_2) \pm 2.575\sqrt{\frac{s_1^2}{n_1} + \frac{s_2^2}{n_2}} = (24.8 - 21.3) \pm 2.575\sqrt{\frac{(7.1)^2}{34} + \frac{(8.1)^2}{41}}$$

$$= 3.5 \pm 4.52 \quad \text{or} \quad (-1.02, 8.02).$$

The difference in mean molt time for "Normal" males versus those "split" from their mates is $(-1.030, 8.030)$ with 99% confidence.

8.49 **a.** A 95% confidence interval for the true percentage of the public favoring drug testing is approximately

$$\hat{p} \pm z_{\alpha/2}\sqrt{\frac{\hat{p}\hat{q}}{n}}$$

or

$$.73 \pm 1.96\sqrt{\frac{(.73)(.27)}{1506}}$$

or

$$.73 \pm .022 = (.708, .752).$$

b. The sampling must be random, and hence each trial (person) independent.

8.50 **a.** Let p_1 = proportion of Americans who ate the recommended amounts of fibrous foods in 1983 and p_2 = proportion of Americans who ate the recommended amounts of fibrous foods in 1992. Then $n_1 = 1250$, $n_2 = 1251$, $\hat{p}_1 = .59$, $\hat{p}_2 = .53$, $\alpha = .02$, and $z_{\alpha/2} = 2.326$. With this information,

$$(\hat{p}_1 - \hat{p}_2) \pm z_{\alpha/2}\sqrt{\frac{\hat{p}_1(1-\hat{p}_1)}{n_1} + \frac{\hat{p}_2(1-\hat{p}_2)}{n_2}}$$

$$= (.59 - .53) \pm 2.326\sqrt{\frac{(.59)(.41)}{1250} + \frac{(.53)(.47)}{1251}} = .06 \pm .046 = (.014, .106)$$

b. Yes, because all of the values in the interval are positive, there is enough evidence to say, with 98% confidence, that the proportion in 1992 is less than the proportion in 1983.

8.51 Let p_1 = proportion of Americans who used seat belts in 1983 and p_2 = proportion of Americans who used seat belts in 1992. Then, $n_1 = 1250$, $n_2 = 1251$, $\hat{p}_1 = .19$, $\hat{p}_2 = .70$, $\alpha = .10$, and $z_{\alpha/2} = z_{.05} = 1.645$.

$$(\hat{p}_1 - \hat{p}_2) \pm z_{\alpha/2}\sqrt{\frac{\hat{p}_1(1-\hat{p}_1)}{n_1} + \frac{\hat{p}_2(1-\hat{p}_2)}{n_2}}$$

$$= (.19 - .70) \pm 1.645\sqrt{\frac{(.19)(.81)}{1250} + \frac{(.70)(.30)}{1251}} = -.51 \pm .028$$

$$= (-.538, -.482)$$

Yes, it is believable, at the 90% confidence level, that the proportion of Americans that used seat belts in 1992 is greater than the proportion in 1983 by as little as .482 or as much as .538.

8.52 Let p_1 = proportion of individuals who used smoke detectors in 1983 and p_2 = proportion of individuals who used smoke detectors in 1992. Then, $n_1 = 1250$, $n_2 = 1251$, $\hat{p}_1 = .67$, $\hat{p}_2 = .90$, $\alpha = .01$, and $z_{\alpha/2} = z_{.005} = 2.576$.

$$(\hat{p}_1 - \hat{p}_2) \pm z_{\alpha/2}\sqrt{\frac{\hat{p}_1(1-\hat{p}_1)}{n_1} + \frac{\hat{p}_2(1-\hat{p}_2)}{n_2}}$$

$$= (.67 - .90) \pm 2.576\sqrt{\frac{(.67)(.33)}{1250} + \frac{(.90)(.10)}{1251}} = -.23 \pm .041$$

$$= (-.271, -.189)$$

8.53 The 98% confidence interval is, since $z_{\alpha/2} = z_{.01} = 2.33$,

$$(\hat{p}_1 - \hat{p}_2) \pm 2.33\sqrt{\frac{\hat{p}_1(1-\hat{p}_1)}{n_1} + \frac{\hat{p}_2(1-\hat{p}_2)}{n_2}}$$

or

$$(.18 - .12) \pm 2.33\sqrt{\frac{(.18)(.82)}{100} + \frac{(.12)(.88)}{100}}$$

or

$$.06 \pm .12 \quad \text{or} \quad (-.06, .18).$$

8.54 **a.** The 95% confidence interval for the difference $\mu_1 - \mu_2$ is approximately

$$\bar{y}_1 - \bar{y}_2 \pm 1.96\sqrt{\frac{s_1^2}{n_1} + \frac{s_2^2}{n_2}} = (11.48 - 13.21) \pm 1.96\sqrt{\frac{(5.69)^2}{252} + \frac{(5.31)^2}{307}}$$

$$= -1.73 \pm .92 \quad \text{or} \quad (-2.65, -.81).$$

b. The 90% confidence interval for the difference $\mu_1 - \mu_2$ is approximately

$$\bar{y}_1 - \bar{y}_2 \pm 1.645\sqrt{\frac{s_1^2}{n_1} + \frac{s_2^2}{n_2}} = (22.05 - 25.96) \pm 1.645\sqrt{\frac{(5.12)^2}{252} + \frac{(5.07)^2}{307}}$$

$$= -3.91 \pm .71 \quad \text{or} \quad (-4.62, -3.20).$$

c. Since both confidence intervals have negative endpoints, neither contains the value $\mu_1 - \mu_2 = 0$. Hence, it is not likely in either case that the means are equal. There appear to be significant differences in the mean scores.

8.55 a. The parameter to be estimated is μ, the mean number of ships passing within ten miles of the proposed power site location per day. The 95% confidence interval is approximately

$$\bar{y} \pm 1.96\left(\frac{s}{\sqrt{n}}\right) = 7.2 \pm 1.96\sqrt{\frac{8.8}{60}} = 7.2 \pm .751 \text{ or } (6.45\,,\, 7.95).$$

b. Now we are interested in the difference between means, $\mu_1 - \mu_2$, for the summer versus the winter months. The interval estimator for the difference between two means is

$$(\bar{y}_1 - \bar{y}_2) \pm z_{\alpha/2}\sqrt{\frac{\sigma_1^2}{n_1} + \frac{\sigma_2^2}{n_2}},$$

which leads to the following approximate 95% confidence interval

$$(\bar{y}_1 - \bar{y}_2) \pm 1.645\sqrt{\frac{s_1^2}{n_1} + \frac{s_2^2}{n_2}} = (7.2 - 4.7) \pm 1.645\sqrt{\frac{8.8}{60} + \frac{4.9}{90}} = 2.5 \pm .738$$

or $(1.762\,,\, 3.238)$.

c. The population used is the difference between the daily mean number of ships sighted in summer months and the mean number sighted in winter months for all summers and winters. One possible problem with the sample of parts **a** and **b** is that the months were not chosen independently or randomly and all of the months were chosen in the same year. Practically, it would be nearly impossible to choose them independently and randomly.

8.56 a. The experiment described in this exercise is multinomial. From Theorem 5.13, we have $V(Y_i) = np_iq_i$ and $\text{Cov}(Y_i, Y_j) = -np_ip_j$. Hence

$$V(Y_1 - Y_2) = V(Y_1) + V(Y_2) - 2\,\text{Cov}(Y_1, Y_2) = np_1q_1 + np_2q_2 + 2np_1p_2$$

b. Since $\hat{p}_1 - \hat{p}_2 = \frac{Y_1 - Y_2}{n}$,

$$V(\hat{p}_1 - \hat{p}_2) = \frac{1}{n^2}V(Y_1 - Y_2) = \frac{p_1q_1}{n} + \frac{p_2q_2}{n} + \frac{2p_1p_2}{n}$$

Using $\hat{p}_i$ to estimate p_i for $i = 1, 2$, we obtain an approximate 95% confidence interval of

$$(\hat{p}_1 - \hat{p}_2) \pm 1.96\sqrt{\frac{\hat{p}_1\hat{q}_1}{n} + \frac{\hat{p}_2\hat{q}_2}{n} + \frac{2\hat{p}_1\hat{p}_2}{n}}$$

or

$$(.06 - .16) \pm 1.96\sqrt{\frac{(.06)(.94)+(.16)(.84)+2(.06)(.16)}{500}} = -.10 \pm .04$$

or $(-.14\,,\, -.06)$.

8.57 Since the $\hat{p}_i$ are independent, $V\left[(\hat{p}_3 - \hat{p}_1) - (\hat{p}_4 - \hat{p}_2)\right] = \sum_{i=1}^{4} \frac{p_iq_i}{n_i}$. If $\hat{p}_i$ is used to estimate p_i, the approximate 95% confidence interval is

$$\left[(\hat{p}_3 - \hat{p}_1) - (\hat{p}_4 - \hat{p}_2)\right] \pm 1.96\sqrt{V\left[(\hat{p}_3 - \hat{p}_1) - (\hat{p}_4 - \hat{p}_2)\right]}$$

or

$$[(.69 - .65) - (.25 - .43)] \pm 1.96\sqrt{\frac{(.65)(.35)}{31} + \frac{(.43)(.57)}{30} + \frac{(.69)(.31)}{26} + \frac{(.25)(.75)}{28}}$$

or

$$.22 \pm .34 = (-.12\,,\, .56).$$

8.58 As in Example 8.9, we need to solve for n in the equation $1.96\sqrt{\frac{pq}{n}} = B$.

a. If $p = .9$ with $B = .05$,

$$1.96\sqrt{\frac{(.9)(.1)}{n}} = .05 \qquad \text{or} \qquad n = 139$$

b. If p is unknown, we use $p = .5$ and

$$1.96\sqrt{\frac{(.5)(.5)}{n}} = .05 \qquad \text{or} \qquad n = 385$$

8.59 $B = 2$, $\sigma = 10$.

$$n = \frac{4\sigma^2}{B^2} = \frac{4(10)^2}{4} = 100.$$

8.60 **a.** We assume $p_1 = p_2 = .75$; $n_1 = n_2 = 1500$.

$$B = 2\sqrt{\frac{2(.75)(.25)}{n}} = .0316.$$

b. For a 90% confidence level, $z_{\alpha/2} = z_{.05} = 1.645$. Assuming equal sample sizes, we consider

$$1.645\sqrt{\frac{2(.75)(.25)}{1500}} = .02$$

$$\sqrt{n} = \frac{1.645\sqrt{2(.75)(.25)}}{.02}$$

$$n = 2536.89 \cong 2537.$$

8.61 From 8.43, $\hat{p} = \frac{2}{3}$. It is given that $B = .02$, $\alpha = .01$, and $z_{\alpha/2} = z_{.005} = 2.576$. Then

$$z_{\alpha/2}\sqrt{\frac{\hat{p}(1-\hat{p})}{n}} = B$$

or

$$2.576\sqrt{\frac{\left(\frac{2}{3}\right)\left(\frac{1}{3}\right)}{n}} = .02.$$

Implying

$$n = \frac{(2.576)^2\left(\frac{2}{3}\right)\left(\frac{1}{3}\right)}{(.02)^2} = 3686.54.$$

Hence we round $n = 3687$.

8.62 This is similar to previous exercises, with $B = .1$ and $\sigma = .5$. The required sample size is obtained by solving

$$.1 = 1.96\left(\frac{.5}{\sqrt{n}}\right) \quad \text{so that} \quad \sqrt{n} = 5(1.96) \quad \text{or} \quad n = 97$$

Notice that water specimens should be selected randomly and not from the same rainfall, in order that all observations be independent.

8.63 There are now two populations of interest and the parameter to be estimated is $\mu_1 - \mu_2$. The bound on the error of estimation is $B = .1$, $\sigma_1^2 = \sigma_2^2 = .25$, and $n_1 = n_2 = n$. With $(1 - \alpha) = .90$, $z_{\alpha/2} = z_{.05} = 1.645$. Hence,

$$1.645\sqrt{\frac{\sigma_1^2}{n_1} + \frac{\sigma_2^2}{n_2}} \leq .1$$

implying

$$1.645\sqrt{\frac{.25+.25}{n}} \leq .1$$

implying

$$\frac{1.645\sqrt{.5}}{.1} \leq \sqrt{n}$$

that is

$$n \geq 135.30.$$

Or $n = 136$ samples should be selected at each location.

8.64 It is given that $n_1 = n_2 = n$, $B = 5$, and $1 - \alpha = 0.9$ so that $z_{\alpha/2} = z_{.05} = 1.645$. From Exercise 8.47, $s_1 = 24.3$ and $s_2 = 17.6$. Using these values to estimate σ_1 and σ_2, the following inequality is to be solved:

$$1.645\sqrt{\frac{\sigma_1^2+\sigma_2^2}{n}} \leq 5$$

implying

$$1.645\sqrt{\frac{(24.3)^2+(17.6)^2}{n}} \leq 5$$

implying

$$\sqrt{n} \geq 9.871$$

that is $n \geq 97.444$. Or, in other words $n_1 = n_2 = 98$.

8.65 Use for estimates of p_1 and p_2, $\hat{p}_1 = .7$ and $\hat{p}_2 = .54$, and note that $z_{.05} = 1.645$. Then

$$1.645\sqrt{\frac{(.7)(.3)+(.54)(.46)}{n}} = .05$$

or $n = 496.17 \cong 497$.

8.66 Assume $n_1 = n_2 = n$ and use the values $\hat{p}_1$ and $\hat{p}_2$ to estimate p_1 and p_2. Then solve for n in

$$s = 1.96\sqrt{\frac{p_1 q_1}{n} + \frac{p_2 q_2}{n}}$$

Since the width of the confidence interval is to be .2, $B = \left(\frac{1}{2}\right)(.2) = .1$ and $z_{.025} = 1.96$ we have

$$.1 = 1.96\sqrt{\frac{(.18)(.82)+(.12)(.88)}{n}} \qquad \text{or} \qquad n = 97.269$$

so that 98 items should be sampled from each line.

8.67 **a.** Assuming equal sample sizes, the estimate of the standard error of the difference is

$$\hat{\sigma}_{\overline{x}_1-\overline{x}_2} = \sqrt{\frac{s_1^2}{n_1} + \frac{s_2^2}{n_2}} = \sqrt{\frac{(2.10)^2+(2.26)^2}{n}} = \frac{3.085}{\sqrt{n}}.$$

Then $z_{.005} = 2.576$ and $B = 1$, so that we have

$$2.576\left(\frac{3.085}{\sqrt{n}}\right) = 1$$

or

$$n = [2.576(3.085)]^2 = 63.16.$$

Therefore we round up to $n =$ 64.

b. For hobbies,

$$\hat{\sigma}_{\overline{x}_1-\overline{x}_2} = \sqrt{\frac{(6.47)^2+(7.46)^2}{n}} = \frac{9.875}{\sqrt{n}}.$$

Then

$$2.576\left(\frac{9.874}{\sqrt{n}}\right) = 1$$

or

$$n = [2.576(9.874)]^2 = 647.09.$$

Collect 648 each of males and females.

c. The activity scores for hobbies have the largest standard deviation of all the activities. Any sample size that meets the criterion for this activity would do better for the other activities. Therefore, a sample of size 648 each of males and females would ensure that 99% confidence intervals to compare males and females in all 7 activities would have widths no larger than 2 units.

8.68 The 95% confidence interval, based on $n - 1 = 20$ degrees of freedom, is

$$\overline{y} \pm t_{.025}\left(\frac{s}{\sqrt{n}}\right) = 26.6 \pm 2.086\left(\frac{7.4}{\sqrt{21}}\right) = 26.6 \pm 3.37 \text{ or } (23.23, 29.97).$$

8.69 Calculate $\Sigma\, y_i = 608$, $\Sigma\, y_i^2 = 37{,}538$, $n = 10$. Then

$$\overline{y} = \frac{608}{10} = 60.8$$

and

$$s^2 = \frac{37{,}538 - \frac{(608)^2}{10}}{9} = 63.5111.$$

The 95% confidence interval based on $n - 1 = 9$ degrees of freedom is

$$\overline{Y} \pm t_{.025}\left(\frac{s}{\sqrt{n}}\right) = 60.8 \pm 2.262\sqrt{\frac{63.5111}{10}} = 60.8 \pm 5.701$$

or (55.099, 66.501).

8.70 **a.** $n = 20$, $\overline{x} = 419$, $s = 57$. Then the 90% confidence interval for the mean SAT scores for urban high school seniors is

$$\overline{y} \pm t_{.05}\left(\frac{s}{\sqrt{n}}\right)$$

where $t_{.05}$ is based on $n - 1 = 19$ degrees of freedom. From the Appendix, this is $t_{.05} = 1.729$. Then the confidence interval is

$$419 \pm 1.729\left(\frac{57}{\sqrt{20}}\right) = 419 \pm 22.04 = (396.96, 441.04).$$

b. The interval does include 422. Thus 422 is a believable value for μ at the 90% confidence level. However, numbers such as 397, 410, and 441, for example, are also believable values for μ.

c. Given $n = 20$, $\overline{x} = 455$, $s = 69$, the 90% confidence interval for the mean mathematics SAT score is

$$\bar{y} \pm t_{.05}\left(\frac{s}{\sqrt{n}}\right) = 455 \pm 1.729\left(\frac{69}{\sqrt{20}}\right) = 455 \pm 26.67 = (428.33, 481.67).$$

The interval does include 474. We would conclude, based on our 90% confidence interval, that the true mean mathematics SAT score is not different from 474.

8.71 **a.** Let μ_1 = mean compartment pressure for all runners under resting condition and μ_2 = mean compartment pressure for all cyclists under resting condition. Then the small-sample 95% confidence interval for $\mu_1 - \mu_2$ is

$$\overline{x}_1 - \overline{x}_2 \pm t_{\alpha/2}\, S_p\sqrt{\frac{1}{n_1} + \frac{1}{n_2}}$$

where $t_{.025} = 2.101$ with $n_1 + n_2 - 2 = 18$ degrees of freedom. Also

$$S_p^2 = \frac{(n_1-1)S_1^2+(n_2-1)S_2^2}{n_1+n_2-2} = \frac{9(3.92)^2+9(3.98)^2}{18} = 15.6034.$$

Then, the interval is

$$14.5 - 11.1 \pm 2.101\sqrt{15.6034\left(\tfrac{1}{10} + \tfrac{1}{10}\right)} = 3.4 \pm 3.7 = (-.3, 7.1).$$

b. Similar to part **a**. Let μ_1 = mean compartment pressure for all runners who exercise at 80% of maximal oxygen consumption and μ_2 = mean compartment pressure for all cyclists who exercise at 80% of maximal oxygen consumption. Then calculate

$$S_p^2 = \frac{(n_1-1)S_1^2+(n_2-1)S_2^2}{n_1+n_2-2} = \frac{9(3.49)^2+9(4.95)^2}{18} = 18.3413.$$

A 90% confidence interval for $\mu_1 - \mu_2$ is

$$(\overline{x}_1 - \overline{x}_2) \pm t_{.05}\sqrt{S_p^2\left(\frac{1}{n_1} + \frac{1}{n_2}\right)}$$

$$= 12.2 - 11.5 \pm 1.734\sqrt{18.3413\left(\tfrac{1}{10} + \tfrac{1}{10}\right)} = .7 \pm 3.32$$

or $(-2.62, 4.02)$

c. Because both intervals contain 0, we cannot conclude that a difference exists in mean compartment pressures between runners and cyclists in either condition.

8.72 Calculate $\bar{y} = \frac{\Sigma y_i}{n} = \frac{37.81}{10} = 3.781$

$$s^2 = \frac{\Sigma y_i^2 - \frac{(\Sigma y_i)^2}{n}}{n-1} = \frac{143.2543 - \frac{(37.81)^2}{10}}{9} = .0327$$

The 95% confidence interval, based on $n - 1 = 9$ degrees of freedom, is

$$\bar{y} \pm t_{.025}\left(\frac{s}{\sqrt{n}}\right) = 3.781 \pm 2.262\sqrt{\frac{.0327}{10}} = 3.781 \pm .129$$

or $(3.652, 3.91)$.

8.73 Calculate

$$S_p^2 = \frac{15(6)^2+19(8)^2}{34} = 51.647$$

The 95% confidence interval for $\mu_1 - \mu_2$ is

$$(\bar{y}_1 - \bar{y}_2) \pm t_{.025}\sqrt{S_p^2\left(\frac{1}{n_1} + \frac{1}{n_2}\right)} = (11 - 12) \pm 1.96\sqrt{51.647\left(\tfrac{1}{16} + \tfrac{1}{20}\right)} = -1 \pm 4.72$$

or $(-5.72, 3.72)$.

8.74 For the $n = 12$ measurements given here, calculate $\Sigma y_i = 108$ and $\Sigma y_i^2 = 1426$. Then

$$\bar{y} = \frac{108}{12} = 9 \qquad \text{and} \qquad s^2 = \frac{1426 - \left[\frac{(108)^2}{12}\right]}{11} = 41.2727$$

The 90% confidence interval is then

$$\bar{y} \pm t_{.05}\left(\frac{s}{\sqrt{n}}\right) = \pm 1.796\sqrt{\frac{41.2727}{12}} = 9 \pm 3.33 \quad \text{or} \quad (5.67, 12.33).$$

8.75 Refer to Exercise 8.74, where $\bar{y}_1 = 9$ and $(n_1 - 1)s_1^2 = \Sigma y_1^2 - \left[\frac{(\Sigma y_1)^2}{n}\right] = 454$. From this,

$$\Sigma y_2 = 10.7 \qquad \Sigma y_2^2 = 65.09$$

$$\Sigma y_2^2 - \frac{(\Sigma y_2)^2}{n_2} = 26.92667 \qquad s_2^2 = \frac{26.92667}{2} = 13.4633$$

a. The 90% confidence interval for μ_2 is

$$\bar{y}_2 \pm t_{.05}\left(\frac{s_2}{\sqrt{n_2}}\right) = 3.57 \pm 2.92\sqrt{\frac{13.4633}{3}}$$

$$= 3.57 \pm 6.19 = (-2.62, 9.76).$$

b. Calculate

$$S_p^2 = \frac{454+26.92667}{13} = 36.9944$$

The 90% confidence interval for $\mu_1 - \mu_2$ is

$$(\overline{y}_1 - \overline{y}_2) \pm t_{.05}\sqrt{S_p^2\left(\frac{1}{n_1}+\frac{1}{n_2}\right)} = (9-3.57) \pm 1.771\sqrt{36.9944\left(\frac{1}{12}+\frac{1}{3}\right)} = 5.43 \pm 6.95$$

or $(-1.52\ ,\ 12.38)$. We must assume that the LC50 measurements are normally distributed and independent and that $\sigma_1^2 = \sigma_2^2$.

8.76 **a.** Let μ_1 = mean verbal score for engineering students and μ_2 = mean verbal score for language/literature students. Then the 95% confidence interval is

$$(\overline{y}_1 - \overline{y}_2) \pm t_{.025}\sqrt{S_p^2\left(\frac{1}{n_1}+\frac{1}{n_2}\right)}$$

where $t_{.025} = 2.048$ with 28 degrees of freedom. Then,

$$S_p^2 = \frac{(n_1-1)S_1^2+(n_2-1)S_2^2}{n_1+n_2-2} = \frac{14(42)^2+14(45)^2}{28} = 1894.5$$

and the confidence interval is

$$446 - 534 \pm 2.048\sqrt{1894.5\left(\frac{1}{15}+\frac{1}{15}\right)} = -88 \pm 32.55 = (-120.55, -55.45)$$

b. Similar to part **a**. Let μ_1 = mean math score for engineering students and μ_2 = mean math score for language/literature students. First, calculate

$$S_p^2 = \frac{(n_1-1)S_1^2+(n_2-1)S_2^2}{n_1+n_2-2} = \frac{14(57)^2+14(52)^2}{28} = 2976.5.$$

Then the interval is

$$548 - 517 \pm 2.048\sqrt{2976.5\left(\frac{1}{15}+\frac{1}{15}\right)} = 31 \pm 40.80 = (-9.80, 71.80).$$

c. The 95% confidence intervals indicate that a significant difference exists in the mean verbal scores for students in engineering and language/literature (since both endpoints of the interval are negative). However, the other interval does not indicate that a significant difference exists in the mean math scores for students in engineering and language/literature, since 0 is in the interval.

d. We assume that the verbal (math) scores for the two groups are randomly and independently selected from two normal distributions with common variance.

8.77 Assume two independent random samples from normal populations with $\sigma_1^2 = \sigma_2^2$. The following calculations are necessary:

	Spring	Summer
$\Sigma\, y_i$	78.1	289.1
$\Sigma\, y_i^2$	1612.15	22,641.47
$\overline{y}$	15.62	72.275
$\Sigma\,(y_i - \overline{y})^2$	392.228	1746.7675
n	5	4

Calculate

$$S_p^2 = \frac{392.228+1746.7675}{7} = 305.57079$$

Then the 95% confidence interval for $\mu_1 - \mu_2$ is

$$(15.62 - 72.275) \pm 2.365\sqrt{S_p^2\left(\frac{1}{5}+\frac{1}{4}\right)} = -56.66 \pm 27.73 = (-84.39\ ,\ -28.93).$$

8.78 We assume $\sigma_1^2 = \sigma_2^2$. Calculate

$$S_p^2 = \frac{3(.001)+4(.002)}{7} = .0016$$

The 95% confidence interval for $\mu_1 - \mu_2$ is

$$(\overline{y}_1 - \overline{y}_2) \pm t_{.025}\sqrt{S_p^2\left(\frac{1}{n_1}+\frac{1}{n_2}\right)} = (.22 - .17) \pm 2.365\sqrt{.0016\left(\frac{1}{4}+\frac{1}{5}\right)} = .05 \pm .063$$

or $(-0.013\ ,\ 0.113)$.

8.79 **a.** Since X and Y are both normally distributed with given means and variances, $2\overline{X} + \overline{Y}$ is normally distributed, with mean $2\mu_1 + \mu_2$ and variance $\frac{4\sigma^2}{n} + \frac{3\sigma^2}{m}$ $= \sigma^2\left(\frac{4}{n}+\frac{3}{m}\right)$. If σ^2 is known, then $2\overline{X} + \overline{Y} \pm 1.96\sigma\sqrt{\frac{4}{n}+\frac{3}{m}}$ is a 95% confidence interval for $2\mu_1 + \mu_2$.

b. Since $\Sigma \frac{(Y_i-\overline{Y})^2}{\sigma^2}$ has a χ^2 distribution with $(n-1)$ degrees of freedom and $\Sigma \frac{(X_i-\overline{X})^2}{3\sigma^2}$ has a χ^2distribution with $(m-1)$ degrees of freedom, the sum
$$\frac{\Sigma (Y_i-\overline{Y})^2+\frac{1}{3}\Sigma (X_i-\overline{X})^2}{\sigma^2}$$
has a χ^2 distribution with $(n+m-2)$ degrees of freedom. Then using Definition 7.2,
$$t = \frac{(2\overline{X}+\overline{Y})-(2\mu_1+\mu_2)}{\sqrt{\hat{\sigma}\left(\frac{4}{n}+\frac{3}{m}\right)}}$$
where
$$\hat{\sigma}^2 = \frac{\Sigma (Y_i-\overline{Y})^2+\left(\frac{1}{3}\right)\Sigma (X_i-\overline{X})^2}{n+m-2}$$
Then, the 95% confidence interval is
$$(2\overline{X}+\overline{Y}) \pm t_{.025}\sqrt{\hat{\sigma}^2\left(\tfrac{4}{n}+\tfrac{3}{m}\right)}.$$

8.80 The appropriate pivotal quantity is
$$T = \frac{(\overline{Y}_1-\overline{Y}_2)-(\mu_1-\mu_2)}{S_p\sqrt{\frac{1}{n_1}+\frac{1}{n_2}}}$$
which has a t distribution with (n_1+n_2-2) degrees of freedom. Select t_α from Table 5, Appendix III such that
$$P(T \geq t_\alpha) = 1-\alpha.$$
Then
$$\begin{aligned}1-\alpha &= P(T \geq t_\alpha)\\ &= P\left[\frac{(\overline{Y}_1-\overline{Y}_2)-(\mu_1-\mu_2)}{S_p\sqrt{\frac{1}{n_1}+\frac{1}{n_2}}} \geq -t_\alpha\right]\\ &= P\left(\overline{Y}_1-\overline{Y}_2+t_\alpha S_p\sqrt{\tfrac{1}{n_1}+\tfrac{1}{n_2}} \geq \mu_1-\mu_2\right).\end{aligned}$$
Thus, $\overline{Y}_1-\overline{Y}_2+t_\alpha S_p\sqrt{\frac{1}{n_1}+\frac{1}{n_2}}$ is a $100(1-\alpha)\%$ upper confidence bound for $\mu_1-\mu_2$.

8.81 It is desired to place a 90% confidence interval on σ^2, the variance of the truck noise emission readings. Indexing $\chi^2_{.05}$ and $\chi^2_{.95}$ with $(n-1)=5$ degrees of freedom in Table 5 yields
$$\chi^2_{.95} = 1.145476 \qquad \text{and} \qquad \chi^2_{.05} = 11.0705.$$
Calculate $\Sigma y_i = 514.4$, $\Sigma y_i^2 = 44{,}103.74$,
$$s^2 = \frac{44{,}103.74-\frac{(514.4)^2}{6}}{5} = .502667.$$
Then the 90% confidence interval for σ^2 is
$$\frac{5(.502667)}{11.0705} < \sigma^2 < \frac{5(.502667)}{1.145476}$$
or
$$.227 < \sigma^2 < 2.194.$$
Intervals constructed in this manner will enclose σ^2 90% of the time in repeated sampling. Hence, we are fairly certain that σ^2 is between .227 and 2.194.

8.82 Calculate $\Sigma y_i = 608$, $\Sigma y_i^2 = 37{,}538$, and
$$(n-1)s^2 = \Sigma y_i^2 - \frac{(\Sigma y_i)^2}{n} = 37{,}538 - \frac{(608)^2}{10} = 571.6$$
Then $\chi^2_{.05} = 16.9190$, $\chi^2_{.95} = 3.32511$, and the 90% confidence interval for σ^2 is
$$\frac{571.6}{16.9190} < \sigma^2 < \frac{571.6}{3.32511}$$
or
$$33.785 < \sigma^2 < 171.90.$$

8.83 **a.** Find a number χ^2 with $(n-1)$ degrees of freedom such that $P\left(\frac{(n-1)s^2}{\sigma^2} \geq \chi^2\right) = 1-\alpha$ from Table 6. That is,
$$1-\alpha = P\left(\frac{(n-1)s^2}{\sigma^2} \geq \chi^2_{1-\alpha}\right) = P\left(\frac{(n-1)s^2}{\chi^2_{1-\alpha}} \geq \sigma^2\right).$$
Then $\frac{(n-1)s^2}{\chi^2_{1-\alpha}}$ is a $100(1-\alpha)\%$ upper confidence bound for σ^2.

b. Similar to part **a**. Find a number χ^2 from Table 6 such that
$$P\left(\frac{(n-1)s^2}{\sigma^2} \le \chi^2\right) = 1-\alpha.$$
Then
$$1-\alpha = P\left(\frac{(n-1)s^2}{\sigma^2} \le \chi^2_\alpha\right) = P\left(\frac{(n-1)s^2}{\chi^2_\alpha} \le \sigma^2\right).$$
Therefore, $\frac{(n-1)s^2}{\chi^2_\alpha}$ is a $100(1-\alpha)\%$ lower confidence bound for σ^2.

8.84 The confidence interval for σ^2 is
$$\left(\frac{(n-1)s^2}{\chi^2_{\alpha/2}}, \frac{(n-1)s^2}{\chi^2_{1-(\alpha/2)}}\right).$$
Since the square root is a one-to-one transformation, a confidence interval for σ is
$$\left(\sqrt{\frac{(n-1)s^2}{\chi^2_{\alpha/2}}}, \sqrt{\frac{(n-1)s^2}{\chi^2_{1-(\alpha/2)}}}\right).$$

8.85 **a.** A $100(1-\alpha)\%$ upper confidence bound for σ is
$$\sqrt{\frac{(n-1)s^2}{\chi^2_{1-\alpha}}}$$
where χ^2_α is chosen such that
$$P\left(\frac{(n-1)s^2}{\sigma^2} \ge \chi^2_\alpha\right) = 1-\alpha.$$
b. A $100(1-\alpha)\%$ lower confidence bound for σ is
$$\sqrt{\frac{(n-1)s^2}{\chi^2_\alpha}}$$
where $\chi^2_{1-\alpha}$ is chosen so that
$$P\left(\frac{(n-1)s^2}{\sigma^2} \le \chi^2_\alpha\right) = 1-\alpha.$$

8.86 Exercise 8.85(a) gives the $100(1-\alpha)\%$ upper confidence bound for σ as
$$\sqrt{\frac{(n-1)s^2}{\chi^2_{1-\alpha}}}$$
where χ^2_α is chosen so that
$$P\left(\frac{(n-1)s^2}{\sigma^2} \ge \chi^2_{1-\alpha}\right) = 1-\alpha.$$
With $(n-1) = 19$ degrees of freedom, $\chi^2_{.99} = 7.6327$. Then
$$(n-1)s^2 = \Sigma\, x_i^2 - \frac{(\Sigma\, x_i)^2}{n} = 92{,}305{,}600 - \frac{(42{,}812)}{20} = 662{,}232.8$$
and the bound is
$$\sqrt{\frac{662{,}232.8}{7.6327}} = 294.55.$$
At the 99% confidence level, the population standard deviation could be less than 150 because the bound is greater than 150. This means that values less than 150 are just as believable as values above 150.

8.87 Calculate
$$\Sigma\, y_i = 56.91 \qquad \Sigma\, y_i^2 = 539.9341 \qquad \Sigma\, y_i^2 - \frac{(y_i)^2}{n} = .14275$$
Thus, $s^2 = .02855$. Then the 90% confidence interval for σ^2 is
$$\frac{(n-1)s^2}{\chi^2_{.05}} < \sigma^2 < \frac{(n-1)s^2}{\chi^2_{.95}} \quad \text{or} \quad \frac{.14275}{11.0705} < \sigma^2 < \frac{.14275}{1.145476} \quad \text{or} \quad (.0129, .1246)$$

8.88 Calculate
$$\Sigma\, y_i = 285 \qquad \Sigma\, y_i^2 = 16823 \qquad \Sigma\, y_i^2 - \frac{(y_i)^2}{n} = 578$$
Then the 99% confidence interval for σ^2 is
$$\frac{(n-1)s^2}{\chi^2_{.005}} < \sigma^2 < \frac{(n-1)s^2}{\chi^2_{.995}} \quad \text{or} \quad \frac{578}{14.8602} < \sigma^2 < \frac{578}{.20699} \quad \text{or} \quad 38.90 < \sigma^2 < 2792.41$$
which means that the 99% confidence interval for σ is $6.24 < \sigma < 52.84$.

8.89 Assume that the measurements are normally distributed. The 90% confidence interval for σ^2 is
$$\frac{(n-1)s^2}{X_U^2} < \sigma^2 < \frac{(n-1)s^2}{X_L^2} \quad \text{or} \quad \frac{11}{7.81473} < \sigma^2 < \frac{11}{.351846} \quad \text{or} \quad 1.408 < \sigma^2 < 31.264$$
The guarantee is not reasonable.

8.90 a. Let p_1 = proportion of survivors in low water group with male parents and p_2 = proportion of survivors in low nutrient group with male parents. Then $n_1 = 578$, $n_2 = 568$, $\hat{p}_1 = \frac{522}{578} = .903$, $\hat{p}_2 = \frac{510}{568} = .898$, $\alpha = .01$, and $z_{\alpha/2} = z_{.005} = 2.576$. The confidence interval is

$$(\hat{p}_1 - \hat{p}_2) \pm z_{\alpha/2}\sqrt{\frac{\hat{p}_1\hat{q}_1}{n_1} + \frac{\hat{p}_2\hat{q}_2}{n_2}}$$

or

$$(.903 - .898) \pm 2.576\sqrt{\frac{(.903)(.097)}{578} + \frac{(.898)(.102)}{568}}$$

or $005 \pm .0456 = (-.0406, .0506)$.

b. Similar to part **a.** Let p_1 = proportion of male survivors in low water group and p_2 = proportion of female survivors in low water group. Then $n_1 = 578$, $n_2 = 510$, $\hat{p}_1 = \frac{522}{578} = .903$, $\hat{p}_2 = \frac{466}{510} = .914$, $\alpha = .01$, and $z_{.005} = 2.576$. Then using the same formula as in part **a**,

$$(.903 - .914) \pm 2.576\sqrt{\frac{(.903)(.097)}{578} + \frac{(.914)(.086)}{510}} = -.011 \pm .045 = (-.056, .034).$$

8.91 We are given that $B = .03$ and $\alpha = .05$. Using the sample statistics from the previous study, we have

$$z_{\alpha/2}\sqrt{\frac{\hat{p}_1\hat{q}_1}{n_1} + \frac{\hat{p}_2\hat{q}_2}{n_2}} = B.$$

Then

$$1.96\sqrt{\frac{(.903)(.097)}{n} + \frac{(.898)(.102)}{n}} = .03$$

implying

$$\frac{(1.96)^2}{(.03)^2}(.179) = n$$

or $764.05 = n$. Therefore we round $n = 765$.

8.92 In this exercise the binomial parameter p is the true fraction of insects killed. Assume that p is approximately equal to .6 and the desired bound is .02. Then

$$z_{\alpha/2}\,\sigma_{\hat{\theta}} = B \qquad \text{or} \qquad 2\sqrt{\frac{pq}{n}} = .02 \qquad \text{or} \qquad 2\sqrt{\frac{(.6)(.4)}{n}} = 0.02$$

Solving for $\sqrt{n}$ and squaring both sides of the inequality yields $n = 2400$ as the necessary sample size.

8.93 a. With $y = 25$ and $n = 400$, the best estimate of p, the proportion of unemployed workers, is $\hat{p} = \frac{25}{400} = .0625$, and a bound on the error of estimation is

$$1.96\sqrt{\frac{pq}{n}} = 1.96\sqrt{\frac{\hat{p}\hat{q}}{n}} = 1.96\sqrt{\frac{(.0625)(.9375)}{400}} = .0237.$$

b. Estimating p by $\hat{p}$ and assuming that the bound on the error of estimation is .02, we have

$$1.96\sqrt{\frac{\hat{p}\hat{q}}{n}} \le .02 \qquad \frac{(1.96)(.24206)}{\sqrt{n}} \le .02 \qquad \sqrt{n} \ge 23.722 \qquad n \ge 562.73$$

so $n = 563$.

8.94 Assume that $\sigma = 400$ and that the desired bound is 50. Then n must be determined so that $z_{\alpha/2}\,\sigma_{\hat{\theta}} \le B$. The population parameter of interest is μ, so that a value of n is determined as follows:

$$1.96\left(\frac{\sigma}{\sqrt{n}}\right) \le 50 \qquad \text{or} \qquad \sqrt{n} \ge 15.68 \qquad \text{or} \qquad n = 245.96 \qquad \text{or} \qquad n = 246$$

8.95 Assume $p = .5$ and the desired bound is .005. Then solving for n in the correct inequality, we have

$$1.96\sigma_{\hat{\theta}} \le B \qquad \text{or} \qquad 1.96\sqrt{\frac{pq}{n}} \le .005 \qquad \text{or} \qquad 1.96\sqrt{\frac{(.5)(.5)}{n}} \le .005$$

or $n = 38{,}416$.

8.96 The objective is to estimate the difference in two binomial parameters p_1 and p_2, using the following information:

Population I (Fraternity)	Population II (Nonfraternity)
$y_1 = 300$	$y_2 = 64$
$n_1 = 500$	$n_2 = 100$

The estimate will be $\hat{p}_1 - \hat{p}_2 = \frac{y_1}{n_1} - \frac{y_2}{n_2} = .60 - .64 = -.04$, and the bound on error will be approximated by

$$2\sqrt{\frac{\hat{p}_1\hat{q}_1}{n_1} + \frac{\hat{p}_2\hat{q}_2}{n_2}} = 2\sqrt{\frac{(.6)(.4)}{500} + \frac{(.64)(.36)}{100}} = .106$$

8.97 Assume that $p = p_1 = p_2 = .6$, $n_1 = n_2 = n$, and that the desired bound is .05. Then

$$2\sqrt{\frac{p_1 q_1}{n_1} + \frac{p_2 q_2}{n_2}} = 2\sqrt{\frac{2pq}{n}}$$

and n is calculated as follows:

$$2\sqrt{\frac{2(.6)(.4)}{n}} = .05$$

or $n = 768$.

8.98 Calculate

$\Sigma y_i = 3975$ $\quad \Sigma y_i^2 = 3{,}160{,}403$ $\quad \overline{y} = 795$ $\quad s^2 = \frac{278}{4} = 69.5$

Then the 90% confidence interval is

$$795 \pm t_{.05,4}\left(\frac{\sqrt{69.5}}{\sqrt{5}}\right) = 795 \pm 2.132(3.728) = 795 \pm 7.95 \text{ or } (787.05\ ,\ 802.95).$$

8.99 Refer to Exercise 8.98. The 90% confidence interval for σ^2 is

$$\frac{(n-1)s^2}{\chi^2_{.05}} < \sigma^2 < \frac{(n-1)s^2}{\chi^2_{.95}} \quad \text{or} \quad \frac{278}{9.48773} < \sigma^2 < \frac{278}{.710721}$$

or

$$29.30 < \sigma^2 < 391.15$$

8.100 The 99% confidence interval for μ is approximately

$$\overline{y} \pm t_{.005}\left(\frac{s}{\sqrt{n}}\right) = 79.47 \pm 2.947\left(\frac{25.25}{\sqrt{16}}\right) = 79.47 \pm 18.60$$

or $60.87 < \mu < 98.07$. Intervals constructed in this manner enclose μ 99% of the time in repeated sampling. Hence, we are fairly certain that this particular interval encloses μ.

8.101 The 90% confidence interval is

$$\overline{y} \pm t_{.05,16}\left(\frac{s}{\sqrt{n}}\right) = 11.3 \pm 1.746\left(\frac{3.4}{\sqrt{17}}\right) = 11.3 \pm 1.44$$

or $9.86 < \mu < 12.74$.

8.102 Assume that the sales are normally distributed. Calculate

$$\overline{y} = \frac{\Sigma y_i}{n} = \frac{22.1}{6} = 3.68$$

$$s^2 = \frac{\Sigma y_i^2 - \left[\frac{(\Sigma y_i)^2}{n}\right]}{n-1} = \frac{99.55 - \left[\frac{(22.1)^2}{6}\right]}{5} = \frac{99.55 - 81.402}{5} = \frac{18.148}{5} = 3.63$$

Then the 90% confidence interval is

$$\overline{y} \pm t_{.05,5}\left(\frac{s}{\sqrt{n}}\right) \quad \text{or} \quad 3.68 \pm 2.015(.778) \quad \text{or} \quad 3.68 \pm 1.57$$

or $2.11 < \mu < 5.25$

8.103 Since n_1 and n_2 are large, the 95% confidence interval for $\mu_1 - \mu_2$ is

$$(\overline{y}_1 - \overline{y}_2) \pm 1.96\sqrt{\frac{s_1^2}{n_1} + \frac{s_2^2}{n_2}} \quad \text{or} \quad (75 - 72) \pm 1.96\sqrt{\frac{(10)^2}{50} + \frac{(8)^2}{45}} \quad \text{or} \quad 3 \pm 3.63$$

or $-.63$ to 6.63.

8.104 Assume the scores are normally distributed with $\sigma_1^2 = \sigma_2^2$. Calculate

$$S_p^2 = \frac{(n_1-1)s_1^2 + (n_2-1)s_2^2}{n_1+n_2-2} = \frac{10(52)+13(71)}{11+14-2} = \frac{1443}{23} = 62.74$$

Then a 95% confidence interval for $\mu_1 - \mu_2$ will be

$$(\overline{y}_1 - \overline{y}_2) \pm t_{.025,23}\sqrt{S_p^2\left(\frac{1}{n_1} + \frac{1}{n_2}\right)} = (64 - 69) \pm 2.069\sqrt{62.74\left(\frac{1}{11} + \frac{1}{14}\right)} = -5 \pm 6.60$$

or $(-11.60\ ,\ 1.60)$.

8.105 Assume the reaction times are normally distributed with $\sigma_1^2 = \sigma_2^2$. Calculate

$$\overline{y}_1 = \frac{\sum_j y_{1j}}{n_1} = \frac{15}{8} = 1.875 \qquad \overline{y}_2 = \frac{21}{8} = 2.625$$

$$S_p^2 = \frac{(n_1-1)s_1^2+(n_2-1)s_2^2}{n_1+n_2-2} = \frac{7(.696)+7(.839)}{14} = .7675$$

A 90% confidence interval for $\mu_1 - \mu_2$ is

$$(\overline{y}_1 - \overline{y}_2) \pm t_{.05,14}\sqrt{S_p^2\left(\frac{1}{n_1}+\frac{1}{n_2}\right)} = (1.875 - 2.625) \pm 1.761\sqrt{.7675\left(\frac{1}{8}+\frac{1}{8}\right)}$$

which yields $-.75 \pm .77$, or $-1.52 < \mu_1 - \mu_2 < .02$.

8.106 It is given that $n = 100$, $\overline{y} = 39.1$, and $s = 17.3$. A 90% confidence interval for μ is

$$\overline{y} \pm 1.645\left(\frac{s}{\sqrt{n}}\right) = 39.1 \pm 1.645\left(\frac{17.3}{\sqrt{100}}\right) = 39.1 \pm 2.846 \text{ or } (36.25\,,\,41.95).$$

Ninety percent of all intervals constructed in this way will include μ.

8.107 It is given that $n = 2300$ and $y = 1914$. Thus, $\hat{p} = \frac{1914}{2300} = .832$ and a 95% confidence interval for p is

$$\hat{p} \pm 1.96\sqrt{\frac{\hat{p}\hat{q}}{n}} = .832 \pm 1.96\sqrt{\frac{(.832)(.168)}{2300}} = .832 \pm .0153$$

or $(.817\,,\,.847)$.

8.108 It is given that $\hat{p} = .67$ and $n = 415$. A 95% confidence interval for p is

$$\hat{p} \pm 1.96\sqrt{\frac{\hat{p}\hat{q}}{n}} = .67 \pm 1.96\sqrt{\frac{(.67)(.33)}{415}} = .67 \pm .045 \text{ or } (.625\,,\,.715).$$

8.109 **a.** From Definition 7.3, the quantity

$$F = \frac{\left[\frac{\frac{(n_1-1)S_1^2}{\sigma_1^2}}{(n_1-1)}\right]}{\left[\frac{\frac{(n_2-1)S_2^2}{\sigma_2^2}}{(n_2-1)}\right]} = \frac{\left(\frac{S_1^2}{\sigma_1^2}\right)}{\left(\frac{S_2^2}{\sigma_2^2}\right)} = \frac{S_1^2}{S_2^2} \times \frac{\sigma_2^2}{\sigma_1^2}$$

has an F distribution with $(n_1 - 1)$ and $(n_2 - 1)$ degrees of freedom.

b. We want

$$P(F_L \le F \le F_U) = 1 - \alpha$$

where F_L and F_U are chosen so that $P(F > F_U) = P(F < F_L) = \frac{\alpha}{2}$. From Table 7, $F_U = F_{\nu_1,\nu_2,\alpha/2}$, where $\nu_i = n_i - 1$, $i = 1, 2$. Then,

$$F_L = \frac{1}{F_{\nu_2,\nu_1,\alpha/2}}.$$

Now, we have

$$P\left(\frac{1}{F_{\nu_2,\nu_1,\alpha/2}} \le \frac{S_1^2}{S_2^2} \times \frac{\sigma_2^2}{\sigma_1^2} \le F_{\nu_1,\nu_2,\alpha/2}\right)$$

$$= P\left(\frac{S_2^2}{S_1^2 F_{\nu_2,\nu_1,\alpha/2}} \le \frac{\sigma_2^2}{\sigma_1^2} \le \frac{S_2^2}{S_1^2} F_{\nu_1,\nu_2,\alpha/2}\right) = 1 - \alpha.$$

Therefore, the $100(1-\alpha)\%$ confidence interval for $\frac{\sigma_2^2}{\sigma_1^2}$ will be

$$\left(\frac{1}{F_{\nu_2,\nu_1,\alpha/2}}\left(\frac{S_2^2}{S_1^2}\right),\, F_{\nu_1,\nu_2,\alpha/2}\left(\frac{S_2^2}{S_1^2}\right)\right).$$

8.110 In Exercise 8.109, the $100(1-\alpha)\%$ confidence interval was derived for $\frac{\sigma_2^2}{\sigma_1^2}$ as

$$\left(\frac{1}{F_{\nu_2,\nu_1,\alpha/2}}\left(\frac{S_2^2}{S_1^2}\right),\, F_{\nu_1,\nu_2,\alpha/2}\left(\frac{S_2^2}{S_1^2}\right)\right).$$

With ten shipments from each supplier, we have $\nu_1 = 9$ and $\nu_2 = 9$. Then $F_{9,9,.025} = 4.03$. Then the confidence interval is

$$\left(\frac{1}{4.03}\left(\frac{.094}{.273}\right),\, 4.03\left(\frac{.094}{.273}\right)\right)$$

and (.085, 1.39) is a 95% confidence interval for the ratio of $\frac{\sigma_2^2}{\sigma_1^2}$.

8.111 Recall that $V(s^2) = 2\sigma^2/(n-1)$.

a. Notice that

$$S'^2 = \frac{\sum(Y_i - \overline{Y})^2}{n} = \frac{(n-1)S^2}{n}$$

Hence

$$V\left(S'^2\right) = \frac{(n-1)^2}{n^2} V\left(S^2\right) = \frac{2(n-1)\sigma^4}{n^2}.$$

b. Since

$$\frac{V\left(S^2\right)}{V\left(S^{2'}\right)} = \frac{2\sigma^4 n^2}{2(n-1)^2\sigma^4} = \left(\frac{n}{n-1}\right)^2 > 1$$

we can say that

$$V\left(S'^2\right) < V\left(S^2\right).$$

8.112

From Exercise 8.1,

$$\text{MSE}\left(S^2\right) = V\left(S^2\right) + \left[E\left(S^2\right) - \sigma^2\right]^2 = V\left(S^2\right) + 0 = \frac{2\sigma^4}{n-1}$$

Similarly,

$$\begin{aligned}\text{MSE}\left(S'^2\right) &= V\left(S'^2\right) + \left[E\left(S'^2\right) - \sigma^2\right]^2 = \frac{(n-1)2\sigma^4}{n^2} + \left(\frac{n-1}{n} \times \sigma^2 - \sigma^2\right)^2 \\ &= \frac{2(n-1)\sigma^4}{n^2} + \frac{\sigma^4}{n^2} = \frac{(2n-1)\sigma^4}{n^2}\end{aligned}$$

Look at

$$\text{MSE}\left(S'^2\right) - \text{MSE}\left(S^2\right) = \frac{(2n-1)\sigma^4}{n^2} - \frac{2\sigma^4}{n-1} = \frac{\sigma^4(1-3n)}{(n-1)n^2}$$

For $n > 1$, the quantity $(1 - 3n)$ will be negative, while σ^4, $n - 1$, and n will be positive. Hence

$$\text{MSE}\left(S'^2\right) - \text{MSE}\left(S^2\right) < 0$$

or

$$\text{MSE}\left(S'^2\right) < \text{MSE}\left(S^2\right).$$

Which indicates that S'^2 may be a better estimator.

8.113

We know that $E\left(S_i^2\right) = \sigma^2$ and $V\left(S_i^2\right) = \frac{2\sigma^4}{n_i-1}$

a. $E\left(S_p^2\right) = \frac{(n_1-1)E\left(S_1^2\right)+(n_2-1)E\left(S_2^2\right)}{n_1+n_2-2} = \frac{n_1+n_2-2}{n_1+n_2-2}\left(\sigma^2\right) = \sigma^2$

so that S_p^2 is an unbiased estimator of σ^2.

b. $V\left(S_p^2\right) = \frac{(n_1-1)^2 V\left(S_1^2\right)+(n_2-1)^2 V\left(S_2^2\right)}{(n_1+n_2-2)^2} = \frac{2\sigma^4(n_1-1)+2\sigma^4(n_2-1)}{(n_1+n_2-2)^2}$

$$= \frac{2\sigma^4}{n_1+n_2-2}$$

8.114

The width of the small-sample confidence interval is $2t_{\alpha/2}\left(\frac{S}{\sqrt{n}}\right)$ and has expected value $2t_{\alpha/2}\left[\frac{E(S)}{\sqrt{n}}\right]$. From Exercise 8.12,

$$E(S) = \frac{\Gamma\left(\frac{n}{2}\right)\sqrt{2}\sigma}{\Gamma\left(\frac{n-1}{2}\right)\sqrt{n-1}}$$

so that

$$E\left(2t_{\alpha/2}\left(\frac{S}{\sqrt{n}}\right)\right) = 2^{3/2}\, t_{\alpha/2}\, \frac{\sigma\Gamma\left(\frac{n}{2}\right)}{\sqrt{n(n-1)}\,\Gamma\left(\frac{n-1}{2}\right)}$$

8.115

The $100(1-\alpha)\%$ confidence interval for σ^2 is given as

$$\frac{(n-1)S^2}{X_U^2} < \sigma^2 < \frac{(n-1)S^2}{X_L^2}$$

so that the midpoint of this interval will be

$$M = \frac{1}{2}\left[\frac{(n-1)S^2}{X_U^2} + \frac{(n-1)S^2}{X_L^2}\right].$$

Now

$$E(M) = \left(\frac{n-1}{2X_U^2} + \frac{n-1}{2X_L^2}\right) E\left(S^2\right) = \frac{(n-1)\sigma^2}{2}\left(\frac{1}{X_U^2} + \frac{1}{X_L^2}\right) \neq \sigma^2.$$

8.116

Consider the quantity $X = Y_p - \overline{Y}$. Since $Y_1, Y_2, \ldots, Y_n, Y_p$ are all independently normally distributed, X is also normally distributed with

$$E(X) = E(Y_p - \overline{Y}) = \mu - \mu = 0$$
$$V(X) = V(Y_p) + V(\overline{Y}) = \sigma^2 + \frac{\sigma^2}{n} = \left(\frac{n+1}{n}\right)\sigma^2.$$

Thus,

$$Z = \frac{Y_p - \overline{Y}}{\sqrt{\left(\frac{n+1}{n}\right)\sigma^2}}$$

has a standard normal distribution. From Theorem 7.3, $\frac{(n-1)S^2}{\sigma^2}$ has a χ^2 distribution with $n-1$ degrees of freedom. Thus,

$$\frac{\frac{Y_p-\overline{Y}}{\sqrt{\left(\frac{n+1}{n}\right)\sigma^2}}}{\sqrt{\frac{(n-1)S^2}{\sigma^2(n-1)}}} = \frac{Y_p-\overline{Y}}{s\sqrt{\frac{n+1}{n}}}$$

has a t distribution (see Definition 7.2). We can now use the same procedure used in Section 8.8 (with Y_p in place of μ and $\frac{n+1}{n}$ in place of $\frac{1}{n}$) to arrive at the confidence interval

$$\overline{Y} \pm t_{\alpha/2}\, s\sqrt{\frac{n+1}{n}}$$

CHAPTER 9 PROPERTIES OF ESTIMATORS AND METHODS OF ESTIMATION

9.1 Refer to Exercise 8.4, where the four necessary variances were calculated. The efficiencies are as follows:

$\hat{\theta}_1$ relative to $\hat{\theta}_5$: $\frac{V(\hat{\theta}_5)}{V(\hat{\theta}_1)} = \frac{\left(\frac{\theta^2}{3}\right)}{\theta^2} = \frac{1}{3}$ $\quad$ $\hat{\theta}_2$ relative to $\hat{\theta}_5$: $\frac{V(\hat{\theta}_5)}{V(\hat{\theta}_2)} = \frac{\left(\frac{\theta^2}{3}\right)}{\left(\frac{\theta^2}{2}\right)} = \frac{2}{3}$

$\hat{\theta}_3$ relative to $\hat{\theta}_5$: $\frac{V(\hat{\theta}_5)}{V(\hat{\theta}_3)} = \frac{\left(\frac{\theta^2}{3}\right)}{\left(\frac{5\theta^2}{9}\right)} = \frac{3}{5}$

9.2 **a.** Since $E(Y_i) = \mu$, $i = 1, 2, \ldots, n$, then

$E(\hat{\mu}_1) = \frac{1}{2}(2\mu) = \mu \qquad E(\hat{\mu}_2) = \frac{1}{4}\mu + \frac{(n-2)\mu}{2(n-2)} + \frac{1}{4}\mu = \mu \qquad E(\hat{\mu}_3) = \frac{n\mu}{n} = \mu$

b. Further,

$V(\hat{\mu}_1) = \frac{1}{4}(2\sigma^2) = \frac{\sigma^2}{2} \quad V(\hat{\mu}_2) = \frac{2}{16}\sigma^2 + \frac{(n-2)\sigma^2}{4(n-2)^2} = \frac{\sigma^2}{8} + \frac{\sigma^2}{4(n-2)} \quad V(\hat{\mu}_3) = \frac{\sigma^2}{n}$

Hence the efficiency of $\hat{\mu}_3$ relative to $\hat{\mu}_1$ is

$$\frac{V(\hat{\mu}_1)}{V(\hat{\mu}_3)} = \frac{\left(\frac{\sigma^2}{2}\right)}{\left(\frac{\sigma^2}{n}\right)} = \frac{n}{2}$$

and the efficiency of $\hat{\mu}_3$ relative to $\hat{\mu}_2$ is

$$\frac{V(\hat{\mu}_2)}{V(\hat{\mu}_3)} = \frac{\frac{\sigma^2}{8} + \frac{\sigma^2}{4(n-2)}}{\frac{\sigma^2}{n}} = \frac{n}{8} + \frac{n}{4(n-2)} = \frac{n^2}{8(n-2)}$$

9.3 For the given uniform distribution we have $E(Y_i) = \theta + \frac{1}{2}$ and $V(Y_i) = \frac{1}{12}$.
We will answer parts **a.** and **b.** at the same time. For $\hat{\theta}_1$ we have,

$$E(\hat{\theta}_1) = E\left(\overline{Y} - \tfrac{1}{2}\right) = \left(\theta + \tfrac{1}{2}\right) - \tfrac{1}{2} = \theta$$

and

$$V(\hat{\theta}_1) = V(\overline{Y}) = \frac{\left(\frac{1}{12}\right)}{n} = \frac{1}{12n}.$$

Now consider $\hat{\theta}_2$. The distribution of $Y_{(n)}$ is given by $g_n(y) = n[F(y)]^{n-1}f(y)$, where

$$F(y) = \int_\theta^y dt = (y - \theta) \qquad \text{and} \qquad f(y) = 1$$

Hence $g_n(y) = n(y-\theta)^{n-1}$ for $\theta \le y \le \theta + 1$. Let $U_{(n)} = Y_{(n)} - \theta$. Then the density of $U_{(n)}$ will be given by (using the transformation method) nu^{n-1} for $0 \le u \le 1$. That is, $U_{(n)}$ is distributed as Beta $(n, 1)$. Then

$$E(\hat{\theta}_2) = E(U_{(n)} + \theta - \tfrac{n}{n+1}) = \tfrac{n}{n+1} + \theta - \tfrac{n}{n+1} = \theta,$$

and,

$$V(\hat{\theta}_2) = V(U_{(n)} + \theta - \tfrac{n}{n+1}) = V(U_{(n)}) = \frac{n}{(n+2)(n+1)^2}.$$

Therefore $\hat{\theta}_1$ and $\hat{\theta}_2$ are unbiased, and the efficiency of $\hat{\theta}_1$ relative to $\hat{\theta}_2$ is

$$\frac{V(\hat{\theta}_2)}{V(\hat{\theta}_1)} = \frac{\frac{n}{(n+2)(n+1)^2}}{\frac{1}{12n}} = \frac{12n^2}{(n+2)(n+1)^2}.$$

9.4 We must first find the $V(\hat{\theta}_1)$ and $V(\hat{\theta}_2)$. Notice that the density of $Y_{(1)}$ is given by

$$nf(y)(1 - F(y))^{n-1} = \left(\tfrac{n}{\theta}\right)\left[1 - \left(\tfrac{y}{\theta}\right)\right]^{n-1} \text{ for } 0 \le y \le \theta.$$

Let $U_{(1)} = Y_{(1)}/\theta$. Then (using the transformation method) the density of $U_{(1)}$is then given by $n(1-u)^{n-1}$ for $0 \le u \le \theta$. That is, $U_{(1)}$ is distributed Beta$(1, n)$. Then

$$E(\theta_1) = E((n+1)Y_{(1)}) = E((n+1)U_{(1)}\theta) = \theta$$

and,

$$V(\theta_{(1)}) = V[(n+1)Y_{(1)}] = V((n+1)U_{(1)}\theta) = (n+1)^2\theta^2 V(U_{(1)}) = \frac{n\theta^2}{n+2}.$$

It was shown in Example 9.1 (or an equivalent calculation as above) yields

$$V\left[\frac{(n+1)Y_{(n)}}{n}\right] = \frac{\theta^2}{n(n+2)}$$

Thus,

$$\text{eff}(\hat{\theta}_1, \hat{\theta}_2) = \frac{V(\hat{\theta}_2)}{V(\hat{\theta}_1)} = \frac{\left(\frac{\theta^2}{n(n+2)}\right)}{\left(\frac{n\theta^2}{n+2}\right)} = \frac{1}{n^2}.$$

9.5 From Exercise 7.10 **b.**, we know that S^2 is unbiased and that $V(S^2) = \frac{2\sigma^4}{n-1}$. Consider $\sigma_2^2 = \left(\frac{1}{2}\right)(Y_1 - Y_2)^2$. Since Y_1 and Y_2 are both normally distributed, the quantity $(Y_1 - Y_2)$ is normal with mean 0 and variance $2\sigma^2$. Finally,

$$Z^2 = \left(\frac{Y_1 - Y_2}{\sqrt{2\sigma^2}}\right)^2 = \frac{(Y_1-Y_2)^2}{2\sigma^2}$$

has by definition a χ^2 distribution with 1 degree of freedom. Thus,

$$E(Z^2) = 1$$

so that

$$E\left[\frac{(Y_1-Y_2)^2}{2}\right] = \sigma^2$$

and

$$V(Z^2) = 2$$

so that

$$V\left[\frac{(Y_1-Y_2)^2}{2}\right] = 2\sigma^4.$$

The efficiency of $\hat{\sigma}_1^2$ relative to $\hat{\sigma}_2^2$ is

$$\frac{V\left(\hat{\sigma}_2^2\right)}{V\left(\hat{\sigma}_1^2\right)} = \frac{2\sigma^4}{\left(\frac{2\sigma^4}{n-1}\right)} = n - 1.$$

This suggest (as our intuition would) that using all of the data is much more efficient than using only two points.

9.6 We have

$$V(\hat{\lambda}_1) = V\left[\frac{1}{2(Y_1+Y_2)}\right] = \left(\frac{2}{4}\right)\lambda = \frac{\lambda}{2}$$

$$V(\hat{\lambda}_2) = V(\overline{Y}) = \frac{\lambda}{n}.$$

Implying

$$\text{eff}(\hat{\lambda}_1, \hat{\lambda}_2) = \frac{2}{n}.$$

This suggest (as our intuition would) that using all of the data is much more efficient than using only two points.

9.7 Since Y is exponential, $E(Y) = \theta$ and $V(Y) = \theta^2$. Then we have $V(\hat{\theta}_1) = \theta^2$, since it was given that $\hat{\theta}_1$ is an unbiased estimator and $\text{MSE}(\hat{\theta}_1) = \theta^2$. Further

$$V(\hat{\theta}_2) = V(\overline{Y}) = \frac{\theta^2}{n}$$

implying

$$\text{eff}(\hat{\theta}_1, \hat{\theta}_2) = \frac{V(\hat{\theta}_2)}{V(\hat{\theta}_1)} = \frac{\left(\frac{\theta^2}{n}\right)}{\theta^2} = \frac{1}{n}.$$

9.8 **a.** For $\hat{\theta} = \overline{Y}$, $V(\hat{\theta}) = V(\overline{Y}) = \frac{\sigma^2}{n}$. The Cramer-Rao lower bound is calculated by using

$$\ln f(y) = c - \frac{(y-\mu)^2}{2\sigma^2}$$

where c is a constant not involving μ. Then we have

$$\frac{\partial \ln f(y)}{\partial \mu} = \frac{y-\mu}{\sigma^2}$$

and

$$\frac{\partial^2 \ln f(y)}{\partial^2 \mu} = -\frac{1}{\sigma^2}.$$

Finally,

$$I(\mu) = \frac{1}{\left[-nE\left(\frac{1}{\sigma^2}\right)\right]} = \frac{\sigma^2}{n},$$

which is equal to $V(\overline{Y})$, so that $\overline{Y}$ is efficient.

b. Similar to part **a** with $V(\overline{Y}) = \frac{\sigma^2}{n} = \frac{\lambda}{n}$ and $p(y) = \frac{\lambda^y e^{-\lambda}}{y!}$. Calculate

$$\ln p(y) = y \ln \lambda - \lambda - \ln y!$$

and

$$\frac{\partial \ln p(y)}{\partial \lambda} = \frac{y}{\lambda} - 1$$

and

$$\frac{\partial^2 \ln p(y)}{\partial^2 \lambda} = -\frac{y}{\lambda^2}.$$

Finally,

$$I(\mu) = \frac{1}{\left[-nE\left(\frac{y}{\lambda^2}\right)\right]} = \frac{\lambda}{n},$$

which equals $V(\overline{Y})$, so that $\overline{Y}$ is efficient for λ.

9.9 Using Theorem 9.1, we have

$$\lim_{n\to\infty} V(\hat{\theta}_1) = \lim_{n\to\infty} \frac{1}{12n} = 0$$

$$\lim_{n\to\infty} V(\hat{\theta}_2) = \lim_{n\to\infty} \frac{n}{(n+2)(n+1)} = \lim_{n\to\infty} \frac{1}{\left[1+\left(\frac{2}{n}\right)\right](n+1)^2} = 0$$

Hence $\hat{\theta}_1$ and $\hat{\theta}_2$ are consistent for θ.

9.10 Notice that $\hat{\sigma}_2^2$ doesn't change with n therefore we wouldn't expect it to be consistent. More precisely, $\lim_{n\to\infty} P\left(\left|\hat{\sigma}_2^2 - \sigma^2\right| > \epsilon\right) = P\left(\left|\hat{\sigma}_2^2 - \sigma^2\right| > \epsilon\right) > 0.$

9.11 From Example 9.2, $\overline{X}$ and $\overline{Y}$ are consistent estimators of μ_1 and μ_2, respectively. Hence, by Theorem 9.2, $\overline{X} - \overline{Y}$ converges in probability to $\mu_1 - \mu_2$.

9.12 Write

$$\frac{\Sigma\left(X_i-\overline{X}\right)^2+\Sigma\left(Y_i-\overline{Y}\right)^2}{2n-2} = \frac{\Sigma\left(X_i-\overline{X}\right)^2}{2n-2} + \frac{\Sigma\left(Y_i-\overline{Y}\right)^2}{2n-2}$$

and consider only $\frac{\Sigma\left(X_i-\overline{X}\right)^2}{2n-2} = \left(\frac{n}{2n-2}\right)\left(\frac{1}{n}\Sigma X_i^2 - \overline{X}^2\right)$. From Example 9.3, the quantity $\left(\frac{1}{n}\right)\Sigma X_i^2 - \overline{X}^2$ converges in probability to σ^2, while $\frac{n}{2n-2} = \frac{1}{2-\left(\frac{2}{n}\right)}$ converges to $\frac{1}{2}$ when $n \to \infty$. Hence $\Sigma\frac{\left(X_i-\overline{X}\right)^2}{2n-2}$ converges in probability to $\frac{\sigma^2}{2}$. A similar argument shows that

$$\frac{\Sigma\left(Y_i-\overline{Y}\right)^2}{2n-2} + \frac{\sigma^2}{2},$$

so that

$$\frac{\Sigma\left(X_i-\overline{X}\right)^2+\Sigma\left(Y_i-\overline{Y}\right)^2}{2n-2} + \sigma^2.$$

9.13 Given $f(y)$, calculate

$$E(Y) = \frac{\theta}{\theta+1}$$

and

$$E\left(Y^2\right) = \frac{\theta}{\theta+2}.$$

Thus

$$V(Y) = \frac{\theta}{\theta+2} - \frac{\theta^2}{(\theta+1)^2} = \frac{\theta}{(\theta+2)(\theta+1)^2}$$

Hence, $E(\overline{Y}) = \frac{\theta}{\theta+1}$. Since $\sigma^2 = V(Y)$ is finite, the law of large numbers (Example 9.2) holds, and $\overline{Y}$ is consistent for $\frac{\theta}{\theta+1}$.

9.14 Let $X_i = 1$ if trial i results in a success and $X_i = 0$ otherwise. Then for $i = 1, 2, \ldots, n$, $X_1, X_2, \ldots, X_n$ constitutes a random sample from a distribution with mean p and variance $pq < \infty$. By the law of large numbers, $\overline{X} = \frac{Y}{n}$ is consistent for p.

9.15 **a.** See problem 9.17 for a very general solution. Here we will approach the problem more statistically. Note that $(Y_{2i} - Y_{2i-1})$ are iid $N(0, 2\sigma^2)$ for $i = 1, ..., k$. Therefore $(Y_{2i} - Y_{2i-1})^2/2\sigma^2 \sim \chi_1^2$ and hence

$$\frac{1}{2k}\sum_{i=1}^{k}(Y_{2i} - Y_{2i-1})^2 = \frac{2\sigma^2}{2k}\sum_{i=1}^{k}(Y_{2i} - Y_{2i-1})^2/2\sigma^2 \sim \frac{2\sigma^2}{2k}\sum_{i=1}^{k}\text{iid } \chi_1^2 = \frac{\sigma^2}{k}\chi_k^2.$$

Thus $\frac{1}{2k}E\left(\sum_{i=1}^{k}(Y_{2i} - Y_{2i-1})^2\right) = \frac{\sigma^2}{k}E(\chi_k^2) = \sigma^2.$

b. Again see problem 9.17 for a very general solution. We will show that the variance of the estimator tends to 0 with k. Using part **a.** we have

$$V\left(\frac{1}{2k}\sum_{i=1}^{k}(Y_{2i}-Y_{2i-1})^2\right)=\frac{\sigma^4}{k^2}V(\chi_k^2)=\frac{\sigma^4 2k}{k^2}=2\sigma^4/k\to 0 \text{ as } k\to\infty.$$

9.16 a. We simply appeal to the result of exercise 9.17 for this answer.

b. Again, we appeal to the result of exercise 9.17 for this answer. All that remains to be shown is that Y_1 has a finite fourth moment. This is easy to see as

$$E(Y_1(Y_1-1)(Y_1-2)(Y_1-3))=\sum_{y_1=4}^{\infty}\frac{1}{(y_1-4)!}\lambda^{y_1}e^{-\lambda}$$

$$=e^{-\lambda}\lambda^4\sum_{y_1=4}^{\infty}\frac{1}{(y_1-4)!}\lambda^{y_1-4}=\lambda^4<\infty.$$

(in general $E\left(\prod_{i=0}^{k-1}(Y_1-i)\right)=\lambda^k$). This result guarantees a finite fourth moment, hence using exercise 9.17 we have the result.

9.17 a. Notice that (for $i=1,...,k$)

$E(Y_{2i}-Y_{2i-1})^2=V(Y_{2i}-Y_{2i-1})+E^2(Y_{2i}-Y_{2i-1})=V(Y_{2i}-Y_{2i-1})=2\sigma^2.$

Then we have

$$\frac{1}{2k}\mathrm{E}\left(\sum_{i=1}^{k}(Y_{2i}-Y_{2i-1})^2\right)=\frac{1}{2k}\sum_{i=1}^{k}E(Y_{2i}-Y_{2i-1})^2=\frac{1}{2k}\sum_{i=1}^{k}2\sigma^2=\sigma^2.$$

b. To prove this we will show that the variance of the estimator tends to 0 as k increase. Notice,

$$V\left(\frac{1}{2k}\sum_{i=1}^{k}(Y_{2i}-Y_{2i-1})^2\right)=\frac{1}{4k^2}\sum_{i=1}^{k}V((Y_{2i}-Y_{2i-1})^2),$$

as each component of the sum is independent (the first summand only involves X_1, X_2, the second only X_3, X_4 and so on). Now note that $V((Y_{2i}-Y_{2i-1})^2)=V((Y_2-Y_1)^2)$ as the X_i are identically distributed. Then we have

$$V\left(\frac{1}{2k}\sum_{i=1}^{k}(Y_{2i}-Y_{2i-1})^2\right)=\frac{1}{4k^2}\sum_{i=1}^{k}V((Y_2-Y_1)^2)=\frac{k}{4k^2}V((Y_2-Y_1)^2).$$

Which will tend to 0 as k tends to infinity, *provided* $V((X_2-X_1)^2)$ *is finite*! Combining this with the unbiasedness (provided in part **a**), we have the result.

c. In part **b** we assumed $V((Y_2-Y_1)^2)<\infty$, that is $E(Y_2-Y_1)^4<\infty$. It is easy to see, that this will be the case only when Y_1 has a finite fourth moment because Y_1 and Y_2 are iid and

$$E(Y_2-Y_1)^4 = \mathrm{E}(Y_1^4)+4E(Y_1^3)E(Y_2) + 6E(Y_1^2)E(Y_2^2)+4E(Y_2^3)E(Y_1)+\mathrm{E}(Y_2^4).$$

9.18 a. We know the distribution of Y_i^2 is Chi squared 1. Then $\sum_{i=1}^{n}Y_i^2$ is Chi squared n. Refer to chapter 6 theorem 6.4.

b. By example 9.2 (the law of large numbers), $W_n=\sum_{i=1}^{n}Y_i^2/\mathrm{n}$ converges to $E(Y_1^2)=1$. To prove this on our own, notice $E(W_n)=\sum_{i=1}^{n}E(Y_i^2)/\mathrm{n}=\sum_{i=1}^{n}1/\mathrm{n}=1$, and

$$V(W_n)=\sum_{i=1}^{n}V(Y_i^2)/\mathrm{n}^2=\sum_{i=1}^{n}2/\mathrm{n}^2=2n/n^2=2/n\to 0 \text{ as } n\to\infty.$$

9.19 a. $E(Y_i)=\mu$

b. $P(|Y_i-\mu|\le 1)=P(-1<Y_i-\mu<1)=P(-1<z<1)=2(.5-.158)=.6826$

c. No, since $\lim_{n\to\infty}.6826\ne 1$

9.20 a. Notice if $\epsilon>\theta$ then ,

$$P(|Y_{(n)}-\theta|\le\epsilon)=P(\theta-\epsilon\le Y_{(n)}\le\theta+\epsilon)=F(\theta+\epsilon)-F(\theta-\epsilon)=1-0=1.$$

If $\epsilon<\theta$ then ,

$$P(|Y_{(n)}-\theta|\le\epsilon)=P(\theta-\epsilon\le Y_{(n)}\le\theta+\epsilon)=F(\theta+\epsilon)-F(\theta-\epsilon)=1-\left(\tfrac{\theta-\epsilon}{\theta}\right)^n.$$

b. Evaluating,
$\lim_{n \to \infty} 1 = 1$ and $\lim_{n \to \infty} 1 - \left(\frac{\theta-\epsilon}{\theta}\right)^n = 1,$ where $\epsilon > 0$
Therefore, for any $\epsilon > 0$, $\lim_{n \to \infty} P(|Y_{(n)} - \theta| \leq \epsilon) = 1$. Hence $Y_{(n)}$ is a consistent estimator for θ.

9.21 $P(|Y_1 - \theta| \leq \epsilon) = F(\theta + \epsilon) - F(\theta - \epsilon) = 1 - \left[\left(1 - \frac{\theta-\epsilon}{\theta}\right)^n\right] = \left(\frac{\theta-\theta+\epsilon}{\theta}\right)^n = \left(\frac{\epsilon}{\theta}\right)^n$
Now,
$$\lim_{n \to \infty} \left(\frac{\epsilon}{\theta}\right)^n = 0 \neq 1.$$
Thus, $Y_{(1)}$ is not a consistent estimator of θ.

9.22 $P(|Y_1 - \beta| \leq \epsilon) = F(\beta + \epsilon) - F(\beta - \epsilon) = 1 - \left(\frac{\beta}{\beta+\epsilon}\right)^{\alpha n} = 1 - 0 = 1$
Thus, $Y_{(1)}$ is a consistent estimator of β.

9.23 $P(|Y_{(n)} - \theta| \leq \epsilon) = F(\theta + \epsilon) - F(\theta - \epsilon) = 1 - \left(\frac{\theta-\epsilon}{\theta}\right)^{\alpha n}$
Now,
$$\lim_{n \to \infty} 1 - \left(\frac{\theta-\epsilon}{\theta}\right)^{\alpha n} = 1 - 0 = 1$$
Thus, $Y_{(1)}$ is a consistent estimator of θ.

9.24 $\overline{Y}$ will converge in probability to $E(Y) = \mu_y$ provided (Y) is finite. Notice
$$E(Y) = \int_0^1 3y^3\, dy = \left(\tfrac{3}{4}\right) y^4\Big]_0^1 = \tfrac{3}{4}$$
and,
$$E(Y^2) = \int_0^1 3y^4\, dy = \tfrac{3}{5} \text{ (this step proves } V(Y) \text{ is finite).}$$
Thus, $\overline{Y}$ converges in probability to $\frac{3}{4}$.

9.25 As $V(Y) = \alpha\beta^2$ is finite we know that $\overline{Y}$ converges in probability to μ_y. Thus $\overline{Y}$ converges in probability to $\alpha\beta$.

9.26 Notice that $E(Y^2) = \int_2^\infty 2\, dy = \infty$, $V(Y)$ is not finite. Hence the law of large numbers cannot be applied to this problem.

9.27 By the law of large numbers, $\overline{X}$ is consistent for λ_1 and $\overline{Y}$ is consistent for λ_2. Using Theorem 9.2, we see that the estimator $\frac{\overline{X}}{\overline{X}+\overline{Y}}$ is consistent for $\frac{\lambda_1}{\lambda_1+\lambda_2}$.

9.28 Because of the Central Limit Theorem (Chapter 7), the quantity $U_n = \frac{\hat{p}-p}{\sqrt{\frac{pq}{n}}}$ converges to a standard normal distribution function. Since $\hat{p}$ is consistent for p (Exercise 9.14), $\hat{q} = 1 - \hat{p}$ is consistent for q and $\hat{p}\hat{q}$ is consistent for pq (Theorem 9.2). This is equivalent to saying that $W_n = \sqrt{\frac{\hat{p}\hat{q}}{pq}}$ converges in probability to 1. Finally, by Theorem 9.3, $\frac{U_n}{W_n} = \frac{\hat{p}-p}{\sqrt{\frac{\hat{p}\hat{q}}{n}}}$ converges to a standard normal distribution.

9.29 The density of the ith sample point is $p^{x_i}(1 - p^{x_i})$ implying
$$L = P(X_1 = x_1, X_2 = x_2, \ldots, X_n = x_n) = p^{\Sigma x_i}(1 - p)^{n-\Sigma x_i}$$
If we consider $g(\Sigma x_i, p) = p^{\Sigma x_i}(1 - p)^{n-\Sigma x_i}$ and $h(X_1, X_2, \ldots, X_n) = 1$, then Theorem 9.4 states that ΣX_i is sufficient for p.

9.30 For this exercise, the likelihood of the sample is
$$L = \frac{1}{\left(\sqrt{2\pi}\right)^n \sigma^n} \exp\left[-\frac{\Sigma(y_i-\mu)^2}{2\sigma^2}\right] = \frac{1}{(2\pi)^{n/2}\sigma^n} \exp\left[-\frac{1}{2\sigma^2}\left(\Sigma y_i^2 - 2\mu\, \Sigma y_i + n\mu^2\right)\right]$$
$$= (2\pi)^{-n/2}\sigma^{-n} \exp\left(-\frac{\Sigma y_i^2}{2\sigma^2}\right) \exp\left(\frac{2\mu n\overline{y}}{2\sigma^2}\right) \exp\left(-\frac{n\mu^2}{2\sigma^2}\right)$$

a. When σ^2 is known, use Theorem 9.4 with
$g(\overline{y}, \mu) = \exp\left(\frac{2\mu n\overline{y} - n\mu^2}{2\sigma^2}\right)$ and $h(y_1, y_2, \ldots, y_n) = (2\pi)^{-n/2}\sigma^{-n} \exp\left(-\frac{\Sigma y_i^2}{2\sigma^2}\right)$
Thus, $\overline{Y}$ is sufficient for μ.

b. When μ is known, use Theorem 9.4 with
$g\left(\Sigma (y_i-\mu)^2, \sigma^2\right) = \sigma^{-n} \exp\left[-\frac{\Sigma (y_i-\mu)^2}{2\sigma^2}\right]$ and $h(y_1, y_2, \ldots, y_n) = \frac{1}{\left(\sqrt{2\pi}\right)^n}$

c. As can be inferred from the text, two statistics U_1 and U_2, are jointly sufficient for θ_1 and θ_2 if and only if the conditional distribution of $Y_1, Y_2, \ldots, Y_n$ given U_1 and U_2 does not depend on θ_1 or θ_2. The factorization criterion states that U_1 and U_2 are jointly sufficient for θ_1 and θ_2 if and only if L (the likelihood) can be factored into two non negative functions,

$$L(y_1, y_2, \ldots, y_n|\theta_1 = \theta_2) = g(u_1, u_2; \theta_1, \theta_2)\, h(y_1, y_2, \ldots, y_n)$$

where $g(u_1, u_2; \theta_1, \theta_2)$ is a function only of u_1, u_2, θ_1, and θ_2, and $h(y_1, y_2, \ldots, y_n)$ is not a function of θ_1 or θ_2. For this exercise, write

$$L = \frac{1}{\left(\sqrt{2\pi}\right)^n \sigma^n} \exp\left[-\frac{\Sigma (y_i-\mu)^2}{2\sigma^2}\right] = \frac{1}{\left(\sqrt{2\pi}\right)^n \sigma^n} \exp\left[-\frac{\Sigma y_i^2 - 2\mu \Sigma y_i + n\mu^2}{2\sigma^2}\right]$$

Take

$$g\left(\Sigma y_i, \Sigma y_i^2, \mu, \sigma^2\right) = \frac{1}{\sigma^n} \exp\left[-\frac{\Sigma y_i^2 - 2\mu \Sigma y_i + n\mu^2}{2\sigma^2}\right]$$

and

$$(y_1, y_2, \ldots, y_n) = \frac{1}{\left(\sqrt{2\pi}\right)^n}$$

and the factorization criterion is satisfied. Hence ΣY_i and ΣY_i^2 are jointly sufficient for μ and σ^2. Then, $\overline{y} = \frac{1}{n} \Sigma y_i$ and

$$\Sigma (y_i - \overline{y})^2 = \Sigma y_i^2 - n\overline{y}^2 = \Sigma y_i^2 - \frac{1}{n} (\Sigma y_i)^2$$

are also jointly sufficient for μ and σ^2.

9.31

Refer to Definition 9.3. Each Y_i has a Poisson distribution with mean λ; hence ΣY_i has a Poisson distribution with mean $n\lambda$. The conditional distribution of $Y_1, Y_2, \ldots, Y_n$ given ΣY_i is

$$P\left(Y_1 = y_1, Y_2 = y_2, \ldots, Y_n = y_n | \Sigma Y_i = x\right) = \frac{p(Y_1=y_1, Y_2=y_2, \ldots, Y_n=y_n)}{P(\Sigma Y_i = x)}$$

$$= \frac{\prod_{i=1}^{n} \frac{e^{-\lambda}\lambda^{y_i}}{y_i!}}{P(\Sigma Y_i = x)} = \frac{\left(\frac{e^{-n\lambda}\lambda^{\Sigma y_i}}{\prod_{i=1}^{n} y_i!}\right)}{\left(\frac{e^{-n\lambda}\lambda^{\Sigma y_i}}{(\Sigma y_i)!}\right)} = \begin{cases} \frac{(\Sigma y_i)!}{\prod_{i=1}^{n} y_i!}, & \text{if } \sum_{i=1}^{n} y_i = x \\ 0, & \text{otherwise} \end{cases}$$

which is independent of λ. Hence ΣY_i is sufficient for λ.

9.32

For the Rayleigh distribution, $f(y) = \left(\frac{2y}{\theta}\right) e^{-y^2/\theta}$, and the likelihood is

$$L = \frac{(2^n y_1 y_2 \cdots y_n)}{e^n} e^{-\Sigma y_i^2/\theta}$$

Take $g\left(\Sigma y_i^2, \theta\right) = \left(\frac{1}{\theta^n}\right) e^{-\Sigma y_i^2/\theta}$ and $h(y_1, y_2, \ldots, y_n) = 2^n \prod_{i=1}^{n} y_i$, and Theorem 9.4 is satisfied. Thus, ΣY_i^2 is sufficient for θ.

9.33

The likelihood is $L = \left(\frac{m^n}{\alpha^n}\right) \left(\prod_{i=1}^{n} y\right)^{m-1} e^{-\Sigma y_i^m/\alpha}$. Let $g\left(\Sigma y_i^m, \alpha\right) = \left(\frac{1}{\alpha^n}\right) e^{-\Sigma y_i^m/\alpha}$ and

$h(y_1, \ldots, y_n) = m^n \left(\prod_{i=1}^{n} y_i\right)^{m-1}$, so that Theorem 9.4 is satisfied. Thus, ΣY_i^m is sufficient for α.

9.34

$f(y|p) = p(1-p)^{y-1}$

$$L(y_1, \ldots, y_n|p) = p^n (1-p)^{\Sigma y_i - n} = p^n (1-p)^{n\overline{y}-n}$$

Let $g(\overline{y}, p) = p^n (1-p)^{n\overline{y}-n}$ and $h(y_1, \ldots, y_n) = 1$, then by Theorem 9.4, $\overline{Y}$ is sufficient for p.

9.35

$L(y_1, \ldots, y_n|\alpha, \theta) = \frac{\alpha^n \Pi y_i^{\alpha-1}}{\theta^{n\alpha}} = g\left(\Pi Y_i, \alpha\right)$

By Theorem 9.4, where $h(y_1, \ldots, y_n) = 1$, $\prod_{i=1}^{n} Y_i$ is sufficient for α.

9.36

$L(y_1, \ldots, y_n|\alpha, \beta) = \alpha^n \beta^{n\alpha} \Pi y_i^{-(\alpha+1)} = g\left(\Pi Y_i, \alpha\right)$

By Theorem 9.4, where $h(y_1, \ldots, y_n) = 1$, $\prod_{i=1}^{n} Y_i$ is sufficient for α.

9.37 The likelihood is $L(y_1, \ldots, y_n|\theta) = a(\theta)^n \Pi\, b(y_i)e^{-c(\theta)\Sigma d(y_i)}$.
Let $g(\Sigma\, d(Y_i), \theta) = a(\theta)^n e^{-c(\theta)\Sigma d(y_i)}$ and $h(y_1, \ldots, y_n) = \Pi b(y_i)$. Then by Theorem 9.4, $\sum_{i=1}^{n} d(Y_i)$ is sufficient for θ.

9.38 The exponential distribution is given by $f(y) = \left(\frac{1}{\theta}\right) e^{-y/\theta}$. Then $L(y_1, \ldots, y_n|\theta) = \frac{1}{\theta^n} e^{-\Sigma y_i/\theta}$ implying by the factorization theorem that $\Sigma\, Y_i$ is sufficient for θ. Then also $\overline{Y}$ is also sufficient for θ. Notice this problem is a special case of 9.37.

9.39 We can write $(y|\alpha, \theta) = \left(\frac{\alpha}{\theta^\alpha}\right) e^{(\alpha-1)\ln y}$. Letting

$$a(\alpha) = \frac{\alpha}{\theta^\alpha}, \qquad b(y) = 1, \qquad c(\alpha) = \alpha - 1 \qquad \text{and} \qquad d(y) = \ln y,$$

we see that $f(y|\alpha, \theta)$ is in the exponential family. It was shown in Exercise 9.38 that $\Sigma\, d(Y_i)$ was a sufficient statistic. Since $d(y) = \ln y$, $\sum_{i=1}^{n} \ln Y_i$ is sufficient for α. It was shown in Exercise 9.35 that $\prod_{i=1}^{n} Y_i$ was sufficient for α. Since $\sum_{i=1}^{n} \ln Y_i$ $= \ln \prod_{i=1}^{n} Y_i$, we have no contradiction.

9.40 We can write $f(y|\alpha, \beta) = \alpha\beta^\alpha e^{-(\alpha+1)\ln y}$. Letting

$$a(\alpha) = \alpha\beta^\alpha, \qquad b(y) = 1, \qquad c(\alpha) = \alpha + 1 \qquad \text{and} \qquad d(y) = \ln y,$$

we see that $f(y|\alpha, \beta)$ is in the exponential family. It was shown in Exercise 9.38 that $\Sigma\, d(Y_i)$ was a sufficient statistic. Since $d(y) = \ln y$, $\sum_{i=1}^{n} \ln Y_i$ is sufficient for α. It was shown in Exercise 9.36 that $\prod_{i=1}^{n} Y_i$ was sufficient for α. Since $\sum_{i=1}^{n} \ln Y_i = \ln \prod_{i=1}^{n} Y_i$, we have no contradiction.

9.41 For the uniform distribution, $f(y_1, y_2, \ldots, y_n|\theta) = \frac{1}{\theta^n}$ for $0 \le y_i \le \theta$. Further, if $U = Y_{(n)}$, then $g_n(u|\theta) = \frac{nu^{n-1}}{\theta^n}$ from Section 6.6 in the text. Hence the conditional distribution of $Y_1, Y_2, \ldots, Y_n$ given U is

$$f((y_1, y_2, \ldots, y_n|u) = \frac{f(y_1, y_2, \ldots, y_n|\theta)}{g_n(u|\theta)} = \frac{1}{nu^{n-1}}$$

which is not dependent on θ. Hence $Y_{(n)}$ is sufficient for θ.
Alternate solution:
Define the indicator function $I(a < y < b)$ by

$$I(a < y < b) = \begin{cases} 1, & \text{if } a < y < b \\ 0, & \text{otherwise} \end{cases}$$

Then we can write the uniform density as $f(y|\theta) = \frac{1}{\theta} I(0 < y < \theta)$. The likelihood is

$$L(y_1, \ldots, y_n|\theta) = \frac{1}{\theta^n} \prod_{i=1}^{n} \mathrm{I}(0 < y_i < \theta) = \frac{1}{\theta^n} \mathrm{I}(0 < y_{(n)} < \theta)$$

Theorem 9.4 is satisfied with

$$g(y_{(n)}, \theta) = \frac{1}{\theta^n} \mathrm{I}(0 < y_{(n)} < \theta) \text{ and } h(y_1, \ldots, y_n) = 1.$$

Thus, $Y_{(n)}$ is sufficient.

9.42 This solution is similar to the alternate solution to Exercise 9.41. The density is $f(y|\theta_1, \theta_2) = \frac{1}{(\theta_2-\theta_1)} \mathrm{I}(\theta_1 < y < \theta_2)$. The likelihood is

$$\begin{aligned} \mathrm{L}(y_1, \ldots, y_n|\theta_1, \theta_2) &= \frac{1}{(\theta_2-\theta_1)} \prod_{1}^{n} \mathrm{I}(\theta_1 < y_i < \theta_2) \\ &= \frac{1}{(\theta_2-\theta_1)^n} \mathrm{I}(\theta_1 < y_{(1)} < \infty)\, \mathrm{I}(-\infty < y_{(n)} \le \theta_2) \end{aligned}$$

Let $g(y_{(1)}, y_{(n)}; \theta_1, \theta_2) = \frac{1}{(\theta_2-\theta_1)^n} \mathrm{I}(\theta_1 < y_{(1)} < \infty)\, \mathrm{I}(-\infty < y_{(n)} < \theta_2)$ and let $h(y_1, \ldots, y_n) = 1$. Thus, $Y_{(1)}$ and $Y_{(n)}$ are jointly sufficient for θ_1 and θ_2.

9.43 Since $g_1(y|\theta) = ne^{-n(y-\theta)}$ from Exercise 6.67, and since $f(y_1, y_2, \ldots, y_n|\theta) = e^{-\Sigma(y_i-\theta)}$, the conditional distribution of $Y_1, Y_2, \ldots, Y_n$ given $Y_{(1)}$ is

$$f(y_1, y_2, \ldots, y_n|y_{(1)}) = \frac{e^{-\Sigma y_i}}{ne^{-ny_{(1)}}}$$

which is independent of θ. Hence $Y_{(1)}$ is sufficient for θ.

Alternate solution:

The density is $f(y|\theta) = e^{-(y-\theta)}\ \mathrm{I}(\theta < y < \infty)$ (see the solutions to Exercises 9.41 and 9.42). The likelihood is

$$L(y_1, \ldots, y_n|\theta) = e^{-\Sigma(y_i-\theta)} \prod_{1}^{n} \mathrm{I}(\theta < y_i < \infty) = e^{-\Sigma y_i} e^{n\theta}\ I(\theta < y_{(1)} < \infty).$$

Let $g(y_{(1)}, \theta) = e^{n\theta}\ I\ (\theta < y_{(1)} < \infty)$ and let $h(y_1, \ldots, y_n) = e^{-\Sigma y_i}$. Thus, $Y_{(1)}$ is sufficient for θ.

9.44 Note that $f(y_1, y_2, \ldots, y_n|\theta) = \begin{cases} \prod_{i=1}^{n} 3y_i^2/\theta^3 \text{ for } 0 \le y_i \le \theta \text{ for } i=1,\ldots,n \\ 0 \text{ otherwise} \end{cases}$. Note also the condition that $0 \le y_i \le \theta$ for $i = 1, \ldots, n$ can be written as $0 \le y_{(1)}$and $y_{(n)} \le \theta$ (if the minimum is greater than 0, then so must be the remainder of the y_i; the same argument applies for the maximum being less than θ). The likelihood can then be written as

$$L(y_1, \ldots, y_n|\theta) = \tfrac{3^n}{\theta^{3n}} \left(\prod_{i=1}^{n} y_i\right) I(0 \le y_{(1)}) I(y_{(n)} \le \theta).$$

Thus with $g(y_{(n)}, \theta) = \frac{3^n}{\theta^{3n}} I(y_{(n)} \ge \theta)$ and $h(y_1, \ldots, y_n) = \left(\prod_{i=1}^{n} y_i\right) I(0 \le y_{(1)})$, we see that, by the factorization theorem, $Y_{(n)}$ is sufficient for θ.

9.45 Note that

$$f(y_1, y_2, \ldots, y_n|\theta) = \begin{cases} \prod_{i=1}^{n} (2\theta^3/\, y_i^2) \text{ for } \theta \le y_i \le \infty \text{ for } i=1,\ldots,n \\ 0 \text{ otherwise} \end{cases}.$$

Note also the condition that $\theta \le y_i \le \infty$ for $i = 1, \ldots, n$ can be written as $\theta \le y_{(1)}$ (if the minimum is greater than θ, then so must be the remainder of the y_i). The likelihood can then be written as

$$L(y_1, \ldots, y_n|\theta) = 2^n \theta^{3n} \prod_{i=1}^{n} (1/\, y_i^2) I(\theta \le y_{(1)}).$$

Thus with $g(y_{(n)}, \theta) = 2^n \theta^{3n} I(\theta \le y_{(1)})$ and $h(y_1, \ldots, y_n) = \prod_{i=1}^{n} (1/\, y_i^2)$ we have that $Y_{(1)}$ is sufficient for θ by the factorization theorem.

9.46 $f(y|\alpha, \theta) = \frac{\alpha y^{\alpha-1}}{\theta^\alpha}\ I(0 \le y \le \theta)$

$$L(y_1, \ldots, y_n|\alpha, \theta) = \left(\tfrac{\alpha}{\theta^\alpha}\right)^n \Pi\, y_i^{\alpha-1}\ I(0 \le y_{(n)} \le \theta)$$

By Theorem 9.4, where $h(y_1, \ldots, y_n) = 1$, we see that $\prod_{i=1}^{n} Y_i$ and $Y_{(n)}$ are jointly sufficient for α and θ.

9.47 Note that

$$f(y|\alpha, \beta) = (\alpha\beta^\alpha)^n\, y^{-(\alpha+1)}\ I(\beta < y)$$

and

$$L(y_1, \ldots, y_n|\alpha, \beta) = (\alpha\beta^\alpha)^n\, \Pi\, y_i^{-(\alpha+1)}\ I(\beta < y_{(1)}).$$

By Theorem 9.4, where $h(y_1, \ldots, y_n) = 1$, we see that $\prod_{i=1}^{n} Y_i$ and $Y_{(1)}$ are jointly sufficient for α and β.

9.48 In Exercise 9.30(b) it was shown that $\sum_{i=1}^{n} (Y_i - \mu)^2$ is sufficient for σ^2 when μ is known.

It is known that $W = \frac{1}{n} \sum_{i=1}^{n} (Y_i - \mu)^2$ is an unbiased estimator for σ^2. Since W is a function of the sufficient statistic, W is an MVUE of σ^2.

9.49 From Example 9.8, $\hat{\sigma}_1^2 = \left(\frac{1}{n-1}\right)\Sigma(X_i - \overline{X})^2$ is the MVUE for σ^2 from the sample $X_1, \ldots, X_n$. Similarly, $\hat{\sigma}_2^2 = \left(\frac{1}{n-1}\right)\Sigma(Y - \overline{Y})^2$ is the MVUE for σ^2 from the sample $Y_1, \ldots, Y_n$. Now,

$$\hat{\sigma}_2^2 = \frac{\Sigma(X_i-\overline{X})^2+\Sigma(Y_i-\overline{Y})^2}{2n-2} = \frac{\hat{\sigma}_1^2}{2} + \frac{\hat{\sigma}_2^2}{2}$$

is unbiased. Also, $V\left(\hat{\sigma}_1^2\right) = V\left(\hat{\sigma}_2^2\right)$. Using the result in Exercise 8.2, we see that $\hat{\sigma}_2$ has minimum variance and is thus an MVUE.

9.50 In Exercise 9.32 it was shown that $\sum_{i=1}^{n} Y_i^2$ is sufficient for θ. Now

$$E(Y^2) = \int_0^\infty \frac{2y^3 e^{-y^2/\theta}}{\theta}\,dy = \int_0^\infty \frac{ze^{-z/\theta}}{\theta}\,dz \qquad \text{with } z = y^2$$
$$= E(z) = \theta.$$

Thus, $E\left(\sum_{i=1}^{n} Y_i^2\right) = n\theta$ and the MVUE is $\hat{\theta} = \frac{1}{n}\sum_{i=1}^{n} Y_i^2$.

9.51 With Y a Poisson random variable with parameter λ, it is necessary to find the MVUE for

$$E(c) = 3E(Y^2) = 3\left(V(Y) + [E(Y^2)]\right) = 3(\lambda + \lambda^2)$$

In Exercise 9.31 it was determined that $\sum_{i=1}^{n} Y_i$ is sufficient for λ and thus for λ^2 and $3(\lambda + \lambda^2)$. If a function of $\sum_{i=1}^{n} Y_i$ that is unbiased for $3(\lambda + \lambda^2)$ can be found, then this function will be MVUE. Note that as $\overline{Y}$ is distributed as Poisson$(n\lambda)$ we can easily calculate

$$E\left(\overline{Y}^2\right) = V(\overline{Y}) + [E(\overline{Y})]^2 = \frac{\lambda}{n} + \lambda^2$$

and

$$E\left(\frac{\overline{Y}}{n}\right) = \frac{1}{n}E(\overline{Y}) = \frac{\lambda}{n}.$$

Then we have

$$\lambda^2 = E\left(\overline{Y}^2\right) - E\left(\frac{\overline{Y}}{n}\right)$$

and

$$\lambda = E(\overline{Y}).$$

$E(c) = 3E\left[\overline{Y}^2 - \left(\frac{\overline{Y}}{n}\right) + \overline{Y}\right]$ and the MVUE is $3\left[\overline{Y}^2 + \overline{Y}\left(1 - \frac{1}{n}\right)\right]$.

9.52 a. The given density can be written in the form of the exponential family by first taking the log and then raising e to the resulting power. This gives

$$f(y|\theta) = e^{\ln\theta + (\theta-1)\ln y} = e^{\ln\theta}e^{(\theta-1)\ln y} = \theta e^{(\theta-1)\ln y} \qquad 0 < y < 1;\ \theta > 0$$

From Exercise 9.37, let $a(\theta) = \theta$, $b(y) = 1$, $c(\theta) = (\theta - 1)$, and $d(y) = -\ln(y)$. The result in Exercise 9.37 showed that $\sum_{i=1}^{n} d(Y_i) = \sum_{i=1}^{n} -\ln(Y_i)$ is sufficient for θ.

b. With $w = -\ln y$, $y = e^{-w}$ and $\left|\frac{dy}{dw}\right| = e^{-w}$. Then

$$f_W(w|\theta) = \theta e^{(\theta-1)\ln y}\left|\frac{dy}{dw}\right| = \theta e^{\theta\ln y}e^{-\ln y}\left|\frac{dy}{dw}\right| = \theta e^{-\theta w}e^{w}e^{-w}.$$

Therefore,

$$f_W(w|\theta) = \begin{cases} \theta e^{-\theta w}, & w > 0;\ \theta > 0 \\ 0, & \text{otherwise} \end{cases}$$

which has the form of an exponential distribution with mean $\frac{1}{\theta}$.

c. Consider the transformation $T_i = 2\theta W_i$.

$$f_T(t) = f_W\left(\frac{t}{2\theta}\right)\frac{d\left(\frac{t}{2\theta}\right)}{dt} = \theta e^{-\theta(t/2\theta)}\left(\frac{1}{2\theta}\right) = \frac{1}{2}e^{-t/2},\ t > 0$$

which is the form of a χ^2 distribution with 2 degrees of freedom. Then, similar to Example 9.10, $\sum_{i=1}^{n} T_i = 2\theta\sum_{i=1}^{w} W_i$ has a χ^2 distribution with $2n$ degrees of freedom.

d. In Exercise 4.90 (similar to 4.89) it was shown that if Y has a χ^2 distribution with ν degrees of freedom, then $E\left(\frac{1}{Y}\right) = \frac{1}{\nu-2}$. Based on this result and the result of part **c**,

$$E\left(\frac{1}{2\theta\sum_{i=1}^{n} W_i}\right) = \frac{1}{2n-2} = \frac{1}{2(n-1)}.$$

e. In parts **a** and **b**, it was shown that $\sum_{i=1}^{n} W_i$ is sufficient for θ while from part **d** we have that

$$E\left(\frac{1}{2\theta\sum_{i=1}^{n} W_i}\right) = \frac{1}{2(n-1)}$$

or

$$E\left(\frac{n-1}{\sum_{i=1}^{n} W_i}\right) = \theta.$$

Thus,

$$\frac{n-1}{\sum_{i=1}^{n} W_i} = \frac{n-1}{-\sum_{i=1}^{n} \ln(Y_i)}$$

is the MVUE for θ.

9.53 In Exercise 9.41, $Y_{(n)}$ was shown to be sufficient for θ. Then, it was shown in Example 9.1 that

$$E(Y_{(n)}) = \int_0^\theta \frac{ny^n}{\theta^n}\,dy = \left(\frac{n}{n+1}\right)\theta$$

Thus, $\left(\frac{n+1}{n}\right)Y_{(n)}$ is unbiased and is the MVUE for θ.

9.54 Calculate

$$E(Y_{(1)}) = \int_\theta^\infty nye^{-n(y-\theta)}\,dy = \int_0^\infty n(z+\theta)e^{-nz}\,dz = \int_0^\infty nze^{-nz}\,dz + \theta\int_0^\infty ne^{-nz}\,dz.$$

Note that the first integral is the expected value of an exponential random variable with mean $\frac{1}{n}$ while the second integral is the integral of an exponential density (with mean $\frac{1}{n}$). Thus,

$$E(Y_{(1)}) = \tfrac{1}{n} + \theta(1) = \theta + \tfrac{1}{n}$$

Thus the statistic $Y_{(1)} - \frac{1}{n}$ is an MVUE for θ, since it is an unbiased function of a sufficient statistic (which was shown in Exercise 9.43).

9.55 **a.** First note that the distribution function corresponding to $f(y)$ is

$$F(y) = \begin{cases} 1 \text{ if } y > \theta \\ y^3/\theta^3 \text{ for } 0 \le y \le \theta \\ 0 \text{ for } y < 0 \end{cases}.$$

Then the density of $Y_{(n)}$ is

$$f_{y_{(n)}}(y) = nf(y)(F(y))^{n-1} = \begin{cases} n3\frac{y^2}{\theta^3}\left(\frac{y^3}{\theta^3}\right)^{n-1} = 3ny^{3n-1}/\theta^{3n} & \text{for } 0 \le y \le \theta \\ 0 & \text{otherwise} \end{cases}.$$

b. We know that the UMVUE will be based on $Y_{(n)}$ as it is a complete sufficient statistic. All we need to do is properly scale it so that it is unbiased. Note

$$E(Y_{(n)}) = \int_0^\theta 3ny^{3n}/\theta^{3n}dy = \left.\frac{3ny^{3n+1}}{(3n+1)\theta^{3n}}\right|_0^\theta = \frac{3n}{(3n+1)}\theta.$$

Thus $\frac{(3n+1)}{3n}Y_{(n)}$ is unbiased for θ. It is then UMVUE as $Y_{(n)}$ is complete sufficient.

9.56 **a.** Exercise 9.30 tells us that $\overline{Y}$ is sufficient for μ when σ is known (in this case $\sigma = 1$). So we know the UMVUE will be based on $\overline{Y}$. To finish the result we need to show that $E(\overline{Y}^2 - \frac{1}{n}) = \mu^2$. Note that $\overline{Y}$ is N(μ, $\frac{1}{n}$). Therefore $E(\overline{Y}^2) = V(\overline{Y}) + E^2(\overline{Y}) = \frac{1}{n} + \mu^2$ which proves the result.

b. Note that

$$\begin{aligned}V(\overline{Y}^2 - \tfrac{1}{n}) &= V(\overline{Y}^2)\\ &= V\left[\left((\overline{Y}-\mu)+\mu\right)^2\right]\\ &= V\left[\left(\sqrt{n}(\overline{Y}-\mu)+\sqrt{n}\mu\right)^2/n\right]\\ &= V\left[\left(Z+\sqrt{n}\mu\right)^2\right]/n^2 \text{ where } Z\sim \text{N}(0,1)\end{aligned}$$

Note also,

$$\begin{aligned}V\left[\left(Z+\sqrt{n}\mu\right)^2\right] &= V(Z^2+2Z\sqrt{n}\mu+n\mu^2)\\ &= V(Z^2+2Z\sqrt{n}\mu)\\ &= V(Z^2)+4n\mu^2V(Z)+4\sqrt{n}\,\mu\text{Cov}(Z^2,Z)\end{aligned}$$

Notice that $\text{Cov}(Z^2,Z) = E(Z^3) - E(Z^2)E(Z) = 0 - 0 = 0$ (if you did not know that $E(Z^3) = 0$, verify this by taking the third derivative of the standard normal mgf). Further, note that $V(Z^2) = 2$ as $Z^2 \sim \chi_1^2$, and $V(Z) = 1$. Then

$$V(\overline{Y}^2 - \tfrac{1}{n}) = [V(Z^2) + 4n\mu^2V(Z)]/n^2 = (2+4n\mu^2)/n^2.$$

An alternative approach would begin by noting that

$$V(\overline{Y}^2 - \tfrac{1}{n}) = V(\overline{Y}^2) = E(\overline{Y}^4) - E^2(\overline{Y}^2) = E(\overline{Y}^4) - (\tfrac{1}{n}+\mu^2)^2.$$

As $\overline{Y}$ is $\text{N}(\mu,\frac{1}{n})$, $E(\overline{Y}^4)$ may be found to be $\frac{3}{n^2}+\frac{6\mu^2}{n}+\mu^4$ by differentiating a $\text{N}(\mu,\frac{1}{n})$ mgf 4 times and evaluating at 0. Finally,

$$E(\overline{Y}^4) - (\tfrac{1}{n}+\mu^2)^2 = \tfrac{3}{n^2}+\tfrac{6\mu^2}{n}+\mu^4 - (\tfrac{1}{n}+\mu^2)^2 = (2+4n\mu^2)/n^2$$

which verifies our previous result.

9.57 **a.** Let $T = 1$ if $Y_1 = 1$, $Y_2 = 0$, and $T = 0$ otherwise. Then
$E(T) = P(Y_1 = 1, Y_2 = 0) = P(Y_1 = 1)P(Y_2 = 0) = p(1-p)$

b. $P(T = 1|W = w) = \frac{P(T=1,\,Y=y)}{P(W=w)} = \frac{P(Y_1=1,\,Y_2=0,\,W=w)}{\binom{n}{w}p^wq^{n-w}} = \frac{P(Y_1=1,Y_2=0,\sum_{i=3}^{n}Y_i=w-1)}{\binom{n}{w}p^wq^{n-w}}$

$= \frac{P(Y_1=1)P(Y_2=0)P(\sum_{i=3}^{n}Y_i=w-1)}{\binom{n}{w}p^wq^{n-w}} = \frac{\binom{n-2}{w-1}p^wq^{n-w}}{\binom{n}{y}p^wq^{n-w}} = \frac{w(n-w)}{n(n-1)}$

Note we used the facts that $\sum_{i=3}^{n}Y_i$ will be distributed as a binomial with $n-2$ trials and success probability p, and that $\sum_{i=3}^{n}Y_i$ will be independent of Y_1 and Y_2.

c. $E(T|W) = (1)P(T = 1|W) = \left(\frac{W}{n}\right)\frac{(n-W)}{(n-1)} = \left(\frac{W}{n}\right)\left(\frac{n}{n-1}\right)\left(\frac{n-W}{n}\right)$
$= \frac{n}{n-1}\left[\left(\frac{W}{n}\right)\left(1-\frac{W}{n}\right)\right]$

Since T is unbiased by (1) and $W = \Sigma Y_i$ is sufficient for p [and hence for $p(1-p)$], then $\left(\frac{n}{n-1}\right)\overline{Y}(1-\overline{Y})$ is the MVUE for $p(1-p)$.

9.58 **a.** i. $L(x_1, x_2, \ldots, x_n|p) = \prod_{i=1}^{n} p^{x_i}(1-p)^{1-x_i} = p^{\Sigma x_i}(1-p)^n(1-p)^{-\Sigma x_i}$

$$= \left(\tfrac{p}{1-p}\right)^{\Sigma x_i}(1-p)^n$$

so that

$$\frac{L(x_1,x_2,\ldots,x_n|p)}{L(y_1,y_2,\ldots,y_n|p)} = \frac{\left(\frac{p}{1-p}\right)^{\Sigma x_i}(1-p)^n}{\left(\frac{p}{1-p}\right)^{\Sigma y_i}(1-p)^n} = \left(\frac{p}{1-p}\right)^{\Sigma x_i - \Sigma y_i}.$$

ii. If $\Sigma x_i = \Sigma y_i$, then the ratio in part (i) equals 1 and is, thus, free from p. If $\Sigma x_i - \Sigma y_i$ is anything other than 0, then the ratio is not free from p.

iii. Letting $g(y_1, y_2, \ldots, y_n) = \sum_{i=1}^{n} y_i$, the result in part (ii) shows that $\sum_{i=1}^{n} Y_i$ is a minimal sufficient statistic for p.

b. i. From Example 9.7,

$$L(y_1, y_2, \ldots, y_n|\theta) = \left(\tfrac{2}{\theta}\right)^n e^{-(1/\theta)\sum_{i=1}^{n} y_i^2} \prod_{i=1}^{n} y_i$$

Then

$$\frac{L(x_1, x_2, \ldots, x_n|\theta)}{L(y_1, y_2, \ldots, y_n|\theta)} = \frac{\left(\frac{2}{\theta}\right)^n \exp\left(-\frac{1}{\theta}\sum_{i=1}^{n} x_i^2\right)\prod_{i=1}^{n} x_i}{\left(\frac{2}{\theta}\right)^n \exp\left(-\frac{1}{\theta}\sum_{i=1}^{n} y_i^2\right)\prod_{i=1}^{n} y_i}$$

$$= \frac{\prod_{i=1}^{n} x_i}{\prod_{i=1}^{n} y_i} \exp\left[-\frac{1}{\theta}\left(\sum_{i=1}^{n} x_i^2 - \sum_{i=1}^{n} y_i^2\right)\right].$$

ii. The ratio in part (i) is free from θ if and only if $\sum_{i=1}^{n} x_i^2 = \sum_{i=1}^{n} y_i^2$. Then, by the Lehmann-Scheffe′ method, $\sum_{i=1}^{n} Y_i^2$ is a minimal sufficient statistic for θ.

9.59 $L(y_1, y_2, \ldots, y_n|\mu_1\sigma^2) = \prod_{i=1}^{n} \frac{1}{\sqrt{2\pi}\sigma} e^{(1/2\sigma^2)(y_i-\mu)^2}$

$$= \frac{1}{(2\pi)^{n/2}\sigma^n} \exp\left[-\frac{1}{2\sigma^2}\sum_{i=1}^{n} (y_i-\mu)^2\right]$$

Then

$$\frac{L(x_1, x_2, \ldots, x_n|\theta)}{L(y_1, y_2, \ldots, y_n|\theta)} = \frac{\frac{1}{(2\pi)^{n/2}\sigma^n} \exp\left[-\frac{1}{2\sigma^2}\sum_{i=1}^{n} (x_i-\mu)^2\right]}{\frac{1}{(2\pi)^{n/2}\sigma^n} \exp\left[-\frac{1}{2\sigma^2}\sum_{i=1}^{n} (y_i-\mu)^2\right]}$$

$$= \exp\left[-\frac{1}{2\sigma^2}\left(\sum_{i=1}^{n} (x_i-\mu)^2 - \sum_{i=1}^{n} (y_i-\mu)^2\right)\right]$$

$$= \exp\left[-\frac{1}{2\sigma^2}\left(\sum_{i=1}^{n} x_i^2 - \sum_{i=1}^{n} y_i^2 - 2\mu\left[\sum_{i=1}^{n} x_i - \sum y_i\right]\right)\right].$$

Let $g(Y_1, Y_2, Y_3, \ldots, Y_n) = \left(\sum_{i=1}^{n} Y_i, \sum_{i=1}^{n} Y_i^2\right)$. Then, the above ratio is free of (μ, σ^2) if and only if $\sum_{i=1}^{n} x_i^2 = \sum_{i=1}^{n} y_i^2$ and $\sum_{i=1}^{n} x_i = \sum_{i=1}^{n} y_i$, i.e., $g(x_1, x_2, x_3, \ldots, x_n)$ $= g(y_1, y_2, y_3, \ldots, y_n)$. Then, by the Lehmann-Scheffe′ method discussed in Exercise 9.58, $\left(\sum_{i=1}^{n} Y_i, \sum_{i=1}^{n} Y_i^2\right)$ jointly form a minimal sufficient statistic for μ and σ^2.

9.60 Since $g_1(U)$ and $g_2(U)$ are unbiased,

$$E(g_1(U)) = \theta \qquad \text{and} \qquad E(g_2(U)) = \theta \qquad \text{so that} \qquad E(g_2(U) - g_2(U)) = 0$$

Since the density of U is complete, $g_1(U) - g_2(U) \equiv 0$ by definition. That is, $g_1(U)$ must equal $g_2(U)$. Thus, there is a unique function of U that is an unbiased estimate of θ.

9.61 To use the method of moments, calculate

$$\mu = \int_{-\infty}^{\infty} yf(y)\,dy = \int_{0}^{1} (\theta+1)y^{\theta+1}\,dy = \left.\frac{(\theta+1)y^{\theta+2}}{(\theta+2)}\right]_0^1 = \frac{\theta+1}{\theta+2}$$

Equating sample and population moments, we obtain the estimator of θ.

$$\overline{Y} = \frac{\hat{\theta}+1}{\hat{\theta}+2} \qquad \text{or} \qquad (\hat{\theta}+2)\overline{Y} = \hat{\theta}+1 \qquad \text{or} \qquad \hat{\theta} = \frac{2\overline{Y}-1}{1-\overline{Y}}$$

Because of the law of large numbers, $\overline{Y}$ is consistent for $\mu = \frac{\theta+1}{\theta+2}$. Then, using Theorem 9.2, we can show that $\hat{\theta}$ is consistent for θ, since $\hat{\theta}$ converges to

$$\frac{2\left(\frac{\theta+1}{\theta+2}\right)-1}{1-\left(\frac{\theta+1}{\theta+2}\right)} = \frac{2(\theta+1)-(\theta+2)}{(\theta+2)-(\theta+1)} = \theta$$

$\hat{\theta}$ is not a function of $-\sum_{i=1}^{n} \ln(Y_i)$ or of $\exp\left[-\left(-\sum_{i=1}^{n} \ln(Y_i)\right)\right] = \prod_{i=1}^{n} Y_i$. This implies that $\hat{\theta}$ is not the MVUE.

9.62 Since $\mu_1' = \lambda$, we equate $\overline{Y} = \Sigma \frac{Y_i}{n}$ to μ_1' and obtain $\hat{\lambda} = \overline{Y}$.

9.63 Since the first population moment is $\mu_1' = \mu$, which does not involve σ^2, we equate the second sample and population moments. That is,

$$m_2' = \frac{\Sigma Y_i^2}{n} \qquad \mu_2' = \sigma^2 + \mu^2 = \sigma^2 + 0 = \sigma^2$$

and the moment estimator of σ^2 is $\hat{\sigma}^2 = \Sigma \frac{Y_i^2}{n}$.

9.64 Similar to Exercise 9.61, with μ unknown. The two equations to be solved are

$$\overline{Y} = \mu$$

and

$$\frac{\Sigma Y_i^2}{n} = \mu_2' = \hat{\sigma}^2 + \hat{\mu}^2.$$

So that

$$\hat{\mu} = \overline{Y}$$

and

$$\hat{\sigma}^2 = \frac{\Sigma Y_i^2}{n} - \overline{Y}^2 = \frac{1}{n}\left[\Sigma\,(Y_i - \overline{Y})^2\right].$$

9.65 Notice for this problem we have one hypergeometric observation, Y. Notice $E(Y) = n\theta/N$. This implies our method of moments estimator is $\hat{\theta} = YN/n$.

9.66 **a.** Calculate

$$\mu_1' = E(Y) = \frac{2}{\theta^2}\int_0^\theta (\theta y - y^2)\,dy = \frac{2}{\theta^2}\left[\frac{\theta y^2}{2} - \frac{y^3}{3}\right]_0^\theta = \frac{\theta}{3}$$

Equating sample and population moments, we have $\frac{\hat{\theta}}{3} = \overline{Y}$, or $\hat{\theta} = 3\overline{Y}$.

b. $\hat{\theta}$ is not a function of a sufficient statistic, since $\hat{\theta}$ is a function of $\overline{Y}$ and, if the likelihood function is examined, $L = \frac{2^n}{e^{2n}} \prod_{i=1}^{n} (\theta - y_i)$, it is found that the likelihood cannot be factored into a function of $\overline{Y}$ and θ $[g(\overline{Y}, \theta)]$ and a function only of the Y_i. Hence $\overline{Y}$ is not a sufficient statistic for θ.

9.67 The density shown in this exercise is that of a beta random variable with $\alpha = \theta$ and $\beta = \theta$. Hence

$$\mu_1' = E(Y) = \frac{\theta}{\theta+\theta} = \frac{\theta}{2\theta} = \frac{1}{2}$$

which does not involve θ and hence is of no use in finding a moment estimator. Thus we equate second sample and population moments. From Exercise 4.98,

$$E\left(Y^2\right) = \frac{(\theta+1)\theta}{2\theta(2\theta+1)} = \frac{\theta+1}{2(2\theta+1)}.$$

Hence with $m_2' = \frac{\Sigma y_i^2}{n}$, we solve for $\hat{\theta}$ in

$$m_2' = \frac{\theta+1}{2(2\theta+1)} \Rightarrow 2m_2'(2\theta+1) = \theta + 1 \Rightarrow \theta\,(4m_2' - 1) = 1 - 2m_2' \Rightarrow \hat{\theta} = \frac{1-2m_2'}{4m_2'-1}.$$

9.68 For a single observation y, the first sample moment is $m_1' = Y$, and since Y has a geometric distribution, $E(Y) = \mu_1' = \frac{1}{p}$. Hence the moment estimator of p is $\hat{p} = \frac{1}{Y}$.

9.69 $\mu_1' = E(Y) = \mu = \frac{3\theta}{2} = \overline{Y}$

Now, $m_1' = \hat{\theta} = \frac{2\overline{Y}}{3}$.

9.70 $E(Y) = \int_0^3 \frac{\alpha y^\alpha}{3^\alpha}\,dy = \left(\frac{\alpha}{3^\alpha}\right)\left(\frac{y^{\alpha+1}}{\alpha+1}\right)\Big]_0^3 = \frac{3\alpha}{\alpha+1} = \overline{Y}.$

$\Rightarrow 3\alpha = \alpha\overline{y} + \overline{y}$

$\Rightarrow (3 - \overline{y})\,\alpha = \overline{y}$

$\Rightarrow \hat{\alpha} = \frac{\overline{Y}}{3-\overline{Y}}.$

9.71 $E(Y) = \int_\beta^\infty \alpha\beta^\alpha y^{-\alpha}\,dy = (\alpha\beta^\alpha)\left(\frac{y^{-\alpha+1}}{-\alpha+1}\right)\Big]_\beta^\infty = \left(\frac{\alpha\beta^\alpha}{-\alpha+1}\right)(y^{-\alpha+1})]_\beta^\infty$

$= 0 - \left[\left(\frac{\alpha\beta^\alpha}{-\alpha+1}\right)(\beta^{-\alpha+1})\right] \qquad \text{if } \alpha > 1$

$= \frac{\alpha\beta}{\alpha-1} \qquad \text{if } \alpha > 1$

is undefined if $\alpha < 1$. That is the methods of moments estimator does not exist for $\alpha < 1$ because the Pareto distribution does not have any moments in that case!

9.72 a. The likelihood function is

$$L = \prod_{i=1}^{n} \frac{\lambda^{y_i} e^{-\lambda}}{y!} = \frac{\lambda^{\Sigma y_i} e^{-n\lambda}}{\prod_{i=1}^{n} y_i!}$$

and

$$\ln L = (\Sigma\, y_i) \ln \lambda - n\lambda - \Sigma \ln y_i!.$$

so that $\left(\frac{d}{d\lambda}\right)[\ln L] = \left(\Sigma\, \frac{y_i}{\lambda}\right) - n$. Equating the derivative to 0, we obtain

$$\frac{\Sigma\, y_i}{\hat{\lambda}} - n = 0$$

or

$$\hat{\lambda} = \frac{\Sigma Y_i}{n} = \overline{Y}.$$

b. Recalling that $E(Y_i) = \lambda$ and $V(Y_i) = \lambda$, we obtain

$$E(\hat{\lambda}) = \frac{\sum_{i=1}^{n} E(Y_i)}{n} = \lambda$$

and

$$V(\hat{\lambda}) = \frac{\sum_{i=1}^{n} V(Y_i)}{n^2} = \frac{\lambda}{n}.$$

c. Since $E(Y_i) = \lambda$ and $V(Y_i) = \lambda < \infty$, the law of large numbers applies and we conclude that $\hat{\lambda}$ converges in probability to λ. Hence $\hat{\lambda}$ is consistent for λ.

d. The MLE of λ was found in part **a** to be $\hat{\lambda} = \overline{Y}$. Then, the MLE for $e^{-\lambda}$ is $e^{-\overline{Y}}$.

9.73 The likelihood is $L(\theta) = f(y_1, y_2, \ldots, y_n|\theta) = \frac{1}{\theta^n} e^{-(\Sigma Y_i)/\theta}$. Then

$$\ln\left[L(\theta)\right] = -n \ln \theta - \frac{\Sigma Y_i}{\theta}.$$

Thus we set

$$\frac{d \ln\left[L(\theta)\right]}{d\theta} = -\frac{n}{\theta} + \frac{\Sigma Y_i}{\theta^2} = 0$$

which implies

$$-n + \frac{\Sigma Y_i}{\theta} = 0$$

or

$$\hat{\theta} = \frac{\Sigma Y_i}{n} = \overline{Y}.$$

Which is the MVUE of θ as shown in Example 9.9.
Assuming that $\theta > 0$, $t(\theta) = \theta^2$ is a one-to-one function of θ. Therefore, $\overline{Y}^2$ is the MLE of θ^2, which is not the MVUE of θ^2 as shown in Example 9.9.

9.74 a. The likelihood function is

$$L = \prod_{i=1}^{n} \frac{r}{\theta} y_i^{r-1} e^{-y_i^r/\theta} = \frac{r^n}{\theta^n} \prod_{i=1}^{n} y^{r-1} e^{-\Sigma y_i^r/\theta} = g(u, \theta) h(y_1, y_2, \ldots, y_n)$$

where

$$u = \sum_{i=1}^{n} y_i^r \qquad g(u, \theta) = \frac{r^n}{\theta^n} e^{-u/\theta} \qquad h(y_1, y_2, \ldots, y_n) = \prod_{i=1}^{n} y_i^{r-1}$$

Hence $\Sigma\, Y_i^r$ is a sufficient statistic for θ.

b. Consider $\ln L = n \ln r - n \ln \theta + (r-1) \sum_{i=1}^{n} \ln y_i - \sum_{i=1}^{n} \frac{y_i^r}{\theta}$ and $\frac{d}{d\theta} \ln L = \frac{-n}{\theta} + \frac{\Sigma y_i^r}{\theta^2}$.

Equating the derivative to 0, the estimator is obtained.

$$\frac{-n}{\hat{\theta}} + \frac{\Sigma\, y_i^r}{\hat{\theta}^2} = 0 \qquad \text{or} \qquad -n\hat{\theta} + \Sigma\, y_i^r = 0 \qquad \text{or} \qquad \hat{\theta} = \frac{\Sigma Y_i^r}{n}.$$

c. The estimator $\hat{\theta}$ given in part **b** is a function of the sufficient statistic. If it is unbiased, or could be adjusted to be unbiased, the MVUE of θ will be obtained. Consider

$$E\left(Y_i^r\right) = \int_0^\infty \frac{r}{\theta} y^{2r-1} e^{-y^r/\theta}\, dy$$

Let $x = y^r$, $dx = r y^{r-1}\, dy$, so that $E\left(Y_i^r\right) = \int_0^\infty \left(\frac{x}{\theta}\right) e^{-x/\theta}\, dx = E(X)$, where X has a gamma distribution with $\alpha = 1$, $\beta = \theta$. Then $E\left(Y_i^r\right) = \theta$ and

$$E(\hat{\theta}) = \frac{\sum_{i=1}^{n} E(Y_i^r)}{n} = \theta$$

Since $\hat{\theta}$ is unbiased for θ, it is the MVUE for θ.

9.75 a. The likelihood function is

$$L = \prod_{i=1}^{n} \frac{1}{(2\theta+1)} I(0 < y_i < 2\theta + 1) = \frac{1}{(2\theta+1)^n} I(0 < y_{(n)} < 2\theta + 1)$$
$$= \frac{1}{(2\theta+1)^n} I(-\tfrac{1}{2} < \tfrac{1}{2}(y_{(n)} - 1) < \theta)$$

Notice that $\frac{1}{(2\theta+1)^n}$ is a decreasing function of θ on the interval $(-\frac{1}{2}, \infty)$. As $-\frac{1}{2} < \frac{1}{2}(y_{(n)} - 1) < \theta$ we know that the maximum must occur at $\hat{\theta} = \left(\frac{1}{2}\right)(Y_{(n)} - 1)$.

b. The variance of the distribution is $\frac{(2\theta+1)^2}{12}$. Since $\theta > 1$, this is a one-to-one function. Thus, the MLE for $\frac{(2\theta+1)^2}{12}$ is

$$\frac{(2\hat{\theta}+1)^2}{12} = \frac{\left[2\left(\frac{1}{2}\right)(Y_{(n)}-1)+1\right]^2}{12} = \frac{Y_{(n)}^2}{12}.$$

9.76 a. As this exercise is a special case of exercise 9.77 **a** (with $\alpha = 2$) we will refer to its results.

$$\hat{\theta} = \left(\frac{\overline{Y}}{2}\right) = \frac{378}{3(2)} = 63.$$

b. From Exercise 9.69 **b**,

$$E(\hat{\theta}) = \theta \qquad V(\hat{\theta}) = \frac{\theta^2}{n\alpha} = \frac{\theta^2}{3(2)} = \frac{\theta^2}{6}$$

c. The bound on the error of estimation is

$$2\sqrt{V(\hat{\theta})} = 2\sqrt{\frac{\theta^2}{6}} = 2\sqrt{\frac{(130)^2}{6}} = 106.14$$

d. The variance of Y is $2\theta^2$. The MLE of θ was found in part **a** to be $\hat{\theta} = 63$. Therefore, the MLE for the variance is $2(63)^2 = 7938$.

9.77 a. The likelihood function, defined as the joint density of $Y_1, Y_2, \ldots, Y_n$ evaluated at $y_1, y_2, \ldots, y_n$, is given by

$$L = \prod_{i=1}^{n} \frac{1}{\Gamma(\alpha)\theta^\alpha} y_i^{\alpha-1} e^{-y_i/\theta} = \frac{1}{[\Gamma(\alpha)]^n \theta^{n\alpha}} e^{-\Sigma y_i/\theta} \prod_{i=1}^{n} y_i^{\alpha-1} = K\left(\frac{1}{\theta^{n\alpha}}\right) e^{-\Sigma y_i/\theta}$$

where K is a constant, independent of θ. Then $\ln L = \ln K - n\alpha \ln \theta - \left(\frac{\Sigma y_i}{\theta}\right)$, and if α is known,

$$\frac{d}{d\theta} \ln L = \frac{\Sigma y_i}{\theta^2} - \frac{n\alpha}{\theta}$$

Equating the derivative to 0, we obtain $\hat{\theta}$.

$$\frac{\Sigma y_i}{\hat{\theta}^2} - \frac{n\alpha}{\hat{\theta}} = 0 \qquad \text{or} \qquad \hat{\theta} = \frac{\Sigma Y_i}{n\alpha} = \frac{\overline{Y}}{\alpha}$$

b. Taking expectations and recalling that $E(Y_i) = \alpha\theta$ and $V(Y_i) = \alpha\theta^2$, we have

$$E(\hat{\theta}) = \frac{\sum_{i=1}^{n} E(Y_i)}{n\alpha} = \frac{n\alpha\theta}{n\alpha} = \theta \qquad \text{and} \qquad V(\hat{\theta}) = \frac{\sum_{i=1}^{n} V(Y_i)}{n^2\alpha^2} = \frac{n\alpha\theta^2}{n^2\alpha^2} = \frac{\theta^2}{n\alpha}$$

c. By the law of large numbers, we know that $\overline{Y}$ is a consistent estimator of $\mu = \alpha\theta$. That is, $\overline{Y}$ converges in probability to $\alpha\theta$. Then, by Theorem 9.2, the quantity $\frac{\overline{Y}}{\alpha} = \hat{\theta}$ converges in probability to $\frac{\mu}{\alpha} = \theta$, so that $\hat{\theta}$ must be a consistent estimator of θ.

d. Using Lehmann and Scheffe''s method, we have

$$\frac{L(x_1, x_2, \ldots, x_n|\alpha, \theta)}{L(y_1, y_2, \ldots, y_n|\alpha, \theta)} = \frac{(\Pi x_i)^{\alpha-1} e^{-\Sigma x_i/\theta}}{(\Pi y_i)^{\alpha-1} e^{-\Sigma y_i/\theta}}$$

In order for this ratio to be free of θ, we need $\Sigma x_i = \Sigma y_i$ so that ΣY_i is the minimal sufficient statistic.

e. Let $U = \sum_{i=1}^{n} Y_i$. The moment-generating function of U is

$$m_U(t) = \prod_{i=1}^{n} m_{Y_i}(t) = \frac{1}{(1-\theta t)^{n\alpha}} = \frac{1}{(1-\theta t)^{1\alpha}}.$$

A random variable that possesses a χ^2 distribution is one whose moment-generating function is $\frac{1}{(1-2t)^k}$, where $2k$ are the degrees of freedom. It is necessary to transform U to obtain a random variable with such a moment-generating function. Consider $X = \frac{2U}{\theta}$, with

$$m_X(t) = m_U\left(\frac{2t}{\theta}\right) = \frac{1}{(1-2t)^{10}}$$

Hence $X = \frac{2U}{\theta}$ has a χ^2 distribution with $2(10) = 20$ degrees of freedom. Using X as a pivotal statistic, write

$$P\left(\chi^2_{.05,\,20} < \frac{2U}{\theta} < \chi^2_{.95,\,.20}\right) = .90$$

or

$$P\left(\frac{2U}{\chi^2_{.95,\,20}} < \theta < \frac{2U}{\chi^2_{.05,\,20}}\right) = .90.$$

Then the 90% confidence interval is $\left[\frac{2\sum_{i=1}^{n} Y_i}{31.41}, \frac{2\sum_{i=1}^{n} Y_i}{10.85}\right]$.

9.78 The likelihood function is

$$L = \prod_{i=1}^{m} \frac{1}{\sqrt{2\pi}\sigma} \exp\left[-\frac{1}{2}\left(\frac{x_i-\mu_1}{\sigma}\right)^2\right] \prod_{i=1}^{n} \frac{1}{\sqrt{2\pi}\theta} \exp\left[-\frac{1}{2}\left(\frac{y_i-\mu_2}{\sigma}\right)^2\right]$$

$$= \frac{1}{(2\pi)^{(m+n)/2}\sigma^{m+n}} \exp\left\{-\frac{1}{2}\left[\sum_{i=1}^{m}\left(\frac{x_i-\mu_1}{\sigma}\right)^2 + \sum_{i=1}^{n}\left(\frac{y_i-\mu_2}{\sigma}\right)^2\right]\right\}$$

and

$$\ln L = \ln K - (m+n)\ln\sigma - \frac{1}{2\sigma^2}\left[\sum_{i=1}^{m}(x_i-\mu_1)^2 + \sum_{i=1}^{n}(y_i-\mu_2)^2\right]$$

Then

$$\frac{d}{d\sigma}\ln L = \frac{-(m+n)}{\sigma} + \frac{1}{\sigma^3}\left[\sum_{i=1}^{m}(x_i-\mu_1)^2 + \sum_{i=1}^{n}(y_i-\mu_2)^2\right]$$

Setting the derivative equal to 0 and solving for $\hat{\sigma}$, we have

$$\frac{m+n}{\hat{\sigma}} = \frac{1}{\hat{\sigma}^3}\left[\sum_{i=1}^{m}(x_i-\mu_1)^2 + \sum_{i=1}^{n}(y_i-\mu_2)^2\right]$$

or

$$\hat{\sigma}^2 = \frac{\sum_{i=1}^{m}(X_i-\mu_1)^2 + \sum_{i=1}^{n}(Y_i-\mu_2)^2}{m+n}.$$

Since μ_1 and μ_2 are unknown, their maximum likelihood estimates must be obtained.

$$\frac{d}{d\mu_1}\ln L = \frac{\sum_{i=1}^{m}(x_i-\mu_1)}{\sigma^2} \qquad \text{and} \qquad \frac{d}{d\mu_2}\ln L = \frac{\sum_{i=1}^{n}(y_i-\mu_2)}{\sigma^2}$$

and, as in Example 9.15 in the text, $\hat{\mu}_1 = \overline{X}$ and $\hat{\mu}_2 = \overline{Y}$. Thus,

$$\hat{\sigma}^2 = \frac{\sum_{i=1}^{m}(X_i-\overline{X})^2 + \sum_{i=1}^{n}(Y_i-\overline{Y})^2}{m+n}$$

9.79 Let p_1, p_2, p_3 be the proportions of voters in the population favoring candidates A, B, and C, respectively. Further, define the random variables n_1, n_2, and n_3 as the number of voters in a random sample of size n who favor candidates A, B, and C, respectively. Note that

$$\sum_{i=1}^{3} p_i = 1 \qquad \text{and} \qquad \sum_{i=1}^{3} n_i = n$$

so that we may write $p_3 = 1 - p_1 - p_2$ and $n_3 = n - n_1 - n_2$. The random variables n_1, n_2, and n_3 follow a multinomial probability distribution (see Section 5.9 of the text), and the likelihood function is

$$L = \frac{n!}{n_1!\,n_2!\,n_3!}\, p_1^{n_1}\, p_2^{n_2}\, (1-p_1-p_2)^{n_3}$$

so that

$$\ln L = \ln K + n_1 \ln p_1 + n_2 \ln p_2 + n_3 \ln(1-p_1-p_2)$$

Differentiating with respect to p_1 and p_2, we have

$$\frac{d\ln L}{dp_1} = \frac{n_1}{p_1} - \frac{n_3}{1-p_1-p_2} \qquad \text{and} \qquad \frac{d\ln L}{dp_2} = \frac{n_2}{p_2} - \frac{n_3}{1-p_1-p_2}$$

Set these two equations equal to 0 and solve simultaneously for $\hat{p}_1$ and $\hat{p}_2$.

$$(*) \quad n_1(1-\hat{p}_1-\hat{p}_2) - n_3\hat{p}_1 = 0 \qquad \text{and} \qquad n_2(1-\hat{p}_1-\hat{p}_2) - n_3\hat{p}_2 = 0$$

Adding the two equations, we have

$(n_1+n_2)(1-\hat{p}_1-\hat{p}_2) = (\hat{p}_1+\hat{p}_2)n_3$ or $\hat{p}_1+\hat{p}_2 = \frac{n_1+n_2}{n}$

Thus, in (*),

$n_1\left[1-\left(\frac{n_1+n_2}{n}\right)\right] = n_3\hat{p}_1$ or $\hat{p}_1 = \frac{(n-n_1-n_2)\theta_1}{n}\left(\frac{1}{n_3}\right) = \frac{n_1}{n}$

Similarly,

$\hat{p}_2 = \frac{n_2}{n}$ and $\hat{p}_3 = 1-\hat{p}_1-\hat{p}_2 = \frac{n_3}{n}$

For the data given in this exercise, $\hat{p}_1 = .30$, $\hat{p}_2 = .38$, and $\hat{p}_3 \equiv .32$. To estimate $p_1 - p_2$, we use $\hat{p}_1 - \hat{p}_2 = \frac{n_1}{n} - \frac{n_2}{n} = -.08$. From Theorem 5.13 of the text, we have $V(n_i) = np_iq_i$ and $\text{Cov}(n_i, n_j) = -np_ip_j$, so that the variance of $\hat{p}_1 - \hat{p}_2$ is

$$V(\hat{p}_1) + V(\hat{p}_2) - 2\,\text{Cov}(\hat{p}_1, \hat{p}_2) = \frac{1}{n^2}V(n_1) + \frac{1}{n^2}V(n_2) - \frac{2}{n^2}\text{Cov}(n_1, n_2)$$
$$= \frac{p_1q_1}{n} + \frac{p_2q_2}{n} + 2\,\frac{p_1p_2}{n}$$

which may be estimated as

$$V(\hat{p}_1 - \hat{p}_2) = \frac{(.30)(.70)}{100} + \frac{(.38)(.62)}{100} + 2\,\frac{(.30)(.38)}{100} = .006736$$

and the approximate bound on the error of estimation is

$$2\sqrt{V(\hat{p}-\hat{p})} = 2\sqrt{.006736} = .1641.$$

9.80 The likelihood function is

$$L = \prod_{i=1}^{n}(\theta+1)y_i^\theta = (\theta+1)^n\prod_{i=1}^{n}y_i^\theta \quad \text{and} \quad \ln L = n\ln(\theta+1) + \theta\sum_{i=1}^{n}\ln y_i$$

so that

$$\frac{d}{d\theta}\ln L = \frac{n}{\theta+1} + \Sigma\ln y_i$$

Equating the derivative to 0, we can find the maximum likelihood estimator,

$$\frac{n}{\hat{\theta}+1} + \Sigma\ln y_i = 0 \quad \text{or} \quad n + (\hat{\theta}+1)\,\Sigma\ln y_i = 0 \quad \text{or} \quad \hat{\theta} = \frac{-n}{\Sigma\ln Y_i} - 1$$

9.81 $P(Y=y) = \binom{2}{y}p^y(1-p)^{2-y}$. Our estimator, $\hat{p}$, must be either 1/4 or 3/4. We choose based on which has the larger likelihood value given the data, Y. It is important to remember in this problem that the likelihood is a function of the parameter p. Therefore we have three possible likelihood functions depending, one for each value of the data, Y.

$L(0,p) = P(Y=0) = (1-p)^2$ implying $\hat{p} = \frac{1}{4}$ as
$L(0,\frac{1}{4}) = (1-\frac{1}{4})^2 > (1-\frac{3}{4})^2 = L(0,\frac{3}{4})$.
$L(1,p) = P(Y=1) = 2p(1-p)$ implying $\hat{p}$ can be either $\frac{1}{4}$ or $\frac{3}{4}$ as
$L(1,\frac{1}{4}) = 2\frac{1}{4}(1-\frac{1}{4}) = 2\frac{3}{4}(1-\frac{3}{4}) = L(1,\frac{3}{4})$.
$L(2,p) = P(Y=2) = p^2$ implying $\hat{p} = \frac{3}{4}$ as $(\frac{1}{4})^2 < (\frac{3}{4})^2$.

Notice the case when $Y=1$ is an instance where the maximum likelihood estimator is not a single unique value!

9.82 Notice under the hypothesis $p_W = p_M = p$ the number of people of our sample who favor the issue is binomial with success probability p and number of trials equal to 200. Then by problem 9.14 we have $\hat{p} = \frac{55}{200}$.

9.83 $L = \left(\frac{1}{2\theta}\right)^n I(0 \le y_{(n)} \le 2\theta) = \left(\frac{1}{2\theta}\right)^n I(0 \le \frac{1}{2}y_{(n)} \le \theta)$. Notice that $\left(\frac{1}{2\theta}\right)^n$ is a decreasing function of θ on $(0,\infty)$. Thus L is a decreasing function of θ on $(\frac{1}{2}y_{(n)},\infty)$. Thus, $\theta = \frac{Y_{(n)}}{2}$.

9.84 **a.** Recall from exercise 9.44 $L(y_1,\ldots,y_n|\theta) = \begin{cases} \frac{3^n}{\theta^{3n}}\left(\prod_{i=1}^{n}y_i\right) & \text{for } y_{(n)} \le \theta \\ 0 & \text{Otherwise} \end{cases}$

That is $L(y_1,\ldots,y_n|\theta) = \frac{c}{\theta^{3n}}$ where $c = 3^n\left(\prod_{i=1}^{n}y_i\right)$ when $\theta \ge y_{(n)}$ (and 0 elsewhere). This is a decreasing function of θ. Therefore it will it will obtain its maximum where θ is smallest, i.e. when $\theta = Y_{(n)}$. Thus the MLE of θ is $Y_{(n)}$.

b. From 9.55 we have $f_{Y_{(n)}}(y) = \begin{cases} 3ny^{3n-1}/\theta^{3n} & \text{for } 0 \le y \le \theta \\ 0 & \text{otherwise} \end{cases}$.

Consider the density of $T = Y_{(n)}/\theta$. Then $\left|\frac{dY}{dT}\right| = \theta$. Let $f_T(t)$ be the density of T. Then

$$f_T(t) = f_{Y_{(n)}}(t\theta)\theta = \begin{cases} 3nt^{3n-1} & \text{for } 0 \le t \le 1 \\ 0 & \text{otherwise} \end{cases}.$$

That is, T is a pivotal quantity.

c. We want to find constants a and b such that $P(a < T < b) = 1 - \alpha$. This will be true if a and b satisfy $P(T < a) = \alpha/2$ and $P(T > b) = \alpha/2$. Notice

$$P(T < a) = \int_0^a 3nt^{3n-1}dt = t^{3n}\Big|_0^a = a^{3n} = \alpha/2 \text{ implying } a = (\alpha/2)^{1/3n}.$$

Further,

$$P(T > b) = \int_b^1 3nt^{3n-1}dt = t^{3n}\Big|_b^1 = 1^{3n} - b^{3n} = \alpha/2$$

implying $b = (1 - \alpha/2)^{1/3n}$. Then

$$1 - \alpha = P(a < T < b) = P(a < \tfrac{Y_{(n)}}{\theta} < b) = P\left(\tfrac{Y_{(n)}}{a} > \theta > \tfrac{Y_{(n)}}{b}\right).$$

Thus the interval is $\left[\frac{Y_{(n)}}{(1-\alpha/2)^{1/3n}}, \frac{Y_{(n)}}{\alpha^{1/3n}}\right]$.

9.85 a. Recall from exercise 9.45 $L(y_1,\ldots,y_n|\theta) = \begin{cases} 2^n\theta^{3n}\prod_{i=1}^{n}(1/y_i^2) & \text{for } y_{(1)} \ge \theta \\ 0 & \text{Otherwise} \end{cases}.$

Which is an increasing function of θ for $y_{(1)} \ge \theta$. Therefore it will obtain its maximum when θ is largest. Thus the MLE for θ is $Y_{(1)}$.

b. First we find the density of $Y_{(1)}$, $f_{Y_{(1)}}(y)$. Note the distribution function of the data is

$$F(y) = \int_\theta^y 2\theta^2/u^3 du = -\theta^2/u^2\Big|_\theta^y = 1 - \theta^2/y^2.$$

Then $f_{Y_{(1)}}(y) = nf(y)(1 - F(y))^{n-1} = \begin{cases} n(2\theta^2/y^3)(\theta^2/y^2)^{n-1} & \text{for } y \ge \theta \\ 0 & \text{otherwise} \end{cases}.$

Let $T = \theta/Y_{(1)}$. Then $\left|\frac{dY}{dT}\right| = \theta/T^2$. Let $f_T(t)$ be the density of T. Then

$$f_T(t) = f_{Y_{(n)}}(\theta/t)\theta/t^2 = \begin{cases} n(2t^3/\theta)(t^2)^{n-1}(\theta/t^2) & \text{for } 0 \le t \le 1 \\ 0 & \text{otherwise} \end{cases}$$

$$= \begin{cases} 2nt^{2n-1} & \text{for } 0 \le t \le 1 \\ 0 & \text{otherwise} \end{cases}$$

so that T is a pivotal quantity (notice t is a beta$(2n, 1)$).

c. We want to find constants a and b such that $P(a < T < b) = 1 - \alpha$. This will be true if a and b satisfy $P(T < a) = \alpha/2$ and $P(T > b) = \alpha/2$. Notice

$$P(T < a) = \int_0^a 2nt^{2n-1}dt = t^{2n}\Big|_0^a = a^{2n} = \alpha/2.$$

Implying $a = (\alpha/2)^{1/2n}$. Further note that

$$P(T > b) = \int_b^1 2nt^{2n-1}dt = t^{2n}\Big|_b^1 = 1 - b^{2n} = \alpha/2.$$

Implying $b = \left(1 - \left(\frac{\alpha}{2}\right)\right)^{1/2n}$. Thus

$$P(a < T < b) = P(a < \tfrac{\theta}{Y_{(1)}} < b) = P(aY_{(1)} < \theta < bY_{(1)}).$$

Hence our interval is $\left[(\alpha/2)^{1/2n}Y_{(1)}, \left(1 - \left(\frac{\alpha}{2}\right)\right)^{1/2n}Y_{(1)}\right]$.

9.86 Let $\beta = t(\theta)$ so that $\theta = t^{-1}(\beta)$. If the maximum of the likelihood is attained when $\theta = \hat{\theta}$, then $L(\theta) \le L(\hat{\theta})$ for all θ. Define $\hat{\beta} = t(\hat{\theta})$ and denote the likelihood as a function of β with $L_1(\beta) \equiv L(t^{-1}(\beta))$. Then, for any β,

$$L_1(\beta) = L(t^{-1}(\beta)) = L(\theta) \le L(\hat{\theta}) = L\left(t^{-1}(\hat{\beta})\right) = L_1(\hat{\beta})$$

So the maximum likelihood estimator of β is $\hat{\beta}$; i.e., the maximum likelihood estimator of $t(\theta)$ is $t(\hat{\theta})$.

9.87 The parameter to be estimated is $R = \frac{p}{1-p}$. Since $\hat{p} = \frac{Y}{n}$ is the maximum-likelihood estimate of p (see Example 9.14), the result proved in Exercise 9.81 implies that $\hat{R} = \frac{\hat{p}}{1-\hat{p}}$ is the maximum-likelihood estimator of R.

9.88 It was shown in Example 9.15 that the maximum-likelihood estimator of σ^2 is s^2. Using this result, we give as the maximum-likelihood estimator for σ to be s.

9.89 From Example 9.18, the MLE for $t(p) = p$ is $\hat{t}(n) = \hat{p} = \frac{Y}{n}$ and $\frac{\partial\, t(p)}{\partial p} = 1$. Also, from Example 9.18,

$$E\left[\frac{-\partial^2 \ln[p(Y|p)]}{\partial p^2}\right] = \frac{1}{p(1-p)}.$$

Then, a $100(1-\alpha)\%$ confidence interval is

$$\hat{p} \pm z_{\alpha/2}\sqrt{\frac{1}{n\left[\frac{1}{p(1-p)}\right]\Big|_{p=\hat{p}}}} = \hat{p} \pm z_{\alpha/2}\sqrt{\frac{\hat{p}(1-\hat{p})}{n}}.$$

This is the confidence interval derived in Section 8.6.

9.90 In Exercise 9.73, $\overline{Y}^2$ was shown to be the MLE of θ^2. Note, $t(\theta) = \theta^2$ and $\frac{\partial\, t(\theta)}{\partial \theta} = 2\theta$. Further we have $f(y|\theta) = \frac{1}{\theta}e^{-y/\theta}$, implying $\ln[f(y|\theta)] = -\ln\theta - \frac{y}{\theta}$. Then

$$\frac{\partial \ln[f(y|\theta)]}{\partial \theta} = -\frac{1}{\theta} + \frac{y}{\theta^2}$$

and

$$\frac{\partial^2 \ln[f(y|\theta)]}{\partial \theta^2} = \frac{1}{\theta^2} - \frac{2y}{\theta^3}.$$

Thus

$$E\left[\frac{-\partial^2 \ln[f(Y|\theta)]}{\partial \theta^2}\right] = -\frac{1}{\theta^2} + \frac{2E(Y)}{\theta^3} = -\frac{1}{\theta^2} + \frac{2\theta}{\theta^3} = \frac{1}{\theta^2}.$$

Then an approximate large-sample $100(1-\alpha)\%$ confidence interval for θ^2 is

$$\begin{aligned} t(\hat{\theta}) \pm z_{\alpha/2}\sqrt{\left(\frac{\left(\frac{\partial\, t(\theta)}{\partial\theta}\right)^2}{nE\left[\frac{-\partial^2 \ln f(Y|\theta)}{\partial\theta^2}\right]}\right)\Bigg|_{\theta=\hat{\theta}}} &= \overline{Y}^2 \pm z_{\alpha/2}\sqrt{\left(\frac{(2\theta)^2}{n\left(\frac{1}{\theta^2}\right)}\right)\Bigg|_{\theta=\overline{Y}}} \\ &= \overline{Y}^2 \pm z_{\alpha/2}\sqrt{\frac{4\overline{Y}^4}{n}} \\ &= \overline{Y} \pm z_{\alpha/2}\left(\frac{2\overline{Y}^2}{\sqrt{n}}\right) \end{aligned}$$

9.91 The MLE for $e^{-\lambda}$ was shown to be $e^{-\overline{Y}}$ in Exercise 9.72. Also, $t(\lambda) = e^{-\lambda}$implying $\frac{\partial\, t(\lambda)}{\partial\lambda} = -e^{-\lambda}$. Note that $p(y|\lambda) = \frac{\lambda^y e^{-\lambda}}{y!}$, thus $\ln[p(y|\lambda)] = y\ln\lambda - \lambda - \ln(y!)$. Then we have

$$\frac{\partial \ln[p(y|\lambda)]}{\partial\lambda} = \frac{y}{\lambda} - 1$$

$$\frac{\partial^2 \ln[p(y|\lambda)]}{\partial\lambda^2} = -\frac{y}{\lambda^2}$$

$$E\left[\frac{-\partial^2 \ln[p(Y|\lambda)]}{\partial\lambda^2}\right] = \frac{E(Y)}{\lambda^2} = \frac{1}{\lambda}.$$

Then, an approximate large-sample $100(1-\alpha)\%$ confidence interval for $e^{-\lambda}$ is

$$t(\hat{\lambda}) \pm z_{\alpha/2}\sqrt{\left(\frac{\left(\frac{\partial\, t(\lambda)}{\partial\lambda}\right)^2}{nE\left[\frac{-\partial^2 \ln f(Y|\lambda)}{\partial\lambda^2}\right]}\right)\Bigg|_{\lambda=\hat{\lambda}}} = e^{-\overline{Y}} \pm z_{\alpha/2}\sqrt{\left(\frac{e^{-2\lambda}}{n\left(\frac{1}{\lambda}\right)}\right)\Bigg|_{\lambda=\overline{Y}}}$$

$$= e^{-\overline{Y}} \pm z_{\alpha/2}\sqrt{\frac{\overline{Y}e^{-2\overline{Y}}}{n}}$$

9.92 **a.** Using the method of moments, calculate

$$\mu_1' = \int_\theta^\infty ye^{-(y-\theta)}\,dy = \int_0^\infty (z+\theta)e^{-z}\,dz = 1+\theta$$

with $z = y - \theta$. Hence the estimator is $1 + \hat{\theta}_1 = \overline{Y}$, or $\hat{\theta}_1 = \overline{Y} - 1$.

b. The likelihood function is

$$L = \prod_{i=1}^{n} e^{-(y_i-\theta)}\ I(\theta \le y_i \le \infty) = e^{n\theta-\Sigma y_i}\ \mathrm{I}(\theta \le y_{(i)} \le \infty)$$

where the indicator function $I(\theta \le y \le \infty)$ equals 1 if $\theta \le y$ and 0 if $\theta > y$. To maximize L, we make θ as large as possible subject to the constraint that $\theta < y_{(1)}$. Thus, $\hat{\theta}_2 = Y_{(1)}$.

c. It is necessary to find $E(\hat{\theta}_i)$ for $i = 1, 2$.

$$E(\hat{\theta}_1) = E(\overline{Y}) - 1 = \mu_1' - 1 = (1+\theta) - 1 = \theta$$

The density of $\hat{\theta}_2$ is given by $g(\hat{\theta}_2) = g(y_{(1)}) = n[1 - F(y)]^{n-1} f(y)$, where

$$F(y) = \int_{-\infty}^{y} f(t)\, dt = \int_{\theta}^{y} e^{-(t-\theta)}\, dt = -e^{-(-\theta)} + 1$$

That is,

$$g(\hat{\theta}_2) = n\left[e^{-(\hat{\theta}_2-\theta)}\right]^{n-1} e^{-(\hat{\theta}_2-\theta)} = ne^{-(\hat{\theta}_2-\theta)n} \qquad \theta \le \theta_2 \le \infty$$

and

$$E(\hat{\theta}_2) = \int_{\theta}^{\infty} n\hat{\theta}_2 e^{-(\hat{\theta}_2-\theta)n}\, d\hat{\theta}_2 = \int_{0}^{\infty} n(z+\theta)e^{-zn}\, dz = \int_{0}^{\infty} nze^{-zn}\, dz$$

$$+\,\theta\int_{0}^{\infty} ne^{-zn}\, dz$$

$$= \tfrac{1}{n}\int_{0}^{\infty} we^{-w}\, dw + \theta\int_{0}^{\infty} e^{-w}\, dw = \tfrac{1}{b} + \theta$$

Notice that $\hat{\theta}_1$ is unbiased, and $\hat{\theta}_2$ can be adjusted to be unbiased by letting $\hat{\theta}_{2a} = \hat{\theta}_2 - \frac{1}{n}$. In order to find the efficiency of $\hat{\theta}_1$ to $\hat{\theta}_{2a}$, the variances must be calculated. Note that

$$V(Y) = \int_{\theta}^{\infty} (y-\theta-1)^2\, e^{-(y-\theta)}\, dy = \int_{0}^{\infty} (z-1)^2\, e^{-z}\, dz$$

which is the variance of an exponential random variable with mean 1. Thus, $V(Y) = 1$. Further

$$V(\hat{\theta}_1) = V(\overline{Y} - 1) = V(\overline{Y}) = \tfrac{V(Y)}{n} = \tfrac{1}{n}$$

and

$$E\left(\hat{\theta}_2^2\right) = \int_{\theta}^{\infty} n\hat{\theta}_2^2 e^{-(\hat{\theta}_2-\theta)/n}\, d\hat{\theta}_2 = \int_{0}^{\infty} n\,(z^2 + 2z\theta + \theta^2)\, e^{-zn}\, dz$$

$$= \tfrac{1}{n^2}\int_{0}^{\infty} w^2 e^{-w}\, dw + \tfrac{2\theta}{n}\int_{0}^{\infty} we^{-w}\, dw + \theta^2\int_{0}^{\infty} e^{-w}\, dw = \tfrac{\Gamma(3)}{n^2} + \tfrac{2\theta}{n} + \theta^2$$

and

$$V(\hat{\theta}_{2a}) = V(\hat{\theta}_2) = \tfrac{2}{n^2} + \tfrac{2\theta}{n} + \theta^2 - \left(\tfrac{1}{n} + \theta\right)^2 = \tfrac{1}{n^2}.$$

Hence the efficiency of $\hat{\theta}_1$ relative to $\hat{\theta}_{2a}$ is

$$\frac{V(\hat{\theta}_{2a})}{V(\hat{\theta}_1)} = \frac{\left(\frac{1}{n^2}\right)}{\left(\frac{1}{n}\right)} = \frac{1}{n}$$

9.93 From Example 9.15, since μ is known, we need to solve

$$\frac{d\ln L}{d\sigma^2} = -\frac{n}{2\sigma^2} + \frac{\Sigma(y_i-\mu)^2}{2\sigma^4} = 0 \qquad \text{or} \qquad \hat{\sigma}^2 = \frac{\Sigma(y_i-\mu)^2}{n}$$

9.94 Using the method given in Exercise 9.49, we construct a random variable T, where $T = 1$ if $Y_1 = 0$ and $T = 0$ otherwise. Then

$$E(T) = P(Y_1 = 0) = \tfrac{\lambda^0 e^{-\lambda}}{0!} = e^{-\lambda}$$

and T is unbiased for $e^{-\lambda}$. Then since ΣY_i is sufficient for λ and $e^{-\lambda}$, we calculate $E(T|\Sigma Y_i)$, which will be the MVUE for $e^{-\lambda}$. Recall that ΣY_i has a Poisson distribution with mean $n\lambda$.

$$E\left(T|\sum_{i=1}^{n} Y_i\right) = P\left(T = 1|\sum_{i=1}^{n} Y_i = x\right) = P\left(Y_1 = 0|\sum_{i=1}^{n} Y_i = x\right)$$

$$= \frac{P\left(Y_1=0, \sum_{i=1}^{n} Y_i=x\right)}{P\left(\sum_{i=1}^{n} Y_i=x\right)} = \frac{P\left(Y_1=0, \sum_{i=2}^{n} Y_i=x\right)}{P\left(\sum_{i=1}^{n} Y_i=x\right)}$$

$$= \frac{e^{-\lambda}\left(e^{-(n-1)\lambda}\frac{[(n-1)\lambda]^x}{x!}\right)}{\frac{e^{-n\lambda}(n\lambda)^x}{x!}} = \frac{e^{-\lambda}e^{\lambda}(n-1)^x}{n^x} = \left(1 - \frac{1}{n}\right)^x = \left(1 - \frac{1}{n}\right)^{\Sigma Y_i}$$

Note that we made use of the fact that, if $Y_i = 0$, the sum $\Sigma\, Y_i = x$ must be obtained on the other $(n-1)$ trials, and that the sum of $(n-1)$ Poisson random variables has a Poisson distribution with mean $(n-1)\lambda$.

9.95 From Exercise 9.77, the MLE for θ is $\overline{Y}$. Hence we use the result of Exercise 9.86 and deduce that the MLE of $e^{-t/\theta} = \overline{F}(t)$ is $e^{-t/\overline{Y}}$.

9.96 **a.** $E(V) = P(Y_1 > t) = \int_t^{\infty} \frac{1}{\theta}\, e^{-y/\theta}\, dy = 1 - F(t) = e^{-t/\theta}$

Hence V is an unbiased estimator of $e^{-t/\theta}$.

b. Since Y_i has a gamma distribution with $\alpha = 1$ and $\beta = \theta$, $\Sigma\, Y_i$ has a gamma distribution with $\alpha = n$ and $\beta = \theta$. (The student may check this result by using moment-generating functions.) Similarly, $U - Y_1 = \sum_{i=2}^{n} Y_i$ has a gamma distribution with $\alpha = n - 1$ and $\beta = \theta$. Then

$$f(u) = \frac{1}{\Gamma(n)\theta^n}\, e^{-u/\theta}\, u^{n-1}$$

Since Y_1 and $U - Y_1 = \sum_{i=1}^{n} Y_i$ are independent, we can write

$$f(y_1, u - y_1) = \left(\frac{1}{\theta}\, e^{-y_1/\theta}\right) \frac{1}{\theta^{n-1}\Gamma(n-1)}\, e^{-(u-y_1)/\theta}(u - y_1)^{n-2}$$

as long as $0 \le y_1 \le u$. Finally, making the transformation to $f(y_1, u)$ with $z = u - y_1$ or $u = z + y_1$, we have $\frac{dz}{du} = 1$ and

$$f(y_1, u) = \frac{1}{\theta^n\Gamma(n-1)}\, e^{-u/\theta}(u - y_1)^{n-2}$$

and

$$f_{Y_1|U}(u) = \frac{f(y_1, u)}{f(u)} = \frac{n-1}{u^{n-1}}\,(u - y_1)^{n-2} \qquad \text{for } 0 \le y_i \le u$$

c. $E(V|U) = P(Y_1 > t|U) = \int_t^u \frac{n-1}{u^{n-1}}\,(u - y_1)^{n-2}\, dy_1 = \int_t^u \frac{n-1}{u}\left(1 - \frac{y_1}{u}\right)^{n-2} dy_1$

$$= -(n-1)\int z^{n-2}\, dz = \left(1 - \frac{y_1}{u}\right)^{n-1}\Big]_t^u = \left(1 - \frac{t}{u}\right)^{n-1}$$

9.97 Let Y_i be the number drawn on the ith draw for $i = 1, 2, \ldots, n$. Then, by definition, $P(Y_i = k) = \frac{1}{N}$ for $k = 1, 2, \ldots, N$, and the Y_i are independent.

a. In order to use the method of moments, calculate

$$\mu_1' = E(Y) = \sum_{k=1}^{N} kP(Y = k) = \sum_{k=1}^{N} k\left(\frac{1}{N}\right) = \frac{N(N+1)}{2N} = \frac{N+1}{2}$$

Then the moment estimator will be

$$\frac{\hat{N}_1+1}{2} = \overline{Y} \qquad \text{or} \qquad \hat{N}_1 = 2\overline{Y} - 1$$

b. First calculate

$$E(Y^2) = \sum_{k=1}^{N} k^2 P(Y = k) = \frac{1}{N}\sum_{k=1}^{N} k^2 = \frac{N(N+1)(2N+1)}{6N} = \frac{(N+1)(2N+1)}{6}$$

so that

$$V(Y) = \frac{(N+1)(2N+1)}{6} - \frac{(N+1)^2}{4} = \frac{(N+1)(N-1)}{12}$$

Now

$$E(\hat{N}_1) = 2E(\overline{Y}) - 1 = 2E(Y) - 1 = (N+1) - 1 = N$$

$$V(\hat{N}_1) = 4V(\overline{Y}) = 4\,\frac{V(Y)}{n} = \frac{4(N^2-1)}{12n} = \frac{1}{3n}\,(N^2 - 1)$$

9.98 **a.** Refer to Exercise 9.91. Since $P(Y_i = k) = \frac{1}{N}$, the likelihood function will be

$$L = P(Y_1 = k_1, Y_2 = k_2, \ldots, Y_n = k_n) = \frac{1}{N^n}\prod_{i=1}^{n} I(y_i \in \{1, 2, \ldots, N\})$$

In order to maximize L, N must be chosen as small as possible subject to the constraint that $y_{(n)} \le N$. Thus, $\hat{N}_2 = \max(Y_1, Y_2, \ldots, Y_n)$.

b. Consider

$$P(\hat{N}_2 \le k) = P[\max(Y_1, Y_2, \ldots, Y_n) \le k] = P(Y_1 \le k)P(Y_2 \le k)\cdots P(Y_n \le k) = \left(\tfrac{k}{N}\right)^n$$

Similarly, $P(\hat{N}_2 \le k-1) = \left(\frac{k-1}{N}\right)^n$. Hence

$$P(\hat{N}_2 = k) = \left(\tfrac{k}{N}\right)^n - \left(\tfrac{k-1}{N}\right)^n = N^{-n}[k^n - (k-1)^n].$$

Then

$$E(\hat{N}_2) = N^{-n}\sum_{k=1}^{N} k[k^n - (k-1)^n] = N^{-n}\sum_{k=1}^{N}[k^{n+1} - (k-1)^{n+1} - (k-1)^n]$$
$$= N^{-n}\left[N^{n+1} - \sum_{k=1}^{N}(k-1)^n\right]$$

Consider

$$\sum_{k=1}^{N}(k-1)^n = 0^n + 1^n + 2^n + \ldots + (N-1)^n$$

For large N, this is approximately the area under the curve $y = x^n$ from $x = 0$ to $x = N$, or

$$\sum_{k=1}^{N}(k-1)^n = \int_0^N x^n\,dx = \frac{N^{n+1}}{n+1} \quad \text{so that} \quad E(\hat{N}_2) = \frac{nN^{n+1}}{N^n(n+1)} = \left(\tfrac{n}{n+1}\right)N$$

and an approximately unbiased estimator will be

$$\hat{N}_3 = \left(\tfrac{n+1}{n}\right)\hat{N}_2.$$

c. $V(\hat{N}_2)$ is given in this exercise. Hence

$$V(\hat{N}_3) = \frac{(n+1)^2}{n^2}V(\hat{N}_2) = \frac{N^2}{n(n+2)}$$

d. From Exercise 9.91, $V(\hat{N}_1) = \left(\frac{1}{3n}\right)(N_2 - 1)$, so that

$$\frac{V(\hat{N}_1)}{V(\hat{N}_3)} = \frac{n(n+2)}{3n}\,\frac{(N^2-1)}{N^2} = \frac{n+2}{3}\left(1 - \frac{1}{N^2}\right)$$

If $n > 1$, $(n+2) > 3$. Moreover, for large N, $1 - \left(\frac{1}{N^2}\right) = 1$. Hence for large N,

$$\frac{V(\hat{N}_1)}{V(\hat{N}_3)} > 1 \quad \text{and} \quad V(\hat{N}_1) > V(\hat{N}_3)$$

9.99

Refer to Exercise 9.98. Using the results of parts **b** and **c**, we have

$$\hat{N}_3 = \frac{n+1}{n}[\max(Y_1, Y_2, \ldots, Y_5)] = \tfrac{6}{5}(210) = 252$$

and an approximate bound on the error of estimation is

$$2\sqrt{V(\hat{N}_3)} = 2\sqrt{\frac{N^2}{n(n+2)}} = 2\sqrt{\frac{(252)^2}{5(7)}} = 2\sqrt{1814.4} = 85.192$$

9.100

a. We will use the large sample properties of MLE's to solve this problem. Recall $\overline{Y}$ is the MLE for λ (exercise 9.72). For our application $t(\lambda) = \lambda$. Thus we must calculate

$$E\left[-\frac{\partial}{\partial\lambda^2}\log(f(y;\lambda)\right] = E\left[-\frac{\partial}{\partial\lambda^2}(-\log(y!) + y\log(\lambda) - \lambda)\right]$$
$$= E\left[-\frac{\partial}{\partial\lambda}\left(\frac{y}{\lambda} - 1\right)\right]$$
$$= E\left[\frac{y}{\lambda^2}\right] = 1/\lambda.$$

Then by the large sample properties of MLE's we have that $\dfrac{\hat{\lambda}-\lambda}{\left(1/nE\left[-\frac{\partial}{\partial\lambda^2}\log(f(y;\lambda)\right]\right)^{1/2}} = \dfrac{\overline{Y}-\lambda}{\sqrt{\overline{Y}/n}}$ has a limiting $\mathrm{N}(0,1)$ distribution.

b. By part **a.** an approximate $(1-\alpha)100\%$ confidence interval for λ would be $\overline{Y} \pm Z_{\alpha/2}\sqrt{\frac{\overline{Y}}{n}}$.

CHAPTER 10 HYPOTHESIS TESTING

10.1 See Definition 10.1.

10.2 The test statistic Y has a binomial distribution with $n = 20$ and p.

a. A Type I error occurs if the experimenter concluded that the drug dosage level induces sleep in less than 80% of the people suffering from insomnia when, in fact, drug dosage level does induce sleep in 80% of insomniacs.

b. $\alpha = P(\text{reject } H_0 | H_0 \text{ true}) = P(Y \leq 12 | p = .8) = .032$, using Table 1, Appendix III.

c. A Type II error would occur if the experimenter concluded that the drug dosage level induces sleep in 80% of the people suffering from insomnia when, in fact, fewer than 80% experience relief.

d. If $p = .6$,

$\beta = P(\text{accept } H_0 | H_0 \text{ false}) = P(Y > 12 | p = .6) = 1 - P(Y \leq 12p = .6) = 1 - .584$
$= .416$

e. If $p = .4$, then

$\beta = P(Y > 12 | p = .4) = 1 - P(Y \leq 12 | p = .4) = 1 - .979 = .021$.

10.3 **a.** With $n = 20$ and $p = .8$, it is necessary to find c such that $\alpha = P(Y \leq c | p = .8)$
$= .01$. From Table 1, Appendix III, this value is $c = 11$.

b. With the rejection region given as $Y \leq 11$,

$\beta = P(Y > 11 | p = .6) = 1 - P(Y \leq 11 | p = .6) = 1 - .404 = .596$

c. $\beta = P(Y > 11 | p = .4) = 1 - P(Y \leq 11 | p = .4) = 1 - .943 = .057$

10.4 **a.** A Type I error occurs if we conclude that the proportion of ledger sheets with errors is larger than .05 when, in fact, the proportion is .05.

b. By the scheme being used, we will reject for the following situations:
(NOTE: NE = no error, E = error)

Sheet 1	Sheet 2	Sheet 3
NE	NE	.
NE	E	NE
E	NE	NE
E	E	NE

Thus, $\alpha = (.95)^2 + 2(.05)(.95)^2 + (.05)^2(.95) = .9025 + .09025 + .002375 = .995125$.

c. A Type II error occurs if we conclude that the proportion of ledger sheets with errors is .05 when, in fact, the proportion is larger than .05.

d. $\beta = P(\text{accept } H_0 \text{ when } H_a \text{ is true}) = P(\text{accepting } H_0 | p = p_a)$
$= 2p_a^2(1 - p_a) + p_a^3$. Since we reject if we observe E,E,E or NE,E,E or E,NE,E,

10.5 Under H_0: $Y_i \sim \text{Uniform}(0, 1)$ and thus $U = Y_1 + Y_2$ has density

$$f_U(u) = \begin{cases} u & 0 \leq u \leq 1 \\ 2 - u & 1 < u \leq 2 \end{cases}$$

Test 1: $P(Y_1 > 0.95) = \int_{0.95}^{1} dy = 0.05$

Test 2: $0.05 = P(U > c) = \int_{c}^{2} (2 - u)du = 2 - 2c + \frac{1}{2}c^2$

Solving the quadratic yields $c = 2 \pm \frac{1}{2}\sqrt{0.4}$. So we will use $c = 2 - \frac{1}{2}(0.652) = 1.684$

10.6 **a.** Let X_1 and X_2 both be binomial with $n = 15$ and $p = 0.10$. By definition,

$$\begin{aligned}
\alpha &= P(\text{type I error}) = P(\text{reject } H_0 \,|H_0 \text{is true}) \\
&= P(\text{reject } H_0 \text{ in stage } 1|H_0 \text{ is true}) + P(\text{reject } H_0 \text{ in stage } 2|H_0 \text{is true}) \\
&= P(X_1 \geq 4) + P(X_1 + X_2 \geq 6 \text{ and } X_1 \leq 3) \\
&= P(X_1 \geq 4) + \sum_{i=0}^{3} P(X_1 + X_2 \geq 6 \text{ and } X_1 = i) \\
&= P(X_1 \geq 4) + \sum_{i=0}^{3} P(X_2 \geq 6 - i)\, P(X_1 = i) \\
&= 1 - P(X_1 \leq 3) + \sum_{i=0}^{3} [1 - P(X_2 \leq 5 - i)][\, P(X_1 \leq i) - P(X_1 \leq i - 1)] \\
&= 1 - 0.944 + (1 - 0.998)(0.206) + \cdots + (1 - 0.816)(0.944 - 0.816). \\
&= 0.0994
\end{aligned}$$

b. This is similar to part **a.** with $p = 0.30$.

c. Let $X_1$and X_2 be binomial with $n = 15$ and $p = 0.30$. By definition,

$$\begin{aligned}
\beta &= P(\text{type II error}) = P(\text{accept } H_0|p = 0.30) \\
&= \sum_{i=0}^{3} P(X_1 = i \text{ and } X_1 + X_2 \leq 5) \\
&= \sum_{i=0}^{3} P(X_2 \leq 5 - i) P(X_1 = i) \\
&= 0.068
\end{aligned}$$

10.7 **a.** Since it is necessary to test a claim that the average amount saved, μ, is \$900, the hypothesis to be tested is two-tailed:

$$H_0\!: \ \mu = 900 \qquad \text{vs.} \qquad H_a\!: \ \mu \neq 900$$

b. The rejection region with $\alpha = .01$ is determined by a critical value of z such that $P\,[|z| > z_0] = .01$

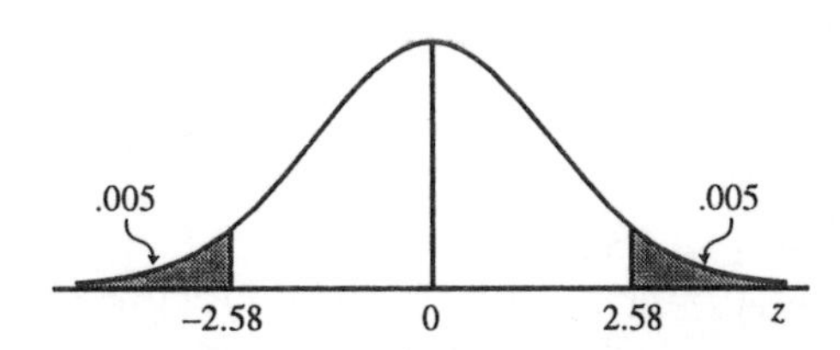

Figure 10.1

This value is $z_0 = 2.58$ (see Figure 10.1) and the rejection region is $|z| > 2.58$.

c. The test statistic is

$$z = \frac{\overline{y}-\mu}{\frac{\sigma}{\sqrt{n}}} \approx \frac{\overline{y}-\mu}{\frac{s}{\sqrt{n}}} = \frac{885-900}{\frac{50}{\sqrt{35}}} = -1.77$$

d. The observed value, $z = -1.77$, does not fall in the rejection region, and H_0 is not rejected. We cannot conclude that the average savings is different than claimed.

10.8 The parameter of interest is μ, the average daily wage of workers in a given company. The objective is to determine whether this company pays inferior wages in comparison to the total industry. Thus the hypothesis to be tested is

$$H_0\!: \ \mu = 13.20 \qquad \text{vs.} \qquad H_a\!: \ \mu < 13.20$$

The best estimator for μ is the sample average, $\overline{y} = 12.20$. The test statistic is

$$Z = \frac{\overline{Y}-\mu}{\frac{\sigma}{\sqrt{n}}}$$

which represents the distance (measured in units of standard deviation) from $\overline{Y}$ to the hypothesized mean μ. Calculating the value of the test statistic using the information contained in the sample, we have

$$z = \frac{\bar{y}-\mu}{\frac{\sigma}{\sqrt{n}}} = \frac{12.20-13.20}{\frac{2.50}{\sqrt{40}}} = \frac{-\sqrt{40}}{2.5} = -2.53$$

The critical value of Z that separates the rejection and non rejection regions will be a value (denoted by z_0) such that $P(Z < z_0) = .01$. That is, $z_0 = -2.326$ (see Figure 10.2). The null hypothesis will be rejected if $z < -2.326$. Note that the observed value of the test statistic falls in the rejection region. Thus the conclusion is to reject the null hypothesis. There is evidence to indicate that this company is paying inferior wages.

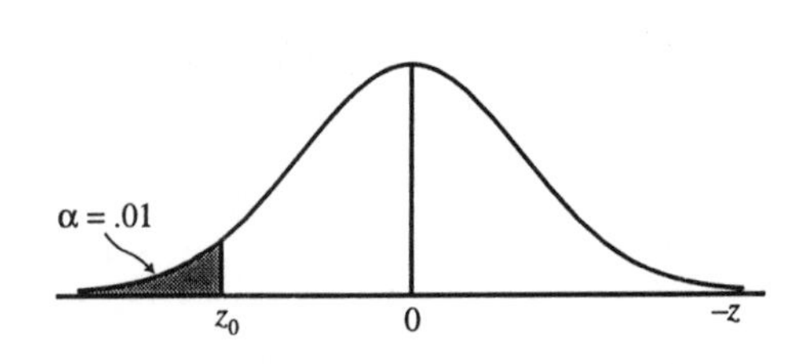

Figure 10.2

10.9 Let μ be the average output voltage. We are to test H_0: $\mu = 130$ vs. H_a: $\mu < 130$ at the $\alpha = 0.05$ level. The test statistic is

$$z = \frac{\bar{y}-\mu_0}{\frac{\sigma}{\sqrt{n}}} = \frac{128.6-130}{\frac{2.1}{\sqrt{40}}} = -4.22$$

where we are using s to estimate σ

RR: Reject H_0 if $z < -z_{.05} = -1.645$.

Conclusion: Reject H_0 at $\alpha = .05$.

That is, there is sufficient evidence at the .05 level to claim that the output voltage is less than 130.

10.10 Let μ be the average hardness index. We are to test H_0: $\mu \geq 64$ vs. H_a: $\mu < 64$. The test statistic is

$$z = \frac{\bar{y}-\mu_0}{\frac{\sigma}{\sqrt{n}}} = \frac{62-64}{\frac{8}{\sqrt{50}}} = -1.77$$

The rejection region is RR: Reject H_0 if $z < -z_{.01} = -2.326$.
Conclusion: Do not reject H_0 at $\alpha = .01$. There is insufficient evidence to reject the manufacturer's claim.

10.11 We are to test H_0: $\mu_1 - \mu_2 = 0$ vs. H_a: $\mu_2 - \mu_2 \neq 0$. The test statistic and rejection region are

$$z = \frac{1.65-1.43}{\sqrt{\frac{(.26)^2}{30}+\frac{(.22)^2}{35}}} = 3.65$$

RR: Reject H_0 if $|z| > 2.575$.
Conclusion: Reject H_0 at $\alpha = .01$. The soils do appear to differ with respect to average shear strength, at the 1% significance level.

10.12 **a.** It might be reasonable to expect that the boys would have a greater interest in sports. Based on this we might choose H_a: $\mu_1 - \mu_2 > 0$, where μ_1 = mean LIC score for sports for boys and μ_2 = mean LIC score for sports for girls.

b. This corresponds to a one-tailed test.

c. The hypothesis of interest is

H_0: $\mu_1 - \mu_2 = 0$ vs. H_a: $\mu_1 - \mu_2 > 0$

and the test statistic is

$$z = \frac{(\bar{y}_1-\bar{y}_2)-0}{\sqrt{\frac{s_1^2}{n_1}+\frac{s_2^2}{n_2}}} = \frac{13.65-9.88}{\sqrt{\frac{(4.82)^2}{252}+\frac{(4.41)^2}{307}}} = 9.56$$

The rejection region, with $\alpha = .01$, is $z > 2.33$ and H_0 is rejected. There is evidence to indicate that the mean for men is greater than that for women.

10.13 **a.–b.** Let μ_i be the mean distance from the release point for each sample $i = 1, 2$. Since there is no prior knowledge as to which mean should be larger, the hypothesis of interest is two-tailed:

H_0: $\mu_1 - \mu_2 = 0$, vs. H_a: $\mu_1 - \mu_2 \neq 0$.

c. The test statistic is

$$z = \frac{\bar{y}_1 - \bar{y}_2}{\sqrt{\frac{s_1^2}{n_1} + \frac{s_2^2}{n_2}}} = \frac{2980-3205}{\sqrt{\frac{(1140)^2+(963)^2}{40}}} = -.954.$$

The rejection region, with $\alpha = .10$, is two-tailed or $|z| > 1.645$. Thus the null hypothesis is not rejected. That is, we conclude there is insufficient evidence to indicate a difference in the two means.

10.14 a. If we define p as the proportion of college students aged 30 years or more, then we test

$$H_0\colon\ p = .25 \qquad \text{vs.} \qquad H_a\colon\ p \neq .25$$

The test statistic is

$$z = \frac{\hat{p} - p_0}{\sqrt{\frac{p_0 q_0}{n}}} = \frac{\frac{98}{300} - .25}{\sqrt{\frac{(.25)(.75)}{300}}} = 3.07$$

and the rejection region, with $\alpha = .05$ is $|z| > 1.96$. H_0 is rejected and we conclude that the 25% figure is not accurate.

b. Yes, the results do give evidence that the columnist's claim is too low.

10.15 Let p be the proportion of adults unable to name an elected official that they admire. Then we want to test

$$\text{H}_0\colon\ p = .5 \qquad \text{vs.} \qquad H_a\colon\ p > .5$$

The test statistic is

$$z = \frac{\hat{p} - p_0}{\sqrt{\frac{p_0(1-p_0)}{n}}} = \frac{.6-.5}{\sqrt{\frac{(.5)(.5)}{1429}}} = 7.56.$$

With $\alpha = .01$, $z_{.01} = 2.326$ and we reject H_0 if $z > 2.326$. H_0 is rejected and we conclude that a majority is unable to name an elected official that they admire.

10.16 Let p be the proportion of Americans with brown eyes. Then the hypothesis of interest is

$$H_0\colon\ p = .45 \qquad \text{vs.} \qquad H_a\colon\ p \neq .45$$

With $\hat{p} = \frac{32}{80} = .40$ and $n = 80$, the test statistic is

$$z = \frac{\hat{p} - p_0}{\sqrt{\frac{p_0 q_0}{n}}} = \frac{.4-.45}{\sqrt{\frac{(.45)(.55)}{80}}} = -.90.$$

The rejection region is two-tailed with $\alpha = .01$, or $|z| > 2.58$ and H_0 is not rejected. There is insufficient evidence to indicate that the proportion of brown-eyed people in the region where the study was performed differs from the value reported in the *Washington Post.*

10.17 The hypothesis of interest is

$$H_0\colon\ p_1 - p_2 = 0 \qquad \text{vs.} \qquad H_a\colon\ p_1 - p_2 \neq 0$$

where p_1 = the proportion of all colleges with an increase in applications in 1992 and p_2 = the proportion of all colleges with an increase in applications in 1991. The test statistic, based on the sample data, will be

$$z = \frac{(\hat{p}_1 - \hat{p}_2) - 0}{\sqrt{\frac{p_1 q_1}{n_1} + \frac{p_2 q_2}{n_2}}}.$$

In order to evaluate the denominator, estimates for p_1 and p_2 must be obtained. Because we are assuming, under H_0, that $p_1 = p_2$, the best estimate for this common value will be (with $\hat{p}_1 = .71$ and $\hat{p}_2 = .50$)

$$\hat{p} = \frac{n_1\hat{p}_1 + n_2\hat{p}_2}{n_1 + n_2} = \frac{875+613}{1232+1225} = .606.$$

The test statistic is, then,

$$z = \frac{\hat{p}_1 - \hat{p}_2}{\sqrt{\hat{p}\hat{q}\left(\frac{1}{n_1} + \frac{1}{n_2}\right)}} = \frac{.71-.50}{\sqrt{(.606)(.394)\left(\frac{1}{1232} + \frac{1}{1225}\right)}} = 10.65.$$

The rejection region, with $\alpha = .01$, is $|z| > 2.58$ and H_0 is rejected. There is evidence of a difference in the proportions between 1991 and 1992.

10.18 Throughout let p_1 be the relevant proportion in 1986 and p_2 be the appropriate proportion in 1991.

a. The hypothesis of interest is

$$H_0\colon\ p_1 - p_2 = 0 \qquad \text{vs.} \qquad H_a\colon\ p_1 - p_2 \neq 0$$

Calculate

$$\widehat{p}_1 = .45, \widehat{p}_2 = .34, \text{ and } \widehat{p} = \frac{n_1\widehat{p}_1+n_2\widehat{p}_2}{n_1+n_2} = \frac{450+340}{1000+1000} = .395$$

The test statistic is then

$$z = \frac{\widehat{p}_1-\widehat{p}_2}{\sqrt{\widehat{p}\widehat{q}\left(\frac{1}{n_1}+\frac{1}{n_2}\right)}} = \frac{.45-.34}{\sqrt{(.395)(.605)\left(\frac{1}{1000}+\frac{1}{1000}\right)}} = 5.03.$$

The rejection region with $\alpha = .05$ is $z > 1.96$ and H_0 is rejected. There is evidence of a difference in the proportion of users in 1986 and 1991.

b. The hypothesis of interest is

$$H_0\colon\ p_1 - p_2 = 0 \qquad \text{vs.} \qquad H_a\colon\ p_1 - p_2 < 0$$

Calculate

$$\widehat{p} = .14, \widehat{p}_2 = .26, \text{ and } \widehat{p} = \frac{n_1\widehat{p}_1+n_2\widehat{p}_2}{n_1+n_2} = \frac{140+260}{1000+1000} = .2$$

The test statistic is then

$$z = \frac{\widehat{p}_1-\widehat{p}_2}{\sqrt{\widehat{p}\widehat{q}\left(\frac{1}{n_1}+\frac{1}{n_2}\right)}} = \frac{.14-.26}{\sqrt{(.2)(.8)\left(\frac{1}{1000}+\frac{1}{1000}\right)}} = -6.71.$$

The rejection region with $\alpha = .05$ is $|z| < -1.645$ and H_0 is rejected. There is evidence to indicate that ibuprofen has significantly increased its market share from 1986 to 1991.

c. Yes. The survey is based on the same samples of 1000 people. Hence, if a person has decreased his use of aspirin, he may have begun using ibuprofen. The tests are not independent.

10.19 The object of this experiment is to make a decision about the binomial parameter p, which is the probability that a customer prefers color A. Hence the null hypothesis will be that a customer has no preference for A, and the alternative will be that he has a preference. The null hypothesis is

$$H_0\colon\ p = P(\text{customer prefers } A) = \tfrac{1}{3},$$

while the alternative hypothesis is

$$H_a\colon\ p > \tfrac{1}{3}.$$

Note that a one-tailed test of hypothesis is implied. The test statistic will be constructed by using $\widehat{p}$, the sample fraction favoring color A. The test statistic is

$$z = \frac{\widehat{p}-p_0}{\sqrt{\frac{p_0q_0}{n}}} = \frac{\left(\frac{400}{1000}\right)-\left(\frac{1}{3}\right)}{\sqrt{\frac{\left(\frac{1}{3}\right)\left(\frac{2}{3}\right)}{1000}}} = 4.47$$

Notice that the value for $\sigma_{\widehat{p}}$ was determined by assuming the null hypothesis to be true, that is, $p = \frac{1}{3}$. The rejection region for this one-tailed test (using $\alpha = .05$) is $z > 1.645$. The test statistic, $z = 4.47$, falls in the rejection region, and hence the null hypothesis is rejected. We conclude that customers have a preference for color A.

10.20 The manufacturer claims that at least 20% of the public prefer her product. In order to test this claim, the following hypothesis is employed:

$$H_0\colon\ p = .2 \qquad \text{vs.} \qquad H_a\colon\ p < .2$$

Rejection of the null hypothesis would imply that the acceptance and rejection regions will be $z = -1.645$, since values of $\widehat{p}$ in the lower tail of the distribution will tend to disprove the null hypothesis (see Figure 10.3). The objective, then, is to determine a value for $\widehat{p}$ such that the corresponding test statistic z will be less than or equal to -1.645. Under the assumption of the null hypothesis, $p = .2$ and

$$\sigma_{\widehat{p}} = \sqrt{\frac{pq}{n}} = \sqrt{\frac{(.2)(.8)}{100}} = .04$$

A value for $\widehat{p}$ must be found so that

$$z = \frac{\widehat{p}-.2}{.04} \le -1.645$$

Solving for $\widehat{p}$ yields $\widehat{p} \le .1342$.

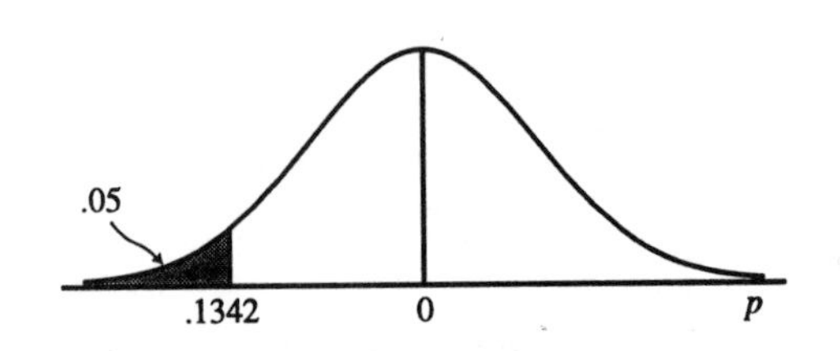

Figure 10.3

10.21 The following conditions must hold in order that the z statistic be appropriate:

(1) $\widehat{\theta}$ is a normally distributed random variable.

(2) $\sigma_{\widehat{\theta}}$ is a known quantity, or n is large, so that a good approximation for $\sigma_{\widehat{\theta}}$ can be obtained from the sample observations.

10.22 **a.** Let p be the proportion of adults who approved of the performance of their congressional representatives. Then we want to test

H_0: $p = .5$ vs. H_a: $p > .5$.

We the test statistic

$$z = \frac{\widehat{p}-p_0}{\sqrt{\frac{p_0q_0}{n}}} = \frac{.56-.5}{\sqrt{\frac{(.5)(.5)}{1429}}} = 4.54.$$

We would reject H_0 if $z > z_{.01} = 2.326$. Therefore, we reject H_0 and say that at the 99% confidence level there is sufficient evidence to indicate that the majority of adults approved of the performance of their congressional representatives.

b. Let p be the proportion of adults who felt that their congressional representatives deserved reelection. Then we want to test

H_0: $p = .5$ vs. H_a: $p > .5$

We use the test statistic

$$z = \frac{\widehat{p}-p_0}{\sqrt{\frac{p_0q_0}{n}}} = \frac{.33-.5}{\sqrt{\frac{(.5)(.5)}{1429}}} = -12.85$$

With $\alpha = .05$, we reject H_0 when $z > 1.645$. Thus there is insufficient evidence, at the 95% confidence level, to indicate that the majority of adults felt that their congressional representatives deserved reelection.

10.23 Two binomial populations are involved. The hypothesis to be tested is

H_0: $p_1 - p_2 = 0$ (i.e., $p_1 = p_2$) vs. H_a: $p_1 - p_2 > 0$ (i.e., $p_1 > p_2$)

The test statistic, based on the sample data, will be

$$Z = \frac{(\widehat{p}_1-\widehat{p}_2)-(p_1-p_2)}{\sqrt{\frac{p_1q_1}{n_1}+\frac{p_2q_2}{n_2}}}$$

In order to evaluate the denominator, estimates for p_1 and p_2 must be obtained, using the assumption, under H_0, that $p_1 - p_2 = 0$. Then the best estimate for this common value is

$$\widehat{p} = \frac{y_1+y_2}{n_1+n_2} = \frac{46+34}{200+200} = .2$$

The test statistic then becomes $z = \frac{\widehat{p}_1-\widehat{p}_2-0}{\sqrt{\widehat{p}\widehat{q}\left[\left(\frac{1}{n}\right)+\left(\frac{1}{n}\right)\right]}} = \frac{.23-.17}{\sqrt{(.2)(.8)\left(\frac{1}{100}\right)}} = 1.5.$

Rejection region: With $\alpha = .05$, the null hypothesis will be rejected if $z > 1.645$.

Conclusion: The test statistic does not fall in the rejection region and we fail to reject the null hypothesis. There is insufficient evidence to support the researcher's belief.

10.24 If μ is the average length of stay the hypothesis of interest is

H_0: $\mu = 5$ vs. H_a: $\mu > 5$

and the test statistic is approximately

$$z = \frac{\overline{y}-\mu_0}{\frac{s}{\sqrt{n}}} = \frac{5.4-5}{\frac{3.1}{\sqrt{500}}} = 2.89$$

The rejection region with $\alpha = .05$ is $z > 1.645$. Hence we reject H_0 and support the agency's hypothesis.

10.25 Let p_1 be the proportion of homeless men currently working and p_2 be the proportion of domiciled men currently working. The hypothesis of interest is

H_0: $p_1 - p_2 = 0$ vs. H_a: $p_1 - p_2 < 0$.

Calculate

$$\widehat{p}_1 = \frac{34}{112} = .30, \ \widehat{p}_2 = .38, \text{ and } \widehat{p} = \frac{x_1+x_2}{n_1+n_2} = \frac{34+98}{112+260} = .355$$

The test statistic is then

$$z = \frac{\widehat{p}_1-\widehat{p}_2}{\sqrt{\widehat{p}\widehat{q}\left(\frac{1}{n_1}+\frac{1}{n_2}\right)}} = \frac{.30-.38}{\sqrt{(.355)(.645)\left(\frac{1}{112}+\frac{1}{260}\right)}} = -1.48.$$

The rejection region with $\alpha = .01$ is $z < -2.326$ and H_0 is not rejected. There is no evidence that the proportion of homeless men working is less than the proportion of domiciled men.

10.26 Let p_1 be the proportion favoring complete protection and p_2 the proportion desiring destruction of the nuisance alligators. The hypothesis to be tested is

$$H_0\colon\ p_1 - p_2 = 0 \qquad \text{vs.} \qquad H_a\colon\ p_1 - p_2 \neq 0$$

and the test statistic is

$$z = \frac{\widehat{p}_1 - \widehat{p}_2}{\sqrt{\left(\frac{\widehat{p}_1\widehat{q}_1}{n}\right)+\left(\frac{\widehat{p}_2\widehat{q}_2}{n}\right)+\left(\frac{2\widehat{p}_1\widehat{p}_2}{n}\right)}} = \frac{-.10}{\sqrt{\frac{(.06)(.94)+(.16)(.84)+2(.06)(.16)}{500}}} = -4.88$$

The rejection region with $\alpha = .01$ is $|z| > 2.58$ and H_0 is rejected. There is a difference between the two proportions.

10.27 For testing

$$H_0\colon\ \mu = 130 \qquad \text{vs.} \qquad H_a\colon\ \mu = 128$$

we reject H_0 if $\frac{\overline{y}-\mu_0}{\frac{\sigma}{\sqrt{n}}} < -1.645$ or, if $\overline{y} < \mu_0 - \frac{1.645\sigma}{\sqrt{n}} = 130 - \frac{1.645(2.1)}{\sqrt{40}} = 129.45$.

If the mean is really 128, then

$$\beta = P\left[\overline{y} > 129.45\right] = P\left[\frac{\overline{y}-\mu_a}{\frac{\sigma}{\sqrt{n}}} > \frac{129.45-128}{\frac{2.1}{\sqrt{40}}}\right] = P[Z > 4.37] \ \leq 0.0000317$$

10.28 For testing $H_0\colon\ \mu \geq 64$ vs. $H_a\colon\ \mu = 60$. We reject H_0 if $\frac{\overline{y}-\mu_0}{\frac{\sigma}{\sqrt{n}}} < -2.33$ or, if

$$\overline{y} < \mu_0 - \frac{2.33\sigma}{\sqrt{n}} = 64 - \frac{2.33(8)}{\sqrt{50}} = 61.36.$$

If the mean is really 60, then

$$\beta = P\left[\overline{y} > 61.36\right] = P\left[\frac{\overline{y}-\mu_a}{\frac{\sigma}{\sqrt{n}}} > \frac{61.36-60}{\frac{8}{\sqrt{50}}}\right] = P[Z > 1.2] = .1151.$$

10.29 In Exercise 10.20 we found that we reject when $\widehat{p} \leq .1342$. Thus

$$\beta = P\left(\widehat{p} > .1342\right).$$

The corresponding z value is

$$z = \frac{.1342-.15}{\sqrt{\frac{(.15)(.85)}{100}}} = -.44$$

Hence

$$\begin{aligned}\beta &= P(Z > -.44)\\ &= P(Z < .44)\\ &= 1 - .3300 = .6700.\end{aligned}$$

10.30 Refer to Figure 10.5, which represents the two probability distributions, one assuming $\mu_{\widehat{p}_1-\widehat{p}_2} = p_1 - p_2 = 0$ and one assuming $\mu_{\widehat{p}_1-\widehat{p}_2} = p_1 - p_2 = .1$. The right curve is the true distribution of the random variable $(\widehat{p}_1 - \widehat{p}_2)$, and consequently any probabilities we wish to calculate concerning the random variable must be calculated as areas under the curve to the right. The object of this exercise is to find a common sample size so that $\alpha = P(\text{reject } H_0|H_0 \text{ true}) = .05$ and $\beta = P(\text{accept } H_0|H_0 \text{ false}) \leq .20$.

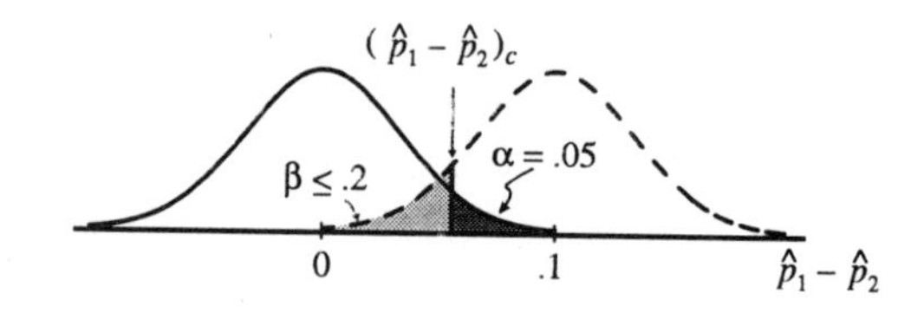

Figure 10.5

For $\alpha = P(\text{reject } H_0|p_1 - p_2 = 0) = .05$, consider the critical value of the random variable $(\widehat{p}_1 - \widehat{p}_2)$ that separates the rejection and acceptance regions. This value will be denoted by $(p_1 - p_2)_c$. Recall that the random variable $z = \frac{y-\mu}{\sigma}$ measures the distance from a particular value y to the mean (in units of standard deviation). Since the z value corresponding to $(\widehat{p}_1 - \widehat{p}_2)_c$ is $z = 1.645$, we have

$$1.645 = \frac{(\widehat{p}_1-\widehat{p}_2)_c - 0}{\sqrt{\frac{p_1q_1}{n_1}+\frac{p_2q_2}{n_2}}} \qquad \text{or} \qquad (\widehat{p}_1 - \widehat{p}_2)_c = 1.645\sqrt{\frac{p_1q_1}{n_1}+\frac{p_2q_2}{n_2}}$$

But $\beta = P[\text{accept } H_0|(p_1 - p_2) = .1]$ equals the area under the curve to the right from $-\infty$ to $(\widehat{p}_1 - \widehat{p}_2)_c$. Since it is required that $\beta = .2$, we must have a z value of $-.84$.

Then

$$-.84 = \frac{(\widehat{p}_1-\widehat{p}_2)_c-.1}{\sqrt{\frac{p_1q_1}{n_1}+\frac{p_2q_2}{n_2}}} \qquad \text{where} \qquad (\widehat{p}_1-\widehat{p}_2)_c = 1.645\sqrt{\frac{p_1q_1}{n_1}+\frac{p_2q_2}{n_2}}$$

Substituting for $(\widehat{p}_1-\widehat{p}_2)_c$, we have

$$-.84 = \frac{1.645\sqrt{\frac{p_1q_1}{n_1}+\frac{p_2q_2}{n_2}}-.1}{\sqrt{\frac{p_1q_1}{n_1}+\frac{p_2q_2}{n_2}}} = 1.645 - \frac{.1}{\sqrt{\frac{p_1q_1}{n_1}+\frac{p_2q_2}{n_2}}}$$

Hence

$$2.485 = \frac{.1}{\sqrt{\frac{p_1q_1}{n_1}+\frac{p_2q_2}{n_2}}}$$

The following two facts will allow us to calculate the appropriate sample size:

(1) $n_1 = n_2 = n$.

(2) The maximum value of $p(1-p)$ will occur when $p = 1-p = .5$. Since values of p_1 and p_2 are unknown, the use of $p = .5$ will provide a valid sample size, although it may be slightly larger than necessary.

Then solving for n, we obtain

$$2.485 = \frac{.1}{\sqrt{(.5)(.5)\left[\left(\frac{1}{n}\right)+\left(\frac{1}{n}\right)\right]}} \qquad \text{or} \qquad 2.485 = \frac{.1}{\left(\frac{.5}{\sqrt{\frac{2}{n}}}\right)} \qquad \text{or} \qquad 12.425 = \sqrt{\frac{n}{2}}$$

or $\sqrt{n} = 17.57$ or $n = 308.76$

Hence, the common sample size for the researcher's test should be $n = 309$.

10.31 Refer to Exercise 10.24. The rejection region, written in terms of $\overline{y}$, is

$$z > 1.645 \qquad \text{or} \qquad \frac{\overline{y}-5}{\left(\frac{3.1}{\sqrt{500}}\right)} > 1.645 \qquad \text{or} \qquad \overline{y} \geq 5.228$$

Then

$$\beta = P(\text{accept } H_0|\mu = 5.5) = P(\overline{y} < 5.228|\mu = 5.5) = P\left[Z < \frac{5.228-5.5}{\frac{3.1}{\sqrt{500}}}\right]$$
$$= P(Z < -1.96) = .025$$

10.32 Using the sample size formula given in this section, we have

$$n = \frac{(z_\alpha+z_\beta)^2\sigma^2}{(\mu_a-\mu_0)^2} = \frac{(2.33+1.645)^2(3.1)^2}{(-.5)^2} = 607.37$$

So, $n = 608$ will provide the desired levels.

10.33 **a.** Let μ_1 be the average manual dexterity score for those that participated in sports and μ_2 the average manual dexterity score for those that didi not participate in sports. The hypothesis of interest is H_0: $\mu_1 - \mu_2 = 0$ vs. H_a: $\mu_1 - \mu_2 > 0$ and the test statistic is given by

$$z = \frac{(\overline{y}_1-\overline{y}_2)-0}{\sqrt{\left(\frac{s_1^2}{n_1}\right)+\left(\frac{s_2^2}{n_2}\right)}} = \frac{32.19-31.68}{\sqrt{\frac{(4.34)^2+(4.56)^2}{37}}} = .49$$

The rejection region with $\alpha = .05$ is $z > 1.645$. Hence H_0 is not rejected. There is insufficient evidence to indicate that $\mu_1 > \mu_2$.

b. The rejection region, written in terms of $\overline{Y}_1 - \overline{Y}_2$, is

$$Z > 1.645 \qquad \text{or} \qquad \frac{\overline{Y}_1-\overline{Y}_2}{\hat{\sigma}_{\overline{Y}_1-\overline{Y}_2}} > 1.645$$

or

$$\overline{Y}_1 - \overline{Y}_2 > 1.645\sqrt{\frac{(4.34)^2+(4.56)^2}{37}} = 1.702$$

Then

$$\beta = P(\text{accept } H_0|\mu_1-\mu_2 = 3) = P\left(\overline{Y}_1 - \overline{Y}_2 < 1.702\right) = P\left[Z < \frac{1.702-3}{\hat{\sigma}_{\overline{Y}_1-\overline{Y}_2}}\right]$$
$$= P(Z < -1.25) = P(Z > 1.25) = .1056.$$

10.34 Using the procedure discussed following Example 10.8 of the text, we can write

$$\alpha = P\left(\overline{Y}_1 - \overline{Y}_2 > k \text{ when } \mu_1-\mu_2 = 0\right) = P\left[Z > \frac{k-0}{\sqrt{\frac{\sigma_1^2+\sigma_2^2}{n}}}\right]$$

so that $z_\alpha = \frac{k\sqrt{n}}{\sqrt{\sigma_1^2+\sigma_2^2}}$. Also

$$\beta = P\left(\overline{Y}_1 - \overline{Y}_2 \le k|\mu_1 - \mu_2 = 3\right) = P\left[Z \le \frac{(k-3)\sqrt{n}}{\sqrt{\sigma_1^2+\sigma_2^2}}\right]$$

so that $-z_\beta = \frac{(k-3)\sqrt{n}}{\sqrt{\sigma_1^2+\sigma_2^2}}$. Eliminating k, we obtain

$$z_\alpha\sqrt{\frac{\sigma_1^2+\sigma_2^2}{n}} = 3 - z_\beta\sqrt{\frac{\sigma_1^2+\sigma_2^2}{n}}$$

Solving for n, we have

$$n = \frac{[2(1.645)]^2\,[(4.34)^2+(4.56)^2]}{3^2} = 47.66$$

or $n = 48$ will provide the given levels of α and β.

10.35 $\overline{y} \pm z_{.005}\frac{s}{\sqrt{n}} = 885 \pm 2.58\left(\frac{50}{\sqrt{35}}\right) = (863.195, 906.805)$

The value $\mu_0 = 900$ is inside the interval. The null hypothesis should not be rejected because 900 is a believable value for the true mean. This is consistent with the conclusion in Exercise 10.7.

10.36 The rejection region is

$$\frac{\widehat{\theta}-\theta_0}{\sigma_{\widehat{\theta}}} > z_\alpha$$

which occurs if and only if

$$\widehat{\theta} - \theta_0 > z_\alpha\sigma_{\widehat{\theta}}$$

which occurs if and only if

$$\widehat{\theta} - z_\alpha\sigma_{\widehat{\theta}} > \theta_0$$

where the left-hand side is the $100(1-\alpha)\%$ lower confidence bound for θ.

10.37 Using Exercise 10.34 a 99% lower confidence bound is given by

$$\widehat{p} - z_{.01}\sqrt{\frac{\widehat{p}(1-\widehat{p})}{n}} = .6 - 2.326\sqrt{\frac{(.6)(.4)}{1429}} = .6 - .03 = .57.$$

Because the interval .57 to 1.0 does not contain .5, the alternative hypothesis of Exercise 10.13 should be accepted at the 99% confidence level. This does not conflict with the answer to Exercise 10.15.

10.38 The rejection region is

$$\frac{\widehat{\theta}-\theta_0}{\sigma_{\widehat{\theta}}} < -z_\alpha$$

which is true if and only if

$$\widehat{\theta} + z_\alpha\sigma_{\widehat{\theta}} < \theta_0.$$

That is, H_0 will be rejected if and only if the $100(1-\alpha)\%$ upper confidence bound for θ, i.e., $\widehat{\theta} + z_\alpha\sigma_{\widehat{\theta}}$, is less than θ_0.

10.39 Using the result of Exercise 10.38, the bound is

$$\overline{y} + z_{.05}\frac{s}{\sqrt{n}} = 128.6 + 1.645\left(\frac{2.1}{\sqrt{40}}\right) = 128.6 + .546 = 129.146.$$

This bound is less than the hypothesized value, i.e., 130. Therefore, the alternative hypothesis should be accepted, which does not conflict with the answer to Exercise 10.9.

10.40 We are to test

$$H_0\colon \mu \ge .6 \qquad \text{vs.} \qquad H_a\colon \mu < .6.$$

The test statistic is

$$z = \frac{\overline{y}-\mu_0}{\frac{\sigma}{\sqrt{n}}} = \frac{.58-.6}{\frac{.11}{\sqrt{120}}} = -1.99$$

The p-value is

$$p\text{-value} = P(Z < -1.99) = .0233$$

Since $.0233 < .10$, we would reject H_0 in a test at level $\alpha = .10$.

10.41 We are to test

$$H_0\colon \mu_1 - \mu_2 = 0 \qquad \text{vs.} \qquad H_a\colon \mu_1 - \mu_2 \ne 0.$$

The test statistic is

$$z = \frac{74-71}{\left(\frac{9^2}{50} + \frac{10^2}{50}\right)^{\frac{1}{2}}} = 1.58.$$

The p-value is

$$p\text{-value} = P(|Z| > 1.58) = 2(.0571) = .1142.$$

Since $.1142 > .05$, we would not reject H_0 in a test at level $\alpha = .05$.

10.42a. Let p_1 and p_2 be the proportions (attending vs. not attending) who were using safety seats 4 to 6 weeks after birth. From the study, $n_1 = 78, \widehat{p}_1 = .96$, $n_2 = 136 - 78 = 58, \widehat{p}_2 = .78$. Then

$$\widehat{p} = \frac{78(.96)+58(.78)}{78+58} = .883.$$

The hypothesis to be tested is

$$H_0\colon\ p_1 - p_2 = 0 \qquad \text{vs.} \qquad H_a\colon\ p_1 - p_2 > 0$$

and the test statistic is

$$z = \frac{\widehat{p}_1-\widehat{p}_2}{\sqrt{\widehat{p}\widehat{q}\left(\frac{1}{n_1}+\frac{1}{n_2}\right)}} = \frac{.96-.78}{\sqrt{(.883)(.117)\left(\frac{1}{78}+\frac{1}{58}\right)}} = 3.23.$$

The rejection region, with $\alpha = .05$, is $z > 1.645$ and H_0 is rejected. There is evidence that the lecture was effective.

b. For the one-tailed test,

$$p\text{-value} = P[z > 3.23] < 0.00135.$$

10.43a. The hypothesis of interest is

$$H_0\colon\ \mu_1 = 3.8 \qquad \text{vs.} \qquad H_a\colon\ \mu_1 < 3.8,$$

where μ_1 is the average drop in FVC for men on the physical fitness program. The test statistic is

$$z = \frac{\overline{y}_1-\mu_1}{\frac{s_1}{\sqrt{n_1}}} = \frac{3.6-3.8}{\frac{1.1}{\sqrt{30}}} = -.996,$$

and from Table 4 the p-value is

$$p\text{-value} = P[z < -1.00] = .1587.$$

b. Since $\alpha = .05$ is smaller than the p-value $= .1587$, H_0 cannot be rejected. We cannot support the contention that the mean decrease in FVC for men is less than 3.8.

c. The hypothesis of interest is

$$H_0\colon\ \mu_2 = 3.1 \qquad \text{vs.} \qquad H_a\colon\ \mu_2 < 3.1,$$

where μ_2 is the average drop in FVC for women on the physical fitness program. The test statistic is

$$z = \frac{\overline{y}_2-\mu_2}{\frac{s_2}{\sqrt{n_2}}} = \frac{2.7-3.1}{\frac{1.2}{\sqrt{30}}} = -1.826,$$

and the p-value is

$$p\text{-value} = P[z < -1.83] = .0336.$$

d. Since $\alpha = .05$ is larger than the p-value, .0336, H_0 can be rejected (at $\alpha = .0336$ or any larger value). The data support the contention that the mean decrease in FVC is less than 3.1 for women on the physical fitness program.

10.44a. The hypothesis to be tested is

$$H_0\colon\ p = .85 \qquad \text{vs.} \qquad H_a\colon\ p > .85$$

where p is the proportion of right-handed executives of large corporations. The test statistic is

$$z = \frac{\widehat{p}-p_0}{\sqrt{\frac{p_0q_0}{n}}} = \frac{.96-.85}{\sqrt{\frac{(.85)(.15)}{300}}} = 5.34.$$

The rejection region, with $\alpha = .01$, is $z > 2.33$ and H_0 is rejected. The percentage of right-handed executives is greater than the proportion of right-handed people in the general population.

b. For a one-tailed test,

$$p\text{-value} = P[z > 5.34] < .001.$$

Hence, H_0 can be rejected for any value of $\alpha > .001$.

10.45 We are to test

$$H_0\colon\ p = .05 \qquad \text{vs.} \qquad H_a\colon\ p < .05$$

with $\widehat{p} = \frac{45}{1124} = .04$.
The test statistic is

$$z = \frac{\widehat{p} - .05}{\sqrt{\frac{(.05)(.95)}{n}}} = \frac{.04 - .05}{\sqrt{\frac{(.05)(.95)}{1124}}} = -1.538.$$

The p-value is

$$p\text{-value} = P(Z < -1.538) = .0618.$$

At level $\alpha = .01$, we would not reject H_0, since $.0618 > .01$.

10.46 We are to test

$$H_0\colon\ \mu_1 - \mu_2 = 0 \qquad \text{vs.} \qquad H_a\colon\ \mu_1 - \mu_2 > 0.$$

The test statistic is

$$z = \frac{6.9 - 5.8}{\sqrt{\frac{(2.9)^2}{35} + \frac{(1.2)^2}{35}}} = 2.074$$

The p-value (attained significance level) is

$$p\text{-value} = P(Z > 2.074) = .0192$$

The company would reject H_0 at the level $\alpha = .05$.

10.47 We wish to test

$$H_0\colon\ p = .60 \qquad \text{vs.} \qquad H_a\colon\ p \neq .60.$$

The sample proportion is $\widehat{p} = \frac{108}{200} = .54$. The test statistic is

$$z = \frac{.54 - .60}{\sqrt{\frac{(.6)(.4)}{200}}} = -1.732.$$

The p-value is

$$p\text{-value} = P\left(|Z| > |-1.732|\right) = 2P(Z > 1.732) = 2(.0418) = .0836.$$

10.48 We wish to test

$$H_0\colon\ p_A = p_B \qquad \text{vs.} \qquad H_a\colon\ p_A > p_B.$$

The sample proportions are $\widehat{p}_A = \frac{37}{50} = .74$ and $\widehat{p}_B = \frac{23}{50} = .46$. The test statistic uses the pooled sample proportion (assuming H_0 to be true) of $\widehat{p} = \frac{37+23}{100} = .6$.

$$z = \frac{\widehat{p}_A - \widehat{p}_B - 0}{\sqrt{\frac{\widehat{p}\widehat{q}}{50} + \frac{\widehat{p}\widehat{q}}{50}}} = \frac{.74 - .46}{\sqrt{\frac{(.6)(.4)}{50} + \frac{(.6)(.4)}{50}}} = 2.858.$$

The p-value is

$$p\text{-value} = P(Z > 2.858) = .0021.$$

Since $.0021 < .05$, we reject H_0 at the level $\alpha = .05$.

10.49 The z test is inappropriate as a test statistic when the sample size is small and σ is unknown. The estimator S can be used as an approximation to σ when n is large (say $n \geq 30$), but the approximation affects the distribution of the test statistic when n is small. Conditions will often be satisfied to permit the use of Student's t when n is small.

10.50 A Student's t test can be employed to test a hypothesis about a single population when the sample has been randomly selected from a normal population. It will work quite satisfactorily for populations that possess mound-shaped frequency distributions resembling the normal distribution.

10.51 The hypothesis to be tested is

$$H_0\colon\ \mu = 800 \qquad \text{vs.} \qquad H_a\colon\ \mu < 800.$$

However, there are only five measurements on which to base the test, and a t statistic must be employed. The test statistic is $T = \frac{\overline{Y} - \mu}{\frac{S}{\sqrt{n}}}$

where

$$\overline{y} = \frac{\sum_i y_i}{n} = \frac{3975}{5} = 795$$

and

$$s^2 = \frac{\sum_i - \frac{\left(\sum_i y_i\right)^2}{n}}{n-1} = \frac{3{,}160{,}403 - \frac{15{,}800{,}625}{5}}{4}$$
$$= 69.5$$

Then
$$t = \frac{\overline{y}-\mu}{\frac{s}{\sqrt{n}}} = \frac{795-800}{\frac{\sqrt{69.5}}{\sqrt{5}}} = \frac{-5}{3.728} = -1.341$$

The rejection region is determined by a t value based on $(n-1) = 4$ degrees of freedom. Indexing $t_{.05}$ in Table 5, the rejection region is $t < -2.132$. Since the observed value of the test statistic does not fall in the rejection region, we do not reject H_0. Refer to Figure 10.6 at the right and notice that the t distribution is similar to the z distribution. The shaded area constitutes the size of the rejection region.

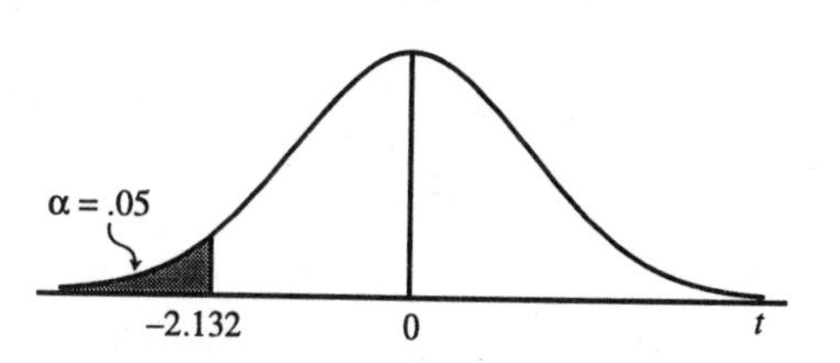

Figure 10.6

To find bounds on the p-value, note in Table 5 that a value of $t = -1.553$ would yield a p-value of .10. The observed value of t is $-1.341 > -1.533$. Thus, p-value > .10.

10.52 The hypothesis to be tested is

$$H_0\!: \mu = 7 \qquad \text{vs.} \qquad H_a\!: \mu \neq 7.$$

The test statistic is
$$t = \frac{\overline{y}-\mu}{\frac{s}{\sqrt{n}}} = \frac{7.1-7.0}{\frac{.12}{\sqrt{10}}} = 2.64$$

To determine the two-tailed rejection region ($\alpha = .10$), Table 5 is used and the rejection region is $|t| > t_{.05,9} = 1.833$. The observed test statistic falls in the rejection region, and the null hypothesis is rejected. There is evidence to suspect that the mean discharge is in excess of 7 ounces.
Note in Table 5 that $2.262 < 2.63 < 2.821$. The p-values associated with 2.262 and 2.821 are $2(.025) = .05$ and $2(.01) = .02$. Thus, $.02 < p\text{-value} < .05$.

10.53 **a.** The hypothesis to be tested is

$$H_0\!: \mu = 45 \qquad \text{vs.} \qquad H_a\!: \mu < 45$$

Calculate
$$\overline{y} = \frac{\sum_{i=1}^{n} y_i}{n} = \frac{712.01}{18} = 39.556$$
$$s^2 = \frac{\sum y_i^2 - \frac{(\sum y_i)^2}{n}}{n-1} = \frac{29{,}040.4275 - \frac{(712.01)^2}{18}}{17} = 50.94593$$

and $s = 7.138$. The test statistic is
$$t = \frac{\overline{y}-\mu_0}{\frac{s}{\sqrt{n}}} = \frac{39.556-45}{\frac{7.138}{\sqrt{18}}} = -3.24\,.$$

With 17 degrees of freedom, the p-value is less than .005. Because this is so small, we reject H_0 and conclude that the average is significantly less than 45 cents.

b. The 95% confidence interval, based on $n - 1 = 17$ degrees of freedom, is
$$\overline{y} \pm t_{.025}\frac{s}{\sqrt{n}} = 39.556 \pm 2.110\left(\frac{7.138}{\sqrt{18}}\right) = 39.556 \pm 3.550$$
or $36.006 < \mu < 43.106$.

10.54 The hypothesis to be tested is

$$H_0\!: \mu = 100 \qquad \text{vs.} \qquad H_a\!: \mu < 100$$

Calculate
$$\overline{y} = \frac{\sum y_i}{n} = \frac{1797.095}{20} = 89.85475$$
$$s^2 = \frac{\sum y_i^2 - \frac{(\sum y_i)^2}{n}}{n-1} = \frac{165{,}697.7081 - \frac{(1797.095)^2}{20}}{19} = 222.115067$$

The test statistic is
$$t = \frac{\overline{y}-\mu}{\frac{s}{\sqrt{n}}} = \frac{89.85475-100}{\sqrt{\frac{222.115067}{20}}} = -3.05$$

The critical value of t with $\alpha = .01$ and $n - 1 = 19$ degrees of freedom is $t_{.01} = 2.539$, and the rejection region is $t < -2.539$. The null hypothesis is rejected and we conclude

that μ is less than 100 DL.
With 19 degrees of freedom, the p-value is less than .005.

10.55a. The hypothesis of interest is

$$H_0\colon\ \mu = 400 \qquad \text{vs.} \qquad H_a\colon\ \mu \neq 400$$

The test statistic is

$$t = \frac{\bar{y}-\mu_0}{\frac{s}{\sqrt{n}}} = \frac{365-400}{\frac{46}{\sqrt{20}}} = -3.4$$

With 19 degrees of freedom, the p-value is less than $2(.005) = .01$. Because the p-value is so small, we reject H_0 and conclude that the mean number of units of vitamin D is not 400.

b. The 95% confidence interval is

$$\bar{y} \pm t_{.025}\frac{s}{\sqrt{n}} = 365 \pm 2.093\left(\frac{46}{\sqrt{20}}\right) = 365 \pm 21.53 = (343.47, 386.53).$$

c. Because the interval in Part **b** does not contain the value 400, we conclude that the true mean is not 400. This conclusion is consistent with the formal test in Part **a.**

10.56 Random samples must be independently drawn from two populations that possess normal distributions with a common variance σ^2. Consequently, it is logical that information in the two sample variances, S_1^2 and S_2^2, should be pooled in order to give the best estimate of the common variance σ^2. In this way all the sample information is being utilized to its best advantage.

10.57 This is similar to Examples 10.14 and 10.15. The hypothesis to be tested is

$$H_0\colon\ \mu_1 - \mu_2 = 0 \qquad \text{vs.} \qquad H_a\colon\ \mu_1 - \mu_2 \doteq 0.$$

We must assume that the data come from two normal populations with a common variance. We obtain an estimate for the common variance σ^2 by calculating

$$s^2 = \frac{(n_1-1)s_1^2+(n_2-1)s_2^2}{n_1+n_2-2} = \frac{10(52)+13(71)}{11+14-2} = \frac{1443}{23} = 62.74$$

The test statistic is

$$t = \frac{\bar{y}_1-\bar{y}_2-D_0}{\sqrt{s^2\left[\left(\frac{1}{n_1}\right)+\left(\frac{1}{n_2}\right)\right]}} = \frac{64-69}{\sqrt{62.74\left[\left(\frac{1}{11}\right)+\left(\frac{1}{14}\right)\right]}} = -1.57$$

The rejection region is $|t| > t_{.025,23} = 2.069$. Since the observed value of the test statistic does not exceed $t = 2.069$ in absolute value, the null hypothesis is not rejected. Note in Table 5 that $1.319 < 1.57 < 1.714$. The p-values associated with 1.319 and 1.714 are 2(.10) and 2(.05). Thus, $.10 < p\text{-value} < .20$.

10.58 The hypothesis to be tested is

$$H_0\colon\ \mu_1 - \mu_2 = 0 \qquad \text{vs.} \qquad H_a\colon\ \mu_1 - \mu_2 > 0.$$

Calculate

$$s^2 = \frac{9(.017)^2+12(.006)^2}{21} = .00014443$$

The test statistic is then

$$t = \frac{\bar{y}_1-\bar{y}_2}{\sqrt{s^2\left[\left(\frac{1}{n_1}\right)+\left(\frac{1}{n_2}\right)\right]}} = \frac{.041-.026}{\sqrt{s^2\left[\left(\frac{1}{10}\right)+\left(\frac{1}{13}\right)\right]}} = 2.97$$

The rejection region, with $\alpha = .05$ and 21 degrees of freedom, is $t > 1.721$, and the null hypothesis is rejected.

10.59 Refer to Exercise 10.58. We are to test

$$H_0\colon\ \mu_1 - \mu_2 \leq .01 \qquad \text{vs.} \qquad H_a\colon\ \mu_1 - \mu_2 > .01.$$

The test statistic is

$$t = \frac{(.041-.026)-.01}{\sqrt{s^2\left[\left(\frac{1}{10}\right)+\left(\frac{1}{13}\right)\right]}} = .989$$

From Table 5, the p-value is greater than .10.

10.60 We are to test

$$H_0\colon\ \mu_1 - \mu_2 = 0 \qquad \text{vs.} \qquad H_a\colon\ \mu_1 - \mu_2 \neq 0.$$

Calculate

$$s^2 = \frac{6(210)^2+9(190)^2}{15} = 39{,}300.$$

The test statistic is

$$t = = \frac{\bar{y}_1-\bar{y}_2}{\sqrt{s^2\left[\left(\frac{1}{n_1}\right)+\left(\frac{1}{n_2}\right)\right]}} = \frac{3250-3240}{\sqrt{s^2\left[\left(\frac{1}{7}\right)+\left(\frac{1}{10}\right)\right]}} = .102.$$

The rejection region is $|t| > 2.131$. We do not reject H_0 when $\alpha = .05$. To find the attained significance level, note, using Table 5, that $.102 < 1.341$. Thus, the p-value is greater than $2(.10) = .20$.

10.61 a. The hypothesis to be tested is

$$H_0\colon\ \mu_1 - \mu_2 = 0 \qquad \text{vs.} \qquad H_a\colon\ \mu_1 - \mu_2 \neq 0.$$

where μ_1 is the average compartment pressure for runners, and μ_2 is the average compartment pressure for cyclists. The pooled estimator of σ^2 is calculated as

$$s_p^2 = \frac{(n_1-1)s_1^2+(n_2-1)s_2^2}{n_1+n_2-2} = \frac{9(3.92)^2+9(3.98)^2}{18} = 15.6034$$

and the test statistic is

$$t = \frac{(\bar{y}_1-\bar{y}_2)-0}{\sqrt{s_p^2\left(\frac{1}{n_1}+\frac{1}{n_2}\right)}} = \frac{14.5-11.1}{\sqrt{15.6034\left(\frac{1}{10}+\frac{1}{10}\right)}} = 1.92$$

The rejection region is two-tailed, based on $n_1 + n_2 - 2 = 18$ degrees of freedom. With $\alpha = .05$, from Table 5 the rejection region is $|t| > t_{.025} = 2.101$. We do not reject H_0; there is insufficient evidence to indicate a difference in the means.
With 18 degrees of freedom, we can say that the p-value is between $2(.025)$ and $2(.05)$; i.e., $.05 < p\text{-value} < .10$.

b. The hypothesis to be tested is

$$H_0\colon\ \mu_1 - \mu_2 = 0 \qquad \text{vs.} \qquad H_a\colon\ \mu_1 - \mu_2 \neq 0.$$

where μ_1 = mean compartment pressure for runners at 80% maximal O_2 consumption and μ_2 = mean compartment pressure for cyclists at 80% maximal O_2 consumption.
First,

$$s_p^2 = \frac{(n_1-1)s_1^2+(n_2-1)s_2^2}{n_1+n_2-2} = \frac{9(3.49)^2+9(4.95)^2}{18} = 18.3413.$$

then the test statistic is

$$t = \frac{(\bar{y}_1-\bar{y}_2)-0}{\sqrt{s_p^2\left(\frac{1}{n_1}+\frac{1}{n_2}\right)}} = \frac{12.2-11.5}{\sqrt{18.3413\left(\frac{1}{10}+\frac{1}{10}\right)}} = .365$$

The rejection region is $|t| > t_{.025} = 2.101$. Fail to reject H_0. We conclude that there is insufficient evidence to indicate a difference in the mean compartment pressure between runners and cyclists at 80% maximal O_2 consumption.
The associated p-value is, with 18 degrees of freedom, greater than $2(.10)$ or the p-value $> .20$.

10.62 The hypothesis of interest is

$$H_0\colon\ \mu = 6 \qquad \text{vs.} \qquad H_a\colon\ \mu < 6,$$

and the test statistic is

$$t = \frac{\bar{y}-\mu_0}{\sqrt{\frac{s^2}{n}}} = \frac{9-6}{\sqrt{\frac{41.2727}{12}}} = 1.62$$

The rejection region is $t < -t_{.05,11} = -1.796$, and the null hypothesis is not rejected. There is insufficient evidence to indicate that μ is less than 6.

10.63 As in Section 10.3, the hypothesis to be tested is

$$H_0\colon\ \mu = 30.31 \qquad \text{vs.} \qquad H_a\colon\ \mu < 30.31.$$

However, there are only 20 measurements on which to base the test, and a t statistic must be employed. The test statistic is

$$t = \frac{\bar{y}-\mu}{\frac{s}{\sqrt{n}}}$$

where

$$\bar{y} = \frac{\sum y_i}{n} = \frac{578.7}{20} = 28.935,$$

$$s^2 = \frac{\sum y_i^2 - \frac{(\sum y_i)^2}{n}}{n-1} = \frac{18{,}462.09-16{,}744.6845}{19} = 90.3898,$$

$$s = \sqrt{90.3898} = 9.507.$$

Then

$t = \frac{\bar{y}-\mu}{\frac{s}{\sqrt{n}}} = \frac{28.935-30.31}{\frac{9.507}{\sqrt{20}}} = -.647.$

The rejection region, with $n-1=19$ degrees of freedom and $\alpha = .05$, is

$t < -t_{.05,19} = -1.729,$

as shown in Figure 10.7. Since the observed value of $t = -.647$ does not fall in the rejection region, H_0 is not rejected. There is no reason to believe that the mean has decreased.

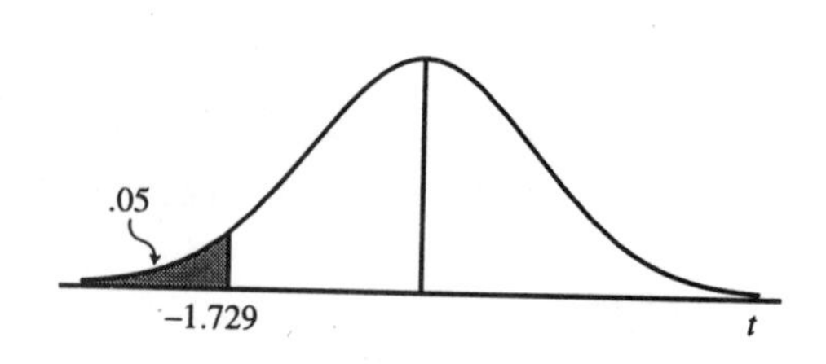

Figure 10.7

10.64 a. If the antiplaque rinse is effective, the plaque buildup should be less for the group using the antiplaque rinse. Hence, the hypothesis to be tested is

$H_0\colon \mu_1 - \mu_2 = 0 \quad \text{vs.} \quad H_a\colon \mu_1 - \mu_2 > 0$

where μ_1 is the mean for the control group and μ_2 is the mean for the antiplaque group.

b. The pooled estimator of σ^2 is

$$s^2 = \frac{(n_1-1)s_1^2+(n_2-1)s_2^2}{n_1+n_2-2} = \frac{6(.32)^2+6(.32)^2}{6+6} = .1024$$

and the test statistic is

$$t = \frac{\bar{y}_1-\bar{y}_2}{\sqrt{s^2\left(\frac{1}{n_1}+\frac{1}{n_2}\right)}} = \frac{1.26-.78}{\sqrt{.1024\left(\frac{2}{7}\right)}} = 2.806.$$

The rejection region, with $\alpha = .05$ and $n_1+n_2-2 = 12$ degrees of freedom, is $t > t_{.05} = 1.782$ and H_0 is rejected. There is evidence to indicate that the rinse is effective.

c. The observed level of significance is

$$p\text{-value} = P[t > 2.806]$$

Since the value $t = 2.806$ falls between two tabled entries for 12 d.f., $t_{.005} = 3.055$ and $t_{.01} = 2.681$, we can conclude that $.005 < p\text{-value} < .01$.

10.65 a. The hypothesis to be tested is

$H_0\colon \mu_1 - \mu_2 = 0 \quad \text{vs.} \quad H_a\colon \mu_1 - \mu_2 \neq 0$

Calculate

$$s_p^2 = \frac{(n_1-1)s_1^2+(n_2-1)s_2^2}{n_1+n_2-2} = \frac{14(42)^2+14(45)^2}{28} = 1894.5$$

and the test statistic is

$$t = \frac{(\bar{y}_1-\bar{y}_2)-0}{\sqrt{s_p^2\left(\frac{1}{n_1}+\frac{1}{n_2}\right)}} = \frac{446-534}{\sqrt{1894.5\left(\frac{1}{15}+\frac{1}{15}\right)}} = -5.54$$

The rejection region is two-tailed, based on $n_1+n_2-2 = 28$ degrees of freedom. From Table 5, the approximate p-value is $p\text{-value} < 2(.005) = .01$. Thus, at the $\alpha = .05$ level there is sufficient evidence to indicate that a difference exists in the mean verbal scores for the two groups.

b. Yes.

c. The hypothesis to be tested is

$H_0\colon \mu_1 - \mu_2 = 0 \quad \text{vs.} \quad H_a\colon \mu_1 - \mu_2 \neq 0$

Calculate

$$s^2 = \frac{(n_1-1)s_1^2+(n_2-1)s_2^2}{n_1+n_2-2} = \frac{14(57)^2+14(52)^2}{28} = 2976.5$$

and the test statistic is

$$t = \frac{\bar{y}_1-\bar{y}_2}{\sqrt{s^2\left(\frac{1}{n_1}+\frac{1}{n_2}\right)}} = \frac{548-517}{\sqrt{2976.5\left(\frac{1}{15}+\frac{1}{15}\right)}} = 1.56$$

The rejection region is two-tailed, based on $n_1+n_2-2 = 28$ degrees of freedom. From Table 5, the approximate p-value is $2(.1) > p\text{-value} > 2(.05)$ or $.2 > p\text{-value} > .1$. There is no evidence of a significant difference in the mean math scores for the two groups.

d. Yes.

10.66a. The force transmitted to a wearer, y, is known to be normally distributed with $\mu = 800$ and $\sigma = 40$. Hence,

$$P[y > 1000] = P\left[z > \frac{1000-800}{40}\right] = P[z > 5] = 0.$$

Hence, it is highly improbable that any particular helmet will transmit a force in excess of 1000 pounds.

b. Since $n = 40$, a large sample z test will be used to test

$$H_0\colon\ \mu = 800 \qquad \text{vs.} \qquad H_a\colon\ \mu > 800.$$

The test statistic is

$$z = \frac{\bar{y}-\mu}{\frac{s}{\sqrt{n}}} = \frac{825-800}{\sqrt{\frac{2350}{40}}} = 3.262$$

and the rejection region with $\alpha = .05$ is $z > 1.645$. H_0 is rejected and we conclude that $\mu > 800$.

c. The hypothesis to be tested is

$$H_0\colon\ \sigma = 40 \qquad \text{vs.} \qquad H_a\colon\ \sigma > 40,$$

and the test statistic is

$$\chi^2 = \frac{(n-1)s^2}{\sigma_0^2} = \frac{39(2350)}{(40)^2} = 57.281.$$

The critical value of χ^2 with $n-1 = 39$ degrees of freedom (approximated with 40 degrees of freedom) is $\chi^2_{.05} = 55.7585$ and the null hypothesis is rejected at the $\alpha = .05$ level of significance. We conclude that σ exceeds 40.

10.67 The hypothesis to be tested is

$H_0\colon\ \sigma^2 = .01$ vs

$H_a\colon\ \sigma^2 > .01.$

The test statistic is

$$\chi^2 = \frac{(n-1)s^2}{\sigma_0^2} = \frac{7(.018)}{.01} = 12.6$$

A one-tailed test is required. Hence, a critical value of χ^2 (denoted by χ^2_c) must be found such that $P(\chi^2 > \chi^2_c) = .05$. Indexing $\chi^2_{.05}$ with $(n-1) = 7$ degrees of freedom (see Table 6), the critical value is found to be $\chi^2_{.05} = 14.07$ (see Figure 10.8).

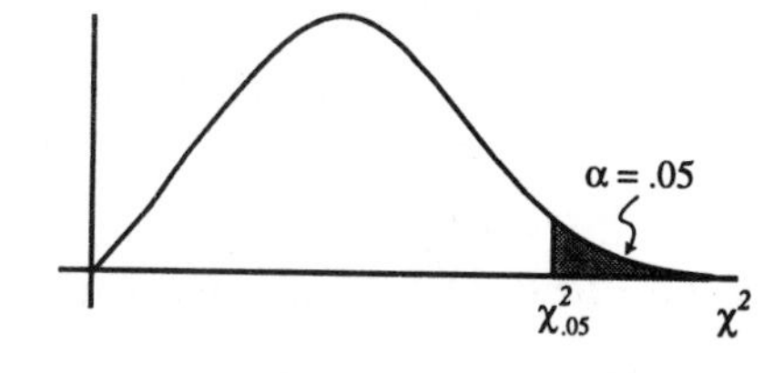

Figure 10.8

The value of the test statistic is not in the rejection region. Consequently, we cannot reject the hypothesis that $\sigma^2 = .01$.

Note in Table 6 that $12.02 < 12.6 \le 14.07$. The p-values associated with 12.02 and 14.07 are .10 and .05. Thus, $.05 < p\text{-value} < .10$.

We must assume that the data (carton weights) are from a normal population.

10.68 In order to employ the F statistic to test a hypothesis concerning the equivalence of two population variances, we must assume that independent random samples have been drawn from two normal populations.

10.69a. The rejection region given is

$$\left\{\frac{S_1^2}{S_2^2} > F^{\nu_1}_{\nu_2,\alpha/2} \text{ or } \frac{S_1^2}{S_2^2} < \left(F^{\nu_2}_{\nu_1,\alpha/2}\right)^{-1}\right\}$$

which, by reversing the second inequality and inverting fractions, becomes

$$\left\{\frac{S_1^2}{S_2^2} > F^{\nu_1}_{\nu_2,\alpha/2} \text{ or } \frac{S_2^2}{S_1^2} > F^{\nu_2}_{\nu_1,\alpha/2}\right\}$$

b.

$$P\left(\frac{S_L^2}{S_S^2} > F^{\nu_L}_{\nu_S,\alpha/2}\right) = P\left\{\frac{S_1^2}{S_2^2} > F^{\nu_1}_{\nu_2,\alpha/2} \text{ or } \frac{S_2^2}{S_1^2} > F^{\nu_2}_{\nu_1,\alpha/2}\right\}$$

$$= P\left\{\frac{S_1^2}{S_2^2} > F^{\nu_1}_{\nu_2,\alpha/2} \text{ or } \frac{S_1^2}{S_2^2} < \left(F^{\nu_1}_{\nu_2,\alpha/2}\right)^{-1}\right\} = \alpha$$

10.70a. Let σ_1^2 = variance of pressure for resting cyclists and σ_2^2 = variance of pressure for resting runners. Then, we are testing

$$H_0\colon\ \sigma_1^2 = \sigma_2^2 \qquad \text{vs.} \qquad H_a\colon\ \sigma_1^2 \ne \sigma_2^2.$$

The test statistic is

$$F = \frac{s_1^2}{s_2^2} = \frac{(3.98)^2}{(3.92)^2} = 1.03$$

The associated p-value is, from Table 7, p-value $> 2(.10) = .20$. Since p-value $> .05$, we do not reject H_0. There is insufficient evidence to claim a difference in the variability of compartment pressure between runners and cyclists who are resting.

b. Let σ_1^2 = variance in pressure for runners at maximal O_2 consumption and let σ_2^2 = variance in pressure for cyclists at maximal O_2 consumption. Then, we are testing

$$H_0\colon \sigma_1^2 = \sigma_2^2 \qquad \text{vs.} \qquad H_a\colon \sigma_1^2 \neq \sigma_2^2.$$

The test statistic is

$$F = \frac{s_1^2}{s_2^2} = \frac{(16.9)^2}{(4.67)^2} = 13.096$$

From Table 7, the p-value $< 2(.005) = .01$. At the $\alpha = .05$ level, there is sufficient evidence to claim a difference in the variability in compartment pressure between runners and cyclists at maximal O_2 consumption.

10.71 It is possible to test the null hypothesis $H_0\colon \sigma_1^2 = \sigma_2^2$ against any one of the three alternative hypotheses:

$$(1)\quad H_a\colon \sigma_1^2 \neq \sigma_2^2 \qquad (2)\quad H_a\colon \sigma_1^2 < \sigma_2^2 \qquad (3)\quad H_a\colon \sigma_1^2 > \sigma_2^2$$

The first alternative would be preferred by the manager of the dairy. He does not know anything about the variability of the two machines and would wish to detect departures from equality of the type $\sigma_1^2 > \sigma_2^2$ or $\sigma_2^2 > \sigma_1^2$. These alternatives are implied in (1). The salesman for company A would prefer that the experimenter select the second alternative. Rejection of the null hypothesis would imply that his machine had smaller variability. Moreover, rejection of the null hypothesis in favor of (2) is more likely than rejection of H_0 in favor of (1) if, in fact, $\sigma_1^2 < \sigma_2^2$. The salesman for company B would prefer the third alternative for a similar reason.

10.72 We wish to test the hypothesis

$$H_0\colon \sigma_1^2 = \sigma_2^2 \qquad \text{vs.} \qquad H_a\colon \sigma_1^2 \neq \sigma_2^2$$

there will be another portion of the rejection region in the lower tail of the distribution. The area to the right of the critical value will represent only $\frac{\alpha}{2}$, and the probability of a Type I error is $2\left(\frac{\alpha}{2}\right) = \alpha$. In this exercise a two-tailed test is employed. The test statistic is, using part **b** of Exercise 10.69,

$$F = \frac{s_1^2}{s_2^2} = \frac{2.96}{1.54} = 1.922$$

For $\alpha = .02$ the critical F, based on $(n_1 - 1) = 15$ and $(n_2 - 1) = 15$ degrees of freedom, is $F_{.01} = 3.52$. The computed value of the test statistic is less than the critical value. Hence the null hypothesis is not rejected. Note that the p-value associated with $F = 1.97$ is $2(.10) = .2$. Since $1.922 < 1.97$, the p-value is greater than .2.

10.73 The hypothesis to be tested is

$$H_0\colon \sigma = .7 \qquad \text{vs.} \qquad H_a\colon \sigma > .7.$$

This is equivalent to testing the hypothesis

$$H_0\colon \sigma^2 = .49 \qquad \text{vs.} \qquad H_a\colon \sigma^2 > .49.$$

Once the value for S^2 is calculated, the test statistic $\chi^2 = \frac{(n-1)S^2}{\sigma_0^2}$ will be used to test the above hypothesis. Then

$$s^2 = \frac{\sum_i y_i^2 - \frac{\left(\sum_i y_i\right)^2}{n}}{n-1} = \frac{497{,}036 - \frac{(1410)^2}{4}}{3} = 3.667$$

and the test statistic is

$$\chi^2 = \frac{(n-1)s^2}{\sigma_0^2} = \frac{3(3.6667)}{.49} = \frac{11}{.49} = 22.45$$

Since $22.45 > 12.8381 = \chi^2_{.005,3}$, the p-value is less than .005.

10.74 The hypothesis of interest is

$$H_0\colon \sigma = 10 \qquad \text{vs.} \qquad H_a\colon \sigma > 10,$$

and the test statistic is

$$\chi^2 = \frac{(n-1)s^2}{\sigma_0^2} = \frac{19(144)}{100} = 27.36$$

The rejection region is $\chi^2 > \chi^2_{.01,19} = 36.1908$ and H_0 is not rejected. The new test is not significantly more variable than the standard.

10.75 Refer to Exercise 10.58. The hypothesis of interest is

$$H_0\colon \sigma_1^2 = \sigma_2^2 \qquad \text{vs.} \qquad H_a\colon \sigma_1^2 > \sigma_2^2$$

and the test statistic is

$$F = \frac{S_1^2}{S_2^2} = \frac{(.017)^2}{(.006)^2} = 8.03$$

The rejection region with $\alpha = .05$ is $F > F^9_{12,.05} = 2.80$ and H_0 is rejected. We conclude that $\sigma_1^2 > \sigma_2^2$.

10.76 Refer to Exercise 10.2. Since the rejection region is $Y \le 12$, the power is $1 - \beta = P(Y \le 12|p)$. To calculate power, use Table 1.

a. power $= P(Y \le 12|p = .4) = .979$
b. power $= P(Y \le 12|p = .5) = .868$
c. power $= P(Y \le 12|p = .6) = .584$
d. power $= P(Y \le 12|p = .7) = .228$

The graph showing p on the horizontal axis and power $= 1 - \beta$ on the vertical axis is omitted here.

10.77 **a.** $\theta = 0.10$, $Y \sim \text{Uniform}(0.10, 1.1)$ so that for Test 1

$$P(Y > 0.95) = \int_{0.95}^{1.1} dy = 1.1 - 0.95 = 0.15.$$

b. $\theta = 0.40 \Rightarrow P(Y > 0.95) = 0.45$
c. $\theta = 0.70 \Rightarrow P(Y > 0.95) = 0.75$
d. $\theta = 1 \Rightarrow P(Y > 0.95) = 1$

10.78 **a.** $P_\theta(Y_1 + Y_2 > 1.684) = P_\theta(Y_1 + Y_2 - 2\theta > 1.684 - 2\theta)$
$= P_\theta(U > 1.684 - 2\theta) = 1 - F_U(1.684 - 2\theta)$

Now U has cdf

$$F_U(u) = \begin{cases} 0 & u < 0 \\ \frac{1}{2}u^2 & 0 \le u \le 1 \\ 2u - \frac{1}{2}u^2 - 1 & 1 < u \le 2 \\ 1 & u > 2 \end{cases}$$

so that

$$\theta = 0.1 \Rightarrow P(U > 1.484) = 0.133$$
$$\theta = 0.4 \Rightarrow P(U > 0.884) = 0.609$$
$$\theta = 0.7 \Rightarrow P(U > 0.284) = 0.960$$
$$\theta = 1 \Rightarrow P(U > -0.316) = 1.$$

10.79 **a.** Refer to Example 10.23 in the text. The uniformly most powerful test is found to be the z test of Section 10.3. That is, reject $H_0\colon \mu = 7$ if

$$Z = \frac{\overline{Y}-7}{\sqrt{\frac{\sigma^2}{20}}} = \frac{\overline{Y}-7}{\sqrt{\frac{5}{20}}} \ge 1.645 \qquad \text{or} \qquad \overline{Y} \ge 1.645\sqrt{.25} + 7 = 7.82$$

b. The power of the test is $1 - \beta = P\left(\overline{Y} > 7.82|\mu\right)$.

For $\mu = 7.5$, $\quad 1 - \beta = P\left(Z > \frac{7.82-7.5}{.5}\right) = P(Z > .64) = .2611$
For $\mu = 8.0$, $\quad 1 - \beta = P\left(Z > \frac{7.82-8}{.5}\right) = P(Z > -.36) = .6406$
For $\mu = 8.5$, $\quad 1 - \beta = P\left(Z > \frac{7.82-8.5}{.5}\right) = P(Z > -1.36) = .9131$
For $\mu = 9.0$, $\quad 1 - \beta = P\left(Z > \frac{7.82-9}{.5}\right) = P(Z > -2.36) = .9909$

c. The graph is omitted here.

10.80 We need to consider $P\left(\overline{Y} > 7.82\right) = P\left[z > \frac{(7.82-8)\sqrt{n}}{\sqrt{5}}\right] = .80.$

$$z = \frac{(7.82-8)\sqrt{n}}{\sqrt{5}} = -.84$$

Implying

$$-.18\left(\sqrt{n}\right) = -1.8783$$

or

$$\sqrt{n} = 10.435$$

or $n = 108.89 \cong 109$.

10.81 Using the sample size formula derived at the end of Section 10.4 in the text, we have

$$n = \frac{(z_\alpha + z_\beta)^2 \sigma^2}{(\mu_a - \mu_0)^2} = \frac{(1.96+1.96)^2 25}{(10-5)^2} = 15.3664$$

or $n = 16$ is the desired sample size.

10.82 We begin by looking at the most powerful test for H_0: $\sigma^2 = \sigma_0^2$ vs. H_a: $\sigma^2 = \sigma_1^2$ for $\sigma_1^2 > \sigma_0^2$. The null hypothesis specifies that $\sigma^2 = \sigma_0^2$, so that

$$L\left(\sigma_0^2\right) = \prod_{i=1}^{n} \frac{1}{\sqrt{2\pi}\,\sigma_0} e^{-(y_i-\mu)^2/2\sigma_0^2} = \frac{1}{\left(\sqrt{2\pi}\right)^n \sigma_0^n} \exp\left[\frac{-\sum(y_i-\mu)^2}{2\sigma_0^2}\right].$$

Similarly,

$$L\left(\sigma_a^2\right) = \frac{1}{\left(\sqrt{2\pi}\right)^n \sigma_1^n} \exp\left[\frac{-\sum(y_i-\mu)^2}{2\sigma_1^2}\right].$$

The most powerful test is

$$\frac{L(\sigma^2)}{L(\sigma_a^2)} = \left(\frac{\sigma_1}{\sigma_0}\right)^n \exp\left[-\frac{\sigma_1^2-\sigma_0^2}{2\sigma_1^2\sigma_0^2}\sum(y_i-\mu)^2\right] \le k.$$

Taking natural logarithms, we have

$$n \ln\left(\frac{\sigma_1}{\sigma_0}\right) - \left(\frac{\sigma_1^2-\sigma_0^2}{2\sigma_1^2\sigma_0^2}\right)\sum(y_i-\mu)^2 \le \ln k$$

or

$$\sum(y_i-\mu)^2 \ge \left[n \ln\left(\frac{\sigma_1}{\sigma_0}\right) - \ln k\right]\left(\frac{2\sigma_1^2\sigma_0^2}{\sigma_1^2-\sigma_0^2}\right) = c.$$

To find the rejection region for a fixed α, write the region as

$$\frac{\sum(y_i-\mu)^2}{\sigma_0^2} \ge \frac{c}{\sigma_0^2} = c'$$

and note that $\frac{\sum(y_i-\mu)^2}{\sigma_0^2}$ has a χ^2 distribution with n degrees of freedom. Since the same rejection region would be used for any $\sigma_1 > \sigma_0$, the test is uniformly most powerful.

10.83 **a.** Under H_0 the likelihood function is

$$L(\theta_0) = \frac{1}{(2\theta_0^3)^4}\left(\prod_{i=1}^{4} y_i^2\right) e^{-\sum y_i/\theta_0}.$$

Under H_a it is

$$L(\theta_a) = \frac{1}{(2\theta_a^3)^4}\left(\prod_{i=1}^{4} y_i^2\right) e^{-\sum y_i/\theta_a}.$$

Using Theorem 10.1, we obtain the most powerful critical region as

$$\frac{L(\theta_0)}{L(\theta_a)} = \frac{\theta_a^{12}}{\theta_0^{12}} \exp\left[-\sum y_i\left(\frac{1}{\theta_0} - \frac{1}{\theta_a}\right)\right] \le k$$

or

$$\exp\left[\sum y_i\left(\frac{1}{\theta_0} - \frac{1}{\theta_a}\right)\right] \le k\left(\frac{\theta_0}{\theta_a}\right)^{12}$$

or

$$-\sum y_i\left(\frac{1}{\theta_0} - \frac{1}{\theta_a}\right) \le \ln k\left(\frac{\theta_0}{\theta_a}\right)^{12}$$

or

$$-\sum y_i \le \frac{\ln k\left(\frac{\theta_0}{\theta_a}\right)^{12}}{\left(\frac{1}{\theta_0}\right)-\left(\frac{1}{\theta_a}\right)}$$

or

$$\sum y_i \ge -k'.$$

If H_0 is true, Y_i has a gamma distribution with $\alpha = 3$ and $\beta = \theta_0$, and $\frac{2Y_i}{\theta_0}$ has a χ^2 distribution with 6 degrees of freedom. Hence $2\left(\sum Y_i\right)\theta_0$ has a χ^2 distribution with 24 degrees of freedom. (Recall the method of moment-generating functions.) The critical region can be written as

$$\frac{2\sum Y_i}{\theta_0} \geq \frac{-2k'}{\theta_0} = k''$$

where k'' is chosen so that the test will have size α.

b. The choice of critical region did not depend on the particular value of θ_a but only upon the fact that $\theta_a > \theta_0$. Hence, for any $\theta > \theta_0$, the above critical region is most powerful and the test given in part **a** is uniformly most powerful for the alternative $\theta > \theta_0$.

10.84 a. The power function is $\int_{.5}^{1} \theta y^{\theta-1} dy = 1 - .5^{\theta}$, this is plotted vs $\theta > 0$.

b. We begin by looking for the most powerful α-level test of $H_0 : \theta = 1$ vs $H_a^* : \theta = \theta_a$ where θ_a is a fixed value that is greater than 1. In this case,

$$\frac{L(1)}{L(\theta_a)} = \frac{1}{\theta_a y^{\theta_a - 1}} \quad \text{for } 0 < y < 1.$$

So that the form of the rejection region is

$$\frac{1}{\theta_a y^{\theta_a - 1}} < k$$

$$\text{or } y > \left(\frac{1}{\theta_a k}\right)^{\frac{1}{\theta_a - 1}}$$

Now set the right hand side equal to k', that is, it is just a constant. Therefore, the most powerful test has the rejection region given by $RR = \{y > k'\}$. Now

$$\alpha = P(Y > k' \text{ when } \theta = 1) = \int_{k'}^{1} dy = 1 - k'$$

$$\Rightarrow k' = 1 - \alpha$$

Thus the rejection region is $RR = \{y > 1 - \alpha\}$ so that neither the test statistic nor the rejection region depends on the value for θ_a. Thus we have found the UMP test for $H_0 : \theta = 1$ vs $H_a : \theta > 1$.

10.85 a. Since the Y_i are iid we have

$$L(\theta) = \prod_{1}^{n} p(y_i|\theta) = \theta^{2N_1}\{2\theta(1-\theta)\}^{N_2}(1-\theta)^{2N_3}$$

b. Applying part **a.** with $\theta_0 < \theta_a$ obtain

$$\begin{aligned}
\frac{L(\theta_0)}{L(\theta_a)} &= \frac{2^{N_2}\theta_0^{2N_1+N_2}(1-\theta_0)^{N_2+2N_3}}{2^{N_2}\theta_a^{2N_1+N_2}(1-\theta_a)^{N_2+2N_3}} \\
&= \left(\frac{\theta_0}{\theta_a}\right)^{2N_1+N_2}\left(\frac{1-\theta_0}{1-\theta_a}\right)^{N_2+2N_3} \\
&= \left(\frac{\theta_0}{\theta_a}\right)^{2N_1+N_2}\left(\frac{1-\theta_0}{1-\theta_a}\right)^{2n-(2N_1+2N_2)} < k
\end{aligned}$$

where $n = N_1 + N_2 + N_3$ is the overall sample size. Thus the rejection region is a function $2N_1 + N_2$, which is the desired result.

c. Our rejection region is

$$RR = \left\{S = 2N_1 + N_2 \middle| \left(\tfrac{\theta_0}{\theta_a}\right)^{S}\left(\tfrac{1-\theta_0}{1-\theta_a}\right)^{2n-S} < k\right\}$$

where k is chosen so that $P(2N_1 + N_2 \in RR|\theta_0) = \alpha$. Notice, however

$$\log\left(\left(\tfrac{\theta_0}{\theta_a}\right)^{S}\left(\tfrac{1-\theta_0}{1-\theta_a}\right)^{2n-S}\right) = S\left[\log\left(\tfrac{\theta_0(1-\theta_a)}{\theta_a(1-\theta_0)}\right)\right] + 2n\log\left(\tfrac{1-\theta_0}{1-\theta_a}\right);$$

which is either strictly increasing or decreasing in S depending on whether the (odds ratio) $\frac{\theta_0(1-\theta_a)}{\theta_a(1-\theta_0)}$ is bigger or smaller than 1. Suppose it is larger than 1, that is $\theta_0 < \theta_a$, then an equivalent rejection would specify

$$RR = \{S = 2N_1 + N_2 | S < k'\}$$

Now finding an α level test is equivalent to finding the value k' so that $P(S < k'|\theta_0) = \alpha$. This can be accomplished by noting that (N_1, N_2, N_3) is a multinomial vector with probabilities $(\theta^2, \theta(1-\theta), (1-\theta)^2)$. Therefore,

for example, the probability that $S = 2$ is the same as the probability that
$N_1 = 0$ $N_2 = 2$ and $N_3 = n - 2$ or $N_1 = 2$ $N_2 = 0$ and $N_3 = n - 1$
which may be calculate from the multinomial distribution.

d. Yes. The test statistic and rejection region do not depend on the actual value of θ_a. Thus we have the UMP test for $H_0 : \theta = \theta_0$ vs $H_a : \theta > \theta_0$.

10.86a. Similar to Exercise 10.83. Let θ_a be a value of $\theta > \theta_0$. Then, using Theorem 10.1, we find the most powerful critical region to be

$$\frac{L(\theta_0)}{L(\theta_a)} = \prod_{i=1}^{n} \frac{\left(\frac{1}{\theta_0}\right) m y_i^{m-1} e^{-y_i^m/\theta_0}}{\left(\frac{1}{\theta_a}\right) m y_i^{m-1} e^{-y_i^m/\theta_a}} = \left(\frac{\theta_a}{\theta_0}\right)^n \exp\left[-\sum y_i^m \left(\frac{1}{\theta_0} - \frac{1}{\theta_a}\right)\right] \le k$$

which is equivalent to

$$-\sum y_i^m \left(\frac{1}{\theta_0} - \frac{1}{\theta_a}\right) \le \ln\left(\frac{\theta_0}{\theta_a}\right)^n k$$

or

$$\sum y_i^m \ge \frac{-\ln\left(\frac{\theta_0}{\theta_a}\right)^{n} k}{\left(\frac{1}{\theta_0}\right) - \left(\frac{1}{\theta_a}\right)} = k'.$$

Note that the inequalities behave as they do because $\theta_a > \theta_0$, which implies that $\left(\frac{1}{\theta_0}\right) - \left(\frac{1}{\theta_a}\right) > 0$. Consider the distribution of $Z = Y^m$. Since $\frac{dz}{dy} = my^{m-1}$, if H_0 is true, then $g(z) = \left(\frac{1}{\theta_0}\right) e^{-z/\theta_0}$ for $z > 0$. That is, Y^m has a gamma distribution with $\alpha = 1$ and $\beta = \theta_0$. Then $\frac{2Y^m}{\theta_0}$ has a χ^2 distribution with 2 degrees of freedom and $\frac{2(\sum Y_i^m)}{\theta_0}$ has a χ^2 distribution with $2n$ degrees of freedom. The critical region can be restated as

$$\frac{2\sum Y_i^m}{\theta_0} \ge \frac{2k'}{\theta_0} = k''$$

where k'' is chosen so that the test has size α.
Since the critical region does not depend upon the particular value of θ_a but only upon the fact that $\theta_a > \theta_0$, the same region will hold for any $\theta > \theta_0$, and hence this is the uniformly most powerful test.

b. If H_0 is true, $\frac{2(\sum Y_i^m)}{100}$ has a χ^2 distribution with $2n$ degrees of freedom. Then

$$\alpha = P\left(\frac{\sum Y_i^m}{50} \ge \chi^2_{.95,2n}\right) = .05.$$

If H_a is true, then $\frac{2(\sum Y_i^m)}{400}$ has a χ^2 distribution with $2n$ degrees of freedom.
Hence

$$\beta = P\left(\frac{2\sum Y_i^m}{100} \le \chi^2_{.05,2n}\right) = P\left(\frac{2\sum Y_i^m}{400} \le \frac{100}{400}\chi^2_{.05,2n}\right) = .05.$$

A value of n is sought such that

$$P\left(\chi^2 \le \tfrac{1}{4}\chi^2_{.05,2n}\right) = .05 \qquad \text{or} \qquad \tfrac{1}{4}\chi^2_{.05,2n} = \chi^2_{.95,2n}$$

Table 6, Appendix III, will show that the degrees of freedom necessary are $2n = 12$, or $n = 6$.

10.87a. $L(\lambda) = \prod_{i=1}^{n} f\left(\frac{y_i}{\lambda}\right) = \frac{\lambda^{\sum y_i} e^{-n\lambda}}{\prod_{i=1}^{n} y_i!}$.

Then, by the Neymann–Pearson Lemma, the test that maximizes the power at θ_a has a rejection region determined by $\frac{L(\lambda_0)}{L(\lambda_1)} < k$ or

$$\frac{\left(\frac{\lambda_0^{\sum y_i} e^{-n\lambda_0}}{\prod_{i=1}^{n} y_i!}\right)}{\left(\frac{\lambda_a^{\sum y_i} e^{-n\lambda_a}}{\prod_{i=1}^{n} y_i!}\right)} < k$$

or

$$\left(\frac{\lambda_0}{\lambda_a}\right)^{\sum y_i} e^{n(\lambda_a - \lambda_0)} < k$$

or

$$\sum_{i=1}^{n} y_i \ln\left(\frac{\lambda_0}{\lambda_a}\right) + n(\lambda_a - \lambda_0) < \ln k$$

or

$$\sum_{i=1}^{n} y_i \ln\left(\frac{\lambda_0}{\lambda_a}\right) < \ln k - n(\lambda_a - \lambda_0)$$

or, since $\lambda_0 < \lambda_a$

$$\sum_{i=1}^{n} y_i > \frac{\ln k - n(\lambda_a - \lambda_0)}{\ln\left(\frac{\lambda_0}{\lambda_a}\right)}$$

or with $k' = \frac{\ln k - n(\lambda_a - \lambda_0)}{\ln\left(\frac{\lambda_0}{\lambda_a}\right)}$

we have $\sum_{i=1}^{n} Y_i > k'$.

b. $\sum_{i=1}^{n} Y_i \sim$ Poisson with mean $n\lambda$. Then, for a given α, the constant k' is the value such that $P\left(\sum_{i=1}^{n} Y_i > k' \text{ when } \lambda = \lambda_0\right) = \alpha$.

c. The form of the rejection region does not depend upon the particular value assigned to λ_a. Therefore, the test derived in part **a** is the uniformly most powerful for the composite hypothesis.

d. The form is similar to that in part **a.** We start with

$$\frac{L(\lambda_0)}{L(\lambda_a)} < k$$

or, since $\lambda_a < \lambda_0$,

$$\sum y_i \ln\left(\frac{\lambda_0}{\lambda_a}\right) < \ln k - n(\lambda_0 - \lambda_a)$$

$$\sum Y_i < \frac{\ln k - n(\lambda_0 - \lambda_a)}{\ln\left(\frac{\lambda_0}{\lambda_a}\right)}$$

or $\sum_{i=1}^{n} Y_i < k'$ with $k' = \frac{\ln k - n(\lambda_0 - \lambda_a)}{\ln\left(\frac{\lambda_0}{\lambda_a}\right)}$.

10.88 $L = \frac{\lambda_2^{\Sigma x_i} \lambda_1^{\Sigma y_j} e^{-n\lambda_1 - m\lambda_2}}{\prod x_i! \prod y_j!}$ where $i = 1, \ldots,$ m and $j = 1, \ldots, n$.

$$L_0 = \frac{\left(2^{\Sigma x_j + \Sigma y_j}\right)\left(e^{-(n+m)2}\right)}{\prod x_i! \prod y_j!}$$

$$L_1 = \frac{\left[\left(\frac{1}{2}\right)^{\Sigma x_i}\right]\left[(3)^{\Sigma y_j}\right]\left(e^{-3m-n/2}\right)}{\prod x_i! \prod y_j!}$$

$$\frac{L_0}{L_1} = \frac{\left(2^{\Sigma x_j + \Sigma y_j}\right)\left(e^{-2n-2m}\right)}{\left[\left(\frac{1}{2}\right)^{\Sigma x_i}\right]\left[(3)^{\Sigma y_j}\right]\left(e^{-3m-n/2}\right)} < K$$

$$\left[4^{\Sigma x_j}\right]\left[\left(\frac{2}{3}\right)^{\Sigma y_j}\right]\left(e^{m-3n/2}\right) < K$$

$$\left[4^{\Sigma x_j}\right]\left[\left(\frac{2}{3}\right)^{\Sigma y_j}\right] < K'$$

Thus, we reject if

$$\left(\sum x_i\right) \ln 4 + \left(\sum y_j\right) \ln\left(\frac{2}{3}\right) < K''.$$

10.89 a. First,

$$L(\theta) = \frac{1}{\theta^n} e^{-\Sigma y_i/\theta}.$$

Then, the rejection region for the most powerful test is

$$\frac{L(\theta_0)}{L(\theta_a)} < k$$

or

$$\frac{\frac{1}{\theta_0^n} e^{-\Sigma y_i/\theta_0}}{\frac{1}{\theta_a^n} e^{-\Sigma y_i/\theta_a}} < k$$

or

$$n \ln\left(\frac{\theta_a}{\theta_0}\right) + \sum_{i=1}^{n} y_i \left(\frac{1}{\theta_a} - \frac{1}{\theta_0}\right) < \ln k$$

or

$$\sum_{i=1}^{n} \frac{y_i}{n} < \left[\frac{\ln k}{n} - \ln\left(\frac{\theta_a}{\theta_0}\right)\right]\left(\frac{\theta_a \theta_0}{\theta_0 - \theta_a}\right)$$

or, finally,

$$\sum y_i < k'$$

for

$$k' = \left[\frac{\ln k}{n} - \ln\left(\frac{\theta_a}{\theta_0}\right)\right]\left(\frac{\theta_a\theta_0}{\theta_0-\theta_a}\right).$$

b. The test derived in part **a** is the uniformly most powerful test for the composite hypothesis since the form of the rejection region does not depend on θ_a.

10.90a. i. $L(p) = \prod_{i=1}^{n} p\left(\frac{y_i}{p}\right) = p^{\Sigma y_i}(1-p)^{n-\Sigma y_i}$

Then,

$$\frac{L(p_0)}{L(p_a)} = \frac{p_0^{\Sigma y_i}(1-p_0)^{n-\Sigma y_i}}{p_a^{\Sigma y_i}(1-p_a)^{n-\Sigma y_i}} = \left(\frac{p_0}{p_a}\right)^{\Sigma y_i}\left(\frac{1-p_0}{1-p_a}\right)^{n}\left(\frac{1-p_0}{1-p_a}\right)^{-\Sigma y_i}$$

$$= \left(\frac{p_0(1-p_a)}{p_a(1-p_0)}\right)^{\Sigma y_i}\left(\frac{1-p_0}{1-p_a}\right)^{n}$$

ii. Then

$$\frac{L(p_0)}{L(p_a)} < k$$

if and only if

$$\left(\frac{p_0(1-p_a)}{p_a(1-p_0)}\right)^{\Sigma y_i}\left(\frac{1-p_0}{1-p_a}\right)^{n} < k$$

if and only if

$$\sum_{i=1}^{n} y_i \ln\left[\frac{p_0(1-p_a)}{p_a(1-p_0)}\right] + n\ln\left(\frac{1-p_0}{1-p_a}\right) < \ln k$$

if and only if

$$\sum_{i=1}^{n} y_i \ln\left[\frac{p_0(1-p_a)}{p_a(1-p_0)}\right] < \ln k - n\ln\left(\frac{1-p_0}{1-p_a}\right)$$

if and only if (since $p_0 < p_a$)

$$\sum_{i=1}^{n} y_i > \frac{\ln k - n\ln\left(\frac{1-p_0}{1-p_a}\right)}{\ln\left[\frac{p_0(1-p_a)}{p_a(1-p_0)}\right]}$$

i.e., if and only if

$$\sum_{i=1}^{n} y_i > k^*$$

for

$$k^* = \frac{\ln k - n\ln\left(\frac{1-p_0}{1-p_a}\right)}{\ln\left[\frac{p_0(1-p_a)}{p_a(1-p_0)}\right]}.$$

iii. Using the Neymann–Pearson *Lemma*, the rejection region for the most powerful test is

$$\sum_{i=1}^{n} Y_i > k^*$$

for some constant k^*.

b. For a given significance level, α, the constant k^* is the value such that

$$P\left\{\sum_{i=1}^{n} Y_i > k^*\right\} = \alpha$$

where $\sum_{i=1}^{n} Y_i$ is distributed binomial with parameters n and p. Because the binomial is a discrete distribution, it may not be possible, for a given α, to find a value for k^* so that $P\left\{\sum_{i=1}^{n} Y_i > k^*\right\}$ is exactly α.

c. The test derived in part **a** is uniformly most powerful for testing $H_0\colon\ p = p_0$ vs. $H_a\colon\ p > p_0$. This is because the rejection region does not depend on the parameter p_a.

10.91a. The density function for y, given θ, is

$$f(y) = \left(\frac{1}{\theta}\right) I(0 < y < \theta)$$

where $I(0 < y < \theta)$ equals 1 if $0 < y < \theta$ and 0 otherwise. The likelihood is

$$L(\theta) = \left(\frac{1}{\theta^n}\right)\prod_{i=1}^{n} I(0 < y_i < \theta) = \left(\frac{1}{\theta^n}\right) I(0 < y_{(n)} < \theta)$$

where $y_{(n)} = \max(y_i)$. Thus, the most powerful test rejects H_0 if

$$\frac{L(\theta_0)}{L(\theta_a)} = \left(\frac{\theta_a}{\theta_0}\right)^n \frac{I(0<y_{(n)}<\theta_0)}{I(0<y_{(n)}<\theta_a)} < k$$

The rejection rule depends on the data only through $y_{(n)}$. (If $\theta_a < y_{(n)} < \theta_0$, the left-hand side of the above inequality is undefined and we know H_0 is true.) We will reject H_0 if $y_{(n)}$ is small.

We will consider the rejection region given by $y_{(n)} < k$, choosing k so that $P(Y_{(n)} < k|\theta = \theta_0) = \alpha$. The density for $Y_{(n)}$ is given as $\frac{ny^{n-1}}{\theta_0^n}$ for $0 \le y \le \theta_0$. Hence

$$\alpha = P(Y_{(n)} < k) = \int_0^k \frac{ny^{n-1}}{\theta_0^n}\, dy = \frac{k^n}{\theta_0^n}$$

so that $k = \theta_0 \alpha^{1/n}$. The most powerful critical region is thus $y_{(n)} < \theta_0 \alpha^{1/n}$.

b. Notice that the choice of critical region depended only upon the fact that $\theta_a < \theta_0$, since this implied rejection of H_0 for small values of $Y_{(n)}$. Therefore, the test in part **a** is uniformly most powerful.

10.92a. As in Exercise 10.91, the test can be based upon $y_{(n)}$. In this case, large values of $Y_{(n)}$ would indicate a large value of θ and hence that $\theta = \theta_a$. Therefore, we reject H_0 if $Y_{(n)} > k$. To determine k, we find

$$\alpha = P(Y_{(n)} > k) = \int_k^{\theta_0} \frac{ny^{n-1}}{\theta_0^n}\, dy = \frac{\theta_0^n - k^n}{\theta_0^n} = 1 - \frac{k^n}{\theta_0^n}$$

Hence $k = \theta_0(1-\alpha)^{1/n} < \theta_0$, so that we would reject H_0 if $Y_{(n)} > \theta_0(1-\alpha)^{1/n}$.

b. As in Exercise 10.91, the test is uniformly most powerful.

c. No, it is not unique. For, suppose we were to pick a rejection region $C = (a, b) + (\theta_0, \infty)$, where (a, b) is any interval in $(0, \theta_0)$ such that $P[Y_{(n)} \in (\text{a, b})] = \alpha$. That is,

$$\alpha = \int_a^b \frac{ny^{n-1}}{\theta_0^n}\, dy = \frac{b^n - a^n}{\theta_0^n}$$

so that

$$b^n - a^n = \text{a}\theta_0^n.$$

The power of this test would be

$$1 - \beta = P[Y_{(n)} \in (\text{a, b})] + P(Y_{(n)} \ge \theta_0) = \int_a^b \frac{ny^{n-1}}{\theta_a^n}\, dy + \int_{\theta_0}^{\theta_a} \frac{ny^{n-1}}{\theta_a^n}\, dy$$

$$\frac{b^n - a^n}{\theta_a^n} + \frac{\theta_a^n - \theta_0^n}{\theta_a^n} = \alpha\left(\frac{\theta_0}{\theta_a}\right)^n + 1 - \left(\frac{\theta_0}{\theta_a}\right)^n = (\alpha - 1)\left(\frac{\theta_0}{\theta_a}\right)^n + 1$$

which is independent of the interval (a, b) chosen to make the test of size α. Note that the power of the test with arbitrary interval (a, b) is exactly the same as the power of the test in part **a**.

10.93 The null hypothesis specifies $\Omega_0 = \{\sigma^2\colon\ \sigma^2 = \sigma_0^2\}$, while $\Omega = \Omega_0 \cup \Omega_a = \{\sigma^2\colon\ \sigma^2 \ge \sigma_0^2\}$. In the restricted space Ω_0, the likelihood function is

$$L(\Omega_0) = \prod_{i=1}^n \frac{1}{(2\pi)^{1/2}\sigma_0} \exp\left[\frac{(y_i-\mu)^2}{2\sigma_0^2}\right]$$

The maximum likelihood estimate of μ is $\overline{Y}$, so that

$$\text{L}(\widehat{\Omega}_0) = \frac{1}{(2\pi)^{n/2}\sigma_0^n} \exp\left[\frac{\sum_{i=1}^n (y_i-\overline{y})^2}{2\sigma_0^2}\right].$$

Now consider the unrestricted space

$$\text{L}(\Omega) = \frac{1}{(2\pi)^{n/2}\sigma^n} \exp\left[\frac{\sum (y_i-\mu)^2}{2\sigma^2}\right].$$

The maximum likelihood estimate of μ is $\widehat{\mu} = \overline{Y}$, while

$$\widehat{\sigma}^2 = \max\left[\sigma_0^2, \widehat{\sigma}^2 = \frac{\sum (Y_i - \overline{Y})^2}{n}\right]$$

and

$$L(\widehat{\Omega}) = \frac{1}{(2\pi)^{n/2}\widehat{\sigma}^2} \exp\left[-\frac{\sum (y_i-\overline{y})^2}{2\widehat{\sigma}^2}\right].$$

The likelihood ratio statistic is

$$\lambda = \frac{L(\widehat{\Omega}_0)}{L(\widehat{\Omega})} = \left(\frac{\widehat{\sigma}^2}{\sigma_0^2}\right)^{n/2} \exp\left[-\frac{\sum(y_i-\overline{y})^2}{2\sigma_0^2} + \frac{\sum(y_i-\overline{y})^2}{2\widehat{\sigma}^2}\right] = 1 \quad \text{if} \quad \widehat{\sigma} \le \sigma_0$$

$$= \left[\frac{\sum(y_i-\overline{y})^2}{n\sigma_0^2}\right]^{n/2} \exp\left[-\frac{\sum(y_i-\overline{y})^2}{2\sigma_0^2}\right] e^{n/2} \quad \text{if} \quad \widehat{\sigma} > \sigma_0$$

Hence the rejection region $\lambda \le k$ is equivalent to

$$g(\chi^2) = (\chi^2)^{n/2} e^{-\chi^2/2} n^{-n/2} e^{n/2} \le k$$

where χ^2 is $\frac{(n-1)s^2}{\sigma_0^2}$, the χ^2 statistic given in Section 10.9.

Note that if $\widehat{\sigma} \le \sigma_0$, $g(\chi^2) = 1$. Further, if $\widehat{\sigma} > \sigma_0$, $g(\chi^2)$ is a monotonically decreasing function of χ^2. Hence the region $\lambda \le k$ is equivalent to $\chi^2 \ge c$, where c is determined so that the test has size α. A rough sketch of $g(\chi^2) = \lambda$ against χ^2 is shown in Figure 10.9.

Figure 10.9

10.94 The hypothesis of interest is H_0: $p_1 = p_2 = p_3 = p_4 = p$ against the alternative that at least one of these equalities is incorrect. In Ω, the likelihood function is

$$L(\Omega) = \prod_{i=1}^{4} \binom{200}{n_i} p_i^{n_i} (1-p_i)^{200-n_i}$$

and the maximum likelihood estimate of p_i is $\widehat{p}_i = \frac{n_i}{200}$.

In the restricted space Ω_0,

$$L(\Omega_0) = \prod_{i=1}^{4} \binom{200}{n_i} p_i^{n_i} (1-p_i)^{200-n_i} = K p^{\sum n_i} (1-p)^{800-\sum n_i}$$

and

$$\ln L = \ln K + \sum n_i \ln p + (800 - \sum n_i) \ln(1-p).$$

One may easily verify that the maximum likelihood estimate of p is $\widehat{p} = \frac{\sum n_i}{800}$.

Then

$$\lambda = \frac{L(\widehat{\Omega}_0)}{L(\widehat{\Omega})} = \frac{\left(\frac{\sum n_i}{800}\right)^{\sum n_i} \left(\frac{800-\sum n_i}{800}\right)^{800-\sum n_i}}{\prod_{i=1}^{4} \left(\frac{n_i}{200}\right)^{n_i} \left(\frac{200-n_i}{200}\right)^{200-n_i}}$$

Since the n_i are large, Theorem 10.2 is applicable, and

$$-2 \ln \lambda = -2\left\{(\sum n_i) \ln\left(\frac{\sum n_i}{800}\right) + (800 - \sum n_i) \ln\left(1 - \frac{\sum n_i}{800}\right) - \sum_{i=1}^{4} \left[n_i \ln\left(\frac{n_i}{200}\right) + (200-n_i) \ln\left(1 - \frac{n_i}{200}\right)\right]\right\}$$

has an approximate χ^2 distribution with 3 degrees of freedom. For this exercise, $n_1 = 76$, $n_2 = 53$, $n_3 = 59$, $n_4 = 48$, and $\sum n_i = 236$, so that $-2 \ln \lambda = -2(5.2676) = 10.54$. The rejection region, for $\alpha = .05$, will be $-2 \ln \lambda > \chi_{.05,3} = 7.81$, and the null hypothesis is rejected. The fraction of voters favoring candidate A is not the same in all four wards.

10.95 Let $X_1, X_2, \ldots, X_n$ be the random sample drawn from population 1, and let $Y_1, Y_2, \ldots, Y_m$ be the random sample from population 2. Assuming that H_0 is true,

$$\frac{\sum(X_i-\overline{X})^2 + \sum(Y_i-\overline{Y})^2}{\sigma_0^2} = \frac{(n-1)S_1^2 + (m-1)S_2^2}{\sigma_0^2} = \chi^2$$

has a χ^2 distribution with $m+n-2$ degrees of freedom. If H_a is true, then S_1^2 and S_2^2 will tend to be larger than σ_0^2 (since they will be estimates of $\sigma^2 > \sigma_0^2$).

Under H_0 the likelihood is

$$L(\widehat{\Omega}_0) = \frac{1}{(2\pi)^{n/2}} \frac{1}{\sigma_0^n} \exp\left\{-\tfrac{1}{2}\chi^2\right\}$$

In the space Ω, the likelihood is maximized either at $\widehat{\sigma} = \sigma_0$ or at $\widehat{\sigma} = \sigma_a$. If $\widehat{\sigma} = \sigma_0$, then $\frac{L(\widehat{\Omega}_0)}{L(\widehat{\Omega})} = 1$. Thus, for $k < 1$, $\frac{L(\widehat{\Omega}_0)}{L(\widehat{\Omega})} \le k$ only if $\widehat{\sigma} = \sigma_a$. In this case,

$$\frac{L(\widehat{\Omega}_0)}{L(\widehat{\Omega})} = \left(\frac{\sigma_a}{\sigma_0}\right)^n \exp\left\{-\tfrac{1}{2}\chi^2 + \tfrac{1}{2}\,\frac{(n-1)S_1^2 + (m-1)S_2^2}{\sigma_a^2}\right\}$$

which is a decreasing function of χ^2. Thus, we reject H_0 if χ^2 is too large. The rejection region is $\chi^2 > \chi_a^2$.

10.96a. In Ω we have three independent random samples from normal populations with different means and different variances. Thus,

$$L(\Omega) = \frac{1}{(2\pi)^{\Sigma n_i/2}\sigma_1^{n_1}\sigma_2^{n_2}\sigma_3^{n_3}} \exp\left[-\frac{1}{2}\sum\left(\frac{x_i-\mu_1}{\sigma_1}\right)\right]^2 \exp\left[-\frac{1}{2}\sum\left(\frac{y_i-\mu_2}{\sigma_2}\right)^2\right]$$
$$\times \exp\left[-\frac{1}{2}\sum\left(\frac{w_i-\mu_3}{\sigma_3}\right)^2\right]$$

The maximum likelihood estimates of μ_1, σ_1^2, μ_2, σ_2^2, μ_3, and σ_3^2 are simply

$$\widehat{\mu}_1 = \overline{X} \qquad \widehat{\sigma}_1^2 = \frac{\sum(X_i-\overline{X})^2}{n_1} \qquad \widehat{\mu}_2 = \overline{Y} \qquad \widehat{\sigma}_2^2 = \frac{\sum(Y_i-\overline{Y})^2}{n_2}$$
$$\widehat{\mu}_3 = \overline{W} \qquad \widehat{\sigma}_3^2 = \frac{\sum(W_i-\overline{W})^2}{n_3}$$

and

$$L(\Omega_0) = \frac{1}{(2\pi)^{\Sigma n_i/2}\sigma^{\Sigma n_i}} \exp\left\{-\frac{1}{2\sigma^2}\left\{\sum(x_i-\mu_1)^2 + \sum(y_i-\mu_2)^2\right.\right.$$
$$\left.\left. + \sum(w_i-\mu_3)^2\right]\right\}$$

The maximum likelihood estimates are

$$\widehat{\mu}_1 = \overline{X} \qquad \widehat{\mu}_2 = \overline{Y} \qquad \widehat{\mu}_3 = \overline{W}$$
$$\widehat{\sigma}^2 = \frac{\sum(X_i-\overline{X})^2 + \sum(Y_i-\overline{Y})^2 + \sum(W_i-\overline{W})^2}{n_1+n_2+n_3} = \frac{n_1\widehat{\sigma}_1^2 + n_2\widehat{\sigma}_2^2 + n_3\widehat{\sigma}_3^2}{n_1+n_2+n_3}$$

and

$$L(\widehat{\Omega}_0) = \frac{1}{(2\pi)^{\Sigma n_i/2}(\widehat{\sigma}^2)^{\Sigma n_i/2}} e^{-\Sigma n_i/2}$$

so that

$$\lambda = \frac{L(\widehat{\Omega}_0)}{L(\widehat{\Omega})} = \frac{(\widehat{\sigma}_1^2)^{n_1/2}(\widehat{\sigma}_2^2)^{n_2/2}(\widehat{\sigma}_3^2)^{n_3/2}}{(\widehat{\sigma}^2)^{(n_1+n_2+n_3)/2}}.$$

b. For large values of n_1, n_2, and n_3, the quantity $-2 \ln \lambda$ will have a χ^2 distribution with 2 degrees of freedom. Hence the rejection region, with $\alpha = .05$, is $-2 \ln \lambda > \chi^2_{.05,2} = 5.99$.

10.97a. In Ω the likelihood function is

$$L(\Omega) = \left(\frac{1}{\theta_1^m}\right) e^{-\Sigma x_i/\theta_1} \left(\frac{1}{\theta_2^n}\right) e^{-\Sigma y_i/\theta_2}$$

The maximum likelihood estimators of θ_1 and θ_2 are $\widehat{\theta}_1 = \overline{X}$ and $\widehat{\theta}_2 = \overline{Y}$, and

$$L(\widehat{\Omega}) = \frac{1}{\overline{X}^m\overline{Y}^n} e^{-(n+m)}$$

In the restricted space Ω_0,

$$L(\Omega_0) = \frac{1}{\theta^{m+n}} e^{-(\Sigma x_i + \Sigma y_i)/\theta},$$

and the maximum likelihood estimate of θ is

$$\widehat{\theta} = \frac{\sum X_i + \sum Y_i}{m+n} = \frac{m\overline{X}+n\overline{Y}}{m+n}$$

so that

$$L(\widehat{\Omega}_0) = \left(\frac{m\overline{X}+n\overline{Y}}{m+n}\right)^{-(m+n)} e^{-(m+n)}$$

Then

$$\lambda = \frac{L(\widehat{\Omega}_0)}{L(\widehat{\Omega})} = \frac{\overline{X}^m\overline{Y}^n}{\left(\frac{m\overline{X}+n\overline{Y}}{m+n}\right)^{m+n}}.$$

b. Notice that X_i has a gamma distribution with $\alpha = 1$ and $\beta = \theta_1$. Hence if H_0 is true, $\frac{2X_i}{\theta_1}$ has a gamma distribution with $\alpha = 1$, $\beta = 2$, which is equivalent to a χ^2 distribution with $\nu = 2$ degrees of freedom. Also, $\frac{2\left(\sum_{i=1}^{m} Y_i\right)}{\theta}$ is the sum of the m independent χ^2 variates and has a χ^2 distribution with $2m$ degrees of freedom. Similarly, $\frac{2\left(\sum_{i=1}^{n} Y_i\right)}{\theta}$ has a χ^2 distribution with $2n$ degrees of freedom. Since X and Y are independent, an F statistic can be formed.

$$\frac{\left(\frac{2\sum X_i}{2m\theta}\right)}{\left(\frac{2\sum Y_i}{2n\theta}\right)} = \frac{\overline{X}}{\overline{Y}}$$

has an F distribution with $2m$ and $2n$ degrees of freedom. Write

$$\lambda = \frac{\overline{X}^m\,\overline{Y}^n}{\left(\frac{m\overline{X}+n\overline{Y}}{m+n}\right)^{m+n}} = \frac{1}{\left[\frac{m\overline{X}+n\overline{Y}}{\overline{X}(m+n)}\right]^m\left[\frac{m\overline{X}+n\overline{Y}}{\overline{Y}(m+n)}\right]^n}$$
$$= \frac{1}{\left[\frac{m}{m+n}+\frac{n}{F(m+n)}\right]^m\left[\left(\frac{m}{m+n}\right)F+\frac{n}{m+n}\right]^n}$$

Note that λ is small if F is either too large or too small.
Then the rejection region, $\lambda \le k$, is equivalent to $F \ge c_1$ and $F \le c_2$, where c_1 and c_2 are chosen so that the test has size α.

10.98 By Theorem 9.4, U is a sufficient statistic if and only if we can factor the likelihood L as

$$L(y_1, y_2, \ldots, y_n|\theta) = g(u,\theta)\,h(y_1, y_2, \ldots, y_n).$$

Now

$$\lambda = \frac{\sup_{\theta\in\Omega_0} L(y_1, y_2, \ldots, y_n|\theta)}{\sup_{\theta\in\Omega} L(y_1, y_2, \ldots, y_n|\theta)2} = \frac{\sup_{\theta\in\Omega_0} g(u,\theta)\,h(y_1, y_2, \ldots, y_n)}{\sup_{\theta\in\Omega} g(u,\theta)\,h(y_1, y_2, \ldots, y_n)} = \frac{\sup_{\theta\in\Omega_0} g(u,\theta)}{\sup_{\theta\in\Omega} g(u,\theta)},$$

since $h(\,\cdot\,)$ does not depend on θ.
Note that this ratio depends on the data only through the sufficient statistic u.

10.99 **a.** Under the null hypothesis, with $\Omega_0 = \{\theta_0\}$, the likelihood is maximized at θ_0. Then, under the alternative hypothesis, with $\Omega = \{\theta_0, \theta_a\}$, the likelihood is maximized at either θ_0 or θ_a. Thus, $L(\widehat{\Omega}_0) = \mathrm{L}(\theta_0)$ and $L(\widehat{\Omega}) = \max\{L(\theta_0), L(\theta_a)\}$, so that

$$\lambda = \frac{L(\widehat{\Omega}_0)}{L(\widehat{\Omega})} = \frac{L(\theta_0)}{\max\{L(\theta_0), L(\theta_a)\}} = \frac{1}{\max\left\{1, \frac{L(\theta_a)}{L(\theta_0)}\right\}}$$

b. First, recognize that

$$\lambda = \frac{1}{\max\left\{1, \frac{L(\theta_a)}{L(\theta_0)}\right\}} = \min\left\{1, \frac{L(\theta_0)}{L(\theta_a}\right\}.$$

Now, as mentioned in Example 10.24, we restrict the attention to $k < 1$. Then

$$\lambda < k$$

if and only if

$$\min\left\{1, \frac{L(\theta_0)}{L(\theta_a}\right\} < k < 1$$

if and only if

$$\frac{L(\theta_0)}{L(\theta_a)} < k.$$

c. These results imply that in the case of both simple null and alternative hypotheses, the likelihood ratio test is equivalent to the most powerful test as given by the Neymann–Pearson Lemma.

10.100 In Ω, $\widehat{\mu}_1 = \overline{Y}_1$, $\widehat{\mu}_2 = \overline{Y}_2$, and

$$\widehat{\sigma}^2 = \frac{\sum(Y_{1j}-\overline{Y}_1)^2+\sum(Y_{2j}-\overline{Y}_2)^2}{n_1+n_2}$$

Then

$$L(\widehat{\Omega}) = \frac{1}{(2\pi\widehat{\sigma})^{(n_1+n_2)/2}}\,e^{-(n_1+n_2)/2}$$

In Ω_0, $\mu_1 = \mu_2$ and σ^2 is unknown. Hence

$$L(\Omega_0) = \frac{1}{(2\pi)^{(n_1+n_2)/2}\sigma^{n_1+n_2}}\exp\left\{-\frac{1}{2\sigma^2}\left[\sum(y_{1j}-\mu)^2+\sum(y_{2j}-\mu)^2\right]\right\}$$

and

$$\ln L = -\left(\frac{n_1+n_2}{2}\right)\ln 1\pi - \left(\frac{n_1+n_2}{2}\right)\ln\sigma^2 - \frac{1}{2\sigma^2}\left[\sum(y_{1j}-\mu)^2+\sum(y_{2j}-\mu)^2\right]$$

Taking derivatives with respect to μ and σ^2, we have

$$\frac{d\ln L}{d\sigma^2} = \frac{-(n_1+n_2)}{2\sigma^2} + \frac{\left[\sum(y_{1j}-\mu)^2+\sum(y_{2j}-\mu)^2\right]}{2\sigma^4} = 0$$
$$\frac{d\ln L}{d\mu} = \frac{\sum(y_{1j}-\mu)+\sum(y_{2j}-\mu)}{\sigma^2} = 0$$

Hence

$$\sum y_{1j} - n_1\mu + \sum y_{2j} - n_2\mu = 0$$

or

$$\widehat{\mu} = \frac{\sum y_{1j}+\sum y_{2j}}{n_1+n_2} = \frac{n_1\overline{y}_1+n_2\overline{y}_2}{n_1+n_2} \quad \text{and} \quad \widetilde{\sigma}^2 = \frac{\sum_i\sum_j(y_{ij}-\widehat{\mu})^2}{n_1+n_2}$$

Finally,

$$L(\widehat{\Omega}_0) = \frac{1}{(2\pi\widehat{\sigma}^2)^{(n_1+n_2)/2}} \qquad \text{and} \qquad \lambda = \left(\frac{\widehat{\sigma}^2}{\widehat{\sigma}^2}\right)^{(n_1+n_2)/2} \le k$$

is the likelihood ratio test.

In order to show that this reduces to the two-sample t test of Section 10.8, define

$$\text{SS}_{Y_1} = \sum (Y_{1j} - \overline{Y}_1)^2 \qquad \text{and} \qquad \text{SS}_{Y_2} = \sum (Y_{2j} - \overline{Y}_2)^2$$

so that

$$\widehat{\sigma}^2 = \frac{\text{SS}_{Y_1} + \text{SS}_{Y_2}}{n_1+n_2}$$

Consider

$$\begin{aligned}(n_1+n_2)\widehat{\sigma}^2 &= \sum\sum (Y_{ij} - \widehat{\mu})^2 = \sum (Y_{1j} - \overline{Y}_1)^2 + n_1(\overline{Y}_1 - \widehat{\mu})^2 + \sum (Y_{2j} - \overline{Y}_2)^2 \\ &\quad + n_2(\overline{Y}_2 - \widehat{\mu})^2 \\ &= \text{SS}_{Y_1} + \text{SS}_{Y_2} + n_1\left(\overline{Y}_1 - \frac{n_1\overline{Y}_1 + n_2\overline{Y}_2}{n_1+n_2}\right)^2 + n_2\left(\overline{Y}_2 - \frac{n_1\overline{Y}_1 + n_2\overline{Y}_2}{n_1+n_2}\right)^2 \\ &= \text{SS}_{Y_1} + \text{SS}_{Y_2} + \frac{n_1n_2(\overline{Y}_1 - \overline{Y}_2)^2}{n_1+n_2}\end{aligned}$$

Now

$$\begin{aligned}\lambda^{2/(n_1+n_2)} &= \frac{\widehat{\sigma}^2}{\widehat{\sigma}^2} = \frac{\text{SS}_{Y_1} + \text{SS}_{Y_2}}{\text{SS}_{Y_1} + \text{SS}_{Y_2} + \left[\frac{n_1n_2(\overline{Y}_1 - \overline{Y}_2)^2}{n_1+n_2}\right]} \\ &= \frac{1}{1 + \frac{(\overline{Y}_1 - \overline{Y}_2)^2}{\{[(\frac{1}{n_1}) + (\frac{1}{n_2})]\, S^2(n_1+n_2-2)\}}} = \frac{1}{1 + \frac{t^2}{n_1+n_2-2}}\end{aligned}$$

where $S^2 = \frac{\text{SS}_{Y_1} + \text{SS}_{Y_2}}{n_1+n_2-2}$. Since we are considering only $\mu_1 > \mu_2$, or equivalently $\overline{Y}_1 - \overline{Y}_2 > 0$, $t = \sqrt{t^2}$ will be positive. Hence, small values of λ imply large positive values of t, and a one-tailed t test is implied.

10.101

Refer to Exercise 10.100, where

$$\lambda^{2/(n_1+n_2)} = \frac{1}{1 + \frac{t^2}{n_1+n_2-2}}$$

In the restricted space Ω_0, $\mu_1 = \mu_2$ so that $t = \sqrt{t^2}$ could be either positive or negative. Hence, small values of λ imply large positive or negative values of t, and a two-tailed t test is implied.

10.102

Using arguments identical to those used in Exercise 10.100, we have

$$L(\widehat{\Omega}) = \frac{1}{(2\pi\widehat{\sigma}^2)^{\sum n_i/2}} e^{-\sum n_i/2} \qquad \text{where} \qquad \widehat{\sigma}^2 = \frac{\sum_{i=1}^{3}\sum_{j=1}^{n_i} (Y_{ij} - \overline{Y}_i)^2}{n_1+n_2+n_3}$$

$$L(\widehat{\Omega}_0) = \frac{1}{(2\pi\widehat{\sigma}^2)^{\sum n_i/2}} e^{-\sum n_i/2} \qquad \text{where} \qquad \widehat{\sigma}^2 = \frac{\sum\sum (Y_{ij} - \widehat{\mu})^2}{n_1+n_2+n_3}$$

and

$$\widehat{\mu} = \frac{n_1\overline{Y}_1 + n_2\overline{Y}_2 + n_3\overline{Y}_3}{n_1+n_2+n_3}.$$

Then $\lambda = \left(\frac{\widehat{\sigma}^2}{\widehat{\sigma}^2}\right)^{\sum n_i/2} \le k$ is the likelihood ratio test. In order to show that this test is equivalent to an exact F test, we refer to results and notation given in Section 13.3 of the text. In particular,

$$(n_1 + n_2 + n_3)\widehat{\sigma}^2 = \text{SSE}$$

and

$$(n_1 + n_2 + n_3)\widehat{\sigma}^2 = \text{TSS} = \text{SST} + \text{SSE}.$$

Then we have the following result:

$$\begin{aligned}\lambda^{2/(n_1+n_2+n_3)} &= \frac{\text{SSE}}{\text{TSS}} = \frac{\text{SSE}}{\text{SSE}+\text{SST}} = \frac{1}{1 + \frac{\text{SST}}{\text{SSE}}} = 1 + \frac{2\text{MST}}{(\sum n_i - 3)\,\text{MSE}} \\ &= \frac{1}{1 + \frac{2F}{(\sum n_i - 3)}}\end{aligned}$$

where $F = \frac{\text{MST}}{\text{MSE}}$ is given in Chapter 13. Thus, λ is small if and only if F is large, and the exact F test is equivalent to the likelihood ratio test.

10.103

a. Test

$$H_0\colon \mu_1 - \mu_2 = 0 \qquad \text{vs.} \qquad H_a\colon \mu_1 - \mu_2 \ne 0$$

where μ_1 = mean nitrogen density of chemical compounds and μ_2 = mean nitrogen density of atmosphere. Then

$$s_p^2 = \frac{(n_1-1)s_1^2 + (n_2-1)s_2^2}{n_1+n_2-2} = \frac{9(.001310)^2 + 8(.000574)^2}{17} = .000001064.$$

The test statistic is

$$t = \frac{\overline{y}_1 - \overline{y}_2}{\sqrt{s_p^2\left(\frac{1}{n_1} + \frac{1}{n_2}\right)}} = \frac{2.29971 - 2.310217}{\sqrt{.000001064\left(\frac{1}{10} + \frac{1}{9}\right)}} = -22.17.$$

The p-value is less than $2(.005) = .010$. Thus, there is sufficient evidence to indicate that a difference exists in the mean mass of nitrogen per flask for chemical compounds and air.

b. The 95% confidence interval is

$$\overline{y}_1 - \overline{y}_2 \pm t_{.025,17}\sqrt{s_p^2\left(\frac{1}{n_1} + \frac{1}{n_2}\right)}$$

or

$$2.29971 - 2.310217 \pm 2.110\sqrt{.000001064\left(\tfrac{1}{10} + \tfrac{1}{9}\right)}$$

or $(-.01151, -.00951)$.

c. Yes, there is sufficient evidence, since the interval does not contain 0.

d. No.

10.104

The hypothesis of interest is

$$H_0\colon\ \mu_1 - \mu_2 = 0 \qquad \text{vs.} \qquad H_a\colon\ \mu_1 - \mu_2 < 0$$

Calculate

$$s^2 = \frac{\sum y_{1i}^2 - \frac{(\sum y_{1i})^2}{n_1} + \sum y_{2i}^2 - \frac{(\sum y_{2i})^2}{n_2}}{n_1 + n_2 - 2} = \frac{.0624 - \frac{(.6)^2}{6} + .1175 - \frac{(.83)^2}{6}}{6 + 6 - 2}$$

$$= \frac{.0024 + .00268}{10} = .0005083$$

Also,

$$\overline{y}_1 = \frac{.6}{6} = .1 \qquad \text{and} \qquad \overline{y}_2 = \frac{.83}{6} = .1383$$

The test statistic is

$$t = \frac{(\overline{y}_1 - \overline{y}_2) - 0}{\sqrt{s^2\left(\frac{1}{n_1} + \frac{1}{n_2}\right)}} = \frac{-.0383}{\sqrt{.0005083\left(\frac{1}{6} + \frac{1}{6}\right)}} = -2.945$$

The one-tailed rejection region with $\alpha = .10$ and 10 d.f. is $t < -t_{.10,10} = -1.372$ and H_0 is rejected. There is sufficient evidence to indicate that $\mu_1 < \mu_2$.

10.105

a. Let $p = P(\text{customer prefers brand } A)$. Then the hypothesis of interest is

$$H_0\colon\ p = .2 \qquad \text{vs.} \qquad H_a\colon\ p > .2.$$

b. It is decided to reject H_0 if $y \geq 92$. Hence

$$\alpha = P(\text{reject } H_0 | H_0 \text{ true}) = P(Y \geq 92 | p = .2)$$

If $p = .2$, then $E(Y) = np = 400(.2) = 80$ and $\sigma = \sqrt{npq} = \sqrt{64} = 8$. Using the normal approximation to the binomial distribution, we have the approximation

$$\alpha = P(Y > 91.5) = P\left(Z > \tfrac{91.5 - 80}{8}\right) = P(Z > 1.44) = .0749$$

10.106

a. $H_0\colon\ \mu = 1100$ vs. $H_a\colon\ \mu < 1100$

b. Rejection region: With $\alpha = .05$, the null hypothesis will be rejected if $z < -1.645$.

c. Test statistic: $\frac{\overline{Y} - \mu}{\frac{\sigma}{\sqrt{n}}}$, which is estimated by

$$z = \frac{1060 - 1100}{\frac{340}{\sqrt{260}}} = \frac{-40}{21.0859} = -1.90$$

Since the observed value of Z falls in the rejection region, we conclude that $\mu < 1100$; that is, there has been a drop in average daily production.

10.107

The test is performed as follows:

(1) $H_0\colon\ \mu_1 - \mu_2 = 0$ vs. $H_a\colon\ \mu_1 - \mu_2 \doteq 0$

(2) Test statistic: $z = \dfrac{\overline{y}_1 - \overline{y}_2}{\sqrt{\left(\frac{s_1^2}{n_1}\right) + \left(\frac{s_2^2}{n_2}\right)}} = \dfrac{118 - 109}{\sqrt{\left(\frac{102}{64}\right) + \left(\frac{87}{64}\right)}} = 5.24$

(3) Rejection region: With $\alpha = .05$, the null hypothesis is rejected if $|z| > 1.96$.

(4) The null hypothesis is rejected, and we conclude that there is a difference in mean stopping time. The p-value is $P\left(|z| > 5.24\right) = 0$

10.108 a. In this exercise we are interested in testing the hypothesis:

H_0: $\sigma_1^2 = \sigma_2^2$ vs. H_a: $\sigma_1^2 > \sigma_2^2$

The test statistic is

$$F = \frac{s_1^2}{s_2^2} = \frac{92{,}000}{37{,}000} = 2.486.$$

The rejection region (a one-tailed rejection region) will be determined by a critical value of F based on $(n_1 - 1) = 49$ and $(n_2 - 1) = 49$ degrees of freedom. With $\alpha = .05$ and interpolating in Table 7, the value $F_{49,49}$ is roughly halfway between $F_{40,40} = 1.69$ and $F_{60,60} = 1.53$. We reject H_0 if $F > F_{49,49} = 1.61$. The observed value of the test statistic falls in the rejection region and we conclude that the "suspect line" possesses a larger variance.

b. We must obtain various critical levels of F from Table 7. We "roughly" interpolate $F_{49,49}$ as halfway between $F_{40,40}$ and $F_{60,60}$.

α	F_a
.05	1.61
.025	1.775
.01	1.975
.005	2.13

Thus, $p = P[F > 2.486] < .005$.

10.109 a. The hypothesis of interest is

H_0: $\sigma_1^2 = \sigma_2^2$ vs. H_a: $\sigma_1^2 \neq \sigma_2^2$

and the test statistic is

$$F = \frac{s_1^2}{s_2^2} = \frac{.273}{.094} = 2.904.$$

The null hypothesis will be rejected if $F > F_{9,9} = 3.18$, with $\alpha = 2(.05) = .10$. Hence, H_0 is not rejected.

b. The 90% confidence interval for σ_B^2 is

$$\frac{(n_2-1)s_2^2}{\chi^2_{.05}} < \sigma_B^2 < \frac{(n_2-1)s_2^2}{\chi^2_{.95}}$$

or

$$\frac{9(.094)}{16.919} < \sigma_B^2 < \frac{9(.094)}{3.32511}$$

or

$$.050 < \sigma_B^2 < .254.$$

Intervals constructed in this manner enclose σ_B^2 90% of the time. Hence, we are fairly certain that σ_B^2 is between .050 and .254.

10.110 The calculations are as follows:

(1) H_0: $\mu_1 - \mu_2 = 0$ vs. H_a: $\mu_1 - \mu_2 \neq 0$.

(2) $s^2 = \frac{\sum(x_i-\overline{x}_1)^2+\sum(x_i-\overline{x}_2)^2}{n_1+n_2-2} = \frac{13.7973-\frac{(11.13)^2}{9}+8.624-\frac{(8.8)^2}{9}}{16} = .0033.$

(3) Test statistic:

$$t = \frac{(\overline{x}_1-\overline{x}_2)-D_0}{\sqrt{s^2\left(\frac{1}{n_1}+\frac{1}{n_2}\right)}} = \frac{1.237-.978}{\sqrt{.0033\left(\frac{1}{9}+\frac{1}{9}\right)}} = \frac{.259}{\sqrt{.0007}} = 9.568.$$

(4) The critical t value is $t_{.05,16} = 1.746$, and the rejection region is $|t| > 1.746$. The calculated t is very large (in the rejection region), and hence the null hypothesis is rejected.

10.111 a. The procedure is as follows:

(1) H_0: $\mu_1 - \mu_2 = 0$ vs. H_a: $\mu_1 - \mu_2 \neq 0$

(2) Calculate

$$\overline{x}_1 = \frac{585}{8} = 73.125, \qquad \overline{x}_2 = \frac{466}{6} = 77.667,$$

$$\sum(x_i - \overline{x}_1)^2 = 42{,}845 - \frac{(585)^2}{8} = 66.875,$$

$$\sum(x_i - \overline{x}_2)^2 = 36{,}246 - \frac{(466)^2}{6} = 53.3333,$$

$$s^2 = \frac{66.875+53.333}{8+6-2} = 10.017.$$

(3) Test statistic:

$$t = \frac{\bar{x}_1 - \bar{x}_2}{\sqrt{s^2\left(\frac{1}{n_1}+\frac{1}{n_2}\right)}} = \frac{-4.542}{\sqrt{2.9217}} = -2.657.$$

(4) The level of significance is $p = 2P[t > 2.657]$ with $8 + 6 - 2 = 12$ degrees of freedom. From Table 5, $t = 2.657$ is between $t_{.025}$ and $t_{.01}$ so that $.02 < p < .05$. Hence, H_0 will be rejected for $\alpha > .02$. If $\alpha = .01$, H_0 would not be rejected. There is evidence of a difference in mean efficiencies.

b. The 90% confidence interval for $\mu_1 - \mu_2$ is

$$(73.125 - 77.667) \pm 1.782\sqrt{10.017\left(\tfrac{1}{8}+\tfrac{1}{6}\right)} = -4.542 \pm 3.046$$

or

$$-7.588 < (\mu_1 - \mu_2) < -1.496.$$

Since 0 is not in the interval there is evidence of a difference at the 0.10 level.

10.112 a. $V(\widehat{\theta}) = a_1^2 V(\overline{X}) + a_2^2 V(\overline{Y}) + a_3^2 V(\overline{W}) = \sigma^2\left(\frac{a_1^2}{n_1} + \frac{a_2^2}{n_2} + \frac{a_3^2}{n_3}\right)$. Thus,

$$\text{SE}(\widehat{\theta}) = \sigma\sqrt{\frac{a_1^2}{n_1} + \frac{a_2^2}{n_2} + \frac{a_3^2}{n_3}}.$$

b. $\widehat{\theta}$ is normally distributed with mean θ and variance $\sigma^2\left(\frac{a_1^2}{n_1} + \frac{a_2^2}{n_2} + \frac{a_3^2}{n_3}\right)$.

c. i. Analogous to the two-sample pooled estimator, $\frac{(n_1+n_2+n_3-3)s_p^2}{\sigma^2}$ has a chi-square distribution with $n_1 + n_2 + n_3 - 3$ degrees of freedom.

ii. Again, similar to Section 8.8 for the two sample case,

$$T = \frac{\widehat{\theta}-\theta}{s_p\sqrt{\frac{a_1^2}{n_1}+\frac{a_2^2}{n_2}+\frac{a_3^2}{n_3}}} = \frac{\frac{\widehat{\theta}-\theta}{\sqrt{\sigma^2\left(\frac{a_1^2}{n_1}+\frac{a_2^2}{n_2}+\frac{a_3^2}{n_3}\right)}}}{\sqrt{\frac{(n_1+n_2+n_3-3)s_p^2}{\sigma^2(n_1+n_2+n_3-3)}}}$$

has a t-distribution with $n_1 + n_2 + n_3 - 3$ degrees of freedom.

d. Refer to Section 8.8 and 10.8. The $(1-\alpha)100\%$ confidence interval for θ is

$$\widehat{\theta} \pm t_{\alpha/2}\, s_p\sqrt{\frac{a_1^2}{n_1} + \frac{a_2^2}{n_2} + \frac{a_3^2}{n_3}}$$

where $t_{\alpha/2}$ comes from the t distribution with $n_1 + n_2 + n_3 - 3$ degrees of freedom.

e. By part **b**, under the null hypothesis,

$$T = \frac{\widehat{\theta}-\theta}{s_p\sqrt{\frac{a_1^2}{n_1}+\frac{a_2^2}{n_2}+\frac{a_3^2}{n_3}}}$$

has a t distribution with $n_1 + n_2 + n_3 - 3$ degrees of freedom. Using this for the test statistic, the rejection region $|t| > t_{\alpha/2}$. See also Section 10.8.

10.113 Let $P = X + Y - W$. Then P has a normal distribution with mean $\mu_1 + \mu_2 - \mu_3$ and variance $(1 + a + b)\sigma^2$. Also, $\overline{P} = \overline{X} + \overline{Y} - \overline{W}$ has a normal distribution with

$$\mu_{\overline{P}} = \mu_1 + \mu_2 - \mu_3 \quad \text{and} \quad \sigma_{\overline{P}}^2 = (1 + a + b)\frac{\sigma^2}{n}$$

Hence

$$\backslash Z = \frac{(\overline{X}+\overline{Y}-\overline{W})-(\mu_1+\mu_2-\mu_3)}{\sqrt{(1+a+b)\left(\frac{\sigma^2}{n}\right)}}$$

is a standard normal variate. Secondly, the quantities

$$\frac{\sum(X_i-\overline{X})^2}{\sigma^2}, \quad \frac{\sum(Y_i-\overline{Y})^2}{a\sigma^2}, \quad \frac{\sum(W_i-\overline{W})^2}{b\sigma^2}$$

are independently distributed as χ^2 variables, each with $(n-1)$ degrees of freedom, so that

$$\frac{1}{\sigma^2}\left[\sum(X_i-\overline{X})^2 + \frac{\sum(Y_i-\overline{Y})^2}{a} + \frac{\sum(W_i-\overline{W})^2}{b}\right]$$

has a χ^2 distribution with $(3n-3)$ degrees of freedom and is independent of $\overline{X}, \overline{Y}$, and $\overline{W}$. A t statistic can now be formed.

$$T = \frac{Z}{\sqrt{\frac{V}{\nu}}} = \frac{(\overline{X}+\overline{Y}-\overline{W})-(\mu_1+\mu_2-\mu_3)}{\left\{\frac{1+a+b}{n(3n-3)\left[\sum(X_i-\overline{X})^2+\left(\frac{1}{a}\right)\sum(Y_i-\overline{Y})^2+\left(\frac{1}{b}\right)\sum(W_i-\overline{W})^2\right]}\right\}^{1/2}}$$

has a Student's t distribution with $(3n-3)$ degrees of freedom. The rejection region for the test is $|t| > t_{\alpha/2,3n-3}$

10.114 From Section 10.3, we have

$$Z = \frac{(\overline{X}-\overline{Y})-(\mu_1-\mu_2)}{\sqrt{\sigma^2\left[\left(\frac{1}{n_1}\right)+\left(\frac{1}{n_2}\right)\right]}}.$$

As in Exercise 10.113, $\frac{\sum(X_i-\overline{X})^2+\sum(Y_i-\overline{Y})^2+\sum(W_i-\overline{W})^2}{\sigma^2}$ has a χ^2 distribution with $(n_1+n_2+n_3-3)$ degrees of freedom. The resulting t statistic is

$$T = \frac{(\overline{X}-\overline{Y})-(\mu_1-\mu_2)}{\sqrt{\left(\frac{1}{n_1}+\frac{1}{n_2}\right)\times\frac{\sum(X_i-\overline{X})^2+\sum(Y_i-\overline{Y})^2+\sum(W_i-\overline{W})^2}{n_1+n_2+n_3-3}}}$$

$$= \frac{(\overline{X}-\overline{Y})-(\mu_1-\mu_2)}{\sqrt{\left(\frac{1}{n_1}+\frac{1}{n_2}\right)\widehat{\sigma}^2}}$$

For the data given in this exercise, the test of hypothesis is as follows:

$H_0\colon\ \mu_1-\mu_2=0$ vs. $H_a\colon\ \mu_1-\mu_2\neq 0$

Calculate

$$\widehat{\sigma}^2 = \frac{36{,}950-\left[\frac{(600)^2}{10}\right]+25{,}850-\left[\frac{(500)^2}{10}\right]+49{,}900-\left[\frac{(700)^2}{10}\right]}{27} = 100$$

and

$$t = \frac{60-50}{\sqrt{\left(\frac{2}{10}\right)(100)}} = 2.326$$

The rejection region with $\alpha=.05$ and 27 degrees of freedom is $|t|>t_{.025,27}=2.052$. Hence the null hypothesis is rejected.

10.115 In Ω we have

$$L(\Omega) = \frac{1}{\theta_1^n}\,e^{-\sum(y_i-\theta_2)/\theta_1}$$

The maximum likelihood estimator (MLE) for θ_2 is $\widehat{\theta}_2=Y_{(1)}$.
To find the MLE of θ_1, consider

$$\ln L = -n\ln\theta_1-\frac{1}{\theta_1}\sum(y_i-\theta_2)$$

and

$$\frac{d\ln L}{d\theta_1} = -\frac{n}{\theta_1}+\frac{1}{\theta_1^2}\sum(y_i-\theta_2)=0$$

or

$$\widehat{\theta}_1 = \frac{\sum(y_i-\widehat{\theta}_2)}{n}$$

In Ω_0, $\theta_1=\theta_{1,0}$ and $\widehat{\theta}_2=Y_{(1)}$ as before. Hence

$$\lambda = \frac{L(\widehat{\Omega}_0)}{L(\widehat{\Omega})} = \left(\frac{\widehat{\theta}_1}{\theta_{1,0}}\right)^n\exp\left[-\frac{\sum(Y_i-Y_{(1)})}{\theta_{1,0}}+\frac{\sum(Y_i-Y_{(1)})}{\widehat{\theta}_1}\right]$$

$$= \left[\frac{\sum(Y_i-Y_{(1)})}{n\theta_{1,0}}\right]^n\exp\left[-\frac{\sum(Y_i-Y_{(1)})}{\theta_{1,0}}+n\right]$$

Values of $\lambda\le k$ will cause rejection of the null hypothesis.

10.116 In Ω we have

$$\widehat{\widehat{\theta}}_1 = \frac{\sum(Y_i-Y_{(1)})}{n} \quad\text{and}\quad \widehat{\theta}_2=Y_{(1)}$$

as in Exercise 10.115. In Ω_0 we have $\theta_2=\theta_{2,0}$ and

$$\widehat{\theta}_1 = \frac{\sum(Y_i-\theta_{2,0})}{n}$$

Hence

$$\lambda = \frac{L(\widehat{\Omega}_0)}{L(\widehat{\Omega})} = \left[\frac{\sum(Y_i-Y_{(1)})}{\sum(Y_i-\theta_{2,0})}\right]^n\exp\left[\frac{-\sum(Y_i-\theta_{2,0})}{\widehat{\theta}_1}+\frac{\sum(Y_i-Y_{(1)})}{\widehat{\widehat{\theta}}_1}\right]$$

$$= \left[\frac{\sum(Y_i-Y_{(1)})}{\sum(Y_i-\theta_{2,0})}\right]^n$$

Small values of λ $(\lambda\le k)$ will cause rejection of the null hypothesis.

CHAPTER 11 LINEAR MODELS AND ESTIMATION BY LEAST SQUARES

11.1 Calculate the following:

$\sum x_i = 0$ $\quad \sum y_i = 7.5$ $\quad \sum x_i y_i = -6$

$\sum x_i^2 = 10$ $\quad \sum y_i^2 = 15.25$ $\quad n = 5$

$S_{xy} = \sum x_i y_i - \frac{1}{n}\left(\sum x_i\right)\left(\sum y_i\right) = -6$

$S_{xx} = \sum x_i^2 - \frac{1}{n}\left(\sum x_i\right)^2 = 10$

Then

$$\widehat{\beta}_1 = \frac{S_{xy}}{S_{xx}} = -\frac{6}{10} = -.6$$

and

$$\widehat{\beta}_0 = \overline{y} - \widehat{\beta}_1\overline{x} = \frac{7.5}{5} - 0 = 1.5.$$

The least squares straight line is $\widehat{y} = 1.5 - .6x$

The observed points and the fitted lines are shown in Figure 11.1.

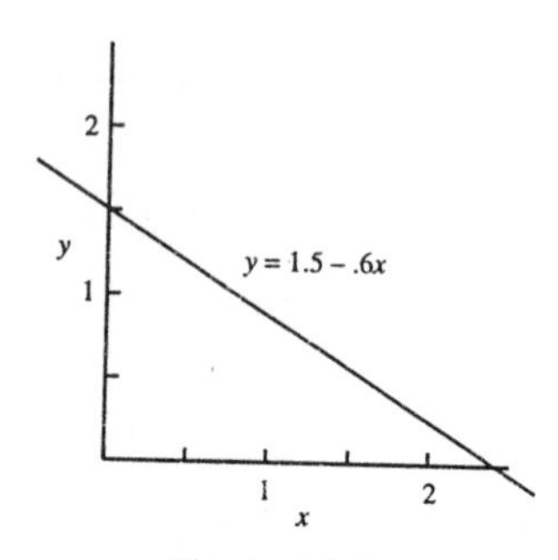

Figure 11.1

11.2 Calculate

$\sum x_i = 720$ $\quad \sum y_i = 721$ $\quad \sum x_i^2 = 106{,}554$

$\sum y_i^2 = 105{,}817$ $\quad \sum x_i y_i = 106{,}155$ $\quad n = 10$

$S_{xy} = \sum x_i y_i - \frac{1}{n}\left(\sum x_i\right)\left(\sum y_i\right) = 54{,}243$

$S_{xx} = \sum x_i^2 - \frac{1}{n}\left(\sum x_i\right)^2 = 54{,}714$

Then

$$\widehat{\beta}_1 = \frac{S_{xy}}{S_{xx}} = \frac{54{,}243}{54{,}714} = .9913916$$

and

$$\widehat{\beta}_0 = \overline{y} - \widehat{\beta}_1\overline{x} = 72.1 - .99(72) = .7198048.$$

The least squares straight line is $\widehat{y} = .72 + .99x$, and the expected change in y for a one-unit change in x is estimated as $\widehat{\beta}_1 = .99$. When $x = 100$, the best estimate of y is $\widehat{y} = .72 + .99(100) = 99.72$.

11.3 Calculate

$\sum x_i = 36$ $\quad \sum y_i = 346.9$ $\quad \sum x_i y_i = 1764.4$

$\sum x_i^2 = 204$ $\quad \sum y_i^2 = 16{,}045.29$ $\quad n = 8$

$S_{xy} = 203.35$ $\quad S_{xx} = 42$

Then

$$\widehat{\beta}_1 = \frac{S_{xy}}{S_{xx}} = \frac{54{,}243}{54{,}714} = 4.84167$$

and

$$\widehat{\beta}_0 = \overline{y} - \widehat{\beta}_1\overline{x} = 43.3625 - 21.7875 = 21.575.$$

The least squares straight line is $\widehat{y} = 21.575 + 4.842x$. The positive slope suggests an increase in sales over time. In particular we expect sales to increase by 4,842 dollars per year.

11.4 Calculate

$\sum x_i = 155.05$ $\quad \sum y_i = 94.48$ $\quad \sum x_i y_i = 3011.3709$

$\sum x_i^2 = 4763.979$ $\qquad \sum y_i^2 = 1993.8156$ $\qquad n = 10$
$S_{xy} = 1546.459$ $\qquad S_{xx} = 2359.929$

Then

$$\widehat{\beta}_1 = \frac{S_{xy}}{S_{xx}} = .65330$$

and

$$\widehat{\beta}_0 = \overline{y} - \widehat{\beta}_1\overline{x} = 9.448 - 10.159039 = -.7124.$$

The least squares straight line is $\widehat{y} = -.712 + .655x$. When $x = 12$, using full accuracy, the estimate of y is $\widehat{y} = -.71 + .655(12) = 7.15$.

11.5 **a.** The data are plotted in Figure 11.2.

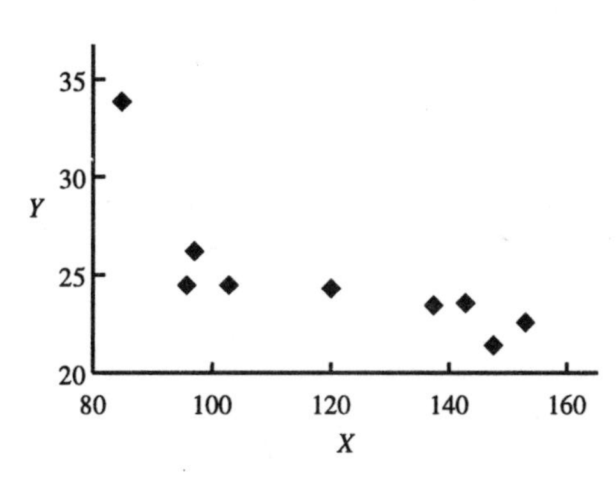

Figure 11.2

b. Calculate

$\sum x_i = 1076$ $\qquad \sum y_i = 216$ $\qquad \sum x_i y_i = 25{,}431$
$\sum x_i^2 = 133{,}336$ $\qquad \sum y_i^2 = 5228$ $\qquad n = 9$
$S_{xy} = -393.0$ $\qquad S_{xx} = 4694.22$

Then

$$\widehat{\beta}_1 = \frac{S_{xy}}{S_{xx}} = -.0837$$

and

$$\widehat{\beta}_0 = \overline{y} - \widehat{\beta}_1\overline{x} = 24 + 10.0092 = 34.0092$$

c. The least squares line is graphed in Figure 11.3.

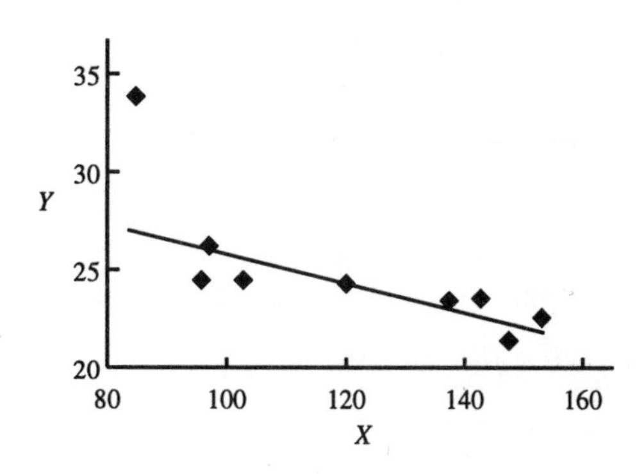

Figure 11.3

d. The estimate of y when $x = 125$ is $\widehat{y} = 34.0092 - .0837(125) = 23.5467$.

11.6 We need to minimize SSE $= \sum_{i=1}^{n} \left[y_i - \widehat{\beta}_1 x_i\right]^2$. Consider

$$\begin{aligned}\frac{d\,\text{SSE}}{d\widehat{\beta}_1} &= -\sum_{i=1}^{n} 2\left[y_i - \widehat{\beta}_1 x_i\right] x_i &= 0\\ &= -2\left[\sum_{i=1}^{n}\left(x_i y_i - \widehat{\beta}_1 x_i^2\right)\right] &= 0\\ \Rightarrow \sum_{i=1}^{n} x_i y_i - \widehat{\beta}_1 \sum_{i=1}^{n} x_i^2 &&= 0\end{aligned}$$

Implying

$$\widehat{\beta}_1 = \frac{\sum_{i=1}^{n} x_i y_i}{\sum_{i=1}^{n} x_i^2}.$$

11.7 Note that $\sum_{i=1}^{n} x_i y_i = 134{,}542$ and $\sum_{i=1}^{n} x_i^2 = 53{,}514$. Thus

$$\widehat{\beta}_1 = \frac{134{,}542}{53{,}514} = 2.514$$

and $\widehat{y}_i = 2.514 x_i$.

11.8 **a.** We calculate:

$\sum_{i=1}^{5} x_i = 102$ $\sum_{i=1}^{5} x_i^2 = 3940$ $\sum_{i=1}^{5} y_i = 64.7$

$\sum_{i=1}^{5} x_i y_i = 894.4$ $\sum_{i=1}^{5} y_i^2 = 949.99$ $n = 5$

$S_{xy} = -425.48$ $S_{xx} = 1859.2$

Implying

$$\widehat{\beta}_1 = \frac{S_{xy}}{S_{xx}} = -.229$$

and

$$\widehat{\beta}_0 = \overline{y} - \widehat{\beta}_1 \overline{x} = \frac{64.7}{5} - (-.229)\left(\frac{102}{5}\right) = 17.611.$$

b. The data and least squares line are plotted in Figure 11.4.

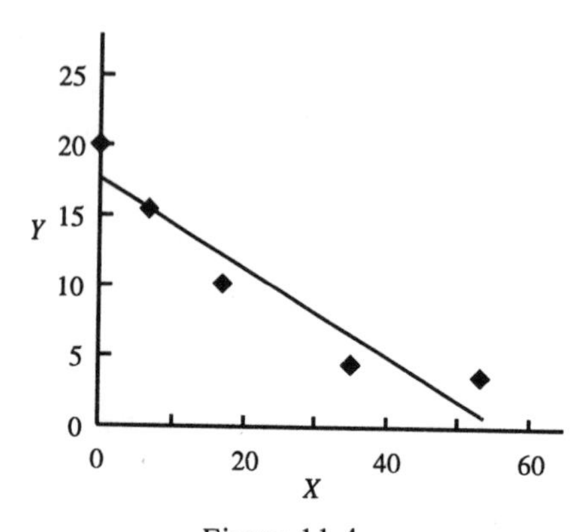

Figure 11.4

c. The estimate of $E(Y)$ when $x = 20$ is $\widehat{y} = 17.611 - .229(20) = 13.031$.

11.9 **a.** Using the model $y = \beta_0 + \beta_1 x + \epsilon$, calculate

$\sum_{i=1}^{14} x_i = 86.48$ $\sum_{i=1}^{14} y_i = 3787$ $\sum_{i=1}^{14} y_i^2 = 1{,}257{,}465$

$\sum_{i=1}^{14} x_i^2 = 732.4876$ $\sum_{i=1}^{5} x_i y_i = 17{,}562.8$ $S_{xy} = -5830.04$

$S_{xx} = 198.29$ $\widehat{\beta}_1 = \frac{S_{xy}}{S_{xx}} = -29.402$

$\widehat{\beta}_0 = \overline{y} - \widehat{\beta}_1 \overline{x} = \frac{3787}{14} - (-29.402)\left(\frac{86.48}{14}\right) = 452.119$

The least squares line is $\widehat{y} = 452.119 - 29.402x$.

b. The graph is omitted.

11.10 a. $\sum x_i = 3.25$ $\quad \sum y_i = 7.55$ $\quad \sum x_i y_i = 2.1825$
$\sum x_i^2 = 1.2625$ $\quad \sum y_i^2 = 6.0725$ $\quad n = 10$
$S_{xy} = -.27125$ $\quad S_{xx} = .20625$
Then

$$\widehat{\beta}_1 = \frac{S_{xy}}{S_{xx}} = -1.31515$$

and

$$\widehat{\beta}_0 = \overline{y} - \widehat{\beta}_1 \overline{x} = 1.182424.$$

b. The graph is omitted but is similar to previous graphs. See Exercise 11.1.

11.11 Since $\widehat{\beta}_0 = \overline{y} - \widehat{\beta}_1 \overline{x}$,

$$\begin{aligned}
\text{SSE} &= \sum \left[y_i - (\overline{y} - \widehat{\beta}_1 \overline{x}) - \widehat{\beta}_1 x_i\right]^2 \\
&= \sum \left[(y_i - \overline{y}) - \widehat{\beta}_1 (x_i - \overline{x})\right]^2 \\
&= \sum (y_i - \overline{y})^2 + \widehat{\beta}_1^2 \sum (x_i - \overline{x})^2 - 2\widehat{\beta}_1 \sum (x_i - \overline{x})(y_i - \overline{y}) \\
&= \sum (y_i - \overline{y})^2 + \widehat{\beta}_1 \times \frac{\sum (x_i - \overline{x})(y_i - \overline{y})}{\sum (x_i - \overline{x})^2} \times \sum (x_i - \overline{x})^2 \ \left(\text{as } \widehat{\beta}_1 = \frac{\sum (x_i - \overline{x})(y_i - \overline{y})}{\sum (x_i - \overline{x})^2}\right) \\
&\quad - 2\widehat{\beta}_1 \sum (x_i - \overline{x})(y_i - \overline{y}) \\
&= \sum (y_i - \overline{y})^2 - \widehat{\beta}_1 \sum (x_i - \overline{x})(y_i - \overline{y}) = S_{yy} - \widehat{\beta}_1 S_{xy}
\end{aligned}$$

11.12 a. Calculate
$\sum x_i = 720$ $\quad \sum y_i = 324$ $\quad \sum x_i y_i = 17{,}540$
$\sum x_i^2 = 49{,}200$ $\quad \sum y_i^2 = 9540$ $\quad n = 12$
$S_{xy} = -1900.0$ $\quad S_{xx} = 6000.0$
Then

$$\widehat{\beta}_1 = \frac{S_{xy}}{S_{xx}} = -.31667$$

and

$$\widehat{\beta}_0 = \overline{y} - \widehat{\beta}_1 \overline{x} = 46.00.$$

b. The least squares line shown in Figure 11.5 is $\widehat{y} = 46.0 - .317x$.

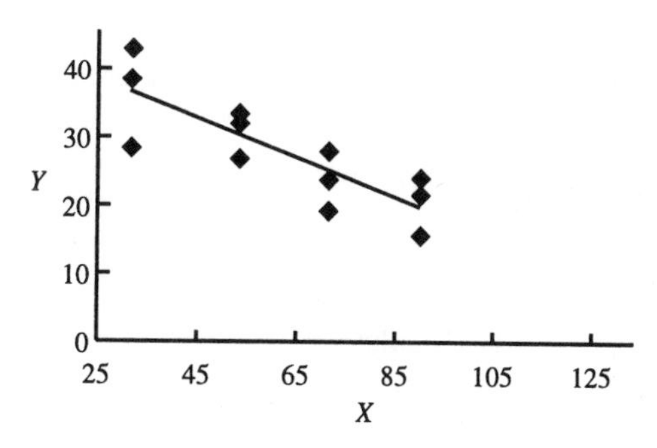

Figure 11.5

c. Refer to Exercise 11.11.

$$\begin{aligned}
\text{SSE} &= S_{yy} - \widehat{\beta}_1 S_{xy} \\
&= 9540 - \frac{(324)^2}{12} + (.31667)\left[17{,}540 - \frac{(720)(324)}{12}\right] = 190.333
\end{aligned}$$

and

$$s^2 = \frac{\text{SSE}}{n-2} = \frac{190.333}{10} = 19.0333.$$

11.13 a. Calculate

$$S_{yy} = \sum (y_i - \overline{y})^2 = 16{,}045.29 - \frac{(346.9)^2}{8} = 1002.8388$$

and

$$S_{xy} = \sum (x_i - \overline{x})(y_i - \overline{y}) = \frac{1626.8}{8} = 203.35.$$

Then

$$\text{SSE} = 1002.8388 - 4.84167(203.35) = 18.2858$$

and

$$s^2 = \frac{\text{SSE}}{6} = 3.0476.$$

b. Using the coding formula given here, we obtain the y values and the corresponding x^* values shown below.

y	27.6	32.5	35.9	39.3	44.2	48.8	55.7	62.9
x^*	−7	−5	−3	−1	1	3	5	7

Then

$\sum x^* = 0$ $\quad\sum y = 346.9$ $\quad\sum x^* y = 406.7$

$\sum x^{*2} = 168$ $\quad\sum y^2 = 16{,}045.29$ $\quad n = 8$

$S_{x^*y} = 406.7$ $\quad S_{x^*x^*} = 168$

so that

$$\widehat{\beta}_1^* = \frac{S_{xy}}{S_{xx}} = 2.42$$

and

$$\widehat{\beta}_0^* = \overline{y} = 43.3625.$$

The fitted model is $\widehat{y} = 43.35 + 2.42x^*$. Calculate $\sum (y_i - \overline{y})^2 = 1002.8388$, as in part **a** , then

$$\text{SSE} = S_{xy} - \widehat{\beta}_1 S_{xx} = 1002.8388 - 2.42(406.7) = 18.2858$$

and

$$s^2 = \frac{\text{SSE}}{6} = 3.0476.$$

as in part **a**. Notice these are the exact same estimates we obtained in part **a** More generally, shifting the X value by a constant does not change SSE.

11.14a. For Exercise 11.4, $S_{yy} = \sum (y_i - \overline{y})^2 = 1101.1686$, $S_{xy} = 1546.459$ and $\text{SSE} = 1101.1686 - (.6552528)(1546.553) = 87.84701$

$$s^2 = \frac{\text{SSE}}{8} = 10.980 \qquad \text{or} \qquad s^2 = 10.98$$

b. Using the coding $x_i^* = x_i - \overline{x}$, we have the new values for x_i^* as given in the table that follows.

x_i^*	23.496	21.996	6.696	1.996	−14.864	−15.054	−12.884
y_i	23.00	22.30	9.40	9.70	.15	.28	.75

x_i^*	−13.144	16.496	−14.734
y_i	.51	28.00	.39

Then

$S_{yy} = 1101.1686$ as in part **a** $\quad S_{x^*y} = 1546.553$

$\widehat{\beta}_1^* = \frac{\sum x_i^* y_i}{\sum x_i^{*2}} = \frac{1546.553}{2360.2388} = .655$

$\text{SSE} = 1101.1686 - (.655)(1546.553) = 37.79$

$s^2 = \frac{\text{SSE}}{8} = 10.97$

as in part **a**.

11.15a. Similar to previous exercises. Calculate

$\sum x_i = 160$ $\quad\sum y_i = 106$ $\quad\sum x_i y_i = 1848$

$\sum x_i^2 = 2880$ $\quad\sum y_i^2 = 1236$ $\quad n = 10$

$S_{xy} = 152.0$ $\quad S_{xx} = 320$ $\quad S_{yy} = 112.4$

Then

$$\widehat{\beta}_1 = \frac{S_{xy}}{S_{xx}} = .475$$

and

$$\widehat{\beta}_0 = \overline{y} - \widehat{\beta}\overline{x} = 10.6 - .475(1.6) = 3.000.$$

b. The least squares line, $\widehat{y} = 3.000 + 4.75x$, is shown in Figure 11.6.

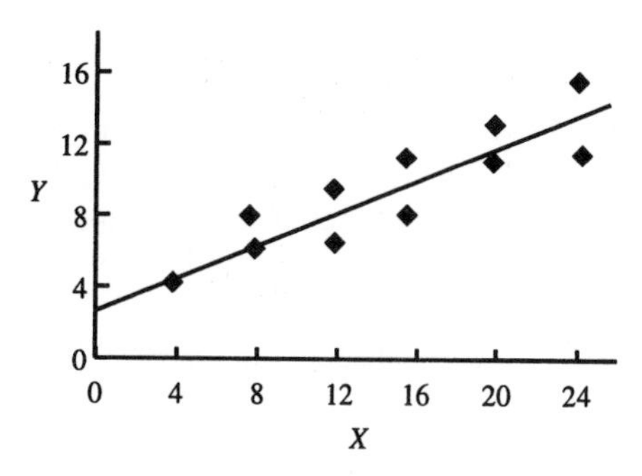

Figure 11.6

c. Calculate $S_{yy} - \widehat{\beta}_1 S_{xy} = 112.4 - .475(152) = 40.2$
and

$$s^2 = \frac{\text{SSE}}{n-2} = \frac{40.2}{8} = 5.025.$$

11.16 The likelihood function is

$$L = \frac{1}{(2\pi)^{n/2}\sigma^n} \prod_{i=1}^{n} \exp\left\{-\frac{1}{2\sigma^2}\left[y_i - E(y_i)\right]^2\right\}$$

$$= K \exp\left[-\frac{1}{2\sigma^2}\sum_{i=1}^{n}(y_i - \beta_0 - \beta_1 x_i)^2\right]$$

so that

$$\ln L = \ln K - \frac{1}{2\sigma^2}\sum_{i=1}^{n}(y_i - \beta_0 - \beta_1 x_i)^2.$$

In order to maximize $\ln L$ with respect to β_0 and β_1, it is necessary to maximize

$$-\frac{1}{2\sigma^2}\sum_{i=1}^{n}(y_i - \beta_0 - \beta_1 x_i)^2$$

which implies choosing $\widehat{\beta}_0$ and $\widehat{\beta}_1$ to minimize $\sum_{i=1}^{n}(y_i - \widehat{\beta}_0 - \widehat{\beta}_1 x_i)^2 = \text{SSE}$. Hence the maximum likelihood estimators of β_0 and β_1 are identical to the least squares estimators.

11.17 Calculate

$$\text{Cov}(\widehat{\beta}_0, \widehat{\beta}_1) = \text{Cov}(\overline{Y} - \widehat{\beta}_1\overline{x}, \widehat{\beta}_1) = \text{Cov}(\overline{Y}, \widehat{\beta}_1) - \overline{x}\,\text{Cov}(\widehat{\beta}_1, \widehat{\beta}_1) = 0 - \overline{x}\,V(\widehat{\beta}_1)$$

$$= \frac{-\overline{x}\sigma^2}{\sum(x_i - \overline{x})^2}$$

This will equal zero if and only if $\overline{x} = 0$ or $\sum x_i = 0$. Since $\widehat{\beta}_1$ and $\widehat{\beta}_0$ are normally distributed, $\text{Cov}(\widehat{\beta}_0, \widehat{\beta}_1) = 0$ if and only if $\widehat{\beta}_1$ and $\widehat{\beta}_0$ are independent.

11.18 Refer to Exercise 11.16, in which the likelihood function is given.

$$\ln L = \ln K' - \frac{n}{2}\ln\sigma^2 - \frac{1}{2\sigma^2}\sum(y_i - \beta_0 - \beta_1 x_i)^2$$

Differentiating with respect to σ^2, we have

$$\frac{\partial \ln L}{\partial \sigma^2} = -\frac{n}{2\sigma^2} + \frac{\sum(y_i - \beta_0 - \beta_1 x_i)^2}{2\sigma^4} = 0 \quad \text{or} \quad \widehat{\sigma}^2 = \frac{\sum(y_i - \beta_0 - \beta_1 x_i)^2}{n}$$

Using the maximum likelihood estimates of β_0 and β_1 (see Exercise 11.16), we have

$$\widehat{\sigma}^2 = \frac{\sum(y_i - \widehat{\beta}_0 - \widehat{\beta}_1 x_i)^2}{n} = \frac{\text{SSE}}{n}$$

11.19 **a.** Refer to Exercise 11.1. Using the information calculated there, we have

$$S_{yy} = \sum_{i=1}^{n} y_i^2 - \frac{1}{n}\left(\sum_{i=1}^{n} y_i\right)^2 = 15.25 - \frac{1}{5}(7.5)^2 = 4.0$$

$$S_{xy} = -6$$

Then

$$\text{SSE} = S_{yy} - \widehat{\beta}_1 S_{xy} = 4 - (-.6)(-6) = .4$$

and

$$s^2 = \frac{\text{SSE}}{n-2} = \frac{.4}{3} = .1333$$

The test of the hypothesis

$$H_0\colon \beta_1 = 0 \qquad \text{vs.} \qquad H_a\colon \beta_1 \neq 0$$

is the familiar t test given in Chapter 10. The test statistic is

$$t = \frac{\widehat{\beta}_1 - \beta_{10}}{s\sqrt{c_{11}}} = \frac{-.6}{\sqrt{.1333}\sqrt{.1}} - 5.20$$

The rejection region for $\alpha = .05$ and 3 degrees of freedom will be all values of t such that $|t| > t_{.025,3} = 3.182$. Since the calculated value of t falls in the rejection region, there is sufficient evidence to conclude that the slope β_1 differs from zero.

b. Using the methods given in Chapter 8, we can obtain a 95% confidence interval for β_1:

$$\widehat{\beta}_1 \pm t_{.025,3}\, s\sqrt{c_{11}} \qquad \text{or} \qquad -.6 \pm 3.182\sqrt{.1333}\,\sqrt{.1} \qquad \text{or} \qquad -.6 \pm .367$$

or

$$[-.967, -.233].$$

11.20 Refer to Exercise 11.9. We wish to test

$$H_0\colon \beta_1 = 0 \qquad \text{vs.} \qquad H_a\colon \beta_1 \neq 0.$$

We have SSE $= S_{yy} - \widehat{\beta}_1 S_{xy} = 233{,}081.5 - 171{,}414.836 = 61{,}667.66$ and

$$s^2 = \frac{61{,}667.66}{12} = 5138.97.$$

The test statistic is

$$t_{12} = \frac{\widehat{\beta}_1 - 0}{s\sqrt{c_{11}}} = \frac{-29.402}{\sqrt{5138.97}\sqrt{.005043}} = -5.775$$

The p-value is $P(|t| > 5.775)$. From Table 5, $P(|t| > 3.055) = 2(.005) = .01$. Thus, the p-value $< .01$. We would reject H_0 when $\alpha = .10$.

11.21 **a.** Refer to Exercise 11.15. We wish to test

$$H_0\colon \beta_1 = 0 \qquad \text{vs.} \qquad H_a\colon \beta_1 \neq 0.$$

To carry out the test we need to estimate the value of $\sigma^2 c_{11} = V(\widehat{\beta}_1)$. From Exercise 11.15, $s^2 = \widehat{\sigma}^2 = 5.025$. From Section 11.4, $V(\widehat{\beta}_1) = \frac{\sigma^2}{\sum (x_i - \overline{x})^2}$. Now

$$\sum (x_i - \overline{x}) = 2880 - \frac{(160)^2}{10} = 320 = \frac{1}{c_{11}}$$

Thus,

$$V(\widehat{\beta}_1) = \frac{\sigma^2}{320} = \frac{5.025}{30}$$

The test statistic is

$$t = \frac{\widehat{\beta}_1}{\sqrt{\frac{s^2}{\sum (x_i - \overline{x})^2}}} = \frac{.475}{\sqrt{\frac{5.025}{320}}} = 3.791$$

The p-value is $P(|t_8| > 3.791) < P(|t_8| > 3.355) = 2(.005) = .01$ (see Table 5).

b. We would reject H_0 when $\alpha = .05$, since the p-value $< .05$.

c. No, there is no reason to believe that the linear trend will continue beyond the region of data collected. For example, we might would expect the rate of increase in the number of mistakes to level off at some point.

d. A 95% confidence interval for β_i is

$$\widehat{\beta}_1 \pm t_{.025}\, s\sqrt{c_{11}} \qquad \text{or} \qquad .475 \pm 2.306\sqrt{\frac{5.025}{320}} \qquad \text{or} \qquad .475 \pm .289$$

We are 95% confident that the change in the number of errors per hour of sleep depravation lies in this interval. Further, he lower endpoint being larger than 0 suggest a statistically significant positive linear increase.

11.22 **a.** $\sum_{i=1}^{6} x_i = 323.4 \qquad \sum_{i=1}^{6} y_i = 42.6 \qquad \sum_{i=1}^{6} x_i y_i = 2495.08$

$\sum_{i=1}^{6} x_i^2 = 19{,}111.95 \qquad \sum_{i=1}^{6} y_i^2 = 326.06 \qquad S_{xy} = 198.94$

$S_{xx} = 1680.69 \qquad S_{yy} = 23.6$

$$\widehat{\beta}_1 = \frac{S_{xy}}{S_{xx}} = \frac{198.94}{1680.69} = .118.$$

$$\widehat{\beta}_0 = \overline{y} - \widehat{\beta}_1 \overline{x} = \frac{42.6}{6} - .118\left(\frac{323.4}{6}\right) = .72.$$

b. SSE $= S_{yy} - \widehat{\beta}_1 S_{xy} = 23.6 - (.118)(198.94) = .125$

and

$$s^2 = \frac{\text{SSE}}{n-2} = \frac{.052}{6-2} = .013$$

A 95% confidence interval for β_1 is

$$\widehat{\beta}_1 \pm t_{.025,4}\, s\sqrt{c_{11}}$$
$$.118 \pm 2.776\sqrt{.013}\ \sqrt{.00059}$$
$$.118 \pm .008$$

c. When $x = 0$, $E(Y) = \beta_0 + \beta_1(0) = \beta_0$. Thus, we must test

H_0: $\beta_0 = 0$ vs. H_a: $\beta_0 \neq 0$.

The test statistic is

$$t = \frac{\widehat{\beta}_0}{s\sqrt{c_{00}}} = \frac{.72}{\sqrt{.013}\ \sqrt{1.895}} = 4.587$$

From Table 5, $3.747 < 4.587 < 4.604$ so that the p-value is between $2(.01)$ and $2(.005)$. Since the p-value $< .05$, we would reject H_0 at the $\alpha = .05$ level.

11.23a. Section 11.4 of the text states the following (assuming that ϵ_i is normally distributed):

(1) $Z = \frac{\widehat{\beta}_i - \beta_i}{\sqrt{\sigma^2 c_{ii}}}$ has a standard normal distribution.

(2) $\frac{[n-(k+1)]S^2}{\sigma^2}$ has a χ^2 distribution with $n - (k+1)$ degrees of freedom.

(3) S^2 and $\widehat{\beta}_i$ are independent.

Thus, constructing the t statistic, we have, from Chapter 7, that

$$T = \frac{2}{\sqrt{\frac{\chi^2}{\nu}}} = \frac{\widehat{\beta}_i - \beta_i}{\sqrt{\left(\frac{S^2\sigma^2}{\sigma^2}\right) c_{ii}}} = \frac{\widehat{\beta}_i - \beta_i}{S\sqrt{c_{ii}}}$$

has a t-distribution with $n - 2$ degrees of freedom.

b. Using the above T as a pivotal statistic, write

$$P(-t_{\alpha/2}) < T < t_{\alpha/2}) = 1 - \alpha$$

or

$$P\left(-t_{\alpha/2} < \frac{\widehat{\beta}_i - \beta_i}{S\sqrt{c_{ii}}} < t_{\alpha/2}\right) = 1 - \alpha$$

or

$$P\big(\widehat{\beta}_i - t_{\alpha/2}\, S\sqrt{c_{ii}} < \beta_i < \widehat{\beta}_i + t_{\alpha/2}\, S\sqrt{c_{ii}}\big) = 1 - \alpha.$$

Then a $(1-\alpha)100\%$ confidence interval for β_i is $\widehat{\beta}_i \pm t_{\alpha/2}\, S\sqrt{c_{ii}}$.

11.24 Restricting ourselves to Ω_0, we find the likelihood function to be

$$L(\Omega_0) = \frac{1}{(2\pi)^{n/2}\sigma^n} \exp\left[-\frac{1}{2\sigma^2}\sum_{i=1}^{n}(y_i - \beta_0)^2\right]$$

One may verify that maximum likelihood estimates of β_0 and σ^2 are

$$\widehat{\beta}_0 = \overline{Y} \quad \text{and} \quad \widehat{\sigma}_0^2 = \frac{\sum_{i=1}^{n}(Y_i - \overline{Y})^2}{n} \text{ so that}$$

$$L(\widehat{\Omega}_0) = \frac{1}{(2\pi)^{n/2}\left(\widehat{\sigma}_0^2\right)^{n/2}} e^{-n/2}.$$

For β_1 in the unrestricted space Ω, the likelihood function is given in the solution to Exercise 11.16, and the maximum likelihood estimates of β_0 and β_1 are the least squares estimates

$$\widehat{\beta}_0 = \overline{Y} - \widehat{\beta}_1\overline{x} \qquad \text{and} \qquad \widehat{\beta}_1 = \frac{S_{xy}}{S_{xx}}$$

The maximum likelihood estimate of σ^2 is $\widehat{\sigma}^2 = \frac{\sum_{i=1}^{n}(y_i - \widehat{\beta}_0 - \widehat{\beta}_1 x_i)^2}{n}$ so that

$$L(\widehat{\Omega}) = \frac{1}{(2\pi)^{n/2}\left(\widehat{\sigma}^2\right)^{n/2}} e^{-n/2}.$$

Hence

$$\lambda^{2/n} = \frac{\widehat{\sigma}^2}{\widehat{\sigma}_0^2} = \frac{\sum_{i=1}^{n}(y_i - \widehat{\beta}_0 - \widehat{\beta}_1 x_i)^2}{S_{yy}} = \frac{\text{SSE}}{S_{yy}}$$

Then

$$\begin{aligned}
S_{yy} &= \sum_{i=1}^{n}\left[y_i - \overline{y} - \widehat{\beta}_1(x_i - \overline{x}) + \widehat{\beta}_1(x_i - \overline{x})^2\right] \\
&= \sum_{i=1}^{n}\left[y_i - \overline{y} - \widehat{\beta}_1(x_i - \overline{x})^2\right] + \widehat{\beta}_1^2 S_{xx} + 2\widehat{\beta}_1 S_{xy} - 2\widehat{\beta}_1^2 S_{xx} \\
&= \text{SSE} + 2\widehat{\beta}_1 S_{xy} - \widehat{\beta}_1^2 S_{xx}
\end{aligned}$$

But $\widehat{\beta}_1 = \frac{S_{xy}}{S_{xx}}$ so that

$$S_{yy} = \text{SSE} + 2\,\frac{S_{xy}^2}{S_{xx}} - \frac{S_{xy}^2}{S_{xx}} = \text{SSE} + \frac{S_{xy}^2}{S_{xx}}.$$

Now

$$\lambda^{2/n} = \frac{\text{SSE}}{\text{SSE}+\left(\frac{S_{xy}^2}{S_{xx}}\right)} = \frac{1}{1+\left(\frac{T^2}{n-2}\right)}$$

where $T^2 = \frac{S_{xy}^2}{S_{xx}\left(\frac{\text{SSE}}{n-2}\right)} = \frac{\beta_1^2 \sum_i (x_i-\overline{x})^2}{\frac{\text{SSE}}{n-2}}$. Hence as λ gets small, T^2 will get large (either positively or negatively), and rejecting H_0 for large or small values of T will be equivalent to rejecting H_0 for small values of λ. Note that

$$T = \frac{\widehat{\beta}_1}{\sqrt{\frac{\widehat{\sigma}^2}{\sum_{i=1}^{n}(x_i-\overline{x})^2}}}$$

is the t test given in Section 11.6.

11.25 We know the following:

(1) $E(\widehat{\beta}_1 - \widehat{\gamma}_1) = \beta_1 - \widehat{\gamma}_1$.

(2) $V(\widehat{\beta}_1 - \widehat{\gamma}_1) = V(\widehat{\beta}_1) + V(\widehat{\gamma}_1) = \sigma^2\left[\frac{1}{S_{xx}} + \frac{1}{S_{cc}}\right]$ (where $S_{cc} = \sum (c_i - \overline{c})^2$))

(3) $\widehat{\beta}_1$ and $\widehat{\gamma}_1$are normally distributed.

Hence under H_0,

$$Z = \frac{((\widehat{\beta}_1-\widehat{\gamma}_1)-0}{\sqrt{\sigma^2\left[\frac{1}{S_{xx}}+\frac{1}{S_{cc}}\right]}}$$

is a standard normal variate. Further, from Section 11.5 we have the following:

(1) $\frac{(n-2)S_1^2}{\sigma^2}$ has a χ^2 distribution with $(n-2)$ degrees of freedom, where $S_1^2 = \frac{\sum(Y_i-\overline{Y})^2}{(n-2)}$.

(2) $\frac{(m-2)S_2^2}{\sigma^2}$ has a χ^2 distribution with $(m-2)$ degrees of freedom $S_2^2 = \frac{\sum(W_i-\overline{W})^2}{(m-2)}$.

(3) $\frac{(n-2)S_1^2+(m-2)S_2^2}{\sigma^2}$ has a χ^2 distribution with $(m+n-4)$ degrees of freedom.

Since S_1^2 is independent of $\widehat{\beta}_1$ and S_2^2 is independent of $\widehat{Y}_1$, the quantity

$$\frac{(n-2)S_1^2+(m-2)S_2^2}{\sigma^2} = \frac{\text{SSE}_1+\text{SSE}_2}{\sigma^2}$$

must be independent of $\widehat{\beta}_1 - \widehat{\gamma}_1$and a t statistic can be constructed.

$$T = \frac{Z}{\sqrt{\frac{\chi^2}{\nu}}} = \frac{\widehat{\beta}_1-\widehat{\gamma}_1}{\sqrt{\frac{\text{SSE}_1+\text{SSE}_2}{m+n-4}\left[\frac{1}{S_{xx}}+\frac{1}{S_{cc}}\right]}}$$

has a t distribution with $(m+n-4)$ degrees of freedom. H_0 will be rejected for very large (negative or positive) values of T.

11.26 a. In order to test

$$H_0\colon\ \beta_1 = 0 \qquad \text{vs.} \qquad H_a\colon\ \beta_1 \neq 0$$

for the two experiments, construct the test statistic

$$t = \frac{\widehat{\beta}_1-0}{\sqrt{V(\widehat{\beta}_1)}}.$$

which for the small-particle catalyst gives $t_1 = \frac{.155}{.0202} = 7.67$ and for the large-particle catalyst gives $t_2 = \frac{.190}{.0193} = 9.84$. The rejection region with $\alpha = .05$ is $|t| > t_{.025,29} = 2.045$ for the first experiment and $|t| > t_{.025,9} = 2.262$ in the second experiment. In each case H_0 is rejected and we conclude that the slopes are significantly different from zero.

b. Using the results of Exercise 11.25, we have

$$\widehat{\sigma}^2 = \frac{\text{SSE}_1+\text{SSE}_2}{m+n-4} = \frac{2.04+1.86}{31+11-4} = .1026$$

Since

$$V(\widehat{\beta}_1) = \frac{\frac{\text{SSE}_1}{n-2}}{S_{xx}}$$

we have

$$S_{xx} = \frac{\frac{\text{SSE}_2}{29}}{(.0202)^2} = 172.3969$$

and

$$S_{cc} = \sum (c_i - \overline{c})^2 = \frac{\frac{\text{SSE}_2}{9}}{(.0193)^2} = 554.82474.$$

and to test H_0: $\beta_1 - \gamma_1 = 0$, the test statistic is

$$t = \frac{.155-.190}{\sqrt{\widehat{\sigma}^2\left[\left(\frac{1}{172.3969}\right)+\left(\frac{1}{554.82474}\right)\right]}} = -1.25$$

The rejection region is $|t| > t_{.025,38} = 1.96$ and the null hypothesis is not rejected. There is no significant difference in the slopes.

11.27 Using the coding $x = \frac{\text{year}-1971.5}{.5}$, we obtain the following calculations:

$\sum x = 0$ $\quad\sum y = 215.9$ $\quad\sum xy = -174.9$

$\sum x^2 = 330$ $\quad\sum y^2 = 4760.43$ $\quad n = 10$

$S_{xy} = -174.9$ $\quad S_{xx} = 330$ $\quad S_{yy} = 99.149$

Then $\widehat{\beta}_1 = \frac{S_{xy}}{S_{xx}} = \frac{-174.9}{330} = -.53$ and

$$\text{SSE} = 99.149 - (-.53)(-174.9) = 6.452$$

and

$$s^2 = \frac{\text{SSE}}{8} = .8065.$$

To test the hypothesis

$$H_0\colon\ \beta_1 = 0 \qquad \text{vs.} \qquad H_a\colon\ \beta_1 < 0$$

we use the test statistic

$$t = \frac{\widehat{\beta}_1}{\sqrt{\frac{s^2}{\sum(x_i-\overline{x})^2}}} = \frac{-.53}{\sqrt{\frac{.8065}{330}}} = -10.72.$$

The rejection region with $\alpha = .05$ is $t < -1.86$ and H_0 is rejected. We conclude that the rate of tuberculosis is decreasing with time.

11.28 **a.** Using the uncoded x's given in Exercise 11.3, we obtain $\widehat{\beta}_1 = 4.8417$ and $S_{xx} = 42$. Then from Exercise 11.13, we obtain $s^2 = 3.0476$. To test the hypothesis

$$H_0\colon\ \beta_1 = 0 \qquad \text{vs.} \qquad H_a\colon\ \beta_1 > 0$$

we use the test statistic

$$t = \frac{\widehat{\beta}_1-0}{\sqrt{\frac{s^2}{S_{xx}}}} = \frac{4.841667}{\sqrt{\frac{3.0476}{42}}} = 17.97$$

The rejection region is $t > t_{.01,6} = 3.143$, and H_0 is rejected. There is evidence of an increase.

b. The 99% confidence interval for β_1 is

$$\widehat{\beta}_1 \pm t_{.005}\sqrt{\frac{s^2}{S_{xx}}} = 4.84 \pm 3.707(.26937) = 4.84 \pm 1.00,$$

or $[3.84, 5.84]$.

11.29 Using the coded x^*'s given in Exercise 11.14, we obtain $\widehat{\beta}_1^* = .655$ and $s^2 = 10.97$. Further

$$\sum(x_i^* - \overline{x^*})^2 = 2360.2388$$

To test H_0: $\beta_1^* = 0$, we use the test statistic

$$t = \frac{.655}{\sqrt{\frac{10.97}{2360.2388}}} = 9.61$$

The rejection region is $|t| > t_{.025,8} = 2.306$, and H_0 is rejected. There is evidence of a linear relationship between X^* and Y (and hence between X and Y).

11.30 Refer to Exercise 11.29, where we computed $t_8 = 9.61$. From Table 5 we see that $P(t_8 > 3.355) = .005$. Thus, the p-value $= P(|t_8| > 9.61) = 2P(t_8 > 9.61) < 2(.005) = .01$.

11.31 $V\left(a_0\widehat{\beta}_0 + a_1\widehat{\beta}_1\right) = \left(\frac{a_0^2\frac{\sum x_i^2}{n}+a_1^2-2a_0a_1\overline{x}}{S_{xx}}\right)$

Letting $a_0 = 1$ and $a_1 = x^*$, we have

$$V\left(\widehat{\beta}_0 + \widehat{\beta}_1 x^*\right) = \frac{\frac{\sum x_i^2}{n}+(x^*)^2-2x^*\overline{x}}{S_{xx}} = \frac{\frac{\sum x_i^2-\frac{1}{n}(\sum x_i)^2}{n}+(x^*)^2-2x^*\overline{x}+\overline{x}^2}{S_{xy}}$$

$$= \frac{\frac{S_{xx}}{n}+(x^*-\overline{x})^2}{S_{xx}} = \frac{1}{n} + \frac{(x^*-\overline{x})^2}{S_{xx}}.$$

Since $(x^* - \overline{x})^2 \geq 0$ for all x^*, $V\left(\widehat{\beta}_0 + \widehat{\beta}_1 x^*\right)$ is minimized for $(x^* - \overline{x})^2 = 0$ or $x^* = \overline{x}$.

11.32 Refer to Exercises 11.13, 11.20, and 11.31. When $x^* = 5$, $\widehat{y} = 452.119 - 29.402(5) = 305.11$. From Exercise 11.31,

$$V(\widehat{y}) = V\left(\widehat{\beta}_0 + \widehat{\beta}_1 x\right) = \left[\frac{1}{14} + \frac{\left(5-\frac{86.48}{14}\right)^2}{732.4876-\left(\frac{(86.48)^2}{14}\right)}\right] s^2 = \left(\frac{1}{14} + \frac{1.3857}{198.288}\right)(5138.97)$$

$= 402.98$

Thus, a 90% confidence interval for $E(Y)$ is

$$\widehat{y} \pm t_{.05,12}\sqrt{V(\widehat{y})} = 305.11 \pm 1.782\sqrt{402.98} = 305.11 \pm 35.773.$$

11.33 From Exercises 11.4 and 11.14, $\widehat{y} = 7.15$ when $x = 12$; $s^2 = 10.97$; $S_{xx} = 2369.929$. With the result of Exercise 11.31, the 95% confidence interval is

$$\widehat{y} \pm t_{.025,8}\sqrt{s^2\left[\frac{1}{n} + \frac{(x^*-\overline{x})^2}{S_{xx}}\right]} = 7.15 \pm 2.306\sqrt{10.97\left[\frac{1}{10} + \frac{(12-15.504)^2}{2359.929}\right]} = 7.15 \pm 2.48$$

or $[4.67,\ 9.63]$.

11.34 Refer to Exercises 11.1 and 11.19, where $s^2 = .1333$, $\widehat{y} = 1.5 - .6x$, and $\overline{x} = 0$.

(1) When $x_0^* = 0$, $\widehat{y} = 1.5$, and the 90% confidence interval for $E(Y)$ is

$$\widehat{y} \pm t_{.05,3}\sqrt{s^2\left[\frac{1}{5} + \frac{(x_0^*-\overline{x}^*)^2}{10}\right]} = 1.5 \pm 2.353\sqrt{.1333\left(\tfrac{1}{5}\right)} = 1.5 \pm .38$$

or $[1.12, 1.88]$.

(2) When $x_0^* = -2$, $\widehat{y} = 1.5 + .6(2) = 2.7$ and the 90% confidence interval is

$$2.7 \pm 2.353\sqrt{.1333\left(\tfrac{1}{5} + \tfrac{4}{10}\right)} = 2.7 \pm .67$$

or $[2.03,\ 3.37]$

(3) When $x_0^* = 2$, $\widehat{y} = 1.5 - 1.2 = .3$, and the 90% confidence interval for $E(Y)$ is

$$.3 \pm 2.353\sqrt{.1333\left(\tfrac{1}{5} + \tfrac{4}{10}\right)}$$

$.3 \pm .67$ or $-.37$ to $.97$

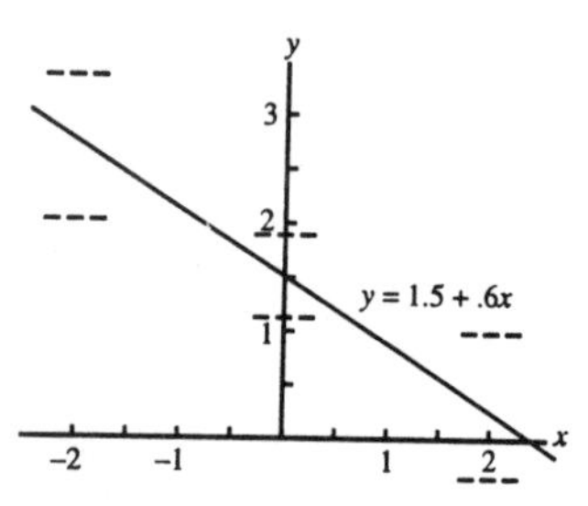

Figure 11.7

These three intervals are shown in Figure 11.7. Notice the interval is shorter when $x_0^* = \overline{x} = 0$.

11.35 Refer to Exercise 11.12. When $x = 65$, $\widehat{y} = 25.395$, and a 95% confidence interval for $E(Y)$ is

$$\widehat{y} \pm t_{.025,10}\sqrt{s^2\left[\frac{1}{12} + \frac{(65-\overline{x})^2}{S_{xx}}\right]} = 25.395 \pm 2.228\sqrt{19.0333\left[\frac{1}{12} + \frac{(65-60)^2}{6000}\right]}$$

or 25.395 ± 2.875.

11.36 Refer to Exercise 11.10. The student may verify that

$S_{xx} = .20625$ $\quad S_{xy} = -.27125$ $\quad S_{yy} = .37225$

$\overline{x} = .325$

Then

$$\text{SSE} = .0155$$

and

$$s^2 = .0019394.$$

When $x = .3$, $\widehat{y} = .7878$, and the 90% confidence interval for $E(Y)$ is

$$.7878 \pm 1.86\sqrt{.0019394}\sqrt{\frac{1}{10} + \frac{(.3-.325)^2}{.20625}} = .7878 \pm .0263$$

or $[.76, .81]$.

11.37 **a.** Using $\widehat{\beta}_0 = \overline{Y} - \widehat{\beta}_1\overline{x}$ and $\widehat{\beta}_1$ as estimators, we have

$$\widehat{\mu}_y = \overline{Y} - \widehat{\beta}_1\overline{x} + \widehat{\beta}_1\mu_x = \overline{Y} + \widehat{\beta}_1(\mu_x - \overline{x})$$

b. Calculate

$$V(\widehat{\mu}_y) = V(\overline{Y}) + (\mu_x - \overline{x})^2 V(\widehat{\beta}_1) = \frac{\sigma^2}{n} + (\mu_x - \overline{x})^2 \times \frac{\sigma^2}{S_{xx}} = \sigma^2\left[\frac{1}{n} + \frac{(\mu_x-\overline{x})^2}{S_{xx}}\right]$$

From Exercise 11.2,

$S_{yy} = 53{,}932.9$ $\quad \text{SSE} = S_{yx} = \widehat{\beta}_1$

$S_{xy} = 53{,}832.9 - (.99139)(54{,}243) = 56.845444$

Thus,

$$\sigma^2 = \frac{56.845444}{8} = 7.1056805$$

and

$$\widehat{\mu}_y = 72.1 + .9913916(74 - 72) = 74.09$$

and

$$V(\widehat{\mu}_y) = 7.1056805\left[\frac{1}{10} + \frac{(74-72)^2}{54{,}714}\right] = .711.$$

Then the two standard deviation error bound would be $2 \times \sqrt{.711} = 1.69$.

11.38 In Exercise 11.31, we showed that

$$V(\widehat{\beta}_0 + \widehat{\beta}_1 x) = \frac{1}{n} + \frac{(x-\overline{x})^2}{S_{xx}}.$$

Hence, the prediction interval can be written as

$$\widehat{y} \pm t_{\alpha/2}\, s\sqrt{1 + \frac{1}{n} + \frac{(x-\overline{x})^2}{S_{xx}}}.$$

as shown in the text. When $x = \overline{x}$, the length of this interval is

$$2t_{\alpha/2}\, s\sqrt{1 + \frac{1}{n}}.$$

which is the minimum value the length can take on. Notice as long as the confidence coefficient stays the same, the width of the interval is controlled by $\frac{(x-\overline{x})^2}{S_{xx}}$, which is minimized at $x = \overline{x}$. I.e. our prediction will be the best at the mean of the independent variable.

11.39 Refer to Exercises 11.3 and 11.13. When $x = 9$, $\widehat{y} = 65.15$, and the 95% prediction interval is

$$65.15 \pm 2.447\sqrt{3.05\left[1 + \frac{1}{8} + \frac{(9-4.5)^2}{42}\right]} = 65.15 \pm 5.42$$

or $[59.73, 70.57]$.

11.40 When the year is 1981, $x = 10$ and $\widehat{y} = 69.99$. The 95% prediction interval is

$$69.99 \pm 2.447\sqrt{3.05\left[1 + \frac{1}{8} + \frac{30.25}{42}\right]} = 69.99 \pm 5.80$$

or $[64.19, 75.79]$.

When the year is 1982, $x = 11$ and $\widehat{y} = 74.83$. The 95% prediction interval is

$$74.83 \pm 2.447\sqrt{3.05\left[1 + \frac{1}{8} + \frac{42.25}{42}\right]} = 74.83 \pm 6.24$$

or $[68.59, 81.07]$.

To answer the final question, we have no reason to expect the linear relationship to hold beyond the limits of experimentation. Therefore we would not feel comfortable predicting at $x = 1988$.

11.41 When $x = 12$, $\widehat{y} = 7.15$, and the 95% prediction interval is

$$7.15 \pm 2.306\sqrt{s^2\left[1 + \frac{1}{10} + \frac{(12-15.504)^2}{2359.929}\right]} = 7.15 \pm 8.03$$

or $[-.86, 15.18]$.

11.42 When $x = 65$, $\widehat{y} = 25.395$. A 95% prediction interval is

$$25.395 \pm 2.228\sqrt{19.0333\left[1 + \frac{1}{12} + \frac{(65-60)^2}{6000}\right]} = 25.395 \pm 10.136$$

11.43 Refer to Exercise 11.10. When $x = .6$, $\widehat{y} = .3933$. To calculate SSE and s^2, refer to Exercise 11.11. The 95% prediction interval is

$$.3933 \pm 2.306\sqrt{.0019394\left[1 + \frac{1}{10} + \frac{(.6-.325)^2}{.20625}\right]} = .3933 \pm .12$$

or $[.27, .51]$.

11.44 Let X denote high temperature and let Y denote peak load. Then

$\sum x_i = 915$ $\quad$ $\sum y_i = 1948$ $\quad$ $\sum x_i y_i = 180{,}798$

$\sum x_i^2 = 84{,}103$ $\quad$ $\sum y_i^2 = 398{,}734$ $\quad$ $n = 10$

$S_{xy} = 2556.0$ $\quad$ $S_{xx} = 380.5$ $\quad$ $S_{yy} = 19{,}263.6$

and $r = \frac{S_{xy}}{S_{xx}S_{yy}} = .944$. The hypothesis to be tested is

$$H_0\colon\ \rho = 0 \qquad \text{vs.} \qquad H_a\colon\ \rho > 0.$$

Since $\rho_0 = 0$, we can use the t statistic as the test statistic

$$t = \frac{r\sqrt{n-2}}{\sqrt{1-r^2}} = \frac{.944\sqrt{8}}{\sqrt{1-(.944)^2}} = 8.0923$$

From Table 7, with $n-2 = 8$ degrees of freedom, the p-value $< .005$. We reject H_0 when $\alpha = .05$.

11.45 From Exercise 11.14, $S_{xy} = 1546.459$ and $S_{yy} = 1101.1686$. Also,

$$S_{xx} = 4763.979 - \left[\frac{(155.04)^2}{10}\right] = 2359.929.$$

Then

$$r = \frac{1546.553}{\sqrt{2360.2388(1101.1686)}} = .9593$$

The hypothesis of interest is

$$H_0\colon\ \rho = 0 \qquad \text{vs.} \qquad H_a\colon\ \rho \neq 0.$$

As in Exercise 11.44, since $\rho_0 = 0$, we use

$$t = \frac{r\sqrt{n-2}}{\sqrt{1-r^2}} = \frac{.9593\sqrt{8}}{\sqrt{1-(.9593)^2}} = 9.608.$$

From Table 7, with 8 degrees of freedom, the p-value $< 2(.005) = .01$. With $\alpha = .01$. we reject H_0 and conclude that the correlation is significantly different from zero.

11.46 **a.** Refer to Exercise 9.56. The method of moments estimator σ_X^2 was shown to be

$$\widehat{\sigma}_X^2 = \frac{1}{n}\sum_{i=1}^{n}(x_i - \overline{x})^2.$$

Therefore, the method of moments estimator for σ_X is

$$\widehat{\sigma}_X = \sqrt{\frac{1}{n}\sum_{i=1}^{n}(x_i - \overline{x})^2}\,.$$

Similarly for σ_Y.

b. $$\widehat{\rho} = \frac{\frac{1}{n}\sum_{i=1}^{n}(x_i-\overline{x})(y_i-\overline{y})}{\sqrt{\frac{1}{n}\sum_{i=1}^{n}(x_i-\overline{x})^2}\sqrt{\frac{1}{n}\sum_{i=1}^{n}(y_i-\overline{y})^2}} = \frac{\frac{1}{n}\sum_{i=1}^{n}(x_i-\overline{x})(y_i-\overline{y})}{\frac{1}{n}\sqrt{\sum_{i=1}^{n}(x_i-\overline{x})^2}\sqrt{\sum_{i=1}^{n}(y_i-\overline{y})^2}}$$

$$= \frac{\sum_{i=1}^{n}(x_i-\overline{x})(y_i-\overline{y})}{\sqrt{\sum_{i=1}^{n}(x_i-\overline{x})^2}\sqrt{\sum_{i=1}^{n}(y_i-\overline{y})^2}} = \frac{S_{xy}}{\sqrt{S_{xx}S_{yy}}}$$

which is the formula for the estimator r.

11.47 We know $\widehat{\beta}_1 = \frac{S_{xy}}{S_{xx}}$, $S = \sqrt{\left(\frac{1}{n-2}\right)\left(S_{yy} - \widehat{\beta}_1 S_{xy}\right)}$, and $r = \frac{S_{xy}}{\sqrt{S_{xx}S_{yy}}} = \widehat{\beta}_1\sqrt{\frac{S_{xx}}{S_{yy}}}$.

With this information,

$$T = \frac{\widehat{\beta}_1 - 0}{\frac{s}{\sqrt{S_{xx}}}} = \frac{\sqrt{S_{xx}}\,\widehat{\beta}_1\left(\sqrt{n-2}\right)}{\sqrt{S_{yy}-\widehat{\beta}_1 S_{xy}}} = \frac{\sqrt{\frac{S_{xx}}{S_{yy}}}\left(\widehat{\beta}_1\right)\sqrt{n-2}}{\sqrt{1-\left(\widehat{\beta}_1\frac{S_{xy}}{S_{yy}}\right)}} = \frac{r\sqrt{n-2}}{\sqrt{1-\left(\frac{S_{xy}^2}{S_{xx}S_{yy}}\right)}} = \frac{r\sqrt{n-2}}{\sqrt{1-r^2}}.$$

11.48 Since the two samples are independent, we form the statistic

$$z = \frac{\frac{1}{2}\ln\left(\frac{1+r_1}{1-r_1}\right) - \frac{1}{2}\ln\left(\frac{1+r_2}{1-r_2}\right) - \frac{1}{2}\ln\left(\frac{1+\rho_1}{1-\rho_1}\right) + \frac{1}{2}\ln\left(\frac{1+\rho_2}{1-\rho_2}\right)}{\sqrt{\left(\frac{1}{n_1-3}\right)+\left(\frac{1}{n_2-3}\right)}}$$

which is approximately normally distributed for large n. The hypothesis of interest is

$$H_0\colon\ \rho_1 = \rho_2 \qquad \text{vs.} \qquad H_a\colon\ \rho_1 \neq \rho_2.$$

Hence

$$z = \frac{\left(\frac{1}{2}\right)\ln\left(\frac{1.9593}{.0407}\right) - \left(\frac{1}{2}\right)\ln\left(\frac{1.85}{.15}\right)}{\sqrt{\left(\frac{1}{7}\right)+\left(\frac{1}{17}\right)}} = 1.52$$

For $\alpha = .05$, H_0 is rejected if $|z| > 1.96$. Therefore, do not reject H_0.

11.49 Refer to Example 11.9 and use the results given there. Then the 90% prediction interval is

$$.979 \pm 2.132(.045)\sqrt{1 + \frac{1}{6} + \frac{(1.5-1.457)^2}{.234}} = .979 \pm .104$$

or $[.875, 1.083]$.

11.50 Using the calculations in Example 11.10, we have

$$r = \frac{S_{xy}}{\sqrt{S_{xx}S_{yy}}} = .9904.$$

11.51 a. Let $W = \ln Y$ so that $E(W) = \ln \alpha_0 - \alpha_1 x = \beta_o + \beta_1 x$ where $\ln \alpha_0 = \beta_o$ and $-\alpha_1 = \beta_1$. After calculating individual values of W for each of the $n = 10$ observations, we have

$\sum x = 55$ $\quad \sum w = 35.505412$ $\quad \sum xw = 194.49729$

$\sum x^2 = 385$ $\quad \sum w^2 = 126.07188$ $\quad n = 10$

$S_{xw} = -.782481$ $\quad S_{xx} = 82.5$ $\quad S_{ww} = .008448$

Hence

$$\widehat{\beta}_1 = \frac{S_{xw}}{S_{xx}} = \frac{-.782481}{82.5} = -.00948$$

and

$$\widehat{\beta}_0 = \overline{w} - \widehat{\beta}_1 \overline{x} = 3.6027.$$

Transforming back to the original variables, we have

$$\widehat{\alpha}_1 = -\widehat{\beta}_1 = +.0095$$

and

$$\widehat{\alpha}_0 = e^{\widehat{\beta}_0} = 36.70.$$

and the prediction equation is $\widehat{y} = 36.70e^{-.0095x}$.

b. In order to find a confidence interval for α_0, we first find a confidence interval for β_0 and then transform the endpoints of the interval. The least squares estimator of β_0 is $\widehat{\beta}_0$, which has variance (given in Section 11.5)

$$V(\widehat{\beta}_0) = \frac{\sigma^2 \sum x_i^2}{n \sum (x_i - \overline{x})^2} = \frac{\sigma^2 \sum x_i^2}{n\, S_{xx}}$$

It is necessary to calculate

$$\text{SSE} = S_{ww} - \widehat{\beta}_1\, S_{xw} = .008448 - (-.00948)(-.782481) = .0010265$$

and

$$s^2 = \frac{\text{SSE}}{8} = .0001283.$$

Then the 90% confidence interval for β_0 is

$$\widehat{\beta}_0 \pm t_{.05,8} \sqrt{V(\widehat{\beta}_0)} = 3.6027 \pm 1.86 \sqrt{.0001283 \left[\frac{385}{10(82.5)}\right]},$$

Yielding the interval $[3.5883, 3.6171]$. Transforming, we have the interval (for α_o)

$$[e^{3.5883}, e^{3.6171}] = [36.17, 37.23].$$

11.52 This is similar to Exercise 11.51. Notice that $\ln(e^{-\alpha_0 X^{\alpha_1}}) = -\alpha_0 X^{\alpha_1}$. Further, $\ln(-\ln(e^{-\alpha_0 X^{\alpha_1}})) = \ln(\alpha_0) + \alpha_1 \ln(X)$. Therefore we would expect $\ln(-\ln(y))$ to be linear in $\ln(X)$. Define $V = \ln(X)$, $Z = \ln(-\ln Y)$, $\beta_o = \ln(\alpha_0)$, and finally, $\beta_1 = \alpha_1$. We fit the model $E[Z] = \beta_o + \beta_1 V$ (notice we are assuming the error remains additive as we take two natural logs, this may not be a valid assumption). To fit the model, we have

$\sum v = -10.15243$ $\quad \sum z = -13.1544$ $\quad \sum vz = 18.521577$

$\sum v^2 = 12.977909$ $\quad \sum z^2 = 28.50516$ $\quad n = 9$

$S_{vz} = 3.6828016$ $\quad S_{vv} = 1.51548$

hence

$$\widehat{\beta}_1 = \frac{S_{vz}}{S_{vv}} = 2.4142$$

and

$$\widehat{\beta}_0 = \overline{z} - \widehat{\beta}_1 \overline{v} = 1.2617.$$

Transforming these estimates, we have

$$\widehat{\alpha}_1 = \widehat{\beta}_1 = 2.4142$$

and

$$\widehat{\alpha}_0 = e^{\widehat{\beta}_0} = 3.5315.$$

Then the predictor is $\widehat{y} = \exp(-3.5315x^{2.4142})$.

11.53 Suppose $E(Y) = 1 - e^{-\beta t}$. Then $E(1 - Y) = e^{-\beta t}$ and $E[\ln(1 - Y)] = -\beta t$. Let $Y^* = \ln(1 - Y)$ and $\beta^* = -\beta$. The new equation is then $E(Y^*) = \beta^* t$ or $Y^* = \beta^* t + \epsilon$, where ϵ is normally distributed with mean 0 and variance σ^2. Using the method of least squares, we must minimize $\text{SSE} = \sum (y_i^* - \widehat{\beta}^* t_i)^2$. Differentiating with respect to $\widehat{\beta}^*$, we have

$$\frac{d\,\text{SSE}}{d\widehat{\beta}^*} = 2\sum t_i(y_i^* - \widehat{\beta}^* t_i) = 0$$
$$\Rightarrow \sum t_i y_i^* - \widehat{\beta}^* \sum t_i^2 = 0$$
$$\Rightarrow \widehat{\beta}^* = \frac{\sum t_i y_i^*}{\sum t_i^2}$$

Note that

$$V(\widehat{\beta}^*) = \frac{1}{(\sum t_i^2)^2}\sum t_i^2\,(\sigma^2) = \frac{\sigma^2}{\sum t_i^2}.$$

Under the assumed model, SSE$/\sigma^2 \sim \chi^2$ with $n-1$ df (the df is $n-1$ rather than $n-2$ as we do not have an intercept). Further we know $\widehat{\beta}^*$ is independent of SSE. Therefore $\frac{\widehat{\beta}^* - \beta^*}{\sqrt{\frac{\sigma^2}{\sum t_i^2}}} \sim$ N(0,1) implying $\frac{\widehat{\beta}^* - \beta^*}{\sqrt{\frac{\text{SSE}}{(n-1)\sum t_i^2}}} = \frac{\widehat{\beta}^* - \beta^*}{\sqrt{\frac{s^2}{\sum t_i^2}}} = \frac{\widehat{\beta}^* + \beta}{\sqrt{\frac{s^2}{\sum t_i^2}}}$ follows a t distribution with $n-1$ df. Therefore a confidence interval for β would be

$$-\widehat{\beta}^* \pm t_{n-1,\alpha/2}\sqrt{\frac{s^2}{\sum t_i^2}}.$$

A much easier (but less rigorous) interval would be as follows; $-\widehat{\beta}^* \pm z_{\alpha/2}\sqrt{\frac{\sigma^2}{\sum t_i^2}}$ is an exact confidence interval for β. Simply estimating σ by s we then have $-\widehat{\beta}^* \pm z_{\alpha/2}\sqrt{\frac{s^2}{\sum t_i^2}}$.

11.54 Using the matrix notation of Section 11.10, write

$$\boldsymbol{X} = \begin{bmatrix} 1 & -2 \\ 1 & -1 \\ 1 & 0 \\ 1 & 1 \\ 1 & 2 \end{bmatrix} \quad \boldsymbol{Y} = \begin{bmatrix} 3 \\ 2 \\ 1 \\ 1 \\ .5 \end{bmatrix} \quad \text{so that} \quad \boldsymbol{X'X} = \begin{bmatrix} 5 & 0 \\ 0 & 10 \end{bmatrix}$$

$$\boldsymbol{X'Y} = \begin{bmatrix} 7.5 \\ -6 \end{bmatrix} \quad (\boldsymbol{X'X})^{-1} = \begin{bmatrix} \frac{1}{5} & 0 \\ 0 & \frac{1}{10} \end{bmatrix}$$

Then

$$\widehat{\boldsymbol{\beta}} = (\boldsymbol{X'X})^{-1}(\boldsymbol{X'Y}) = \begin{bmatrix} 1.5 \\ -.6 \end{bmatrix}$$

so that the least squares straight line is $\widehat{y} = 1.5 - .5x$. The observed points and the fitted line were shown in Figure 11.1.

11.55 $$\boldsymbol{X} = \begin{bmatrix} 1 & -1 \\ 1 & 0 \\ 1 & 1 \\ 1 & 2 \\ 1 & 3 \end{bmatrix} \quad \boldsymbol{Y} = \begin{bmatrix} 3 \\ 2 \\ 1 \\ 1 \\ .5 \end{bmatrix} \quad \boldsymbol{X'Y} = \begin{bmatrix} 7.5 \\ 1.5 \end{bmatrix} \quad \boldsymbol{X'X} = \begin{bmatrix} 5 & 5 \\ 5 & 15 \end{bmatrix}$$

The student may verify that

$$(\boldsymbol{X'X})^{-1} = \begin{bmatrix} .3 & -.1 \\ -.1 & .1 \end{bmatrix} \quad \text{and} \quad \widehat{\boldsymbol{\beta}} = (\boldsymbol{X'X})^{-1}(\boldsymbol{X'Y}) = \begin{bmatrix} 2.1 \\ -.6 \end{bmatrix}$$

so that the least squares line is

$\widehat{y} = 2.1 - .6x$

The line is shown in Figure 11.8. Notice that for this exercise $\boldsymbol{X'X}$ was no longer diagonal and hence the calculation of $(\boldsymbol{X'X})^{-1}$ was a bit more tedious.

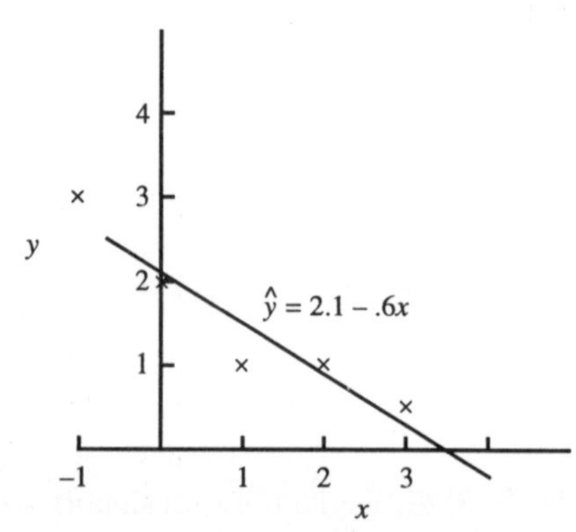

Figure 11.8

11.56 It is necessary to fit the linear model

$$y = \beta_0 + \beta_1 x + \beta_2 x^2$$

using the method of least squares. For this exercise,

$$X = \begin{bmatrix} 1 & -3 & 9 \\ 1 & -2 & 4 \\ 1 & -1 & 1 \\ 1 & 0 & 0 \\ 1 & 1 & 1 \\ 1 & 2 & 4 \\ 1 & 3 & 9 \end{bmatrix} \quad Y = \begin{bmatrix} 1 \\ 0 \\ 0 \\ -1 \\ -1 \\ 0 \\ 0 \end{bmatrix} \quad X'Y = \begin{bmatrix} -1 \\ -4 \\ 8 \end{bmatrix} \quad X'X = \begin{bmatrix} 7 & 0 & 28 \\ 0 & 28 & 0 \\ 28 & 0 & 196 \end{bmatrix}$$

The student may verify, using the procedures given in Appendix I, that

$$(X'X)^{-1} = \begin{bmatrix} .33333 & 0 & -.047619 \\ 0 & .035714 & 0 \\ -.047619 & 0 & .011905 \end{bmatrix}$$

and $\quad \widehat{\beta} = \begin{bmatrix} -.714285 \\ -.142857 \\ .142859 \end{bmatrix}$

The fitted parabola is

$$\widehat{y} = -.71 - .14x + .14x^2.$$

The fitted parabola is shown in Figure 11.9.

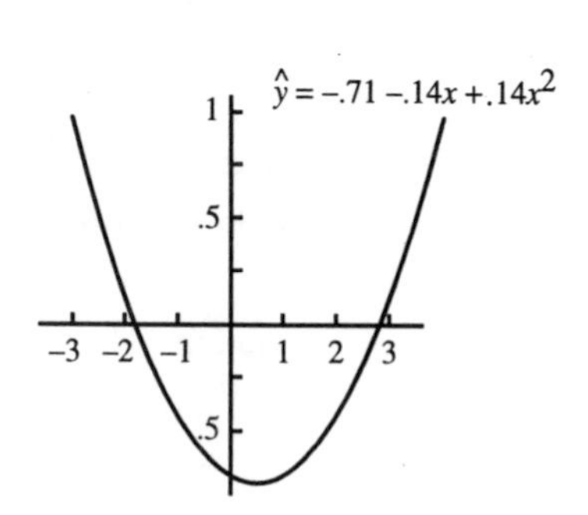

Figure 11.9

11.57a. Using the model $y = \beta_0 + \beta_1 x + \epsilon$, calculate

$$(X'X) = \begin{bmatrix} 10 & 0 \\ 0 & 330 \end{bmatrix} \quad X'Y = \begin{bmatrix} 299.9 \\ 458.3 \end{bmatrix}$$

$$\widehat{\beta} = (X'X)^{-1}(X'Y) = \begin{bmatrix} 29.99000 \\ 1.3887879 \end{bmatrix}$$

and the least squares line is

$$\widehat{y} = 29.99 + 1.39x.$$

b. Using the model $y = \beta_0 + \beta_1 x + \beta_2 c^2 + \epsilon$, calculate

$$X'X = \begin{bmatrix} 10 & 0 & 330 \\ 0 & 330 & 0 \\ 330 & 0 & 19{,}338 \end{bmatrix} X'Y = \begin{bmatrix} 299.9 \\ 458.3 \\ 8220.7 \end{bmatrix}$$

$$\widehat{\beta} = (X'X)^{-1}(X'Y) = \begin{bmatrix} 36.54 \\ 1.59 \\ -.20 \end{bmatrix}.$$

Hence the least squares line is

$$\widehat{y} = 36.54 + 1.59x - .20x^2.$$

11.58a. For Exercise 11.2,

$$X = \begin{pmatrix} 1 & 10 \\ 1 & 12 \\ 1 & 9 \\ 1 & 27 \\ 1 & 47 \\ 1 & 112 \\ 1 & 36 \\ 1 & 241 \\ 1 & 59 \\ 1 & 167 \end{pmatrix} \quad Y = \begin{pmatrix} 9 \\ 14 \\ 7 \\ 29 \\ 45 \\ 109 \\ 40 \\ 238 \\ 60 \\ 7 \end{pmatrix} \quad X'X = \begin{pmatrix} 10 & 720 \\ 720 & 106{,}554 \end{pmatrix}$$

Then

$$X'Y = \begin{pmatrix} 721 \\ 106,155 \end{pmatrix} \text{ implying } \widehat{\beta} = (X'X)^{-1}(X'Y) = \begin{pmatrix} .7198048 \\ .9913916 \end{pmatrix}.$$

Thus

$$\text{SSE} = Y'Y - \widehat{\beta}' X^{*\prime} Y = 105{,}817 - 105{,}760.155 = 56.845.$$

and

$$s^2 = \tfrac{\text{SSE}}{8} = 7.1056805.$$

b. Using the coded x_i^* values, we obtain $x_i - \overline{x}$ as shown below.

x_i^*	−62	−60	−63	−45	−25	40	−36	169	−13	95
y_i	9	14	7	29	45	109	40	238	60	70

Then

$$(X^{*\prime}X^*) = \begin{bmatrix} 10 & 0 \\ 0 & 54{,}714 \end{bmatrix} \quad X^{*\prime}Y = \begin{bmatrix} 721 \\ 54{,}243 \end{bmatrix} \quad \widehat{\beta}^* = \begin{bmatrix} 72.1 \\ .9913916 \end{bmatrix}$$

$$\text{SSE} = Y'Y - \widehat{\beta}^{*\prime} X^{*\prime} Y = 105{,}817 - 105{,}760.155 = 56.845$$

and

$$s^2 = \tfrac{\text{SSE}}{8} = 7.1057$$

(exactly the same values as in part **a.**)

11.59 Since the vector a_i is a vector of k 0's and one 1 (in the jth position), we can write

$$\beta_i = a_i'\beta \qquad \text{and} \qquad \widehat{\beta}_i = a_i'\widehat{\beta}$$

in vector notation. Then with $U = a_i'\widehat{\beta}$,

$$E(\widehat{\beta}_i) = E\left(a_i'\widehat{\beta}\right) = a_i'\, E(\widehat{\beta}) = a_i'\beta = \beta_i$$

and

$$V(\widehat{\beta}_i) = V\left(a_i'\widehat{\beta}\right) = [a_i'(X'X)^{-1}a_i]\,\sigma^2.$$

But

$$a_i'(X'X)^{-1}a_i = [c_{0i} \quad c_{1i} \quad c_{2i} \quad \cdots \quad c_{ki}] \begin{bmatrix} 0 \\ 0 \\ \cdot \\ \cdot \\ 1 \\ \cdot \\ \cdot \\ 0 \end{bmatrix} = c_{ii}$$

so that $V(\widehat{\beta}_i) = c_{ii}\sigma^2$.

11.60 **a.** From Exercise 11.57, $\widehat{\beta}_2 = -.19839$, $c_{22} = .0001184712$. Then

$$\text{SSE} = Y'Y - \widehat{\beta}' X'Y = 245.68562$$

using the matrices given in Exercise 11.57. The hypothesis to be tested is

$$H_0\colon\ \beta_2 = 0 \qquad \text{vs.} \qquad H_a\colon\ \beta_2 \neq 0$$

and the test statistic is

$$t = \frac{\widehat{\beta}_2 - 0}{\sqrt{s^2(c_{22})}} = \frac{-.19839}{\sqrt{\left(\frac{\text{SSE}}{7}\right) c_{22}}} = -3.08$$

The rejection region with $\alpha = .10$, and 7 degrees of freedom is $|t| > 1.895$, and H_0 is rejected. There is evidence of a quadratic effect.

b. The 90% confidence interval for β_2 is

$$\widehat{\beta}_2 \pm t_{.05}\sqrt{s^2 c_{22}} = -.19839 \pm 1.895(.064456) = -.20 \pm .12$$

or $[-.32, -.08]$.

11.61 If the minimum value is to occur at $x_0 = 1$, then this implies $\beta_1 + 2\beta_2 = 0$. To test this claim, let $a' = [0 \quad 1 \quad 2]$, so that $U = a'\widehat{\beta} = \widehat{\beta}_1 + 2\widehat{\beta}_2$ and the hypothesis to be tested is

$$H_0\colon\ E(U) = 0 \qquad \text{vs.} \qquad H_a\colon\ E(U) \neq 0.$$

From Exercise 11.56, we have $\widehat{\beta}$ and $(X'X)^{-1}$, with $s^2 = .14285$. Then

$$a'(X'X)^{-1}a = [-.095238 \quad .035714 \quad .02381] \begin{bmatrix} 0 \\ 1 \\ 2 \end{bmatrix} = .083334$$

and $U = \hat{\beta}_1 + 2\hat{\beta}_2 = .142861$. The test statistic is

$$t = \frac{U - E(U)}{\sqrt{s^2\,[a'(X'X)^{-1}a]}} = \frac{.142861}{\sqrt{(.14285)(.08333)}} = 1.31$$

The rejection region with $\alpha = .05$ is $|t| > 2.776$, and H_0 is not rejected.

11.62 **a.** Defining $x_1 = \frac{T_1 - 60}{10}$, $x_2 = \frac{P - 15}{5}$, $x_3 = \frac{C - 1.5}{.5}$, $x_4 = \frac{T_2 - 150}{50}$ yields the desired result. Notice we are simply subtracting the midpoint (average) and dividing by the distance to the midpoint For example the midpoint for T_1 is $(70 + 50)/2 = 60$ and each point is $(70 - 50)/2 = 10$ from the midpoint.

b. The matrix solutions are

$$Y = \begin{bmatrix} 22.2 \\ 19.4 \\ 22.1 \\ 14.2 \\ 24.5 \\ 24.1 \\ 19.6 \\ 12.7 \\ 24.4 \\ 25.2 \\ 23.5 \\ 19.3 \\ 25.9 \\ 28.4 \\ 16.5 \\ 16.0 \end{bmatrix} \qquad X = \begin{bmatrix} 1 & -1 & -1 & -1 & 1 \\ 1 & -1 & 1 & -1 & 1 \\ 1 & 1 & -1 & -1 & 1 \\ 1 & 1 & 1 & -1 & 1 \\ 1 & -1 & -1 & 1 & 1 \\ 1 & -1 & 1 & 1 & 1 \\ 1 & 1 & -1 & 1 & 1 \\ 1 & 1 & 1 & 1 & 1 \\ 1 & -1 & -1 & -1 & -1 \\ 1 & -1 & 1 & -1 & -1 \\ 1 & 1 & -1 & -1 & -1 \\ 1 & 1 & 1 & -1 & -1 \\ 1 & -1 & -1 & 1 & -1 \\ 1 & -1 & 1 & 1 & -1 \\ 1 & 1 & -1 & 1 & -1 \\ 1 & 1 & 1 & 1 & -1 \end{bmatrix} \qquad X'Y = \begin{bmatrix} 338 \\ -50.2 \\ -19.4 \\ -2.6 \\ -20.4 \end{bmatrix}$$

$$X'X = \begin{bmatrix} 16 & 0 & 0 & 0 & 0 \\ 0 & 16 & 0 & 0 & 0 \\ 0 & 0 & 16 & 0 & 0 \\ 0 & 0 & 0 & 16 & 0 \\ 0 & 0 & 0 & 0 & 16 \end{bmatrix}$$

$$(X'X)^{-1} = \begin{bmatrix} \frac{1}{16} & 0 & 0 & 0 & 0 \\ 0 & \frac{1}{16} & 0 & 0 & 0 \\ 0 & 0 & \frac{1}{16} & 0 & 0 \\ 0 & 0 & 0 & \frac{1}{16} & 0 \\ 0 & 0 & 0 & 0 & \frac{1}{16} \end{bmatrix} \qquad \beta = \begin{bmatrix} 21.125 \\ -3.1375 \\ -1.2125 \\ -.1625 \\ -1.275 \end{bmatrix}$$

and the fitted model is

$$\hat{y} = 21.125 - 3.1375x_1 - 1.2125x_2 - .1625x_3 - 1.275x_4$$

c. Calculate

$$\text{SSE} = Y'Y - \hat{\beta}'X'Y = 7446.52 - 7347.7075 = 98.8125$$

and

$$s^2 = \frac{\text{SSE}}{n - [k+1]} = \frac{98.8125}{16 - 5} = 8.98.$$

The test of

$$H_0\colon \beta_i = 0 \qquad \text{vs.} \qquad H_a\colon \beta_i \neq 0$$

for $i = 1, 2, 3, 4$, will be based on the test statistic

$$t_i = \frac{\hat{\beta}_i}{s\sqrt{c_{ii}}} = \frac{4\hat{\beta}_i}{\sqrt{8.98}} = \frac{\hat{\beta}_i}{.7493}.$$

For $\alpha = .01$, each null hypothesis will be rejected, $i = 1, 2, 3, 4$, if $|t| > t_{.005,11} = 3.106$. The test statistics are

$$t_1 = \frac{-3.1375}{.7493} = -4.19 \qquad t_3 = \frac{-.1625}{.7493} = -.22$$
$$t_2 = \frac{-1.2125}{.7493} = -1.62 \qquad t_4 = \frac{-1.275}{.7493} = -1.70$$

Hence for $i = 1$, the null hypothesis is rejected, while for $i = 2, 3, 4$, the null hypothesis is not rejected.

For β_1, the p-value is less than .01 since $|t| > t_{.005,11} = 3.106$. For β_2 and β_4, $|t|$ is between $t_{.05,11}$ and $t_{.10,11}$; thus $.1 < p\text{-value} < .2$. For β_3, $|t|$ is less than $1.363 = t_{.10,11}$; thus the p-value $> .2$.

11.63 When $T_1 = 50$, $P = 20$, $C = 1$, and $T_2 = 200$, we have $x_1 = -1$, $x_2 = 1$, $x_3 = -1$, and $x_4 = 1$, so that

$$\boldsymbol{a}' = [1 \quad -1 \quad 1 \quad -1 \quad 1]$$

and the

$$\boldsymbol{a}'(\boldsymbol{X}'\boldsymbol{X})^{-1}\boldsymbol{a} = \tfrac{5}{16} = .3125$$

The estimate of $E(Y)$ at this particular setting is

$$\widehat{y} = 21.125 + 3.1375 - 1.2125 + .1625 - 1.275 = 21.9375$$

and the confidence interval is (based on 11 df)

$$\widehat{y} \pm t_{\alpha/2}\, s\sqrt{\boldsymbol{a}'(\boldsymbol{X}'\boldsymbol{X})^{-1}\boldsymbol{a}} = 21.94 \pm 1.796\sqrt{8.98}\,\sqrt{.3125} = 21.94 \pm 3.01$$

or $[18.93, 24.95]$.

11.64 If the year is 1977, $x = 11$ and $\widehat{y} = 36.54 + 1.39(11) - .20(121) = 27.63$ and

$$\boldsymbol{a}' = [1 \quad 11 \quad 121].$$

Then

$$\boldsymbol{a}'(\boldsymbol{X}'\boldsymbol{X})^{-1}\boldsymbol{a} = [-.243744 \quad .0333 \quad .0104167]\begin{bmatrix} 1 \\ 11 \\ 121 \end{bmatrix} = 1.3833$$

and

$$\text{SSE} = \boldsymbol{Y}'\boldsymbol{Y} - \widehat{\boldsymbol{\beta}}'\boldsymbol{X}'\boldsymbol{Y} = 245.68$$

and

$$s^2 = \tfrac{245.68}{7} = 35.0978.$$

Then the 98% prediction interval is

$$27.63 \pm 2.998\sqrt{35.0978(1 + 1.3833)} = 27.63 \pm 27.42,$$

or $[.21, 55.05]$.

11.65 Refer to Exercises 11.62 and 11.63. For the given levels

$\widehat{y} = 21.9375$ and $\boldsymbol{a}'(\boldsymbol{X}'\boldsymbol{X})^{-1}\boldsymbol{a} = .3135$ and $s^2 = 8.98$.

Hence the 90% prediction interval is (based on 11 df)

$$21.94 \pm 1.796\sqrt{8.98(1 + .3135)} = 21.94 \pm 6.17$$

or $[15.77, 28.11]$.

11.66 From Exercise 11.27, the complete model is $Y = \beta_0 + \beta_1 x + \epsilon$ with $\text{SSE}_c = 6.452$ (8 d.f.) The reduced model is $Y = \beta_0 + \epsilon$ with $\widehat{\beta}_0 = \overline{Y}$ and $\text{SSE}_R = \sum(Y_i - \widehat{Y})^2 = \sum(Y_i - \overline{Y})^2 = 99.149$ (9 d.f.). The test statistic for testing H_0: $\beta_1 = 0$ is

$$F = \frac{S_3^2}{S_2^2} = \frac{\frac{\text{SSE}_R - \text{SSE}_c}{9-8}}{\frac{\text{SSE}_c}{8}} = \frac{92.697}{\frac{6.452}{8}} = 114.94$$

The rejection region with $\alpha = .05$ is $F_{1,8} = 5.32$, and H_0 is rejected. There is evidence to indicate that $\beta_1 \neq 0$.

11.67 First, note that with $k = 1$, $\text{SSE}_R = S_{yy}$. Then, $g = 0$, and $\text{SSE}_c = S_{yy} - \widehat{\beta}_1 S_{xy}$.

$$F = \frac{(\text{SSE}_R - \text{SSE}_c)(k-g)}{\frac{\text{SSE}_c}{n-(k+1)}} = \frac{S_{yy} - S_{yy} + \widehat{\beta}_1 S_{xy}}{\frac{\text{SSE}_c}{n-2}} = \frac{\widehat{\beta}_1 S_{xy}}{\frac{\text{SSE}_c}{n-2}} = \frac{\frac{\widehat{\beta}_1 S_{xy}}{S_{xy}}}{\frac{s^2}{S_{xx}}} = \frac{\widehat{\beta}_1^2}{\frac{s^2}{S_{xx}}}$$
$$= t^2$$

In Exercise 11.66, $F = 114.94$, and in Exercise 11.27, $t = -10.72$. Thus, $t^2 = 1$ which equals F (except for roundoff error).

11.68 a. Refer to Exercise 11.57. For the complete model, $Y = \beta_0 + \beta_1 x + \beta_2 x^2 + \epsilon$, $\text{SSE}_c = 245.69$ (see Exercise 11.60) with 7 degrees of freedom. For the reduced model, $Y = \beta_0 + \beta_1 x + \epsilon$,

$$\text{SSE}_R = \boldsymbol{Y'Y} - \beta_1 \boldsymbol{X'Y} = 10.208.67 - \begin{bmatrix} 29.99 & 1.3879 \end{bmatrix} \begin{bmatrix} 299.9 \\ 458.3 \end{bmatrix} = 578.6$$

with 8 d.f. The test statistic for testing H_0: $\beta_2 = 0$ is

$$F = \frac{\frac{\text{SSE}_R - \text{SSE}_c}{8-7}}{\frac{\text{SSE}_c}{7}} = \frac{332.91}{\frac{245.69}{7}} = 9.49$$

The rejection region with $\alpha = .05$ is $F > F_{1,7} = 5.59$, and H_0 is rejected. There is evidence that $\beta_2 \neq 0$. These results do agree with the results of problem 11.60.

b. For the reduced model, $Y = \beta_0 + \epsilon$, $\text{SSE}_R = \sum (y_i - \overline{y})^2 = 1214.669$ with 9 d.f. Then

$$F = \frac{\frac{\text{SSE}_R - \text{SSE}_c}{9-7}}{\frac{\text{SSE}_c}{7}} = \frac{\frac{968.98}{2}}{\frac{245.69}{7}} = 13.80$$

The rejection region with $\alpha = .05$ is $F > F_{2,7} = 4.74$, and the null hypothesis, H_0: $\beta_1 = \beta_2 = 0$, is rejected.

11.69 The hypothesis of interest is H_0: $\beta_1 - \beta_4 = 0$. For the complete model,

$Y = \beta_0 + \beta_1 x_1 + \beta_2 x_2 + \beta_3 x_3 + \beta_4 x_4 + \epsilon$, $\text{SSE}_c = 98.8125$ with 11 d.f.

For the reduced model, the 2nd and 5th columns of the $\boldsymbol{X}$ matrix are deleted, so that

$$(\boldsymbol{X'X})^{-1} = \begin{bmatrix} \frac{1}{16} & 0 & 0 \\ 0 & \frac{1}{16} & 0 \\ 0 & 0 & \frac{1}{16} \end{bmatrix} \qquad \boldsymbol{X'Y} = \begin{bmatrix} 338 \\ -19.4 \\ -2.6 \end{bmatrix}$$

Then

$$\text{SSE} = \boldsymbol{Y'Y} - \widehat{\boldsymbol{\beta}}' \boldsymbol{X'Y} = 7446.52 - 7164.195 = 282.325$$

with 13 d.f. The test statistic is

$$F = \frac{\frac{\text{SSE}_R - \text{SSE}_c}{2}}{\frac{\text{SSE}_c}{11}} = 10.21$$

The rejection region with $\alpha = .05$ is $F > F_{2,11} = 3.98$, and H_0 is rejected. There is reason to believe that either T_1 or T_2 or both affect the yield.

11.70 We need to carry out an F test with test statistic

$$F = \frac{\frac{\text{SSE}_R - \text{SSE}_c}{3}}{\frac{\text{SSE}_c}{24-6}} = \frac{\frac{465.134 - 152.177}{3}}{\frac{152.177}{18}} = 12.34$$

Refer to Table 7. We have $12.34 > 5.92 = F_{.005}$, thus the p-value $< .005$.

11.71 Refer to Example 11.13. For the reduced model, $S^2 = \frac{326.623}{8} = 40.83$,

$$(\boldsymbol{X'X})^{-1} = \begin{bmatrix} \frac{1}{11} & 0 & 0 \\ 0 & \frac{1}{11} & 0 \\ 0 & 0 & \frac{1}{11} \end{bmatrix} \qquad \boldsymbol{a}' = \begin{bmatrix} 1 & 1 & -1 \end{bmatrix}$$

Then $\widehat{y} = \boldsymbol{a}'\widehat{\boldsymbol{\beta}} = 93.73 + 4.00 - 7.35 = 90.38$, $\boldsymbol{a}'(\boldsymbol{X'X})^{-1}\boldsymbol{a} = .31620$, and the 95% confidence interval for $E(Y)$ is

$$\widehat{y} \pm t_{.025}\sqrt{s^2 \boldsymbol{a}'(\boldsymbol{X'X})^{-1}\boldsymbol{a}} = 90.38 \pm 2.306\sqrt{40.83(.32620)} = 90.38 \pm 8.42$$

or $[81.96, 98.80]$.

11.72 The hypothesis of interest is

$$H_0\colon \beta_i = 0 \qquad \text{vs.} \qquad H_a\colon \beta_i \neq 0$$

for $i = 3, 4, 5$, and the test statistics are

$$t_3 = \frac{\widehat{\beta}_3}{\sqrt{s^2 c_{33}}} = \frac{-.88}{\sqrt{\left(\frac{77.948}{5}\right)(.15)}} = -.58$$

$$t_4 = \frac{\widehat{\beta}_4}{\sqrt{s^2 c_{44}}} = \frac{-4.66}{\sqrt{s^2(.15)}} = -3.05$$

$$t_5 = \frac{\widehat{\beta}_5}{\sqrt{s^2 c_{55}}} = \frac{5.00}{\sqrt{s^2(.25)}} = 2.53$$

The rejection region for each test is $|t| > t_{.005,5} = 4.032$. Hence we conclude that none of the three intercepts is significantly different from zero.

11.73a. Calculate

$\sum x = -2682.8$ $\sum y = 6.826$ $\sum xy = -1847.0073$
$\sum x^2 = 720{,}039.3$ $\sum y^2 = 5.6326$ $n = 10$
$S_{xy} = 15.728$ $S_{xx} = 297.716$ $S_{yy} = .9732$

Then

$$\widehat{\beta}_1 = \frac{S_{xy}}{S_{xx}} = -.053 \quad \text{and} \quad \widehat{\beta}_0 = \frac{6.826}{10} - \widehat{\beta}_1\left(\frac{-2682.8}{10}\right) = -13.54$$

The least squares line is $\widehat{y} = -13.54 - .053x$.

b. Note that

$$\text{SSE} = S_{yy} - \widehat{\beta}_1 S_{xy} = .9732 - (-.053)(-15.72802) = .14225$$

and

$$s^2 = \frac{.14225}{8} = .01778$$

Also,

$$V(\widehat{\beta}_1) = \frac{\sigma^2}{\sum(x_i - \overline{x})^2} = \frac{s^2}{297.716}. \text{ Thus,}$$

$$t = \frac{\widehat{\beta}_1}{\sqrt{V(\widehat{\beta}_1)}} = \frac{-.053}{\sqrt{\frac{.01778}{297.716}}} = -6.84$$

With $\alpha = .01$, we reject H_0 if $t < -2.896$. Thus, we reject H_0.

c. With $x = -273$, $\widehat{y} = -13.54 + .053(.273) = .93$. A 95% prediction interval is

$$\widehat{y} \pm t_{.025}\, s\sqrt{1 + \frac{1}{n} + \frac{(x_0 - \overline{x})^2}{\sum(x_i - \overline{x})^2}}$$

or

$$.93 \pm 2.306\sqrt{.01778}\sqrt{1 + \frac{1}{10} + \frac{(273 - 268.28)^2}{297.716}}$$

or $.93 \pm .33$.

11.74a. Calculate

$\sum x = 4.951$ $\sum y = 118.76$ $\sum xy = 59.207$
$\sum x^2 = 2.489$ $\sum y^2 = 1415.704$ $n = 10$
$S_{xy} = .4089$ $S_{xx} = .0378$ $S_{yy} = 1413.25$

Thus,

$$\widehat{\beta}_1 = \frac{S_{xy}}{S_{xx}} = 10.817$$

and

$$\widehat{\beta}_0 = 11.876 - 10.817(.4951) = 6.52.$$

b. $\text{SSE} = S_{yy} - \widehat{\beta}_1 S_{xy} = 1413.25 - 10.817(.4089) = .8871$

and

$$s^2 = \frac{.8871}{8} = .1109$$

The test statistic is

$$t = \frac{10.817}{\sqrt{\frac{.1109}{.0378}}} = 6.32$$

Since $6.32 > 2.306 = t_{.025}$, we reject H_0.

c. When $x = .59$, $\widehat{y} = 6.52 + 10.817(.59) = 12.902$. A 90% confidence interval for $E(Y)$ is

$$12.902 \pm 1.860\sqrt{.1109}\sqrt{\frac{1}{10} + \frac{(.59 - .4951)^2}{.0378}} = 12.902 \pm .36.$$

11.75a.

$$X = \begin{bmatrix} 1 & -3 & 5 & -1 \\ 1 & -2 & 0 & 1 \\ 1 & -1 & -3 & 1 \\ 1 & 0 & -4 & 0 \\ 1 & 1 & -3 & -1 \\ 1 & 2 & 0 & -1 \\ 1 & 3 & 5 & 1 \end{bmatrix} \qquad Y = \begin{bmatrix} 1 \\ 0 \\ 0 \\ 1 \\ 2 \\ 3 \\ 3 \end{bmatrix} \qquad X'Y = \begin{bmatrix} 10 \\ 14 \\ 10 \\ -3 \end{bmatrix}$$

$$X'X = \begin{bmatrix} 7 & 0 & 0 & 0 \\ 0 & 28 & 0 & 0 \\ 0 & 0 & 84 & 0 \\ 0 & 0 & 0 & 6 \end{bmatrix} \qquad (X'X)^{-1} = \begin{bmatrix} \frac{1}{7} & 0 & 0 & 0 \\ 0 & \frac{1}{28} & 0 & 0 \\ 0 & 0 & \frac{1}{84} & 0 \\ 0 & 0 & 0 & \frac{1}{6} \end{bmatrix}$$

$$\widehat{\beta} = (X'X)^{-1}X'Y = \begin{bmatrix} 1.4285 \\ .5000 \\ .1190 \\ -.5000 \end{bmatrix}$$

and the fitted model is $\widehat{y} = 1.4825 + .5000x_1 + .1190x_2 - .5000x_3$.

b. When $x_1 = 1$, $x_2 = -3$, and $x_3 = -1$, the predicted value of y is

$$\widehat{y} = 1.4285 + .5000 - .3570 + .5000 = 2.0715$$

whereas the observed response at this setting was $y = 2$. The difference appears because the former is a predicted value based on a model fit using all of the data where as the latter is an observed response.

c. Calculate

$$\text{SSE} = Y'Y - \widehat{\beta}'X'Y = 24 - 23.9757 = .0243$$

and

$$s^2 = \frac{\text{SSE}}{n-4} = \frac{.0243}{3} = .008.$$

In order to test the hypothesis

$$H_0\colon \beta_3 = 0 \qquad \text{vs.} \qquad H_a\colon \beta_3 \neq 0,$$

we use the test statistic

$$t = \frac{\widehat{\beta}_3 - \beta_3}{s\sqrt{c_{33}}} = \frac{-.5000}{\sqrt{.008\left(\frac{1}{6}\right)}} = \frac{-.5000}{.0365} = -13.7$$

The rejection region, with $\alpha = .05$ and 3 degrees of freedom, will be $|t| > t_{.025,3} = 3.182$, and the null hypothesis is rejected.

d. Refer to Section 11.8 of the text and write

$$\widehat{Y} = U = \sum_{i=0}^{k} a_i \widehat{\beta}_i$$

In this exercise, $\widehat{Y} = U = \widehat{\beta}_0 + \widehat{\beta}_1 - 3\widehat{\beta}_2 - \widehat{\beta}_3$ and $a' = [1 \quad 1 \quad -3 \quad -1]$. A 95% confidence interval for $E(Y)$ is given by

$$\widehat{Y} \pm t_{\alpha/2} S\sqrt{a'(X'X)^{-1}a}$$

where

$$a'(X'X)^{-1}a = \begin{bmatrix} \frac{1}{7} & \frac{1}{28} & -\frac{3}{84} & -\frac{1}{6} \end{bmatrix} \begin{bmatrix} 1 \\ 1 \\ -3 \\ -1 \end{bmatrix} = .45238$$

Hence the 95% confidence interval is

$$2.0715 \pm 3.182\sqrt{.008}\,\sqrt{.45238} = 2.07 \pm .19$$

e. Refer to Section 11.9. The 95% prediction interval for Y will be

$$\widehat{y} \pm t_{\alpha/2}\, s\sqrt{1 + a'(X'X)^{-1}a} = 2.07 \pm 3.182\sqrt{.008}\,\sqrt{1.45238}$$

or $2.07 \pm .34$.

11.76 Symmetric spacing about the origin simplifies the $X'X$ matrix and produces zero entries in the off-diagonal elements, making the matrix easier to invert.

11.77 It is known that $V(\widehat{\beta}_1) = \frac{\sigma^2}{\sum_{i=1}^{n} (X_i - \overline{X})^2}$. This will be minimized when $\sum_{i=1}^{n} (X_i - \overline{X})^2$ is maximized. However, $\sum_{i=1}^{n} (X_i - \overline{X})^2$ will be maximized when all the X_i are as far away from $\overline{X}$ as possible. That is, take $\frac{n}{2}$ data points at $X = -9$ and $\frac{n}{2}$ data points at $X = 9$.

11.78 Suppose $n = 10$. Based on the spacing used in Exercise 11.77, the ten values of X will be $x = -9, -9, -9, -9, -9, 9, 9, 9, 9, 9$, and

$$\sum_{i=1}^{n} (x_i - \overline{x})^2 = \sum_{i=1}^{n} x_i^2 = 810$$

so that

$$V(\widehat{\beta}_1) = \frac{\sigma^2}{810}.$$

If equal spacing is employed, then the ten values of X will be $x = -9, -7, -5, -3, -1, 1, 3, 5, 7, 9$, and

$$\sum_{i=1}^{n}(x_i-\overline{x})^2=\sum_{i=1}^{n}x_i^2=330$$

so that

$$V(\widehat{\beta}_1^*)=\frac{\sigma^2}{330}.$$

The relative efficiency of $\widehat{\beta}_1^*$ to $\widehat{\beta}_1$ is

$$\frac{V(\widehat{\beta}_1)}{V(\widehat{\beta}_1^*)}=\frac{330}{810}=\frac{11}{27}$$

11.79 a.
$$\boldsymbol{Y}=\begin{bmatrix}8.0\\9.0\\9.1\\10.2\\10.4\\10.0\\10.3\\12.2\\12.6\\13.9\end{bmatrix}\quad \boldsymbol{X}=\begin{bmatrix}1&0&-2&0\\1&0&-1&0\\1&0&0&0\\1&0&1&0\\1&0&2&0\\1&1&-2&-2\\1&1&-1&-1\\1&1&0&0\\1&1&1&1\\1&1&2&2\end{bmatrix}\quad \boldsymbol{X}'\boldsymbol{Y}=\begin{bmatrix}105.7\\59.0\\16.1\\10.1\end{bmatrix}$$

$$\boldsymbol{X}'\boldsymbol{X}=\begin{bmatrix}10&5&0&0\\5&5&0&0\\0&0&20&10\\0&0&10&10\end{bmatrix}$$

The student may verify that

$$(\boldsymbol{X}'\boldsymbol{X})^{-1}=\begin{bmatrix}.2&-.2&0&0\\-.2&.4&0&0\\0&0&.1&-.1\\0&0&-.1&.2\end{bmatrix}\quad \text{and}\quad \widehat{\boldsymbol{\beta}}=\begin{bmatrix}9.34\\2.46\\.60\\.46\end{bmatrix}$$

b. The observed points and the two growth lines are shown in Figure 11.10. They are

$$\widehat{y}_A=\widehat{\beta}_0+\widehat{\beta}_2x_2=9.34=.60x_2$$

and

$$\widehat{y}_B=(\widehat{\beta}_0+\widehat{\beta}_1)+(\widehat{\beta}_2+\widehat{\beta}_3)x_3$$
$$=11.80+1.01x_2$$

c. If we are interested in the growth of bacteria A, then $x_1=0$, $x_2=0$, $x_1x_2=0$, and the prediction is

$$\widehat{y}=\widehat{\beta}_0=9.34$$

For bacteria B, $x_1=1$, $x_2=0$, $x_1x_2=0$ and

$$\widehat{y}=\widehat{\beta}_0+\widehat{\beta}_1=9.34+2.46=11.80$$

The observed growths for bacteria A and B were 9.1 and 12.2, respectively.

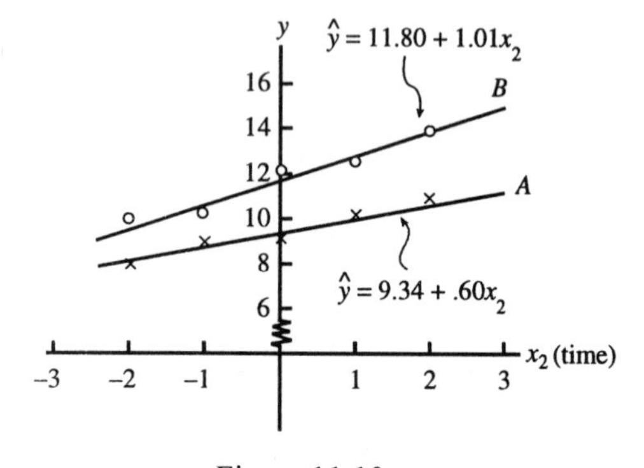

Figure 11.10

d. Calculate

$$\text{SSE}=\boldsymbol{Y}'\boldsymbol{Y}-\widehat{\boldsymbol{\beta}}'\boldsymbol{X}'\boldsymbol{Y}=1146.91-1146.179=.731$$

and

$$s^2=\frac{\text{SSE}}{n-(k+1)}=\frac{.731}{6}=.121833.$$

Refer to Figure 11.10. The difference in the rates of growth for the two types of bacteria is represented by the difference in the slopes of the two lines. Since for bacteria A the model is $y=\beta_0+\beta_2x_2+\epsilon$ and for bacteria B the model is $y=(\beta_0+\beta_1)+(\beta_2+\beta_3)x_2+\epsilon$, the growth rates will be different only if $\beta_3\neq 0$. Hence it is necessary to test

$$H_0\colon\ \beta_3=0 \qquad \text{vs.} \qquad H_a\colon\ \beta_3\neq 0$$

using the test statistic

$$t=\frac{\widehat{\beta}_3-\beta_3}{s\sqrt{c_{33}}}=\frac{.41}{\sqrt{.121833}\sqrt{.175}}=2.63$$

With $\alpha = .05$ and 6 degrees of freedom, the rejection region will be $|t| > t_{.025,6} = 2.447$. The null hypothesis is rejected, and there is sufficient evidence to detect a difference in the growth rates.

e. It is given that $x_1 = 1$ and $x_2 = 1$, so that $\widehat{y} = 12.81$,

$$\boldsymbol{a}' = \begin{bmatrix} 1 & 1 & 1 & 1 \end{bmatrix}$$

and

$$\boldsymbol{a}'(\boldsymbol{X}'\boldsymbol{X})^{-1}\boldsymbol{a} = \begin{bmatrix} 0 & \frac{1}{5} & 0 & \frac{1}{10} \end{bmatrix} \begin{bmatrix} 1 \\ 1 \\ 1 \\ 1 \end{bmatrix} = \frac{3}{10}.$$

A 90% confidence interval for $E(Y)$ at this particular setting will be

$$\widehat{y} \pm t_{\alpha/2}\, s\sqrt{\boldsymbol{a}'(\boldsymbol{X}'\boldsymbol{X})^{-1}\boldsymbol{a}} = 12.81 \pm 1.943\sqrt{.121833}\,\sqrt{\tfrac{3}{10}} = 12.81 \pm .37.$$

f. Refer to part **e**. A 90% prediction interval for growth of bacteria B at time $x_2 = 1$ is

$$\widehat{y} \pm t_{\alpha/2}\, s\sqrt{1 + \boldsymbol{a}'(\boldsymbol{X}'\boldsymbol{X})^{-1}\boldsymbol{a}} = 12.81 \pm 1.943\sqrt{.121833}\,\sqrt{1 + .3}$$

or $12.81 \pm .78$.

11.80

The F statistic is

$$F = \frac{\frac{\text{SSE}_1 - \text{SSE}_2}{2}}{\frac{\text{SSE}_2}{200-5}} = \frac{\frac{795.23 - 783.9}{2}}{\frac{783.9}{195}} = 1.41$$

The critical value is $F_{.05} = 3.00 > 1.41$, so we do not reject H_0: salary is not dependent on sex.

11.81

Let

$$\boldsymbol{Y}' = \begin{bmatrix} Y_1 & Y_2 & Y_3 & \cdots & Y_n \end{bmatrix} \quad \text{and} \quad \boldsymbol{1}' = \begin{bmatrix} 1 & 1 & 1 & \cdots & 1 \end{bmatrix}$$

be two $1 \times n$ vectors. Then we can write

$$\overline{Y} = \begin{bmatrix} \frac{1}{n} & \frac{1}{n} & \cdots & \frac{1}{n} \end{bmatrix} \begin{bmatrix} Y_1 \\ Y_2 \\ Y_3 \\ \vdots \\ Y_n \end{bmatrix} = \frac{1}{n}\boldsymbol{1}'\boldsymbol{Y}$$

In matrix form, the equation of interest is $Y = \boldsymbol{x}'\widehat{\boldsymbol{\beta}}$, where

$$\boldsymbol{x}' = \begin{bmatrix} 1 & x_1 & x_2 & \cdots & x_k \end{bmatrix}$$

and

$$\widehat{\boldsymbol{\beta}}' = \begin{bmatrix} \widehat{\beta}_0 & \widehat{\beta}_1 & \cdots & \widehat{\beta}_k \end{bmatrix}.$$

Suppose $Y = \overline{Y}$. Then

$$\overline{Y} = \boldsymbol{x}'\widehat{\boldsymbol{\beta}} = \boldsymbol{x}'(\boldsymbol{X}'\boldsymbol{X})^{-1}\boldsymbol{X}'\boldsymbol{Y}$$

implying

$$\frac{1}{n}\boldsymbol{1}'\boldsymbol{Y}\boldsymbol{Y}' = \boldsymbol{x}'(\boldsymbol{X}'\boldsymbol{X})^{-1}\boldsymbol{X}'\boldsymbol{Y}\boldsymbol{Y}'.$$

Which must then imply

$$\frac{1}{n}\boldsymbol{1}' = \boldsymbol{x}'(\boldsymbol{X}'\boldsymbol{X})^{-1}\boldsymbol{X}'$$

or

$$\frac{1}{n}\boldsymbol{1}'\boldsymbol{X} = \boldsymbol{x}'(\boldsymbol{X}'\boldsymbol{X})^{-1}\boldsymbol{X}'\boldsymbol{X}$$

or

$$\frac{1}{n}\boldsymbol{1}'\boldsymbol{X} = \boldsymbol{x}'.$$

That is

$$\boldsymbol{x}' = \begin{bmatrix} 1 & \overline{x}_1 & \overline{x}_2 & \cdots & \overline{x}_k \end{bmatrix}.$$

That is, the point $(\overline{x}_1, \overline{x}_2, \ldots, \overline{x}_k, \overline{Y})$ satisfies the equation $Y = \boldsymbol{x}'\widehat{\boldsymbol{\beta}}$, so that the least squares prediction line must pass through this point.

11.82

a. Use the coding

$$X_1^* = \frac{X_1 - 65}{15}$$

and

$$X_2^* = \frac{X_2 - 200}{100}.$$

Then $X_1 = 50, 80$ correspond to $X_1^* = -1, 1$, while $X_2 = 100, 200, 300$ correspond to $X_2^* = -1, 0, 1$. Now

$$\boldsymbol{Y} = \begin{bmatrix} 21 \\ 23 \\ 26 \\ 22 \\ 23 \\ 28 \end{bmatrix} \quad \boldsymbol{X}^* = \begin{bmatrix} 1 & -1 & -1 & 1 \\ 1 & -1 & 0 & 0 \\ 1 & -1 & 1 & 1 \\ 1 & 1 & -1 & 1 \\ 1 & 1 & 0 & 0 \\ 1 & 1 & 1 & 1 \end{bmatrix} \quad \boldsymbol{X}^{*\prime}\boldsymbol{X}^* = \begin{bmatrix} 6 & 0 & 0 & 4 \\ 0 & 6 & 0 & 0 \\ 0 & 0 & 4 & 0 \\ 4 & 0 & 0 & 4 \end{bmatrix}$$

(columns of $\boldsymbol{X}^*$: $x_0 \quad x_1 \quad x_2 \quad x_2^2$)

$$(\boldsymbol{X}^{*\prime}\boldsymbol{X}^*)^{-1} = \begin{bmatrix} \frac{1}{2} & 0 & 0 & -\frac{1}{2} \\ 0 & \frac{1}{6} & 0 & 0 \\ 0 & 0 & \frac{1}{4} & 0 \\ -\frac{1}{2} & 0 & 0 & \frac{3}{4} \end{bmatrix} \quad \boldsymbol{X}^{*\prime}\boldsymbol{Y} = \begin{bmatrix} 143 \\ 3 \\ 11 \\ 97 \end{bmatrix} \quad \widehat{\boldsymbol{\beta}}^* = \begin{bmatrix} 23 \\ .5 \\ 2.75 \\ 1.25 \end{bmatrix}$$

Finally, the least squares equation is

$$\begin{aligned}\widehat{y} &= 23 + .5\left(\tfrac{x_1-65}{15}\right) + 2.75\left(\tfrac{x_2-200}{100}\right) + 1.25\left(\tfrac{x_2-200}{100}\right)^2 \\ &= 20.33 + .0333x_1 - .0225x_2 + .000125x_2^2\end{aligned}$$

b. The hypothesis of interest, H_0: $\beta_3 = 0$, is equivalent to a test of H_0: $\beta_3^* = 0$ since $\beta_3 = \left(\frac{1}{100}\right)^2 \beta_3^*$. Using the coded calculations, then, we have

$$\text{SSE} = \boldsymbol{Y}'\boldsymbol{Y} - \widehat{\boldsymbol{\beta}}^{*\prime}\boldsymbol{X}'\boldsymbol{Y} = 3443 - 3442 = 1$$

and

$$s^2 = \tfrac{\text{SSE}}{n-4} = .5.$$

The test statistic is

$$t = \frac{\widehat{\beta}_3^*}{\sqrt{c_{33}s^2}} = \frac{1.25}{\sqrt{\left(\frac{1}{2}\right)\left(\frac{3}{4}\right)}} = 2.042$$

The rejection region with $\alpha = .05$ is $|t| > 4.303$, and H_0 is not rejected. The quadratic temperature effect is not significant.

c. In order to test H_0: $\beta_2 = \beta_3 = 0$, or, equivalently, H_0: $\beta_2^* = \beta_3^* = 0$, complete and reduced models are fitted. For the complete model, $\text{SSE}_2 = 1$ with 2 d.f. [from part **b**]. For the reduced model, columns 3 and 4 of the $\boldsymbol{X}^*$ matrix are omitted and

$$(\boldsymbol{X}^{*\prime}\boldsymbol{X}^*)^{-1} = \begin{bmatrix} \frac{1}{6} & 0 \\ 0 & \frac{1}{6} \end{bmatrix}$$

$$\boldsymbol{X}^*\boldsymbol{Y} = \begin{bmatrix} 143 \\ 3 \end{bmatrix}$$

so that

$$\text{SSE}_1 = 3443 - \begin{bmatrix} 23.8333 & .5 \end{bmatrix} \begin{bmatrix} 143 \\ 3 \end{bmatrix} = 33.33.$$

with 4 d.f. The test statistic is then

$$F = \frac{\frac{\text{SSE}_1 - \text{SSE}_2}{2}}{\frac{\text{SSE}_2}{2}} = \frac{\frac{33.33-1}{2}}{\frac{1}{2}} = 32.33$$

The rejection region with $\alpha = .05$ is $F > F_{2,2} = 19.00$, and the null hypothesis is rejected. Temperature does affect yield.

11.83a. Let $S_{xy} = \sum (x_i - \overline{x})(y_i - \overline{y})$. Write

$$\widehat{\beta}_1 = \frac{S_{xy}}{S_{xx}} = \frac{S_{xy}}{\sqrt{S_{xx}S_{yy}}}\sqrt{\frac{S_{yy}}{S_{xx}}} = \text{r}\sqrt{\frac{S_{yy}}{S_{xx}}}$$

b. If X and Y have a bivariate normal distribution then the conditional distribution of Y_i, given $X_i = x_i$, will be normal with mean

$$E(Y_i|X_i = x_i) = \mu_y + \rho\,\tfrac{\sigma_y}{\sigma_x}(x_i - \mu_x)$$

and variance

$$V(Y_i|X_i = x_i) = \sigma_y^2(1-\rho^2)$$

(see exercise 5.109). Redefine $\beta_1 = \rho\left(\frac{\sigma_y}{\sigma_x}\right)$, $\beta_0 = \mu_y - \beta_1\mu_x$. Notice if $\rho = 0,\ \beta_1 = 0$.
In Exercise 11.29 we showed in general that

$$T = \frac{\widehat{\beta}_i - \beta_i}{S\sqrt{c_{ii}}}$$

has a t distribution with $n-(k+1)$ degrees of freedom. In particular, for $k=1$ and $i=1$,

$$\frac{\widehat{\beta}_1 - \beta_1}{S\sqrt{\frac{1}{S_{xx}}}}$$

has a t distribution with $(n-2)$ degrees of freedom. Finally, if $\rho = 0$, then $\beta_1 = 0$ and the t statistic becomes

$$T = \frac{\widehat{\beta}_1}{S\sqrt{\frac{1}{S_{xx}}}} = \frac{\widehat{\beta}_1}{\sqrt{\frac{\text{SSE}}{(n-2)}}\sqrt{\frac{1}{S_{xx}}}} = \frac{\widehat{\beta}_1\sqrt{(n-2)S_{xx}}}{\sqrt{(1-r^2)\,S_{yy}}}$$

since

$$\text{SSE} = (1-r^2)\,S_{yy}.$$

c. Since $\widehat{\beta}_1 = r\sqrt{\frac{S_{yy}}{S_{xx}}}$, the t statistic in part **b** becomes

$$T = \frac{r\sqrt{n-2)}}{\sqrt{1-r^2}}$$

and it has a t distribution with $(n-2)$ degrees of freedom. Since the t distribution depends only on $\nu = n-2$ and not on the particular value of X_i, the distribution of T will be the same no matter what values of X_i are observed. Hence T has the same distribution unconditionally.

11.84 Calculate

$\sum x = 1480$	$\sum y = 1572.3$	$\sum xy = 290{,}946.62$
$\sum x^2 = 273{,}866.54$	$\sum y^2 = 309{,}109.89$	$n = 8$
$S_{xx} = 66.54$	$S_{yy} = 93.97875$	$S_{xy} = 71.12$

Then

$$r = \frac{S_{xy}}{\sqrt{S_{xx}S_{yy}}} = \frac{71.12}{\sqrt{(66.54)(93.97895)}} = .8994.$$

Note from Exercise 11.83 that

$$\widehat{\beta}_1 = r\sqrt{\frac{S_{yy}}{S_{xx}}} = .8994\sqrt{\frac{93.97875}{66.54}} = 1.06883$$

and

$$\text{SSE} = (1-r^2)\,S_{yy} = [1-(.8994)^2]\,(93.97875) = 17.957$$

Thus,

$$s^2 = \frac{17.957}{6} = 2.99.$$

Testing H_0: $\rho = 0$ is equivalent to testing H_0: $\beta_1 = 0$. The test statistic is

$$t = \frac{\widehat{\beta}_1}{s\sqrt{S_{xx}}} = \frac{1.06883}{\sqrt{\frac{2.99}{66.54}}} = 5.04$$

Since $5.03 > 3.707$, the p-value is less than $2(.005) = .01$.

11.85 The necessary computations are

$\sum x = 741$	$\sum y = 108$	$\sum xy = 10{,}016.3$
$\sum x^2 = 68{,}789$	$\sum y^2 = 1459.34$	$S_{xx} = 153.875$
$S_{yy} = 1.34$	$S_{xy} = 12.8$	

a. $r = \frac{12.8}{\sqrt{153.875(1.34)}} = .89$

b. We need to test

$$H_0\colon \beta_1 = 0 \qquad \text{vs.} \qquad H_a\colon \beta_1 \neq 0.$$

Now

$$\widehat{\beta}_1 = \frac{12.8}{153.875} = .0832$$

Also,

$$\text{SSE} = S_{yy} - \widehat{\beta}_1 S_{xy} = 1.34 - .0832(12.8) = .275$$

so

$$s^2 = \frac{.275}{6} = .046.$$

The test statistic is

$$t = \frac{.0832}{\sqrt{\frac{.046}{153.875}}} = 4.81$$

For $\alpha = .05$ the rejection region is $|t| > t_{.025,6} = 2.447$. Thus, we reject H_0.
Since $t = 4.81 > 3.707$, the p-value is less than $2(.005) = .01$.

CHAPTER 12 CONSIDERATIONS IN DESIGNING EXPERIMENTS

12.1 Suppose that $\sigma_1^2 = 9$, $\sigma_2^2 = 25$, $n = 90$. Then $V(\overline{Y}_1 - \overline{Y}_2)$ will be minimized when

$$n_1 = \left(\frac{\sigma_1}{\sigma_1+\sigma_2}\right) n = \left(\frac{3}{3+5}\right) 90 = 33.75 \text{ or } 34 \qquad \text{and} \qquad n_2 = 90 - 34 = 56$$

12.2 Refer to Exercise 12.1. If $n_1 = 34$ and $n_2 = 56$, then

$$\sigma_{\overline{Y}_1 - \overline{Y}_2} = \sqrt{\frac{\sigma_1^2}{34} + \frac{\sigma_2^2}{56}} = \sqrt{\frac{9}{34} + \frac{25}{56}} = \sqrt{.7111}$$

In order to achieve this same bound with $n_1 = n_2 = n$, we must have

$$\sqrt{\frac{9}{n} + \frac{25}{n}} = \sqrt{.7111} \qquad \text{or} \qquad \frac{34}{n} = .7111 \qquad \text{or} \qquad n = \frac{34}{.7111} = 47.8$$

Hence, in order to acquire the same amount of information as that implied by the proportional allocation of Exercise 12.1, it is necessary to use $n_1 + n_2 = 48 + 48 = 96$ experimental units.

12.3 From Exercise 12.1 we know that $n_1 = \left(\frac{3}{8}\right) n$ and $n_2 = \left(\frac{5}{8}\right) n$, where n is the combined sample size. The length of a 95% confidence interval for $(\mu_1 - \mu_2)$ is twice the half width, or

$$2(1.96)\sqrt{\frac{\sigma_1^2}{n_1} + \frac{\sigma_2^2}{n_2}}$$

This width must be equal to two units. Hence

$$2 = 2(1.96)\sqrt{\frac{9}{\left(\frac{3}{8}\right)n} + \frac{25}{\left(\frac{5}{8}\right)n}} \qquad \text{or} \qquad 1 = (1.96)^2\left[\frac{\left(\frac{45}{8}\right)+\left(\frac{75}{8}\right)}{\left(\frac{15}{64}\right)n}\right]$$

or

$$1 = (1.96)^2\left[\frac{120(8)}{15n}\right] \Rightarrow n = 245.9 \Rightarrow n_1 = 93 \text{ and } n_2 = 154.$$

12.4 Refer to Exercise 12.3. If $n_1 = n_2$, then the equation to be solved is

$$2 = 2(1.96)\sqrt{\frac{9}{n_1} + \frac{25}{n_1}}$$

or

$$1 = \frac{34(1.96)^2}{n_1}$$

or

$$n_1 = 130.6$$

or $n_1 = 131$. The total sample size is $n_1 + n_2 = 2(131) = 262$.

12.5 Refer to Section 12.2. The maximum quantity of information, or the minimum value for $\sigma_{\widehat{\beta}_1}$, occurs when the six data points are equally divided, with half located at $x = 2$ and half located at $x = 5$. Notice that this allocation of rats provides no information on curvature of the response curve, and it is necessary that the experimenter be almost certain that the response is linear. Again, we are employing the principle of signal amplification, since we are concerned with the selection of treatments (dosages) and the number of experimental units (rats) to be assigned to each dosage.

12.6 A $100(1-\alpha)\%$ confidence interval for β_1 when σ is known is

$$\widehat{\beta}_1 \pm Z_{\alpha/2}\frac{\sigma}{\sqrt{\sum_i (x_i - \overline{x})^2}}$$

where x_i are the dose levels for $i = 1, 2, \ldots, 6$. Suppose that method 1 is used to select the six dose levels. Then $x_1 = x_2 = x_3 = 2$ and $x_4 = x_5 = x_6 = 5$. Hence the half width of the confidence interval will be

$$Z_{\alpha/2}\frac{\sigma}{\sqrt{\sum x_i^2 - \frac{(\sum x_i)^2}{6}}} = Z_{\alpha/2}\frac{\sigma}{\sqrt{87 - \frac{(21)^2}{6}}} = Z_{\alpha/2}\frac{\sigma}{\sqrt{13.5}}$$

If method 2 is used, then $x_1 = 2$, $x_2 = 2.6$, $x_3 = 3.2$, $x_4 = 3.8$, $x_5 = 4.4$, $x_6 = 5.0$, and the half width of the confidence interval is

$$Z_{\alpha/2} \frac{\sigma}{\sqrt{79.8 - \frac{(21)^2}{6}}} = Z_{\alpha/2} \frac{\sigma}{\sqrt{6.3}}$$

Hence method 2 produces a larger confidence interval; in fact, the ratio of the two half widths is

$$R = \frac{\frac{\sigma Z_{\alpha/2}}{\sqrt{6.30}}}{\frac{\sigma Z_{\alpha/2}}{\sqrt{13.50}}} = \sqrt{\frac{13.5}{6.3}} = 1.46$$

The confidence interval produced by method 2 is 1.46 times as large as the interval produced by method 1.

In order to obtain the same amount of information as implied by the optimal assignment, it would be necessary to take repeated measurements at each of the six dose levels. Suppose we take n measurements at each of the six dose levels. Then the reader may verify that the half width of the confidence interval is

$$\frac{\sigma Z\alpha/2}{\sqrt{n \sum_{i=1}^{6} (x_i - \overline{x})^2}} = \frac{\sigma Z\alpha/2}{\sqrt{6.30n}}$$

To obtain the same half width as obtained by using method 1, we need

$$\frac{\sigma Z\alpha/2}{\sqrt{6.30n}} = \frac{\sigma Z\alpha/2}{\sqrt{13.50}} \qquad \text{or} \qquad \sqrt{13.50} = \sqrt{6.30n} \qquad \text{or} \qquad n = 2.14$$

Hence, slightly more than twice as many observations are required.

12.7 Refer to Exercise 12.5. If all the experimental units are assigned to the points $x = 2$ and $x = 5$, the experimenter has no way to determine whether or not the response function is truly linear over the experimental region. By assigning one or two points at $x = 3.5$, the experimenter can check for curvature in the response function.

12.8 Spreading the x's out as much as possible makes it less likely that the assumption that the variance of the error term ϵ does not depend on the value of the independent variable x.

12.9 **a.** Because the two halves come from the same sample, the percentage of iron ore in each should be similar. If both methods work reasonably well, then each will be measuring similar amounts. Therefore, the observations will be positively correlated.

b. Either analysis would have worked. However, the paired analysis requires fewer assumptions, removes the unwanted variability caused by the inherent differences in the ore samples, and produces a smaller standard error because of the positive correlation discussed in part **a.**

12.10 **a.** A paired-difference test is used to test the hypothesis

$$H_0\colon \mu_1 - \mu_2 = 0 \qquad \text{vs.} \qquad H_a\colon \mu_1 - \mu_2 = 0.$$

The differences are shown below:

d_i		
$-.03$	$\sum d_i = -.13$	Then
$.01$	$\sum d_i^2 = .0073$	$\overline{d} = \frac{-.13}{6} = -.02167$
$-.06$		
$-.01$	$n = 6$	
$.01$		$s_d^2 = \frac{.0073 - \frac{(-.13)^2}{6}}{5} = .0008967$
$-.05$		

The test statistic is

$$t = \frac{\overline{d} - 0}{\frac{s_d}{\sqrt{n}}} = \frac{-.02167}{\sqrt{\frac{.0008967}{6}}} = -1.773$$

and the rejection region with $\alpha = .05$ and $n - 1 = 5$ degrees of freedom is $|t| > t_{.025,5} = 2.571$. H_0 is not rejected.

b. The p-value is

$$p\text{-value} = 2P[t > 1.772].$$

From Table 5, $t = 1.772$ falls between $t_{.10}$ and $t_{.05}$. Hence, $2(.05) < p < 2(.10)$ or $.10 < p < .20$.

c. The 95% confidence interval for $\mu_1 - \mu_2$ is

$$\overline{d} \pm t_{.025}\sqrt{\frac{s_d^2}{n}} \quad \text{or} \quad -.02167 \pm 2.57\sqrt{\frac{.0008967}{6}} \quad \text{or} \quad -.022 \pm .031$$

or

$$-.053 < \mu_1 - \mu_2 < .009.$$

12.11 a. $\sigma^2_{(\overline{Y}_1 - \overline{Y}_2)}$ equals $\sigma^2_{\overline{D}}$ if $(2\rho\sigma_1\sigma_2) = 0$. This will happen if $\rho = 0$, i.e., if the two samples are uncorrelated.

b. $\sigma^2_{(\overline{Y}_1 - \overline{Y}_2)}$ is less than $\sigma^2_{\overline{D}}$ if $2\rho\sigma_1 s_2 < 0$. This happens if $\rho < 0$, i.e., if the two samples are negatively correlated.

c. It would be better to implement the paired experiment if the samples are positively correlated and the independent samples experiment when the samples are uncorrelated or negatively correlated. Note that if the samples are uncorrelated and the variances are equal, then independent samples yield about double the degrees of freedom.

12.12 We will use a paired-difference test to test

$$H_0\colon \mu_D = 0 \quad \text{vs.} \quad H_a\colon \mu_D > 0,$$

where

$$d = \text{Procedure 1} - \text{Procedure II}.$$

The differences are shown below.

d_i		
-2	$\sum d_i = -2$	Then
1	$\sum d_i^2 = 28$	$\overline{d} = -.333$
-3		
-2	$n = 6$	
1		$s_D^2 = \frac{28 - \frac{(-2)^2}{6}}{5} = 5.466$
3		

The test statistic is

$$t = \frac{\overline{d} - 0}{\frac{s_D}{\sqrt{n}}} = \frac{-.333}{\sqrt{\frac{5.466}{6}}} = -.35$$

The p-value is

$$p\text{-value} = P(t_5 > -.35) > .10$$

(see Table 5). Since the p-value $> .05$, we would not reject H_0 at the $\alpha = .05$ level.

12.13 a. A paired-difference test is used since the samples are not independent. The hypothesis of interest is

$$H_0\colon \mu_A - \mu_B = 0 \quad \text{vs.} \quad H_a\colon \mu_A - \mu_B \neq 0.$$

The differences are

$d_i = 2, 1, 0, 3, -1, 2, 4, 1.$

Then

$$\sum d_i = 12 \quad \text{and} \quad \sum d_i^2 = 36$$

$$\overline{d} = \frac{\sum d_i}{n} = 1.5 \quad \text{and} \quad s_D^2 = \frac{\sum d_i^2 - \frac{(\sum d_i^2)}{n}}{n-1} = \frac{36-18}{7} = 2.571$$

The test statistic is

$$t = \frac{\overline{d} - 0}{\frac{s_D}{\sqrt{n}}} = \frac{1.5}{\sqrt{\frac{2.571}{8}}} = 2.65$$

The rejection region is based on a t distribution with $n - 1 = 7$ degrees of freedom, and is $|t| > 2.365$ with $\alpha = .05$. The null hypothesis is rejected, and we conclude that there is a difference between the two machines.

b. Note that the variation among technicians is very high (some high and some low), while the variation with technicians between machines is not. Hence pairing is important to screen out the variation among technicians.

c. The population of differences is normally distributed with variance σ_D^2. A random sample of n differences has been selected from this population.

12.14 a. Let D = 1978 reading − 1972 reading. Then the 15 values of d_i to be used in a paired-difference test are

−3, −6.1, −2, −4, −2.5, −8.9, −1.2, −4.2, −8.8, −3.3,
−2.3, −3.7, −2.5, 1.8, −7.5

Calculate

$\sum d_i = -58.2$ $\qquad \overline{d} = -3.88$ $\qquad n = 15$

$s_D^2 = \frac{343.8 - \frac{(58.2)^2}{15}}{14} = 8.427$

To test

H_0: $\mu_D = 0$ vs. H_a: $\mu_D < 0$

the test statistic is

$$t = \frac{-3.88}{\sqrt{\frac{8.427}{15}}} = -5.176$$

We see from Table 5 that, since $t < -2.977$, the p-value is less than .005. If $\alpha = .01$, we would reject H_0.

b. A 95% confidence interval for μ_D is

$$\overline{D} \pm t_{.025}\sqrt{\frac{s_D^2}{n}} = -3.88 \pm 2.145\sqrt{\frac{8.427}{15}} = -3.88 \pm 1.608$$

We are highly confident that the true difference in the means for 1978 and 1972 $(\mu_{1978} - \mu_{1972})$ is inside this interval.

c. Using just the 1972 data, we have $\overline{y} = \frac{553.9}{15} = 36.926$

$$s_y^2 = \frac{21{,}026.13 - \frac{(553.9)^2}{15}}{14} = 40.889$$

A 95% confidence interval for μ_{1972} is

$$36.926 \pm 2.145\sqrt{\frac{40.889}{15}} \quad \text{or} \quad 36.926 \pm 3.541$$

d. Using only the 1978 data, we have

$$\overline{x} = \frac{495.7}{15} = 33.046$$

$$s_x^2 = \frac{16{,}878.47 - \frac{(495.7)^2}{15}}{14} = 35.517$$

A 95% confidence interval for μ_{1978} is

$$33.046 \pm 2.145\sqrt{\frac{35.517}{15}} \quad \text{or} \quad 33.046 \pm 3.301$$

e. In parts **a** and **b**, we assumed that the differences were normally distributed with constant variance. In parts **c** and **d**, we assumed that the individual measurements were normally distributed with constant variances σ_Y^2 for part **c** and σ_X^2 for part **d**.

12.15 a. $E(Y_{ij}) = \mu_i + E(U_j) + E(\epsilon_{ij}) = \mu_i$.

b. The Y_{ij} are not normally distributed because the sum of a uniform random variable and a normal random variable is not normal.

c.

$$\begin{aligned}\text{Cov}(Y_{1j}, Y_{2j}) &= \text{Cov}(\mu_1 + u_j + \epsilon_{ij}, \mu_2 + u_j + \epsilon_{2j}) \\ &= \text{Cov}(u_j, u_j) + \text{Cov}(\epsilon_{1j}, \epsilon_{2j}) + \text{Cov}(u_j, \epsilon_{2j}) + \text{Cov}(u_j, \epsilon_{2j}) \\ &= V(u_j) + 0 + 0 + 0 = \tfrac{1}{3}\end{aligned}$$

d. $D_j = Y_{1j} - Y_{2j} = \mu_1 + u_j + \epsilon_{1j} - (\mu_2 + u_j + \epsilon_{2j}) = \mu_1 - \mu_2 + \epsilon_{1j} - \epsilon_{2j}$.

Since ϵ_{1j} and ϵ_{2j} are independent normals, D_j is also normal. Finally, for $j \neq j'$,

$$\begin{aligned}\text{Cov}(D_j, D_{j'}) &= \text{Cov}(\mu_1 - \mu_2 + \epsilon_{1j} - \epsilon_{2j}, \mu_1 - \mu_2 + \epsilon_{1j'} - \epsilon_{2j'}) \\ &= \text{Cov}(\epsilon_{1j} - \epsilon_{2j}, \epsilon_{1j'} - \epsilon_{2j'}) = 0\end{aligned}$$

since the ϵ_{ij} are independent.

Thus, the D_j are independent, normally distributed random variables.

e. Any choice of non normal distribution for the u_j will work as long as $E(u_j) = 0$ and $V(u_j) = \sigma_u^2 > 0$.

12.16 Use Table 12 and see Section 12.4 of the text.

12.17 Use Table 12 and see Section 12.4 of the text.

12.18 **a.** There are six treatments. One treatment would be catalyst A combined with the first temperature setting.

b. After assigning the n experimental units to the treatments, the experimental units are numbered from 1 to n. Then use the tables to randomly select numbers until all experimental units have been selected.

12.19 Randomization avoids the possibility of bias introduced by a nonrandom selection of sample elements. Also, it provides a probabilistic basis for the selection of a sample (see Section 12.4).

12.20 Factors are independent experimental variables.

12.21 A treatment is a specific combination of factor levels.

12.22 An independent variable could quite possibly be a factor in one experiment and a nuisance variable in another. Consider the enzyme growth experiment of Section 12.5. Suppose that we were interested in the differences in the samples with regard to enzyme growth. In this case, "samples" would be a factor instead of a nuisance variable.

12.23 The parts of the experimental design that would increase the accuracy of the experiment are (a) the selection of treatments and (b) choice of the number of experimental units to be assigned to each treatment. The part of the design that would decrease the impact of extraneous sources of variability is the method of assigning treatments to the experimental units.

12.24 By designing the experiment in this way, it is possible to isolate unwanted variation due to rats. Hence the random experimental error is reduced, and the amount of information in the experiment is increased. The principle of noise reduction is being employed, since we are concerned with the manner in which the treatments (dosages) should be applied to the experimental units. Notice that the experimental design described in this exercise is a randomized block design.

12.25 The treatments are assigned so that each treatment appears in each row and column exactly once. Hence, for this exercise the assignment of treatments is as follows

B	A	C
C	B	A
A	C	B

12.26 A quantity of information in a sample pertinent to a specific population parameter is the width of the confidence interval that could be constructed from the sample.

12.27 A random sample is a sample in which every possible sample from the population has an equal probability of being selected.

12.28 From Section 12.5, the choice of factor levels and the allocation of the experimental units to the treatments as well as the total number of experimental units affect the total quantity of information. Randomization and blocking control these factors.

12.29 Given the model proposed in this exercise, we have the following:

a. $E(Y_{ij}) = E(\mu_i + P_j + \epsilon_{ij}) = \mu_i + E(P_j) + E(\epsilon_{ij}) = \mu_i + 0 + 0 = \mu_i$

b. $\overline{Y}_i$ is the mean of the n observations receiving treatment i. That is,

$$\overline{Y}_i = \frac{\sum_{j=1}^{n} Y_{ij}}{n} \quad \text{and} \quad E(\overline{Y}_i) = \frac{\sum_{j=1}^{n} E(Y_{ij})}{n} = \frac{n\mu_i}{n} = \mu_i$$

Further,

$$V(\overline{Y}_i) = \frac{\sum_{j=1}^{n} V(Y_{ij})}{n^2} = \frac{\sum_{j=1}^{n} V(\mu_i+P_j+\epsilon_{ij})}{n^2} = \frac{\sum_{j=1}^{n} [V(P_j)+V(\epsilon_{ij})]}{n^2}$$
$$= \frac{n\,(\sigma_P^2+\sigma^2)}{n^2} = \frac{\sigma_P^2+\sigma^2}{n}$$

Notice that P_j and ϵ_{ij} are independent random variables for all i and j.

c. Let $\overline{D} = \overline{Y}_1 - \overline{Y}_2$. Then

$$E(\overline{D}) = E(\overline{Y}_1) - E(\overline{Y}_2) = \mu_1 - \mu_2$$

The variance of $\overline{D}$ can be most easily seen by noting that

$$\overline{D} = \frac{\sum_{j=1}^{n}(Y_{1j}-Y_{2j})}{n} = \frac{\sum_{j=1}^{n}[\mu_1-\mu_2+(P_j-P_j)+(\epsilon_{1j}-\epsilon_{2j})}{n}$$

$$= \frac{n}{n}(\mu_1-\mu_2) + \frac{\sum_{j=1}^{n}\epsilon_{1j}}{n} - \frac{\sum_{j=1}^{n}\epsilon_{2j}}{n} = (\mu_1-\mu_2) + \overline{\epsilon}_1 - \overline{\epsilon}_2$$

where $\overline{\epsilon}_i$ is the mean of the random errors associated with experimental units receiving treatment i. Then

$$V(\overline{D}) = V(\overline{\epsilon}_1) + V(\overline{\epsilon}_2) = \frac{V(\epsilon_{1j})}{n} + \frac{V(\epsilon_{2j})}{n} = \frac{2\sigma^2}{n}$$

Finally, since $\overline{D}$ is a linear combination of the Y_{ij}, which are normally distributed, $\overline{D}$ is normally distributed.

12.30 From Exercise 12.29, the quantity $U = \frac{D-(\mu_1-\mu_2)}{\sqrt{\frac{2\sigma^2}{n}}}$ has a normal distribution with zero mean and unit variance. Further, consider D_j, which has a normal distribution with mean $\mu_1 - \mu_2$ and variance $2\sigma^2$. From Theorem 7.3,

$$V = \frac{\sum(D_i-\overline{D})^2}{2\sigma^2} = \frac{(n-1)S_D^2}{2\sigma^2}$$

has a χ^2 distribution with $(n-1)$ degrees of freedom. Forming the t statistic, assuming $\mu_1 - \mu_2 = 0$, then

$$T = \frac{U}{\sqrt{\frac{V}{\nu}}} = \frac{\overline{D}\sqrt{\frac{2\sigma^2}{n}}}{\sqrt{(n-1)\frac{S_D^2}{2\sigma^2(n-1)}}} = \frac{\overline{D}\sqrt{n}}{S_D}$$

12.31 In this situation the model must be written as $Y_{ij} = \mu_i + P_{ij} + \epsilon_{ij}$ since the pair effect will differ from one of the $2n$ observations to another. That is, for a fixed pair j, the pair effect P_{ij} may be different for the first member of the pair (P_{1j}) and the second (P_{2j}). Hence

$$\overline{D} = \frac{\sum_{j=1}^{n}(Y_{1j}-Y_{2j})}{n} = \frac{\sum_{j=1}^{n}[\mu_1-\mu_2+(P_{1j}-P_{2j})+(\epsilon_{1j}-\epsilon_{2j})}{n}$$

and

$$V(\overline{D}) = \frac{1}{n^2}\sum_{j=1}^{n}[V(P_{1j}) + V(P_{2j}) + V(\epsilon_{1j}) + V(\epsilon_{2j})] = \frac{2\sigma_p^2+2\sigma^2}{n} > \frac{2\sigma^2}{n}$$

Notice that $V(\overline{D})$ is larger for the completely randomized design, since the unwanted variation due to pairs is not being eliminated.

12.32 a. We expect there to be a positive correlation because jobs that are estimated to take a long time are more likely to actually take a long time.

b. The hypothesis of interest is

$$H_0\colon \mu_1 - \mu_2 = 0 \qquad \text{vs.} \qquad H_a\colon \mu_1 - \mu_2 < 0.$$

where μ_1 is the average estimated CPU time and μ_2 is the average actual CPU time. The differences $y_1 - y_2$ are shown below along with the calculation of $\overline{d}$ and s_D^2.

d_i
.04
−.12
−.04
−.08
.04
−.11
.11
−.30
−.11

$\sum d_i = -.69$

$\sum d_i^2 = .1719$

$n = 11$

Then

$$\overline{d} = \frac{\sum d_i}{n} = \frac{-.69}{11} = -.062727,$$

$$s_D^2 = \frac{.1719 - \frac{(-.69)^2}{11}}{10} = .0128618.$$

The test statistic is then

.02 $\qquad t = \frac{\bar{d}-0}{\frac{s_D}{\sqrt{n}}} = \frac{-.062727}{\sqrt{\frac{.0128618}{11}}} = -1.834.$

$-.14$

The rejection region with $\alpha = .10$ is $t < -t_{10,.10} = -1.372$, and H_0 is rejected. There is reason to believe that the customer tends to underestimate CPU time.

c. From Table 5,

$$p\text{-value} = P[t < -1.834] = P[t > 1.834]$$

Since $t = 1.834$ falls between the two tabulated values, $t_{10,.05}$ and $t_{10,.025}$, $.025 < p\text{-value} < .05$.

d. A 90% confidence interval for μ_d is

$$\bar{d} \pm t_{10,.05}\frac{s_D}{\sqrt{n}} = -.063 \pm 1.812\sqrt{\frac{.0128618}{11}} = -.063 \pm .062$$

or

$$-.125 < (\mu_1 - \mu_2) < -.001.$$

12.33 The differences along with the calculation of $\bar{d}$ and s_D^2 are shown below.

d_i		Then
80	$\sum d_i = 450$	
90	$\sum d_i^2 = 69{,}900$	$\bar{d} = \frac{450}{4} = 112.5$
230		
50	$n = 4$	
		$s_D^2 = \frac{69{,}900 - \frac{(450)^2}{4}}{3} = 6425$

The 90% confidence interval is

$\bar{d} \pm t_{3,.05}\frac{s_D}{\sqrt{n}}$

$112.5 \pm 2.353\sqrt{\frac{6425}{4}}$ or 112.5 ± 94.104 or $18.196 < (\mu_1 - \mu_2) < 206.804$

12.34 a. Each subject was presented with both signs in random order. If his reaction time in general is high, both responses will be high; if his reaction time in general is low, both responses will be low. The large variability from subject to subject will mask the variability due to the difference in sign types. The paired-difference design will eliminate the subject to subject variability.

b. The hypothesis of interest is

$$H_0\colon \mu_1 - \mu_2 = 0 \qquad \text{vs.} \qquad H_a\colon \mu_1 - \mu_2 \neq 0$$

The differences along with the calculation of $\bar{d}$ and s_D^2 follow.

d_i		Then
122	$\sum d_i = 1069$	
141	$\sum d_i^2 = 126{,}561$	$\bar{d} = \frac{\sum d_i}{n} = \frac{1069}{10} = 106.9$
97		
107	$n = 10$	
37		
56		$s_D^2 = \frac{126{,}561 - \frac{(1069)^2}{10}}{9} = 1364.9889$
110		
146		
104		
149		

The test statistic is

$$t = \frac{\bar{d}-0}{\frac{s_D}{\sqrt{n}}} = \frac{106.9}{\sqrt{\frac{1364.9889}{10}}} = 9.150$$

The rejection region with $\alpha = .05$ and $n - 1 = 9$ d.f. is $|t| > t_{.025} = 2.262$, and H_0 is rejected.

c. The observed level of significance is

$$P[|t| > 9.150] = 2P[t > 9.150] < 2(.005) = .01$$

d. The 95% confidence interval is

$$\overline{d} \pm t_{.025}\frac{s_D}{\sqrt{n}} = 106.9 \pm 2.262\sqrt{\frac{1364.9889}{10}} = 106.9 \pm 26.428$$

or

$$80.472 < (\mu_1 - \mu_2) < 133.328.$$

12.35 There will be nk_1 points at the setting $x = -1$, nk_2 points at $x = 0$, and nk_3 points at $x = 1$. Hence the design matrix will be

$$\boldsymbol{X} = \begin{bmatrix} 1 & -1 & 1 \\ 1 & -1 & 1 \\ \vdots & \vdots & \vdots \\ 1 & -1 & 1 \\ 1 & 0 & 0 \\ 1 & 0 & 0 \\ \vdots & \vdots & \vdots \\ 1 & 0 & 0 \\ 1 & 1 & 1 \\ 1 & 1 & 1 \\ \vdots & \vdots & \vdots \\ 1 & 1 & 1 \end{bmatrix}$$

and

$$\boldsymbol{X}'\boldsymbol{X} = \begin{bmatrix} n & n(k_3 - k_1) & n(k_1 + k_3) \\ n(k_3 - k_1) & n(k_1 + k_3) & n(k_3 - k_1) \\ n(k_1 + k_3) & n(k_3 - k_1) & n(k_1 + k_3) \end{bmatrix}$$
$$= n\begin{bmatrix} 1 & b & a \\ b & a & b \\ a & b & a \end{bmatrix} = nA$$

where $a = k_1 + k_3$ and $b = k_3 - k_1$. The inverse of this matrix can be shown to be

$$\frac{1}{n\,|A|}\begin{bmatrix} a^2 - b^2 & 0 & b^2 - a^2 \\ 0 & a - a^2 & ab{-}b \\ b^2 - a^2 & ab - b & a - b^2 \end{bmatrix}$$

with

$$|A| = a^2 - b^2 - b(ab - ab) + \text{a}\,(b^2 - a^2) = (a - 1)\,(b^2 - a^2) = -k_2\,(b^2 - a2)$$
$$= -k_2\,(k_3^2 - 2k_1k_3 + k_1^2 - k_1^2 - k_3^2 - 2k_1k_3) = 4k_1k_2k_3$$

Hence

$$V(\widehat{\beta}_2) = \frac{\sigma^2\,(a - b^2)}{n4k_1k_2k_3} = \frac{\sigma^2\,[k_1 + k_2 - (k_3 - k_1)^2]}{4nk_1k_2k_3}$$

We must minimize

$$Q = \frac{(k_1+k_3)-(k_3-k_1)^2}{4k_1k_2k_3} = \frac{(k_1+k_3)-[(k_1+k_3)^2-4k_1k_3]}{4k_1k_2k_3}$$
$$= \frac{(k_1+k_3)(1-k_1-k_3)}{4k_1k_2k_3} + \frac{4k_1k_3}{4k_1k_2k_3} = \frac{k_1+k_3}{4k_1k_3} + \frac{1}{k_2} = \frac{k_1+k_3}{4k_1k_3} + \frac{1}{1-k_1-k_3}$$

Differentiating with respect to k_1 and k_2 and setting the resulting equations equal to zero implies

$$(^*)\quad 4k_1^2 = (1 - k_1 - k_3)^2$$
$$4k_3^2 = (1 - k_1 - k_3)^2$$

Hence, since k_1, k_2, k_3 are positive, $k_1 = k_3$. From (*), noting that $1 - 2k_1 = 1 - k_1 - k_3 = k_2 > 0$,

$$4k_1^2 = (1 - 2k_1)^2$$
$$2k_1 = 1 - 2k_1$$
$$k_1 = \tfrac{1}{4}$$

It follows that $k_1 = k_3 = \frac{1}{4}$ and $k_2 = 1 - k_1 - k_3 = \frac{1}{2}$.

CHAPTER 13 THE ANALYSIS OF VARIANCE

13.1 **a. & c.** The hypothesis to be tested is

$$H_0\colon \mu_1 = \mu_2 \qquad \text{vs.} \qquad H_a\colon \mu_1 \neq \mu_2$$

The analysis of variance F test will be used. A completely randomized design has been employed. We assume that independent random samples of size $n_1 = n_2 = 8$ have been drawn from two normal populations with means μ_1 and μ_2, respectively, and with common variance σ^2. The object of the analysis of variance is to partition the total sum of squares into two parts, the sum of squares for treatment and error. Then $F = \frac{\text{MST}}{\text{MSE}}$. will provide a test statistic to test the null hypothesis H_0: $\mu_1 = \mu_2$ against the alternative that this equality does not hold. In order to perform the analysis of variance, we must calculate the following quantities:

$$\bar{y}_1 = 1.875 \qquad \bar{y}_2 = 2.625 \qquad \sum_{ij} y_{ij}^2 = 94 \qquad \sum_{ij} y_{ij} = 36$$

Using the formulas given in this section, calculate

$$\text{SST} = \frac{n_1}{2}(\bar{y}_1 - \bar{y}_2)^2 = \frac{8}{2}(1.875 - 2.625)^2 = 4(.5625) = 2.25$$

$$\text{TSS} = \sum_i \sum_j y_{ij}^2 - \frac{\left(\sum_i \sum_j y_{ij}\right)^2}{16} = 94 - \frac{(36)^2}{16} = 94 - 81 = 13$$

Then

$$\text{SSE} = \text{TSS} - \text{SST} = 13 - 2.25 = 10.75$$

For the test of the hypothesis of equality of means, the appropriate mean squares are

$$\text{MST} = \frac{\text{SST}}{1} = 2.25$$

and

$$\text{MSE} = \frac{\text{SSE}}{2n_1 - 2} = .7679$$

and

$$F = \frac{\text{MST}}{\text{MSE}} = \frac{2.25}{.7679} = 2.93$$

Notice that the critical value for rejection with $\alpha = .05$, based on 1 and 14 degrees of freedom, is $F = 4.60$. Hence the null hypothesis is not rejected. There is not sufficient evidence to conclude that there is a difference in mean reaction times for the two stimuli, using a 5% significance level (p-value $> .10$).

b. The value MSE $= .7679$ is the same as s^2. The t statistic is

$$t = \frac{\bar{y}_1 - \bar{y}_2}{\sqrt{s^2\left[\left(\frac{1}{n_1}\right) + \left(\frac{1}{n_2}\right)\right]}} = \frac{1.875 - 2.625}{\sqrt{.7679\left[\left(\frac{1}{8}\right) + \left(\frac{1}{8}\right)\right]}} = -1.712$$

Since $|-1.712| < 2.145 = t_{.025,14}$, we do not reject H_0. Note that $t^2 = F$, i.e., $(-1.712)^2 = 2.93$.

13.2 **a.** $\bar{y} = \frac{15(446) + 15(534)}{30} = 490$

$$\text{MST} = \frac{\text{SST}}{1} = n_1 \sum_{i=1}^{2} (\bar{y}_i - \bar{y})^2 = 15\left[(534 - 490)^2 + (446 - 490)^2\right] = 58{,}080$$

$$\text{MSE} = \frac{\text{SSE}}{2n_1 - 2} = \frac{(n_1 - 1)s_1^2 + (n_2 - 1)s_2^2}{2n_1 - 2} = \frac{14(42)^2 + 14(45)^2}{28} = 1894.5$$

Then the test statistic is

$$F = \frac{58{,}040}{1894.5} = 30.64$$

Since $F > F_{.005} = 9.28$ with 1 and 28 numerator and denominator degrees of freedom, respectively, the p-value $< .005$. We would reject at $\alpha = .05$ and conclude that there is a difference in mean verbal SAT scores for high school students who intend to major in engineering and language/literature.

b. F in part **a** is the square of the t statistic obtained in Exercise 10.63.

c. We assume that independent random samples have been drawn from two normal populations of verbal SAT scores for high school students who intend to major in engineering and language/literature. We also assume that the two populations have a common variance of σ^2.

13.3 See Section 13.3 of the text.

13.4 Recall that in Chapter 10 we calculated

$$S^2 = \frac{(n_1-1)S_1^2+(n_2-1)S_2^2}{n_1+n_2-2}$$

$$= \frac{\left[\sum_j Y_{1j}^2 - \frac{1}{n_1}\left(\sum_j Y_{1j}\right)^2\right] + \left[\sum_j Y_{2j}^2 - \frac{1}{n_2}\left(\sum_j Y_{2j}\right)^2\right]}{n_1+n_2-2}$$

This procedure is now extended in order to obtain a pooled sum of squares of error using the four samples of Example 13.2. The calculations are as follows:

$$(n_1-1)s_1^2 = \sum_j y_{1j}^2 - \frac{\left(\sum_j y_{1j}\right)^2}{6} = 34{,}686 - 34{,}352.667 = 333.333$$

$$(n_2-1)s_2^2 = \sum_j y_{2j}^2 - \frac{\left(\sum_j y_{2j}\right)^2}{7} = 43{,}361 - 43{,}057.285 = 303.715$$

$$(n_3-1)s_3^2 = \sum_j y_{3j}^2 - \frac{\left(\sum_j y_{3j}\right)^2}{6} = 30{,}563 - 30{,}104.1667 = 458.833$$

$$(n_4-1)s_4^2 = \sum_j y_{4j}^2 - \frac{\left(\sum_j y_{4j}\right)^2}{4} = 30{,}901 - 30{,}800.25 = 100.75$$

Then

$$s^2 = \frac{\sum_{i=1}^{4}(n_i-1)s_i^2}{\sum_{i=1}^{4} n_i - 4} = \frac{1196.631}{19} = 62.98$$

Notice that SSE has the same value (1196.631) as was obtained by subtraction in Example 13.2.

13.5 Since W has a χ^2 distribution with r degrees of freedom, the moment-generating function is

$$M_W(t) = (1-2t)^{-r/2} = E\left(e^{tw}\right) = E\left(e^{t(u+v)}\right) = E\left(e^{tu}e^{tv}\right)$$
$$= E\left(e^{tu}\right)E\left(e^{tv}\right) \quad \text{since } u \text{ and } v \text{ are independent}$$
$$= E\left(e^{tu}\right)(1-2t)^{-s/2} \quad \text{since } v \text{ has a } \chi^2 \text{ distribution with } s \text{ degrees of freedom.}$$

Therefore,

$$E\left(e^{tu}\right) = \frac{(1-2t)^{-r/2}}{(1-2t)^{-s/2}} = (1-2t)^{-(r-s)/2}$$

By Theorem 6.1, u has a χ^2 distribution with $r-s$ degrees of freedom.

13.6 **a.** From Theorem 7.3,

$$\frac{(n_i-1)S_i^2}{\sigma^2}$$

has a χ^2 distribution with n_i degrees of freedom. Thus, from Exercise 6.45 and the fact that the S_i are independent,

$$\frac{\sum_{i=1}^{k}(n_i-1)S_i^2}{\sigma^2} = \frac{\text{SSE}}{\sigma^2}$$

has a χ^2 distribution with

$$(n_1-1)+(n_2-1)+\ldots+(n_k-1) = n_1+\ldots+n_k-k = n-k$$

degrees of freedom.

b. Since a random sample is taken from each of the k populations, the observations within each population are independent. Also, since the samples are independent, two observations from different populations are independent. Thus, all the Y_{ij} are independent, normally distributed random variables assumed to have common variance σ^2.

If H_0: $\mu_1 = \mu_2 = \ldots = \mu_k$ holds, and μ denotes the common value of the k

means, then all the Y_{ij} have mean μ and variance σ^2.
Notice that

$$\frac{\text{Total SS}}{n-1} = \frac{1}{n-1}\sum_{i=1}^{k}\sum_{j=1}^{n_i}(Y_{ij} - \overline{Y})^2$$

is the sample variance for this random sample of size n. Thus,

$$\frac{\text{Total SS}}{\sigma^2} = \frac{(n-1)S^2}{\sigma^2},$$

which, by Theorem 7.3, has a χ^2 distribution with $n-1$ degrees of freedom under H_0: $\mu_1 = \ldots = \mu_k$.

c. Since Total SS = SST + SSE,

$$\frac{\text{Total SS}}{\sigma^2} = \frac{\text{SST}}{\sigma^2} + \frac{\text{SSE}}{\sigma^2}$$

Recall Exercise 13.5: $w = u + v$. Let

$W = \frac{\text{Total SS}}{\sigma^2}$ and $U = \frac{\text{SST}}{\sigma^2}$ and $V = \frac{\text{SSE}}{\sigma^2}$

From part **a**, V has a χ^2 distribution with $n-k$ degrees of freedom.
From part **b**, under H_0: $\mu_1 = \ldots = \mu_k$, W has a χ^2 distribution with $n-1$ degrees of freedom.
Also, SST and SSE are independent as described in the problem.
Thus, from Exercise 13.5, under H_0: $\mu_1 = \ldots = \mu_k$, $\frac{\text{SST}}{\sigma^2}$ has a χ^2 distribution with $(n-1)-(n-k) = k-1$ degrees of freedom.

d. From part **a**, $\frac{\text{MSE}}{\sigma^2} = \frac{\text{SSE}}{\sigma^2(n-k)}$ is a χ^2 random variable divided by its degrees of freedom.
From part **c**, under H_0: $\mu_1 = \ldots = \mu_k$, $\frac{\text{MST}}{\sigma^2} = \frac{\text{SST}}{\sigma^2(k-1)}$ is a χ^2 random variable divided by its degrees of freedom.
From part **c**, $\frac{\text{MST}}{\sigma^2}$ and $\frac{\text{MSE}}{\sigma^2}$ are independent.
Thus, from Definition 7.3, under H_0: $\mu_1 = \ldots = \mu_k$,

$$F = \frac{\text{MST}}{\text{MSE}} = \frac{\frac{\text{MST}}{\sigma^2}}{\frac{\text{MSE}}{\sigma^2}}$$

has an F distribution with $k-1$ and $n-k$ numerator and denominator degrees of freedom, respectively.

13.7

This is similar to Example 13.2 in the text. The null hypothesis is H_0: $\mu_1 = \mu_2 = \mu_3 = \mu_4$ against the alternative that at least one of the above equalities does not hold. The following calculations are necessary.

(1) $\text{CM} = \frac{\left(\sum_i\sum_j y_{ij}\right)^2}{12} = 58.08$

(2) $\text{TSS} = \sum_i\sum_j y_{ij}^2 - \text{CM} = 58.115 - \text{CM} = .035$

(3) $\text{SST} = \sum_i \frac{y_{i\cdot}^2}{n_i} - \text{CM} = \frac{(6.75)^2+(6.50)^2+(6.50)^2+(6.65)^2}{3} - \text{CM} = 58.095 - \text{CM}$
$= .015$

There are four treatments and hence 3 degrees of freedom for treatments, $(n-1) = 11$ total degrees of freedom, and $(11-3) = 8$ degrees of freedom for error. The ANOVA table is shown below. In order to determine whether or not the four types of concrete differ in average strength, we run an F test. The critical value of F with $\alpha = .05$ and 3 and 8 degrees of freedom is $F = 4.07$. The computed value of F is

$$F = \frac{\text{MST}}{\text{MSE}} = \frac{.0050}{.0025} = 2$$

Source	d.f.	SS	MS
Treatments	3	.015	.0050
Error	8	.020	.0025
Total	11	.035	

which does not exceed the critical value. Therefore, the null hypothesis of equality cannot be rejected and thus there is insufficient evidence to suggest that the strengths of the four types of concrete are significantly different.

13.8

This is similar to Exercise 13.7. The design is completely randomized with three treatments, containing five, three, and three measurements, respectively. The analysis is as follows:

(1) $\text{CM} = \frac{\left[\sum_i \sum_j y_{ij}\right]^2}{n} = \frac{(850)^2}{11} = 64,145,4545$

(2) $\text{TSS} = \sum_i \sum_j y_{ij}^2 - \text{CM} = 65,286 - \text{CM} = 1140.5455$

(3) $\text{SST} = \sum_i \frac{Y_i^2}{n_i} - \text{CM} = \frac{(380)^2}{5} + \frac{(199)^2}{3} + \frac{(261)^2}{3} - \text{CM} = 641.87883$

(4) $\text{SSE} = \text{TSS} - \text{SST} = 498.6667$

The ANOVA table is shown below. Finally, the F test to detect a difference in mean student response is

$F = \frac{\text{MST}}{\text{MSE}} = 5.148$

Source	d.f.	SS	MS
Treatments	2	641.8788	320.939
Error	8	498.6667	62.333
Total	10	1140.5455	

The critical value of F for $\alpha = .05$, with 2 and 8 degrees of freedom, is $F = 4.46$. Because the computed value of F exceeds the critical value, $F = 4.46$, we reject H_0 when $\alpha = .05$. Also, $4.46 < 5.148 < 6.06$, so $.025 < p\text{-value} < .05$.

13.9

H_0: $\mu_1 = \mu_2 = \mu_3 = 0$ vs. H_a: One or more of μ_i's differ

a. $y_{1\cdot} = 14(.93) = 13.02$; $y_{2\cdot} = 14(1.21) = 16.94$; $y_{3\cdot} = 14(.92) = 12.88$

b. $\text{Total} = y_{1\cdot} + y_{2\cdot} + y_{3\cdot} = 42.84$

c. $\text{CM} = \frac{(42.84)^2}{42} = 43.6968$

d. $\text{SST} = \frac{(13.02)^2+(16.94)^2+(12.88)^2}{14} - \text{CM} = .7588$

e. $s_1^2 = 14(.04)^2 = .0224$; $s_2^2 = 14(.03)^2 = .0126$; $s_3^2 = 14(.04)^2 = .0224$

f. & g. $\text{SSE} = \sum_{i=1}^{p} (n_i - 1)s_1^2 = 13(.0224) + 13(.0126) + 13(.0224) = .7462$

h.

Source	df	SS	MS	F
Treatments	2	.7588	.3794	19.83
Error	39	.7462	.019133	
Total	41			

i. The test statistic is

$F = \frac{\text{MST}}{\text{MSE}} = \frac{.3794}{.019133} = 19.83$

and the rejection region with 2 and 39 df is approximately $F > F_{.05} = 3.23$. The null hypothesis is rejected and there is a difference between the means.

$p\text{-value} < 0.005$.

13.10

a. Calculate

(1) $\text{CM} = \frac{\left[\sum_i \sum_j y_{ij}\right]^2}{n} = \frac{573,306.4}{36} = 15,925.178$

(2) $\text{TSS} = \sum_i \sum_j y_{ij}^2 - \text{CM} = 16,160.397 - \text{CM} = 235.219$

(3) $\text{SST} = \sum_i \frac{y_{i\cdot}^2}{n_i} - \text{CM} = \frac{(287.58)^2+(245.56)^2+(224.03)^2}{12} - \text{CM} = 174.106$

(4) $\text{SSE} = \text{TSS} - \text{SST} = 61.113$

The ANOVA table is shown below. The F statistic is

$F = \frac{\text{MST}}{\text{MSE}} = 47.007$

Source	d.f.	SS	MS
Treatments	2	174.106	87.053
Error	33	61.113	1.852
Total	35	235.219	

From Table 7, $F = 47.007 > 6.266$, so the p-value $< .005$. We reject H_0 when $\alpha = .01$. Note that the value 6.266 was found by interpolation $\left[6.266 = 6.35 - \left(\frac{3}{10}\right)(6.35 - 6.07)\right]$.

b. We must assume that we have normally distributed data with a common variance. Also we assume that the values from low, medium, and high concentrations of acetonitrile are normally distributed with common variance.

13.11 Using the pooled sum of squares formula for SSE, we have

$$\text{SSE} = \sum (n_i - 1)s_i^2 = 44(.7)^2 + 101(.64)^2 + 17(.9)^2 = 76.6996$$

The treatment totals are

$$y_{1\cdot} = 45(4.59) = 206.55 \qquad y_{2\cdot} = 102(4.88) = 497.76 \qquad y_{3\cdot} = 18(6.24) = 112.32$$

so that

$$\text{CM} = \frac{[\sum\sum y_{ij}]^2}{n} = \frac{(816.63)^2}{165} = 4041.725$$

$$\text{SST} = \sum \frac{y_{i\cdot}^2}{n_i} - \text{CM} = \sum n_i \bar{y}_{i\cdot}^2 - \text{CM} = 45(4.59)^2 + 102(4.88)^2 + 18(6.24)^2 - \text{CM}$$
$$= 4078.010 - 4041.725 = 36.285$$

The ANOVA table is shown below. Then

$$F = \frac{\text{MST}}{\text{MSE}} = \frac{18.143}{.4735} = 38.316$$

Source	d.f.	SS	MS
Treatments	2	36.285	18.143
Error	162	76.6996	0.4735
Total	164		

From Table 7, $F = 38.316 > 7.88 = F_{.005}$ for 2 and ∞ degrees of freedom. Thus, the p-value is less than .005. We reject H_0 at the $\alpha = .05$ level (p-value $< .005$).

13.12 This is similar to Exercise 13.11. Calculate

$$\text{SSE} = 9(.014)^2 + 9(.008)^2 + 9(.017)^2 = .004941$$

Treatment totals are $y_{1\cdot} = .32$, $y_{2\cdot} = .22$, and $y_{3\cdot} = .41$, so that

$$\text{SST} = \frac{(.32)^2 + (.22)^2 + (.41)^2}{10} - \frac{(.95)^2}{30} = .03189 - .0300833 = .0018067$$

The ANOVA table is shown below. Then

$$F = \frac{\text{MST}}{\text{MSE}} = 4.94$$

Source	d.f.	SS	MS
Treatments	2	.0018067	.009033
Error	27	.004941	.000183
Total	29		

which is compared to $F_{.05} = 3.35$ and 2 and 27 degrees of freedom. H_0 is rejected; there are significant differences.

13.13 We have a completely randomized design with four treatments. The analysis is as follows:

(1) $\text{CM} = \frac{\left[\sum_i \sum_j y_{ij}\right]^2}{n} = \frac{(110.6)^2}{19} = 643{,}8084$

(2) $\text{TSS} = \sum_i \sum_j y_{ij}^2 - \text{CM} = 652.26 - 643.8084 = 8.4516$

(3) $\text{SST} = \sum_i \frac{y_{i\cdot}^2}{n_i} - \text{CM} = \frac{(30.4)^2}{5} + \frac{(32.2)^2}{5} + \frac{(23.9)^2}{5} + \frac{(24.1)^2}{4} - 643.808 = 7.8361$

(4) $\text{SSE} = \text{TSS} - \text{SST} = 8.452 - 7.836 = .616$

The ANOVA table is shown below. The F statistic is

$$F = \frac{\text{MST}}{\text{MSE}} = \frac{2.61203}{.04103} = 63.66$$

Source	d.f.	SS	MS
Treatments	2	7.8361	2.61203
Error	15	.6155	.04103
Total	18	8.4516	

The critical value for F with $\alpha = .005$ and with 3 and 15 degrees of freedom is $F = 6.48 < 63.66$. Thus, the p-value $< .005$. We conclude that there is a difference in mean dissolved oxygen content for the four locations.

13.14 This is similar to previous exercises. The analysis is as follows:

(1) $\text{CM} = \frac{(1161)^2}{40} = 33{,}698.025$

(2) $\text{TSS} = 34{,}701 - \text{CM} = 1002.975$

(3) $\text{SST} = \frac{(309)^2 + \ldots + (282)^2}{10} - \text{CM} = 33{,}765.5 - \text{CM} = 67.475$

(4) $\text{SSE} = \text{TSS} - \text{SST} = 935.5$

The ANOVA table is shown below. The F statistic is

$$F = \frac{\text{MST}}{\text{MSE}} = \frac{22.4917}{25.9861} = .87$$

Source	d.f.	SS	MS
Treatments	3	67.475	22.4917
Error	36	935.5	25.9861
Total	39	1002.975	

which does not fall in the rejection region with $\alpha = .05$ ($F > F_{.05} = 2.87$ by interpolation). The null hypothesis is not rejected and we cannot conclude that there is a difference in mean increase among the four groups.

13.15

$$E\left(\overline{Y}_{i\cdot}\right) = E\left(\frac{1}{n_i}\sum_{j=1}^{n_i} Y_{ij}\right) = E\left(\frac{1}{n_i}\sum_{j=1}^{n_i} (\mu + \tau_i + \epsilon_{ij})\right)$$

$$= E\left(\frac{1}{n_i}\sum_{j=1}^{n_i} (\mu_i + \epsilon_{ij})\right) \qquad \text{since } \mu_i = \mu + \tau_i,$$

$$= \frac{1}{n_i}\sum_{j=1}^{n_i} E(\mu_i) + \frac{1}{n_i}\sum_{j=1}^{n_i} E(\epsilon_{ij})$$

$$= \mu_i + 0 = \mu_i, \qquad \text{the mean of } i^{th} \text{ population.}$$

Similarly,

$$V\left(\overline{Y}_{i\cdot}\right) = V\left(\frac{1}{n_i}\sum_{j=1}^{n_i} Y_{ij}\right) = V\left(\mu + \tau_i + \frac{1}{n_i}\sum_{j=1}^{n_i} \epsilon_{ij}\right) = V\left(\frac{1}{n_i}\sum_{j=1}^{n_i} \epsilon_{ij}\right)$$

$$= \frac{1}{n_i^2}\left(\sum_{j=1}^{n_i} V(\epsilon_{ij}) + 2\sum_{j<j'}\sum \text{Cov}\left(\epsilon_{ij}, \epsilon'_{ij}\right)\right)$$

$$= \frac{1}{n_i^2}\left(\sum_{j=1}^{n_i} \sigma^2 + 0\right) \qquad \text{since the } \epsilon_{ij} \text{ are independent,}$$

$$= \frac{\sigma^2}{n_i}$$

13.16 a.

$$E\left(\overline{Y}_{i\cdot} - \overline{Y}_{i'\cdot}\right) = E\left(\overline{Y}_{i\cdot}\right) - E\left(\overline{Y}_{i'\cdot}\right)$$

$$= \mu_i - \mu_{i'} \qquad \text{from Exercise 13.15}$$

$$= \mu + \tau_i - (\mu + \tau_{i'})$$

$$= \tau_i - \tau_{i'} \qquad \text{the difference in the effects of treatments } i \text{ and } i'.$$

b.

$$V\left(\overline{Y}_{i\cdot} - \overline{Y}_{i'\cdot}\right) = V\left(\overline{Y}_{i\cdot}\right) + V\left(\overline{Y}_{i'\cdot}\right) - 2\,\text{Cov}\left(\overline{Y}_{i\cdot}, \overline{Y}_{i'\cdot}\right)$$

$$= \frac{\sigma^2}{n_i} + \frac{\sigma^2}{n_{i'}} - 0 \qquad \text{from Exercise 13.15 and since } Y_{i\cdot} \text{ and } Y_{i'\cdot} \text{ are independent.}$$

$$= \sigma^2\left(\frac{1}{n_i} + \frac{1}{n_{i'}}\right)$$

13.17 a Recall τ_i is defined as $\mu_i - \mu$.
Thus, $\tau_1 = \tau_2 = \ldots = \tau_k = 0$ can be written as

$$\mu_1 - \mu = \mu_2 - \mu = \ldots = \mu_k - \mu = 0$$

Now, $\mu_i - \mu = 0 \Rightarrow \mu_i = \mu$ for all $i = 1, \ldots, k$.
This implies $\mu_1 = \ldots = \mu_k$.
Conversely, if $\mu_1 = \ldots = \mu_k$,

$$\mu + \tau_1 = \ldots = \mu + \tau_k.$$

This simplifies to $\tau_1 = \ldots = \tau_k$.
Then since $\sum_{i=1}^{k} \tau_i = 0$, $\tau_1 = \ldots = \tau_k = 0$.

b. Consider $\mu_i = \mu + \tau_i$ and $\mu_{i'} = \mu + \tau_{i'}$
If $\mu_i \neq \mu_{i'}$, then $\mu + \tau_i \neq \mu + \tau_{i'}$, which simplifies to $\tau_i \neq \tau_{i'}$.
Since $\sum_{j=1}^{k} \tau_j = 0$, at least one $\tau_j \neq 0$ (actually at least two).

Conversely, let $\tau_i \neq 0$. As we know $\sum_{j=1}^{k} \tau_j = 0$ we know there must be an i' such that $\tau_{i'} \neq \tau_i$. Recall that

$$\mu_i = \mu + \tau_i$$
$$\mu_{i'} = \mu + \tau_{i'}$$

Thus, $\mu_i \neq \mu_{i'}$.

13.18a. Using only the data on Technique 1, the confidence interval for the mean test score is

$$\overline{Y}_{1\cdot} \pm t_{.025} \frac{S_1}{\sqrt{n_1}},$$

where

$$S_1^2 = \frac{1}{n_1 - 1} \sum_{j=1}^{n_1} \left(Y_{ij} - \overline{Y}_{1\cdot}\right)^2 = \frac{1}{5}(333.33) = 66.67,$$

and $t_{.025}$ is based upon $n_1 - 1 = 5$ degrees of freedom. This gives

$$75.67 \pm (2.571)\frac{\sqrt{66.67}}{\sqrt{6}} \qquad \text{or} \qquad 75.67 \pm 8.57 = (67.1, 84.24)$$

b. The interval in part **a** is longer than the interval given in Example 13.3, with lengths 17.14 and 13.56, respectively.

c. The major reason the interval in part **a** is larger is the loss of 14 degrees of freedom resulting from excluding data on Techniques 2, 3, and 4 when estimating σ^2. The $t_{.025}$ value is 2.571 for the interval in part **a** based on 5 degrees of freedom compared to 2.093 in Example 13.3 based on 19 degrees of freedom.

13.19a. The 95% confidence interval is

$$\left(\overline{Y}_{1\cdot} - \overline{Y}_{4\cdot}\right) = t_{.025}\, S_{14} \sqrt{\tfrac{1}{6} + \tfrac{1}{4}},$$

where

$$S_{14}^2 = \frac{(n_1-1)S_1^2 + (n_4-1)S_4^2}{n_1+n_4-2} = \frac{5(66.67)+3(33.58)}{8} = 54.26,$$

and $t_{.025}$ is based upon 8 degrees of freedom. This gives

$$-12.08 \pm (2.306)\sqrt{54.26}\left(\sqrt{\tfrac{1}{6} + \tfrac{1}{4}}\right) \quad \text{or} \quad -12.08 \pm 10.96 = (-23.04, -1.12),$$

suggesting $\mu_4 > \mu_1$.

b. The interval in part **a** is longer than the interval given in Example 13.4, with lengths 21.92 and 21.46, respectively.

c. The major reason the interval in part **a** is longer is the loss of 11 degrees of freedom resulting from excluding data on Techniques 2 and 3 when estimating σ^2. The $t_{.025}$ value is 2.306 based on 8 degrees of freedom compared to 2.093 based on 19 degrees of freedom.

13.20a. We would expect confidence intervals computed using data from all k samples, as in this section, to be shorter than the corresponding intervals that make use of only one or two of the samples. This is because the value $\frac{t\alpha}{2}$ is based on a greater number of degrees of freedom $(n-k)$ instead of $n_i - 1$ or $n_i + n_{i'} - 2$, respectively.

b. It is possible that a 95% confidence interval for the mean of a single population based only on the sample taken from that population to be shorter than the 95% confidence interval for the same population mean computed in this section. This could occur if the estimate of σ^2, based on a single population (S_i^2), is sufficiently smaller than the estimate of σ^2 based on all k populations (MSE) to offset the decrease in degrees of freedom.

13.21 Need to consider

$$\left(\overline{Y}_{1\cdot} - \overline{Y}_{2\cdot}\right) \pm (t_{.025,39})s\sqrt{\frac{1}{n_1} + \frac{1}{n_2}}$$

or

$$s = \sqrt{\text{MSE}} = .1383.$$

We give as our 95% estimate interval

$$(.93 - 1.21) \pm 1.96(.1383)\sqrt{\tfrac{2}{14}}$$

or

$$-.28 \pm .102 = (-.382, -.178).$$

At the 95% confidence level, we would conclude that there is a significant difference between the mean bone densities for the two groups of women since the confidence interval formed contains all negative values. This suggests that $\mu_2 > \mu_1$.

13.22 a. A 90% confidence interval for μ_A is obtained as in Chapter 8, where $S^2 = \text{MSE}$. Thus, estimating μ_A by $\overline{Y}_A$, the confidence interval becomes

$\overline{y}_A \pm t_{.05,8}\frac{s}{\sqrt{n_A}}$ or $\frac{6.75}{3} \pm 1.86\frac{.05}{\sqrt{3}}$ or $2.25 \pm .05$ or $[2.20, 2.30]$

Hence we are 90% confident that the interval, 2.20 to 2.30, encloses μ_A.

b. Similarly, a 95% confidence interval for $\mu_A - \mu_B$ is

$$(\overline{y}_A - \overline{y}_B) \pm t_{.025,8}\, s\sqrt{\frac{1}{n_A} + \frac{1}{n_B}} = (2.25 - 2.166) \pm 2.306(.05)\sqrt{\frac{1}{3} + \frac{1}{3}}$$

or

$$.0833 \pm .094 = [-.01, .18]$$

and we are 95% confident that $\mu_A - \mu_B$ will lie between $-.01$ and $.18$.

13.23 Confidence intervals are calculated as in Exercise 13.22.

a. $\overline{y}_A \pm t_{.025,8}\frac{s}{\sqrt{n_A}}$ or $76 \pm 2.306\frac{\sqrt{62.233}}{\sqrt{5}}$ or 76 ± 8.142

or

$$[67.86, 84.14].$$

b. $\overline{y}_B \pm t_{.025,8}\frac{s}{\sqrt{n_B}}$ or $66.33 \pm 2.306\frac{\sqrt{62.233}}{\sqrt{3}}$ or 64.33 ± 10.51

or

$$[55.82, 76.84].$$

c. $(\overline{y}_A - \overline{y}_B) \pm t_{.025,8}\, s\sqrt{\frac{1}{n_A} + \frac{1}{n_B}}$ or $9.667 \pm 2.306(7.895)\sqrt{\frac{1}{3} + \frac{1}{5}}$

or

$$9.667 \pm 13.295 = [-3.628, 22.962]$$

13.24 a. A 95% confidence interval for μ_L is

$$\overline{y}_L \pm t_{.025,33}\frac{s}{\sqrt{n_L}} = 23.965 \pm 1.96\sqrt{\frac{1.852}{12}} = 23.965 \pm .77$$

b. A 90% confidence interval for $\mu_L - \mu_M$ is

$$(\overline{y}_L - \overline{y}_M) \pm t_{.05,33}\frac{s}{\sqrt{\frac{1}{n_L} + \frac{1}{n_M}}}$$

or

$$(23.965 - 20.463) \pm 1.645\sqrt{1.852}\sqrt{\frac{1}{12} + \frac{1}{12}} = 3.502 \pm .914$$

13.25 a. A 95% confidence interval for μ_B is

$$\overline{y}_B \pm t_{.025,162}\frac{s}{\sqrt{n_B}} = 6.24 \pm 1.96\sqrt{\frac{.4735}{18}} = 6.24 \pm .318$$

b. A 95% confidence interval for $\mu_s - \mu_L$ is

$$(\overline{y}_s - \overline{y}_L) \pm t_{.05,162}\frac{s}{\sqrt{\frac{1}{n_s} + \frac{1}{n_L}}}$$

or

$$(4.59 - 4.58) \pm 1.96\sqrt{.4735}\sqrt{\frac{1}{45} + \frac{1}{102}} \quad \text{or} \quad -.29 \pm .241$$

c. Probably not. The driving habits of people vary from town to town. Thus, the vehicles sampled do not represent a random sample of vehicles from all towns.

13.26 a. A summary of the data from this completely randomized design follows.

$y_{1\cdot} = 83.90$	$y_{2\cdot} = 98.65$	$y_{3\cdot} = 87.00$	$y_{4\cdot} = 76.37$
$n_1 = 7$	$n_2 = 7$	$n_3 = 7$	$n_4 = 7$
$\overline{y}_{1\cdot} = 11.986$	$\overline{y}_{2\cdot} = 14.093$	$\overline{y}_{3\cdot} = 12.429$	$\overline{y}_{4\cdot} = 10.910$

$\text{CM} = \frac{(345.92)^2}{28} = 4273.5945$

$\text{TSS} = 4370.6264 - 4273.5945 = 97.0319$

$\text{SST} = \frac{(83.9)^2}{7} + \frac{(98.65)^2}{7} + \frac{(87)^2}{7} + \frac{(76.37)^2}{7} - \frac{(\sum x)^2}{28} = 36.7497$

SSE = TSS − SST = 60.2822
MST = $\frac{\text{SST}}{3}$ = 12.2499
MSE = $\frac{\text{SSE}}{24}$ = 2.5118

Let the mean wear for treatments A, B, C, and D be denoted by μ_1, μ_2, μ_3, and μ_4, respectively. We test

H_0: $\mu_1 - \mu_2 = \mu_3 = \mu_4$ vs. H_a: At least two of the means differ

The test statistic is

$$F = \frac{\text{MST}}{\text{MSE}} = 4.88$$

Since $F = 4.88 > 3.01 = F_{.05}$, we reject H_0 at the .05 level. There is a difference in mean wear among the four treatments.

b. A 99% confidence interval for $(\mu_2 - \mu_3)$ is

$$(14.093 - 12.429) \pm (2.797)\sqrt{2.5118}\sqrt{\tfrac{1}{7} + \tfrac{1}{7}} \quad \text{or} \quad 1.664 \pm 2.3695$$

or

$$(-.7055, 4.0335)$$

There is insufficient evidence of a difference between μ_2 and μ_3.

c. A 90% confidence interval for μ_1 is

$$11.986 \pm 1.711\sqrt{\tfrac{2.5118}{7}} \quad \text{or} \quad 11.986 \pm 1.025 \quad \text{or} \quad (10.961, 13.011)$$

13.27 a. We wish to test H_0: $\mu_1 = \mu_2 = \mu_3 = \mu_4$ against the general alternative. Calculate

CM = $\frac{(49)^2}{16}$ = 150.0625
TSS = 183 − CM = 32.9375
SST = $\frac{(8)^2+(12)^2+(13)^2+(16)^2}{4}$ − CM = 8.1875

The ANOVA table is shown below. Then

$$F = \frac{\text{MST}}{\text{MSE}} = 1.32$$

Source	d.f.	SS	MS
Treatments	3	8.1875	2.7292
Error	12	24.7500	2.0625
Total	15	32.9375	

which is compared to $F_{.05} = 3.49$ with 3 and 12 degrees of freedom. We do not reject H_0.

b. $(\overline{y}_{4\cdot} - \overline{y}_{1\cdot}) \pm t_{.025,12}\sqrt{s^2\left(\frac{1}{n_4} + \frac{1}{n_1}\right)}$ or $2 \pm 2.179\sqrt{2.0625\left(\frac{2}{4}\right)}$ or 2 ± 2.21

or

$$-.21 < \mu_4 - \mu_1 \leq 4.21.$$

Intervals constructed in this manner will enclose $(\mu_4 - \mu_1)$ 95% of the time in repeated sampling. Hence we are fairly confident that this particular interval encloses $(\mu_4 - \mu_1)$.

13.28 $\overline{y}_{3\cdot} \pm t_{.025,27}\sqrt{s^2\left(\frac{1}{n_3}\right)}$ or $.041 \pm 2.052\sqrt{\frac{.000183}{10}}$ or $.041 \pm .009$

or [.032, .050].

13.29 $(\overline{y}_{2\cdot} - \overline{y}_{3\cdot}) \pm t_{\alpha/2,15}\, s\sqrt{\frac{1}{n_2} + \frac{1}{n_3}}$ or $(6.44 - 4.78) \pm 2.131\, s\sqrt{\frac{1}{5} + \frac{1}{5}}$

or

$$1.66 \pm 2.131(.2025)\sqrt{\tfrac{2}{5}} \quad \text{or} \quad 1.66 \pm .273 \quad \text{or} \quad [1.39,\ 1.93]$$

13.30 The estimator for $\left(\frac{1}{2}\right)(\mu_1 + \mu_2) - \mu_4$ is $\left(\frac{1}{2}\right)(\overline{y}_{1\cdot} + \overline{y}_{2\cdot})$, which has variance

$$\left(\tfrac{1}{4}\right)\left[\left(\tfrac{\sigma^2}{n_1}\right) + \left(\tfrac{\sigma^2}{n_2}\right)\right] + \tfrac{\sigma^2}{n_4}.$$

The 95% confidence interval is

$$\left[\left(\tfrac{1}{2}\right)(\overline{y}_{1\cdot} + \overline{y}_{2\cdot}) - \overline{y}_{4\cdot}\right] \pm t_{.025,15}\sqrt{s^2\left(\frac{1}{4n_1} + \frac{1}{4n_2} + \frac{1}{n_4}\right)}$$

or

$$[6.26 - 6.025] \pm 2.131\sqrt{.04103\left(\tfrac{1}{20} + \tfrac{1}{20} + \tfrac{1}{4}\right)} \quad \text{or} \quad .235 \pm .255$$

or $[-.020,\ .490]$.

13.31a. The 90% confidence interval for $(\mu_1 - \mu_4)$ is

$$(\overline{y}_{1\cdot} - \overline{y}_{4\cdot}) \pm t_{.05,36}\sqrt{\text{MSE}\left(\tfrac{1}{n_1} + \tfrac{1}{n_4}\right)} = (30.9 - 28.2) \pm 1.645\sqrt{25.986\left(\tfrac{2}{10}\right)}$$

or

$$2.7 \pm 3.75 \quad \text{or} \quad [-1.05, 6.45]$$

b. A 90% confidence interval for μ_2 is

$$\overline{y}_{2\cdot} \pm t_{.05,36}\sqrt{\tfrac{\text{MSE}}{n_2}} = 27.5 \pm 1.645\sqrt{\tfrac{25.9861}{10}} = 27.5 \pm 2.65$$

or $[24.85,\ 30.15]$.

13.32 See Sections 12.3 and 13.7.

13.33a.

$$\tfrac{1}{bk}\sum_{i=1}^{k}\sum_{j=1}^{b} E(Y_{ij}) = \tfrac{1}{bk}\sum_{i=1}^{k}\sum_{j=1}^{b}(\mu + \tau_i + \beta_j) = \tfrac{1}{bk}\left(bk\mu + b\sum_{i=1}^{k}\tau_i + k\sum_{j=1}^{k}\beta_j\right)$$
$$= \tfrac{1}{bk}(bk\mu + 0 + 0) = \mu$$

b. μ is the overall mean.

13.34

$$\overline{Y}_{i\cdot} = \tfrac{1}{b}\sum_{j=1}^{b} Y_{ij} = \tfrac{1}{b}\sum_{j=1}^{b}(\mu + \tau_i + \beta_j + \epsilon_{ij}) = \tfrac{1}{b}\left(b\mu + \sum_{j=1}^{b}\tau_i + \sum_{j=1}^{b}\beta_j + \sum_{j=1}^{b}\epsilon_{ij}\right)$$
$$= \tfrac{1}{b}\left(b\mu + b\tau_i + 0 + \sum_{j=1}^{b}\epsilon_{ij}\right) = \mu + \tau_i + \tfrac{1}{b}\sum_{j=1}^{b}\epsilon_{ij}$$

$$E(\overline{Y}_{i\cdot}) = E\left(\mu + \tau_i + \tfrac{1}{b}\sum_{j=1}^{b}\epsilon_{ij}\right) = E(\mu) + E(\tau_i) + \tfrac{1}{b}\sum_{j=1}^{b}E(\epsilon_{ij}) = \mu + \tau_i + 0$$
$$= \mu + \tau_i = \mu_i,$$

which by definition of unbiasedness shows that $\overline{Y}_{i\cdot}$ is an unbiased estimator.

$$V(\overline{Y}_{i\cdot}) = V\left(\tfrac{1}{b}\sum_{j=1}^{b}(\mu + \tau_i + \beta_j + \epsilon_{ij})\right) = \tfrac{1}{b^2}V\left(\sum_{j=1}^{b}\epsilon_{ij}\right)$$
$$= \tfrac{1}{b^2}\left(\sum_{j=1}^{b}V(\epsilon_{ij}) + 2\sum_{j<j'}\sum \text{Cov}(\epsilon_{ij}, \epsilon_{ij'})\right)$$
$$= \tfrac{1}{b^2}(b\sigma^2 + 0) \qquad \text{since the } \epsilon_{ij} \text{ are independent}$$
$$= \tfrac{\sigma^2}{b}.$$

13.35a. $E(\overline{Y}_{i\cdot} - \overline{Y}_{i'\cdot}) = E(\overline{Y}_{i\cdot}) - E(\overline{Y}_{i'\cdot}) = \mu_i - \mu_{i'} = (\mu + \tau_i) - (\mu + \tau_{i'}) = \tau_i - \tau_{i'}$.

b.

$$V(\overline{Y}_{i\cdot} - \overline{Y}_{i'\cdot}) = V(\overline{Y}_{i\cdot}) + V(\overline{Y}_{i\cdot}) - 2\,\text{Cov}(\overline{Y}_{i\cdot}, \overline{Y}_{i'\cdot})$$
$$= \tfrac{\sigma^2}{b} + \tfrac{\sigma^2}{b} - 0 \qquad \text{from Exercise 13.34 and since } \overline{Y}_{i\cdot}, \overline{Y}_{i'\cdot} \text{ are independent.}$$
$$= \tfrac{2\sigma^2}{b}$$

To further motivate the independence of $\overline{Y}_{i\cdot}$ and $\overline{Y}_{i'}$ notice that

$$\overline{Y}_{i\cdot} = \mu + \tau_i + \tfrac{1}{b}\sum_j \epsilon_{ij}$$

and

$$\overline{Y}_{i'\cdot} = \mu + \tau_i + \tfrac{1}{b}\sum_j \epsilon_{i'j}$$

are functions of different ϵ's.

13.36a.

$$\overline{Y}_{\cdot j} = \tfrac{1}{k}\sum_{i=1}^{k} Y_{ij} = \tfrac{1}{k}\sum_{i=1}^{k}(\mu + \tau_i + \beta_j + \epsilon_{ij}) = \tfrac{1}{k}\left(k\mu + \sum_{i=1}^{k}\tau_i + k\beta_j + \sum_{i=1}^{k}\epsilon_{ij}\right)$$
$$= \mu + \tfrac{1}{k}\sum_{i=1}^{k}\tau_i + \beta_j + \tfrac{1}{k}\sum_{i=1}^{k}\epsilon_{ij} = \mu + \beta_j + \tfrac{1}{k}\sum_{i=1}^{k}\epsilon_{ij}$$

Thus,

$$E(\overline{Y}_{\cdot j}) = E\left(\mu + \beta_j + \tfrac{1}{k}\sum_{i=1}^{k} \epsilon_{ij}\right) = E(\mu) + E(\beta_j) + \tfrac{1}{k}\sum_{i=1}^{k} E(\epsilon_{ij})$$
$$= \mu + \beta_j + 0 = \mu + \beta_j = \mu_j$$
$$V(\overline{Y}_{\cdot j}) = V\left(\mu + \beta_j + \tfrac{1}{k}\sum_{i=1}^{k} \epsilon_{ij}\right) = \tfrac{1}{k^2}\left(\sum_{i=1}^{k} V(\epsilon_{ij}) + 2\sum_{i<i'}\sum \text{Cov}\,(\epsilon_{ij}, \epsilon_{i'j})\right)$$
$$= \tfrac{1}{k^2}\,(k\sigma^2 + 0) \qquad \text{since } \epsilon_{ij} \text{ are independent}$$
$$= \tfrac{\sigma^2}{k}$$

b. $E(\overline{Y}_{\cdot j} - \overline{Y}_{\cdot j'}) = E(\overline{Y}_{\cdot j}) - E(\overline{Y}_{\cdot j'}) = \mu_j - \mu_{j'} = (\mu + \beta_j) - (\mu + \beta_{j'}) = \beta_j - \beta_{j'}$

c. $V(\overline{Y}_{\cdot j} - \overline{Y}_{\cdot j'}) = V(\overline{Y}_{\cdot j}) + V(\overline{Y}_{\cdot j'}) - 2\sum_{j<j'}\sum \text{Cov}\,(\overline{Y}_{\cdot j}, \overline{Y}_{\cdot j'})$

$= \frac{\sigma^2}{k} + \frac{\sigma^2}{k} + 0$ from part a and since $Y_{\cdot j}$, $Y_{\cdot j'}$ are independent.

$= \frac{2\sigma^2}{k}$

13.37

a. A summary of the data is

$Y_{1\cdot} = 9.32$ $\quad Y_{2\cdot} = 9.45$ $\quad Y_{\cdot 1} = 2.27$
$Y_{\cdot 2} = 3.45$ $\quad Y_{\cdot 3} = 2.14$ $\quad Y_{\cdot 4} = 3.73$
$Y_{\cdot 5} = 2.93$ $\quad Y_{\cdot 6} = 4.35$

$\sum_{i=1}^{2}\sum_{j=1}^{6} y_{ij} = 18.77$

$\sum_{i=1}^{2}\sum_{j=1}^{6} y_{ij}^2 = 31.1013$

$\text{CM} = \frac{(18.77)^2}{12} = 29.3594$

$\text{Total SS} = 31.1013 - 29.3594 = 1.7419$

$\text{SST} = \frac{(9.32)^2+(9.45)^2}{6} - 29.3594 = .0014$

$\text{SSB} = \frac{(2.27)^2+(2.45)^2+(2.14)^2+(3.73)^2+(2.93)^2+(4.25)^2}{2} - 29.3594$
$= 1.7382$

$\text{SSE} = 1.7419 - .0014 - 1.7382 = .0023$

The ANOVA table is

Source	df	SS	MS
Computer	1	.0014	.0014
Program	5	1.7382	.3476
Error	5	.0023	.00045
Total	11	1.7419	

To test H_0: $\mu_1 = \mu_2$, we use

$F = \frac{\text{MST}}{\text{MSE}} = 3.05$

Since $3.05 < 6.61 = F_{.05}$ with 1 and 5 degrees of freedom, we fail to reject H_0. Thus, we see no evidence of a difference in mean CPU time between computer 1 and computer 2.

This decision is the same as the one reached in Exercise 12.10**a**.

b. The p-value is greater than .10. This is consistent with Exercise 12.10**b**.

c. Except for round off, $S_D^2 = 2\text{MSE}$.

13.38

Using the formulas given in this section, calculate $\text{TSS} = 674 - 588 = 86$,

$\text{SSB} = \frac{(20)^2+(36)^2+(28)^2}{4} - \text{CM} = 32 \quad \text{SST} = \frac{(21)^2+\ldots+(18)^2}{3} - \text{CM} = 42$

The remaining calculations are shown in the ANOVA table at the right. Then, for a test of difference among chemical means, the test statistic is

Source	d.f.	SS	MS
Treatments	3	42	14
Blocks	2	32	16
Error	6	12	2
Total	11	86	

$F = \frac{\text{MST}}{\text{MSE}} = \frac{14}{2} = 7$

which for $\alpha = .05$ is compared to $F_{.05} = 4.76$ with 3 and 6 degrees of freedom. H_0 is rejected; there is evidence to suggest a difference among chemical means. Note that $6.6 < F < 9.78$, so $.01 < p\text{-value} < .025$.

13.39 The factor of interest is "soil preparation," and the blocking factor is "locations." A randomized block design is employed and the analysis of variance is as shown below.

(1) $\text{CM} = \frac{(162)^2}{12} = 2187$

(2) $\text{TSS} = 2298 - \text{CM} = 111$

(3) $\text{SST (preparations)} = \frac{8900}{4} - \text{CM} = 38$

(4) $\text{SSB (locations)} = \frac{6746}{3} - \text{CM} = 61.67$

The ANOVA table is shown at the right.

Source	d.f.	SS	MS
Blocks	3	61.67	20.56
Treatments	2	38.00	19.00
Error	6	11.33	1.89
Total	11	111.00	

a. The F statistic to detect a difference owing to soil preparations is

$$F = \frac{\text{MST}}{\text{MSE}} = 10.05$$

Then the critical value of F, based on 2 and 6 degrees of freedom, is $F_{.05} = 5.14$, and we conclude that there is a significant effect owing to soil preparation.

b. The F statistic to detect a difference owing to locations is

$$F = \frac{\text{MSB}}{\text{MSE}} = 10.88$$

Then the critical value of F, based on 3 and 6 degrees of freedom, is $F_{.05} = 4.76$, and we conclude that there is evidence to suggest a significant effect owing to locations.

13.40 Calculate

$\text{CM} = \frac{(52.333)^2}{20} = 136.937$

$\text{SST} = \frac{549.558}{4} - \text{CM} = .452$

$\text{TSS} = 138.603 - \text{CM} = 1.666$

$\text{SSB} = \frac{689.943}{5} - \text{CM} = 1.052$

The ANOVA table is shown at the right.

Source	d.f.	SS	MS
Treatments	4	.452	.113
Blocks	3	1.052	.3507
Error	12	.162	.0135
Total	19	1.666	

a. The F statistic to detect a difference due to varieties has 4 and 12 degrees of freedom:

$$F = \frac{\text{MST}}{\text{MSE}} = 8.37$$

Since $8.37 > 6.52$, the p-value $< .005$, and we reject H_0 at the $\alpha = .01$ level.

b. The F statistic for blocks, with 3 and 12 degrees of freedom, is

$$F = \frac{\text{MSB}}{\text{MSE}} = 25.97$$

Since $25.97 > 3.49 = F_{.05}$, we reject H_0: there is evidence to suggest a significant difference due to blocks.

13.41 Using a randomized block design with locations as blocks, recall that $\text{CM} = 150.0625$, $\text{TSS} = 32.9375$, and $\text{SST} = 8.1875$. Then

$$\text{SSB} = \frac{(15)^2+(8)^2+(14)^2+(12)^2}{4} - \text{CM} = 7.1875$$

SSE will be obtained by subtraction; see the ANOVA table. To test for a difference in treatment means, the test statistic is

Source	d.f.	SS	MS
Treatments	3	8.1875	2.729
Blocks	3	7.1875	2.396
Error	9	17.5625	1.95139
Total	15	32.9375	

$$F = \frac{\text{MST}}{\text{MSE}} = 1.40$$

which is compared to $F_{.05} = 3.86$ with 3 and 9 degrees of freedom. There is no evidence of significant treatment differences. Notice that the test for block differences yields $F = \frac{\text{MST}}{\text{MSE}} = 1.23$, which is not significant. That is, the evidence supports the hypothesis that blocking was not worthwhile.

13.42 The most straightforward implementation is to calculate block-treatment means or totals. Then perform the analysis of variance on the means or totals.

$$\text{TSS} = \sum_{ijl} y_{ijl}^2 - \text{CM} \qquad \text{CM} = \frac{\left(\sum_{ijl} y_{ijl}\right)}{2bk} \qquad \text{SSB} = \frac{\sum Y_{\cdot j}^2}{2k} - \text{CM}$$

$$\text{SST} = \frac{\sum Y_{i\cdot}^2}{2b} - \text{CM}$$

and SSE is obtained by subtraction. The degrees of freedom are the same for treatments $(k-1)$ and blocks $(b-1)$, while error degrees of freedom increases to $(2bk-1)-(b-1)-(k-1) = 2bk-b-k-1$.

13.43 A summary of the data is

$y_{1\cdot} = 497.7$	$y_{2\cdot} = 531.3$	$y_{3\cdot} = 491.3$
$k = 3$	$y_{\cdot 1} = 211.1$	$y_{\cdot 2} = 202.7$
$y_{\cdot 3} = 233.1$	$y_{\cdot 4} = 218.1$	$y_{\cdot 5} = 220.5$
$y_{\cdot 6} = 205.3$	$y_{\cdot 7} = 229.5$	$b = 7$
$\sum y = 1520.3$	$\sum y^2 = 110{,}587.13$	

$\text{CM} = \frac{(1520.3)^2}{21} = 110{,}062.48$

$\text{TSS} = 110{,}587.13 - \frac{(1520.3)^2}{21} = 524.650$

$\text{SST} = \frac{(497.7)^2}{7} + \frac{(531.3)^2}{7} + \frac{(491.3)^2}{7} - \frac{(1520.3)^2}{21} = 131.901$

$\text{SSB} = \frac{(211.1)^2}{3} + \ldots + \frac{(229.5)^2}{3} - \frac{(1520.3)^2}{21} = 268.290$

$\text{SSE} = 524.650 - 131.901 - 268.290 = 124.459.$

The ANOVA table is given on the right. To test $H_0\colon\ \mu_1 = \mu_2 = \mu_3$, we use

Source	d.f.	SS	MS
Treatments	2	131.901	65.9505
Blocks	6	268.90	44.8167
Error	12	124.459	10.3716
Total	20	524.65	

$F = \frac{\text{MST}}{\text{MSE}} = 6.36$

Since $6.36 > 3.89 = F_{.05}$ with 2 and 12 degrees of freedom, we reject H_0.

13.44 The company has designed a randomized block experiment since the same shipment is handled by each carrier. The shipments form the blocks and the carriers are the treatments

$y_{1\cdot} = 73.3$	$y_{2\cdot} = 81.5$	$y_{3\cdot} = 81.4$
$\overline{y}_{1\cdot} = 14.66$	$\overline{y}_{2\cdot} = 16.3$	$\overline{y}_{3\cdot} = 16.28$
$y_{\cdot 1} = 49.2$	$y_{\cdot 2} = 46.8$	$y_{\cdot 3} = 46.3$
$y_{\cdot 4} = 48.8$	$y_{\cdot 5} = 45.1$	$b = 5$
$b = 5$	$k = 3$	$n = 15$
$\sum y = 236.2$	$\sum y^2 = 3732.62$	

$\text{TSS} = 3732.62 - \frac{(236.2)^2}{15} = 13.2573$

$\text{SST} = \frac{(73.3)^2}{5} + \frac{(81.5)^2}{5} + \frac{(81.4)^2}{5} - \frac{(236.2)^2}{15} = 8.8573$

$\text{SSB} = \frac{(49.2)^2}{3} + \frac{(46.8)^2}{3} + \ldots + \frac{(45.1)^2}{3} - \frac{(236.2)^2}{15} = 3.9773$

$\text{SSE} = \text{TSS} - \text{SST} - \text{SSB} = .4227$

The ANOVA table is given on the right. To test $H_0\colon\ \mu_1 = \mu_2 = \mu_3$, we obtain an F statistic with 2 and 8 degrees of freedom.

Source	d.f.	SS	MS
Treatments	2	8.8573	4.4286
Blocks	6	3.9773	.9943
Error	8	.4227	.0528
Total	14	13.2573	

$F = \frac{\text{MST}}{\text{MSE}} = 83.88$

Since $83.88 > 9.60 = F_{.005}$, the p-value $< .005$.

13.45 Some preliminary results will be necessary in order to obtain the solution.

(1) $E\left(Y_{ij}^2\right) = V(Y_{ij}) + [(E(Y_{ij})]^2 = \sigma^2 + (\mu + \tau_i + \beta_j)^2$

(2) With $\overline{Y}_{\cdot\cdot} = \frac{\sum_i \sum_j Y_{ij}}{bk} = \mu + \frac{\sum_i \sum_j \epsilon_{ij}}{bk}$, then

$$E\left(\overline{Y}_{\cdot\cdot}\right) = \mu \qquad V\left(\overline{Y}_{\cdot\cdot}\right) = \frac{1}{(bk)^2} \sum_i \sum_j V(\epsilon_{ij}) = \frac{\sigma^2}{bk} \qquad E\left(\overline{Y}_{\cdot\cdot}^2\right) = \frac{\sigma^2}{bk} + \mu^2$$

(3) With $\overline{Y}_{\cdot j} = \frac{\sum_i Y_{ij}}{k} = \mu + \beta_j = \frac{\sum_i \epsilon_{ij}}{k}$, then

$$E\left(\overline{Y}_{\cdot j}\right) = \mu + \beta_j \qquad V\left(\overline{Y}_{\cdot j}\right) = \frac{\sigma^2}{k} \qquad E\left(\overline{Y}^2_{\cdot j}\right) = \frac{\sigma^2}{k} + (\mu + \beta_j)^2$$

(4) $E\left(\overline{Y}^2_{i\cdot}\right) = V\left(\overline{Y}_{i\cdot}\right) + \left[E\left(\overline{Y}_{i\cdot}\right)\right]^2 = \frac{\sigma^2}{b} + (\mu + \tau_i)^2$

a. $E(\text{MST}) = \frac{b}{k-1} E\left[\sum_i \left(\overline{Y}_{i\cdot} - \overline{Y}_{\cdot\cdot}\right)^2\right] = \frac{b}{k-1} E\left[\sum_i \overline{Y}^2_{i\cdot} - k\overline{Y}^2_{\cdot\cdot}\right]$

which can be seen by expanding $\sum_i \left(\overline{Y}_{i\cdot} - \overline{Y}_{\cdot\cdot}\right)^2$. Then

$$\begin{aligned} E(\text{MST}) &= \frac{b}{k-1}\left[\sum_i E\left(\overline{Y}^2_{i\cdot}\right) - kE\left(\overline{Y}^2_{\cdot\cdot}\right)\right] \\ &= \frac{b}{k-1}\left[\sum_i \left(\frac{\sigma^2}{b} + \mu^2 + 2\mu\tau_i + \tau^2 i\right) - k\left(\frac{\sigma^2}{bk} + \mu^2\right)\right] \\ &= \frac{b}{k-1}\left[\frac{(k-1)\sigma^2}{b} + \sum_i \tau_i^2\right] = \sigma^2 + \frac{b}{k-1}\sum_i \tau_i^2 \end{aligned}$$

b.

$$\begin{aligned} E(\text{MSB}) &= \frac{k}{b-1}\left[\sum_j E\left(\overline{Y}^2_{\cdot j}\right) - bE\left(\overline{Y}^2_{\cdot\cdot}\right)\right] \\ &= \frac{k}{b-1}\left[\sum_j \left(\frac{\sigma^2}{k} + \mu^2 + 2\mu\beta_j + \beta^2 j\right) + \frac{b\sigma^2}{bk} - b\mu^2\right] \\ &= \frac{k}{b-1}\left[\frac{(b-1)\sigma^2}{k} + \sum_j \beta_j^2\right] = \sigma^2 + \frac{k}{b-1}\sum_j \beta_j^2 \end{aligned}$$

c. Recall that $\text{TSS} = \sum_i \sum_j Y_{ij}^2 - bk\overline{Y}^2_{\cdot\cdot}$, so that

$$\begin{aligned} E(\text{TSS}) &= \sum_i \sum_j \left(\sigma^2 + \mu 2 + \tau_i^2 + \beta_j^2\right) - bk\left(\frac{\sigma^2}{bk} + \mu^2\right) \\ &= (bk-1)\sigma^2 + b\sum_i \tau_i^2 + k\sum_j \beta_j^2 \end{aligned}$$

By the additivity property,

$$\begin{aligned} E(\text{SSE}) &= E(\text{TSS}) - E(\text{SST}) - E(\text{SSB}) \\ &= (bk-1)\sigma^2 - (k-1)\sigma^2 - (b-1)\sigma^2 + b\sum_i \tau_i^2 - b\sum_i \tau_i^2 \\ &\quad + k\sum_j \beta_j^2 - k\sum_j \beta_j^2 \\ &= (bk - k - b + 1)\sigma^2 \end{aligned}$$

Finally, since

$$\text{MSE} = \frac{\text{SSE}}{bk-k-b+1} \qquad \text{then} \qquad E(\text{MSE}) = \sigma^2$$

13.46 The 95% confidence interval for the difference in mean response for CPU time between computer 1 and computer 2 is

$$\left(\overline{Y}_{1\cdot} - \overline{Y}_{2\cdot}\right) \pm t_{.025}\sqrt{\text{MSE}}\sqrt{\tfrac{2}{b}},$$

where $t_{.025}$ is based upon 5 degrees of freedom, yielding

$$(1.553 - 1.575) \pm (2.571)\sqrt{.00045}\sqrt{\tfrac{2}{6}} = -.022 \pm .031 = (-.053, .009)$$

This interval is the same as the interval in Exercise 12.10(c).

13.47 $\left(\overline{Y}_A - \overline{Y}_B\right) \pm t_{.025,6}\sqrt{\text{MSE}\left(\frac{2}{b}\right)}$ or $(7-5) \pm 2.447\sqrt{2\left(\frac{2}{3}\right)}$ or 2 ± 2.83 or $[-.83, 4.83]$.

13.48 $\left(\overline{Y}_B - \overline{Y}_A\right) \pm t_{.05,6}\sqrt{\text{MSE}}\sqrt{\frac{2}{b}}$ or $\left(\frac{64}{4} - \frac{50}{4}\right) \pm 1.943\sqrt{1.89}\sqrt{\frac{2}{4}}$ or 3.5 ± 1.89 or $[1.61, 5.39]$.

13.49 $\left(\overline{Y}_B - \overline{Y}_D\right) \pm t_{.025,12}\sqrt{\text{MSE}}\sqrt{\frac{2}{b}}$ or $\left(\frac{10.756}{4} - \frac{10.175}{4}\right) \pm 2.179\sqrt{.0135}\sqrt{\frac{2}{4}}$ or $.145 \pm .179$.

13.50 $\left(\overline{Y}_4 - \overline{Y}_1\right) \pm t_{.025,9}\sqrt{\text{MSE}\left(\frac{2}{b}\right)}$ or $2 \pm 2.262\sqrt{1.95139\left(\frac{2}{4}\right)}$ or 2 ± 2.23

or $-.23$ to 4.23, which differs very little from the interval obtained without blocking.

13.51 $(\overline{Y}_{1\cdot} - \overline{Y}_{2\cdot}) \pm t_{.055,12}\sqrt{\text{MSE}}\sqrt{\frac{2}{b}}$ or $(71.1 - 75.9) \pm 3.055\sqrt{10.3716}\sqrt{\frac{2}{7}}$
or -4.8 ± 5.259.

13.52 Refer to Exercise 13.7. The difference in mean strengths is to be estimated to within .02 ton. The problem is to determine the number of observations per treatment necessary to achieve this bound. Thus, we must have

$$2\sigma_{\overline{Y}_{1\cdot}-\overline{Y}_{2\cdot}} \le .02 \quad \text{or} \quad 2\sigma\sqrt{\tfrac{2}{n}} \le .02$$

The best estimate available for σ^2 is $s^2 = \text{MSE} = .0025$. Hence,

$$2(.05)\sqrt{\tfrac{2}{n}} \le .02 \quad \text{or} \quad \sqrt{n} \ge \tfrac{.1\sqrt{2}}{.02}$$

and $n \ge \frac{2}{.04}$ and $n \ge 50$ will give the desired bound.
Thus, the total number of observations required in the entire experiment is

$$4n \ge 4(50) = 200.$$

13.53 It is necessary to have $\frac{2\sigma}{\sqrt{n_A}} \le 10$. Estimating σ with $\sqrt{\text{MSE}}$, we solve

$$\tfrac{2\sqrt{62.333}}{\sqrt{n_A}} \le 10 \quad \text{or} \quad n_A \ge 2.49$$

Hence, $n = 3$ observations are necessary.

13.54 It is necessary to have

$$2\sigma_{\overline{Y}_{A\cdot}-\overline{Y}_{B\cdot}} \le 20 \quad \text{or} \quad 2\sigma\sqrt{\tfrac{2}{n}} \le 20$$

Using $\sigma = \sqrt{\text{MSE}}$, solve

$$2\sqrt{62.333}\sqrt{\tfrac{2}{n}} \le 20 \quad \text{or} \quad n \ge 1.24$$

Hence, $n = 2$ observations are necessary.
This means that the total number of observations required in the entire experiment is $3n = 6$.

13.55 It is necessary to have

$$2\sigma_{\overline{Y}_{i\cdot}-\overline{Y}_{i'\cdot}} \le 1 \quad \text{or} \quad 2\sigma\sqrt{\tfrac{2}{b}} \le 1$$

Using $\sigma = \sqrt{\text{MSE}}$, solve

$$2\sqrt{1.89\left(\tfrac{2}{b}\right)} \le 1 \quad \text{or} \quad b \ge 15.12$$

Hence, $b = 16$ locations must be used.
Thus, $3b = 48$ is the total number of observations required in the entire experiment.

13.56 It is necessary to have $2\sigma_{\overline{T}_1-\overline{T}_4} \le .5$ or $2\sigma\sqrt{\frac{2}{b}} \le .5$. Using $\sigma = \sqrt{\text{MSE}}$, solve

$$2\sqrt{1.95139\left(\tfrac{2}{b}\right)} \le .5 \quad \text{or} \quad b \ge 62.44$$

Hence, $b = 63$ locations are needed.

13.57 There are three intervals to construct, so each interval should have confidence coefficient $1 - \left(\frac{.05}{3}\right) = .9833$. Since MSE $= .4735$ with 162 degrees of freedom, we may use the Z table. Now, $1 - .9833 = .01667$ and $\frac{.01667}{2} = .00833$. Thus, we will use $Z_{.0083} = 2.39$. The confidence intervals are of the form

$$(\overline{y}_{i\cdot} - \overline{y}_{j\cdot}) \pm 2.39\sqrt{\text{MSE}\left(\tfrac{1}{n_i} + \tfrac{1}{n_j}\right)}$$

For the pairs (i, j) of $(1, 2)$, $(1, 3)$, and $(2, 3)$, they are

$$(1, 2):\ -.29 \pm 2.39\sqrt{.4735\left(\tfrac{1}{45} + \tfrac{1}{102}\right)} \quad \text{or} \quad -.29 \pm .294$$

$$(1, 3):\ -1.65 \pm 2.39\sqrt{.4735\left(\tfrac{1}{45} + \tfrac{1}{18}\right)} \quad \text{or} \quad -1.65 \pm .459$$

$$(2, 3):\ -1.36 \pm 2.39\sqrt{.4735\left(\tfrac{1}{102} + \tfrac{1}{18}\right)} \quad \text{or} \quad -1.36 \pm .420$$

The simultaneous coverage rate of intervals constructed in this manner is .95.

13.58 In this case three pair wise comparisons can be made. Indeed, the reader of the report has implicitly done this in choosing the largest pair wise difference for study. Thus, the Bonferroni technique should be used, with $m = 3$.

13.59 There are three intervals to construct, so that each interval should have confidence coefficient $1 - \left(\frac{\alpha}{3}\right) = 1 - \left(\frac{.10}{3}\right) = .97$. Since no value of t is given with area $\frac{.1}{(2)(3)} = .01667$ to its right, we choose to use $t_{.01}$ so that the overall confidence coefficient will be .94 rather than .85, which would occur if $t_{.025}$ were used. The three intervals all have half width $3.143\sqrt{1.89\left(\frac{2}{4}\right)} = 3.06$. The three intervals are shown below.

$(1,2)$:	-3.5 ± 3.06	or	-6.56 to $-.44$
$(1,3)$:	$.5 \pm 3.06$	or	-2.56 to 3.56
$(2,3)$:	4.0 ± 3.06	or	$.94$ to 7.06

13.60 As we want an overall confidence level of .95 and there are three intervals, we use $t_{.05/(2)(3)}$. This is close enough $t_{.01}$. The width of each interval is then

$$t_{.01,9} = \sqrt{\text{MSE}\left(\tfrac{2}{b}\right)} = 2.821\sqrt{1.95139\left(\tfrac{2}{4}\right)} = 2.79$$

The intervals are shown below.

$(1,4)$:	-2 ± 2.79	or	-4.79 to $.79$
$(2,4)$:	-1 ± 2.79	or	-3.79 to 1.79
$(3,4)$:	$-.75 \pm 2.79$	or	-3.54 to 2.04

13.61 **a.** $\beta_0 + \beta_3$ is the mean response to treatment A in block III.

b. $\beta_3 =$ difference in mean responses to chemicals A and D in block III.

13.62 **a.** The complete model is

$$Y_{ij} = \beta_0 + \beta_1 x_1 + \beta_3 x_2 + \epsilon$$

where

$$x_1 = \begin{cases} 1, & \text{if method } A \\ 0, & \text{otherwise} \end{cases}$$

and

$$x_2 = \begin{cases} 1, & \text{if method } B \\ 0, & \text{otherwise} \end{cases}.$$

Then for the complete model,

$$\boldsymbol{Y} = \begin{bmatrix} 73 \\ 83 \\ 76 \\ 68 \\ 80 \\ 54 \\ 74 \\ 71 \\ 79 \\ 95 \\ 87 \end{bmatrix} \qquad \boldsymbol{X} = \begin{bmatrix} 1 & 1 & 0 \\ 1 & 1 & 0 \\ 1 & 1 & 0 \\ 1 & 1 & 0 \\ 1 & 1 & 0 \\ 1 & 0 & 1 \\ 1 & 0 & 1 \\ 1 & 0 & 1 \\ 1 & 0 & 0 \\ 1 & 0 & 0 \\ 1 & 0 & 0 \end{bmatrix} \qquad \boldsymbol{X'X} = \begin{bmatrix} 11 & 5 & 3 \\ 5 & 5 & 0 \\ 3 & 0 & 3 \end{bmatrix}$$

$$(\boldsymbol{X'X})^{-1} = \begin{bmatrix} \frac{1}{3} & -\frac{1}{3} & -\frac{1}{3} \\ -\frac{1}{3} & \frac{8}{15} & \frac{1}{3} \\ -\frac{1}{3} & \frac{1}{3} & \frac{2}{3} \end{bmatrix} \qquad \widehat{\boldsymbol{\beta}} = (\boldsymbol{X'X})^{-1}\boldsymbol{X'Y} = \begin{bmatrix} 87 \\ -11 \\ -20.67 \end{bmatrix}$$

$$\text{SSE}_c = \boldsymbol{Y'Y} - \widehat{\boldsymbol{\beta}}'\boldsymbol{X'Y} = 65{,}286 - 54{,}787.33 = 498.67$$

with $11 - 3 = 8$ degrees of freedom. The reduced model is $Y_{ij} = \beta_0 + \epsilon$, and the $\boldsymbol{X}$ matrix becomes

$$\boldsymbol{X} = \begin{bmatrix} 1 \\ 1 \\ 1 \\ 1 \\ 1 \\ 1 \\ 1 \\ 1 \\ 1 \\ 1 \\ 1 \end{bmatrix} \quad \text{and} \quad \boldsymbol{X'X} = \begin{bmatrix} 11 \end{bmatrix} \quad (\boldsymbol{X'X})^{-1} = \begin{bmatrix} \frac{1}{11} \end{bmatrix} \quad \widehat{\boldsymbol{\beta}} = \begin{bmatrix} 76.3636 \end{bmatrix}$$

$$\text{SSE}_R = 65{,}286 - 64{,}145.455 = 1140.5455$$

with $11 - 1 = 10$ degrees of freedom. Then

$$s_2^2 = \frac{498.67}{8} = 62.3333$$

and

$$s_3^2 = \frac{\text{SSE}_R - \text{SSE}_c}{10-8} = \frac{641.8788}{2} = 320.9394$$

and $F = \frac{s_3^2}{s_2^2} = 5.15$,

which is compared to $F_{.05} = 4.46$ with 2 and 8 degrees of freedom. There is evidence of a difference in treatment means.

b. The hypothesis of interest is

$H_0\colon\ \mu_A - \mu_B = 0$ vs. $H_a\colon\ \mu_A - \mu_B \neq 0.$

With $\frac{\text{SSE}_2}{B}$ used to estimate σ^2, the test statistic is

$$t = \frac{\bar{y}_A - \bar{y}_B}{\sqrt{\text{MSE}_2\left(\frac{1}{5}+\frac{1}{3}\right)}} = \frac{76-66.33}{\sqrt{62.333\left(\frac{8}{15}\right)}} = 1.68$$

The rejection region is $|t| > t_{.025,8} = 2.306$, and the null hypothesis is not rejected. There is not significant evidence of a difference between A and B.

c. For part **a**, note in Table 7 that $4.46 < 5.15 < 6.06$. Thus, $.025 < p\text{-value} \leq .05$. For part **b**, note in Table 5 that $1.397 \leq 1.68 \leq 1.860$. Thus, $2(.05) < p\text{-value} \leq 2(.1)$.

13.63 The complete model is

$$Y_{ij} = \beta_0 + \beta_1 x_1 + \beta_3 x_2 + \beta_3 x_3 + \beta_4 x_4 + \beta_5 x_5 + \epsilon$$

where

$$x_1 = \begin{cases} 1, & \text{if block 1} \\ 0, & \text{otherwise} \end{cases} \quad x_2 = \begin{cases} 1, & \text{if block 2} \\ 0, & \text{otherwise} \end{cases} \quad x_3 = \begin{cases} 1, & \text{if treatment 1} \\ 0, & \text{otherwise} \end{cases}$$

$$x_4 = \begin{cases} 1, & \text{if treatment 2} \\ 0, & \text{otherwise} \end{cases} \quad x_5 = \begin{cases} 1, & \text{if treatment 3} \\ 0, & \text{otherwise} \end{cases}$$

Then for the complete model,

$$\boldsymbol{Y} = \begin{bmatrix} 5 \\ 3 \\ 8 \\ 4 \\ 9 \\ 8 \\ 13 \\ 6 \\ 7 \\ 4 \\ 9 \\ 8 \end{bmatrix} \qquad \boldsymbol{X} = \begin{bmatrix} 1 & 1 & 0 & 1 & 0 & 0 \\ 1 & 1 & 0 & 0 & 1 & 0 \\ 1 & 1 & 0 & 0 & 0 & 1 \\ 1 & 1 & 0 & 0 & 0 & 0 \\ 1 & 0 & 1 & 1 & 0 & 0 \\ 1 & 0 & 1 & 0 & 1 & 0 \\ 1 & 0 & 1 & 0 & 0 & 1 \\ 1 & 0 & 1 & 0 & 0 & 0 \\ 1 & 0 & 0 & 1 & 0 & 0 \\ 1 & 0 & 0 & 0 & 1 & 0 \\ 1 & 0 & 0 & 0 & 0 & 1 \\ 1 & 0 & 0 & 0 & 0 & 0 \end{bmatrix}$$

$$\boldsymbol{X}'\boldsymbol{X} = \begin{bmatrix} 12 & 4 & 4 & 3 & 3 & 3 \\ 4 & 4 & 0 & 1 & 1 & 1 \\ 4 & 0 & 4 & 1 & 1 & 1 \\ 3 & 1 & 1 & 3 & 0 & 0 \\ 3 & 1 & 1 & 0 & 3 & 0 \\ 3 & 1 & 1 & 0 & 0 & 3 \end{bmatrix} \qquad (\boldsymbol{X}'\boldsymbol{X})^{-1} = \begin{bmatrix} \frac{1}{2} & -\frac{1}{4} & -\frac{1}{4} & -\frac{1}{3} & -\frac{1}{3} & -\frac{1}{3} \\ -\frac{1}{4} & \frac{1}{2} & \frac{1}{4} & 0 & 0 & 0 \\ -\frac{1}{4} & \frac{1}{4} & \frac{1}{2} & 0 & 0 & 0 \\ -\frac{1}{3} & 0 & 0 & \frac{2}{3} & \frac{1}{3} & \frac{1}{3} \\ -\frac{1}{3} & 0 & 0 & \frac{1}{3} & \frac{2}{3} & \frac{1}{3} \\ -\frac{1}{3} & 0 & 0 & \frac{1}{3} & \frac{1}{3} & \frac{2}{3} \end{bmatrix}$$

$$\boldsymbol{X}'\boldsymbol{Y} = \begin{bmatrix} 84 \\ 20 \\ 36 \\ 21 \\ 15 \\ 30 \end{bmatrix} \qquad \widehat{\boldsymbol{\beta}} = \begin{bmatrix} 6 \\ -2 \\ 2 \\ 1 \\ -1 \\ 4 \end{bmatrix}$$

and $\text{SSE}_2 = 674 - 662 = 12$, with $12 - 6 = 6$ degrees of freedom. The reduced model is $Y_{ij} = \beta_0 + \beta_1 x_1 + \beta_2 x_2 + \epsilon$, with x_1 and x_2 as defined in the complete model. The $\boldsymbol{X}$ matrix has columns 4, 5, and 6 deleted, and

$$\boldsymbol{X}'\boldsymbol{X} = \begin{bmatrix} 12 & 4 & 4 \\ 4 & 4 & 0 \\ 4 & 0 & 4 \end{bmatrix} \qquad (\boldsymbol{X}'\boldsymbol{X})^{-1} = \begin{bmatrix} \frac{1}{4} & -\frac{1}{4} & -\frac{1}{4} \\ -\frac{1}{4} & \frac{1}{2} & \frac{1}{4} \\ -\frac{1}{4} & \frac{1}{4} & \frac{1}{2} \end{bmatrix} \qquad \boldsymbol{X}'\boldsymbol{Y} = \begin{bmatrix} 84 \\ 20 \\ 36 \end{bmatrix}$$

$$\widehat{\boldsymbol{\beta}} = \begin{bmatrix} 7 \\ -2 \\ 2 \end{bmatrix}$$

and $\text{SSE}_1 = 674 - 620 = 54$ with $12 - 3 = 9$ degrees of freedom. Then

$$F = \frac{S_3^2}{S_2^2} = \frac{\frac{54-12}{3}}{\frac{12}{6}} = 7.00$$

which is compared to $F_{.05} = 4.76$ with 3 and 6 degrees of freedom. H_0 is rejected; there is a difference in the treatment means.

13.64 The complete model is

$$Y_{ij} = \beta_0 + \beta_1 x_1 + \beta_3 x_2 + \beta_3 x_3 + \beta_4 x_4 + \beta_5 x_5 + \epsilon$$

where x_1, x_2, and x_3 are dummy variables for blocks and x_4 and x_5 are dummy variables for treatments. Then we have the following matrices:

$$\boldsymbol{Y} = \begin{bmatrix} 11 \\ 15 \\ 10 \\ 13 \\ 17 \\ 15 \\ 16 \\ 20 \\ 13 \\ 10 \\ 12 \\ 10 \end{bmatrix} \qquad \boldsymbol{X} = \begin{bmatrix} 1 & 1 & 0 & 0 & 1 & 0 \\ 1 & 1 & 0 & 0 & 0 & 1 \\ 1 & 1 & 0 & 0 & 0 & 0 \\ 1 & 0 & 1 & 0 & 1 & 0 \\ 1 & 0 & 1 & 0 & 0 & 1 \\ 1 & 0 & 1 & 0 & 0 & 0 \\ 1 & 0 & 0 & 1 & 1 & 1 \\ 1 & 0 & 0 & 1 & 0 & 1 \\ 1 & 0 & 0 & 1 & 0 & 0 \\ 1 & 0 & 0 & 0 & 1 & 0 \\ 1 & 0 & 0 & 0 & 0 & 1 \\ 1 & 0 & 0 & 0 & 0 & 0 \end{bmatrix}$$

$$(\boldsymbol{X}'\boldsymbol{X})^{-1} = \begin{bmatrix} \frac{1}{2} & -\frac{1}{3} & -\frac{1}{3} & -\frac{1}{3} & -\frac{1}{4} & -\frac{1}{4} \\ -\frac{1}{3} & \frac{2}{3} & \frac{1}{3} & \frac{1}{3} & 0 & 0 \\ -\frac{1}{3} & \frac{1}{3} & \frac{2}{3} & \frac{1}{3} & 0 & 0 \\ -\frac{1}{3} & \frac{1}{3} & \frac{1}{3} & \frac{2}{3} & 0 & 0 \\ -\frac{1}{4} & 0 & 0 & 0 & \frac{1}{2} & \frac{1}{4} \\ -\frac{1}{4} & 0 & 0 & 0 & \frac{1}{4} & \frac{1}{2} \end{bmatrix} \qquad \boldsymbol{X}'\boldsymbol{Y} = \begin{bmatrix} 162 \\ 36 \\ 45 \\ 49 \\ 50 \\ 64 \end{bmatrix} \qquad \widehat{\boldsymbol{\beta}} = \begin{bmatrix} 9.17 \\ 1.33 \\ 4.33 \\ 5.67 \\ .5 \\ 4.0 \end{bmatrix}$$

and $\text{SSE}_2 = 2298 - 2286.6667 = 11.3333$, with $12 - 6 = 6$ degrees of freedom. For the reduced model, columns 2, 3, and 4 are omitted and

$$X'X = \begin{bmatrix} 12 & 4 & 4 \\ 4 & 4 & 0 \\ 4 & 0 & 4 \end{bmatrix} \quad (X'X)^{-1} = \begin{bmatrix} \frac{1}{4} & -\frac{1}{4} & -\frac{1}{4} \\ -\frac{1}{4} & \frac{1}{2} & \frac{1}{4} \\ -\frac{1}{4} & \frac{1}{4} & \frac{1}{2} \end{bmatrix} \quad X'Y = \begin{bmatrix} 162 \\ 50 \\ 64 \end{bmatrix}$$

$$\widehat{\beta} = \begin{bmatrix} 12.5 \\ .5 \\ 4.0 \end{bmatrix}$$

and $\text{SSE}_1 = 2298 - 2225 = 73$ with $12 - 3 = 9$ degrees of freedom. Then

$$F = \frac{S_3^2}{S_2^2} = \frac{\frac{73-11.333}{3}}{\frac{11.333}{6}} = 10.88$$

which is compared to $F_{.05} = 4.76$ with 3 and 6 degrees of freedom. H_0 is rejected; there is a difference due to locations.

13.65 a. Experimental units are patches of skin, while the three people act as blocks.

b. Incorporate the given sums of squares into an ANOVA table, as shown at the right. Then

Source	d.f.	SS	MS
Treatments	2	1.18	.59
Blocks	2	.78	.39
Error	4	2.24	.56
Total	8	4.2	

$$F = \frac{\text{MST}}{\text{MSE}} = \frac{.59}{.56} = 1.05$$

which is compared to $F_{.05} = 6.94$ with 2 and 4 degrees of freedom. H_0 is not rejected; there is no evidence to suggest a significant difference in treatment means.

13.66 Refer to Exercise 13.7. In that exercise we calculated CM = 58.08, TSS = .035, and and SST = .015. Then

$$\text{SSB} = \frac{\sum_i Y_j^2}{n_B} - \text{CM} = \frac{(8.9)^2+(8.6)^2+(8.9)^2}{4} - \text{CM} = 58.095 - \text{CM} = .015$$

The ANOVA table is shown at the right. Note that the addition of SSB to the analysis causes a reduction to SSE. The test for a difference in response owing to the type of sand is essentially a test to see whether or not blocks are significant. We calculate

Source	d.f.	SS	MS
Treatments	3	.015	.005000
Blocks	2	.015	.007500
Error	6	.005	.000833
Total	11	.035	

$$F = \frac{\text{MSB}}{\text{MSE}} = 9$$

The critical value of F is 5.14, and we can conclude that the data suggest "type of sand" is important.

Similarly, the test for difference in mean strength for mixes will be

$$F = \frac{\text{MST}}{\text{MSE}} = 6$$

which is compared to a critical F with 3 and 6 degrees of freedom. This value is $F_{.05} = 4.76$. Hence we conclude that mixes are indeed an important factor. Notice that by eliminating the extraneous factor (sand pits) by blocking, SSE is significantly reduced and it is possible to conclude that the mixes are significantly different without changing the value of α, the probability of falsely rejecting H_0.

13.67 A 95% confidence interval for $(\mu_A - \mu_B)$ is given by

$$(\overline{y}_A - \overline{y}_B) \pm t_{.025,6}\, s\sqrt{\frac{1}{n_A} + \frac{1}{n_B}} = (2.25 - 2.166) \pm 2.447\sqrt{.000833}\,\sqrt{\frac{1}{3} + \frac{1}{3}}$$

or

$$.08 \pm .06 \qquad \text{or} \qquad [.02\,,\,.14]$$

Thus, in repeated sampling, confidence intervals constructed in this manner will enclose the mean difference $(\mu_A - \mu_B)$ 95% of the time. Notice that this confidence interval is not as wide as the interval found in Exercise 13.22(b). That is, reducing SSE by accounting for blocks has reduced the width of the confidence intervals. In other words, accounting for the block effect has improved the precision of our results.

13.68 a. The experiment is designed to compare two drugs, each at three levels. Thus there are $3(3) = 9$ treatments that are of interest to the experimenter. The first treatment is defined as the combination "first level of drug 1 and first level of drug

2"; the second is defined as the combination "second level of drug 1 and first level of drug 2"; and so on for all nine treatments. Denote these nine treatments by $A, B, \ldots, H, I$. The layout of the design is shown in the table below. One response, denoted by y_i, $i = 1, \ldots, 9$, is measured for each treatment. It is not a randomized block design.

Treatment	A	B	C	D	E	F	G	H	I
Response	y_1	y_2	y_3	y_4	y_5	y_6	y_7	y_8	y_9

b. The second design is similar to the first, except that two patients are randomly assigned to each treatment. Thus a completely randomized design is employed, with nine treatments and two responses per treatment. The layout of the design is shown in the table below.

Treatment	A	B	C	D	E	F	G	H	I
Response	y_{11}	y_{21}	y_{31}	y_{41}	y_{51}	y_{61}	y_{71}	y_{81}	y_{91}
	y_{12}	y_{22}	y_{32}	y_{42}	y_{52}	y_{62}	y_{72}	y_{82}	y_{92}

13.69 a. This is similar to Exercise 13.45. It is necessary to have

$$2\sigma_{\overline{Y}} \le 10 \qquad \text{or} \qquad 2\frac{\sigma}{\sqrt{n}} \le 10 \qquad \text{or} \qquad \frac{2(20)}{\sqrt{n}} \le 10$$

so that $\sqrt{n} \ge 4$ and $n \ge 16$.

b. When 16 patients are assigned to each of the 9 treatments, there are $(n_1 + n_2 + \ldots + n_9 - 9)$ degrees of freedom for estimating σ^2. In this case there are $16(9) - 9 = 135$ degrees of freedom.

c. With 16 replications for each treatment, an approximate half width of the confidence interval for the difference in mean response for two treatments is

$$2\sigma\sqrt{\frac{1}{n_1} + \frac{1}{n_2}} = 2(20)\sqrt{\frac{1}{16} + \frac{1}{16}} = 14.14$$

13.70 A randomized block design is employed and the analysis of variance is as follows:

$$(1)\ \text{CM} = \frac{\left[\sum_i\sum_j y_{ij}\right]^2}{n} = \frac{(180.4)^2}{9} = 3616.017$$

$$(2)\ \text{TSS} = \sum_i\sum_j y_{ij}^2 - \text{CM} = 3637.82 - \text{CM} = 21.802$$

$$(3)\ \text{SST (cars)} = \frac{\sum_j Y_{j\cdot}^2}{n_T} - \text{CM} = \frac{10{,}894.46}{3} - \text{CM} = 3631.486 - \text{CM} = 15.4696$$

$$(4)\ \text{SSB (gas)} = \frac{\sum_i Y_{\cdot i}^2}{n_B} - \text{CM} = \frac{10{,}852.08}{3} - \text{CM} = 3617.36 - \text{CM} = 1.343$$

The ANOVA table is shown at the right.

Source	d.f.	SS	MS
Treatments	2	15.4696	7.7348
Blocks	2	1.3430	.6715
Error	4	4.9894	1.2470
Total	8	21.8020	

a. For a test of the null hypothesis that there is no difference in mean gasoline consumption for the three cars, an F test is used, where

$$F = \frac{\text{MST}}{\text{MSE}} = 6.2027$$

The critical value of F, based on 2 and 4 degrees of freedom, is $F_{.05} = 6.94$. The computed value of F is less than 6.94, and hence the null hypothesis of equality cannot be rejected.

b. Similarly, the F test used to test the null hypothesis of no difference due to brand of gasoline is

$$F = \frac{\text{MSB}}{\text{MSE}} = .53849$$

The critical value of F is, again, 6.94. The conclusion is that brand of gasoline does not affect mileage.

13.71 Refer to Exercise 13.70. If the experiment is run as a completely randomized design, the block factor (brand of gasoline) has a negligible effect on the response. Hence the sum of squares for blocks (which was calculated in Exercise 13.61) will not be removed from the sum of squares for error. The calculations for SST, TSS, and CM remain the same, and the ANOVA table is as shown below.

a. For a test of significance of treatments, the test statistic is

Source	d.f.	SS	MS
Treatments	2	15.4696	7.7348
Error	6	6.3324	1.0554
Total	8	21.8020	

$F = \frac{\text{MST}}{\text{MSE}} = 7.3287$

The critical value of F, based on 2 and 6 degrees of freedom, is 5.14. Notice that the observed value of the test statistic exceeds the critical value of F. Hence we reject the null hypothesis of no difference between treatments and conclude that cars have a significant effect on gasoline mileage.

b. Notice that we are able to reject the null hypothesis in this case but were unable to do so in Exercise 13.70. The reason for this is that the drop in SSE (due to the isolation of block sum of squares in the first case) was not sufficient to compensate for the loss of degrees of freedom for estimating σ^2. The loss of 2 degrees of freedom caused the critical value of F to be substantially increased, and the information gained by blocking was not sufficient to overcome this loss.

c. The randomized block design randomly assigns treatments to experimental units within each block, while the completely randomized design randomly assigns the treatments to all experimental units. Thus, when we have data from an RBD, the randomization is wrong for the completely randomized design and the results do not fit the model.

13.72

a. We have independent random samples of sizes 4, 5, 5, and 3 from plans 1, 2, 3, and 4, respectively. Hence, this qualifies as a completely randomized design.

b. The necessary computations follow:

$y_{1\cdot} = 107$ $\quad y_{2\cdot} = 134$ $\quad y_{3\cdot} = 162$ $\quad y_{4\cdot} = 94$

$\sum y = 497$ $\quad \sum y^2 = 14{,}713$ $\quad \text{CM} = \frac{(497)^2}{17} = 1452.941$

$\text{TSS} = 14{,}713 - \text{CM} = 183.059$

$\text{SST} = \frac{(107)^2}{4} + \frac{(134)^2}{5} + \frac{(162)^2}{5} + \frac{(94)^2}{3} - \text{CM} = 117.642$

$\text{SSE} = \text{TSS} - \text{SST} = 65.417$

The ANOVA table is given at the right. The test statistic is

Source	d.f.	SS	MS
Treatments	3	117.642	39.214
Error	13	65.417	5.032
Total	16	183.059	

$F = \frac{\text{MST}}{\text{MSE}} = 7.79$

which is compared to $F_{.01} = 5.74$ with 3 and 13 degrees of freedom. We reject H_0: there is significant evidence of a difference.

c. A 95% confidence interval for the difference between plan 1 and plan 3 is

$$\left(\tfrac{107}{4} - \tfrac{162}{5}\right) \pm 2.160\sqrt{65.417}\sqrt{\tfrac{1}{4} + \tfrac{1}{5}} = -5.65 \pm 11.72$$

13.73

a. A rearrangement of the results gives

Levels of Digitalis	Dogs 1	2	3	4
A	1342	1140	1029	1150
B	1608	1387	1296	1319
C	1881	1698	1549	1579

The analysis is as follows:

$$(1)\ \text{CM} = \frac{\left[\sum_i \sum_j y_{ij}\right]^2}{n} = \frac{(16{,}978)^2}{12} = 24{,}021{,}040.333$$

$$(2)\ \text{TSS} = \sum_i \sum_j y_{ij}^2 - \text{CM} = 24{,}724{,}722 - 24{,}021{,}040.333 = 703{,}681.667$$

$$(3)\ \text{SST (digitalis)} = \frac{\sum_i Y_{i\cdot}^2}{n_T} - \text{CM} = \frac{(4661)^2}{4} + \frac{(5610)^2}{4} + \frac{(6707)^2}{4} - 24{,}021{,}040$$
$$= 524{,}177{,}167$$

(4) $\text{SSB (dogs)} = \frac{\sum_j Y_j^2}{n_B} - \text{CM}$

$= \frac{(4831)^2}{3} + \frac{(4225)^2}{3} + \frac{(3874)^2}{3} + \frac{(4048)^2}{3} - 24{,}021{,}040.33$

$= 173{,}415$

(5) $\text{SSE} = \text{TSS} - \text{SST} - \text{SSB} = 6089.5$

The ANOVA table is shown below.

Source	d.f.	SS	MS	F
Treatments	2	524,177.167	262,088.58	258.237
Blocks	3	173,415.00	57,805.00	56.95
Error	6	6,089.5	1,014.9167	
Total	11	703,681.667		

b. There are $n - k - b + 1 = 12 - 3 - 4 + 1 = 6$ degrees of freedom associated with SSE.

c. To test the null hypothesis that there is no difference in mean uptake of calcium for the three levels of digitalis, we use an F test, where

$$F = \frac{\text{MST}}{\text{MSE}} = \frac{262{,}088.58}{1014.9167} = 258.237$$

The critical value of F, based on 2 and 6 degrees of freedom, is $F_{.05} = 5.14$. Since the test statistic is in the rejection region, we reject the null hypothesis and conclude that at least one of the levels of digitalis causes a different level of calcium in the heart muscle of dogs.

d. To test the null hypothesis that there is no difference in the mean uptake in calcium for the four heart muscles, we use an F test where

$$F = \frac{\text{MSB}}{\text{MSE}} = \frac{57{,}805.00}{1014.9167} = 56.95.$$

The critical value of F, based on 3 and 6 degrees of freedom, is $F_{.05} = 4.76$. Since the test statistic is in the rejection region, we reject the null hypothesis and conclude that at least one of the heart muscles differs in mean calcium uptake.

e. The standard deviation of the difference between the mean calcium uptake for two levels of digitalis is

$$s\sqrt{\frac{1}{n_i} + \frac{1}{n_j}} = \sqrt{\text{MSE}}\sqrt{\frac{1}{4} + \frac{1}{4}} = (1014.9167)\sqrt{.5} = 22.53.$$

f. A 95% confidence interval for the difference in mean response between treatments A and B is

$$(\bar{y}_A - \bar{y}_B) \pm t_{\alpha/2,6}\, s\sqrt{\frac{1}{n_A} + \frac{1}{n_B}} = (1165.25 - 1402.5) \pm (2.447)(22.53)$$

or

$$-237.25 \pm 55.13 \qquad \text{or} \qquad [-292.38, -182.12].$$

13.74

We need $2\sigma\sqrt{\frac{1}{b} + \frac{1}{b}} \le 20$. From Exercise 13.73, $\sigma = \sqrt{\text{MSE}} = \sqrt{1015.0833}$. Thus, we need

$$2\sqrt{1015.0833}\sqrt{\frac{2}{b}} \le 20 \qquad \text{or} \qquad \sqrt{b} \ge \frac{2\sqrt{1015.0833}\sqrt{2}}{20} = 4.505 \qquad \text{or} \qquad b \ge 20.3$$

Thus, we need $b = 21$.

13.75

a. The design is completely randomized with five treatments, containing 4, 7, 6, 5, and 5 measurements, respectively. The analysis is as follows:

(1) $\text{CM} = \frac{(20.6)^2}{27} = 15.717$

(2) $\text{TSS} = 17{,}500 - \text{CM} = 1.783$

(3) $\text{SST} = \frac{(2.5)^2}{4} + \frac{(4.7)^2}{7} + \frac{(6.4)^2}{6} + \frac{(4.6)^2}{5} + \frac{(2.4)^2}{5} - \text{CM} = 1.212$

The ANOVA table is shown at the right. The F test to detect differences in mean reaction time to five stimuli is

Source	d.f.	SS	MS
Treatments	4	1.212	.303
Error	22	.571	.02596
Total	26	1.783	

$$F = \frac{\text{MST}}{\text{MSE}} = 11.68$$

The critical value of F for $\alpha = .05$, with 4 and 22 degrees of freedom, is $F = 2.82$, and we conclude that there is a significant difference owing to treatments. Since $F = 11.68 > 5.02$, the p-value is less than .005.

b. The hypothesis of interest is

$$H_0\colon\ \mu_A - \mu_D = 0 \qquad \text{vs.} \qquad H_a\colon\ \mu_A - \mu_D \neq 0,$$

and the test statistic is

$$t = \frac{\bar{y}_A - \bar{y}_D}{s\sqrt{\frac{1}{n_A} + \frac{1}{n_D}}} = \frac{.625 - .920}{\sqrt{.02596}\sqrt{\frac{1}{4} + \frac{1}{5}}} = -2.73$$

The critical value of t, with $\alpha = .05$ and 22 degrees of freedom, is $t_{.025,22} = 2.074$ and the rejection region is $|t| > 2.074$. Hence the null hypothesis is rejected, and we conclude that there is a difference between stimuli A and D. Since $2.508 < |t| < 2.819$, $2(.005) < p\text{-value} < 2(.01)$.

13.76 A randomized block design is used with people as blocks and stimuli as treatments. Calculate

$$\text{CM} = \frac{(16.4)^2}{20} = 13.448$$
$$\text{TSS} = 14.46 - \text{CM} = 1.012$$
$$\text{SST (stimuli)} = \frac{56.94}{4} - \text{CM} = .787$$
$$\text{SSB} = \frac{67.94}{5} - \text{CM} = .14$$

The ANOVA table is shown at the right. The F test for treatments is

Source	d.f.	SS	MS
Treatments	4	.787	.197
Blocks	3	.140	.047
Error	12	.085	.0071
Total	19	1.012	

$$F = \frac{\text{MST}}{\text{MSE}} = 27.7$$

The critical F based on 4 and 12 degrees of freedom is $F_{.05} = 3.25$, and the null hypothesis of no difference is rejected.

13.77 We will construct four confidence intervals, so each interval should have confidence coefficient $1 - \left(\frac{.05}{4}\right) = .9875 = .99$. We will use the t value $t_{.005,12} = 3.055$. The intervals all have half width $3.055\sqrt{.0135}\sqrt{\frac{2}{4}} = .251$. The intervals for the various varieties are

$\mu_A - \mu_D$: $.320 \pm .251$ $\qquad$ $\mu_B - \mu_D$: $.145 \pm .251$

$\mu_C - \mu_D$: $.023 \pm .251$ $\qquad$ $\mu_E - \mu_D$: $-.124 \pm .251$

13.78 Following the procedure used in Section 13.4 of the text, write

$$\begin{aligned}
\text{TSS} &= \sum_{j=1}^{b}\sum_{i=1}^{k} (y_{ij} - \bar{y}_{..})^2 \\
&= \sum_{j=1}^{b}\sum_{i=1}^{k} (y_{ij} - \bar{y}_{i.} + \bar{y}_{i.} - \bar{y}_{.j} + \bar{y}_{.j} - \bar{y}_{..} + \bar{y}_{..} - \bar{y}_{..})^2 \\
&= \sum_{j=1}^{b}\sum_{i=1}^{k} (\bar{y}_{.j} - \bar{y}_{..})^2 + \sum_j\sum_i (\bar{y}_{i.} - \bar{y}_{..})^2 \\
&\quad + \sum_j\sum_i (y_{ij} - \bar{y}_{i.} - \bar{y}_{.j} + \bar{y}_{..})^2 + 2\sum_j\sum_i (\bar{y}_{.j} - \bar{y}_{..})(\bar{y}_{i.} - \bar{y}_{..}) \\
&\quad + 2\sum_j\sum_i (\bar{y}_{.j} - \bar{y}_{..})(y_{ij} - \bar{y}_{i.} - \bar{y}_{.j} + \bar{y}_{..}) \\
&\quad + 2\sum_j\sum_i (\bar{y}_{i.} - \bar{y}_{..})(y_{ij} - \bar{y}_{i.} - \bar{y}_{.j} + \bar{y}_{..}) \\
(^*) &= k\sum_{j=1}^{b} (\bar{y}_{.j} - \bar{y}_{..})^2 + b\sum_{i=1}^{k} (\bar{y}_{i.} - \bar{y}_{..})^2 + \text{SSE} \\
&\quad + 2\sum_{j=1}^{b} (\bar{y}_{.j} - \bar{y}_{..}) \sum_{i=1}^{k} (\bar{y}_{i.} - \bar{y}_{..}) \\
&\quad + 2\sum_{j=1}^{b} (\bar{y}_{.j} - \bar{y}_{..}) \sum_{i=1}^{k} (y_{ij} - \bar{y}_{i.} - \bar{y}_{.j} + \bar{y}_{..}) \\
&\quad + 2\sum_{j=1}^{k} (\bar{y}_{i.} - \bar{y}_{..}) \sum_{j=1}^{b} (y_{ij} - \bar{y}_{i.} - \bar{y}_{.j} + \bar{y}_{..})
\end{aligned}$$

Note that

$$(1)\ \sum_{j=1}^{b} (\overline{y}_{.j} - \overline{y}_{..}) = \sum_{j} \left(\frac{\sum_i y_{ij}}{k} - \frac{\sum_j \sum_i y_{ij}}{bk} \right) = \frac{\sum_j \sum_i y_{ij}}{k} - \frac{b \sum_j \sum_i y_{ij}}{bk} = 0$$

$$(2)\ \sum_{i} (y_{ij} - \overline{y}_{.j} - \overline{y}_{i.} + \overline{y}_{..}) = \sum_{i} \left(y_{ij} - \frac{\sum_j y_{ij}}{b} - \frac{\sum_i y_{ij}}{k} + \frac{\sum_j \sum_i y_{ij}}{bk} \right)$$

$$= \sum_i y_{ij} - \frac{\sum_j \sum_i y_{ij}}{b} - \sum_i y_{ij} + \frac{\sum_j \sum_j y_{ij}}{b} = 0$$

An expansion similar to (2) will show that

$$\sum_j (y_{ij} - \overline{y}_{.j} - \overline{y}_{i.} + \overline{y}_{..}) = 0$$

Returning to (*), whose last three terms are zero, we have

$$\text{TSS} = \text{SSB} + \text{SST} + \text{SSE}.$$

13.79a. Y_{ij} and $Y_{ij'}$ are independent if $\text{Cov}(Y_{ij}, Y_{ij'}) = 0$. Refer to Section 5.10 on the bivariate normal distribution. Using Theorem 5.12,

$$\begin{aligned}\text{Cov}(Y_{ij}, Y_{ij'}) &= \text{Cov}(\mu + \tau_i + \beta_j + \epsilon_{ij}, \mu + \tau_i + \beta_{j'} + \epsilon_{ij'}) \\ &= \text{Cov}(\beta_j + \epsilon_{ij}, \beta_{j'} + \epsilon_{ij'}) \\ &= \text{Cov}(\beta_j, \beta_{j'}) + \text{Cov}(\epsilon_{ij}, \beta_{j'}) + \text{Cov}(\beta_j, \epsilon_{ij'}) + \text{Cov}(\epsilon_{ij}, \epsilon_{ij'}) \\ &= 0 + 0 + 0 + 0 = 0\end{aligned}$$

Since the β_j's are independent, the ϵ_{ij}'s are independent, and the β_i and ϵ_{ij} are independent of each other.

$$\begin{aligned}\text{Cov}(Y_{ij}, Y_{i'j'}) &= \text{Cov}(\mu + \tau_i + \beta_j + \epsilon_{ij}, \mu + \tau_{i'} + \beta_{j'} + \epsilon_{i'j'}) \\ &= \text{Cov}(\beta_j + \epsilon_{ij}, \beta_{j'} + \epsilon_{i'j'}) \\ &= \text{Cov}(\beta_j, \beta_{j'}) + \text{Cov}(\epsilon_{ij}, \beta_{j'}) + \text{Cov}(\beta_j, \epsilon_{i'j'}) \\ &\quad + \text{Cov}(\epsilon_{ij}, \epsilon_{i'j'}) \\ &= 0 + 0 + 0 + 0 = 0\end{aligned}$$

b.
$$\begin{aligned}\text{Cov}(Y_{ij}, Y_{i'j}) &= \text{Cov}(\mu + \tau_i + \beta_j + \epsilon_{ij}, \mu + \tau_{i'} + \beta_j + \epsilon_{i'j}) \\ &= \text{Cov}(\beta_j + \epsilon_{ij}, \beta_j + \epsilon_{i'j}) \\ &= \text{Cov}(\beta_j, \beta_j) + \text{Cov}(\epsilon_{ij}, \beta_{i'j}) + \text{Cov}(\beta_j, \epsilon_{i'j}) + \text{Cov}(\epsilon_{ij}, \beta_j) \\ &= V(\beta_j) + 0 + 0 + 0 = \sigma_B^2.\end{aligned}$$

c. When $\sigma_B^2 = 0$, $\text{Cov}(Y_{ij}, Y_{i'j}) = 0$.

13.80a.
$$E(Y_{ij}) = E(\mu + \tau_i + \beta_j + \epsilon_{ij}) = E(\mu) + E(\tau_i) + E(\beta_j) + E(\epsilon_{ij}) = \mu + \tau_i + 0 + 0$$
$$= \mu + \tau_i$$
$$V(Y_{ij}) = V(\mu + \tau_i + \beta_j + \epsilon_{ij}) = V(\beta_j) + V(\epsilon_{ij}) + \text{Cov}(\beta_j, \epsilon_{ij}) = \sigma_B^2 + \sigma_\epsilon^2 + 0$$
$$= \sigma_B^2 + \sigma_\epsilon^2$$

b.
$$\begin{aligned}E(\overline{Y}_{i.}) &= E\left(\tfrac{1}{b} \sum_{j=1}^{b} Y_{ij} \right) = E\left(\tfrac{1}{b} \sum_{j=1}^{b} (\mu + \tau_i + \beta_j + \epsilon_{ij}) \right) \\ &= E\left(\mu + \tau_i + \tfrac{1}{b} \sum_{j=1}^{b} \beta_j + \tfrac{1}{b} \sum_{j=1}^{b} \epsilon_{ij} \right) \\ &= E(\mu) + E(\tau_i) + \tfrac{1}{b} \sum_{j=1}^{b} E(\beta_j) + \tfrac{1}{b} \sum_{j=1}^{b} E(\epsilon_{ij}) = \mu + \tau_i\end{aligned}$$

Thus, $\overline{Y}_{i.}$ is an unbiased estimator for $\mu + \tau_i$, the mean response to treatment i.

$$\begin{aligned}V(\overline{Y}_{i.}) &= V\left(\mu + \tau_i + \tfrac{1}{b} \sum_{j=1}^{b} \beta_j + \tfrac{1}{b} \sum_{j=1}^{b} \epsilon_{ij} \right) \\ &= \tfrac{1}{b^2} \sum_{j=1}^{b} V(\beta_j) + \tfrac{1}{b} \sum_{j=1}^{b} V(\epsilon_{ij}), \quad \text{since we have independence everywhere} \\ &= \tfrac{\sigma_B^2}{b} + \tfrac{\sigma_\epsilon^2}{b} = \tfrac{\sigma_B^2 + \sigma_\epsilon^2}{b}\end{aligned}$$

c.
$$\begin{aligned}E(\overline{Y}_{i.} - \overline{Y}_{i'.}) &= E(\overline{Y}_{i.}) - E(\overline{Y}_{i'.}) \\ &= (\mu + \tau_i) - (\mu + \tau_{i'}) \quad \text{from part } \mathbf{b} \\ &= \tau_i - \tau_{i'}\end{aligned}$$

d. $V(\overline{Y}_{i\cdot} - \overline{Y}_{i'\cdot})$

$$= V\left(\mu + \tau_i + \tfrac{1}{b}\sum_{j=1}^{b}\beta_j + \tfrac{1}{b}\sum_{j=1}^{b}\epsilon_{ij} - \left(\mu + \tau_i + \tfrac{1}{b}\sum_{j=1}^{b}\beta_j + \tfrac{1}{b}\sum_{j=1}^{b}\epsilon_{i'j}\right)\right)$$

$$= V\left(\tfrac{1}{b}\sum_{j=1}^{b}\epsilon_{ij} - \tfrac{1}{b}\sum_{j=1}^{b}\epsilon_{i'j}\right) = \tfrac{1}{b^2}\sum_{j=1}^{b}V(\epsilon_{ij}) + \tfrac{1}{b^2}\sum_{j=1}^{b}V(\epsilon_{ij})$$

$$= \tfrac{\sigma_\epsilon^2}{b} + \tfrac{\sigma_\epsilon^2}{b} = \tfrac{2\sigma_\epsilon^2}{b}$$

13.81a. $\overline{Y}_{\cdot j} = \tfrac{1}{k}\sum_{i=1}^{k}Y_{ij} = \tfrac{1}{k}\sum_{i=1}^{k}(\mu + \tau_i + \beta_j + \epsilon_{ij}) = \mu + \tfrac{1}{k}\sum_{i=1}^{k}\tau_i + \beta_j + \tfrac{1}{k}\sum_{i=1}^{k}\epsilon_{ij}$

$$= \mu + \beta_j + \overline{\epsilon}_{\cdot j}$$

$$E(\overline{Y}_{\cdot j}) = E(\mu + \beta_j + \epsilon_{\cdot j}) = \mu + E(\beta_j) + E(\overline{\epsilon}_{\cdot j}) = \mu + 0 + 0 = \mu$$

$$V(\overline{Y}_{\cdot j}) = V(\mu + \beta_j + \epsilon_{\cdot j}) = \mu + V(\beta_j) + \tfrac{1}{k^2}\sum_{i=1}^{k}V(\epsilon_{ij}) = \sigma_B^2 + \tfrac{\sigma_\epsilon^2}{k}$$

b. $E(\text{MST}) = \sigma^2 + \left(\tfrac{b}{k-1}\right)\sum_{i=1}^{k}\tau_i^2$ as calculated in Exercise 13.45, since the block effects cancel when we are making treatment comparisons.

c. $E(\text{MSB}) = k\left[\dfrac{\sum_{j=1}(\overline{Y}_{\cdot j} - \overline{Y})^2}{b-1}\right]$

$= k\left[\text{an unbiased estimator of } V(\overline{Y}_{\cdot j})\right]$

$= kV(\overline{Y}_{\cdot j})$

$= k\left[\sigma_B^2 + \tfrac{\sigma_\epsilon^2}{k}\right]$ from part **a**

$= \sigma_\epsilon^2 + k\sigma_B^2$

d. $E[\text{MSE}] = \sigma_\epsilon^2$ (similar to derivation in Exercise 13.45 **c**)

13.82a. $\widehat{\sigma}_\epsilon^2 = \text{MSE}$

b. $\widehat{\sigma}_\beta^2 = \frac{\text{MSB}-\text{MSE}}{k}$ (Notice this estimate may actually be less than 0!)

13.83a. The vector $\boldsymbol{AY}$ can be displayed as

$$\boldsymbol{AY} = \begin{bmatrix} \frac{\sum_i Y_i}{\sqrt{n}} \\ \frac{Y_1 - Y_2}{\sqrt{2}} \\ \frac{Y_1 + Y_2 - 2Y_3}{\sqrt{2\cdot 3}} \\ \vdots \\ \frac{[(Y_1 + Y_2 + \ldots + Y_{n-1}) - (n-1)Y_n]}{\sqrt{n(n-1)}} \end{bmatrix} = \begin{bmatrix} \sqrt{n}\,\overline{Y} \\ U_1 \\ U_2 \\ \vdots \\ U_{n-1} \end{bmatrix}$$

where $U_1, U_2, \ldots, U_{n-1}$ are linear functions of $Y_1, Y_2, \ldots, Y_n$. Then

$$\sum_{i=1}^{n} Y_i^2 = Y'Y = Y'A'AY = n\overline{Y}^2 + \sum_{i=1}^{n-1} U_i^2.$$

b. Write $L_i = \sum_{j=1}^{n} a_{ij}Y_j$ to be a linear function of $Y_1, Y_2, \ldots, Y_n$. Then two such linear functions, say L_i and L_k, will be pair wise orthogonal if and only if $\sum_{j=1}^{n} a_{ij}a_{kj} = 0$, and hence L_i and L_k will be independent if the Y_j are normal (see Exercise 5.81). Let $L_1, L_2, \ldots, L_n$ be the n linear functions defined by $\sqrt{n}\,\overline{Y}$, $U_1, U_2, \ldots, U_{n-1}$. The constants a_{ij}, $j = 1, 2, \ldots, n$, are the elements of the ith row of the matrix $\boldsymbol{A}$. Moreover, if any two rows of the matrix $\boldsymbol{A}$ are multiplied together, the result is zero. Consider the row vectors

$$\boldsymbol{a}_1' = \begin{bmatrix} \frac{1}{\sqrt{n}} & \frac{1}{\sqrt{n}} & \cdots & \frac{1}{\sqrt{n}} \end{bmatrix}$$

and

$$\boldsymbol{a}_i' = \begin{bmatrix} \frac{1}{\sqrt{i(i-1)}} & \frac{1}{\sqrt{i(i-1)}} & \cdots & \frac{-(i-1)}{\sqrt{i(i-1)}} & 0 & \ldots & 0 \end{bmatrix}$$

for $i = 2, \ldots, n$,

$$a_1' a_i = \frac{i-1}{\sqrt{n}\sqrt{i(i-1)}} - \frac{i-1}{\sqrt{n}\sqrt{i(i-1)}} = 0$$
$$a_i' a_j = \frac{j-1}{\sqrt{i(i-1)(j)(j-1)}} - \frac{j-1}{\sqrt{i(i-1)(j)(j-1)}} = 0$$

Thus, $L_1, L_2, \ldots, L_n$ are independent linear functions of $Y_1, Y_2, \ldots, Y_n$.

c. $\sum_{i=1}^{n} (Y_i - \overline{Y})^2 = \sum_{i=1}^{n} Y_i^2 - n\overline{Y}^2 = n\overline{Y}^2 + \sum_{i=1}^{n-1} U_i^2 - n\overline{Y}^2 = \sum_{i=1}^{n-1} U_i^2$

Since U_i is independent of $\sqrt{n}\,\overline{Y}$ for $i = 1, 2, \ldots, n-1$, so is U_i^2 and also $\sum_{i=1}^{n-1} U_i^2$.

Thus,

$$\sum_{i=1}^{n-1} U_i^2 = \sum_{i=1}^{n} (Y_i - \overline{Y})^2$$

is independent of $\overline{Y}$.

d. Write

$$\frac{\sum_{i=1}^{n} (Y_i-\mu)^2}{\sigma^2} = \frac{\sum_{i=1}^{n} (Y_i-\overline{Y}+\overline{Y}-\mu)^2}{\sigma^2} = \frac{\sum_{i=1}^{n} (Y_i-\overline{Y})^2}{\sigma^2} + \frac{n(\overline{Y}-\mu)^2}{\sigma^2} = X_1 + X_2$$

X_1 and X_2 are independent from part **c**.

(1) Since Y_i are normal, $i = 1, 2, \ldots, n$, $\frac{(Y_i-\mu)}{\sigma}$ is standard normal, and $\left(\frac{Y_i-\mu}{\sigma}\right)^2$ has a χ^2 distribution with 1 degree of freedom.

(2) Since $Y_1, Y_2, \ldots, Y_n$ are independent,

$$\sum_{i=1}^{n} \left(\frac{Y_i-\mu}{\sigma}\right)^2 = \frac{\sum_{i=1}^{n} (Y_i-\mu)^2}{\sigma^2} = X_1 + X_2$$

has a χ^2 distribution with n degrees of freedom.

(3) $\overline{Y}$ has a normal distribution with mean μ and variance $\frac{\sigma^2}{n}$, so that $\sqrt{n}\,\frac{\overline{Y}-\mu}{\sigma}$ has a standard normal distribution and $X_2 = \frac{n(\overline{Y}-\mu)^2}{\sigma^2}$ has a χ^2 distribution with 1 degree of freedom. Now consider the distribution of X_1 using moment-generating functions. Since X_1 and X_2 are independent,

$$m_{X_1+X_2}(t) = m_{X_1}(t) m_{X_2}(t)$$

or

$$\frac{1}{(1-2t)^{n/2}} = m_{X_1}(t) \left[\frac{1}{(1-2t)^{1/2}}\right]$$

or

$$m_{X_1}(t) = \frac{1}{(1-2t)^{(n-1)/2}}$$

and X_1 is evidently distributed as a χ^2 random variable with $(n-1)$ degrees of freedom.

13.84 a. For the completely randomized design,

$$\text{SSE} = \sum_{i=1}^{k} \sum_{j=1}^{n_i} (Y_{ij} - \overline{Y}_i)^2$$

Consider a particular treatment 1. Then the sample mean $\overline{Y}_1$ will be independent of the sample sum of squares, $\sum_{j=1}^{n_1} (Y_{1j} - \overline{Y}_1)^2$, from Exercise 13.74(c).

Furthermore, the rest of the terms in SSE do not involve $\overline{Y}_1$, since

$$\text{SSE} = \sum_{i=1}^{l-1} \sum_{j=1}^{n_i} (Y_{ij} - \overline{Y}_i)^2 + \sum_{j=1}^{n_1} (Y_{1j} - \overline{Y}_1)^2 + \sum_{i=l+1}^{k} \sum_{i=1}^{n_i} (Y_{ij} - \overline{Y}_i)^2$$

Therefore, $\overline{Y}_1$ must be independent of SSE.

b. With $n_1 = n_2 = \ldots = n_p$, and assuming H_0 to be true, we have

$$\text{MST} = \frac{n\sum_{i=1}^{k} (\overline{Y}_i-\overline{Y})^2}{k-1}$$

and

$$\text{MSE} = \frac{\sum_{i=1}^{k} \sum_{j=1}^{n} (Y_{ij}-\overline{Y}_i)^2}{nk-k}.$$

Consider $(k-1)\text{MST} = n\sum_{i=1}^{k} (\overline{Y}_i - \overline{Y})^2$. If H_0 is true, and $\mu_1 = \mu_2 = \ldots = \mu_k$

$= \mu$, then $\overline{Y}_1, \overline{Y}_2, \ldots, \overline{Y}_k$ is a random sample of p observations from a normal population with mean μ and variance $\frac{\sigma^2}{n}$, where $\sigma^2 = V(Y_{ij})$. Hence, by Exercise 13.83(d),

$$\frac{\sum_{i=1}^{n} (\overline{Y}_i - \overline{Y})^2}{\frac{\sigma^2}{n}} = \frac{(k-1)\text{MST}}{\sigma^2}$$

has a χ^2 distribution with $(k-1)$ degrees of freedom.

Consider now, for a fixed value of i, the quantity

$$\sum_{j=1}^{n} (Y_{ij} - \overline{Y}_i)^2$$

$Y_{i1}, Y_{i2}, \ldots, Y_{in}$ is a random sample of n observations from a normal distribution with mean $\mu_i = \mu$ and variance σ^2. Hence, from Exercise 13.74(d),

$$\frac{\sum_{i=1}^{n} (Y_{ij} - \overline{Y}_i)^2}{\sigma^2}$$

has a χ^2 distribution with $(n-1)$ degrees of freedom. Then

$$\frac{(nk-k)\text{MSE}}{\sigma^2} = \frac{\sum_{i=1}^{k}\sum_{j=1}^{n} (Y_{ij} - \overline{Y}_i)^2}{\sigma^2}$$

is a sum of k independent χ^2 variates and hence has a χ^2 distribution with $k(n-1) = nk - k$ degrees of freedom.

If MSE and MST are independent, then

$$\frac{\frac{\text{MST}}{\sigma^2}}{\frac{\text{MSE}}{\sigma^2}} = \frac{\text{MST}}{\text{MSE}}$$

has an F distribution with $(k-1)$ and $(nk-k)$ degrees of freedom. But MST is a function only of $\overline{Y}_i$, $i = 1, 2, \ldots, k$, since we can write

$$\text{MST} = \frac{n}{k-1} \sum_{i=1}^{k} \left[\overline{Y}_i \frac{\sum_{i=1}^{k} \overline{Y}_i}{k} \right]^2$$

and $\overline{Y}_i$ is independent of SSE and MSE. Hence MSE and MST are independent and $\frac{\text{MST}}{\text{MSE}}$ has the desired F distribution.

CHAPTER 14 ANALYSIS OF CATEGORICAL DATA

14.1 One thousand cars were each classified according to the lane that they occupied (1 through 4). The objective is to determine whether or not some lanes were preferred over others. This is a multinomial experiment with $k = 4$ cells. If no lane is preferred over another, the probability that a car will be driven in lane i, $i = 1, 2, 3, 4$, is $\frac{1}{4}$. The null hypothesis is then

$$H_0\colon\ p_1 = p_2 = p_3 = p_4 = \tfrac{1}{4}$$

Notice that the hypothesis to be tested is a test of specified numerical values for the probabilities rather than a test of their relationship to one another. Hence no degrees of freedom are lost for estimating cell probabilities. The test statistic is

$$X^2 = \sum_{i=1}^{k} \frac{[n_i - E(n_i)]^2}{E(n_i)}$$

which, when n is large, will possess an approximate chi-square distribution in repeated sampling. The values of n_i are the actual counts observed in the experiment, and

$E(n_i) = np_i = 1000\left(\frac{1}{4}\right) = 250$

A table of observed and expected cell counts is shown at the right. Then

	Lane 1	Lane 2	Lane 3	Lane 4
n_i	294	276	238	192
$E(n_i)$	250	250	250	250

$$X^2 = \frac{(294-250)^2}{250} + \frac{(276-250)^2}{250} + \frac{(238-250)^2}{250} + \frac{(192-250)^2}{250}$$

$$= \frac{6120}{250} = 24.48$$

To obtain the rejection region for this test, the degrees of freedom associated with X^2 must be determined. The number of degrees of freedom is equal to the number of cells, k, less 1 degree of freedom for each linearly independent restriction placed on $n_1, n_2, \ldots, n_k$. For this example, $k = 4$ and one degree of freedom is lost because of the restriction that $\sum_i n_i = n$. Hence X^2 has $(k-1) = (4-1) = 3$ degrees of freedom and the appropriate upper-tailed rejection region is

$$X^2 \geq X^2_{3,.005} = 7.81$$

Thus the conclusion is to reject the null hypothesis, with a probability of error equal to $\alpha = .05$. Remember that a one-tailed test is employed, using the upper-tail values of X^2, because large deviations of the observed cell counts will tend to contradict H_0. Hence we will reject the null hypothesis when X^2 is large. Since $24.48 > 12.8381 = X^2_{3(.005)}$, the p-value is less than .005.

14.2 If the frequency of occurrence of a heart attack is the same for each day of the week, then when a heart attack occurs, the probability that it falls in one cell (day) is the same as for any other cell (day). Hence,

$$H_0\colon\ p_1 = p_2 = \ldots = p_7 = \tfrac{1}{7}$$

vs.

$$H_a\colon \text{ at least one } p_i \text{ is different from the others}$$

or equivalently,

$$H_a\colon\ p_i \neq p_j \text{ for some pair } i \neq j.$$

Since $n = 200$,

$$E(n_i) = np_i = 200\left(\tfrac{1}{7}\right) = 28.571429$$

and the test statistic is

$$X^2 = \frac{(24-28.571429)^2}{28.571429} + \ldots + \frac{(29-28.571429)^2}{28.571429} = \frac{103.71429}{28.571429} = 3.63.$$

The degrees of freedom for this test of specified cell probabilities are $k - 1 = 7 - 1 = 6$, and the upper-tailed rejection region is

$$X^2 > \chi^2_{6,.05} = 12.59.$$

and H_0 is not rejected. There is insufficient evidence to indicate a difference in frequency of occurrence from day to day.

14.3 **a.** Let p denote the true proportion of heart attacks occurring on Monday. The hypothesis to be tested is

$$H_0\colon\ p = \tfrac{1}{7} \qquad \text{vs.} \qquad H_a\colon\ p > \tfrac{1}{7}$$

From Section 8.3 and 10.3, the observed value of the test statistic is

$$z = \frac{\widehat{p} - p_0}{\sqrt{\frac{p_0(1-p_0)}{n}}} = \frac{.18 - \frac{1}{7}}{\sqrt{\frac{\left(\frac{1}{7}\right)\left(\frac{6}{7}\right)}{200}}} = 1.50.$$

From Table IV, Appendix 3, we see that $P(z > 1.645) = .05$. Hence, we take $z > 1.645$ as the rejection region. Since the observed value of z is not in the rejection region, there is not sufficient evidence of heart attacks being more likely to occur on Monday than any other day of the week.

b. The test is suggested by the data. Our hypothesis should test a question of interest formulated before the experiment is conducted. Data is then gathered to support or refute the given hypothesis. That is, we should apply the scientific method.

c. Monday is popularly known as the most stressful workday of the week; one has a long five days until the weekend. One might wish to investigate if this extra stress results in a disproportionate amount of heart attacks.

14.4 The hypothesis of interest is

$$H_0\colon \text{ ratio is } 9:3:3:1 \qquad \text{vs.} \qquad H_a\colon \text{ ratio is not } 9:3:3:1$$

which can be alternatively stated as

$$H_0\colon\ p_1 = \tfrac{9}{16},\ p_2 = \tfrac{3}{16},\ p_3 = \tfrac{3}{16},\ p_4 = \tfrac{1}{16} \qquad \text{vs.}$$

H_a: at least one of these equalities is incorrect

The expected and observed cell counts are shown in the table below.

n_i	56	19	17	8
$E(n_i)$	56.25	18.75	18.75	6.25

Then

$$X^2 = \frac{(56-56.25)^2}{56.25} + \frac{(19-18.75)^2}{18.75} + \frac{(17-18.75)^2}{18.75} + \frac{(8-6.25)^2}{6.25} = .658$$

The degrees of freedom are $k - 1 = 4 - 1 = 3$ and, with $\alpha = .05$, the critical value for rejection will be $X^2_{3,.05} = 7.81$. Hence the null hypothesis is not rejected. We do not have sufficient evidence to conclude that the ratio is not $9:3:3:1$.

14.5 Similar to previous exercises. The null hypothesis to be tested is

$$H_0\colon\ p_1 = .69;\ p_2 = .21;\ p_3 = .07;\ p_4 = .03$$

against the alternative that at least one of these probabilities is incorrect. The observed and expected cell counts are shown below.

User	1	2	3	4
n_i	102	32	12	4
$E(n_i)$	103.5	31.5	10.5	4.5

The test statistic is

$$X^2 = \frac{(102-103.5)^2}{103.5} + \ldots + \frac{(4-4.5)^2}{4.5} = .2995$$

and the p-value with $k - 1 = 3$ d.f. is p-value $> .95$. The null hypothesis is not rejected and we cannot conclude that the figures given are inaccurate.

14.6 The hypothesis of interest is

$$H_0\colon\ p_1 = \tfrac{5}{10},\ p_2 = \tfrac{2}{10},\ p_3 = \tfrac{2}{10},\ p_4 = \tfrac{1}{10} \qquad \text{vs.} \qquad H_a\colon \text{ at least one equality does not hold}$$

The observed and expected cell counts are shown in the table below.

n_i	48	18	21	13
$E(n_i)$	50	20	20	10

Then

$$X^2 = \frac{(48-50)^2}{50} + \ldots + \frac{(13-10)^2}{10} = 1.23$$

The rejection region with $\alpha = .05$ and d.f. $= k - 1 = 3$ is $X^2 \geq 7.81$. The null hypothesis is not rejected; there is no reason to suspect that the ratio is incorrect.

14.7 This is similar to Example 14.2. Using $\overline{y}$ to estimate the Poisson parameter λ, calculate

$$\overline{y} = \frac{\sum y_i f_i}{n} = \frac{0(56)+1(104)+\ldots+10(2)+11(0)+19(1)}{400} = \frac{976}{400} = 2.44$$

The expected cell counts are the estimated as

$$\widehat{E}(n_i) = n\widehat{p}_i = 400\,\frac{e^{-2.44}(2.44)^{y_i}}{y_i!}$$

Notice that when $Y = 7$, the expected cell count drops below 5. Hence the final group is $Y \geq 7$. The observed and estimated expected cell counts are shown in the table at the right.

No. of Colonies	n_i	$\widehat{p}_i$	$\widehat{E}(n_i)$
0	56	.087	34.86
1	104	.2127	85.07
2	80	.2595	103.73
3	62	.2110	84.41
4	42	.1287	51.49
5	27	.0628	25.13
6	9	.0255	10.22
7 or more	20		$400 - 394.96 = 5.04$

Then

$$X^2 = \frac{(56-34.86)^2}{34.86} + \ldots + \frac{(20-5.04)^2}{5.04} = 69.42.$$

The rejection region, based on $k - 2 = 6$ degrees of freedom (see Example 14.2), is $X^2 \geq 12.59$, and the null hypothesis is rejected. The data do not fit the Poisson distribution.

14.8 This is similar to Exercise 14.7. Calculate

$$\widehat{\lambda} = \overline{y} = \frac{0(296)+1(74)+\ldots+8(1)}{414} = 483.09$$

The observed and estimated cell counts are given in the table at the right.

y	n_i	$\widehat{p}_i$	$\widehat{E}(n_i)$
0	296	.6169	255.38
1	74	.298	123.38
2	26	.072	29.80
3 or more	18	.0131	5.44

Then

$$X^2 = \frac{(296-255.38)^2}{255.38} + \ldots + \frac{(18-5.44)^2}{5.44} = 55.71$$

The rejection region with $k - 2 = 2$ degrees of freedom is $X^2 \geq 5.99$, and the null hypothesis is rejected. The data do not come from a Poisson distribution.

14.9 **a.** The table of estimated expected cell counts is

	JAS Score		
3-year follow-up	Less than -5	-5 to 5	Greater than 5
Died	17.09	16.24	15.67
Alive	162.91	154.76	149.33

The test statistic is

$$X^2 = \frac{(21-17.09)^2}{17.09} + \frac{(17-16.24)^2}{16.24} + \ldots + \frac{(154-149.33)^2}{149.33} = 2.56$$

with $(r-1)(c-1) = 2$ d.f. Since

$$X^2 = 2.56 < 5.99 = \chi^2_{.05},$$

we do not reject H_0 at the $\alpha = .05$ level. There is insufficient evidence to indicate a dependence between mortality rate and level of Type A behavior.

b. Since $X^2 = 2.56 < 4.61 = \chi^2_{.10}$, $p > .10$. Thus the results are not significant.

14.10 **a.** The hypothesis of independence between attachment pattern and child care time is tested using the chi-square statistic. The contingency table, including column and row totals and the estimated expected cell counts in parentheses, follows.

Pattern	0–3 hours	4–19 hours	20–54 hours	Total

Secure	24	35	5	64
	(24.09)	(30.97)	(8.95)	
Anxious	11	10	8	29
	(10.91)	(14.03)	(4.05)	
Total	35	45	13	93

The test statistic is

$$X^2 = \frac{(24-24.09)^2}{24.09} + \frac{(35-30.97)^2}{30.97} + \ldots + \frac{(8-4.05)^2}{4.05} = 7.267$$

and the rejection region is $X^2 > \chi^2_{2,.05} = 5.99$ with 2 d.f., H_0 is rejected. There is evidence of a dependence between attachment pattern and child care time.

b. The value $X^2 = 7.267$ is between $\chi^2_{.05}$ and $\chi^2_{.025}$ so that $.025 < p\text{-value} < .05$. The results are significant.

14.11 a. $X^2 = \sum_{j=1}^{c}\sum_{i=1}^{r} \frac{[n_{ij}-E(\hat{n}_{ij})]^2}{E(\hat{n}_{ij})} = \sum_{j=1}^{c}\sum_{i=1}^{r}\left[\frac{\left(n_{ij}-\frac{r_ic_j}{n}\right)^2}{\frac{r_ic_j}{n}}\right]$

$$= n\sum_{j=1}^{c}\sum_{i=1}^{r}\left[\frac{n_{ij}^2-\frac{2n_{ij}r_ic_j}{n}+\frac{r_i^2c_j^2}{n^2}}{r_ic_j}\right]$$

$$= n\left[\sum_{j=1}^{c}\sum_{i=1}^{r}\frac{n_{ij}^2}{r_ic_j} - 2\sum_{j=1}^{c}\sum_{i=1}^{r}\frac{n_{ij}}{n} + \sum_{j=1}^{c}\sum_{i=1}^{r}\frac{r_ic_j}{n^2}\right]$$

$$= n\left[\sum_{j=1}^{c}\sum_{i=1}^{r}\frac{n_{ij}^2}{r_ic_j} - 2 + \frac{\left(\sum_{j=1}^{c} c_j\right)\left(\sum_{i=1}^{r} r_i\right)}{n^2}\right]$$

$$= n\left[\sum_{j=1}^{c}\sum_{i=1}^{r}\frac{n_{ij}^2}{r_ic_j} - \frac{2n}{n} + \frac{n^2}{n^2}\right] = n\left[\sum_{j=1}^{c}\sum_{i=1}^{r}\frac{n_{ij}^2}{r_ic_j} - 1\right]$$

b. When every entry in the contingency is multiplied by the same $k > 0$,

$$X^2 = kn\left[\sum_{j=1}^{c}\sum_{i=1}^{r}\frac{(kn_{ij})^2}{(kr_i)(kc_j)} - 1\right] = kn\left[\sum_{j=1}^{c}\sum_{i=1}^{r}\frac{n_{ij}^2}{r_ic_j} - 1\right]$$

Thus, if the pattern of responses is the same, then the X^2 will be increased k times.

14.12 The data are analyzed as a 2×6 contingency table with estimated expected cell counts shown in parentheses.

	Type of Activity						
Sex	Walking	Cycling	Aerobics	Running	Calisthenics	Swimming	Total
Male	60	85	28	113	79	179	544
	(81.9)	(81.9)	(81.9)	(81.9)	(81.9)	(81.9)	
Female	106	81	138	55	89	90	559
	(84.1)	(84.1)	(84.1)	(84.1)	(84.1)	(84.1)	
Total	166	16	166	168	168	269	1103

The test statistic is

$$X^2 = \frac{(60-81.9)^2}{81.9} + \frac{(85-81.9)^2}{81.9} + \ldots + \frac{(90-84.1)^2}{84.1} = 135.62$$

using calculator accuracy on the expected cell counts. The rejection region with $\alpha = .05$ and 5 d.f. is $X^2 > 11.0705$, and H_0 is rejected. There is a difference in the proportion according to sex. The value $X^2 = 135.62$ is greater than $16.75 = \chi^2_{5,.005}$, so that the p-value $< .005$.

14.13 a.-b. The MINITAB printouts below are used to analyze the data for the two contingency tables. The observed values of the test statistics are $X^2 = 19.043$ and $X^2 = 60.139$, for faculty and student responses, respectively. The rejection region, with $(3)(2) = 6$ d.f., is $X^2 > 16.81$ with $\alpha = .01$, and H_0 is rejected for both cases.

```
MTB > CHISQ C1-C3
Expected counts are printed below observed counts

        C1        C3        C3      Total
```

1	4	0	0	4
	1.53	1.47	1.00	
2	15	12	3	30
	11.50	11.00	7.50	
3	2	7	7	16
	6.13	5.87	4.00	
4	2	3	5	10
	3.83	3.67	2.50	
Total	23	22	15	60
ChiSq=	3.968 +	1.467 +	1.000 +	
	1.065 +	0.091 +	2.700 +	
	2.786 +	0.219 +	2.250 +	
	0.877 +	0.121 +	2.500 =	19.043

df = 6
7 cells with expected counts less than 5.0
MTB > CHISQ C4-C6
Expected counts are printed below observed counts

	C4	C5	C6	Total
1	19	6	2	27
	6.88	9.56	10.57	
2	19	41	27	87
	22.16	30.80	34.04	
3	3	7	31	41
	10.44	14.52	16.04	
4	0	3	3	6
	1.53	2.12	2.35	
Total	41	57	63	161
ChiSq=	21.379 +	1.325 +	6.944 +	
	0.449 +	3.377 +	1.457 +	
	5.303 +	3.891 +	13.943 +	
	1.528 +	0.361 +	0.181 =	60.139

df = 6
3 cells with expected counts less than 5.0

c. Expected values of some cell counts are less than 5. Thus, the χ^2 approximation to the test statistic X^2 may not be valid.

14.14 The table of observed and estimated expected cell counts is given below.

	Age		
	1	2	3
Low	8(13.16)	12(13.67)	21(14.17)
High	18(12.84)	15(13.33)	7(13.83)

Then

$$X^2 = \frac{(8-13.16)^2}{13.16} + \ldots + \frac{(7-13.83)^2}{13.83} = 11.18$$

and the rejection region is $X^2 > \chi_2^2 = 5.99$ with $\alpha = .05$. The null hypothesis is rejected. The two classifications are dependent.

14.15 **a.** The estimated expected cell counts are calculated as

$$\widehat{E}(n_{ij}) = \frac{r_i c_j}{n},$$

and are shown in parentheses in the table below.

	Life Threatening Complications		
Results	No	Yes	Total
Negative	166 (151.69)	1 (15.31)	135
Positive	260 (274.31)	42 (27.69)	302
Total	426	43	469

Then

$$X^2 = \sum_{ij} \frac{\left[n_{ij}-\widehat{E}(n_{ij})\right]^2}{\widehat{E}(n_{ij})} = \frac{(166-151.69)^2}{151.69} + \ldots + \frac{(42-27.69)^2}{27.69} = 22.94$$

With $\alpha = .05$ and $1(1) = 1$ d.f., a one-tailed rejection region is found using Table 5 to be $X^2 > \chi^2_{.05} = 3.84$.

Since $X^2 = 22.87$, H_0 is rejected. The probability of having life-threatening myocardial infarctions is dependent upon the electrocardiogram results.

b. The value $X^2 = 22.87$ is greater than $7.88 = \chi^2_{.005}$, so that the p-value $< .005$.

14.16

The data are rearranged into a 2×2 contingency table, and the estimated expected cell counts are calculated as usual (see the table at the right).

	B	$\overline{B}$	Total
A	48(45.54)	18(20.46)	66
$\overline{A}$	21(23.46)	13(10.54)	34
Total	69	31	100

Then

$$X^2 = \frac{(48-45.54)^2}{45.54} + \ldots + \frac{(13-10.54)^2}{10.54} = 1.26$$

The rejection region based on 1 degree of freedom is $X^2 \geq 3.84$, and we conclude that the two defects are independent.

14.17

Three different contingency tables are given and the tests proceed as in previous exercises. Tables are provided for each situation.

a.

20 (13.44)	4 (10.56)
8 (14.56)	18 (11.44)

$X^2 = 13.99$; rejection region is $X^2 \geq 3.84$; reject H_0: species segregate.

b.

4 (10.56)	20 (13.44)
18 (11.44)	18 (14.56)

$X^2 = 13.99$; rejection region is $X^2 \geq 3.84$; reject H_0: species are overly mixed.

c.

20 (18.24	4 (5.76)
18 (19.76)	8 (6.24)

$X^2 = 1.36$; rejection region is $X^2 \geq 3.84$; do not reject H_0.

14.18

a. This is similar to previous exercises. The observed and estimated expected cell counts are given in the table below.

	Treated	Untreated	Total
Improved	117(95.5)	74(95.5)	191
Not Improved	83(104.5)	126(104.5)	209
Total	200	200	400

Then

$$X^2 = \frac{(117-95.5)^2}{95.5} + \ldots + \frac{(126-104.5)^2}{104.5} = 18.53$$

The degrees of freedom associated with the test have been shown to be $(r-1)(c-1)$ and the rejection region will be $X^2 \geq \chi^2_{1,.05} = 3.84$. Notice that the observed value of the test statistic exceeds the critical value. Hence the conclusion is to reject the null hypothesis; that is, we conclude that the serum is effective.

b. Consider the treated and untreated patients as comprising random samples of 200 each, drawn from two populations (i.e., a sample of 200 treated patients and a sample of 200 untreated patients). Let p_1 be the probability that a treated patient improves and let p_2 be the probability that an untreated patient improves. Then the hypothesis to be tested is

$$H_0\colon\ p_1 - p_2 = 0 \qquad \text{vs.} \qquad H_a\colon\ p_1 - p_2 \neq 0.$$

With the procedure described in Section 10.3 for testing a hypothesis about the difference between two binomial parameters, the following estimators are calculated:

$$\widehat{p}_1 = \frac{y_1}{n_1} = \frac{117}{200}$$
$$\widehat{p}_2 = \frac{y_2}{n_2} = \frac{74}{200}$$

$$\widehat{p} = \frac{y_1+y_2}{n_1+n_2} = \frac{117+74}{400} = .4775.$$

The test statistic is

$$z = \frac{\widehat{p}_1-\widehat{p}_2-0}{\sqrt{\widehat{p}\widehat{q}\left[\left(\frac{1}{n_1}\right)+\left(\frac{1}{n_2}\right)\right]}} = \frac{.215}{\sqrt{(.4775)(.5225)(.01)}} = 4.3$$

and the rejection region for $\alpha = .05$ is $|z| \geq 1.96$. Again the test statistic falls in the rejection region. We reject the hypothesis of no difference and conclude that the serum is effective.

c. Since $18.53 > 7.88 = \chi^2_{1,.005}$, the p-value is less than .005.

14.19 Refer to Section 10.3. The two-tailed z test was used to test a hypothesis

$$H_0\colon\ p_1 - p_2 = 0 \qquad \text{vs.} \qquad H_a\colon\ p_1 \neq p_2.$$

The test statistic was

$$Z = = \frac{\widehat{p}_1-\widehat{p}_2}{\sqrt{\widehat{p}\widehat{q}\left[\left(\frac{1}{n_1}\right)+\left(\frac{1}{n_2}\right)\right]}}$$

and

$$Z^2 = \frac{(\widehat{p}_1-\widehat{p}_2)^2}{\widehat{p}\widehat{q}\left(\frac{n_1+n_2}{n_1n_2}\right)} = \frac{n_1n_2(\widehat{p}_1-\widehat{p}_2)^2}{(n_1+n_2)\widehat{p}\widehat{q}}.$$

Notice that

$$\widehat{p} = \frac{y_1+y_2}{n_1+n_2} = \frac{n_1\widehat{p}_1+n_2\widehat{p}_2}{n_1+n_2}$$

Now consider the chi-square test statistic used in Exercise 14.17. The hypothesis to be tested is

H_0: Independence of classification vs. H_a: dependence of classification

That is, the null hypothesis asserts that the percentage of patients who show improvement is independent of whether or not they have been treated with the serum. If the null hypothesis is true, then $p_1 = p_2$. Hence the two tests are designed to test the same hypothesis.

In order to show that Z^2 is equivalent to X^2, it is necessary to rewrite the chi-square test statistic in terms of quantities $\widehat{p}_1$, $\widehat{p}_2$, n_1, and n_2.

(1) Consider n_{11}, the observed number of treated patients who have improved. Since $\widehat{p}_1 = \frac{n_{11}}{n_1}$, we have $n_{11} = n_1\widehat{p}_1$. Similarly, $n_{21} = n_1\widehat{q}_1$, $n_{12} = n_2\widehat{p}_2$, $n_{22} = n_2\widehat{q}_2$.

(2) The estimated expected cell counts are calculated under the assumption that the null hypothesis is true. Consider

$$\widehat{E}(n_{11}) = \frac{r_1c_1}{n} = \frac{(n_{11}+n_{12})(n_{11}+n_{21})}{n_1+n_2} = \frac{(y_1+y_2)(n_{11}+n_{21})}{n_1+n_2} = n_1\widehat{p}$$

Similarly, $\widehat{E}(n_{21}) = n_1\widehat{q}$, $\widehat{E}(n_{12}) = n_2\widehat{p}$, and $\widehat{E}(n_{22}) = n_2\widehat{q}$. The table of observed and estimated expected cell counts is shown below.

	Treated	Untreated
Improved	$n_1\widehat{p}_1\ (n_1\widehat{p})$	$n_2\widehat{p}_2\ (n_2\widehat{p})$
Not Improved	$n_1\widehat{q}_1\ (n_1\widehat{q})$	$n_2\widehat{q}_2\ (n_2\widehat{q})$

Then

$$\begin{aligned}
X^2 &= \sum_i\sum_j \frac{\left[n_{ij}-\widehat{E}(n_{ij})\right]^2}{\widehat{E}(n_{ij})} \\
&= \frac{n_1^2(\widehat{p}_1-\widehat{p})^2}{n_1\widehat{p}} + \frac{n_1^2(\widehat{q}_1-\widehat{q})^2}{n_1\widehat{q}} + \frac{n_2^2(\widehat{p}_2-\widehat{p})^2}{n_2\widehat{p}} + \frac{n_2^2(\widehat{q}_2-\widehat{q})^2}{n_2\widehat{q}} \\
&= \frac{n_1(\widehat{p}_1-\widehat{p})^2}{\widehat{p}} + \frac{n_1\left[(1-\widehat{p}_1)-(1-\widehat{p})\right]^2}{\widehat{q}} + \frac{n_2(\widehat{p}_2-\widehat{p})^2}{\widehat{p}} \\
&\quad + \frac{n_2\left[(1-\widehat{p}_2)-(1-\widehat{p})\right]^2}{\widehat{q}} \\
&= \frac{(1-\widehat{p})n_1(\widehat{p}_1-\widehat{p})^2+n_1\widehat{p}(\widehat{p}_1-\widehat{p})^2}{\widehat{p}\widehat{q}} + \frac{(1-\widehat{p})n_2(\widehat{p}_2-\widehat{p})^2+n_2\widehat{p}(\widehat{p}_2-\widehat{p})^2}{\widehat{p}\widehat{q}} \\
&= \frac{n_1(\widehat{p}_1-\widehat{p})^2}{\widehat{p}\widehat{q}} + \frac{n_2(\widehat{p}_2-\widehat{p})^2}{\widehat{p}\widehat{q}}
\end{aligned}$$

Substituting for $\widehat{p}$, we obtain

$$\begin{aligned}
X^2 &= \frac{n_1}{\widehat{p}\widehat{q}}\left(\frac{n_1\widehat{p}_1+n_2\widehat{p}_1-n_1\widehat{p}_1-n_2\widehat{p}_2}{n_1+n_2}\right)^2 + \frac{n_2}{\widehat{p}\widehat{q}}\left(\frac{n_1\widehat{p}_2+n_2\widehat{p}_2-n_1\widehat{p}_1-n_2\widehat{p}_2}{n_1+n_2}\right)^2 \\
&= \frac{n_1n_2(n_1+n_2)(\widehat{p}_1-\widehat{p}_2)^2}{\widehat{p}\widehat{q}(n_1+n_2)^2} = \frac{n_1n_2(\widehat{p}_1-\widehat{p}_2)^2}{\widehat{p}\widehat{q}(n_1+n_2)}
\end{aligned}$$

Note that X^2 is identical to Z^2, as defined at the beginning of the exercise.

14.20a. The data are analyzed as a contingency table as shown below.

Groups	Memory Lapses Decrease	No Decrease	Total
Lecithin	37 (30.246)	4 (10.754)	41
Placebo	8 (14.754)	12 (5.246)	20
Total	45	16	61

The test statistic is

$$X^2 = \frac{(37-30.246)^2}{30.246} + \frac{(4-10.754)^2}{10.754} + \ldots + \frac{(12-5.246)^2}{5.246} = 17.54$$

The rejection region with $\alpha = .05$ and 1 d.f. is $X^2 > 3.84$, and H_0 is rejected. There is evidence that a decrease in memory lapses depends on whether a subject has been on a daily regimen of lecithin.

b. The hypothesis to test is

$$H_0\colon\ p_1 - p_2 = 0 \qquad \text{vs.} \qquad H_a\colon\ p_1 \neq p_2$$

where

p_1 = proportion of individuals who take lecithin that report decrease in memory = lapse.

p_2 = proportion of individuals taking placebo that report decrease in memory = lapse.

The test statistic is

$$Z = \frac{(\hat{p}_1 - \hat{p}_2)}{\sqrt{\frac{\hat{p}_1\hat{q}_1}{n_1} + \frac{\hat{p}_2\hat{q}_2}{n_2}}} = \frac{.90 - .4}{\sqrt{\frac{(.90)(.10)}{41} + \frac{(.4)(.6)}{20}}} = 4.19$$

Using $\alpha = .05$, the rejection region is $Z > 1.96$. Since $z = 4.19 > 1.96$, we reject H_0.

14.21 The contingency table, including column and row totals and the estimated expected cell counts, follows.

Age Group	More	Less	Same	Total
35–54	90 (65.00)	18 (39.00)	92 (96.00)	200
55+	40 (65.00)	60 (39.00)	100 (96.00)	200
Total	130	78	192	400

The test statistic is

$$X^2 = \frac{(90-65)^2}{65} + \frac{(18-39)^2}{39} + \ldots + \frac{(100-96)^2}{96} = 42.179$$

using computer accuracy. The rejection region with $\alpha = .01$ and 2 d.f. is $X^2 > 9.21$. Hence H_0 is rejected. The investing pattern of the baby-boomer group differs from that of the older group.

14.22 The objective of the experiment is to determine whether the number of defective and non defective buttons is affected by the particular machine on which they are produced. Hence buttons have been classified according to machine and whether they were defective or nondefective. A sample of 400 buttons was selected from each machine.

a. The complete contingency table, with column totals fixed at 400, is shown below.

	Machine			Total
Number Defective	16	24	9	49
Number Non defective	384	376	391	1151
Total	400	400	400	1200

Recall that a contingency table is an example of a multinomial experiment. Each of the $n = 1200$ items must fall in one of the six cells shown in the table. (Note that it is necessary to add the cells corresponding to the nondefectives.) The test procedure is identical to that used in previous exercises.

The estimated expected cell counts are

$$\widehat{E}(n_{11}) = \widehat{E}(n_{12}) = \widehat{E}(n_{13}) = 16.33$$
$$\widehat{E}(n_{21}) = \widehat{E}(n_{22}) = \widehat{E}(n_{23}) = 383.67$$

The rejection region, based on $(r-1)(c-1) = 2$ degrees of freedom, will be $X^2 \geq \chi^2_{2,.05} = 5.99$. Then the calculated value of the test statistic, $X^2 = 7.19$ exceeds the critical value, $\chi^2 = 5.99$. Hence the null hypothesis is rejected. Note that the first three sums (i.e., the sums associated with the first row) produce a value of 6.89, and the hypothesis can be rejected at this point.

b. The procedure for the likelihood ratio test is identical to that of Exercise 10.75. The hypothesis of interest is H_0: $p_1 = p_2 = p_3 = p$ against the alternative that at least one of these equalities is incorrect. In Ω, the likelihood function is

$$L(\Omega) = \prod_{i=1}^{3} \binom{400}{n_i} p_{p_i}^{n_i} (1-p_i)^{400-n_i}$$

and the maximum likelihood estimate of p_i is $\widehat{p}_i = \frac{n_i}{400}$.
In the restricted space Ω_0,

$$L(\Omega_0) = \prod_{i=1}^{3} \binom{400}{n_i} p^{n_i} (1-p)^{400-n_i} = Kp^{\Sigma n_i}(1-p)^{1200-\Sigma n_i}$$

and

$$\ln L = \ln K + \sum n_i \ln P + (1200 - \sum n_i) \ln (1-p)$$

where K is a quantity not involving p.
The student may verify that the maximum likelihood estimate of p is $\widehat{p} = \frac{\sum n_i}{1200}$.
Then

$$\lambda = \frac{L(\widehat{\Omega}_0)}{L(\widehat{\Omega})} = \frac{\left(\frac{\sum n_i}{1200}\right)^{\Sigma n_i} \left(\frac{1200-\sum n_i}{1200}\right)^{1200-\Sigma n_i}}{\prod_{i=1}^{3} \left(\frac{n_i}{400}\right) \left(\frac{400-n_i}{400}\right)^{400-n_i}}.$$

Since the n_i are large, Theorem 10.2 is applicable and

$$-2 \ln \lambda = -2\left\{\sum n_i \ln\left(\frac{\sum n_i}{1200}\right) + (1200 - \sum n_i) \ln\left(1 - \frac{\sum n_i}{1200}\right) - \sum_{i=1}^{3} \left[n_i \ln\left(\frac{n_i}{400}\right) + (400 - n_i) \ln\left(1 - \frac{n_i}{400}\right)\right]\right\}$$

has an approximate χ^2 distribution with 2 degrees of freedom. For this exercise $n_1 = 16$, $n_2 = 24$, $n_3 = 9$, and $\sum n_i = 49$, so that $-2 \ln \lambda = -2(-3.689) = 7.378$. The rejection region, for $\alpha = .05$, will be $-2 \ln \lambda > \chi^2_{2,.05} = 5.99$, and the null hypothesis is rejected.

14.23 Similar to previous exercises, except that observed cell counts must be obtained from the given information as

$$n_i = \frac{(\text{number of samples})_i \times (\text{percentage})_i}{100}.$$

The approximate observed and estimated expected cell counts are shown in the following table.

	Age							
	1	2	3	4	5	6	7	Total
Have nodules	23 (54.80)	25 (19.72)	35 (30.14)	18 (11.83)	52 (34.79)	159 (157.77)	11 (13.95)	323
No nodules	366 (334.20)	115 (120.28)	179 (183.86)	66 (72.17)	195 (212.21)	961 (962.23)	88 (85.05)	1970
Total	389	140	214	84	247	1120	99	2293

The test statistic is

$$X^2 = \frac{(23-54.80)^2}{54.80} + \ldots + \frac{(88-85.05)^2}{85.05} = 38.429$$

and the rejection region with $\alpha = .05$ and $(r-1)(c-1) = 6$ degrees of freedom is

$$X^2 \geq 12.59.$$

The null hypothesis is rejected and we conclude that age and probability of finding nodules are dependent.

14.24 The contingency table, including column and row totals and the estimated expected cell counts, follows.

	Annoyance Level			
Age Group	High	Moderate	None	Total
18-34	18 (19.50)	15 (17.00)	17 (13.50)	50
35-54	21 (19.50)	19 (17.00)	10 (13.50)	50
Total	39	34	27	100

The test statistic is

$$X^2 = \frac{(18-19.50)^2}{19.50} + \ldots + \frac{(10-13.50)^2}{13.50} = 2.516$$

The rejection region with $\alpha = .01$ and 2 d.f. is $X^2 > 9.21$, and H_0 is not rejected. There is no significant difference in the proportions for the two age groups. $X^2 = 2.516 \le 4.61 = \chi^2_{.10}$ based on 2 d.f., so that the p-value $> .1$.

14.25 **a.** We wish to test a hypothesis of equivalence among the proportions of residents with lung disease in the four areas, which implies a null hypothesis of independence between the row and column classifications. The estimated expected and observed cell counts are given below.

	City A	City B	Non urban Area 1	Non urban Area 2	Total
Number with	34 (28.75)	42 (28.75)	21 (28.75)	18 (28.75)	115
Number without	366 (371.25)	358 (371.25)	379 (371.25)	382 (371.25)	1485
Total	400	400	400	400	1600

$$X^2 = \frac{(34-28.75)^2}{28.75} + \ldots + \frac{(382-371.25)^2}{371.25} = 14.19$$

The rejection region with $(r-1)(c-1) = 3$ degrees of freedom and $\alpha = .05$ is $X^2 \ge 7.81$, and the null hypothesis is rejected. There is a dependence between the proportion of people with lung disease and the location (i.e., there is a difference in the proportions with lung disease for the four locations).

b. Cigarette smokers probably should have been excluded from the sample. Cigarette smoking is a major contributor to lung disease, and a greater proportion of smokers in one or more of the locations would affect the proportions with lung disease. The number of people who smoke is an unknown variable in the four locations. It is quite possible that more people who live in urban areas smoke, owing perhaps to increased advertising or tension in the cities. The result may then indicate a dependence between lung disease and location that is not caused by air pollution but by an increased proportion of people in certain locations who smoke.

14.26 Refer to Exercise 14.25. Using a large-sample confidence interval for $p_1 - p_2$ given in Chapter 8, calculate

$$\widehat{p}_1 = \frac{34}{400} = .085$$

and

$$\widehat{p}_2 = \frac{42}{400} = .105.$$

The interval is

$$(\widehat{p}_1 - \widehat{p}_2) \pm 1.96\sqrt{\frac{\widehat{p}_1\widehat{q}_1}{n_1} + \frac{\widehat{p}_2\widehat{q}_2}{n_2}}$$

or

$$-.02 \pm 1.96\sqrt{\frac{(.085)(.915)+(.105)(.895)}{400}}$$

or $-.02 \pm .041 = [-.061, .021]$.

14.27 The table of estimated cell counts is shown below.

	Rhode Island	Colorado	California	Florida
Participate	63.62	78.63	97.88	97.88
Do Not Participate	131.38	162.37	202.12	202.12

The observed value of the test statistic is calculated to be $X^2 = 21.51$. Since the degrees of freedom associated with X^2 are $(r-1)(c-1) = 1(3) = 3$, the rejection region is $X^2 \geq \chi^2_{3,.01} = 11.3449$. Hence the null hypothesis of independence is rejected.

14.28 See Section 5.9 of the text.

14.29 This is similar to previous exercises. The estimated expected cell counts are given in the table below.

	Student	Faculty	Administration
Favor	237.44	114.16	50.40
Oppose	153.56	73.84	32.60

The rejection region, based on $(r-1)(c-1) = 1(2) = 2$ degrees of freedom, is $X^2 \geq \chi^2_{2,.05} = 5.99$. The observed value of the test statistic is $X^2 = 6.18$, which falls in the rejection region, and the null hypothesis of independence is rejected. Since $5.99 < 6.18 < 7.38$, $.025 < p\text{-value} < .05$.

14.30 **a.** The contingency table with estimated expected cell counts in parentheses is shown in the MINITAB printout below.

Expected counts are printed below observed counts

	C1	C3	C3	Total
1	43	48	9	100
	43.50	50.50	6.00	
2	44	53	3	100
	43.50	50.50	6.00	
Total	87	101	12	200
ChiSq =	0.006 +	0.124 +	1.500 +	
	0.006 +	0.124 +	1.500 =	3.259

The test statistic is

$$X^2 = \frac{(43-43.5)^2}{43.5} + \frac{(48-50.5)^2}{50.5} + \ldots + \frac{(3-6.00)^2}{6.00} = 3.259$$

The observed value of X^2 is less than $\chi^2_{2,.10}$, so that the p-value $> .10$. and H_0 is not rejected. There is no evidence of a difference due to gender.

b. The contingency table with estimated expected cell counts in parentheses is shown in the MINITAB printout below.

Expected counts are printed below observed counts

	C1	C3	C3	C4	Total
1	4	42	41	13	100
	3.50	45.00	38.00	13.50	
2	3	48	35	14	100
	3.50	45.00	38.00	13.50	
Total	7	90	76	27	200
ChiSq =	0.071 +	0.200 +	0.237 +	0.019 +	
	0.071 +	0.200 +	0.237 =	0.019 =	1.054

df = 3

2 cells with expected counts less than 5.0

The test statistic is

$$X^2 = \frac{(4-3.5)^2}{3.5} + \frac{(42-45)^2}{45} + \ldots + \frac{(14-13.5)^2}{13.5} = 1.054$$

The observed value of X^2 is less than $\chi^2_{2,.10}$, so that the p-value $> .10$, and H_0 is not rejected. There is no evidence of a difference due to gender.

c. Notice that the computer printout in part **b** warns that 2 cells have expected cell counts less than 5. This is a violation of the assumptions necessary for this test, and results should thus be viewed with caution.

14.31 a. The data are analyzed as a 5×3 contingency table with estimated expected cell counts shown in parentheses.

Level	0 to 1	2	3 to 6	Total
1	3	21	20	44
	(11.00)	(14.67)	(18.12)	
2	2	12	9	23
	(5.75)	(7.67)	(9.47)	
3	26	20	31	77
	(19.25)	(25.67)	(31.71)	
4	13	10	18	41
	(10.25)	(13.67)	(16.88)	
5	7	5	7	19
	(4.75)	(6.33)	(7.82)	
Total	51	68	84	204

The test statistic is

$$X^2 = \frac{(3-11.00)^2}{11.00} + \frac{(21-14.67)^2}{14.67} + \ldots + \frac{(7-7.82)^2}{7.82} = 20.513$$

using calculator accuracy.

b. The derived value, $X^2 = 20.513$, lies between $\chi^2_{.01}$ and $\chi^2_{.005}$ with 8 d.f. Hence $.005 < p\text{-value} < .01$. The author's value is correct.

c. Based on the p-value observed in part **b**, H_0 is rejected. The level of angina is dependent on the level of coronary artery obstruction.

14.32 a. Similar to previous exercises. The table of observed and estimated expected cell counts is shown below.

	Participant	Nonparticipant	Total
Less than once	4	16	20
a week	(6.75)	(13.25)	
More than once a	15	20	35
week but less	(11.82)	(23.18)	
than daily			
Daily	7	15	22
	(7.43)	(14.57)	
Total	26	51	77

The test statistic is

$$X^2 = \frac{(4-6.75)^2}{6.75} + \frac{(16-13.25)^2}{13.25} + \ldots + \frac{(15-14.57)^2}{14.57} = 3.025$$

using calculator accuracy. The rejection region with $(r-1)(c-1) = 2$ d.f. and $\alpha = .05$ is $X^2 > 5.99$ and H_0 is not rejected. There is no evidence of a difference due to organized sports.

b. From Table 5, the observed value of $X^2 = 3.025$ is less than $\chi^2_{.10}$, so that the p-value $> .10$. The calculated results agree with the published results.

14.33 For a chi-square goodness-of-fit test of the given data, it is necessary that the values n_i and $E(n_i)$ be known for each of the five cells. The n_i (the number of measurements falling in the ith cell) are given. However, $E(n_i) = np_i$ must be calculated. Remember that p_i is the probability that a measurement falls in the ith cell. The hypothesis to be tested will be

H_0: the experiment is binomial vs. H_a: the experiment is not binomial

Let Y = number of successes and p = probability of success on a single trial. Then assuming the null hypothesis to be true, we have

$p_0 = P(Y=0) = \binom{4}{0} p^0(1-p)^4$ $\qquad$ $p_3 = P(Y=3) = \binom{4}{3} p^3(1-p)^1$

$p_1 = P(Y = 1) = \binom{4}{1} p^1(1-p)^3$ $\qquad$ $p_4 = P(Y = 4) = \binom{4}{4} p^3(1-p)^0$

$p_2 = P(Y = 2) = \binom{4}{2} p^2(1-p)^2$

Hence, once an estimate for p is obtained, the expected cell frequencies can be calculated by using the above probabilities. Note that each of the 100 experiments consists of four trials and hence the complete experiment involves a total of 400 trials. The maximum likelihood estimator of p is $p = \frac{Y}{n}$ (as in Chapter 10). Thus,

$$\widehat{p} = \frac{y}{n} = \frac{\text{number of successes}}{\text{number of trials}} = \frac{0(11)+1(17)+2(42)+3(21)+4(9)}{400} = \frac{1}{2}$$

The experiment consisting of four trials was repeated 100 times. There is a total of 400 trials in which the result "no successes in four trials" was observed 11 times, the result "one success in four trials" was observed 17 times, and so on. Then

$p_0 = \binom{4}{0}\left(\frac{1}{2}\right)^0\left(\frac{1}{2}\right)^4 = \frac{1}{16}$ $\qquad$ $p_3 = \binom{4}{3}\left(\frac{1}{2}\right)^3\left(\frac{1}{2}\right)^1 = \frac{4}{16}$

$p_1 = \binom{4}{1}\left(\frac{1}{2}\right)^1\left(\frac{1}{2}\right)^3 = \frac{4}{16}$ $\qquad$ $p_4 = \binom{4}{4}\left(\frac{1}{2}\right)^4\left(\frac{1}{2}\right)^0 = \frac{1}{16}$

$p_2 = \binom{4}{2}\left(\frac{1}{2}\right)^2\left(\frac{1}{2}\right)^2 = \frac{6}{16}$

The observed and expected cell frequencies are shown in the table below.

	y				
	0	1	2	3	4
n_i	11	17	42	21	9
$E(n_i)$	6.25	25.00	37.50	25.00	6.25

The test statistic is

$$X^2 = \frac{(11-6.25)^2}{6.25} + \ldots + \frac{(9-6.25)^2}{6.25} = 8.56$$

In order to set up the rejection region, we must first determine the degrees of freedom associated with the test statistic. Two restrictions are placed upon the cell counts:

(1) $n_1 + n_2 + \ldots + n_5 = 100$.

(2) The binomial parameter p is estimated by using a linear combination of the n_i.

The number of degrees of freedom is equal to the number of cells (k) less 1 degree of freedom for each independent linear restriction placed on the cell frequencies. Therefore, $(k-2) = 5-2 = 3$, and the critical value of $\chi^2 = 7.81$, using $\alpha = .05$. The test statistic falls in the rejection region, and we conclude that the experiment does not fulfill the properties of a binomial experiment. The p-value is between .025 and .05.

14.34a. The likelihood is

$$L(\theta) = (-1)^n \frac{1}{\prod\limits_{i=1}^{n} \ln(1-\theta)} \frac{\theta^{\sum_{i=1}^{n} y_i}}{\prod\limits_{i=1}^{n} y_i}.$$

So,

$$\ln L = -\sum_{i=1}^{n} \ln(\ln(1-\theta)) + \sum_{i=1}^{n} y_i(\ln\theta) + k$$

where k is a quantity not involving θ. Setting the derivative with respect to θ of $\ln L$ equal to zero yields

$$\sum_{i=1}^{n} \frac{1}{\ln(1-\widehat{\theta})}\left(\frac{1}{1-\widehat{\theta}}\right) + \sum_{i=1}^{n} \frac{y_i}{\widehat{\theta}} = 0$$

or

$$\frac{-n}{(1-\widehat{\theta})\ln(1-\widehat{\theta})} = \frac{\sum_{i=1}^{n} y_i}{\widehat{\theta}}$$

implying

$$\overline{y} = \frac{\widehat{\theta}}{-(1-\widehat{\theta})\ln(1-\widehat{\theta})}.$$

b. $\overline{y} = \frac{1(359)+2(146)+3(57)+4(41)+5(26)+6(17)+7(29)}{675} = \frac{1421}{675} = 2.105$

To find $\widehat{\theta}$, we must solve

$$2.105 = \frac{-\widehat{\theta}}{(1-\widehat{\theta})\ln(1-\widehat{\theta})}$$

Proceeding by trial and error, and using interpolation, we get $\widehat{\theta} = .7385$.

$\widehat{p}(1) = \frac{-\widehat{\theta}}{\ln(1-\widehat{\theta})} = .5506$ $\qquad$ $\widehat{p}(2) = \frac{-\widehat{\theta}^2}{2\ln(1-\widehat{\theta})} = .2033$

$\widehat{p}(3) = \frac{-\widehat{\theta}^3}{3\ln(1-\widehat{\theta})} = .1001$ $\qquad$ $\widehat{p}(4) = \frac{-\widehat{\theta}^4}{4\ln(1-\widehat{\theta})} = .0554$

$\widehat{p}(5) = \frac{-\widehat{\theta}^5}{5\ln(1-\widehat{\theta})} = .0328$ $\widehat{p}(6) = \frac{-\widehat{\theta}^6}{6\ln(1-\widehat{\theta})} = .0202$

$\widehat{p}(\geq 7) = 1 - \sum_{i=1}^{6} \widehat{p}(i) = .0376$

We get the table of estimated expected values by multiplying these values by $\sum y_i = 675$:

	y						
	1	2	3	4	5	6	≥ 7
$E(n_i)$	371.66	137.23	67.57	37.40	22.14	13.64	25.38

The value of the test statistic is

$$X^2 = 4.99$$

which is less than $11.07 = \chi^2_{5,.05}$. We do not reject H_0.

14.35 In order to find the maximum likelihood estimator of p_i, the probability of falling in row i, consider row i as a single cell with r_i observations falling in this cell. Then the variables $r_1, r_2, \ldots, r_r$ follow a multinomial distribution with parameters n, $p_1, p_2, \ldots, p_r$. Hence the likelihood function is

$$L = \frac{n!}{r_1! r_2! \cdots r_r!} p_1^{r_1} p_2^{r_2} \cdots p_r^{r_r} = K \prod_{j=1}^{r} p_j^{r_j}$$

so that

$$\ln L = \ln K + \sum_{j=1}^{r} r_j \ln p_j$$

with

$$\sum_{j=1}^{r} p_j = 1,$$

where K does not involve any of the p_i.

Notice that because of the above restriction, we may write $p_r = 1 - \sum_{j=1}^{r-1} p_j$ and that p_r is really a function of p_i for $i = 1, 2, \ldots, r-1$. Also, $r_r = n - \sum_{j=1}^{r-1} r_j$. Hence

$$\ln L = \ln K + \sum_{j=1}^{r-1} r_j \ln p_j + \left(n - \sum_{j=1}^{r-1} r_j\right) \ln \left(1 - \sum_{j=1}^{r-1} p_j\right)$$

Now

$$\frac{d(\ln L)}{dp_i} = \frac{r_i}{p_i} - \frac{n - \sum_{j=1}^{r-1} r_j}{1 - \sum_{j=1}^{r-1} p_j} \qquad \text{for} \qquad i = 1, 2, \ldots, r-1$$

Setting these $(r-1)$ equations equal to zero, we have, for $i = 1, 2, \ldots, r-1$,

$$(*) \qquad r_i \left(1 - \sum_{j=1}^{r=1} \widehat{p}_j\right) = \widehat{p}_i \left(n - \sum_{j=1}^{r-1} r_j\right)$$

In order to solve the $(r-1)$ equations simultaneously, add them together to obtain

$$\sum_{i=1}^{r-1} r_i \left(1 - \sum_{j=1}^{r-1} \widehat{p}_j\right) = \sum_{i=1}^{r-1} \widehat{p}_i \left(n - \sum_{j=1}^{r-1} r_j\right)$$

or

$$n \sum_{j=1}^{r-1} \widehat{p}_j = \sum_{j=1}^{r-1} r_j$$

or

$$\sum_{j=1}^{r-1} \widehat{p}_j = \frac{1}{n} \left(\sum_{j=1}^{r-1} r_j\right).$$

Substituting in (*), we have

$$r_i\left(1-\frac{1}{n}\sum_{j=1}^{r-1} r_j\right)=\widehat{p}_i\left(n-\sum_{j=1}^{r-1} r_j\right)\ \Rightarrow\ \widehat{p}_i=\frac{r_i\left(1-\frac{1}{n}\sum_{j=1}^{r-1} r_j\right)}{n-\sum_{j=1}^{r-1} r_j}=\frac{r_i}{n}.$$

14.36 The model states that for this particular multinomial experiment with $k=3$, the hypothesized cell probabilities are $p_1=p^2$, $p_2=2p(1-p)$, $p_3=(1-p)^2$ for some parameter p. Hence the likelihood function is

$$L=\frac{n!}{n_1!\,n_2!\,n_3!}(p^2)^{n_1}[2p(1-p)]^{n_2}(1-p)^{2n_3}=\frac{n!\,2^{n_2}}{n_1!\,n_2!\,n_3!}p^{2n_1+n_2}(1-p)^{2n_3+n_2}$$

a. If p is unspecified, an estimate for p must be used.

$$\ln L=\ln K+[2n_1+n_2)\ln p+(n_2+2n_3)\ln(1-p)$$

implying

$$\frac{d(\ln L)}{dp}=\frac{2n_1+n_2}{p}-\frac{n_2+2n_1}{1-p}.$$

Putting $\frac{d(\ln L)}{dp}=0$ and solving for $\widehat{p}$, we find the maximum likelihood estimator for p.

$$\frac{2n_1+n_2}{p}=\frac{n_2+2n_3}{1-p}$$

or

$$\widehat{p}=\frac{2n_1+n_2}{2(n_1+n_2+n_3)}.$$

For this experiment, $n_1=30$, $n_2=40$, and $n_3=30$, so that

$$\widehat{p}=\frac{60+40}{2(100)}=\frac{1}{2}$$

The estimated expected cell counts are

$$\widehat{E}(n_1)=n\widehat{p}_1=n\widehat{p}^2=25$$
$$\widehat{E}(n_2)=n\widehat{p}_2=2n\widehat{p}(1-\widehat{p})=50$$
$$\widehat{E}(n_3)=n\widehat{p}_3=n(1-\widehat{p})^2=25$$

The test statistic is

$$X^2=\frac{(30-25)^2}{25}+\frac{(40-50)^2}{50}+\frac{(30-25)^2}{25}=4$$

The degrees of freedom associated with the test statistic are $(k-1-1)=3-2=1$, since

(i) $\sum_{i=1}^{k} n_i=n$ and (ii) $\frac{2n_1+n_2}{2(n_1+n_2+n_3)}=\widehat{p}$

Hence the critical value of X^2 is $\chi^2_{1,.05}=3.84$, and the null hypothesis is rejected. It does not appear that the model is correct.

b. If the null hypothesis specifies that $p=\frac{1}{2}$, there is no need to obtain an estimate. The expected cell counts are $E(n_1)=25$, $E(n_2)=50$, and $E(n_3)=25$, so that $X^2=4$ as in part **a**. However, the degrees of freedom are now $(k-1)=2$ and the critical value of χ^2 is 5.99. The null hypothesis is not rejected, and we cannot conclude that the data contradict the hypothesis.

14.37 The problem describes a multinomial experiment with $k=4$ cells. Under the null hypothesis, the four cell probabilities are

$$p_1=\frac{p}{2}\qquad p_2=\frac{p^2}{2}+pq\qquad p_2=\frac{q}{2}\qquad p_4=\frac{q^2}{2}$$

where p is unspecified, but $p+q=1$. In order to use the chi-square goodness-of-fit test for this genetic model, we must first obtain an estimate of p. We will use a maximum likelihood estimate for p, with the likelihood function given by

$$L=\frac{n!}{n_1!\,n_2!\,n_3!\,n_4!}\left(\frac{p}{2}\right)^{n_1}\left(\frac{p^2}{2}+pq\right)^{n_2}\left(\frac{q}{2}\right)^{n_3}\left(\frac{q^2}{2}\right)^{n_4}$$
$$==\frac{n!}{\prod_{i=1}^{4} n_i\,2^{\Sigma n_i}}p^{n_1}(p^2+2pq)^{n_2}(1-p)^{n_3}(1-p)^{2n_4}$$
$$=kp^{n_1}[p(2-p)]^{n_2}(1-p)^{n_3+2n_4}=Kp^{n_1+n_2}(2-p)^{n_2}(1-p)^{n_3+2n_4}$$

so that

$$\ln L=\ln K+(n_1+n_2)\ln p+n_2\ln(2-p)+(n_3+2n_4)\ln(1-p)$$

where K does not depend on p. Now

$$\frac{d(\ln L)}{dp}=\frac{n_1+n_2}{p}-\frac{n_2}{2-p}-\frac{n_3+2n_4}{1-p}$$

Setting the derivative equal to zero and solving for $\widehat{p}$, we obtain

$$(n_1+n_2)(2-\widehat{p})(1-\widehat{p})-n_2\widehat{p}(1-\widehat{p})-(n_3+2n_4)(\widehat{p})(2-\widehat{p})=0$$

implying

$$(n_1+2n_2+n_3+2n_4)\widehat{p}^2-(3n_1+4n_2+2n_3+4n_4)\widehat{p}+2(n_1+n_2)=0$$

That is, $\widehat{p}$ is the root (between 0 and 1) of a quadratic equation in the form

$$a\widehat{p}^2+b\widehat{p}+c=0,$$

with

$$a=n_1+2n_2+n_3+2n_4=3040,$$
$$b=-(3n_1+4n_2+2n_3+4n_4)=-6960,$$
$$c=2(n_1+n_2)=3824.$$

The solution will be

$$\widehat{p}=\frac{-b-\sqrt{b^2-4ac}}{2a}=\frac{6960-\sqrt{1,941,760}}{6080}=.9155.$$

Now the estimated cell probabilities, estimated expected cell counts, and the value of the test statistic can be obtained. (See the table below.)

$\widehat{p}_i$	$\widehat{E}(n_i)$	n_i
$\frac{\widehat{p}}{2}=.45775$	915.50	880
$\left(\frac{\widehat{p}}{2}\right)+\widehat{p}\widehat{q}=.49643$	992.86	1032
$\frac{\widehat{q}}{2}=.04225$	84.50	80
$\frac{\widehat{q}^2}{2}=.00357$	7.14	8

Then

$$X^2=\frac{(880-915.5)^2}{915.5}+\ldots+\frac{(8-7.14)^2}{7.14}=3.26$$

The degrees of freedom are $(k-1-1)=2$, since:

$$\sum_{i=1}^{k} n_i=n$$

and

(2) the observed counts, n_i, were used to calculate $\widehat{p}$.

For $\alpha=.05$, the critical value of X^2 is $\chi^2_{2,.05}=5.99$, and the hypothesized model is not rejected.

14.38 Suppose that we have two multinomial experiments, each with k cells. The cell counts and cell probabilities are n_i and p_i, m_i and p_i^*, respectively, and

$$\sum_{i=1}^{k} p_i=\sum_{i=1}^{k} p_i^*=1$$

The hypothesis of interest is H_0: $p_1=p_1^*$, $p_2=p_2^*$, ... , $p_k=p_k^*$. If H_0 is true, then $p_i=p_i^*$, so that two expected cell counts could be obtained if we could estimate p_i, the probability of falling in cell i for each of the two experiments. In general, the likelihood function can be written as

$$L=f(n_1,n_2,\ldots,n_k,m_1,m_2,\ldots,m_k)$$
$$=\frac{n!}{\prod_{j=1}^{k} n_j!}\prod_{j=1}^{k} p_j^{n_j}\frac{m!}{\sum_{j=1}^{k} m_j!}\prod_{i=1}^{n} p_j^{*m_j}=K\prod_{j=1}^{k} p_j^{n_j}p_j^{*n_j}$$

Under H_0:

$$L=K\prod_{j=1}^{k} p_j^{n_j+m_j}$$

and

$$\ln L=\ln K+\sum_{j=1}^{k}(n_j+m_j)\ln p_j.$$

Maximizing $\ln L$ subject to the restrictions $\sum_{j=1}^{k} p_i=1$, we obtain, as in Exercise 14.32,

$$\widehat{p}_i=\frac{n_i+m_i}{n+m}\qquad\text{for}\qquad i=1,2,3,\ldots,k-1$$

and

$$\widehat{p}_k=1-\sum_{i=1}^{k-1} p_i.$$

Then

$$\widehat{E}(n_i) = n\widehat{p}_i = n\left(\frac{n_i+m_i}{n+m}\right) \text{ and } \widehat{E}(m_i) = m\widehat{p}_i = m\left(\frac{n_i+m_i}{n+m}\right)$$

for $i = 1, 2, 3, \ldots, k$ and the test statistic is

$$X^2 = \sum_{i=1}^{k} \frac{\left[n_i - n\left(\frac{n_i+m_i}{n+m}\right)\right]^2}{n\left(\frac{n_i+m_i}{n+m}\right)} = \sum_{i=1}^{k} \frac{\left[m_i - m\left(\frac{n_i+m_i}{n+m}\right)\right]^2}{m\left(\frac{n_i+m_i}{n+m}\right)}$$

X^2 will have an approximate chi-square distribution with degrees of freedom of $(2k-2)-(k-1) = k-1$ since there are $2k$ cells. Two degrees of freedom are lost since

$$\sum_{i=1}^{k} n_i = n \qquad \text{and} \qquad \sum_{i=1}^{k} m_i = m$$

and $(k-1)$ cell probabilities have been estimated using the observed cell counts n_i and m_i. Hence a rejection region for the test will be based on $(k-1)$ degrees of freedom.

14.39 In this exercise there are four binomial experiments performed, one at each of four dosage levels. Let n_i = number of survivors for dose i, $p_i = P$(insect survives at dose i), and $q_i = P$(insect dies at dose i). The hypothesis of interest is

H_0: $p_1 = 1+\beta, p_2 = 1+2\beta, p_3 = 1+3\beta, p_4 = 1 = 4\beta$

Notice that this automatically implies that $q_1 = -\beta, q_2 = -2\beta, q_3 = -3\beta, q_4 = -4\beta$. It is necessary to obtain an estimate of β, which can be done by using the method of maximum likelihood.

$$L = \prod_{i=1}^{4} \binom{1000}{n_i} (1+i\beta)^{n_i}(-i\beta)^{1000-n_i} = K\prod_{i=1}^{4} (1+i\beta)^{n_i}(\beta)^{1000-n_i}$$

implying

$$\ln L = \ln K + \sum_{i=1}^{4} n_i \ln(1+i\beta) + \sum_{i=1}^{4} (1000-n_i) \ln \beta,$$

where K does not involve β. Then

$$\frac{d(\ln L)}{d\beta} = \sum_{i=1}^{4} \frac{in_i}{1+i\beta} + \frac{1}{\beta}\sum_{i=1}^{4} (1000-n_i)$$

Putting $\frac{d(\ln L)}{d\beta} = 0$ and expanding, we obtain a quartic equation in $\widehat{\beta}$.

$$\left(4000 - \sum_{i=1}^{4} n_i\right)(1+\widehat{\beta})(1+2\widehat{\beta})(1+3\widehat{\beta})(1+4\widehat{\beta}) + n_1\widehat{\beta}(1+2\widehat{\beta})(1+3\widehat{\beta})(1+4\widehat{\beta})$$
$$+ 2n_2\widehat{\beta}(1+\widehat{\beta})(1+3\widehat{\beta})(1+4\widehat{\beta}) + 3n_3\widehat{\beta}(1+\widehat{\beta})(1+2\widehat{\beta})(1+4\widehat{\beta})$$
$$+ 4n_4\widehat{\beta}(1+\widehat{\beta})(1+2\widehat{\beta})(1+3\widehat{\beta})$$
$$= 0$$

or

$$\left(4000 - \sum_{i=1}^{4} n_i\right) + \left[10\left(4000 - \sum_{i=1}^{4} n_i\right) + n_1 + 2n_2 + 3n_3 + 4n_4\right]\widehat{\beta}$$
$$+ \left[35\left(4000 - \sum_{i=1}^{4} n_i\right) + 9n_1 + 16n_2 + 21n_3 + 24n_4\right]\widehat{\beta}^2$$
$$+ \left[50\left(4000 - \sum_{i=1}^{4} n_i\right) + 26n_1 + 38n_2 + 42n_3 + 44n_4\right]\widehat{\beta}^3$$
$$+ \left[24\left(4000 - \sum n_i\right) + 24\sum_{i=1}^{4} n_i\right]\widehat{\beta}^4 = 0$$

Substituting $n_1 = 820$, $n_2 = 650$, $n_3 = 310$, and $n_4 = 50$, we obtain

$$217 + 2495\widehat{\beta} + 10{,}144\widehat{\beta}^2 + 16{,}974\widehat{\beta}^3 + 9600\widehat{\beta}^4 = 0$$

This equation can best be solved iteratively, using a "guessed" value for $\widehat{\beta}$ based on a graph of p_i against D_i. Recall that β is the slope of the line $p = 1 + \beta D$, which is a line with intercept 1. Using the four points available to us, a good guess for the slope of the line would be $\beta = -.2$ (see Figure 14.1). We use this as an initial estimate and solve the equation until a value is found such that $f(\widehat{\beta}) = 0$.

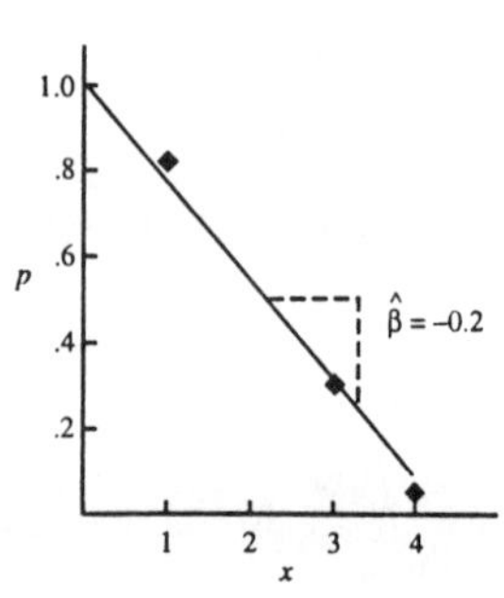

Figure 14.1

$\widehat{\beta}$	$-.2$	$-.22$	$-.225$	$-.226$	$-.228$	$-.230$	$-.232$	$-.233$
$f(\widehat{\beta})$	3.328	.819	.424	.355	.226	.110	.004	$-.040$

Hence, our estimate is

$$\widehat{\beta} = -.232,$$

so that

$$\widehat{p}_1 = 1 + \widehat{\beta} = .768 \qquad \widehat{p}_3 = 1 + 3\widehat{\beta} = .304$$
$$\widehat{p}_2 = 1 + 2\widehat{\beta} = .536 \qquad \widehat{p}_4 = 1 + 4\widehat{\beta} = .072$$

and the table of observed and estimated expected cell counts is as shown below.

	Dosage			
	1	2	3	4
Survived	820 (768)	650 (536)	320 (304)	50 (72)
Died	180 (232)	350 (464)	690 (696)	950 (928)

The calculated test statistic is

$$X^2 = 74.8$$

and the rejection region based on a chi-square variable with $(8 - 4 - 1) = 3$ degrees of freedom is

$$X^2 \geq \chi^2_{3,.05} = 7.81$$

The null hypothesis is rejected and the hypothesis $p = 1 + \beta D$ is contradicted.
Note: There are 8 cells, but 5 restrictions:
(1) $p_i + q_i = 1$ for $i = 1, 2, 3, 4$.
(2) Estimate β using n_i.

CHAPTER 15 NONPARAMETRIC STATISTICS

15.1 It is necessary that α (the probability of rejecting the null hypothesis when it is true) take values between $\alpha = .01$ and $\alpha = .15$. Assuming the null hypothesis to be true, the two populations are identical and consequently $p = P(A$ exceeds B for a given pair of observations) is $\frac{1}{2}$.

The binomial probability distribution was discussed in Chapter 3. In particular, it was noted that the distribution of the random variable Y is symmetrical about the mean np when $p = \frac{1}{2}$. For example, with $n = 25$, $P(Y = 0) = P(Y = 25)$. That is,

$$P(Y = 0) = \binom{25}{0}\left(\tfrac{1}{2}\right)^0\left(\tfrac{1}{2}\right)^{25} = \left(\tfrac{1}{2}\right)^{25}$$

and

$$P(Y = 25) = \binom{25}{25}\left(\tfrac{1}{2}\right)^{25}\left(\tfrac{1}{2}\right)^0 = \left(\tfrac{1}{2}\right)^{25}$$

Similarly, $P(Y = 1) = P(Y = 24)$, and so on. Hence is it necessary to determine a rejection region such that $.005 \le \frac{\alpha}{2} \le .075$, where $P(Y \le a) = \frac{\alpha}{2}$ is given in Table 1. Indexing $n = 25$, $p = \frac{1}{2}$ in Table 1, it is found that the critical values of Y that determine the lower tail of the desired rejection region are $Y = 6$, $Y = 7$, and $Y = 8$. The reader may verify that the corresponding values for the upper tail of the rejection region are $Y = 19$, $Y = 18$, and $Y = 17$. The significance levels and corresponding rejection regions are shown in the table below.

$\frac{\alpha}{2}$	α	Rejection Region
.007	.014	$Y \le 6; Y \ge 19$
.022	.044	$Y \le 7; Y \ge 18$
.054	.108	$Y \le 8; Y \ge 17$

15.2 **a.** A sign test is employed to test the following hypothesis:

$$H_0\colon\ p = \tfrac{1}{2} \quad \text{vs.} \quad H_a\colon\ p \ne \tfrac{1}{2}$$

where

$p = P(\text{school } A \text{ exceeds school } B \text{ in test score})$

The test statistic is M, the number of times school A exceeds school B in test score. Then M will be equivalent to the number of plus signs observed in the table shown at the right. Notice that $m = 7$. The p-value is found using Table 1:

Twin Pair	Sign $(A - B)$
1	+
2	+
3	−
4	+
5	+
6	−
7	+
8	+
9	−
10	+

$$P(M \ge 7 \text{ or } M \le 3) = 2P(M \le 3)$$
$$= 2(.172) = .344$$

Since $.344 > .05$, we do not reject H_0.

b. Consider the one-tailed test of hypothesis as follows:

$$H_0\colon\ p = \tfrac{1}{2} \quad \text{vs.} \quad H_a\colon\ p > \tfrac{1}{2}$$

This alternative will imply that school A is superior to school B, since, if this alternative is true, we would expect p (the probability that school A exceeds school B in test scores) to be greater than $\frac{1}{2}$. For this alternative hypothesis, the p-value is

$$P(M \ge 7) = 1 - P(M \le 6) = 1 - .828 = .172.$$

Hence we fail to reject H_0.

15.3 Let $p = P(\text{judge favors mixture } B)$ and M = number of judges favoring mixture B. The hypothesis of interest is

$$H_0\colon\ p = \tfrac{1}{2} \quad \text{vs.} \quad H_a\colon\ p \ne \tfrac{1}{2},$$

with $n = 10$. The observed value of M is m = 2. The p-value is

$$2P(M \le 2) = 2(.055) = .11 > .05.$$

There is no significant difference between the tastes of A and B.

15.4

a. Let $p = P$(high elevation exceeds low elevation) and $M =$ number of nights during which high elevation exceeds low elevation. Then

$$H_0\colon\ p = \tfrac{1}{2} \qquad \text{vs.} \qquad H_a\colon\ p > \tfrac{1}{2}$$

and $n = 10$. Large values of M will tend to favor H_a, and an upper-tailed rejection region is used. The observed value of M is $m = 9$. The p-value is

$$P(M \geq 9) = 1 - P(M \leq 8) = 1 - .989 = .011.$$

The data supports H_a.

b. Extremal variables, such as the minimum temperatures in this example, often have skewed distributions, making the assumptions of the t test invalid.

15.5

a. The hypothesis to be tested is

$$H_0\colon\ p = \tfrac{1}{2} \qquad \text{vs.} \qquad H_a\colon\ p \neq \tfrac{1}{2}$$

where $p = P$(response for stimulus 1 exceeds that for stimulus 2). Then the test statistic is M, the number of times the response for stimulus 1 exceeds that for stimulus 2. Again, denote a positive difference by a plus sign and a negative difference by a minus sign. Then M will be equivalent to the number of plus signs observed. These signs are

Subject	1	2	3	4	5	6	7	8	9
Sign	−	−	+	−	−	−	−	+	−

and $m = 2$. The next step is to select a rejection region such that $\alpha = P\left(\text{reject } H_0 | p = \tfrac{1}{2}\right)$ is close to .05. Using the rejection region $M = 0, 1, 8, 9$, we find that

$$\alpha = P\left(M = 0, 1, 8, 9 | p = \tfrac{1}{2}\right) = \binom{9}{0}\left(\tfrac{1}{2}\right)^0\left(\tfrac{1}{2}\right)^9 + \binom{9}{1}\left(\tfrac{1}{2}\right)^1\left(\tfrac{1}{2}\right)^8 + \binom{9}{8}\left(\tfrac{1}{2}\right)^8\left(\tfrac{1}{2}\right)^1 + \binom{9}{9}\left(\tfrac{1}{2}\right)^9\left(\tfrac{1}{2}\right)^0 = .04$$

Using the large rejection region $M = 0, 1{,}2{,}7{,}8{,}9$ we find that

$$\alpha = P\left(M = 0, 1, 2, 7, 8,\ 9 | p = \tfrac{1}{2}\right) = .180.$$

Therefore we see that $M = 0, 1, 8, 9$ is the correct rejection region to have .05 level test. Examining the test statistic ($m = 2$), we fail to reject the null hypothesis.

b. The two samples are not random and independent. Rather, the experiment has been conducted in a paired manner, and a paired-difference analysis is used. The differences and the associated t test are given below.

d_i	d_i^2
−.9	.81
−1.1	1.21
1.5	2.25
−2.6	6.76
−1.8	3.24
−2.9	8.41
−2.5	6.25
2.5	6.25
−1.4	1.96

(1) $H_0\colon\ \mu_1 - \mu_2 = 0$ vs. $H_a\colon\ \mu_1 - \mu_2 \neq 0$

(2) $\bar{d} = \frac{\sum_i d_i}{n} = \frac{-9.2}{9} = -1.022$

$$s^2 = \frac{\sum_i d_i^2 - \left(\sum_i d_i / n\right)^2}{n-1} = \frac{37.14 - 9.404}{8} = \frac{27.736}{8} = 3.467$$

(3) Test statistic: $t = \frac{\bar{d} - 0}{\frac{s_d}{\sqrt{n}}} = \frac{-1.022}{1.86373} = -1.65$

(4) Rejection region: The critical value of t for a two-tailed test, based on 8 degrees of freedom, will be $t_{.025,8} = 2.306$, and the rejection region is $|t| > 2.306$. The test statistic does not fall in the rejection region. Hence the null hypothesis cannot be rejected.

15.6

Let $p = P(B$ exceeds $A)$ and $M =$ number of technicians for which B exceeds A with $n = 7$ (since one tied pair is deleted). A two-tailed rejection region is needed to test

$$H_0\colon\ p = \tfrac{1}{2} \qquad \text{vs.} \qquad H_a\colon\ p \neq \tfrac{1}{2}.$$

The observed value of M is 1. The p-value is

$$P(M \leq 1 \text{ or } M \geq 6) = \left[\binom{7}{0} + \binom{7}{1} + \binom{7}{6} + \binom{7}{7}\right] \left(\tfrac{1}{2}\right)^7 = .125$$

Since $.125 > .10$, H_0 is not rejected.

15.7 **a.** Since two of the pairs are tied, $n = 10$. Let $p = P(\text{before exceeds after})$. Then

$$H_0\colon p = \tfrac{1}{2} \qquad \text{vs.} \qquad H_a\colon p > \tfrac{1}{2}.$$

Large values of M will favor H_a. The observed value of the test statistic is $m = 9$. The p-value is $P(M \geq 9) = 1 - .989 = .011$. H_0 is not rejected at the $\alpha = .01$ level. Note: the p-value is larger than α, although by only a small amount.

b. The number of accidents is a small integer and hence probably not well approximated by a continuous distribution (such as the t distribution).

15.8 **a.** Define d_i to be the difference between the math score and the art score for a particular student. The differences, along with their ranks (according to absolute magnitude), are shown in the table at the right. Then the rank sum for positive differences is $T^+ = 14$ and the rank sum for negative differences is $T^- = 106$. Consider the minimum rank sum.

d_i	Rank $\lvert d_i \rvert$	d_i	Rank $\lvert d_i \rvert$
−31	14.5	7	7
−31	14.5	−9	10.5
−6	4.5	−2	2
−11	12.5	−8	9
−9	10.5	−1	1
−7	7	−6	4.5
7	7	−3	3
−11	12.5		

Indexing $n = 15$ in Table 9, we see that $T = 14 < 16$, so the p-value $< .01$. The null hypothesis of no difference between math and art scores is rejected.

b. The hypothesis of interest is

H_0: population distributions for A and B are identical

vs.

H_a: population distributions for A and B differ in location

Under the null hypothesis of no difference, we would expect that positive and negative differences of equal absolute magnitude will occur with equal probability. Hence, the rank sum of the negative differences should be nearly equal to the rank sum of the positive differences.

15.9 **a.** Define d_i to be the difference between school A and school B (i.e., $A - B$). The differences, along with their ranks (according to absolute magnitude) are shown in the table at the right. Then the rank sum for positive differences is $T^+ = 49$, and the rank sum for negative differences is $T^- = 6$. Consider the minimum rank sum, $T = 6$. Indexing $n = 10$ in Table 9, $5 < T < 8$ so $.02 < p\text{-value} < .05$. When $\alpha = .05$, the null hypothesis is rejected.

d_i	Rank $\lvert d_i \rvert$
28	13
5	4
−4	3
15	9
12	7
−2	1
7	5
9	6
−3	2
13	8

b. If a one-tailed test of hypothesis is desired, the p-value is between .01 and .025. We therefore conclude that school A is superior to school B. Notice that the rank-sum test leads to rejection of the null hypothesis but the sign test does not. The rank-sum test utilizes more information than the sign test and is more powerful in detecting departures from the null hypothesis.

15.10 The differences, with their ranks (according to absolute magnitude) are given in the table at the right. Then $T^- = 1$ and $T^+ = 54$. Hence $T = 1$. Using Table 9 we see that $T^{-1} = 1 < 3$, so that the p-value $< .005$ for this one-tailed test.

d_i	Rank $\lvert d_i \rvert$	d_i	Rank $\lvert d_i \rvert$
1.1	6	.7	4
1.3	8.5	1.1	6
2.8	10	.6	3
−.1	1	1.3	8.5
.5	2	1.1	6

15.11 The experimenter's design was paired using people as blocks. The Wilcoxon rank-sum test is the appropriate test for this experiment, with the null hypothesis that the populations follow the same distributions. The data are shown below.

Subject	Normal	Stress	Difference $(N - S)$	Rank
1	126	130	−4	5
2	117	118	−1	1
3	115	125	−10	7
4	118	120	−2	2
5	118	121	−3	3.5
6	128	125	+3	3.5
7	125	130	−5	6
8	120	120	0	–

The rank sum for positive values is $T^+ = 3.5$, and for negative values $= 24.5$. For $n = 7$ (one tie) and for a one-tailed test, we use the rank sum $T^* = 3.5$. Since $2 < T^* < 4$, $.025 < p\text{-value} < .05$.

15.12 **a.** The paired data are given in the exercise. The differences, along with their ranks (according to absolute magnitude), are shown in the following table.

d_i	+1	+2	−1	+1	+3	+1	−1	+3	−2	+3	+1	0
Rank $\lvert d_i \rvert$	3.5	7.5	3.5	3.5	10	3.5	3.5	10	7.5	10	2.5	–

Let $p = P[A$ exceeds B for a given intersection] and M = number of intersections at which A exceeds B. The hypothesis to be tested is

$$H_0\colon\ p = \tfrac{1}{2} \qquad \text{vs.} \qquad H_a\colon\ p \neq \tfrac{1}{2}$$

Various two-tailed rejection regions are tried in order to find a region with $\alpha \approx .05$. These are shown in the following table.

Rejection Region	α
$M \leq 1$; $M \geq 10$	.012
$M \leq 2$; $M \geq 9$	.066
$M \leq 3$; $M \geq 8$	.226

We choose to reject H_0 if M ≤ 2 or $M \geq 9$ with $\alpha = .066$. Since $= 8$, H_0 is not rejected. There is insufficient evidence to indicate a difference between the two methods.

b. To use the Wilcoxon signed-rank test, we use the ranks of the absolute differences shown in the table above. Then $T^+ = 51.5$ and $T^- = 14.5$ with $n = 11$. Indexing $n = 11$ and $\alpha = .05$ in Table 8, the lower portion of the two-tailed rejection region is $T \leq 11$ and H_0 is not rejected, as in part **a**.

15.13 The ranked differences are

Difference untreated-treated	Absolute value of difference	Rank of absolute value
0	0	(eliminated)
2	2	6
2	2	6
1	1	2.5
−1	1	2.5
1	1	2.5
4	4	9
−2	2	6
5	5	10
0	0	(eliminated)
3	3	8
−1	1	2.5

$T^+ =$ sum of positive ranks $= 44$ $\qquad$ $T^- =$ sum of negative ranks $= 11$

The Wilcoxon test is

H_0: The probability distributions of the number of cavities are the same for the two populations.

H_a: The distribution of the number of cavities for the untreated teeth is shifted to the right of the distribution of the number of cavities for the treated teeth.

The test statistic is $T^- = 11$. Using Table 9, since $T^- = 11 \geq 11$, we reject H_0 at the level $\alpha = .05$.

15.14 The ranked differences are

1972–1978	Absolute difference	Rank
3	3	7
6.1	6.1	12
2	2	3
4	5	10
2.5	2.5	5.5
8.9	8.9	14
.8	.8	1
4.2	4.2	11
9.8	9.8	15
3.3	3.3	8
2.3	2.3	4
3.7	3.7	9
2.5	2.5	5.5
−1.8	1.8	2
7.5	7.5	13

$T^+ = 118$ $\qquad$ $T^- = 2$

For the one-tailed test, the test statistic is $T^- = 2$. From Table 9, $T^- < 16$, so the p-value $< .005$. We reject H_0 at the $\alpha = .01$ level.

15.15 Let ξ be the median of a random variable Y with distribution function $F(y)$. By definition, then,

$$P(Y > \xi) = P(Y < \xi) = \tfrac{1}{2}$$

It is necessary to test

$$H_0\colon \xi = \xi_0 \qquad \text{vs.} \qquad H_a\colon \xi \neq \xi_0.$$

Notice that, if H_0 is true,

$$P(Y < \xi_0) = \tfrac{1}{2}.$$

a. Instead of defining $d_i = X_i - Y_i$, we define $d_i = Y_i - \xi_0$ and let M be the number of negative differences. If H_0 is true, $p = P(\text{negative difference}) = \frac{1}{2}$, and M will have a binomial distribution with n trials and $p = \frac{1}{2}$. If M is too large or too small, H_0 will be rejected. The rejection region for M will be obtained by using the binomial distribution of Chapter 3.

b. We now make use of the magnitude of the differences as well as their sign. Define

$d_i = Y_i - \xi_0$

so that if H_0 is true, $P(d_i < 0) = \frac{1}{2}$. Rank the d_i according to absolute magnitude and define T^- to be the sup of the negative ranks. T^- will have a distribution identical to the Wilcoxon signed-rank statistic given in Section 15.4. If T^- is too large or too small (using an appropriate rejection region), the null hypothesis will be rejected.

15.16 Use the results of Exercise 15.15 with $\xi_0 = 15{,}000$. The differences, $d_i = Y_i - 15{,}000$, are given in the table at the right.

d_i	Rank $\lvert d_i \rvert$	d_i	Rank $\lvert d_i \rvert$
−200	2	3500	7
1900	5	5000	10
3000	6	4200	9
4100	8	100	1
−1800	4	1500	3

a. Using the sign test with $p = P(Y < 15{,}000) = \frac{1}{2}$ under the null hypothesis, a lower-tailed rejection region is used. The observed value of M is $m = 2$. The p-value is $P(M \leq 2) = .055$ (use Table 1).

b. $T^- = 6, T^+ = 49$, and $T = 6$. Using a one-tailed test, the p-value is between .01 and .025. Thus, H_0 is rejected.

15.17 The hypothesis to be tested is

H_0: the population distributions for plastics 1 and 2 are the same

H_a: the population distributions differ in location

We rank the 12 observations in order of magnitude across groups. The data, with corresponding ranks, are shown in the table at the right.

Plastic 1	Plastic 2
15.3 (2)	21.2 (9)
18.7 (6)	22.4 (11)
22.3 (10)	18.3 (5)
17.6 (4)	19.3 (8)
19.1 (7)	17.1 (3)
14.8 (1)	27.7 (12)

The two possible values for U are

$$U_A = n_1 n_2 + \frac{n_1(n_1+1)}{2} - W_A = 36 + \frac{6(7)}{2} - W_A = 57 - 30 = 27$$

$$U_B = n_1 n_2 + \frac{n_2(n_2+1)}{2} - W_B = 36 + \frac{6(7)}{2} - 48 = 9$$

Since we have agreed to use the smaller value of U as a test statistic, a lower-tailed rejection region must be determined so that α is close to .10. Notice that the hypothesis to be tested is actually two-tailed. That is, both large and small values of U will tend to contradict the null hypothesis. Hence, although we will only consider the area below some critical value of U (denoted by U_0) in determining α for the test, there is a similar area in the upper tail of the distribution. Thus the area below the critical U is actually $\frac{\alpha}{2}$ and must be doubled to obtain the value for α.

Referring to Table 8 and indexing $n_1 = n_2 = 6$, a rejection region is determined such that

$$P(U \leq U_0) = \frac{\alpha}{2} = .05$$

Hence we will reject if $U \leq 7$, with $\frac{\alpha}{2} = .0465$ or $\alpha = .093$. The minimum of U_1 and U_2 is $U = 9$, and H_0 is not rejected. We cannot conclude that the populations differ in location.

15.18 a. The data with their corresponding ranks are given in the table at the right. Notice that the observations corresponding to ranks 9 and 10 and 13 and 14 were tied. Hence an average rank was assigned to both (9.5 and 13.5, respectively).

	A	B
	6.1 (1)	9.1 (16)
	9.2 (17)	8.2 (8)
	8.7 (12)	8.6 (11)
	8.9 (13.5)	6.9 (2)
	7.6 (5)	7.5 (4)
	7.1 (3)	7.9 (7)
	9.5 (18)	8.3 (9.5)
	8.3 (9.5)	7.8 (6)
	9.0 (15)	8.9 (13.5)
	14.8 (1)	27.7 (12)
Rank Sum	94	77

Then the two possible values for U are

$$U_A = n_1n_2 + \frac{n_1(n_1+1)}{2} - W_A = 126 - 94 = 32$$

$$U_B = n_1n_2 + \frac{n_2(n_2+1)}{2} - W_B = 126 - 77 = 49$$

Thus, $\mu = 32$. Referring to Table 8 and indexing $n_1 = n_2 = 9$, the p-value is $2P(U \leq 32) = 2(.2447) = .4894$.

b. Since the data are presented in an unpaired manner (that is, the two samples are independent and random), the procedure is identical to that used in Section 10.6 of the text. The analysis is as follows:

(1) $H_0\colon\ \mu_1 - \mu_2 = 0$ vs. $H_a\colon\ \mu_1 - \mu_2 \neq 0$

(2) $$s^2 = \frac{\sum_j y_{1j}^2 - \left[\frac{\left(\sum_j y_{1j}\right)^2}{n_1}\right] + \sum_j y_{2j}^2 - \left[\frac{\left(\sum_j y_{2j}\right)^2}{n_2}\right]}{n_1+n_2-2}$$

$$= \frac{625.06 - \left[\frac{(74.4)^2}{9}\right] + 599.22 - \left[\frac{(73.2)^2}{9}\right]}{16} = .8675$$

(3) Test statistic: $$t = \frac{(\bar{y}_1 - \bar{y}_2) - D_0}{\sqrt{s^2\left(\frac{1}{n_1} + \frac{1}{n_2}\right)}} = \frac{.1333}{\sqrt{.8675\left(\frac{1}{9}+\frac{1}{9}\right)}} = .30$$

(4) Since $|.3| < 1.337$, the p-value $> 2(.1) = .2$.

15.19 If the alternative hypothesis is true, we would expect the batteries from plant A to fail later than the batteries from plant B. Hence the observations from plant A will be ranked near the end of the sequence and the U statistic (i.e., the number of observations from plant A that precede each observation from plant B) will be small. Hence, small values of U will tend to contradict the null hypothesis, and a lower-tailed rejection region is desired. Remember that the test statistic U has expected value $E(U) = \frac{n_1n_2}{2}$ and variance $V(U) = \frac{n_1n_2(n_2+n_2+1)}{12}$ when the null hypothesis is true. Also, for large values of n_1 and n_2, the quantity

$$Z = \frac{U - E(U)}{\sigma_\mu}$$

will be approximately normal with mean 0 and variance 1. Once this z value has been calculated, the null hypothesis will be rejected if $z < -1.645$ (cf. Chapter 10). The following data are available: $n_1 = n_2 = 15$, $W_A = 276$, and $W_B = 189$. Notice that observations are presented in order of failure, so that

$$W_A = 1 + 5 + 7 + 8 + 13 + 15 + 20 + 21 + 23 + 24 + 25 + 27 + 28 + 29 + 30 = 276$$

The value of U is

$$U_A = n_1n_2 + \frac{n_1(n_1+1)}{2} - W_A = 345 - 276 = 69$$

Also,

$$E(U) = \frac{n_1n_2}{2} = \frac{225}{2} = 112.5$$

and

$$V(U) = \frac{n_1 n_2(n_1+n_2+1)}{12} = \frac{1}{12}(15)(15)(31) = 581.25$$

Thus,

$$z = \frac{U - E(U)}{\sigma_\mu} = \frac{69 - 112.5}{\sqrt{581.25}} = \frac{-43.5}{24.1} = -1.80$$

The null hypothesis is rejected since $z = -1.80$ falls in the rejection region.

15.20 Although $n_2 = 3$, we use a large-sample approximation (Table 8 stops at 10, and $n_1 = 12$). The data with corresponding ranks are shown in the table at the right. Then

DDT	Diazinon
16 (13)	7.8 (9)
5 (6.5)	1.6 (2)
21 (15)	1.3 (1)
19 (14)	
10 (12)	
5 (6.5)	
8 (10)	
2 (3.5)	
7 (8)	
2 (3.5)	
4 (5)	
9 (11)	

$$U_A = 12(3) + \frac{12(13)}{2} - 108 = 6$$

$$U_B = 12(3) + \frac{3(4)}{2} - 12 = 30$$

$$E(U) = \frac{12(3)}{2} = 18$$

and

$$V(U) = \frac{12(3)(16)}{12} = 48$$

The test statistic is

$$z = \frac{U - E(U)}{\sqrt{V(U)}} = \frac{6 - 18}{\sqrt{48}} = -1.732$$

The p-value is $P(|z| > 1.732) = 2(.0418) = .0836$. Since $.0836 < .10$, the null hypothesis is rejected. There is a difference in the locations of the two populations.

15.21 To test for a difference in location, a two-tailed Mann-Whitney U test is used. The data, with corresponding ranks, are shown in the following table.

Sample 1	Sample 2
235 (10)	180 (3.5)
225 (9)	169 (1)
190 (8)	180 (3.5)
188 (7)	185 (6)
	178 (2)
	182 (5)

Calculate

$$U_A = n_1 n_2 + \frac{n_1(n_1+1)}{2} - W_A = 4(6) + \frac{4(5)}{2} - 34 = 0$$

$$U_B = n_1 n_2 + \frac{n_2(n_2+1)}{2} - W_B = 4(6) + \frac{6(7)}{2} - 21 = 24$$

The test statistic is $U = 0$ and the rejection region, found in Table 8, is $U \leq 3$ with $\alpha = 2(.0333) = .0666$. The null hypothesis is rejected. There is a difference in the distributions of wing stroke frequencies.

The p-value can be found in Table 8. Since the test is two-tailed, the observed level of significance is $2P[U \leq 0] = 2(.0048) = .0096$.

15.22 Similar to previous exercises. The data with corresponding ranks are shown in the table at the right. With $n_1 = n_2 = 12$, the two possible values for U are

$$U_A = 144 + \frac{12(13)}{2} - 89.5 = 132.5$$

$$U_B = 144 + \frac{12(13)}{2} - 210.5 = 11.5$$

Since n_1 and n_2 are greater than 10, the large-sample approximation can be used. Since it is necessary to detect a shift of the B observations to the right of the A observations, we let $U = U_A$.

A	B
84 (3)	140 (10)
128 (9)	184 (14.5)
168 (12)	368 (19)
92 (4.5)	96 (6)
184 (14.5)	480 (23)
92 (4.5)	188 (16)
76 (2)	480 (23)
104 (7)	244 (18)
72 (1)	440 (21)
180 (13)	380 (20)
144 (11)	480 (23)
120 (8)	196 (17
$W_A = 89.5$	$W_B = 210.5$

Calculate

$$z = \frac{U_A - \frac{n_1 n_2}{2}}{\sqrt{n_1 n_2 \left(\frac{n_1+n_2+1}{12}\right)}} = \frac{132.5-72}{\sqrt{300}} = 3.49$$

and reject H_0 if $z > z_{.05} = 1.645$. The null hypothesis is rejected and we conclude that rats in population B tend to survive longer than rats in population A.

15.23 **a.** The necessary computations follow:

$T_1 = 209$ $\quad T_2 = 199$ $\quad T_3 = 343$

$\sum y = 751$ $\quad \sum y^2 - 51{,}889$

$\text{TSS} = 51{,}889 - \frac{(751)^2}{15} = 14{,}288.933$

$\text{SST} = \frac{(209)^2}{5} + \frac{(199)^2}{5} + \frac{(343)^2}{5} - \frac{(751)^2}{15} = 2586.1333$

$\text{SSE} = \text{TSS} - \text{SST} = 11{,}702.8$

We test

$H_0\colon \mu_A = \mu_B = \mu_C$ vs. $H_a\colon$ At least two means differ

The test statistic is

$$F = \frac{\frac{2586.1333}{2}}{\frac{11{,}702.8}{12}} - 1.33$$

Since $F < 3.89 = F_{.05}$, we do not reject H_0. There is not sufficient evidence to indicate a significant difference among the means for the three brands. The analysis above is valid if the three population probability distributions are normal and the three population variances are equal. The normality assumption is probably not valid since life length data usually follow an exponential distribution (skewed to the right).

b. A composite ranking of the data is

A	B	C
5	8	14
7	4	3
2	12	15
13	1	11
9	10	6
$R_1 = 36$	$R_2 = 35$	$R_3 = 49$

We test

H_0: The population probability distributions of life lengths are identical for the three brands.

H_a: At least two of the probability distributions differ in location.

The test statistic is

$$H = \frac{122}{15(16)}\left[\frac{(36)^2}{5} + \frac{(35)^2}{5} + \frac{(49)^2}{5}\right] - 3(16) = 1.22$$

Since $H < \chi^2_{2,.05} = 5.99$, we do not reject H_0. There is not enough evidence to conclude that the brands of magnetron tubes tend to differ in length of life under stress.

15.24a. The ranked data are given in the test. The rank sums are

$R_1 = 103.5$ $\quad R_2 = 83$ $\quad R_3 = 44.5$

A test of identical distributions of recovery times is

H_0: The population probability distributions of recovery times for the three types of flu are identical.

H_a: At least two of the types of flu have probability distributions with different locations.

The test statistic is

$$H = \frac{12}{21(22)}\left[\frac{(103.5)^2}{7} + \frac{(83)^2}{7} + \frac{(44.5)^2}{7}\right] - 3(22) = 6.66$$

Since $\chi^2_{2,.05} < 6.66 < \chi^2_{2,.025}$, the p-value is between .025 and .05. The data suggest at least one of the flu types requires a longer recovery time than another.

b. We use the Mann-Whitney U test to detect a difference in locations. A composite ranking of the recovery times for Victoria A and Russian is

Victoria A	Russian
13	7
4.5	1
14	7
11	3
9.5	4.5
12	2
7	9.5
$T_A = 71$	$T_B = 34$

We test

H_0: The probability distributions of recovery times for the Victoria A and the Russian flu are identical.

H_A: The probability distributions differ in location.

The test statistic is

$U = 7(7) + \frac{7181}{2} - 71 = 6.$

From Table 8, the p-value is $2P(U \leq 6) = 2(.0087) = .0174$. There is sufficient evidence to claim that the recovery times differ for Victoria A and Russian flu.

15.25 A composite ranking of the data is

$38°F$	$42°F$	$46°F$	$50°F$
16	3	2	6.5
18.5	14	22	8.5
4.5	21	14	1
8.5	4.5	10.5	12
10.5	20	18.5	14
	6.5	17	
$R_1 = 58$	$R_2 = 69$	$R_3 = 84$	$R_4 = 42$

We test

H_0: The population probability distributions of weights are identical for the four water temperatures.

H_a: At least two of the probability distributions differ in location.

The test statistic is

$$H = \frac{12}{(22)(23)}\left[\frac{(58)^2}{5} + \frac{(69)^2}{6} + \frac{(84)^2}{6} + \frac{(42)^2}{5}\right] - 3(23) = 2.03$$

Since $H < \chi^2_{3,.1} = 6.25139$, we do not reject H_0. There is insufficient evidence to detect a difference among the weight distributions.

15.26 A composite ranking of the data is

A	B	C
3.5	27	1.5
11	21	7
19	17	12
15	30	8
16	23	10
1.5	25	3.5
13	29	6
18	26	14
24	22	9
20	28	5
$R_1 = 141$	$R_2 = 248$	$R_3 = 76$

We test

H_0: The probability distributions of percentage of plants with weevil damage are identical for the three chemicals.

H_a: The probability distributions differ in location.

The test statistic is

$$H = \frac{12}{30(31)}\left[\frac{(141)^2}{10} + \frac{(248)^2}{10} + \frac{(76)^2}{10}\right] - 3(31) = 19.47$$

Since $19.47 > \chi^2_{2,.005} = 10.5966$, the p-value is less than .005.

15.27 Expanding H, we have

$$\begin{aligned}
H &= \frac{12}{n(n+1)} \sum \left[n_i \left(\overline{R}_i^2 - 2\overline{R}_i \frac{(n+1)}{2} + \frac{(n+1)^2}{4}\right)\right] \\
&= \frac{12}{n(n+1)} \sum \left[n_i \left(\frac{R_i^2}{n_i^2} - (n+1)\frac{R_i}{n_i} + \frac{(n+1)^2}{4}\right)\right] \\
&= \frac{12}{n(n+1)} \sum \frac{R_i^2}{n_i} - \frac{12}{n} \sum R_i + \frac{3(n+1)}{n} \sum n_i \\
&= \frac{12}{n(n+1)} \sum \frac{R_i^2}{n_i} - \frac{12n(n+1)}{2n} + \frac{3(n+1)(n)}{n} \\
&= \frac{12}{n(n+1)} \sum \frac{R_i^2}{n_i} - 3(n+1)
\end{aligned}$$

15.28 Using the hint given, we need not worry about the population in which a rank falls; we need only consider the 15 possible pairings of ranks. The statistic H is

$$H = \frac{12}{6(7)} \sum \frac{R_i^2}{2} - 3(7) = \frac{\sum R_i^2 - 147}{7}$$

The possible pairings are given below, along with the value of H for each.

Pairing			H
(1, 2)	(3, 4)	(5, 6)	32/7
(1, 2)	(3, 5)	(4, 6)	26/7
(1, 2)	(3, 6)	(5, 6)	24/7
(1, 3)	(2, 4)	(5, 6)	26/7
(1, 3)	(2, 5)	(4, 6)	18/7
(1, 3)	(2, 6)	(4, 5)	14/7
(1, 4)	(2, 3)	(5, 6)	24/7
(1, 4)	(2, 5)	(3, 6)	8/7
(1, 4)	(2, 6)	(3, 5)	6/7
(1, 5)	(2, 3)	(4, 6)	14/7
(1, 5)	(2, 4)	(3, 6)	6/7
(1, 5)	(2, 6)	(3, 4)	2/7
(1, 6)	(2, 3)	(4, 5)	8/7
(1, 6)	(2, 4)	(3, 5)	2/7
(1, 6)	(2, 5)	(3, 4)	0/7

Thus, the null distribution of H is

h	$p(h)$
0	1/15
2/7	2/15
6/7	2/15
8/7	2/15
2	2/15
18/7	1/15
24/7	2/15
26/7	2/15
32/7	1/15

15.29 A summary of the ranked data (ranked within each metal) is

Metal	I	II	III
1	2	1	3
2	3	1	2
3	1.5	3	1.5
4	3	1	2
5	3	1	2
6	2	1	3
7	1.5	1.5	3
8	1	2	3
9	3	1	2
10	3	1	2

Thus, $R_1 = 23$, $R_2 = 13.5$, and $R_3 = 23.5$. We test

H_0: The probability distributions for the amount of corrosion are the same for all three types of sealer.

H_a: At least two of the probability distributions differ in location.

The test statistic is

$$F_r = \frac{12}{10(3)(4)}[(23)^2 + (13.5)^2 + (23.5)^2] - 3(10)(4) = 6.35$$

Since $F_r > \chi^2_{2,.05} = 5.99$, we reject H_0 with $\alpha = .05$. There is sufficient evidence of a difference in the abilities of the sealers to prevent corrosion.

15.30 A summary of the ranked data is

	Sprays		
Ear	A	B	C
1	2	3	1
2	2	3	1
3	1	3	2
4	3	2	1
5	2	1	3
6	1	3	2
7	2.5	2.5	1
8	2	3	1
9	2	3	1
10	2	3	1

Thus, $R_1 = 19.5$, $R_2 = 26.5$, and $R_3 = 14$.

The Friedman test is

H_0: The probability distributions of the amount of aflatoxin are identical for the three sprays.

H_a: At least two of the probability distributions differ in location.

The test statistic (with 2 degrees of freedom) is

$$F_r = \frac{12}{10(3)(4)}[(19.5)^2 + (26.5)^2 + (14)^2] - 3(10)(4) = 7.85$$

Since $7.38 < 7.85 < 9.21$, the p-value is between .01 and .025.

15.31 a. To carry out the Friedman test, we need the rank sums, R_i, for each model. We can find these by adding the ranks given for each model. Thus for model A, $R_1 = 8(15) = 120$. For model B, $R_2 = 4 + 2(6) + 7 + 8 + 9 + 2(14) = 68$, etc. The R_i values are 120, 68, 37, 61, 31, 87, 100, 34, 32, 62, 85, 75, 30, 71, 67. Thus,

$$\sum R_i^2 = 71{,}948$$

and

$$F_r = \frac{12}{8(15)(16)}(71{,}948) - 3(8)(16) = 65.675$$

Since $F_r > \chi^2_{14,.005} = 31.32$, the p-value is less than .005. We reject the hypothesis that the 15 distributions are identical.

b. The highest (best) rank given to model H is lower than the lowest (worst) rank given to model M. Thus, the value of the test statistic is $m = 0$. Since $P(M = 0) = \left(\frac{1}{2}\right)^8 = \frac{1}{256}$, the p-value is $2\left(\frac{1}{256}\right) = \frac{1}{128}$.

c. To use the sign test we must know, for each judge, whether the judge preferred model H or model M. We do not have this information since the rankings "overlap," unlike the situation in part **b**.

15.32 We rank the skin irritation scores from low to high for each rat:

	Chemical		
Rat	A	B	C
1	3	2	1
2	3	2	1
3	2	3	1
4	1	3	2
5	1	2	3
6	1	3	2
7	2	3	1
8	2	1	3

Thus, $R_1 = 15$, $R_2 = 19$, and $R_3 = 14$. The complete test is

H_0: The probability distributions of skin irritation scores are identical for the three chemicals.

H_a: At least two of the distributions differ in location.

The test statistic is

$$F_r = \frac{12}{8(3)(4)}\left[(15)^2 + (19)^2 + (14)^2\right] - 3(8)(4) = 1.75$$

Since $F_r < \chi^2_{2,.01} = 9.21034$, we do not reject H_0. There is insufficient evidence to conclude that the chemicals cause differing degrees of irritation.

15.33 If $k = 2$ and $b = n$, then

$$F_r = \frac{2}{n}\left(R_1^2 + R_2^2\right) - 9n$$

If $R_1 = 2n - M$ and $R_2 = n + M$, then

$$\begin{aligned} F_r &= \frac{2}{n}\left[(2n - M)^2 + (n + M)^2\right] - 9n \\ &= \frac{2}{n}\left[\left(4n^2 - 4nM + M^2\right) + \left(n^2 + 2nM + M^2\right) - 4.5n^2\right] \\ &= \frac{2}{n}\left(-.5n^2 - 2nM + 2M^2\right) \\ &= \frac{4}{n}\left(M^2 - nM - \frac{1}{4}n^2\right) \\ &= \frac{4}{n}\left(M - \frac{n}{2}\right)^2 \end{aligned}$$

The Z statistic from Section 15.3 is

$$Z = \frac{M - \frac{n}{2}}{\left(\frac{1}{2}\right)\sqrt{n}} = \frac{2}{\sqrt{n}}\left(M - \frac{n}{2}\right)$$

so

$$Z^2 = \frac{4}{n}\left(M - \frac{n}{2}\right)^2 = F_r$$

15.34 Using the hints given, we have

$$F_r = \frac{12b}{k(k+1)} \sum \left(\overline{R}_i^2 - 2\overline{R}_i\overline{R} + \overline{R}^2\right) = \frac{12b}{k(k+1)} \sum \left(\frac{R_i^2}{b^2} - \frac{(k+1)R_i}{b} + \frac{(k+1)^2}{4}\right)$$

$$= \frac{12b}{k(k+1)} \sum \frac{R_i^2}{b^2} - \frac{12}{k}\frac{bk(k+1)}{2} + \frac{12b(k+1)k}{4k} = \frac{12}{bk(k+1)} \sum R_i^2 - 3b(k+1)$$

15.35 This is similar to Exercise 15.28. As in that exercise, we need only worry about the 3! possible rank pairings. They are listed below, with the R_i values and F_r. When $b = 2$ and $k = 3$,

$$F_r = \frac{\sum R_i^2}{2} - 24.$$

Block 1	Block 2	R_i
1	1	2
2	2	4
3	3	6

$F_r = 4$

Block 1	Block 2	R_i
1	1	2
2	3	5
3	2	5

$F_r = 3$

Block 1	Block 2	R_i
1	2	3
2	1	3
3	3	6

$F_r = 3$

Block 1	Block 2	R_i
1	2	3
2	3	5
3	1	4

$F_r = 1$

Block 1	Block 2	R_i
1	3	4
2	1	3
3	2	5

$F_r = 1$

Block 1	Block 2	R_i
1	3	4
2	2	4
3	1	4

$F_r = 0$

Thus, $P(F_r = 0) = P(F_r = 4) = \frac{1}{6}$ and $P(F_r = 1) = P(F_r = 3) = \frac{1}{3}$

15.36 Use Table 10, indexing the row marked (5, 5). Since the table is cumulative, successive entries can be subtracted to obtain individual probabilities.

a. $P(R = 2) = P(R \le 2) = .008$, since the minimum value of R is 2

b. $P(R \le 3) = .040$

c. $P(R \le 4) = .167$

15.37 The runs test is used to test the null hypothesis of randomness. Either a large or a small number of runs indicates non randomness, and a two-tailed test is used. The data are shown below:

$$W, W, W, W, B, W, W, W, B, B, W, B, B$$

There are $n_1 = 5$ blacks hired and $n_2 = 8$ whites hired. The number of runs observed is $R = 6$. From Table 10, the p-value is $2P(R \le 6) = 2(.347) = .694$. We do not reject the null hypothesis of randomness. The data do not suggest a nonrandom racial selection in the hiring of the union's members.

15.38 The hypothesis to be tested is

$$H_0\text{: no contagion (randomly diseased)}$$

vs.

$$H_a\text{: contagion (non randomly diseased)}$$

The test statistic is $R = 5$, the number of runs observed. Since contagion would be indicated by a grouping of diseased trees, a small number of runs will tend to support the alternative hypothesis. Hence, using $n_1 = n_2 = 5$ in Table 10, the p-value is .357. We cannot conclude that there is evidence of contagion.

15.39 **a.** In this exercise it is necessary to calculate $P(R \le 11)$, where $n_1 = 11$ and $n_2 = 23$. Since it is known that the quantity

$$z = \frac{R - E(R)}{\sigma_R}$$

is approximately normally distributed for large n_1 and n_2 (say, $n_1 \ge 10, n_2 \ge 10$), we may use the normal approximation to calculate $P(R \le 11)$. The first step is to determine the z value corresponding to an R value of 11. Note that

$$E(R) = \frac{2n_1n_2}{n_1+n_2} + 12 = \frac{2(11)(23)}{11+23} + 1 = 15.88$$

$$V(R) = \frac{2n_1n_2(2n_1n_2-n_1-n_2)}{(n_1+n_2)^2(n_1+n_2-1)} = \frac{2(11)(23)[2(11)(23)-11-23]}{(11+23)^2(11+23-1)} = \frac{238,832}{38,148}$$

$$= 6.2607$$

$$\sigma_R = \sqrt{6.2607} = 2.50$$

Hence the corresponding z value will be

$$z = \frac{R-E(R)}{\sigma_R} = \frac{11-15.88}{2.50} = -1.95$$

Thus,

$$P(R \leq 11) \approx P(Z \leq -1.95) = P(Z \geq 1.95) = .0256$$

b. The hypothesis to be tested is

H_0: randomness of occurrence vs. H_a: non randomness of occurrence

The test statistic is R, the number of runs observed. The reader may verify that the observed value of R is $R = 11$, with $n_1 = 11$ and $n_2 = 23$. Using a large-sample approximation, the standardized test statistic is

$$z = \frac{R-E(R)}{\sigma_R} = -1.95 \quad \text{(calculated above)}$$

Since an unusually small or unusually large number of runs would imply a non randomness of defectives, a two-tailed test is employed and the rejection region is $z < -1.96$ or $z > 1.96$. Since the test statistic, $z = -1.95$, does not fall in the rejection region, the null hypothesis is not rejected. Hence, there is not sufficient evidence of a non randomness of defectives.

15.40a. The hypothesis to be tested is

H_0: process is stable (random fluctuation) vs. H_a: process is unstable (nonrandom fluctuation)

and the test statistic is R, the number of runs observed. The mean of the 16 measurements is calculated to be

$$\bar{y} = \frac{\sum_i y_i}{n} = \frac{1082.7}{16} = 67.67$$

The measurements are classified as A if they lie above the mean and as B if they fall below. The sequence of runs generated by using this procedure is

$$\underline{A\,A\,A\,A\,A}\;\underline{B\,B\,B\,B\,B\,B}\;\underline{A}\;\underline{B}\;\underline{A}\;\underline{B}\;\underline{A}$$

Notice that the observed value of the test statistic is $R = 7$, with $n_1 = 8$, $n_2 = 8$. Nonrandom fluctuation will be implied by a small number of runs. (That is, the process would be higher than the mean for a period of time and then would drop below the mean for a longer period of time. This would imply a lack of stability in the process.) Consulting Table 10, with $n_1 = 8$ and $n_2 = 8$, the p-value is $P(R \leq 7) = .214$. Hence the null hypothesis is not rejected.

b. If the time period is divided into two equal parts, the data appear as shown in the table at the right. The hypothesis to be tested (using the t test described in Section 10.6) is

$$H_0\colon \mu_1 - \mu_2 - 0 \quad \text{vs.} \quad H_a\colon \mu_1 - \mu_2 \neq 0$$

Then

Group I	Group II
68.2	65.3
71.6	64.2
69.3	67.6
71.6	68.6
70.4	66.8
65.0	68.9
63.6	66.8
64.7	70.1

$$\bar{y}_1 = \frac{\sum_i y_{1j}}{n_i} = \frac{544.4}{8} = 68.05 \qquad \bar{y}_2 = \frac{\sum_j y_{2j}}{n_2} = \frac{538.30}{8} = 67.29$$

$$s^2 = \frac{\sum_j y_{1j}^2 - \left[\frac{\left(\sum_j y_{1j}\right)^2}{n_1}\right] + \sum_j y_{2j}^2 - \left[\frac{\left(\sum_j y_{2j}\right)^2}{n_2}\right]}{n_1+n_2-2}$$

$$= \frac{37{,}119.06 - \left[\frac{(544.4)^2}{8}\right] + 36{,}247.15 - \left[\frac{(538.30)^2}{8}\right]}{14} = \frac{72.64+26.29}{14} = 7.066$$

The test statistic is

$$t = \frac{(\bar{y}_1 - \bar{y}_2) - D_0}{\sqrt{s^2\left(\frac{1}{n_1}+\frac{1}{n_2}\right)}} = \frac{68.05-67.29}{\sqrt{7.066\left(\frac{1}{8}+\frac{1}{8}\right)}} = \frac{.76}{1.328} = .57$$

Using a two-tailed rejection region (with $\alpha = .10$), the critical value of t is $t_{.05,14} = 1.761$, and the rejection region is $|t| > 1.761$. The null hypothesis cannot be rejected.

15.41 Let A represent an observation from population A, and let B represent an observation from population B. Now referring to Exercise 15.18, the observations are arranged according to rank, and the population from which they were drawn is noted. Using the ranks obtained in Exercise 15.18 to arrange the observations, the sequence of runs is as follows:

$$\underline{A}\ \underline{B}\ \underline{A}\ \underline{B}\ \underline{A}\ \underline{B\ B\ B}\ \underline{A}\ \underline{B\ B}\ \underline{A\ A}\ \underline{B}\ \underline{A}\ \underline{B}\ \underline{A\ A}$$

Notice that the 9th and 10th and the 13th and 14th letters in the sequence represent the two pairs of tied observations. If the tied observations were reversed in the sequence of runs, we would still obtain $R = 13$. Hence the order of the tied observations is irrelevant.

Consider the alternative situation that asserts that the two distributions are not identical. If the alternative is true, we would expect a small number of runs because most of the measurements for population A will fall below those for population B (or vice versa). Hence small values for R will tend to contradict the null hypothesis. A one-tailed test of hypothesis is employed with a lower-tailed rejection region. Table 10 is then used to find the p-value. For $n_1 = n_2 = 9$ and $R = 13$, the p-value is $P(R \leq 13) = .956$. The null hypothesis is not rejected. This is the same conclusion that was reached when the Mann-Whitney U test was employed.

15.42 Refer to Exercise 15.19. The test statistic is R, the number of runs observed. If the alternative is true, and the experimental batteries have greater mean life, then most of the observations from plant B will be smaller than those for plant A. Consequently, the expected number of runs will be small.

To use a large-sample test, the values for R, $E(R)$, and $V(R)$ must be determined. Then when n_1 and n_2 are large, the quantity $Z = \frac{R-E(R)}{\sigma_R}$ will be approximately normally distributed with mean 0 and variance 1. Since we are interested in a one-tailed test of hypothesis (that is, a small number of runs will tend to contradict the null hypothesis), the rejection region will be $z \leq -1.645$.

$$E(R) = \frac{2n_1n_2}{n_1+n_2} + 1 = \frac{2(15)(16)}{30} + 1 = 16$$

$$V(R) = \frac{2n_1n_2(2n_1n_2-n_1-n_2)}{(n_1+n_2)^2(n_1+n_2-1)} = \frac{2(15)(15)[2(15)(15)-15-15]}{(30)^2(29)} = \frac{210}{29} = 7.24137$$

$$\sigma_R = \sqrt{7.24137} = 2.69$$

The reader may verify that $R = 15$, and the standardized test statistic is

$$z = \frac{R-E(R)}{\sigma_R} = \frac{15-16}{2.69} = -.37$$

Hence, the null hypothesis is not rejected, and there is insufficient evidence to indicate a difference between the distributions of the two types of batteries.

15.43 The ranks of the two variables are shown below.

Leaf	Rank, x	Rank, y
1	10.5	12
2	5.5	7.5
3	7.5	9
4	7.5	6
5	4	4.5
6	9	10
7	2	3
8	5.5	4.5
9	1	1
10	12	11
11	10.5	7.5
12	3	2

Calculate

$\sum x_i y_i = 636.25$ $\qquad \sum x_i^2 = 648.5$ $\qquad \sum y_i^2 = 649$

$n = 12$ $\qquad \sum x_i = 78$ $\qquad \sum y_i = 78$

Then

$S_{xy} = 636.25 - \frac{(78)^2}{12} = 129.25$ $\qquad S_{xx} = 648.5 - \frac{(78)^2}{12} = 141.5$

$S_{yy} = 649 - \frac{(78)^2}{12} = 142$

and

$$r_s = \frac{S_{xy}}{\sqrt{S_{xx}S_{yy}}} = \frac{129.25}{\sqrt{141.5(142)}} = .912$$

To test for correlation with $\alpha = .05$, index .025 in Table 11, Appendix III, and the rejection region is $|r_s| \geq .591$. The null hypothesis is rejected and we conclude that there is a correlation between the two variables.

15.44 The two sets of ranks are

Food Portion	Days $= x_i$	Taste $= y_i$
1	2	11
2	5	7.5
3	1	12
4	11	3
5	7	6
6	10	2
7	3	10
8	9	9
9	4	7.5
10	12	1
11	6	5
12	8	4

Calculate

$\sum x_i y_i = 381.5$ $\qquad \sum x_i = 78$ $\qquad \sum y_i = 78$

$\sum x^2 = 650$ $\qquad \sum y_i^2 = 649.5$ $\qquad n = 12$

Thus

$$r_s = \frac{12(381.5)-(78)(78)}{\text{r}[12(650)-(78)^2][12(649.5)-(78)^2]} = -.879$$

From Table 11, since $|-.879| > .780$, the p-value is less than $2(.005) = .01$. We reject H_0 at the $\alpha = .05$ level.

15.45 The objective is to determine whether or not test scores are correlated with interview ratings. Hence a Spearman rank correlation coefficient may be used to test for a relation between two ranked variables. Since the first variable (interview rating) is already in ranked form, we need only rank the second variable (test score). This variable will be ranked from low to high. The ranks (x_i and y_i) are shown in the table below.

Subject	x_i	y_i	Subject	x_i	y_i
1	8	5	6	1	9
2	5	6	7	4	10
3	10	2.5	8	7	4
4	3	7	9	9	1
5	6	2.5	10	2	8

a. $r_s = \frac{n\sum_i x_i y_i - \left(\sum_i x_i\right)\left(\sum_i y_i\right)}{\sqrt{\left[n\sum_i x_i^2 - \left(\sum_i x_i\right)^2\right]\left[n\sum_i y_i^2 - \left(\sum_i y_i\right)^2\right]}}$

$= \frac{(10)(233)-(55)(56)}{\sqrt{[10(385)-(55)^2][10(384.5)-(55)^2]}}$

$= \frac{-695}{\sqrt{825(820)}} = \frac{-695}{822.5} = -.845$

b. The hypothesis to be tested is

H_0: no correlation between interview rank and test score

H_a: negative correlation

and the test statistic will be the Spearman rank correlation coefficient, r_s. For a one-tailed test with $\alpha = .05$ and $n = 10$, the critical value for rejection is $-.564$ (see Table 11). The test statistic falls in the rejection region. Hence the null hypothesis is rejected. There is evidence of a significant negative correlation between the two ranked variables. The p-value $= P(r_s \le -.845) < .005$.

15.46 The two variables (rating and distance) are ranked from low to high, and the results are shown in the table below.

Voter	x_i	y_i	Voter	x_i	y_i
1	7.5	3	7	6	4
2	4	7	8	11	2
3	3	12	9	1	10
4	12	1	10	5	9
5	10	8	11	9	5.5
6	7.5	11	12	2	5.5

a. $r_s = \frac{n\sum_i x_i y_i - \left(\sum_i x_i\right)\left(\sum_i y_i\right)}{\sqrt{\left[n\sum_i x_i^2 - \left(\sum_i x_i\right)^2\right]\left[n\sum_i y_i^2 - \left(\sum_i y_i\right)^2\right]}}$

$= \frac{(12)(422.5)-(78)(78)}{\sqrt{[12(649.5)-(78)^2][120(649.5)-(78)^2]}}$

$= \frac{5070-6084}{7794-6084} = \frac{-1014}{1710} = -.593$

b. The hypothesis of interest is

H_0: no correlation vs. H_a: negative correlation

Consulting Table 11 for $\alpha = .05$, the critical value of r_s, denoted by r_0, is $-.497$. Since the value of the test statistic, r_s, is less than the critical value, the null hypothesis is rejected. There is evidence of a significant negative correlation between rating and distance.

15.47 The ranks for the two variables of interest (x_i and y_i, corresponding to math and art, respectively) are shown in the table below.

Student	x_i	y_i	Student	x_i	y_i
1	1	5	9	10.5	6
2	3	11.5	10	12	15
3	2	1	11	13.5	11.5
4	4	2	12	6	7
5	5	3.5	13	13.5	10
6	7.5	8.5	14	15	14
7	7.5	3.5	15	10.5	8.5
8	9	13			

Then

$$r_s = \frac{15(1148.5)-120(120)}{\sqrt{[15(1238.5)-(120)^2]^2}} = .6768$$

Consulting Table 11 for $\alpha = .10$, the critical value of r_s is .441, and the rejection region is $|r_s| > .441$. (Notice that the α for a two-tailed test requires doubling the probabilities tabulated in Table 11.) Since the calculated value of r_s falls in the rejection region, H_0 is rejected, and we conclude that there is a correlation between math and art scores.

15.48 **a.** The data are ranked separately according to the variables x and y.

Rank x	7	6	5	4	1	12	8	3	2	11	10	9
Rank y	7	8	4	5	2	10	12	3	1	6	11	9

Since there were no tied observations, the simpler formula for r_s is used, and

$$r_s = 1 - \frac{6\sum d_i^2}{n(n^2-1)} = 1 - \frac{6[(0)^2+(-2)^2+\ldots+(0)^2]}{12(143)} = 1 - \frac{6(54)}{1716} = .811$$

b. To test for positive correlation with $\alpha = .05$, index .025 in Table 9 and the rejection region is $r_s \geq .497$. Hence, H_0 is rejected; there is a correlation between x and y (p-value $< .005$).

15.49 The ranks for the two variables of interest are given below, along with $d_i = x_i - y_i$.

y_i	2	3	1	4	6	8	5	10	7	9
x_i	2	3	1	4	6	8	5	10	7	9
d_i	0	0	0	0	0	0	0	0	0	0

Using the alternative formula given in this section, we have

$$r_s = 1 - \frac{6\sum d_i^2}{n(n^2-1)} = 1 - 0 = 1$$

Using Table 11, $1 > .794$, so the p-value $< .005$.

15.50 Let y denote the rank of the y observations and x denote the rank of the x observations. These ranks are shown below.

y_i	9	8	6	7	1	2	5	4	10	3
x_i	10	9	7	6	2	1	5	4	8	3
d_i	-1	-1	-1	1	-1	1	0	0	2	0

Then

$$\sum d_i^2 = 10$$

and

$$r_s = 1 - \frac{6(10)}{10(99)} = .9394$$

For a two-tailed test with $\alpha = .10$, the hypothesis of no correlation will be rejected if $r_s \geq .564$. Hence there is evidence of significant correlation.

15.51 **a.** Since we are interested in a difference in recovery rates, let

$p = P$(recovery rate for A exceeds B at a given hospital)
$M =$ number of times A exceeds B

The hypothesis to be tested is

$$H_0\colon p = \tfrac{1}{2} \quad \text{vs.} \quad H_a\colon p \neq \tfrac{1}{2},$$

and the data are shown in the following table.

Hospital	A	B	Sign of $(A - B)$
1	75.0	85.4	$-$
2	69.8	83.1	$-$
3	85.7	80.2	$+$
4	74.0	74.5	$-$
5	69.0	70.0	$-$
6	83.3	81.5	$+$
7	68.9	75.4	$-$
8	77.8	79.2	$-$
9	72.2	85.4	$-$
10	77.4	80.4	$-$

Various rejection regions are tried in order to find $\alpha = .10$. (Use Table 1,

Appendix III).

Rejection Region	α
$M = 0, M = 10$	.002
$M \leq 1, M \geq 9$	.022
$M \leq 2, M \geq 8$	.110

Using the rejection region $M \leq 2$ or $M \geq 8$, the null hypothesis is rejected since the observed value of M is $m = 2$. We conclude that a difference does exist in the recovery rates for the two drugs.

b. In the above analysis we made no assumptions concerning the underlying distributions of the data. To use the t test, we must be able to assume normality of the distributions and equal variances for the two populations. Since the observations given above are percentages, their distributions may be almost mound-shaped, but the variances will not be equal.

15.52 Let $p = P$(gourmet A's rating exceeds gourmet B's for a given meal) and $M =$ a number of meals for which gourmet A exceeds B. The hypothesis to be tested is

$$H_0\colon\ p = \tfrac{1}{2} \qquad \text{vs.} \qquad H_a\colon\ p \neq \tfrac{1}{2}.$$

The sign test will be used with M as the test statistic. Notice that for this exercise, $n = 17$, since a tied rating was given to meals 7, 14, and 20. The value of the test statistic is $m = 8$. Various two-tailed rejection regions are tried in order to find a region with $\alpha = .05$. The calculations are shown below.

Rejection Region	$\alpha = P\left(\text{reject } H_0 \vert p = \frac{1}{2}\right)$
$M \leq 2, M \geq 15$	$2\left[\binom{17}{0} + \binom{17}{1} + \binom{17}{2}\right]\left(\frac{1}{2}\right)^{17} = .00235$
$M \leq 4, M \geq 13$	$2\left[\binom{17}{0} + \binom{17}{1} + \binom{17}{2} + \binom{17}{3} + \binom{17}{4}\right]\left(\frac{1}{2}\right)^{17} = .04904$

Since the second region gives $\alpha = .05$, we choose to reject H_0 if $M \leq 4$ or $M \geq 13$. Since the observed value of M is $m = 8$, the null hypothesis is not rejected. There is insufficient evidence to indicate a difference between the two gourmets.

15.53 For the Wilcoxon signed-rank test, the differences and the ranks of their absolute values are given below for $n = 17$ differences.

d_i	Rank $\vert d_i \vert$	d_i	Rank $\vert d_i \vert$
−2	10.5	−3	15
−1	4.5	3	15
3	15	3	10.5
1	4.5	−2	10.5
−1	4.5	−1	4.5
3	15	1	4.5
−1	4.5	−2	10.5
−3	15	1	4.5
1	4.5		

Then $T^+ = 73.5$ and $T^- = 79.5$. With $\alpha = .05$ and $n = 17$, the lower portion of the rejection region is $T \leq 35$ (see Table 9). Since the observed value of T is $T = 73.5$, the null hypothesis is not rejected, as in Exercise 15.52.

15.54 Having made no assumptions concerning the underlying distribution of the populations, we cannot use a parametric test of means; rather we use the nonparametric Mann-Whitney U test to test the equivalence of the population distributions. The rank sums are $W_A = 126$ and $W_B = 45$. With $n_1 = n_2 = 9$,

$$U_A = n_1 n_2 + \frac{n_1(n_1+1)}{2} - T_A = 9(9) + \frac{9(10)}{2} - 126 = 0$$

$$U_B = n_1 n_2 + \frac{n_2(n_2+1)}{2} - T_B = 9(9) + \frac{9(10)}{2} - 45 = 81$$

For $n_1 = n_2 = 9$ in Table 8, the lower tail of the two-tailed rejection region is $U \leq 18$ with $\alpha = 2(.0252) = .0504$. The observed value of U is $U = 0$, and the null hypothesis is rejected. We conclude that the deaf children do differ from the hearing children in eye movement rate.

15.55 The data along with corresponding ranks are shown in the table at the right. Then

$$U_A = 8^2 + \frac{8(9)}{2} - 53.5 = 46.5$$

$$U_B = 8^2 + \frac{8(9)}{2} - 82.5 = 17.5$$

Stimuli 1	Stimuli 2
1 (2.5)	4 (16)
3 (12.5)	2 (7)
2 (7)	3 (12.5)
1 (2.5)	3 (12.5)
2 (7)	1 (2.5)
1 (2.5)	2 (7)
3 (12.5)	3 (12.5)
2 (7)	3 (12.5)
$W_1 = 53.5$	$W_2 = 82.5$

Consulting Table 8 with $n_1 = n_2 = 8$, the hypothesis of no difference will be rejected if $U \le 13$ with $\alpha = 2(.0249) = .0498$. We choose the minimum value of U_i, $U = 17.5$, and H_0 is not rejected. There is no evidence of a difference in location for the two distributions. This is the same conclusion as reached in Exercise 13.1.

15.56 **a.** The measurements are ordered according to magnitude and ranked from the "outside in," as described in this exercise. The resulting ranks are shown in the table at the right. The hypothesis to be tested is

$$H_0\colon \sigma_A^2 = \sigma_B^2 \quad \text{vs.} \quad H_a\colon \sigma_A^2 > \sigma_B^2$$

Instrument	Response	Rank
A	1060.21	1
B	1060.24	3
A	1060.27	5
B	1060.28	7
B	1060.30	9
B	1060.32	8
A	1060.34	6
A	1060.36	4
A	1060.40	2

and the test statistic is the Mann-Whitney U. If the alternative hypothesis is true (that is, the variance for instrument A is greater than the variance for instrument B), then the measurements for instrument A should be very low and very high in the sequence of measurements. Hence they will be assigned the lower ranks, and the sum of ranks for the A observations will be small. A one-tailed test of hypothesis is required, with α near .05. Calculating the U values, we obtain

$$U_1 = n_1n_2 + \frac{n_1(n_1+1)}{2} - W_A = 5(4) + \frac{5(6)}{2} - 18 = 17$$

$$U_2 = n_1n_2 + \frac{n_2(n_2+1)}{2} - W_B = 20 + 10 - 27 = 3$$

The rejection region will be $U \le 3$ with $\alpha = .056$. The test statistic falls in the rejection region, and hence the null hypothesis is rejected.

b. Calculate

$$s_1^2 = \frac{\sum_j y_{1j}^2 - \left[\frac{\left(\sum_j y_{1j}\right)^2}{n_1}\right]}{n_1 - 1} = \frac{.0230}{4} = .00575$$

$$s_2^2 = \frac{\sum_j y_{2j}^2 - \left[\frac{\left(\sum_j y_{2j}\right)^2}{n_2}\right]}{n_2 - 1} = \frac{.0035}{3} = .00117$$

Then the test statistic is

$$F = \frac{s_1^2}{s_2^2} = \frac{.00575}{.00117} = 4.914$$

The critical value of F (with 4 and 3 degrees of freedom) is $F_{.05} = 9.12$. The null hypothesis is not rejected.

15.57 Designate the five observations from samples 1 and 2 as A and B, respectively. When the $(n_1 + n_2) = 10$ observations are ordered according to their magnitude, the U statistic is obtained by counting the number of observations in sample A that precede each observation in sample B. Then

$$P(U \le 2) = P(U = 0) + P(U = 1) + P(U = 2)$$

The only sample point associated with $U = 0$ is

$$B\ B\ B\ B\ B\ A\ A\ A\ A\ A$$

because there are no A's preceding any of the B observations. In order to obtain the probability of observing this sample point, we proceed as follows:

(1) The total number of permutations of the 10 observations is 10!.

(2) The number of different arrangements of the five A observations is 5!. Similarly, the number of different arrangements of the five B observations is 5!.

(3) The total number of distinct arrangements of the five A's and five B's will be $\frac{10!}{5!\,5!}$. Hence

$$P(U = 0) = \frac{1}{\frac{10!}{5!\,5!}}$$

The only sample point associated with $U = 1$ is

$$B\ B\ B\ B\ A\ B\ A\ A\ A\ A$$

where only one A observation precedes a B observation. Then

$$P(U = 1) = \frac{1}{\frac{10!}{5!\,5!}}$$

Finally, the event $U = 2$ will occur when one of the following two sample points occurs:

$B\ B\ B A\ B\ B\ A\ A\ A\ A$ or $B\ B\ B\ B\ A\ A\ B\ A\ A\ A$

and

$$P(U = 2) = \frac{2}{\frac{10!}{5!\,5!}}.$$

Then

$$P(U \le 2) = P(U = 0) + P(U = 1) + P(U = 2) = \frac{1+1+2}{\frac{10!}{5!\,5!}} = \frac{4(5!)(5!)}{10!}$$

$$= \frac{1}{63} = .0159$$

15.58 Let Y be the number of positive differences and let T be the rank sum of the positive differences. Then

$$P(T \le 2) = P(T = 0) + P(T = 1) + P(T = 2)$$

Note that a particular value of T is the union of four mutually exclusive events ($Y = 0$ and the rank sum is T, $Y = 1$ and the rank sum is T, $Y = 2$ and the rank sum is T, $Y = 3$ and the rank sum is T). Then, assuming the null hypothesis to be true,

$$p = P(\text{observation } A \text{ exceeds observation } B) = P(\text{a difference is positive}) = \tfrac{1}{2}$$

and Y possesses the binomial probability distribution

$$p(y) = \binom{3}{y}\left(\tfrac{1}{2}\right)^y\left(\tfrac{1}{2}\right)^{3-y} \qquad y = 0, 1, 2, 3$$

Consider the three pairs of observations and rank the differences according to absolute magnitude. Let d_1, d_2, and d_3 denote the ranked differences. The table below presents the possible algebraic signs of these differences and the corresponding values of Y and T, respectively.

d_1	d_2	d_3	y	T
+	+	+	3	6
−	+	+	2	5
+	−	+	2	4
+	+	−	2	3
−	−	+	1	3
−	+	−	1	2
+	−	−	1	1
−	−	−	0	0

(1) $P(T = 0) = P(Y = 0, T = 0) + P(Y = 1, T = 0)$

$$+ P(Y = 2, T = 0) + P(Y = 3, T = 0)$$

$$= P(Y = 0)P(T = 0|Y = 0) + P(Y = 1)P(T = 0|Y = 1)$$

$$+ P(Y = 2)P(T = 0|Y = 2) + P(Y = 3)P(T = 0|Y = 3)$$

Note that when $Y > 0$, $T > 0$. Hence

$$P(T = 0) = P(Y = 0)P(T = 0|Y = 0) + 0 + 0 + 0 = \binom{3}{0}\left(\tfrac{1}{2}\right)^0\left(\tfrac{1}{2}\right)^3(1) = \tfrac{1}{8}$$

(2) The value $T = 1$ occurs only when $Y = 1$. Hence

$$P(T = 1) = P(Y = 0, T = 1) + P(Y = 1, T = 1) + P(Y = 2, T = 1) + P(Y = 3, T = 1)$$
$$= 0 + P(Y = 1)P(T = 1|Y = 1) + 0 + 0$$
$$= \binom{3}{1}\left(\tfrac{1}{2}\right)^1\left(\tfrac{1}{2}\right)^2\left(\tfrac{1}{3}\right) = \left(\tfrac{3}{8}\right)\left(\tfrac{1}{3}\right) = \tfrac{1}{8}$$

Note that $P(T = 1|Y = 1) = \frac{1}{3}$ because there are three possible values of T, which occur with equal probability when $Y = 1$.

(3) The value $T = 2$ occurs only when $Y = 1$. Hence

$$P(T = 2) = P(Y = 0, T = 2) + P(Y = 1, T = 2) + P(Y = 2, T = 2) + P(Y = 3, T = 2)$$
$$= 0 + P(Y = 1)P(T = 2|Y = 1) + 0 + 0$$
$$= \binom{3}{1}\left(\tfrac{1}{2}\right)^1\left(\tfrac{1}{2}\right)^2\left(\tfrac{1}{3}\right) = \left(\tfrac{3}{8}\right)\left(\tfrac{1}{3}\right) = \tfrac{1}{8}$$

Again, $P(T = 2|Y = 1)$ is $\frac{1}{3}$ because any one of three values for T occur with equal probability when $Y = 1$. Then

$$P(T \le 2) = P(T = 0) + P(T = 1) + P(T = 2) = \tfrac{1}{8} + \tfrac{1}{8} + \tfrac{1}{8} = \tfrac{3}{8} = .375$$

15.59 A composite ranking of the data is

Line 1	Line 2	Line 3
19	14	2
16	10	15
12	5	4
20	13	11
3	9	1
18	17	8
21	7	6
$R_1 = 109$	$R_2 = 75$	$R_3 = 47$

We test

H_0: The probability distributions of the production figures are the same for all three production lines.

H_a: At least two of the probability distributions differ in location.

The test statistic is

$$H = \frac{12}{21(22)}\left[\frac{(109)^2}{7} + \frac{(75)^2}{7} + \frac{(47)^2}{7}\right] - 3(22) = 7.154$$

Since $H > \chi^2_{2,.05} = 5.99147$, we reject H_0 with $\alpha = .05$. The data provide sufficient evidence to indicate a difference in location for the three sets of production figures.

15.60 **a.** A composite ranking of the data is

Supervisor			
I	II	III	IV
21.5	16.5	14.5	1
20	5.5	12.5	7
21.5	8.5	8.5	3.5
18.5	12.5	18.5	10.5
16.5	10.5	5.5	2
	14.5		3.5
$R_1 = 98$	$R_2 = 68$	$R_3 = 59.5$	$R_4 = 27.5$

We test

H_0: The probability distributions are identical for the four types of personalities.

H_a: At least two of the types of personalities have probability distributions with different locations.

The test statistic is

$$H = \frac{12}{22(23)}\left[\frac{(98)^2}{5} + \frac{(68)^2}{6} + \frac{(59.5)^2}{5} + \frac{(27.5)^2}{6}\right] - 3(23) = 14.61$$

Since $H > \chi^2_{3,.05} = 7.81473$, we reject H_0 with $\alpha = .05$. There is sufficient evidence to indicate that one or more of the supervisors tend to receive higher ratings than the others.

b. We use the Mann-Whitney U test to detect a difference in location. A composite ranking of the ratings for supervisor I and supervisor III is

Supervisor I	Supervisor III
9.5	4
8	3
9.5	2
6.5	6.5
5	1
$W_A = 38.5$	$W_B = 16.5$

We test

H_0: The probability distributions of the ratings of the personality types represented by supervisors I and II are identical.

H_a: The probability distributions differ in location.

The test statistic is $U = (5)(5) + \frac{5(6)}{2} - 38.5 = 1.5$

The p-value is less than $2P(U \leq 2) = 2(.0159) = .0318$. Thus, we reject H_0 with $\alpha = .05$.

15.61 The ranked data are best analyzed using the Friedman test with people representing the blocks. The rank sums are $R_1 = 19$, $R_2 = 21.5$, $R_3 = 27.5$, $R_4 = 32$.

We test

H_0: The probability distributions are identical for the four items.

H_a: At least two of the probability distributions differ in location.

The test statistic is

$$F_r = \frac{12}{10(4)(5)}[(19)^2 + (21.5)^2 + (27.5)^2 + (32)^2] - 3(10)(5) = 6.21$$

Since $F_r < \chi^2_{3,.05} = 7.81473$, we do not reject H_0. There is insufficient evidence to indicate that one or more of the items are preferred to the others.

15.62 We use the Friedman test with groups as blocks. The ranks and rank sums are

Group	Lecture	Demonstration	Teaching Machine
1	1	2	3
2	1.5	1.5	3
3	1	3	2
4	2	1	3
5	3	1	2
6	1	2	3
	$R_1 = 9.5$	$R_2 = 10.5$	$R_3 = 16$

We test

H_0: The probability distributions are identical for the three methods.

H_a: At least two probability distributions differ in location.

The test statistic is

$$F_r = \frac{12}{6(3)(4)}[(9.5)^2 + 910.5)^2 + (16)^2] - 3(6)(4) = 4.08$$

Since $4.08 < 4.605 = \chi^2_{2,.10}$, the p-value is greater than .10.

15.63 The necessary probability can be read directly from Table 10, Appendix III, or can be obtained by using the results given in Section 15.9 of the text. We must obtain the probability of observing exactly Y_1 S runs and Y_2 F runs, where $Y_1 + Y_2 = \text{R}$. Note that

$$P[y_1, y_2] = \frac{\binom{n_1-1}{y_1-1}\binom{n_2-1}{y_2-1}}{\binom{n_1+n_2}{n_1}} = \frac{\binom{7}{y_1-1}\binom{7}{y_2-1}}{\binom{16}{8}}$$

(1) Consider $P(R = 2)$. The event $R = 2$ will occur when $y_1 = 1$ and $y_2 = 1$, with either the S elements or F elements beginning the sequence. Then

$$P(R = 2) = 2P(Y_1 = 1, Y_2 = 1) = \frac{2\binom{7}{0}\binom{7}{0}}{\binom{16}{8}} = \frac{2}{12{,}870}$$

(2) The event $R = 3$ will occur when we observe $y_1 = 1$ S run and $y_2 = 2$ F runs, or when we observe $y_1 = 2$ S runs and $y_2 = 1$ F run. Note that if there are two F runs and one S run, there is only one possible ordering because the F's must commence the sequence. Then

$$P(R=3) = P(Y_1=1, Y_2=2) + P(Y_1=2, Y_2=1) = \frac{\binom{7}{1}\binom{7}{0}}{\binom{16}{8}} + \frac{\binom{7}{0}\binom{7}{1}}{\binom{16}{8}}$$
$$= \frac{14}{12{,}870}$$

(3) Similarly,

$$P(R=4) = 2P(Y_1=2, Y_2=2) = \frac{2\binom{7}{1}\binom{7}{1}}{\binom{16}{8}} = \frac{98}{12{,}870}$$

(4) $P(R=5) = P(Y_1=3, Y_2=2) + P(Y_1=2, Y_2=3) = \frac{\binom{7}{2}\binom{7}{1}}{\binom{16}{8}} + \frac{\binom{7}{1}\binom{7}{2}}{\binom{16}{8}} = \frac{294}{12{,}870}$

(5) $P(R=6) = 2P(Y_1=3, Y_2=3) = \frac{2\binom{7}{2}\binom{7}{2}}{\binom{16}{8}} = \frac{882}{12{,}870}$

Then

$$P(R \le 6) = P(R=2) + P(R=3) + P(R=4) + P(R=5) + P(R=6) = \frac{1290}{12{,}870}$$
$$= .100$$

15.64 As in Exercise 15.57, designate the five observations from sample 1 as A's and the five from sample 2 as B's and arrange them in order of magnitude. The Wilcoxon W statistic is obtained by summing the ranks for the A's. Notice that the rank of a particular A is given by the position in the sequence that it occupies. Also, the minimum value of W is $W = 1+2+3+4+5 = 15$, which will occur if the following sequence is observed:

$$A\,A\,A\,A\,A\,B\,B\,B\,B\,B$$

From Exercise 15.57, the total number of distinct sequences is

$$\binom{10}{5} = \frac{10!}{5!\,5!} = 252$$

Hence, $P(W=15) = \frac{1}{252}$. The value $W = 16$ will occur only if the following sequence is observed:

$$A\,A\,A\,A\,B\,A\,B\,B\,B\,B$$

since $W = 1+2+3+4+6 = 16$, so that $P(W=16) = \frac{1}{252}$. Finally, the value $W = 17$ will occur in two ways:

$$A\,A\,A\,B\,A\,A\,B\,B\,B\,B \qquad \text{or} \qquad A\,A\,A\,A\,B\,B\,A\,B\,B\,B$$

so that $P(W=17) = \frac{2}{252}$. Then $P(W \le 17) = \frac{4}{252} = \frac{1}{63}$. This is the same result as obtained in Exercise 15.57, since $W = 15$ is equivalent to $U = 0$, $W = 16$ to $U = 1$, and $W = 17$ to $U = 2$.

15.65 We have n_1 A observations, $A_1, A_2, \ldots, A_{n_1}$, and n_2 B observations, $B_1, B_2, \ldots, B_{n_2}$. The Mann-Whitney U statistic is defined as

$$U = \sum_{i=1}^{n_2} U_i$$

where u_i is the number of A observations preceding the ith B. Write $B_{(i)}$ to be the ith B in the combined sample when it has been ranked from smallest to largest, and write $R[B_{(i)}]$ to be the rank of the ith ordered B in the total ranking of the A's and B's. Then u_i is the number of A observations preceding $B_{(i)}$. We know that there are $(i-1)$ B's preceding $B_{(i)}$, and that there are $R[B_{(i)}] - 1$ A's and B's preceding $B_{(i)}$. Hence there are $u_i = R[B_{(i)}] - i$ A's before $B_{(i)}$. Then

$$U = \sum_{i=1}^{n_2} u_i = \sum_{i=1}^{n_2} [R(B_{(i)}) - i] = \sum_{i=1}^{n_2} R[B_{(i)}] - \sum_{i=1}^{n_2} i = W_B - \frac{n_2(n_2+1)}{2}$$

since the first term is simply the sum of the B ranks, which is, by definition, W_B. The second term is the sum of the first n_2 integers, whose value is as given. Now if we let $n_1 + n_2 = N$, and define Z_i to be the ith observation in the combined sample of A's and B's, where $i = 1, 2, \ldots, N$, we can write

$$W_A + W_B = \sum_{i=1}^{N} R(Z_i) = \sum_{i=1}^{N} i = \frac{N(N+1)}{2}$$

so that

$$W_B = \frac{N(N+1)}{2} - W_A$$

and

$$U = \frac{N(N+1)}{2} - \frac{n_2(n_2+1)}{2} - W_A = \frac{N^2+N-n_2^2-n_2}{2} - W_A$$
$$= \frac{n_1^2+2n_1n_2+n_2^2+n_1+n_2-n_2^2-n_2}{2} - W_A$$
$$= n_1n_2 + \frac{n_1(n_1+1)}{2} - W_A$$

15.66 We will use the notation of Exercise 15.65, finding first the mean and variance of W_A.

Notice that

$$W_A = \sum_{i=1}^{n_1} R(A_i) = \sum_{i=1}^{N} X_i$$

where

$$X_i = \begin{cases} R(z_i), & \text{if } z_i \text{ is from sample } A \\ 0, & \text{if } z_i \text{ is from sample } B \end{cases}.$$

Now if H_0 is true,

$$E(X_i) = R(z_i)P[X_i = R(z_i)] + 0 \cdot P(X=0) = R(z_i)\tfrac{n_1}{N}$$
$$E(X_i^2) = R(z_i)^2\tfrac{n_1}{N}$$
$$V(X_i) = R(z_i)^2\tfrac{n_1}{N} - R(z_i)^2\left(\tfrac{n_1}{N}\right)^2 = R(z_i)^2\left[\tfrac{n_1(N-n_1)}{N^2}\right]$$
$$E(X_iX_j) = R(z_i)R(z_j)P[X_i = R(z_i) \text{ and } X_j = R(z_j)] = R(z_i)R(z_j)\left(\tfrac{n_1}{N}\right)\left(\tfrac{n_1-1}{N-1}\right)$$

so that

$$\text{Cov}(X_i, X_j) = E(X_iX_j) - E(X_i)E(X_j) = R(z_i)R(z_j)\left[\tfrac{n_1(n_1-1)}{N(N-1)} - \tfrac{n_1^2}{N^2}\right]$$
$$= R(z_i)R(z_j)\left[\tfrac{-n_1(N-n_1)}{N^2(N-1)}\right]$$

Now

$$E(W_A) = \sum_{i=1}^{N} E(X_i) = \tfrac{n_1}{N}\sum_{i=1}^{N} R(z_i) = \tfrac{n_1}{N}\left[\tfrac{N(N+1)}{2}\right] = \tfrac{n_1(N+1)}{2}$$
$$V(W_A) = V\left(\sum_{i=1}^{N} X_i\right) = \sum_{i=1}^{N} V(X_i) + \sum_{i \neq j}^{N}\sum^{N} \text{Cov}(X_i, X_j)$$
$$= \tfrac{n_1(N-n_1)}{N^2}\sum_{i=1}^{N} R(z_i)^2 - \tfrac{n_1(N-n_1)}{N^2(N-1)}\left[\sum_{i=1}^{N}\sum_{j=1}^{N} R(z_i)R(z_j) - \sum_{i=1}^{N} R(x_i)^2\right]$$
$$= \tfrac{n_1(N-n_1)}{N^2}\left[\tfrac{N(N+1)(2N+1)}{6}\right] - \tfrac{n_1(N-n_1)}{N^2(N-1)}\left\{\left[\sum_{i=1}^{N} R(z_i)\right]^2 - \sum_{i=1}^{N} R(z_i)^2\right\}$$
$$= \frac{2n_1(N-n_1)(N+1)(2N+1)}{12N} - \frac{n_1(N-n_1)}{N^2(N-1)}\left[\frac{N^2(N+1)^2}{4} - \frac{N(N+1)(2N+1)}{6}\right]$$
$$= \frac{n_1n_2(n_1+n_2+1)}{12}\left[\frac{4N+2}{N} - \frac{(3N+2)(N-1)}{N(N-1)}\right]$$
$$= \frac{n_1n_2(n_1+n_2+1)}{12}$$

Recall from Exercise 15.65 that

$$U = n_1n_2 + \frac{n_1(n_1+1)}{2} - W_A$$

Then

$$E(U) = n_1n_2 + \frac{n_1(n_1+1)}{2} - E(W_A) = \frac{n_1n_2}{2}$$

and

$$V(U) = V(W_A) = \frac{n_1n_2(n_1+n_2+1)}{12}.$$

15.67 In order to obtain T, the Wilcoxon signed-rank statistic, the differences d_i are calculated and ranked according to absolute magnitude. The rank sum of the positive or negative differences is then calculated. We will work, for the time being, with T^+ and write, as in Exercise 15.66,

$$T^{+} = \sum_{i=1}^{n} X_i$$

where

$$X = \begin{cases} R(d_i), & \text{if } d_i \text{ is positive} \\ 0, & \text{if } d_i \text{ is negative} \end{cases}$$

If H_0 is true and the two populations are identical, then the probability of observing a positive difference is $\frac{1}{2}$, and

$$E(X_i) = E(d_i)P[X_i = R(d_i)] = \tfrac{1}{2}\, R(d_i)$$
$$E\left(X_i^2\right) = R(d_i)^2 P[X_i = R(d_i)] = \tfrac{1}{2}\, R(d_i)^2$$
$$E(X_i X_j) = R(d_i)R(d_j)P[X_i = R(d_i), X_j = R(d_j)] = \tfrac{1}{4}\, R(d_i)R(d_j)$$

Then

$$V(X_i) = \tfrac{1}{2}\, R(d_i)^2 - \tfrac{1}{4}\, R(d_i)^2 = \tfrac{1}{4}\, R(d_i)^2$$
$$\text{Cov}\,(X_i, X_j) = \tfrac{1}{4}\, R(d_i)R(d_j) - \tfrac{1}{4}\, R(d_i)R(d_j) = 0$$

Finally,

$$E(T^{*}) = \sum_{i=1}^{n} E(X_i) = \tfrac{1}{2} \sum_{i=1}^{n} R(d_i) = \left(\tfrac{1}{2}\right) \tfrac{n(n+1)}{2} = \tfrac{n(n+1)}{4}$$
$$V(T^{*}) = \sum_{i=1}^{n} V(X_i) = \tfrac{1}{4} \sum_{i=1}^{n} R(d_i)^2 = \left(\tfrac{1}{4}\right) \tfrac{n(n+1)(2n+1)}{6} = \tfrac{n(n+1)(2n+1)}{24}$$

Had we calculated the mean and variance of

$$T^{-} = \sum_{i=1}^{n} Y_i$$

where

$$Y_i = \begin{cases} R(d_i), & \text{if } d_i \text{ is negative} \\ 0, & \text{if } d_i \text{ is positive} \end{cases}.$$

we would find that $E(T^{-}) = E(T^{+})$ and $V(T^{-}) = V(T^{+})$, since the probability of observing a negative difference is also $\frac{1}{2}$ under H_0.

15.68 Notice that X_i is, in fact, the rank of the ith X, and Y_i is the rank of the ith Y. Hence

$$\sum_{i=1}^{n} X_i = \sum_{i=1}^{n} Y_i = \tfrac{n(n+1)}{2} \quad \text{and} \quad \sum_{i=1}^{n} X_i^2 = \sum_{i=1}^{n} Y_i^2 = \tfrac{n(n+1)(2n+1)}{6}.$$

Write $d_i = X_i - Y_i$. Then

$$\sum_{i=1}^{n} d_i^2 = \sum_{i=1}^{n} (X_i^2 - 2X_iY_i + Y_i^2) = \tfrac{2n(n+1)(2n+1)}{6} - 2\sum_{i=1}^{n} X_iY_i$$

or

$$\sum_{i=1}^{n} X_iY_i = \tfrac{n(n+1)(2n+1)}{6} - \tfrac{1}{2}\sum_{i=1}^{n} d_i^2$$

Now

$$r_s = \frac{n\sum_{i=1}^{n} X_iY_i - \left(\sum_{i=1}^{n} X_i\right)\left(\sum_{i=1}^{n} Y_i\right)}{\sqrt{\left[n\sum_{i=1}^{n} X_i^2 - \left(\sum_{i=1}^{n} X_i\right)^2\right]\left[n\sum_{i=1}^{n} Y_i^2 - \left(\sum_{i=1}^{n} Y_i\right)^2\right]}} = \frac{\frac{n^2(n+1)(2n+1)}{6} - \frac{n}{2}\sum_{i=1}^{n} d_i^2 - \frac{n^2(n+1)^2}{4}}{\frac{n^2(n+1)(2n+1)}{6} - \frac{n^2(n+1)^2}{4}}$$

$$= \frac{\frac{n^2(n+1)(n-1)}{12} - \frac{n}{2}\sum_{i=1}^{n} d_i^2}{\frac{n^2(n+1)(n-1)}{12}} = 1 - \frac{\frac{n}{2}\sum_{i=1}^{n} d_i^2}{\frac{n^2(n^2-1)}{12}} = 1 - \frac{6\sum_{i=1}^{n} d_i^2}{n(n^2-1)}$$